OKANAGAN UNIV/COLLEGE LIBRARY

S0-AVL-095

QE 571 .B66 1987b
Principles of sedimentology
Boggs, Sam

214708

DATE DUE

SEP 17 1996	
OCT 03 1996	
NOV 23 1996	
MAY 18 2000	
FEB 9 2004	

BRODART   Cat. No. 23-221

Principles of
Sedimentology and
Stratigraphy

Principles of Sedimentology and Stratigraphy

Sam Boggs, Jr.
UNIVERSITY OF OREGON

MERRILL PUBLISHING COMPANY
A BELL & HOWELL COMPANY
COLUMBUS   TORONTO   LONDON   MELBOURNE

To Sumiko, Barbara, Steve, and Cindy

Cover Photo: Cross-bedded and jointed Navajo sandstone, Zion National Park, Utah

Published by Merrill Publishing Company
A Bell & Howell Company
Columbus, Ohio 43216

This book was set in Melior

Developmental Editor: Jennifer Knerr
Production Coordinator: Mary Harlan
Art Coordinator: Mark D. Garrett
Cover Designer: Cathy Watterson
Text Designer: Cynthia Brunk

Copyright © 1987 by Merrill Publishing Company. All rights reserved. No part of this book may be reproduced in any form, electronic or mechanical, including photocopy, recording, or any information storage and retrieval system, without permission in writing from the publisher. "Merrill Publishing Company" and "Merrill" are registered trademarks of Merrill Publishing Company.

Library of Congress Catalog Card Number: 86–61423
International Standard Book Number: 0–675–20487–9
Printed in the United States of America
    2   3   4   5   6   7   8   9—91   90   89   88   87

Preface

The fields of sedimentology and stratigraphy differ in some important respects, but they are so closely related that they are logically treated as a unified subject in an introductory textbook on sedimentary rocks. Krumbein and Sloss's classic textbook *Stratigraphy and Sedimentation*, last published in 1963, provides one of the few such integrated approaches to the study of sedimentary rock. *Stratigraphy and Sedimentation* remains a useful reference book for many aspects of sedimentology and stratigraphy; however, significant new data have been generated since the 1960s and important new concepts, such as magnetostratigraphy and seismic stratigraphy, have developed. Because no subsequent textbook has given an equally comprehensive, integrated treatment of sedimentary rock, a modern book is needed that brings together these new concepts and data and presents them, along with established principles of sedimentology and stratigraphy, in a format suitable for undergraduate majors in geology. This book was written to meet that need. It is the outgrowth of material presented in my own courses over a period of years. It is aimed primarily at undergraduate students, and it is written at a level that requires only basic preparation in physical and historical geology, mineralogy, physics, chemistry, and mathematics.

The choice and emphasis of subjects in the book has necessarily been influenced by space limitations. No single book of reasonable length can cover all aspects of sedimentology and stratigraphy in great depth. I have chosen throughout to emphasize fundamental *principles* rather than extensive discussion of case histories, nonetheless I have included selected case histories to emphasize important concepts.

I have drawn heavily in this book on the published work of other authors in an effort to bring together the most up-to-date results of modern research in sedimentology and stratigraphy. I have attempted in this synthesis to arrive at a reasonable compromise between breadth and depth, but this compromise has required presenting some material in simplified form with minimum supporting data. For the benefit of readers who may wish more detailed information on some subjects covered in the book, a list of additional readings is included at the end of each chapter.

The book is divided into seven parts. Part One introduces the subject of sedimentology and stratigraphy. These disciplines are defined and a brief review of their development and application in the study of Earth history and resource exploitation is presented. Part Two discusses the fundamental processes of weathering, transportation, and deposition, which act collectively to form sedimentary rocks. Part Three describes the textures and structures of sedimentary rocks, and Part Four deals with their composition and classification. Part Five describes the diagenetic processes that act during sediment burial to alter sediment and bring about lithification to sedimentary rock. Part Six treats the interpretation of ancient depositional environments and the characteristics of sedimentary rocks formed in the major sedimentary environments. The close relationship between sedimentology and stratigraphy is clearly demonstrated in this section. Part Seven covers the principles of stratigraphy: stratification, stratigraphic relations, stratigraphic nomenclature, age determination, and correlation.

I wish to thank the following individuals who reviewed various portions of the manuscript and gave me their critical comments: Brian H. Baker, University of Oregon; Charles W. Byers, University of Wisconsin—Madison; H. Edward Clifton, U.S. Geological Survey; Joseph E. Goebel, Indiana State University—Evansville; Gordon G. Goles, University of Oregon; Ralph E. Hunter, U.S. Geological Survey; Charlotte J. Mehrtens, University of Vermont; Jeffrey Mount, University of California—Davis; and William N. Orr, University of Oregon. I wish also to thank my wife, Sumiko, for her unstinting support during preparation of the manuscript.

Contents

PART SIX
SEDIMENTARY ENVIRONMENTS: THE LINK
BETWEEN SEDIMENTOLOGY AND STRATIGRAPHY

10 Principles of Environmental Interpretation and Classification   305

11 Continental Environments   341

12 Marginal-Marine Environments   395

13 The Marine Environment   453

PART ONE
Introduction

Sedimentary rocks form at the Earth's surface by a variety of low-temperature, low-pressure processes. These rocks are composed either of particles derived from preexisting rocks by weathering and erosion or of crystalline materials precipitated from seawater or freshwater by chemical and biochemical processes. Sedimentary rocks cover approximately three-fourths of the surface area of the continents and an even larger percentage of the ocean basins. Together with metasedimentary rocks, they form a thin shell, punctuated here and there by bodies of igneous rock, that envelops Earth. The average thickness of sedimentary rocks is about 1800 meters (6000 feet) on the continents and 240 meters (800 feet) in the oceans (Blatt, 1970); however, the thickness of sedimentary rocks in some ancient sedimentary basins ranges to 13 kilometers (8 miles) (Pettijohn, 1975).

Sedimentary rocks are characterized especially by their distinctive layers, which reflect differences in sediment particle size and mineral composition. Many sedimentary rocks are distinguished also by unique suites of fossils. Sedimentary rocks are of particular interest to geologists because a record of Earth history dating back almost four billion years is locked up and preserved within these strata. It is the study of this reservoir of Earth history that constitutes the sciences of sedimentology and stratigraphy. Many sedimentary rocks also have economic importance as sources of fossil fuels, valuable minerals, and building materials.

Sedimentology is defined as the scientific study of sedimentary rocks and the processes by which they formed; that is, the classification, origin, and interpretation of sediments (Bates and Jackson, 1980). Sedimentology is often incorrectly referred to as sedimentation. The term **sedimentation** is correctly used to describe the processes of sediment accumulation. Two related but more specialized fields closely akin to sedimentology are sedimentary petrography and sedimentary petrology. **Sedimentary petrography** is the description and classification of sedimentary rocks, especially by means of microscopic study. **Sedimentary petrology** is the study of the composition, characteristics, and origin of sediments and sedimentary rocks.

It is difficult to draw a sharp distinction between sedimentology and **stratigraphy,** which is defined simply as the science of rock strata. In the broadest sense of this definition, stratigraphy covers all aspects of the study of sedimentary rocks. In fact, this definition may be interpreted to apply to all stratified rocks, including layered igneous and metamorphic rocks. Thus, stratigraphers who deal with sedimentary rocks cover much the same subject field as sedimentologists; however, stratigraphers are more concerned with age relationships of strata, successions of beds, local and worldwide correlation of strata, and stratigraphic order and chronological arrangement of beds in the geologic column. Two specialized fields of stratigraphy that have developed since about the 1960s are **seismic stratigraphy,** the study of stratigraphic relationships and depositional facies as interpreted from seismic data, and **magnetostratigraphy,** which is based on study of the natural remanent magnetic fields in sedimentary rocks.

The first chapter of this book begins with a brief history of sedimentology and stratigraphy to give readers some insight into their development as sciences. In this chapter, I also describe some of the modern tools and techniques used in studying sedimentary rocks and discuss in general terms the applications of sedimentologic and stratigraphic study to interpretation of Earth history and to exploitation of economic resources. The remaining chapters are devoted to detailed descriptions and discussions of the processes that form sedimentary rocks; the physical, chemical, and biological properties of the rocks that result from these processes; and interpretation of these properties and stratigraphic relationships in terms of Earth history.

1

Development and Application of Sedimentology and Stratigraphy

1.1 INTRODUCTION

The study of sedimentary rocks in some form can be traced back at least to the sixteenth century. The gradual evolution of sedimentology and stratigraphy as sciences constitutes one of the most fascinating chapters in the overall history of the earth sciences. The development of sedimentologic and stratigraphic study has been described by several workers, including Dunbar and Rodgers (1957), Weller (1960), Krumbein and Sloss (1963), Pettijohn (1975), and Friedman and Sanders (1978). From these and other accounts we can place the beginning of sedimentology and stratigraphy at about A.D. 1500 with the observations of Leonardo da Vinci on the fossils in sedimentary rocks of the Italian Apennines. Da Vinci deduced that the fossils were the remains of ancient organisms and concluded that the shells visible in the rocks belonged to animals that lived in a sea that once covered the area. Very little additional study of sedimentary rocks appears to have taken place until about the middle of the seventeenth century, when Nicolas Steno began investigating the fossil-bearing strata around Rome. On the basis of this study, Steno made the first known attempt to place strata in some kind of positional order. In 1669 he postulated that in any sequence of flat-lying strata the oldest layers are at the bottom and the youngest at the top, a concept referred to as the principle of **superposition.** He also proposed the principle of **original horizontality,** which states that beds are always deposited initially in a nearly horizontal position, even though they may later be found dipping steeply. These principles are still considered fundamental to stratigraphy. About the same time that Steno made his studies in Italy, Robert Hooke in England initiated use of the microscope to study fossils. Hooke also apparently suggested the possibility of using fossils to make chronologic comparisons of sedimentary rocks, although such comparisons were not actually attempted until much later.

3

From this rather modest beginning in the sixteenth and seventeenth centuries, study and understanding of sedimentary strata have continued to grow, although progress at times has been slow and somewhat erratic. A very generalized discussion of some of the important stages in the gradual evolution of sedimentology and stratigraphy into modern sciences is presented in the following section.

1.2 DEVELOPMENT OF SEDIMENTOLOGY AND STRATIGRAPHY AS SCIENCES

Organization of Sedimentary Rocks into Stratigraphic Successions

As interest in sedimentary rocks gradually increased into the eighteenth century, it became apparent to serious workers that systematic study of rock strata required organization of the strata into some kind of stratigraphic sequence. The most important examples of early attempts at stratigraphic organization are those made by Giovanni Arduino, an Italian professor and provincial director of mines, and Johann Gottlob Lehman, a German professor of mineralogy. Arduino (1714–1795) divided all rocks into four groups: (1) **primary** mountains composed of rocks containing metallic ores but devoid of fossils; (2) **secondary** mountains consisting of stratified and well-lithified rocks containing fossils but without ore deposits; (3) **tertiary** low mountains consisting of fossiliferous but unconsolidated gravels, sand, and clays with associated volcanic rocks; and (4) **alluvium** consisting of earth and rocky materials washed down from mountains and overlying the other kinds of rocks. Here we see introduced the concept of stratigraphic ordering by relative age, as implied by the terms primary, secondary, and tertiary. Lehmann (1719–1767) recognized three classes of mountains very similar to those of Arduino: (1) primitive mountains composed of crystalline rocks devoid of fossils and unstratified or poorly bedded; (2) layered mountains, or secondary mountains, consisting of well-bedded strata with fossils and containing material eroded from older rocks; and (3) mountains composed of loosely consolidated surficial sands and gravels called alluvium.

Arduino and Lehmann did not know the actual ages of the strata in their groups and may have grouped together rocks of widely different ages. Nonetheless, their efforts at organization were important steps in developing the concept of relative age as a basis for ordering stratigraphic successions. The Tertiary even survived as a name to become part of modern stratigraphic terminology. The term Quarternary, which is used today for one of the geologic systems of Cenozoic rocks, is based on Arduino's fourth category of rocks. Quaternary was introduced into the geologic literature by Nesnoyers in France in 1829 as a specific term to parallel primary, secondary, and tertiary.

The Geologic Cycle and Uniformitarianism

The emergence of geology as a modern science began in the late eighteenth century with the work of James Hutton (1727–1797). Hutton, a Scottish physician and gentleman farmer, was the first worker to recognize and describe the cyclic behavior of earth processes and materials. He visualized tectonic uplift, erosion, sediment transport, and deposition as parts of a continuous cycle, repeated throughout geologic time. He wrote in 1788 that "the result, therefore, of our present enquiry is that we find no vestige of a beginning, no prospect of an end." This concept was labeled by later workers as the **geologic cycle.** On the basis of his observations of the cyclic behavior of geologic processes and his penetrating ideas concerning the significance of the rock record, Hutton has been credited by many subsequent workers with conceiving the **principle of uniformitarianism.** This principle, also sometimes called **actualism,** is commonly expressed to mean that the processes that shaped Earth throughout geologic time were the same as those observable today. It has often been stated simply as "The present is the key to the past."

Uniformitarianism became one of the guiding principles of geologic philosophy and has exerted great influence on geologic thinking since Hutton's time. Unfortunately, controversy exists regarding the exact meaning and usefulness of the term. Shea (1982) challenges the commonly held concept that the present is the key to the past and sets forth what he calls twelve fallacies of uniformitarianism. He argues that the geologic literature is riddled with false and misleading statements about uniformitarianism and that uniformitarianism consists only of the **scientific approach** to the study of nature. That is, as scientists we must follow the rule of simplicity—the rule of choosing first the simplest hypothesis that fits the relevant observations and that also leads to least complexity in overall theory. Shea suggests also that Hutton was not the first to propose uniformitarianism in the sense of proposing scientifically reasonable interpretations of geologic phenomena. Whether or not we eventually abandon the term uniformitarianism, Hutton's place in the history of geology is secure. He was the first scientist to present the ideas embodied in the concept of uniformitarianism in such a way that, to quote Bushman (1983, p. 313), they "illuminated the rock record as it had never been illuminated before, and geologists were able to see and interpret things so effectively that geology underwent its greatest development and became established as an important field of science."

Birth of Biostratigraphy and Stratigraphic Correlation

William Smith (1769–1839), an English surveyor and engineer, is given credit for initiating the science of biostratigraphy. In his work as a canal builder, Smith discovered that different layers of strata are characterized by unique assemblages of fossils. He displayed remarkable insight for his time by initiating the use of fossils for correlation of sedimentary strata from one area to another. He demonstrated the practical importance of the principle of superposition. Through his study of the relationship of fossils and rock strata, he also laid the foundation for development of the **law of faunal succession**—the formal statement of the principle that fossil organisms succeed each other in the stratigraphic record in an orderly, recognizable fashion. In recognition of his pioneer work in biostratigraphy, Smith is sometimes referred to as the "father of stratigraphy."

Smith's work was followed in 1842 by introduction of the concept of the biologic **stage.** A French paleontologist named Alcide d'Orbigny conceived stages as major subdivisions of strata, each stage systematically following the other and each containing a characteristic assemblage of fossils; he believed that his stages had worldwide extent and could be recognized everywhere. Following closely on the heels of the stage concept, a German geologist named Albert Oppel introduced in 1856 the concept of the biologic **zone,** or biozone. Oppel visualized biozones as small-scale stratigraphic units that include all the strata deposited during the existence of specific fossil organisms. He based his zones on the overlapping stratigraphic ranges of these organisms. Stratigraphic range is the stratigraphic interval between the first and last appearance of a fossil species in the stratigraphic record. Using the overlapping ranges of species, Oppel found that he could subdivide stages and delineate the boundaries between small-scale rock units on the basis of fossil content, irrespective of the lithology of the fossil-bearing beds. Furthermore, these zones could be correlated over wide distances.

The pioneer work of Smith, d'Orbigny, and Oppel laid the foundation for developing a standard worldwide stratigraphic column. The stratigraphic succession gradually unfolded with study of well-exposed sections of strata in different areas of Europe. By the early part of the twentieth century, a composite **standard stratigraphic column** covering the entire rock record had been constructed and its subdivisions, called systems, named and defined. Relative ages were initially assigned to the systems on the basis of their fossil content. With the development of radiochronologic methods for estimating absolute ages of rocks, absolute ages have gradually been assigned to the boundaries of the systems.

Development of Petrographic Microscopy

An English scientist named Henry Clifton Sorby initiated microscopic study of rocks around 1850 through his work on limestones. Sorby's work paved the way for development of the science of petrography. **Petrography** is that branch of geology dealing with the description and systematic classification of rocks of all types, especially by means of microscopic examination of thin slices of rock called **thin sections.** Students of sedimentary rocks failed to follow up on the momentum begun by Sorby and largely allowed geologists who study igneous rocks to develop the science of petrography; however, interest in sedimentary petrography resurged in the early part of the twentieth century. Today, thin-section petrography remains a standard and essential tool for study of sedimentary rocks.

The Geologic Revolution—Seafloor Spreading and Global Plate Tectonics

For a century following Sorby's introduction of thin-section microscopic techniques, progress in sedimentology and stratigraphy was relatively slow. New stratigraphic, paleontologic, and sedimentologic data were collected and analyzed during this period, and knowledge of sedimentary rock characteristics steadily grew. One searches in vain, however, for evidence of significant new discoveries or important new concepts that brought advances in understanding of the magnitude of those made by Hutton and Smith. This period is distinguished particularly by the appearance in print of several classic stratigraphic and sedimentological syntheses that brought together the cumulative knowledge of the time. Particularly noteworthy are Lyell's *Principles of Geology* (1833), Grabau's *Principles of Stratigraphy* (1913), Twenhofel's (with others) *Treatise on Sedimentation* (1926) and *Principles of Sedimentation* (1939). Cayeux's *Introduction à l'étude Petrographique des Roches Sédimentaires*, published in 1931 (in French), and *Les Roches Sédimentaires de Frances: Roches Carbonatées*, published in 1935, are also classic examples of outstanding syntheses produced during this period.

About the middle of the twentieth century, stirrings of fresh creativity and imagination became noticeable in published sedimentologic and stratigraphic studies. Furthermore, an undercurrent of excitement began to be felt in the geologic community as a whole. The late 1950s and 1960s ushered in a new era in geologic study—an era marked by significant increase in research activity and data gathering by all types of earth scientists. This period of renewed and intensified research soon led to discoveries that brought about an almost quantum jump in understanding of Earth history and rapid advancement in ideas and concepts in every branch of the earth sciences, including sedimentology and stratigraphy. This renaissance in geology came about for many reasons, but four seem particularly important: (1) greater availability of funds to support research and development of new tools and techniques for field and laboratory study; (2) expanded geophysical exploration, particularly in the ocean basins; (3) deep coring of the ocean basins under the Deep Sea Drilling Project, initiated by a group of United States research institutions; and (4) an accelerated pace in exploration, drilling, and research by energy companies.

Improved Funding for Research. The greater availability of research funds in the United States beginning about 1960 was made possible by grants from the National Science Foundation and other federal agencies and private foundations. These funds provided new research opportunities for scientists in all fields of the earth sciences. They also made possible the development of new or improved research tools that facilitated the process of data gathering and opened up new avenues of investigation. Much credit for the rapid advances made in geology since the 1960s must go to development of these new tools and techniques. The result of this research impetus in the fields of sedimentology and stratigraphy has been an enormous increase in new data pertaining to the physical, chemical, and biological properties of sedimentary rocks and the distribution, ages, and stratigraphic relations of these rocks.

Geophysical Exploration and Deep-Sea Drilling. Exploration of the ocean basins intensified in the 1950s with use of geophysical techniques involving magnetic, seismic, and gravity surveying. These geophysical surveys and other research efforts brought rapid advances in knowledge and understanding of tectonic relationships and depositional settings in the oceans. They generated a massive body of data that led to significant new ideas about the tectonic evolution of continents and ocean basins. These fresh insights paved the way in the late 1950s and early 1960s for the birth of one of the most far-reaching concepts in geologic philosophy—**seafloor spreading** and **global plate tectonics.** This concept envisions Earth's crust as a rigid layer broken into several distinct segments, or plates. These crustal plates move slowly about with respect to one another, sliding over a deeper, plastic layer beneath the crust. They spread apart along mid-ocean ridges, where new crustal rocks are generated by volcanism. Plates move together or converge in major deep-sea trenches, where one plate may thrust beneath another to form a **subduction zone** (Fig. 2.1). The implications of seafloor spreading and plate tectonics have revolutionized all branches of the earth sciences and dramatically changed many previously held ideas about the thickness and age of oceanic rocks, the tectonic setting in which sediments accumulate, and the processes by which sedimentary rocks become tectonically emplaced.

One of the most significant research spinoffs spawned by the plate tectonics revolution was initiation of a drilling program to recover cores of sedimentary and volcanic rock from the deep ocean floor. This program, commonly referred to as the Deep Sea Drilling Project, was initiated in 1963 as the Joint Oceanographic Institutes Deep Earth Sampling Program (J.O.I.D.E.S.), the work of a group of United States planning institutions. A research drilling vessel called the *Glomar Challenger* (Fig. 1.1) was commissioned by the National Science Foundation and was launched in 1968. The *Glomar Challenger* was capable of taking incremental cores up to 1000 m in cumula-

FIGURE 1.1   The drilling ship
Glomar Challenger, previously
used for coring and drilling in the
Deep Sea Drilling Project. The *Glomar Challenger* has now been re-
tired from service. (Photo courtesy
of National Science Foundation.)

tive length from a single hole while floating above the ocean floor in water as much as
6000 m deep. Prior to development of this capability to do rotary drilling and coring
in deep water, the longest cores that could be taken in deep water by piston coring
were about 18 m.

The Deep Sea Drilling Project was financed mainly by the National Science
Foundation until 1975, but subsequently received support from the United Kingdom,
France, West Germany, Japan, and Russia. The *Glomar Challenger* drilled almost 600
holes and recovered cores having a cumulative length exceeding 35 mi (56 km). It
steamed more than 300,000 mi (483,000 km), a distance greater than the distance from
Earth to the moon. The cores taken by the *Glomar Challenger*, together with other data
such as seismic records, have furnished firsthand information about the ages, thicknesses, and character of oceanic rocks. These data provide spectacular confirmation of
seafloor spreading (Warme et al., 1981). The *Glomar Challenger* has now completed
its mission, but the wealth of research data that it generated is still being evaluated.

Research by Energy Companies. Spurred by the need to increase the world's fossil
fuel resources, energy companies have been extremely active since World War II in
exploration and research. Through exploration and development activities that involve
subsurface investigation by drilling and coring, seismic prospecting, and magnetic and
gravity surveying, these companies have accumulated vast quantities of data on the
properties and stratigraphic relationships of sedimentary rocks. Most major petroleum
companies also have research laboratories, where scientists carry out a wide range of
basic and applied research studies. Energy company geologists have made particularly
important contributions to knowledge of subsurface stratigraphy through study of seismic reflection characteristics. More than any other group of scientists, petroleum geologists and geophysicists have been responsible for developing the field of seismic
stratigraphy to a full-blown science since the mid-1960s. These geologists have added
many new data about depositional environments and the physical and biological properties of sedimentary rocks. They have likewise made significant contributions to understanding of the organic geochemistry of sedimentary rocks. Although data obtained
by energy companies are commonly treated as confidential and proprietary, much of
this information is eventually released through talks and publications by the companies' research scientists.

1.3 NEW TOOLS AND TECHNIQUES

As mentioned, much of the credit for rapid post-1950s advances in geologic knowledge must go to improvements in instrumentation and in techniques for geophysical, geochemical, and geologic study of rocks. Many of these techniques have direct application to sedimentology and stratigraphy. In addition to the deep-sea coring techniques described, significant advances have been made in marine magnetic and seismic surveying, underwater photography and sonar imagery, high-magnification electron microscopy, and development of rapid techniques for chemical and grain-size analyses of sediments. A few of these methods are described briefly in the following sections.

Magnetic and Seismic Surveying

Volcanic rocks become magnetized in the Earth's magnetic field at the time of their extrusion and cooling. They can retain this magnetism—called remanent magnetism—for hundreds of millions of years. The magnetic characteristics of seafloor volcanic rocks can be detected and mapped by remote-sensing techniques that use magnetometers towed on the water surface behind an oceanographic vessel. Reversal patterns of remanent magnetism mapped in this way reflect reversals in Earth's magnetic field in the geologic past. The discovery of reversal patterns of magnetic striping on the sea floor (Fig. 1.2) was one of the keys that helped unlock the secrets of seafloor spreading. Magnetic reversals can be detected also in cores of sediment and sedimentary rock recovered from ocean basins or on land. The use of magnetic polarity reversals in sediments to subdivide and correlate stratigraphic units is called magnetostratigraphy. Magnetostratigraphy has found numerous applications in geology, as described in Chapter 15.

FIGURE 1.2 Pattern of magnetic anomalies on the Reykjanes Ridge south of Iceland. The black areas represent seafloor volcanic rocks with normal remanent magnetism; the white areas represent rocks with reversed remanent magnetism. (From Vine, F. J., 1966, Spreading of the ocean floor: New evidence: Science, v. 154. Fig. 2, p. 1407, reprinted by permission of American Association for the Advancement of Science, Washington, D.C.)

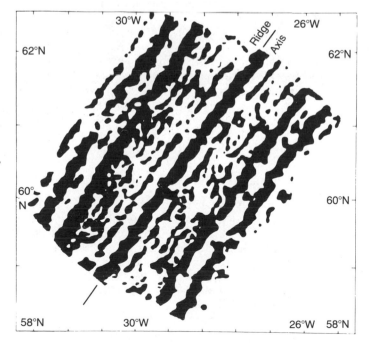

Seismic surveying is a technique for mapping the structure of subsurface formations by detection of sound signals reflected from buried strata. Improvements in techniques for both on-land and marine seismic surveying have made possible the generation of new data that have greatly aided understanding of subsurface stratigraphic relationships. Changes in on-land seismic techniques have included development of methods other than dynamite explosions for generating sound signals; improvements in signal detectors and detector layout patterns; and new methods of filtering, enhancing, and displaying seismic data. In marine surveying, the development of multichannel systems for generating sound signals and improved floating streamer cables for deploying signal detectors, as well as introduction of sophisticated satellite navigation systems for accurate positioning at sea, have brought great improvements in the quality and usefulness of marine seismic records. Hundreds of thousands of kilometers of high-quality seismic profiles have been generated since the 1950s, giving a new dimension to our knowledge of the stratigraphy, structure, and tectonic evolution of the ocean basins. Seismic methods also provide an essential tool for oil and gas exploration on land and in offshore areas. Improvements in the quality and detail of seismic records played an important part in the development of the science of seismic stratigraphy (Chapter 16). Figure 1.3 shows an example of a modern seismic record.

Underwater Photography and Sonar Imagery

Detailed, closeup photography of sediment textures and small seabed structures such as ripple marks and organic markings (Fig. 1.4) is now possible in water of virtually any depth by use of automatic exposure cameras of various types. Side-scan sonar (radar) techniques allow rapid surveying of larger areas of the seafloor and detection of larger bedforms such as sand waves (Fig. 1.5). These tools have added greatly to understanding sediment distribution patterns on the seafloor and the nature of bedload sediment-transport processes in the ocean.

Electron Microscopy

Development of the scanning electron microscope and the transmission electron microscope, which permit magnification up to 70,000 times or more, greatly enhanced study of very fine-size particles that cannot be examined effectively with a standard petrographic microscope. The great depth of field of these instruments and their ability to examine opaque objects have rendered the electron microscope particularly effective in studies of clay minerals, microfossils (Fig. 1.6), microrelief features on quartz grains, and textural patterns of cements and matrix in sedimentary rocks.

Chemical Analyses of Rocks

Several tools that do not require wet-chemical methods are now available for chemical analysis of rocks. These tools permit rapid and accurate analysis of sedimentary rocks and other rocks at very moderate cost. The **electron microprobe,** for example, allows chemical analysis of small areas (on the order of one micron in size) of a rock. Rapid analysis of bulk rock samples for major elements and some trace elements can be done by **x-ray fluorescence** methods and **atomic absorption spectrophotometry.** A comparatively new technique called **inductively coupled argon plasma emission spectrophotometry** can apparently analyze for a greater number of trace elements and has better detection limits than x-ray fluorescence and atomic absorption methods. **Instrumental**

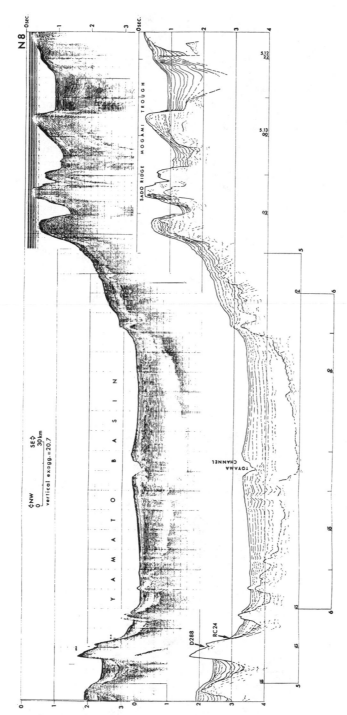

FIGURE 1.3 Example of a modern seismic record (top) and the geologists' interpretation of the record (bottom). This is a marine seismic profile run across the Toyama Deep Sea Channel in the Sea of Japan. The distinctly layered part of the record indicates the presence of well-bedded sedimentary rocks. The indistinct basal part of the record probably represents volcanic basement rocks. (After Tamaki, K., F. Murakami, K. Nishimura, and E. Honza, 1979, Continuous seismic reflection profiling, in E. Honza (ed.), Geological investigation of the Japan Sea, April–June 1978 (GH78-2 Cruise): Geol. Survey Japan Cruise Rept 13. Plate between p. 50 and p. 51, reprinted by permission of Geological Survey of Japan.)

FIGURE 1.4 Bottom photograph showing surface trails of a large bathyal sand dollar. Photograph taken at a depth of 684 m in Suruga Bay off central Japan. (From Okada, H., S. Ohta, and N. Niitsuma, 1980, Lebensspuren photographed on the deep-sea floor of Suruga Bay, central Japan: Inst. Geosciences Bull., v. 5. Fig. 1, pl. 22, reprinted by permission of Institute of Geosciences, Shizuoka, Japan. Photo courtesy of H. Okada and S. Ohta.)

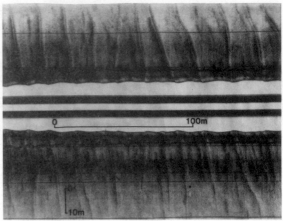

FIGURE 1.5 Side-scan sonar sonograph of straight- to sinuous-crested sandwaves at 23 m depth in the northeast Chukchi Sea. The sandwaves range in height from 1.0 to 1.3 m and have a wave length of approximately 38 m. Photograph courtesy of R.L. Phillips, U.S. Geological Survey.

FIGURE 1.6 Example of a microfossil photographed under the scanning electron microscope. This specimen is a calcareous planktonic foraminifer, *Globigerina bulloides*, ×100. (Specimen furnished by R.A. Linder, photograph by M.B. Shaffer.)

neutron activation analysis is a technique particularly well suited to analysis of trace elements. The **mass spectrometer** is used to measure abundances of carbon, oxygen, and other isotopes. These tools and techniques are now making chemical analysis of sedimentary rocks almost routine. The greater availability and lower cost of chemical analyses have increased interest among sedimentologists in sedimentary geochemistry.

Cathode Luminescence

Petrographic microscopy has now been supplemented by a technique called cathode luminescence. This technique is based on activating various parts of a thin rock slab with an electron beam that excites certain ions and produces luminescence. Inhomogeneities of particles and cements are revealed by differences in luminescence that reflect differences in concentrations of various excitor and inhibitor ions. Thus, many small-scale textures show up more clearly than they do in polarized light. Cathode luminescence has proven particularly useful for studying cements in sandstones and carbonate rocks.

Rapid Grain-Size Analysis

Determining the grain size of sediment by traditional sieving and pipette methods is a laborious, time-consuming process. Automatic-recording settling tubes and balances for analysis of sediment grain size by sedimentation techniques are now readily available (Fig. 1.7). These instruments make possible rapid analysis of large numbers of samples and have a precision roughly equal to that of sieving and pipette methods. This capability for rapid grain-size analysis is particularly important in oceanographic studies, where many thousands of surface samples from the seafloor may have to be analyzed.

1.4 SEDIMENTARY ROCKS AND EARTH HISTORY

The ultimate objective of all geologic study is to further understanding of the origin and evolution of Earth through time. All earth materials, whether igneous, metamorphic, or sedimentary, hold clues to Earth history. Sedimentary rocks and the fossils they contain are particularly important because they provide significant information about Earth's past geography, climates, depositional environments, life forms, and ocean composition. The fossils in sedimentary rocks also provide a means for determining the relative ages of sedimentary rocks, making possible the organization of these layered rocks into a meaningful stratigraphic succession. The following paragraphs explore a few of the ways that study of sedimentary rocks has contributed and continues to contribute to knowledge of Earth history.

Paleogeography and Paleoclimatology

Paleogeography is the study and description of the physical geography of the geologic past. It entails historical reconstruction of the patterns of Earth's surface, or of a given area, at a particular time in the geologic past and comparison with reconstructions for other times in order to determine the successive changes in geographic patterns through time. It is the science that tells us literally how the face of Earth has changed with time. Paleogeography thus involves, among other things, interpretation of the

FIGURE 1.7 Large automated settling tube for rapid size analysis of coarse-grained sediment; located in the Department of Geology, Portland State University, Oregon. An electronic signal from the settling tube, which is approximately 2.5 m high and 20 cm wide, is fed through the microprocessor (on table to left) to a small desk computer that calculates grain-size statistics and plots graphical representations of grain-size distributions.

changing relationships of continents and oceans. On a global scale, great strides have now been made in interpreting the changing relative positions of ancient continental masses by applying the principles of seafloor spreading and plate tectonics. On a smaller regional scale, geologists can study the characteristics of ancient sedimentary rocks and their stratigraphic relationships to reconstruct ancient sedimentary environments and ecological conditions. This knowledge allows them to fix the approximate positions of shorelines at various times in the geologic past and to map advances and retreats of the ocean throughout geologic time. Interpretation of ancient depositional environments involves study of the textures, structures, fossils, and other properties of sedimentary rocks as a basis for deducing ancient environmental conditions and processes (Chapter 10). **Paleoecology** is the science of the relationships between ancient organisms and their environment and is an integral part of environmental analysis.

Paleogeography also involves interpretation of the relative positions of major ocean basins and uplifted sediment source areas. The former presence of vanished continental highlands can be deduced from the nature of the sediments shed from these mountains and deposited in adjacent ocean basins as the mountains were lowered by erosion. For example, a chain of ancestral "Rocky Mountains" is believed to have arisen across the Colorado region of western North America in Pennsylvanian time. This ancient mountain chain is no longer present, but its former existence is postulated on the basis of minerals, rock fragments, and sedimentary structures preserved in Pennsylvanian-age sediments deposited in sedimentary basins adjacent to the mountains. Composition of the existing rocks allows geologists to interpret the probable composition of the ancestral mountain chain. In addition, regional grain-size distribution patterns and paleocurrent indicators in the basins allow geologists to reconstruct the approximate location of the mountains, because these features show the flow direction of currents that carried sediment away from the mountains. Paleocurrent indicators are sedimentary structures such as cross-bedding that indicate the direction of flow of depositing currents.

Paleoclimatology is the study of ancient climates. Paleoclimate analysis is based on identification of paleoclimatic indicators in sedimentary rocks. These indicators include such features as poorly sorted tillites, which suggest glacial climates; distinctive fossils such as palm leaves and corals that indicate warm climates; and distinctive lithologies that point toward deposition under special climatic conditions. For example, windblown sands and evaporite deposits such as gypsum suggest deposition in arid to semiarid desert climates. Extensive coal beds, on the other hand, suggest moist climatic conditions under which lush swamp vegetation flourished. These very simple examples are intended only to emphasize the point that paleoclimate analysis depends almost entirely on analysis of sedimentary rocks and the minerals, textures, and fossils they contain.

Successions of Ancient Life Forms

As noted, William Smith's pioneer work on fossils initiated the science of biostratigraphy and laid the foundation upon which later scientists built to establish the law of faunal succession. Fossil organisms succeed each other in the rock record in a definite and recognizable order, each geologic formation having a total aspect of life different from that in the formations above and below it. The orderly succession of fossils found in sedimentary strata thus provides a means of correlating strata and organizing them by relative age. This succession also makes it possible for a geologist or paleontologist to determine the relative ages of fossiliferous strata anywhere by reference to the standard stratigraphic column. Furthermore, it furnishes strong corroborating evidence for the principle of **organic evolution**—the theory that life forms on Earth have developed gradually from one or a few simple organisms to more complex organisms.

Composition of the Ancient Ocean and Atmosphere

Even clues to evolution of the Earth's atmosphere and ocean are locked up in certain types of sedimentary rocks. For example, differences in the relative degree of oxidation of iron-bearing minerals in older and younger Precambrian rocks may indicate changes in the relative levels of oxygen in the early atmosphere. Variations in relative abundance of sulfur isotopes, particularly $^{34}S/^{32}S$ ratios, in ancient evaporite and shale deposits provide evidence of the sulfur isotope composition of the ancient ocean and thus the sulfate and sulfide composition of the ocean. The presence of thick, widespread deposits of salt, gypsum, or other evaporites in stratigraphic sequences of various ages not only furnishes information about past climates, but also suggests temporary changes in salinity of the ocean, at least locally. Changes in salinity are inferred because deposition of large quantities of evaporites may locally deplete the ocean in salts, causing a temporary episode of lowered salinity. These examples are highly generalized, but they illustrate how we can use the chemical composition and mineralogy of sedimentary rocks to illuminate Earth history.

1.5 PRACTICAL APPLICATIONS

This book is aimed primarily at developing understanding of the basic principles of sedimentology and stratigraphy. Nonetheless, many of the principles presented here have practical applications also. A good example of the practical side of sedimentology and stratigraphy is seen in the fossil fuel industries. Almost all of the world's oil and gas and all of its coal occur in sedimentary rocks. Successful exploration for oil and gas requires the services of trained geologists who can evaluate the sedimentological characteristics and stratigraphic relationships of subsurface formations and identify favorable reservoir and trapping conditions for petroleum. Knowledge of rock properties such as porosity and permeability, organic geochemistry, age, and stratigraphic relationships thus plays a significant role in the search for new fossil fuels. The development of the field of seismic stratigraphy as a tool for petroleum exploration is an outstanding example of the practical application of stratigraphic principles. The value of applied sedimentology and stratigraphy to industry is clearly indicated by the fact that energy companies have long been the major employers of geologists, particularly sedimentologists, stratigraphers, and paleontologists.

Sedimentology and stratigraphy have applications also in the mineral industries. Certain types of ores, including uranium, vanadium, manganese, iron, lead, zinc, and copper, may become localized in sedimentary deposits of particular environments, such as fluvial or reef environments. The geologist's knowledge and ability to interpret ancient sedimentary environments on the basis of physical and biological characteristics of sedimentary units thus become of paramount importance in this type of exploration. Exploration for commercial deposits of phosphate rock, salt, gypsum, and other nonmetallic mineral deposits is equally dependent upon knowledge of environments and stratigraphy. Other examples of practical applications of sedimentological and stratigraphic principles include exploration for groundwater resources (which occur primarily but not exclusively in sedimentary rocks) and application to engineering problems. Engineering problems involving sediment transport in estuaries and other nearshore regions, shore-zone erosion, silting up of reservoirs, stream-channel control, highway and dam construction, and foundation evaluation are but a few examples. Somewhat more esoteric applications of sedimentological principles include study of the influence of sedimentation on concrete aggregates; study of clay minerals in the field of ceramics; evaluation of the quality of foundry sands; and use of sedimentary rocks as storage sites for water, natural gas, and radioactive and chemical wastes.

ADDITIONAL READINGS

Ager, D. V., 1981, The nature of the stratigraphical record, 2nd ed.: John Wiley & Sons, New York, 122 p.

Brenchley, P. J., and B. P. J. Williams (eds.), 1985, Sedimentology: Recent developments and applied aspects: Geol. Soc. Spec. Pub. 12, Blackwell Scientific Publications, Oxford, 320 p.

Conklin, B. A., and J. E. Conklin (eds.), Stratigraphy: Foundations and concepts: Benchmark Papers in Geology, v. 82, Van Nostrand Reinhold, New York, 365 p.

Dunbar, C. O., and J. Rodgers, 1957, Principles of stratigraphy: John Wiley & Sons, New York, 356 p.

Friedman, G. M., and J. E. Sanders, 1978, Principles of sedimentology, Chap. 1: John Wiley & Sons, New York, 792 p.

Hallam, A., 1973, A revolution in the earth sciences: From continental drift to plate tectonics: Oxford University Press, London, 127 p.

Holland, H. D., and A. F. Trendall (eds.), 1984, Patterns of change in Earth evolution: Springer-Verlag, New York, 432 p.

Krumbein, W. C., and L. L. Sloss, 1963, Stratigraphy and sedimentation, 2nd ed.: W. H. Freeman, San Francisco, 660 p.

Trask, P. D. (ed.), 1950, Applied sedimentation: John Wiley & Sons, New York, 707 p.

Warme, J. E., R. G. Douglas, and E. L. Winterer (eds.), 1981, The Deep Sea Drilling Project: A decade of progress: Soc. Econ. Paleontologists and Mineralogists Spec. Pub. 32, 564 p.

Weller, J. M., 1960, Stratigraphic principles and practices, Chaps. 1, 2: Harper & Brothers, New York, 725 p.

PART TWO
Principles of
Sedimentology

2

Sedimentary Cycles and Weathering

2.1 INTRODUCTION

Sedimentary rocks are composed of grains derived either directly or indirectly from preexisting rocks. Some of these grains are individual mineral crystals; others are composite grains made up of aggregates of crystals bonded or cemented together (Table 2.1). Individual crystals range in size from a few microns to a few millimeters. Composite grains range from millimeter-scale particles to cobbles and boulders tens to hundreds of centimeters in diameter.

Some grains that form sedimentary rocks originate as resistant **residues** of an older generation of igneous, metamorphic, or sedimentary rocks subsequently broken down by weathering. The residues include both individual grains of minerals such as quartz and feldspar and larger fragments, or clasts, of older rocks. Mineral grains and clasts can also originate as **pyroclastic particles** through explosive volcanism. Pyroclastic particles may or may not undergo further breakdown by weathering. Other grains, such as fine-size iron oxides and clay minerals, originate at the weathering site as **secondary minerals,** formed by chemical recombination and crystallization of elements released from rocks undergoing weathering. These particulate weathering residues, pyroclastic particles, and secondary weathering minerals are collectively called **terrigenous grains;** that is, grains derived from the land. Because most terrigenous grains are composed in part of silica, they are often referred to also as **siliciclastic grains.** Terrigenous grains originate mainly outside depositional basins and are thus considered to be **extrabasinal** constituents, although some pyroclastic particles may be generated within depositional basins. Terrigenous grains are transported as solids to depositional basins, where they form the primary source materials for siliciclastic sedimentary rocks such as sandstones and shales.

TABLE 2.1   Principal constituents of sedimentary rocks

Grain type	Origin	Example
Terrigenous grains		
Weathering residues	Chemical and physical weathering of igneous, metamorphic, and sedimentary rocks	Quartz, feldspars, rock fragments
Pyroclastic particles	Explosive volcanism	Volcanic rock fragments, pumice, glass, feldspars
Secondary minerals	Crystallization at the weathering site	Iron oxides, clay minerals, fine-size quartz
Intrabasinal precipitates		
Single crystals	Chemical and biochemical precipitation	Carbonate minerals, chert, iron-bearing minerals, evaporites, phosphorites
Composite grains	Chemical and biochemical precipitation, mechanical reworking and transport	Oolites, pellets, fossil shells, limeclasts
Organic residues	Partial decomposition of plant and animal organic matter	Organic residues in coals, disseminated fine organic matter in sediments

Not all sedimentary rocks are composed of siliciclastic grains derived from the land. Some sediment grains originate within depositional basins by chemical or biochemical precipitation of minerals from water. The dissolved constituents from which these **intrabasinal** precipitates form are derived from older rocks by subaerial and subaqueous weathering. Thus, the precipitates are the indirect products of rock weathering. They may be either individual fine-size crystals of minerals such as calcite or gypsum or composite grains such as calcium carbonate pellets, oolites, and organic skeletal remains. Intrabasinal precipitates may consist of both silicate minerals and nonsilicate minerals; no general name, analogous to the term siliciclastic used for terrigenous grains, is in general use for such precipitates. Some precipitates are deposited in situ; others may undergo transportation as solids within the depositional basin before final deposition. These chemically or biochemically formed intrabasinal grains are the source materials of sedimentary rocks such as limestones, cherts, and evaporites.

The final type of constituent that may make up sedimentary rocks occurs in coals and other carbonaceous rocks. These rocks are special types of sedimentary rocks composed in part of nonskeletal organic materials. Thus, they are enriched in organic carbon. They contain high concentrations of organic residues derived by partial decomposition of plant and animal remains. The organic matter in coals originates from plants and is deposited in continental environments without significant transport. Organic-rich shales and limestones, called sapropelitic rocks, contain fine-size particulate organic matter derived from plant or animal sources either on land or in the ocean. This fine organic debris commonly undergoes some transport before deposition.

Sedimentologists can often identify the type of source rock from which terrigenous sedimentary particles were derived by studying the particles preserved in ancient sedimentary rocks. Such studies show that sediment source rocks are very diverse. For example, a piece of sandy conglomerate may contain clasts of several different kinds of sedimentary, igneous, or metamorphic rocks as well as many kinds of individual mineral grains. Any type of rock exposed to weathering, including sedimentary rock, can break down to yield sediment and soluble constituents. The fact that one generation of sedimentary rocks can provide source materials for another suggests that the processes involved in the formation of sedimentary rocks are repeated in cyclic fashion, a deduction first made by Hutton in 1788. The processes of weathering, coupled with those of transport and subsequent deposition of sediment and dissolved constituents, account for movement of rock materials from uplifted areas of the Earth's surface to depressed areas on the surface. Sediment is later buried beneath the Earth's surface and lithified through diagenetic processes (Chapter 9) to form sedimentary rock. Deeply buried sedimentary rock may subsequently be uplifted by tectonic forces to create elevated areas standing well above sea level, setting the stage for a new cycle of weathering, transport, and deposition. This cyclic behavior of earth materials is called the **sedimentary cycle.**

2.2 SEDIMENTARY CYCLE

The sedimentary cycle is extremely complex and is an imperfect cycle in the sense that not all sedimentary rocks are recycled to weathering sites on continents. Understanding the complexities of the cycle requires that the formation of sedimentary rocks be considered in the light of modern plate tectonics theory. The concept of seafloor spreading and plate tectonics is now familiar to most geology students; however, I will briefly review those aspects of the theory that relate to the sedimentary cycle. The basic tectonic framework of the continents and oceans as interpreted from plate tectonics concepts is illustrated in Figure 2.1. Mid-ocean ridges are sites where new oceanic crust is generated from magma brought upward from the asthenosphere. This new basaltic crust is welded onto the old ocean crust along the axes of the ridges. The ocean crust is broken up by ridges, trenches, and major fault zones into several rigid plates. These plates move away from mid ocean-ridge spreading centers at velocities of a few millimeters to a few tens of millimeters per year, presumably driven by convection in the mantle. Where the leading edge of an oceanic plate meets a continent or a volcanic island arc, the plate may be thrust beneath the continent or arc to form a **subduction zone.** Subduction zones are the sites of the deep ocean trenches. If an oceanic plate meets another plate head on rather than thrusting beneath it, a **collision zone** develops instead. Plate collisions result in deformation and uplift of crustal rocks and appear to be an important process in mountain building.

Fine-grained pelagic sediment, volcanic ash, and biogenic oozes are deposited on oceanic crustal plates as they are rafted slowly away from mid-ocean ridges. Sediment deposited on plates at or near a ridge in a large ocean basin may require 150–200 million years to move from the ridge to a subducting trench. During transport across the ocean basin on top of the ocean crust, older sediment is progressively covered by a blanket of younger sediment that is deposited more or less continuously throughout the deep ocean basins at rates ranging from about 1 to 20 mm per 1000 years. This process causes total sediment thickness to increase away from the ridges and also accounts for the fact that the oldest sedimentary rocks in the ocean basins are generally located farthest from the mid-ocean ridges. Thus, the sediment or sedimentary rock on top of a spreading ocean plate can, when it reaches a trench, range in age from as much as 150–200 million years at the base to a few tens of years at the top.

The fine-grained oceanic sediment that enters a trench on the subducting oceanic plate may come in contact within the trench with coarser terrigenous sediment derived from the continental arc. Some scientists believe that as the subducting plate descends

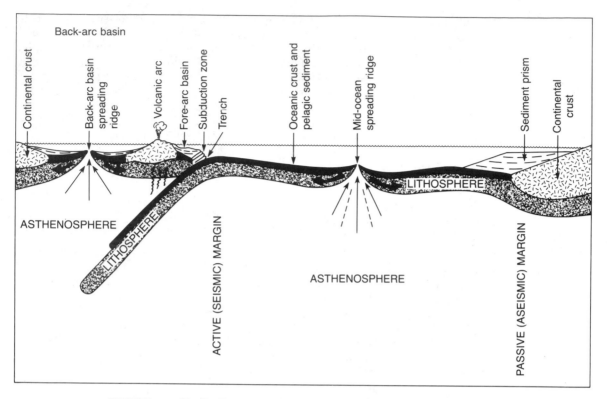

FIGURE 2.1   Idealized tectonic setting for the accumulation and burial of sedimentary rocks in the ocean.

into a trench some of the oceanic sediment is scraped off the plate. This offscraped sediment together with terrigenous sediment in the trench is plastered onto the over-riding plate in the inner trench wall to form an **accretionary wedge** of sediment dipping in the landward direction. Such an accretionary wedge may eventually be uplifted above sea level and exposed at the surface as isostatic readjustment or tectonic movement takes place. In this way, sediments deposited in deep ocean basins may eventually be exposed on land to begin a new cycle of weathering.

On the other hand, some or all of the oceanic sediment may remain attached to the ocean plate as it descends and be carried to great depths in the subduction zone. No sedimentary rock older than about 200 million years has been recovered from the ocean basins seaward of trenches, presumably because all older sedimentary rock within the ocean basins has already been stuffed down a subduction zone in this manner. As the downgoing slab of crustal rock and sediment heats up at depth, some of the sediment may melt and subsequently rise as magma to form plutonic bodies or to nourish the volcanic arc. The remaining part of the crustal slab with its sediment cover is carried to greater depths and incorporated into the mantle. It may eventually move laterally at these depths, as part of the asthenosphere, to a spreading ridge and there rise as magma to form a new generation of oceanic crustal rocks.

Sediments deposited on continental crust in the fore-arc or back-arc regions of continents or island arcs (Fig. 2.1), or on passive continental margins without trenches, have a different history of burial and uplift. They are not carried down subduction zones, but they can be buried to great depths owing to basin subsidence. Some sedimentary rocks deposited in these settings are tectonically uplifted after burial and lithification before they have been significantly altered by metamorphism. Others, in some fore-arc and back-arc settings, become metamorphosed before uplift, owing possibly to rising heat that is generated along a subducting slab under a continental margin. Sedimentary rocks that become deeply buried in fore-arc regions may become partly melted and thereby furnish some of the magma that forms the plutonic igneous bodies commonly found in the roots of fore-arc fold belts.

Thus, sedimentary rock buried in depositional basins or subducted in trenches may eventually be brought upward toward the Earth's surface in one form or another. Depending upon its location within the tectonic setting (Fig. 2.1), this rock may subsequently be exposed by subaerial or submarine weathering and erosion to act as a source for a new generation of sediment grains. Alternatively, through subduction and melting it may become part of new ocean crust and thus enter into a new cycle of seafloor spreading.

The time required for a complete sedimentary cycle obviously depends upon the rates of erosion on the continents and the subsequent history of burial and uplift. These factors in turn depend upon the loci of deposition within the plate tectonics regime. Stated somewhat differently, the time required for sediment recycling is a function of rates of erosion, rates of sedimentation and subsidence, rates of seafloor spreading (particularly for oceanic sediment), and rates of uplift. All of these rates differ geographically on Earth at the present time and have undoubtedly varied throughout geologic time. An excellent discussion of the various factors that affect the rates of sediment recycling is given by Blatt et al. (1980). These authors conclude that the rate of erosional decline of sedimentary volumes with age is so fast that most sediment is derived from the erosion of sedimentary rock and that sedimentary materials must have been recycled, on the average, from three to ten times since the late Precambrian.

The preceding discussion indicates that tectonic processes play an important part in the sedimentary cycle; however, our immediate concern at this point is with the details of surficial processes and burial changes that collectively influence the formation of sedimentary rocks. We can consider for convenience that the processes of sedimentation begin when uplifted source rocks are subjected to **weathering,** as illustrated in Figure 2.2, to yield particulate residues and dissolved constituents. These residues and dissolved materials undergo **transportation** to sedimentary basins and ultimately **deposition.** Subsequently, the deposits are buried and altered by **diagenesis** to produce lithified sedimentary rock. We will examine each of these major processes in succeeding chapters, beginning in this chapter with weathering processes.

2.3   WEATHERING PROCESSES AND PRODUCTS

Breakdown of rocks by weathering is an essential part of the sedimentary cycle. Weathering involves chemical, physical, and biological processes, although chemical processes are by far the most important. Details of these processes may be found in several books devoted to weathering and soil formation; for example, those by Birkland (1974), Bohn et al. (1979), Carroll (1970), Drever (1985), Loughnan (1969), Ollier (1969), and Keller (1955). Only a brief summary of weathering processes is presented

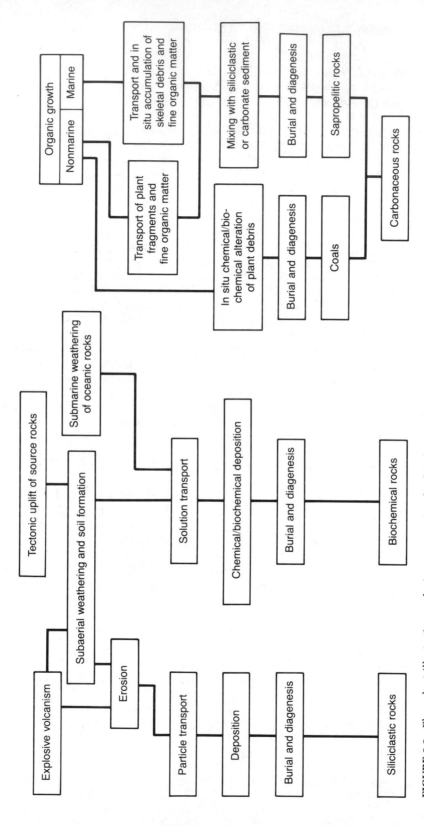

FIGURE 2.2 Flow chart illustrating geologic processes involved in formation of principal types of sedimentary rocks.

26

here to illustrate how weathering acts to decompose and disintegrate exposed rocks, producing particulate residues and dissolved constituents. These weathering products are the source materials of sedimentary rocks.

It is important in studying sedimentary processes to understand how weathering attacks source rocks and what remains after weathering to be transported as sediment and dissolved constituents to depositional basins. The ultimate composition of a terrigenous sedimentary rock bears a strong relationship to the composition of the source rock, but it is clear from the study of residual soil profiles that both the mineral composition and bulk chemical composition of soils may differ greatly from those of the bedrock on which they form. Some minerals in the source rock are destroyed completely during weathering, whereas more chemically resistant or stable minerals are loosened from the fabric of the decomposing and disintegrating rock and accumulate as residues. During this process, new minerals such as ferric oxides and clay minerals may form in situ in the soils from chemical elements released during breakdown of the source rocks. Thus, soils are composed of survival assemblages of minerals and rock fragments derived from the parent rocks plus any new minerals formed at the weathering site. Soil composition is governed not only by the parent-rock composition but also by the nature, intensity, and duration of the weathering process.

In the following section we examine the principal processes of subaerial weathering and discuss the nature of the particulate residues and dissolved constituents that result from weathering. We also look briefly at the processes of submarine weathering. Submarine weathering includes both the interaction of seawater with hot oceanic rocks along mid-ocean ridges—a process that leaches important amounts of chemical constituents from hot crustal rocks—and low-temperature alteration of volcanic rocks and sediments on the ocean floor.

Subaerial Weathering Processes

Physical Weathering

Physical weathering is the process by which rocks are broken into smaller fragments through a variety of causes, but without significant change in chemical or mineralogical composition. Except in extremely cold or very dry climates, physical weathering occurs together with chemical weathering and it is difficult to separate their effects.

Frost wedging, caused by freezing and thawing of water in rock fractures, is the most important physical weathering process in climates where recurring freezing and thawing take place. Water increases in volume by about 9 percent when it changes to ice, creating enough pressure in tortuous rock fractures to crack most types of rock. This process commonly tends to produce large, angular blocks of rock, but may also cause granular disintegration of coarse-grained rocks such as granites.

Alternate expansion and contraction of rock surfaces as a result of diurnal changes in temperature was suggested by early workers as a cause of weakening of bonds along grain boundaries and subsequent flaking off of rock fragments or dislodging of mineral grains. The quantitative importance of this postulated process is still being debated. Griggs (1936) carried out experiments in the laboratory over a temperature range of 110°C to simulate the equivalent of 244 years of heating and cooling; he showed that in the absence of water little disintegration occurred. Ollier (1969) points out, however, that Griggs' laboratory specimens were unconfined and could expand in all directions, whereas a patch of rock on the surface of a boulder is confined by neighboring patches and can expand outward only. Such a confined area of rock is more likely to fracture under repeated stress. Ollier suggested also that small heating and cooling stresses maintained for longer periods of time than the 15-minute heating and cooling cycles in Griggs' experiments might lead to permanent strain. Thus, high temperatures and temperature changes from day to night may cause fracturing of rocks by thermal changes alone (Kerr et al., 1984). High temperatures in desert environments tend to promote weathering caused by crystallization of salts in fractures (Sperling and Cooke, 1980). Growth of salt crystals in saline solutions that have access to rock fractures generates internal pressures that force the cracks apart.

Release of overburden pressure owing to erosion of overlying strata causes the development of rock fractures that are nearly parallel to the topographic surface. These fractures divide the rock into a series of layers, or sheets; hence this process of crack formation is called **sheeting.** These layers increase in thickness with depth and may exist for several tens of meters below the Earth's surface. Sheeting is most conspicuous in homogeneous rocks such as granite, but may occur also in layered rocks.

Other processes that may contribute to physical weathering under certain conditions include volume increases owing to hydration of clay minerals or other minerals, volume changes owing to alteration of minerals such as biotite and plagioclase to clay minerals, alternate wetting and drying of rocks, growth of plant roots in the cracks of rocks, plucking of mineral grains and rock fragments from rock surfaces by lichens as they expand and contract in response to wetting and drying, and burrowing and ingestion of soils and loosened rock materials by worms or other organisms.

The grain size of the particulate rock materials that result from physical weathering is a function of the thoroughness of the weathering process, but ultimately is determined by the grain size and degree of cementation of the parent rock. Coarse-grained parent rocks such as granites tend to yield grains of individual minerals upon disintegration, whereas physical weathering of fine-grained sedimentary, volcanic, or metamorphic rocks is more likely to produce rock fragments as disintegration products (Boggs, 1968; Carroll, 1970).

Chemical Weathering

Weathering Processes. Chemical weathering involves changes that can alter both the chemical and mineralogical composition of rocks. Minerals in the rocks are attacked by water, oxygen, and carbon dioxide of the atmosphere, causing some components of the minerals to dissolve and be removed in solution. Other mineral constituents recombine in situ and crystallize to form new mineral phases. These chemical changes, along with changes caused by physical weathering processes, disrupt the fabric of the weathered rock, producing a loose residue of resistant grains and secondary minerals. Water plays a dominant role in every aspect of chemical weathering. Because some water is present in almost every environment, chemical weathering processes are commonly far more important than physical weathering processes, even in very arid climates. Nevertheless, owing to the low temperatures of the weathering environment, chemical weathering occurs very slowly. The principal processes of chemical weathering are listed and briefly described in Table 2.2 along with selected examples of new minerals formed in situ during the weathering processes.

Hydrolysis is an extremely important chemical reaction between silicate minerals and water, leading to breakdown of the silicate minerals and releasing metal cations and silica. If aluminum is present in the minerals undergoing weathering, clay minerals may form as a byproduct of hydrolysis. Thus, orthoclase feldspar can break

TABLE 2.2 Principal processes of chemical weathering

Name of process	Nature of process	Examples	Principal types of rock materials affected
Hydrolysis	Reaction between H^+ and OH^- ions of water and the ions of silicate minerals, yielding soluble cations, silicic acid and clay minerals (if Al present)	$Mg_2SiO_4 + 4H_2O \rightarrow 2Mg^{+2} + 4OH^- + H_4SiO_4$ (silicic acid) (forsterite) $2KAlSi_3O_8 + 2H^+ + 9H_2O \rightarrow H_4Al_2Si_2O_9 + 4H_4SiO_4 + 2K^+$ (orthoclase) aq (kaolinite) (silicic acid) aq $2NaAlSi_3O_8 - 2H^+ + 9H_2O \rightarrow H_4Al_2Si_2O_9 + 4H_4SiO_4 + 2Na^+$ (albite) aq (kaolinite) (silicic acid)	Silicate minerals
Hydration and dehydration	Gain or loss of water molecules from a mineral, resulting in formation of a new mineral	$CaSO_4 \cdot 2H_2O \rightleftharpoons CaSO_4 + 2H_2O$ (gypsum) (anhydrite) $Fe_2O_3 + H_2O \rightleftharpoons 2FeOOH$ (hematite) (goethite)	Evaporites Ferric oxides
Oxidation	Loss of an electron from an element (commonly Fe or Mn) in a mineral, resulting in the formation of oxides or, if water is present, hydroxides	$4FeSiO_3 + O_2 \rightarrow 2Fe_2O_3 - 4SiO_2$ (pyroxene) (hematite) (quartz) $MnSiO_3 + \frac{1}{2}O_2 + 2H_2O \rightarrow MnO_2 + H_4SiO_4$ (rhodonite) $2FeS_2 + 15/2\, O_2 + 4H_2O \rightarrow Fe_2O_3 + 4SO_4^{2-} + 8H^+$ (pyrite) (hematite)	Iron and manganese-bearing silicate minerals, sulfur
Solution	Dissolution of soluble minerals, commonly in the presence of CO_2, to yield cations and anions in solution	$H_2O + CO_2 + CaCO_3 \rightleftharpoons Ca^{2-} + 2HCO_3^-$ [carbonation] (calcite) (bicarbonate) $CaSO_4 \cdot 2H_2O \rightarrow Ca^{2+} + SO_4^{2-} + 2H_2O$ [direct solution] (gypsum)	Carbonate rocks Evaporites
Ion exchange	Exchange of ions, principally cations, between solutions and minerals	Na-clay + $H^+ \rightarrow$ H-clay − Na^+	Clay minerals
Chelation	Bonding of metal ions to organic molecules having ring structures	Metal ions (cations) + chelating agent [excreted by lichens] $\rightarrow H^+$ ions + chelate [in solution]	Silicate minerals

29

down to yield kaolinite or illite, albite can decompose to kaolinite or montmorillonite, and so forth, as illustrated by the reactions in Table 2.2. Most of the silica set free during hydrolysis goes into solution as silicic acid (H_4SiO_4); however, some of the silica may separate as colloidal or amorphous SiO_2 and be left behind during weathering to combine with aluminum to form clay minerals or crystallize into minute grains of quartz (Krauskopf, 1979). Hydrolysis is the primary process by which silicate minerals decompose during weathering.

Hydration is the process by which water molecules are added to a mineral to form a new mineral. Common examples of hydration are the addition of water to hematite to form goethite or to anhydrite to form gypsum. Hydration is accompanied by volume changes that may lead to physical disruption of rocks. Under some conditions, hydrated minerals may lose their water, a process called **dehydration,** and be converted to the anhydrous forms, with accompanying decrease in mineral volume. Dehydration is relatively uncommon in the weathering environment because some water is generally present in this environment.

Oxidation of iron in silicate minerals such as biotite and pyroxenes, owing to oxygen dissolved in water, is also an important weathering process because of the abundance of iron in the common rock-forming silicate minerals. Loss of electrons from iron during oxidation causes loss of other cations from crystal lattices to maintain electrical neutrality. Cation loss leaves vacancies in the crystal lattice that either bring about the collapse of the lattice or make the mineral more susceptible to attack by other weathering processes (Birkland, 1974). Oxidation of manganese minerals to form oxides and silicic acid or other soluble products is a less important but common weathering process. Another element that oxidizes during weathering is sulfur. For example, pyrite (FeS_2) is oxidized to form hematite (Fe_2O_3), with release of soluble sulfate ions.

Solution of highly soluble minerals such as calcite, dolomite, and gypsum owing to exposure to rainwater, or meteoric water, during weathering can result in decomposition of these minerals. If carbon dioxide is dissolved in the rainwater through interaction with atmospheric or soil CO_2—the usual case in the weathering environment—the solubilizing ability of water is enhanced, particularly for carbonate minerals. Simple solution of this type is an important process only in moderately wet climates where carbonate rocks or evaporites are present near the surface.

Ion exchange is a weathering process that is particularly important in alteration of one type of clay mineral to another. It is the reaction between ions in solution and those held in a mineral; for example, the exchange of sodium for calcium. Most ion exchange takes place between cations, but anion exchange also occurs.

Chelation involves the bonding of metal ions to organic substances to form organic molecules having a ring structure. Chelation, or organic complexing, performs the dual role during weathering of removing cations from mineral lattices and also of keeping the cations in solution until they are removed from the weathering site. Chelated metal ions will remain in solution under pH conditions and at concentration levels at which nonchelated ions would normally be precipitated. A good example of natural chelation is provided by lichens, which, by secreting organic chelating agents, cause an increase in the rate of chemical weathering on the rock surfaces on which they grow. In addition to the role of plant organic matter as a chelating agent, plants also enhance chemical weathering processes by retaining soil moisture and by acidifying waters by release of CO_2 and various types of organic acids during decay.

Rates of Chemical Weathering. Chemical weathering proceeds at different rates depending upon the climate and the mineral composition and grain size of the rocks. Weathering processes are more rapid in humid, hot climates than in cold or very dry climates, but the influence of temperature on weathering rate is difficult to quantify. A general qualitative rule is that chemical reaction rates are approximately doubled with a 10°C increase in temperature. The rate of weathering of silicate rocks of a given grain size is related to the relative stabilities of the common rock-forming silicate minerals. Chemical stability refers to the resistance of minerals to alteration or destruction by chemical processes. The order of relative stability of the most important mafic and felsic minerals is shown in Table 2.3. Readers will recognize this order as the same as that in which minerals crystallize in Bowen's reaction series. Minerals that crystallize at high temperatures (e.g., olivine) have the greatest degree of disequilibrium with surface weathering temperatures and thus tend to be less stable than minerals that crystallize at lower temperatures (e.g., quartz). Furthermore, the high-temperature minerals are bonded with weaker ionic or ionic-covalent bonds, whereas quartz is bonded with strong covalent bonds.

Owing to the preponderance of low-stability minerals in basic igneous rocks, these rocks tend to weather faster than acid igneous rocks of the same grain size. Thus, gabbro weathers faster than granite, and basalt weathers faster than rhyolite. Fine-grained basic igneous rocks such as basalt may, however, weather more slowly than coarse-grained granitic rocks (Birkland, 1974). There is no rule of weathering susceptibility that can be applied generally to sedimentary rocks. Rates of weathering of these rocks are a function of the mineralogy and the amount and type of cement in the rocks. Limestones, for example, weather rapidly by solution in wet climates, whereas quartz-rich sandstones cemented with silica cement weather very slowly under these same conditions.

Finally, it should be mentioned that, depending upon climatic conditions and vegetative cover, rates of weathering have probably varied throughout geologic time. Prior to the development of land plants in early Paleozoic time, absence of plant cover to hold soil moisture and contribute organic acids probably slowed rates of chemical weathering while contributing to increased rates of physical erosion.

TABLE 2.3 Relative stability of the common silicate minerals during weathering

Mafic minerals	Felsic minerals	
Olivine		
	Ca-plagioclase	
Pyroxene		
	Ca-Na plagioclase	
Amphibole	Na-Ca plagioclase	Increasing stability
	Na-plagioclase	
Biotite		
	K-feldspar, muscovite	
	Quartz	

Source: Goldich, S. S., 1938, A study of rock weathering: Jour. Geology, v. 46, pp. 17–58.

Products of Subaerial Weathering

Subaerial weathering generates three types of weathering products: (1) source-rock residues consisting of chemically resistant minerals and rock fragments; (2) secondary
minerals formed in situ by chemical recombination and crystallization, largely as a
result of hydrolysis and oxidation; and (3) soluble constituents released from parent
rocks by hydrolysis and solution (Table 2.4). Until they are removed by erosion, residues and secondary minerals accumulate at the weathering site to form a soil mantle
composed of particles of various compositions and of grain sizes ranging from clay to
gravel. Grain size and composition depend upon the grain size and composition of the
parent rock and upon the nature and intensity of the weathering process. These characteristics of the weathering environment are in turn functions of the climate, topography, and duration of the weathering process.

Source Rock Residues. The residual particles in young soils developed on igneous or
metamorphic rocks may include, in addition to rock fragments, assemblages of immature minerals with low chemical stability; for example, biotite, pyroxenes, hornblende,
and calcic plagioclase. Mature soils, developed after more prolonged or intensive
weathering of these rocks, tend to contain only the most stable minerals: quartz, muscovite, and perhaps potassium feldspars. Because the silicate minerals that make up
siliciclastic sedimentary rocks such as sandstones have already passed through a
weathering cycle before the siliciclastic rocks were deposited, the weathering products
of these rocks tend to be depleted in easily weathered minerals. Thus, even young
soils developed on siliciclastic sedimentary rocks may have assemblages of mature
minerals. Weathering of limestones by solution produces thin soils composed of the
fine-size insoluble silicate and iron oxide residues of these rocks.

Secondary Minerals. Secondary minerals developed at the weathering site are dominantly clay minerals, ferric oxides or hydroxides, and aluminum hydroxides. Clay
minerals formed in immature soils under only moderately intense chemical weathering conditions may be illites or smectites. More prolonged and intense leaching conditions lead to formation of kaolinite. Under extremely intense chemical weathering

TABLE 2.4 Principal kinds of products formed by subaerial weathering processes and the types of sedimentary
rocks ultimately formed from these products

Weathering process	Type of weathering product	Example	Ultimate depositional product
Physical weathering	Particulate residues	Silicate minerals such as quartz and feldspar; all types of rock fragments	Sandstones, conglom-erates, mudrocks
Chemical weathering Hydrolysis	Secondary minerals	Clay minerals; fine quartz	Mudrocks; mud ma-trix
	Soluble constituents	Silicic acid; K^+, Na^+, Mg^{2+}, Ca^{2+}, etc.	Chert, limestones, evaporites, etc.
Oxidation	Secondary minerals	Fine-grained SiO_2 minerals; fer-ric oxides	Mudrocks; mud ma-trix
	Soluble constituents	Silicic acid, SO_4^{2-}, etc.	Chert, evaporites, etc.
Solution	Soluble constituents	Bicarbonate, SO_4^{2-}, Ca^{2+}, Mg^{2+}, etc.	Limestones, evapor-ites, etc.

conditions, aluminum hydroxides such as gibbsite and diaspore are formed. These latter clay minerals are aluminum ores. The common secondary iron minerals include goethite, limonite, and hematite. Comparison of the chemical composition of unweathered silicate rocks with that of the weathering products of these rocks shows a net loss owing to weathering of all major cations except aluminum and iron (Krauskopf, 1979). Although considerable silica is lost during weathering as soluble silicic acid, loss of Mg, Ca, Na, and K is comparatively much greater. Therefore, the relative abundance of silica, aluminum, and ferric iron in the particulate weathering residues of silicate rocks is greater than in the parent source rocks.

Soluble Materials. Soluble materials extracted from parent rocks by chemical weathering processes are removed from the weathering site in surface water or soil groundwater more or less continuously throughout the weathering process. Ultimately these soluble products make their way into rivers and are carried to the ocean. The most abundant inorganic constituents of rivers, representing the principle soluble products of weathering, are, in order of decreasing abundance: HCO_3^- (bicarbonate), Ca, H_4SiO_4 (silicic acid), SO_4^{2-} (sulfate), Cl, Na, Mg, and K (Garrels and McKenzie, 1971). These constituents are the raw materials from which chemically and biochemically deposited rocks such as limestones and cherts are formed in the oceans.

Submarine Weathering Processes and Products

Geologists have long recognized that sediments and rocks on the seafloor are altered by reaction with seawater, a process called **halmyrolysis**. Halmyrolysis includes alteration of clay minerals from one type to another, formation of glauconite from feldspars and micas, and formation of phillipsite (a zeolite mineral) and palagonite (altered volcanic glass) from volcanic ash. Dissolution of the siliceous and calcareous tests of organisms may also be considered a type of submarine weathering. Prior to the 1970s, submarine weathering processes had not received a great deal of research, and it was not recognized that they might have a significant effect on the overall chemical composition of the oceans. Our concept of the importance of submarine weathering has changed dramatically since the middle 1970s because recent studies of volcanic rocks and weathering processes on the seafloor show that submarine weathering of basalts, particularly on mid-ocean ridges, is an extremely important chemical phenomenon. This process results in both widespread hydration of basalts and changes in composition of seawater owing to ion exchange during the reaction of seawater with basalt.

Alteration of oceanic rocks occurs both at low temperatures (less than 20°C) and at higher temperatures (ranging up to 350°C). Low-temperature alteration takes place as seawater percolates through fractures and voids in the uppermost part of the ocean crust, perhaps extending to depths of 2–5 km. Olivine and interstitial glass in the basalts are replaced by smectite clay minerals, and further alteration may lead to formation of zeolite minerals and chlorite. As a result of these changes, an exchange of elements between rock and water takes place, and large volumes of seawater become fixed in the oceanic crust in hydrous clay minerals and zeolites.

Scientists have only very recently become aware, as a result of the discovery in 1977 of submarine thermal springs along the Galapagos Rift (Corliss et al., 1979), that large-scale hydrothermal activity is taking place along the crests of mid-ocean ridges. Since that initial discovery, scientists using deep submersible vehicles have located additional hot springs along other ridge crests. These springs originate from the activity of seawater in areas of active or recent volcanism along the ridge crests. Seawater enters the ocean crust along fractures or other voids and comes in contact with hot volcanic rock. The heated water then flows out into the ocean through vents on the ocean floor and mixes with the overlying water. On the top of the East Pacific Rise investigators have found spectacular vents composed of sulfide, sulfate, and oxide deposits up to 10 m tall and discharging plumes of hot solutions. These vents, or chimneys, called **black smokers** or **white smokers** depending upon the color of water in the plumes (McDonald et al., 1980), discharge water ranging in temperature up to 350°C. When these hot solutions mix with seawater of ambient temperature, they precipitate various minerals, particularly pyrite and chalcopyrite, to build sulfide deposits around the vents (Fig. 2.3).

Reactions between hot basalt and seawater may play a major role in regulating the chemical composition of seawater. Magnesium and sulfate ions, for example, are removed from seawater during this exchange, whereas some other elements such as calcium, lithium, rubidium, barium, manganese, silicon, and potassium are enriched in the seawater. The magnitude of hydrothermal alteration of basalts along mid-ocean ridges is still being investigated; however, Edmond (1980) suggests that a volume of seawater equivalent to the entire ocean could be circulated through crustal rocks along ridges every eight million years or so. This flow rate is only about 0.5 percent of that of all the world's rivers, but the concentration of some elements in these hot waters is 100 to 1000 times greater than in the average river. Therefore, the fluxes of dissolved material out of and into ridge axes could be comparable to those derived from continental weathering. Clearly, we must revise our previous ideas about the dominant role of continental weathering in supplying the salts of the ocean.

FIGURE 2.3   A multiple-orifice black smoker issuing from its constructional chimney of chalcopyrite–sphalerite–anhydrite on the East Pacific Rise. The temperature of the water issuing from the chimney is 350°C. The "smoke" is caused by the presence of fine-grained sulfide precipitates that form by reaction of the hot waters with cold, ambient seawater. (Photo by Fred Spiess. Courtesy of Scripps Institution of Oceanography, University of California, San Diego.)

ADDITIONAL READINGS

Birkland, P. W., 1974, Pedology, weathering and geomorphological research. Oxford University Press, New York, 285 p.

Bohn, H. L., B. L. McNeal, and G. A. O'Connor, 1979, Soil chemistry: John Wiley & Sons, New York, 329 p.

Carroll, D., 1970, Rock weathering: Plenum Press, New York, 203 p.

Drever, J. I. (ed.), 1985, The chemistry of rock weathering: D. Reidel, Hingham, Mass., 336 p.

Loughnan, F. C., 1969, Chemical weathering of silicate minerals: Elsevier, New York, 154 p.

Ollier, C., 1969, Weathering: Elsevier, New York, 304 p.

Rona, P. A., and R. P. Lowell (eds.), 1980, Seafloor spreading centers: Hydrothermal systems: Benchmark Papers in Geology, v. 56, Dowden, Hutchinson and Ross, Stroudsburg, Pa., 424 p.

3

Transport and Deposition of Siliciclastic Sediment

3.1 INTRODUCTION

The weathering residues and pyroclastic particles discussed in Chapter 2 are ultimately eroded from highlands and transported to depositional basins where they may undergo additional transport before final deposition. Sediment transport occurs either by processes that involve direct fluid flow of air, water, or ice or by gravity-driven processes that commonly involve the presence of water. During gravity-flow transport, fluids may act as a support mechanism and lubricant for moving sediment, and some sediment gravity flows such as debris flows behave like fluids. Study of sediment transport thus requires some understanding and application of the principles of fluid flow. The fundamental laws of fluid dynamics are complex when applied to fluid flow alone. These complexities are greatly magnified when particles are entrained in the flow, as during sediment transport. The problem of understanding sediment transport by fluid flow is further magnified by the fact that geologists are primarily interested in understanding conditions that existed in the past. They must attempt to interpret flow conditions based upon only depositional products of sediment transport, long after the fluid flows or sediment gravity flows themselves have disappeared. In short, they must infer from the preserved characteristics of the sediment the nature of flows that took place millions of years ago. Thus, geologists seek to find a relationship between physical properties of sedimentary rocks, such as sedimentary structures and texture, and parameters of fluid flow, such as velocity and water depth, that lend insight into depositional mechanisms and environments. This task is a formidable one and requires application of principles and knowledge gained from theoretical studies, experimental laboratory research, and study of sediment transport and depositional processes in modern environments.

In this chapter we investigate sediment-transport processes by first examining some of the properties of fluids and the basic concepts of fluid flow. We will then consider the problems involved in entrainment and transport of particles by fluid-flow and gravity-flow processes. Fluid mechanics is a subject of particular importance to engineers, and much of the theory of fluid flow has been developed for application to engineering problems. Unfortunately, many of these principles of fluid dynamics are not readily applicable to the more pragmatic problem of understanding sediment transport. For example, engineers are usually not greatly concerned with such aspects of fluid flow as the decay of flow velocity, which is of special interest to geologists studying sediment deposition. No attempt is made here to give a comprehensive review of fluid mechanics. Only those concepts of fluid flow that are important to understanding sediment transport and deposition are discussed, and these concepts are presented in very simplified form. Introduction of some terminology peculiar to fluid dynamics is unavoidable, but new terminology is kept to a minimum. The emphasis in this chapter is on flow processes as they relate to sediment transport. I have not attempted to explore fully the relationships between fluid-flow parameters and the physical properties of sedimentary rocks. Only a very general discussion of the characteristics of fluid-flow deposits and sediment gravity-flow deposits is given. The relationship between flow parameters and sediment properties cannot be adequately developed until sedimentary textures and structures have been discussed. Therefore, this relationship will be examined further in Chapters 5 and 6 and in subsequent chapters of the book, as appropriate.

3.2 FUNDAMENTALS OF FLUID FLOW

Fluids are substances that change shape easily and continuously as external forces are applied. Thus, they have negligible resistance to shearing forces. **Shearing forces** set up conditions within a body that create a tendency for parts of the body to slide over other parts along a series of parallel shear planes. Natural fluids include crude petroleum, natural gas, air, and water; however, air, water, and water containing various amounts of suspended sediment are the fluids of primary geological interest. The basic physical properties of these fluids are **density** and **viscosity.** Differences in these properties markedly affect the ability of fluids to erode and transport sediment.

The fluid density (ρ), defined as mass per unit fluid volume, affects the magnitude of forces that act within a fluid and on the bed as well as the rate at which particles fall or settle through a fluid. Density particularly influences the movement of fluids downslope under the influence of gravity. Density increases with decreasing temperature of a fluid. The density of water (0.998 g/mL at 20°C) is more than 700 times greater than that of air. This density difference influences the relative abilities of water and air to transport sediment. Water can transport particles of much larger size than those transported by wind.

Fluid viscosity describes the ability of fluids to flow. Fluids with low viscosity flow readily, and vice versa. For example, air has very low viscosity and ice has very high viscosity. To appreciate the significance of viscosity, imagine a simple experiment in which a fluid is trapped between two parallel plates. The lower plate is stationary and the upper plate is moving over it with a constant velocity (V). The fluid can be thought of as forming parallel sheets between the plates. As the upper plate moves over the lower, the fluid in between is put in motion with a velocity that varies linearly from zero at the lower plate to velocity (V) at the upper plate (Fig. 3.1). The shearing force per unit area needed to produce a given rate of shearing, or a given velocity gradient normal to the shear planes, is determined by the viscosity. **Dynamic viscosity** (μ) is thus the measure of resistance of a substance to change in shape taking place at finite speeds during flow. It is force per unit velocity gradient and is defined as the ratio of shear stress (τ) to the rate of deformation (du/dy) sustained across the fluid:

$$\mu = \frac{\tau}{du/dy} \tag{3.1}$$

Shear stress is the shearing force per unit area (e.g., dynes/cm^2) exerted across the shearing surface at some point in a fluid. It acts on the fluid parallel to the surface of the fluid body. Shear stress is generated at the boundary of two moving fluids and is a function of the extent to which a slower moving mass retards a faster moving one. The velocity gradient du/dy is the rate of change of the local fluid velocity u in the direction y normal to the shearing surface. Thus, as a faster moving layer moves over a slower moving layer, the shear stress is the force that produces a change in velocity (du) relative to height (dy) (Fig. 3.1). The greater the viscosity, the greater the shear stress must be to produce the same rate of deformation. Shear stress at the bed is also referred to as **shear velocity.** When written as shear velocity it is expressed in velocity units such as cm/sec. Shear stress, or shear velocity, is a very important factor in sediment transport. It plays a critical role in the erosion or entrainment of sediment on a stream bed or ocean floor and the continued downcurrent transport of sediment.

Viscosity decreases with temperature; thus, a given fluid flows more readily at higher temperatures. Because both density and dynamic viscosity strongly affect fluid behavior, fluid dynamicists commonly combine the two into a single parameter called **kinematic viscosity** (ν), which is the ratio of dynamic viscosity to density:

$$\nu = \frac{\mu}{\rho} \tag{3.2}$$

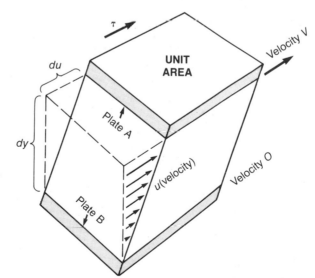

FIGURE 3.1 Geometric representation of the factors that determine fluid viscosity. A fluid is enclosed between two rigid plates, A and B. Plate A moves at a velocity (V) relative to Plate B. A shear force (τ) acting parallel to the plates creates a steady-state velocity profile, shown by the inclined line, where fluid velocity (u) is proportional to the length of the arrows. The shear stress may be thought of as the force that produces a change in velocity (du) relative to height (dy) as one fluid layer slides over another. The ratio of shear stress to du/dy is the viscosity (μ).

Kinematic viscosity is an important factor in determining the extent to which fluid flows exhibit turbulence.

Types of Fluids

Air and water are the only fluids of importance in sediment transport; water, however, can display variable properties as a fluid medium if it contains substantial concentrations of sediment or is frozen into ice. Because these fluid properties affect the way fluids flow and transport sediment, it is important to understand the behavior of various types of fluids. Depending upon the extent to which dynamic viscosity (μ) changes with shear or strain rate, three general types of fluids can be defined. **Newtonian fluids** have no strength and do not undergo a change in viscosity as the shear rate increases. Thus, ordinary water, which does not change viscosity as it is stirred or agitated, is a Newtonian fluid. Additional resistance to flow in water does arise during turbulence owing to movement of eddies, which absorb energy. This resistance is called eddy viscosity, but it does not represent a change in the dynamic viscosity. **Non-Newtonian fluids** have no strength, but show variable viscosity (μ) with change in shear or strain rate. Water containing dispersions of sand in concentrations greater than about 30 percent by volume—or even lower concentrations of cohesive clay—behaves as a non-Newtonian fluid. Therefore, highly water-saturated, noncompacted muds display non-Newtonian behavior. Such muds may flow very sluggishly at low flow velocities, but display much less viscous flow at higher velocities.

Some extremely concentrated dispersions of sediment may behave as plastic substances, which have an initial strength that must be overcome before yield occurs. If the plastic material behaves as a substance with constant viscosity after the yield strength is exceeded, it is called a **Bingham plastic.** Debris flows in which large cobbles or boulders are supported in a matrix of interstitial fluid and fine sediment are examples of natural substances that behave as Bingham plastics. Water with dispersed sediment and other plastic materials such as ice that behave as substances with variable viscosity after yield strength is exceeded and they start to flow are called **pseudoplastics. Thixotropic substances,** a special type of pseudoplastic, have strength until sheared. Shearing destroys their strength; the substances behave like a fluid (commonly non-Newtonian) until allowed to rest a short while, after which strength is regained. Freshly deposited muds commonly display thixotropic behavior. Shearing resulting from earthquake tremors, for example, can cause liquefaction and failure of such muds. Such momentary liquefaction may result in downslope movement of sediment that otherwise would not undergo transport. It may also lead to formation of certain kinds of deformation structures. Differences in behavior of Newtonian fluids, non-Newtonian fluids, and plastic substances in response to shear stress are illustrated in Figure 3.2.

Laminar vs. Turbulent Flow

Fluids in motion display two modes of flow depending upon the velocity and viscosity of the fluids. Experiments with dyes show that a thin stream of dye injected into a slowly moving, unidirectional fluid will persist as a straight, coherent stream of nearly constant width. This type of movement is **laminar flow.** It can be visualized as a series of parallel sheets or filaments by which movement is occurring on a molecular scale owing to constant vibration and translation of the fluid molecules (Fig. 3.3A,B). If velocity of flow is increased or viscosity of the fluid decreased, the dye stream is no

FIGURE 3.2 Rates of deformation vs. shear stress for fluids and plastics. (After Blatt, H., G. V. Middleton, and R. Murray, Origin of sedimentary rocks, 2nd ed., © 1980, Fig. 5.26, p. 187. Reprinted by permission of Prentice-Hall, Englewood Cliffs, N. J.)

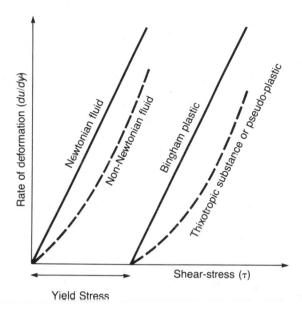

longer maintained as a coherent stream but breaks up and becomes highly distorted. It moves as a series of constantly changing and deforming masses in which there is sizeable transport of fluid perpendicular to the mean direction of flow (Fig. 3.3C). This type of flow is called **turbulent flow** because of the transverse movement of these masses of fluid. Turbulence is thus an irregular or random component of fluid motion. Highly turbulent water masses are referred to as **eddies.** Most flow of water and air under natural conditions is turbulent, although flow of ice and of mud-supported debris (non-Newtonian fluids) is laminar.

The upward motion of water particles in turbulent water masses slows the fall of settling particles and thus decreases their settling velocity. Also, fluid turbulence tends to increase the effectiveness of fluid masses in eroding and entraining particles from a sediment bed. Because of the significance of turbulence in sediment transport, it is important to develop a fuller understanding of this property. Velocity measured over a period of time at a particular point in a laminar flow is constant. By contrast, velocity measured at a point in turbulent flow tends toward an average value when measured over a long period of time, but varies from instant to instant about this

FIGURE 3.3 Schematic representation of laminar vs. turbulent fluid flow: A. Laminar flow over a smooth stream bed. B. Laminar flow over a spherical particle on a smooth bed. C. Turbulent flow over a smooth bed. The arrows indicate flow paths of the fluid.

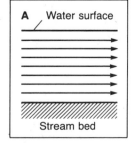
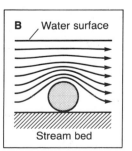
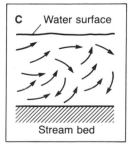

average value. As we shall see, a calculated variable called the Reynolds number can be used to predict the boundary conditions separating laminar and turbulent flow. Turbulent flow resists distortion to a much greater degree than laminar flow. Thus, a fluid undergoing turbulent flow appears to have a higher viscosity than the same fluid undergoing laminar flow. As mentioned, this apparent viscosity, which varies with the character of the turbulence, is called **eddy viscosity.** Eddy viscosity results from turbulent momentum transfer and is the rate of exchange of fluid mass between adjacent water bodies. It is necessary in dealing with fluids undergoing turbulence to rewrite the equation for shear stress to include a term for eddy viscosity. Thus, for laminar flow, shear stress is given by the relationship

$$\tau = \mu \frac{du}{dy} \tag{3.3}$$

but for turbulent flow

$$\tau = (\mu + \eta)\frac{du}{dy} \tag{3.4}$$

where η is eddy viscosity, which is commonly several orders of magnitude higher than dynamic viscosity.

Velocity Profiles and Bed Roughness

Because of the greater shear stress required to maintain a particular velocity gradient in turbulent flow, both the vertical velocity profile above the bed and the velocity profile in a flow channel, as observed from above, have a different shape than laminar-flow velocity profiles (Fig. 3.4). Owing to variations in flow velocity during turbulent flow, the shape of the turbulent-flow vertical profile is determined by time-averaged values of velocity. Under conditions of turbulent flow, laminar or near-laminar flow occurs only very near the bed. The exact shape of the turbulent profile depends upon the nature of the bed over which the flow takes place. For smooth beds, there is a thin layer close to the bed boundary where molecular viscous forces dominate. Molecular adhesion causes the fluid immediately at the boundary to remain stationary. Successive overlying layers of fluid slide relative to those beneath at a rate dependent upon the fluid viscosity (Fig. 3.1). Flow within this thin boundary layer tends toward laminar, although it is characterized by streaks of faster and slower moving fluid and is not truly laminar. This layer is the **viscous sublayer,** sometimes called the **laminar sublayer.** Over a very rough or irregular bed such as coarse sand or gravel, the viscous sublayer is destroyed by these irregularities, which extend through the layer into the turbulent flow. The flow of fluid over a boundary is thus affected by the roughness of the boundary. Obstacles on the bed generate eddies at the boundary of a flow; the larger and more abundant the obstacles the more turbulence is generated. The presence or absence of a viscous sublayer may be an important factor in initiating grain movement.

Boundary Shear Stress

As a fluid flows across its bed, a stress that opposes the motion of the fluid exists at the bed surface. This stress is called the **boundary shear stress** (τ_0) to differentiate it from fluid shear stress (τ) and is defined as force per unit area parallel to the bed; that

FIGURE 3.4 Comparison of vertical velocity profiles for (A) laminar, and (B) turbulent flow in a wide channel. Velocities in the turbulent profile are time-averaged values. C. General form of laminar and turbulent velocity profiles as observed from above. (After Collinson, J. D., and D. B. Thompson, 1982, Sedimentary structures. Fig. 3.3, p. 22, reprinted by permission of George Allen & Unwin, London; and Leeder, M. R., 1982, Sedimentology: Process and product. Fig. 5.8, p. 53, reprinted by permission of George Allen & Unwin, London.)

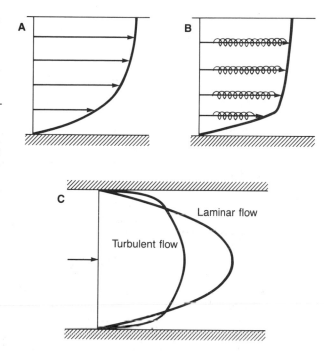

is, tangential force per unit area of surface. It is a function of the specific gravity of the fluid, slope of the bed, and water depth. Boundary shear stress is expressed as

$$\tau_0 = \gamma R_h S \tag{3.5}$$

where γ is specific gravity of the fluid, R_h is hydraulic radius (cross-sectional area divided by wetted perimeter), and S is the slope. The boundary shear stress is also a function of velocity of flow, a complex mathematical relationship not shown here. It tends to increase as velocity increases, although not in a direct way.

Because boundary shear stress is determined by the force that a flow is able to exert on the sediment bed and is related to flow velocity, it is an extremely important variable in determining the erosion and transport of sediment on the bed below a flow. Equation 3.5 indicates that boundary shear stress increases directly with increase in specific gravity of the moving fluid, increasing diameter and depth of the stream channel, and increasing slope of the stream bed. Other factors being equal, we would thus expect to see greater boundary shear stress developed, and greater ability to erode and transport sediment, in water flows than in air flows, in larger stream channels than in smaller channels, and in high-gradient streams than in low-gradient streams.

Reynolds Number

The fundamental differences in laminar and turbulent flow arise from the ratio of inertial forces that tend to cause fluid turbulence and viscous forces that tend to suppress turbulence. Inertial forces are related to the scale and velocity of fluids in motion. Viscous forces arise from the viscosity of the fluid, and these forces resist deformation of the fluid. The relationship of inertial to viscous forces can be shown mathematically by a dimensionless value called the **Reynolds number** (R_e), which is expressed as

$$R_e = \frac{UL\rho}{\mu} = \frac{UL}{\nu} \tag{3.6}$$

where U is the mean velocity of flow, L is some length (commonly water depth) that characterizes the scale of flow, and ν is kinematic viscosity. When viscous forces dominate, as in highly concentrated mudflows, Reynolds numbers are small and flow is laminar. Very low flow velocity or shallow depth also produces low Reynolds numbers and laminar flow. When inertial forces dominate and flow velocity increases, as in the atmosphere and most flow in rivers, Reynolds numbers are large and flow is turbulent. Thus, most flow under natural conditions is turbulent. Note from equation 3.6 that an increase in viscosity can have the same effects as a decrease in flow velocity or flow depth. The transition from laminar flow to turbulent flow takes place above a critical value of Reynolds number, commonly between 500 and 2000, that will depend upon boundary conditions such as channel depth and geometry. Thus, under a given set of boundary conditions, the Reynolds number can be used to predict whether flow will be laminar or turbulent and to derive some idea of the magnitude of turbulence. Because the Reynolds number is dimensionless, it is of particular value when used to compare scaled-down models of flow systems with natural flow systems. If the length value in equation 3.6 is replaced by grain diameter, the Reynolds number becomes an important parameter in evaluating sediment erosion and entrainment from the bed, a subject discussed subsequently in this chapter.

Froude Number

In addition to the effects of fluid viscosity and inertial forces, gravity forces also play an important role in fluid flow because gravity influences the way in which a fluid transmits surface waves. The velocity (celerity) with which small gravity waves move in shallow water is given by the expression $\sqrt{gd}$ in which g is gravitational acceleration and D is water depth. The ratio between inertial and gravity forces is the **Froude number** (F_r), which is expressed as

$$F_r = \frac{U}{\sqrt{gD}} \tag{3.7}$$

where U is again the mean velocity of flow and D is depth, in the case of water flowing in an open channel. The Froude number, like the Reynolds number, is a dimensionless value.

When the Froude number is less than 1, the velocity at which waves move is greater than flow velocity and waves can travel upstream. Flow under these conditions is called tranquil, streaming, or subcritical. If the Froude number is greater than 1, waves cannot be propagated upstream and flow is said to be rapid, shooting, or supercritical. Thus, Froude number can be used to define the critical velocity of water (but not air) at which flow at a given depth changes from tranquil to rapid, or vice versa. The Froude number also has a relationship to flow regimes, which are defined by characteristic bedforms such as ripples, that develop during fluid flow over a sediment bed. This relationship is discussed further in Chapter 6.

3.3 PARTICLE TRANSPORT BY FLUIDS

Having established some of the fundamentals of fluid behavior during flow of fluids alone, we are now at a point where we can consider the more complicated processes of transport of sediment by fluid flow. Transport of sediment by fluid flow involves two fundamental steps: (1) erosion and entrainment of sediment from the bed and (2) subsequent, sustained downcurrent or downwind movement of sediment along or above the bed. The term **entrainment** refers to the processes involved in lifting resting grains from the bed or otherwise putting them in motion. More energy is commonly required to initiate particle movement than to keep particles in motion after entrainment. Thus, a great deal of experimental and theoretical work has been directed toward study of the conditions necessary for particle entrainment. Once particles are lifted from the sediment bed into the overlying water or air column, the rate at which they fall back to the bed—the settling rate, or settling velocity—is an important factor in determining how far the particles travel downcurrent before they again come to rest on the sediment bed. Like particle entrainment, the settling velocity of particles has been studied extensively. We will now examine some of these fundamental aspects of particle transport by fluids, beginning with a look at the factors involved in entrainment of sediments by a moving body of fluid.

Particle Entrainment

Entrainment by Currents. As the velocity and shear stress of a fluid moving over a sediment bed increase, a critical point is reached at which grains begin to move downcurrent. Commonly the smallest and lightest grains move first. As shear stress increases, larger grains are put into motion until finally grain motion is common everywhere on the bed. This **critical threshold** for grain movement is a direct function of several variables, including boundary shear stress; fluid viscosity; and particle size, shape, and density. Indirectly, it is also a function of the velocity of flow, which varies as the logarithm of the distance above the bottom.

To understand the problems involved in lifting particles from the bed and initiating motion, let us consider the opposing forces that come into play as a fluid moves across its bed. As shown in Figure 3.5A, forces owing to gravity act downward to resist motion and hold particles against the bed. The gravity forces result from the weight of the particles and are aided in resisting grain movement by frictional resistance between particles. Fine, clay-size particles have added resistance to movement owing to cohesiveness caused by electrochemical bonds between these small grains. The motive forces that must be generated by fluid flow to overcome the resistance to movement imposed by these retarding factors include a **drag force** that acts parallel to the bed and is related to the boundary shear stress and a **lift force** due to the Bernoulli effect of fluid flow over projecting grains. The drag force (F_D) depends upon the boundary shear stress (τ_0) and the drag exerted on each grain exposed to this stress. Thus

$$F_D = \frac{\tau_0}{N} \tag{3.8}$$

where N is the number of exposed grains per unit area. The hydraulic lift force known as the Bernoulli effect is caused by the convergence of fluid streamlines over a pro-

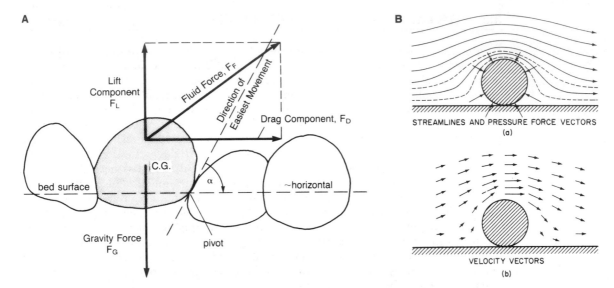

FIGURE 3.5 A. Forces acting during fluid flow on a grain resting on a bed of similar grains. B. Flow pattern of fluid moving over a grain, illustrating the lift forces generated owing to the Bernoulli effect: (a) streamlines and the relative magnitude of pressures acting on the surface of the grain, (b) direction and relative velocity of velocity vectors; higher velocities occur where streamlines are closer together. (A, after Middleton, G. V., and J. B. Southard, 1978, Mechanisms of sediment movement: Eastern Section, Soc. Econ. Paleoutologists and Mineralogists Short Course No. 3. Fig. 6.1, p. 6.3, reprinted by permission of SEPM, Tulsa, Okla. B, from Blatt, H., G. V. Middleton, and R. Murray, Origin of sedimentary rocks, 2nd ed., © 1980, Fig. 4.9, p. 107. Reprinted by permission of Prentice-Hall, Englewood Cliffs, N. J.)

jecting grain. The Bernoulli effect results from an increase in flow velocity in the zone where the streamlines converge over the grain. The velocity increase causes pressure to decrease above the grain. Hydrostatic pressure from below then tends to push the grain up off the bed into this low-pressure zone (Fig. 3.5B). The drag force and lift force combine to produce the total fluid force, represented by the fluid-force vector in Figure 3.5A. The fluid force must be large enough to overcome the gravity and frictional forces for grain movement to occur.

The preceding discussion is greatly simplified and generalized, and a number of factors complicate calculation of critical thresholds of grain movement under natural conditions. These factors include variations in shape, size, and sorting of grains; bed roughness, which controls the presence or absence of a viscous sublayer; and cohesion of small particles. Because of these complicating factors, the critical conditions for particle entrainment must be determined experimentally. Two widely used plots that show experimentally derived threshold graphs for initiation of grain movement are the **Hjulström curve** and the **Shields diagram.**

In the Hjulström curve (Fig. 3.6), the velocity at which grain movement begins as flow velocity increases above the bed is plotted against mean grain size (grain diameter). This diagram shows the critical velocity for movement of quartz grains on a plane bed at a water depth of one meter. The curve separates the graph into two fields. Points above the graph indicate the conditions under which grains are in motion and points below indicate no motion. Note from this figure that critical entrainment veloc-

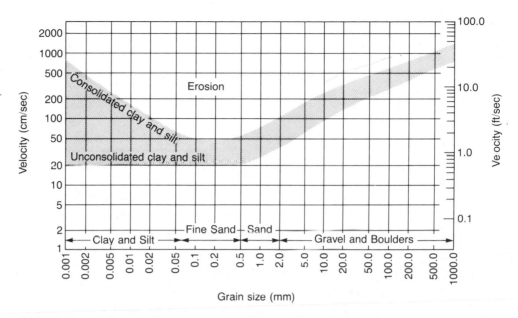

FIGURE 3.6 Hjulström's diagram, as modified by Sundborg, showing the critical current velocity required to move quartz grains on a plane bed at a water depth of 1 m. The shaded area indicates the scatter of experimental data, and the increased width of this area in the finer grain sizes shows the effect of sediment cohesion and consolidation on the critical velocity required for sediment entrainment. (After Sundborg, A., 1956, The River Klaralven, a study of fluvial processes: Geografiska Annaler, Ser. A, v. 38. Fig. 16, p. 197, reprinted by permission.)

ity for grains larger than about 0.5 mm increases gradually with increasing mean grain size, whereas the entrainment velocity for grains smaller than 0.05 mm increases with decreasing grain size. This seemingly anomalous behavior at smaller grain sizes is apparently due mainly to increasing cohesion of finer sized particles, making them more difficult to erode than larger, noncohesive particles.

The Shields diagram (Fig. 3.7) is widely used by sedimentologists and is well established by experimental work. However, it is more complex and difficult to understand than the Hjulström curve because it involves two dimensionless relationships. In this diagram, a value called **dimensionless shear stress** (τ^*) is used instead of flow velocity as a measure of critical shear, and the mean grain size parameter used in Hjulström's diagram is replaced by **grain Reynolds number** (R_{eg}), another dimensionless quantity. The dimensionless bed shear stress is given by

$$\tau^* = \frac{\tau_0}{(\gamma_s - \gamma_f)d} \tag{3.9}$$

where γ_s is specific gravity of the particles, γ_f is specific gravity of the fluid, and d is particle diameter. The value of dimensionless shear stress is the ratio of the bed shear stress to the product of the specific weight (immersed weight) of the grain and its diameter; that is, the weight of a layer of grains underlying the unit area of bed. The value increases with increasing bed shear stress and increasing velocity and decreases with increasing density and size of the particles. Unlike the separate velocity and

FIGURE 3.7 Shields diagram, showing the relationship between two dimensionless parameters: grain Reynolds number (R_{eg}) and dimensionless shear stress (τ^*). (From Pettijohn, F. J., P. E. Potter, and R. Siever, 1973, Sand and sandstone. Fig. 9.10, p. 341, used by permission of Springer-Verlag, New York.)

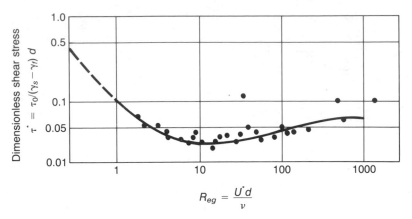

$$R_{eg} = \frac{U^{\cdot} d}{\nu}$$

Grain Reynolds number

grain-size parameters in Hjulström's diagram, the dimensionless shear stress thus incorporates shear stress (velocity), grain size, and grain and fluid density into a single term. An increase in dimensionless shear stress indicates either an increase in flow velocity and shear stress or a decrease in grain size or density.

The grain Reynolds number (R_{eg}) differs from the ordinary Reynolds number previously discussed. The length, or water depth, value (L) of the ordinary Reynolds number is replaced by particle diameter (d) and the flow velocity (U) by friction velocity (U^*), a measure of turbulent eddying. The grain Reynolds number in effect is a measure of turbulence at the grain–fluid boundary. It is thus expressed as

$$R_{eg} = \frac{U^* d}{\nu} \tag{3.10}$$

The grain Reynolds number is clearly not the same thing as mean grain size; however, it can be seen from Figure 3.7 that the grain Reynolds number increases with increasing grain size if friction velocity and kinematic viscosity remain constant. Thus, an increase in grain Reynolds number means either an increase in grain size, an increase in friction velocity and turbulence, or a decrease in kinematic viscosity.

The Shields diagram is more difficult to interpret than Hjulström's diagram, but, as in the Hjulström diagram, points above the curve indicate that grains on the bed are fully in motion and points below indicate no motion. Beginning of motion is determined by the dimensionless shear stress, which increases with increasing bed shear stress under a given set of conditions for grain density, fluid density, and grain size. The critical dimensionless shear stress required to initiate grain movement thus depends upon the grain Reynolds number, which in turn is a function of grain size, kinematic viscosity, and turbulence. Note from the Shields diagram that the dimensionless bed shear stress increases slightly with increasing grain Reynolds number above about 5–10, although it remains mainly between 0.03 and 0.05. At lower Reynolds numbers, the value increases steadily to a value of 0.1 or higher. This greater rate of increase at lower Reynolds numbers is related to the presence of the viscous sublayer. When the bed is composed of small particles on the order of fine sand or smaller, a smooth boundary to flow results; the particles lie entirely within the viscous sublayer, where flow is essentially nonturbulent and instantaneous velocity variations are less than in the lowermost part of the overlying turbulent boundary layer. For coarser particles, the viscous sublayer is so thin that the grains project through the layer into the turbulent flow. The reason that sedimentologists use the Shields diagram in preference to the simpler Hjulström diagram is that the former has more general application. For example, it can be used for wind as well as water and for a variety of conditions in water. By contrast, the Hjulström diagram is valid only for water in which fluid and grain density and dynamic viscosity are constant, as in freshwater streams in a given season during average flow (Pettijohn et al., 1973).

Several complicating factors not covered in the Hjulström and Shields diagrams make prediction of the onset of grain movement difficult. Instantaneous fluctuations in boundary shear stress may arise from local eddies or from wave action superimposed on current flow and cause some particles to move before the general onset of grain movement. Fine muds and silts may not erode as predicted by the Hjulström and Shields curves owing to the tendency of such cohesive materials to be removed as chunks or aggregates of grains rather than as individual grains. Entrainment of grains by wind action can be strongly affected by impact of moving grains hitting the bed. At a value of wind velocity below the critical velocity needed to initiate grain movement, grain motion can be started and propagated downwind by throwing grains onto the bed, a process referred to as seeding. This lower threshold for grain movement is called the **impact threshold.**

Entrainment by Waves. A special problem arises in evaluating the threshold of sediment movement under the action of orbital waves in nearshore areas of the ocean. In shallow water the passage of waves over the ocean surface sets water in motion near the seabed; the water may have sufficient velocity to entrain and shift sediment about on the seafloor. The origin of this near-bottom water motion is related to the orbital motion generated by waves in near surface water. If an observer watches the movement of an object floating on a lake or in the ocean during the passage of a wave, it is apparent that the object simply bobs up and down as the wave passes. It undergoes no noticeable forward motion. Numerous studies have established that water moves in orbital paths during the passage of waves (Fig. 3.8). Experiments with dyes and glass beads show that water moves forward in the crest of a wave, then downward, and finally backward under the trough and upward as the wave passes. Water particles retrace this orbit with each passing wave, returning to nearly their original position after the wave passes. Actually, a slight net forward movement of water particles occurs in shallow water during each orbit. Some forward movement takes place owing to a small "time asymmetry" in the velocity with which water moves forward under the foreshortened wave crest compared to the velocity of backward movement under the longer trough (Clifton and Dingler, 1984).

Orbital motion of water dies out downward at a depth equal to about one-half the wave length. Therefore, in deep water the orbital motion is unimpeded by the bottom, and orbits are nearly circular. As waves move into shallow water where depth is less than one-half the wave length, the bottom begins to interfere with orbital motion and thus to affect the shape of the orbits. By the time that waves reach very shallow water, where depth is less than about $\frac{1}{20}$ the wave length, the motion of the particles is strongly affected by interaction with the bottom and the orbits become much more elliptical. They become progressively flatter downward until near bottom they are essentially linear, generating a to-and-fro motion as waves pass (Fig. 3.8). This motion

FIGURE 3.8 Orbital motion of water particles caused by passage of waves in (A) deep water and (B) shallow water. Note that in shallow water the orbits become both smaller and flatter with depth, until near the bottom they are essentially flat. (After Gross, M. G., Oceanography, 3rd ed., © 1982, Fig. 8.2, p. 214. Reprinted by permission of Prentice-Hall, Englewood Cliffs, N. J.)

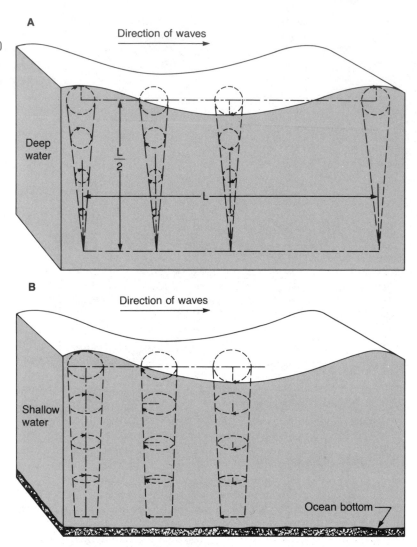

produces bidirectional flow of water along the seafloor as each wave passes over the surface. The velocity of this bottom flow is referred to as the **orbital velocity.** Orbital velocity varies directly as a function of the magnitude of the orbital diameter and indirectly as a function of the wave period, which is the time required for passage of one wave length. Owing to the time asymmetry referred to above, the orbital velocity is commonly greater in one direction than the other. This difference in velocity becomes particularly important when only the stronger velocity flow exceeds the threshold of movement for grains, resulting in net transport of grains in one direction. The factors that influence the threshold for grain movement are discussed in the following paragraph.

Komar and Miller (1975) show that the threshold of grain movement under orbital waves is a function of grain diameter, wave period, and orbital velocity of the waves. Orbital velocity is in turn a function of wave height, water depth, wave period, and wave length. This relationship is shown by

$$u_t = \frac{\pi d_o}{T} = \frac{\pi H}{T \sin h \, (2\pi h/L)} \qquad (3.11)$$

in which u_t is the threshold orbital velocity, d_o is orbital diameter of the wave motion, H is wave height, L is wave length, T is wave period, and h is water depth. Note from this equation that, as mentioned, the threshold orbital velocity varies directly with the magnitude of the orbital diameter and inversely with the magnitude of the wave period. Komar and Miller (1975) used equation 3.11 to determine the threshold values for movement of grains ranging from 0.01 to 100 mm (Fig. 3.9) under waves having periods ranging from 1 to 15 sec. Note the consistent, though nonlinear, increase in threshold orbital velocity with increasing grain diameter. Presumably this relationship holds only for noncohesive sediment. Komar and Miller used the threshold values in Figure 3.9 to calculate the depths at which waves with a period of 15 sec and heights up to 6 m would set sediment in motion on the seafloor. Their calculations showed that particles in the size range of coarse silt (0.05 mm) to medium sand (0.3 mm) can be moved by orbital wave motion to depths of 100 m and more. Waves of longer period could entrain sediment at even greater depths. The movement of sediment under the influence of orbital waves has important implications for both transport of sediments in the nearshore zone and formation of sedimentary structures such as oscillation rip-

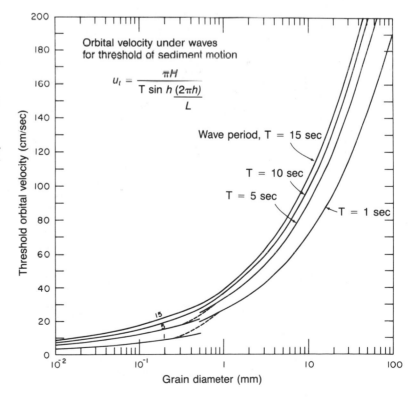

FIGURE 3.9 The near-bottom critical orbital velocity (u_t) required to initiate motion of quartz grains by waves having various wave periods. The orbital velocity in turn is related to wave height (H), water depth (h), and wave length (L). (After Komar, P. O., and M. C. Miller, 1975, Sediment threshold under oscillatory waves: Proc. 14th Conf. on Coastal Engineering. Fig. 7, p. 772, reprinted by permission of American Society of Civil Engineers.)

ples. The relationship between sediment entrainment and sedimentary structures is explored further in Chapter 6.

Settling Velocity

As soon as grains are lifted above the bed during the entrainment process, they begin to fall back to the bed. The distance that they travel downcurrent before again coming to rest on the bed depends upon the drag force exerted by the current and the settling velocity of the particles. A particle initially accelerates as it falls through a fluid, but acceleration gradually decreases until a steady rate of fall, called the terminal **fall velocity,** is achieved. For small particles, terminal fall velocity is reached very quickly. The rate at which particles settle after reaching fall velocity is a function of the viscosity of the fluid and the size, shape, and density of the particles. The settling rate is determined by the interaction of upwardly directed forces—owing to buoyancy of the fluid and viscous resistance (drag) to fall of the particles through the fluid—and downwardly directed forces arising from gravity. The drag force exerted by the fluid on a falling grain has been determined for spherical grains to be proportional to the density of the fluid (ρ_f), the diameter (d) of the grains, and the fall velocity (V) as given by the relationship

$$C_D \, \pi \frac{d^2}{4} \frac{\rho_f V^2}{2} \tag{3.12}$$

where C_D is a drag coefficient that depends upon the grain Reynolds number and the particle shape. The upward force resulting from buoyancy of the fluid is given by

$$\frac{4}{3}\pi \left(\frac{d}{2}\right)^3 \rho_f g \tag{3.13}$$

where ρ_f is fluid density and g is gravitational acceleration. The downward force due to gravity is given by

$$\frac{4}{3}\pi \left(\frac{d}{2}\right)^3 \rho_s g \tag{3.14}$$

where ρ_s is particle density. As the particle stops accelerating and achieves fall velocity, the drag force of the liquid on the falling particle is equal to the downward force due to gravity minus the upward force resulting from buoyancy of the liquid. Thus

$$C_D \pi \frac{d^2}{4} \frac{\rho_f V^2}{2} = \frac{4}{3}\pi \left(\frac{d}{2}\right)^3 \rho_s g - \frac{4}{3}\pi \left(\frac{d}{2}\right)^3 \rho_f g \tag{3.15}$$

Rearranging terms, this relationship can be expressed in terms of fall velocity (V) as

$$V^2 = \frac{4gd}{3C_D} \frac{(\rho_s - \rho_f)}{\rho_f} \tag{3.16}$$

For slow laminar flow at low concentrations of particles and low grain Reynolds numbers, C_D has been determined to equal $24/R_{eg}$ (Rouse and Howe, 1953, p. 182). Substituting this value ($24/U^* d/\mu/\rho_f$) for C_D, then

$$V = \frac{1}{18} \frac{(\rho_s - \rho_f)\, gd^2}{\mu} \tag{3.17}$$

which is **Stokes Law** of settling, with particle size expressed as diameter in centimeters. The law, formulated by Stokes in 1845, is often simplified in textbooks to

$$V = CD^2 \tag{3.18}$$

where C is a constant equaling $(\rho_s - \rho_f)g/18\mu$ and D is the diameter of particles (spheres) expressed in centimeters. Values of C have been calculated for a range of common laboratory temperatures (Galehouse, 1971); thus settling velocity (V) can be calculated quickly for any value of particle diameter (D).

Experimental determination of particle fall velocity shows that Stokes Law accurately predicts settling velocity of particles only for particles less than about 0.2 mm in diameter. Larger particles have fall velocities lower than those predicted by Stokes Law, apparently owing to inertial (turbulent) effects caused by the increased rates of fall of these larger grains. The fall velocity is also decreased by decrease in temperature, which increases viscosity; decrease in particle density; and decrease in the sphericity of the particles. Most natural particles are not spheres, and departure from spherical shape decreases fall velocity. Fall velocity is also decreased by increasing concentration of suspended sediment in the fluid; this increases the apparent viscosity and density of the fluid.

Sediment Loads and Transport Paths

Once sediment has been eroded and put into motion, the transport path that it takes during further sustained downcurrent movement is a function of the settling velocity of the particle and the magnitude of the current velocity and turbulence. Under a given set of conditions, the sediment load may consist entirely of very coarse particles, entirely of very fine particles, or of mixtures of coarse and fine particles. Coarse sediment such as sand and gravel moves on or very close to the bed during transport and is considered to constitute the **bedload.** Finer material carried higher up in the main flow above the bed makes up the **suspended load.**

Bedload Transport. Particles larger than sand size are commonly transported as part of the bedload in essentially continuous contact with the bed. This type of transport is called **traction** transport and may include rolling of large or elongated grains, sliding of grains over or past each other, and creep. Creep results when grains are pushed a short distance along the bed in a downcurrent direction by impact of other moving grains. **Saltation** is a type of bedload transport in which grains, particularly sand-size grains, tend to move in intermittent contact with the bed. Saltating grains move by a series of jumps or hops, rising off the bed at a steep angle ($\sim 45°$) to a height of a few grain diameters and then falling back along a shallow descent path of about 10 degrees. This asymmetric saltation path may be interrupted by turbulence or by collision with another grain (Fig. 3.10). Saltation is a particularly prevalent mode of transport of fine sand by wind; the impact of saltating grains is primarily responsible for creep in eolian sands. Saltation transport may be thought of as intermediate between traction transport and suspension transport, but it is included here as part of bedload transport because most saltating grains remain relatively close to the bed during movement.

FIGURE 3.10 Schematic illustration of grain paths during bedload, suspension, and saltation transport. (From Leeder, M. R., Bedload dynamics: Grain-grain interactions in water flows: Earth surface processes, v. 4. © 1979 by John Wiley & Sons, Ltd. Fig. 5, p. 237, reprinted by permission of John Wiley & Sons, Ltd., Chichester, Sussex, England.)

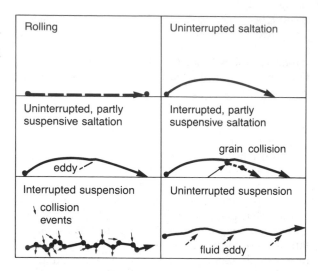

Suspended Load Transport. As flow strength of a current increases, the intensity of turbulence increases close to the bed. Particle trajectories become longer, more irregular, and higher up from the bed than the trajectories of saltating grains. Upward components of fluid motion resulting from turbulence increase to the point that they balance downward gravitational forces on the particles, allowing the particles to stay suspended above the bed far longer than could be predicted from their settling velocities in nonturbulent water. If the lift forces arising from turbulence are erratic and do not continuously maintain this balance—a common occurrence during transport of fine-to-medium sand—the grains may drop back from time to time onto the bed. This behavior is called **intermittent suspension.** Intermittent suspension differs from saltation because the suspended particles tend to be carried higher above the bed and remain off the bed for longer periods of time. Smaller particles have settling velocities that may be so low that they remain in nearly **continuous suspension** and are carried along at almost the same velocity as the fluid flow.

Transport of sediment in suspension is common both by wind movement and stream flow. A special type of intermittent suspension transport takes place in the ocean within cloudy layers of near-bottom water called **nepheloid layers,** first named and surveyed by Ewing and Thorndike (1965). A nepheloid layer is a turbid body of suspended sediment that may reach heights of several hundred meters above the seafloor. It is more dense than the ambient water but not dense enough to sink rapidly. Thus, sediment may remain suspended in such a layer for a long period of time. Most of the material of a nepheloid layer consists of very fine clay particles derived initially from land. Some of this material reached the nepheloid layer by settling through the water column from the surface waters. Most of it is probably fine sediment resuspended from the ocean floor owing to erosion of the seabed by strong bottom currents, or it is fine material injected into the water column by turbidity currents or other mechanisms. Fine sediment may remain in suspension in the nepheloid layer for periods ranging from days to weeks in the lowest 15 m of the water column and from weeks to months in the lowest 100 m (Kennett, 1982). Nepheloid layers in the modern ocean extend seaward for hundreds of kilometers and to water depths of 6000 m or more. Because of the relatively short residence times indicated above for fine sediment in this layer, sediment transported long distances from shore in a nepheloid layer must have been deposited on the ocean floor and resuspended many times.

Washload and Dustload. Much of the sediment load undergoing continuous suspension transport is composed of fine, clay-size particles with very low settling velocities. In rivers, this sediment is derived either from upstream source areas or by erosion of the bank, rather than from the stream bed, and is called the **washload.** Rivers have the capacity to transport large washloads even at very low velocities of flow. Because the washload travels in continuous suspension at about the same velocity as the water, it is transported rapidly through river systems. Similar suspended loads carried by the wind are called **dustloads.** Upward diffusion in unstable, buoyant air masses at an advancing front have been known to carry dust clouds rapidly to heights of hundreds or even thousands of meters during volcanic eruptions. Material carried to such great heights may remain in suspension for long periods of time and subsequently be spread over a very broad area, including the ocean basins (Prospero, 1981). In fact, the very fine-grained component of deep-sea pelagic sediments is believed to be largely of windblown origin.

Ice-transport Load. Transport of sediment by ice is a type of fluid flow transport, although ice flows very slowly as a high-viscosity, non-Newtonian, pseudoplastic. Rates of flow range from 1×10^{-6} cm/sec to 2×10^{-3} cm/sec (J. R. L. Allen, 1970a). Glaciers advance or retreat depending upon the balance between rates of accumulation of snow in the upper reaches of the glacier and rates of ablation (melting) of ice in the lower reaches. Flow of ice is laminar, and flow velocity is greatest near the top and center of the glacier. Velocity decreases toward the walls and floor, although not necessarily to zero. Sediment is entrained by glaciers by quarrying and abrasion by ice as a glacier erodes its bed and by falling or sliding of material from the valley walls. Much of the sediment transported by a glacier is carried along the bottom and sides. Some of this sediment is transported in contact with the valley walls and floor and is responsible for much of the abrasion. Part of the remaining load is carried on the upper surface of the glacier and part is carried within. The internal load is derived either from the joining of ice streams from two or more valleys or by washing or falling of material from the surface into crevasses.

Deposits of Fluid Flows

Water and air are responsible for most sediment transport by fluid flow; ice, however, may account for local transport of large volumes of sediment and particles of very large size. Sediment entrainment and transport by the various processes discussed stops, and deposition occurs, when local hydrologic conditions change sufficiently to cause decrease in bed shear stress to the point that it is no longer adequate to initiate and sustain particle movement. This decrease in bed shear stress is caused fundamentally by decrease in flow velocity. Flow velocity and shear stress may decrease below the critical level required for sediment transport owing to a variety of causes. In the case of water transport, these causes include decrease in the slope of the bed, increase in bed roughness, and loss of water volume. Decrease in wind velocity may result from increase in bed roughness or from changes in surface topography and weather conditions. Deposition from glaciers is brought about on land when the glaciers either become stagnant or retreat owing to decrease in snow accumulation rates or increase in melting rates. Glaciers that run out to sea and calve to form icebergs eventually melt and drop their loads on the seafloor.

Sediment deposition may be temporary or permanent. For example, sediment deposited in river channels and in very nearshore environments such as point bars and beaches can be reentrained and subjected to continued transport as seasonal or longer range changes in the hydrologic regimen take place. In fact, river sediment may be deposited and reentrained numerous times before finally reaching a depositional basin in the ocean. On the other hand, some river sediment, lake sediment, and wind-transported sediment may become deposited in continental settings and be preserved for long periods of time to become part of the geologic record. The great bulk of sediment undergoing transportation ultimately finds its way into ocean basins, where it is eventually deposited below wave base and more or less permanently immobilized until buried.

Sediments deposited by normal fluid flow of water or wind are commonly characterized by layers or beds of various thickness, scarcity of vertical grain-size grading, grain-size sorting ranging from poor to excellent depending upon depositional conditions, and the presence of a variety of sedimentary structures. Sediments deposited from traction currents commonly preserve sedimentary structures such as cross-beds, ripple marks, and pebble imbrication that display directional features from which the direction of ancient fluid flow can be determined. Sediments deposited from suspension lack these flow structures and are characterized instead by fine laminations. Wind is competent to transport and deposit particles in the size range of sand to dust (clay) only. By contrast, the grain size of sediment deposited by water may range from clay size to cobbles or boulders tens to hundreds of centimeters in diameter. These variations in grain size reflect the wide range of energy conditions of wind and water that prevail under natural conditions and the variations in relative competence of wind and water to initiate and sustain sediment transport. Ice does not behave as a Newtonian fluid and because of its much greater viscosity is capable of transporting particles of enormous size as well as particles of the smallest sizes. Sediments deposited by glaciers are characteristically poorly layered and extremely poorly sorted, with particles ranging from meter-size boulders to clay-size grains.

This very brief description of fluid-flow deposits is given here to illustrate the relationship between flow processes and the characteristics of the resulting sedimentary deposits. The textures and structures of sedimentary rocks are discussed in detail in Chapters 5 and 6; the characteristics of sediments deposited by fluid flows in different sedimentary environments are described in Chapters 11–13.

3.4 PARTICLE TRANSPORT BY SEDIMENT GRAVITY FLOWS

Introduction

In the preceding section, we examined sediment transport resulting from the interaction of moving fluids and sediment. During fluid-flow transport, the fluids (water, wind, ice) move in various ways under the action of gravity and the sediment is simply carried along with and by the fluid. Sediment can also be transported independently of fluid flow by the effect of gravity acting directly on the sediment. In this type of transport, fluids may play a role in reducing internal friction and in supporting grains but are not primarily responsible for downslope movement of the sediment. Movement of sediment under the influence of gravity creates the flow, and flow stops when the sediment load is deposited.

Sediment transport owing to the direct action of gravity can occur in both subaerial and subaqueous environments. Gravity transport under submarine conditions has the greatest geological significance. A spectrum of gravity movements exists, ranging from those in which sediment is moved en masse and in which fluids act mainly to reduce internal friction by lubricating the grains to those in which transport is on a grain-by-grain basis and fluids play an important role in supporting the sediment during transport. Gravity mass movements can be grouped into rock falls, slides, and sediment gravity flows, as shown in Table 3.1. **Rock fall** involves free fall of blocks or clasts from cliffs or steep slopes. **Slides** are en masse movements of rock or sediment owing to shear failure that takes place with little accompanying internal deformation of the mass. **Sediment gravity flows** are more "fluid" types of movement in which

TABLE 3.1 Major types of mass-transport processes, their mechanical behavior, and transport and sediment support mechanisms

Mass transport processes			Mechanical behavior	Transport mechanism and sediment support
Rockfall			Elastic	Freefall and subordinate rolling of individual blocks or clasts along steep slopes
Slide	Glide			Shear failure along discrete shear planes with little internal deformation or rotation
	Slump			Shear failure accompanied by rotation along discrete shear surfaces with little internal deformation
			— Plastic limit —	
Sediment gravity flow	Mass flow	Debris flow / Mud flow	Plastic	Shear distributed throughout sediment mass; strength principally from cohesion due to clay content; additional matrix support possibly from buoyancy
		Grain flows: Inertial / Viscous	— Liquid limit —	Cohesionless sediment supported by dispersive pressure; flow in inertial (high-concentration) or viscous (low-concentration) regime; steep slopes usually required
	Fluidal flow	Liquefied flow	Viscous fluid	Cohesionless sediment supported by upward displacement of fluid (dilatance) as loosely packed structure collapses, settling into a more tightly packed framework; slopes in excess of 3° required
		Fluidized flow		Cohesionless sediment supported by the forced upward motion of escaping pore fluid; thin (<10 cm) and short-lived
		Turbidity current		Supported by fluid turbulence

Source: Nardin, T. R., F. J. Hein, D. S. Gorsline, and B. D. Edwards, 1979a, A review of mass movement processes, sediments, and acoustical characteristics, and contrasts in slope and base-of-slope systems versus canyon-fan-basin flow systems, *in* L. J. Doyle and O. R. Pilkey (eds.), Geology of continental slopes: Soc. Econ. Paleontologists and Mineralogists Spec. Pub. 27. Table 1, p. 64, reprinted by permission of SEPM, Tulsa, Okla.

breakdown in grain packing occurs and internal deformation of the sediment mass is intense.

Sediment gravity flows are of particular interest because they are capable of rapidly transporting large quantities of sediment, including very coarse sediment, even into very deep water in the oceans. Gravity flows that occur in subaerial environments can be considered in a broad sense to include pyroclastic flows and base surges resulting from volcanic eruptions, grain flow of dry sand down the slip face of sand dunes, and both volcanic and nonvolcanic debris flows and mudflows, in which large particles are transported in a slurrylike matrix of finer material. Subaqueous sediment gravity flows also include grain flows and debris flows, as well as turbidity currents and fluidized and liquefied sediment flows.

Sediment gravity flows can occur only when grains become separated and dispersed to the point that internal friction and cohesiveness are sufficiently reduced to lower the strength of the sediment mass below the critical point required for gravity to initiate movement. Four types of dispersive and support mechanisms that can achieve this reduction in internal strength have been identified; from these mechanisms four principal types of sediment gravity flows are recognized (Fig. 3.11): (1) **Turbidity currents** are gravity flows in which sediment is supported by the upward component of fluid turbulence. The presence of this suspended sediment in the flow causes its density to increase above that of the ambient water, resulting in downslope flow. Flow can occur quite rapidly, even on very low slopes. (2) **Fluidized and liquefied flows** are concentrated dispersions of grains in which the sediment is supported either by the upward flow of pore water escaping from between the grains as they settle downward by gravity, or by pore water that is forced upward by injection from below. Liquefaction can occur by sudden shocking of the sediment mass, greatly reducing friction between the grains. These flows can move rapidly down relatively gentle slopes (3–10 degrees). (3) **Grain flows** are dispersions of cohesionless sediment in

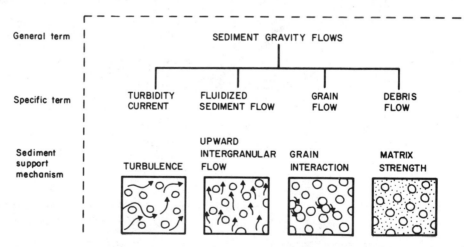

FIGURE 3.11 Principal types of sediment gravity flows and the types of interactions between fluids and grains that keep sediment supported during transport. (After Middleton, G. V., and M. A. Hampton, Subaqueous sediment transport and deposition by sediment gravity flows, in D. H. Stanley and D. J. P. Swift (eds.), Marine sediment transport and environmental management. © 1976 by John Wiley & Sons, Inc. Fig. 1, p. 198, reprinted by permission of John Wiley & Sons, Inc., New York.)

which the sediment is supported by dispersive pressures owing to direct grain-to-grain collisions or close approaches. Flow can occur rapidly under both subaerial and subaqueous conditions, especially on steep slopes that approach the angle of repose for the sediment. (4) **Debris flows** and **mudflows** are slurrylike flows in which large grains, ranging up to boulder size, are supported in a matrix of fine sediment and interstitial water that has enough cohesive strength to prevent larger particles from settling out, but not enough strength to prevent flow. Debris flows can occur on gentle or steep slopes in subaerial or subaqueous environments.

These four mechanisms of gravity transport are best thought of as end members of a spectrum of gravity-flow processes. One type of process may grade into another under some conditions. For example, submarine mudflows may change into turbidity currents downslope with additional mixing and dilution by water. We will now examine each of these major sediment gravity-flow processes in greater detail.

Turbidity Currents

Density currents are currents that are generated by gravity acting on differences in density between adjacent bodies of fluids. Density differences may arise from salinity or temperature variations or from sediment suspended in the fluid. A turbidity current is a special type of density current that flows downhill along the bottom of an ocean or lake because of density contrasts with the ambient water caused by sediment suspended in the water owing to turbulence. Turbidity currents can be generated experimentally in the laboratory by sudden release of muddy, dense water into the end of a sloping flume filled with less dense, clear water. They have been observed to occur under natural conditions in lakes where muddy river water enters the lakes. There is strong geologic evidence to support the supposition that they have also occurred throughout geologic time in the marine environment on continental margins. In this setting, they appear to originate mainly in or near the heads of submarine canyons, although they may occur on the continental shelf in areas where submarine canyons are absent.

Turbidity currents can be generated by two mechanisms. Those that occur on the continental margins appear to be **surges,** or spasmodic turbidity currents, that are initiated by some short-lived catastrophic event such as earthquake-triggered massive sediment slumping or storm waves acting on a continental shelf. Such an event creates intense turbulence in the water overlying the ocean floor, resulting in extensive erosion and entrainment of sediment, which is rapidly thrown into suspension. The sediment then remains suspended, supported in the water column by turbulence. This process generates a dense, turbid cloud that moves downslope, eroding and picking up more sediment as it increases in speed. Middleton and Hampton (1976) suggest that surge flows develop into three main parts as they move away from the source: (1) the **head,** which is about twice as thick as the rest of the flow and in which turbulence is intense; (2) the **body,** of almost uniform thickness, in which nearly steady, uniform flow occurs; and (3) the **tail,** where the flow thins abruptly and becomes more dilute. The head is overhanging and is divided transversly into lobes and clefts (Fig. 3.12). According to Middleton and Hampton, the head of the turbidity current moves with a velocity (V) given by

$$V = 0.7 \sqrt{\frac{\Delta\rho}{\rho} g d_2}$$ (3.19)

FIGURE 3.12 Postulated structure
of the head and body of a turbidity
current advancing into deep water.
The tail is not shown. (After Allen,
J. R. L., 1985, Loose boundary hydraulics and fluid mechanics: Selected advances since 1961, *in* P. J.
Brenchley and D. J. P. Swift (eds.),
Sedimentology: Recent developments and applied aspects. Fig. 8,
p. 20, reprinted by permission of
Blackwell Scientific Publications
Limited, Oxford.)

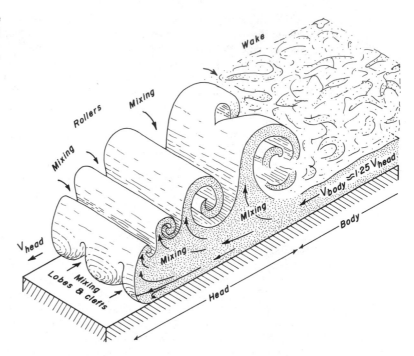

where ρ is the density of the ambient water, $\Delta\rho$ is the difference in density of the
ambient water and the turbid flow, d_2 is the thickness of the head, and g is gravitational acceleration. Note from this relationship that flow velocity of the head is independent of the slope of the ocean floor. The body of a turbidity current in deep water
flows about 25 percent faster than the head, causing the current to consume itself in
the process of mixing with the ambient water (J. R. L. Allen, 1985).

The second type of turbidity current arises from steady uniform flow of a dense
fluid moving below a less dense fluid in a channel with a sloping floor. Such currents
have been observed to flow along the sloping bottoms of lakes where sediment-laden
rivers flow into the lakes. It is postulated that they may also occur on a continental
shelf where muddy rivers discharge onto the shelf; however, they are less likely to
occur in this setting because the density contrast between muddy river water and
ocean water is less than that between muddy river water and freshwater. The velocity
of steady uniform flows ($\overline{u}$) is given by

$$\overline{u} = \sqrt{\frac{8g}{f_0 + f_1}\left(\frac{\Delta\rho}{\rho}\right)dS} \qquad \textbf{(3.20)}$$

where d is the thickness of the flow, S is the slope of the bottom, f_0 is the frictional
resistance at the bottom of the flow, and f_1 is the frictional resistance at the upper
interface of the flow in contact with the overlying water layer (Middleton and Hampton, 1976). The velocity of steady-flow turbidity currents, including the body of surge
flows, is sensitive to the slope over which flow takes place, although flows may occur
on slopes as low as 1 degree (Kersey and Hsü, 1976).

Once sediment is suspended owing to turbulence generated by the initial event, the turbidity current, without additional energy input, continues to flow for some time under the action of gravity. Flow will stop when the sediment–water mixture that produces the density contrast with the ambient water is exhausted by settling of the suspended load. Rapid deposition of coarser particles from suspension appears to occur in regions near the source because of early decay of the extremely intense turbulence generated by the initial event. As the flow continues to move forward, the remaining coarser material will be progressively concentrated in the head of the flow—denser fluid must be continuously supplied to the head to replace that lost to eddies that break off from the head and rejoin the body of the flow. Owing to differences in turbulence in the head and body, the head may be a region of erosion while deposition is taking place from the body.

Theoretically, much of the sediment remaining in suspension after initial deposition of coarse material in the proximal area can be maintained in suspension for a very long time in a state of dynamic equilibrium called **autosuspension** (Bagnold, 1962; Pantin, 1979; Parker, 1982). A condition of autosuspension is presumably maintained because turbulence continues to be generated in the bottom of the flow owing to gravity-generated downslope flow of the turbidity current over the bed. Thus, loss of energy by friction of the flow with the bottom is compensated for by gravitational energy. The distance that turbidity currents can travel in the ocean is not known from unequivocal evidence. A presumed turbidity current triggered by the 1929 Grand Banks earthquake off Nova Scotia appears to have traveled south across the floor of the Atlantic for a distance of several hundred kilometers at a velocity of at least 7 m/sec (based on timing of breaks in submarine telegraph cables). If sediment was indeed transported over this distance by a turbidity current, such transport constitutes supporting evidence that autosuspension actually works. The Grand Banks event has been interpreted by some scientists to have been a grain flow or fluidized sediment flow rather than a turbidity current flow.

The velocity of a turbidity current eventually diminishes owing to flattening of the canyon slope, overbank flow of the current along a submarine channel, or spreading of the flow over the flat ocean floor at the base of the slope. As the flow slows, turbulence generated along the sole of the flow also diminishes and the current gradually becomes more dilute owing to mixing with ambient water around the head and along the upper interface. The remaining sediment carried in the head eventually settles out, causing the head to sink and dissipate. The exact process by which deposition takes place from various parts of a turbidity current is still not thoroughly understood, although it seems clear from experimental results that deposition does not occur in all parts of the current at the same time. As mentioned, for example, the head may be a region of potential erosion at the same time deposition is taking place from the body behind the head. Sediment that is deposited very rapidly from some part of the flow, such as the head, may undergo little or no subsequent traction transport before being quickly buried. On the other hand, in more distal parts of the flow or in areas where the head overflows the channel, a period of scouring by the head may be followed by slow deposition from the body and tail, during which process additional traction transport of the deposited sediment takes place. Final deposition from the tail may take place after movement of the current is too weak to produce traction transport.

Depending upon position within the turbidity flow and the initial amount of sediment put into suspension by the flow, a turbidity current may contain either high concentrations of sediment or relatively low concentrations. Two principal types of turbidity currents, based on suspended particle concentration, can be considered: **low-**

density flows, containing less than about 20–30 percent grains, and **high-density flows,** containing greater concentrations (Lowe, 1982). Low-density flows are made up largely of clay, silt, and fine- to medium-grained sand-size particles that are supported in suspension entirely by turbulence. High-density flows may include coarse-grained sands and pebble- to cobble-size clasts as well as fine sediment. Support of coarse particles during flow is provided by turbulence aided by hindered settling resulting from high particle concentrations and the buoyant lift provided by the interstitial mixture of water and fine sediment. For example, the heads of turbidity currents may be high-density flows, whereas the tails may be dilute, low-density flows. As we shall see, different types of deposits form from high-density and low-density flows.

Fluidized and Liquefied Flows

Loosely packed, cohesionless sediment such as sand can become temporarily liquefied owing to a sudden shock, or series of shocks, that causes the grains to momentarily lose contact with each other and become suspended in their own pore fluid. Grain contact may also be lost if a fluid is introduced into the base of a mass or column of cohesionless sediment and injection is continued until the grains are pushed apart, with their weight being supported by the rising fluid. This process is called **fluidization.** Once the cohesionless sediment has become liquefied or fluidized, it loses its strength and behaves like a high-viscosity fluid that can flow quite rapidly down slopes as low as 3 degrees.

Flow can occur only as long as grain dispersion is maintained. As soon as the grains settle out of the fluid and reestablish grain-to-grain contact, the flowing layer will "freeze up" and stop moving. "Freezing" begins at the base of the flow; a surface of settled grains rises through the dispersion (Fig. 3.13) at a rate determined by the settling velocity of the particles. The time required for settling to occur is on the order of hours for thick, fine-grained flows (Lowe, 1976); therefore, fluidized and liquefied flows may travel short, though potentially important, distances before deposition occurs. The upward movement of pore waters through the settling grains as deposition occurs leads to formation of a number of fluid escape structures such as dish structures. Some fluidized and liquefied flows may become turbulent as the flowing sediment mass is accelerated downslope and thus change into turbidity currents.

Grain Flows

Grain flow is the movement of cohesionless sediments down steep slopes owing to sudden loss of internal shear strength of the sediment. Grain flow begins when traction processes cause cohesionless sediment, commonly sand, to be piled up beyond the critical angle of repose. This angle is a function of grain packing and grain shape and tends to be greatest in deposits with angular grains of low sphericity. When the angle of repose for a particular sediment is exceeded, avalanching occurs; flow quickly begins when the internal shear stresses owing to gravity exceed the internal shear strength of the sediment. The **dispersive pressures** needed to force the grains apart and keep them suspended during flow are provided not by fluid but by grain-to-grain collisions and close encounters as the failed mass of sediments moves down a slope. During the interaction of grains, dispersive pressure is the force normal to the plane of shearing and tends to expand or disperse the grains in that direction. Bagnold (1956) suggested that the relation between the shear stress (T) acting on grains and the dispersive pressure (P) is

FIGURE 3.13 Schematic representation of grain settling and water expulsion during deposition of sand from a liquefied flow. (After Allen, J. R. L., and N. L. Banks, 1972, An interpretation and analysis of recumbent-folded deformed cross bedding: Sedimentology, v. 19. Fig. 3, p. 267, reprinted by permission of Elsevier Science Publishers, Amsterdam.)

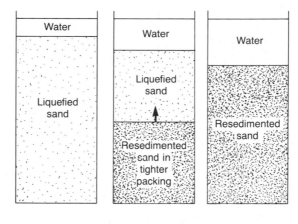

$$T/P = \tan a \qquad\qquad (3.21)$$

where a is the angle of internal friction. The minimum slope on which sustained grain flow is possible is about 30 degrees (Blatt et al., 1980). Although dispersion, or dilation, of sand grains is achieved and maintained during flow primarily by grain collisions, dispersion may be aided under some conditions by upward flow of pore fluids as grains settle or possibly by buoyancy of a dense mud matrix. Grain flow is similar to fluidized and liquefied flow in many respects and may, in fact, grade into these flows. In contrast to fluidized and liquefied flows, grain flow can occur under subaerial conditions as well as subaqueous conditions.

Grain flow is a common occurrence on the lee slopes of sand dunes. Flows of cohesionless sand have also been observed and photographed in the ocean as they moved down steep slopes in submarine canyons (Shepard, 1961; Dill, 1966; Shepard and Dill, 1966). Grain flows over the floors of Norwegian fjords are reported to be responsible for breaking submarine telephone cables. Grain flows may be of limited geological significance because of the steep slopes required to initiate flow, although it has been suggested that grain flow may accompany turbidity currents on less steep slopes, moving beneath but independently of the turbidity currents (Friedman and Sanders, 1978). Deposition of grain-flow sediment occurs quickly and en masse by sudden "freezing" owing primarily to reduction of slope angle.

A grain-flow origin has been suggested by some workers for very thick, almost massive sandstone beds; however, Lowe (1976) concludes that the deposits of a single grain flow cannot be thicker than a few centimeters for sand-size grains. Reverse grading—that is, grading from fine size to coarse size upward—that occurs in some sandstones has been attributed to grain-flow processes. Reverse grading is assumed to occur during grain flow as a result of smaller particles filtering down through larger particles while they are in the dispersed state, a process called **kinetic sieving.**

Debris Flows and Mud Flows

Debris flows and mudflows are sediment gravity flows that behave as Bingham plastics; that is, they have a yield strength that must be overcome before flow begins. They consist of mixtures of fine sand and clay particles forming a muddy matrix that has sufficient cohesive strength and buoyancy to support very large boulders. Subaerial flows occur under many climatic conditions, but are particularly common in arid and semiarid regions, where they are usually initiated after heavy rainfalls. **Lahars** are debris flows composed largely of volcanic particles that become water saturated during heavy rains that accompany volcanic eruptions or of particles from melting of ice and snow that accumulate on volcanic cones between eruptions.

After the yield strength of a debris flow has been overcome owing to water saturation, and movement begins, the flow may continue to move over slopes as low as 1 or 2 degrees (Curray, 1966). Debris flows are believed to occur also in subaqueous environments, possibly as a result of mixing at the downslope ends of subaqueous slumps. As subaqueous debris flows move rapidly downslope and are diluted by mixing with more water, their strength is reduced and they may pass into turbidity currents. Deposition of the entire mass of debris flows and mudflows occurs quickly. When the shear stress owing to gravity no longer exceeds the yield strength of the base of the flow, the mass "freezes" and stops moving.

Deposits of Sediment Gravity Flows

Important differences exist between the deposits of sediment gravity flows and the sediments deposited by normal fluid flows. Sediment deposition by fluid flow takes place on a grain-by-grain basis and may be either rapid or slow depending upon the sizes of the particles and the depositional conditions. Under the conditions that exist at many depositional sites, such as beach and eolian environments, grains may be deposited and reentrained many times before final deposition occurs, leading to improved sorting and increased rounding of the grains. Because of the prevalence of traction during fluid flow, sediments deposited by fluid flows commonly display a variety of sedimentary structures such as fine laminations, cross-bedding, and ripple marks, as mentioned. By contrast, sediments transported by gravity-flow processes are deposited rapidly, commonly by mass emplacement. Little or no opportunity exists for reworking of the sediment, with the exception of some turbidity current deposits, which can be reworked to some extent during subsequent traction transport. The deposits of sediment gravity flows tend to be poorly sorted and the grains poorly rounded and, with the exception of turbidity current deposits, show few internal current-generated structures. The characteristics of the principal types of sediment gravity flow deposits are illustrated in Figure 3.14.

Turbidity current deposits, commonly called **turbidites,** are of two basic types. Turbidites deposited from high-density flows with high sediment concentrations tend to be thick bedded and coarse grained. They typically have relatively poor grading and few internal laminations, and basal scour marks are either poorly developed or absent. Some thick-bedded turbidites may grade upward to finer grained deposits that display traction structures such as laminations and small-scale cross-bedding. In the uppermost part of the flow units, the sediments may consist of very fine-grained, nearly homogeneous muds deposited from the tail of the flow. The deposits of more dilute, low-concentration turbidity current flows are thin bedded and fine grained at the base with good vertical size grading, well-developed laminations, and small-scale cross-bedding. Scour marks may be present on the soles or bottoms of the beds (Fig. 3.14A).

An ideal turbidite sequence, called a **Bouma sequence,** has been proposed. It consists of five structural units (Fig. 3.15) that include the characteristics of both types of turbidites. These structural subdivisions presumably record the decay of flow strength of a turbidity current with time and the progressive development of different sedimentary structures and bedforms in adjustment to different flow regimes (Chapter 6) as current flow velocity wanes. Most turbidites do not contain all of these structural

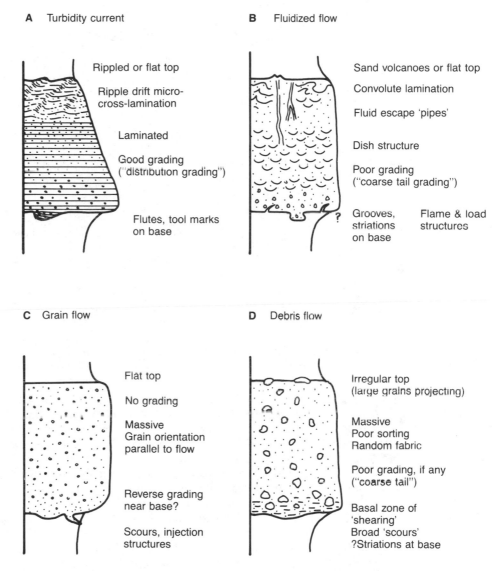

A Turbidity current

Rippled or flat top

Ripple drift micro-cross-lamination

Laminated

Good grading ("distribution grading")

Flutes, tool marks on base

B Fluidized flow

Sand volcanoes or flat top

Convolute lamination

Fluid escape 'pipes'

Dish structure

Poor grading ("coarse tail grading")

Grooves, striations on base   Flame & load structures

C Grain flow

Flat top

No grading

Massive
Grain orientation parallel to flow

Reverse grading near base?

Scours, injection structures

D Debris flow

Irregular top (large grains projecting)

Massive
Poor sorting
Random fabric

Poor grading, if any ("coarse tail")

Basal zone of 'shearing'
Broad 'scours'
?Striations at base

FIGURE 3.14 Comparison of sedimentary structures in different types of sediment gravity flow deposits. (After Middleton, G. V., and M. A. Hampton, Subaqueous sediment transport and deposition by sediment gravity flows, in D. H. Stanley and D. J. P. Swift (eds.), Marine sediment transport and environmental management. © 1976 by John Wiley & Sons, Inc. Fig. 9, p. 213, reprinted by permission of John Wiley & Sons, Inc., New York.)

units. Thick-bedded, coarse-grained turbidites tend to show well-developed A and B units, but C through E units are commonly poorly developed or absent. Thin-bedded, finer grained turbidites commonly display well-developed C–E units and poorly developed or absent A and B units.

It has frequently been suggested that thick-bedded turbidites are deposited in areas proximal to the turbidite source and thin-bedded turbidites in more distal re-

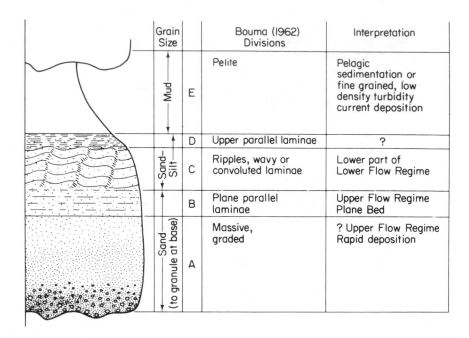

Grain Size		Bouma (1962) Divisions	Interpretation
Mud	E	Pelite	Pelagic sedimentation or fine grained, low density turbidity current deposition
Sand-Silt	D	Upper parallel laminae	?
	C	Ripples, wavy or convoluted laminae	Lower part of Lower Flow Regime
	B	Plane parallel laminae	Upper Flow Regime Plane Bed
Sand (to granule at base)	A	Massive, graded	? Upper Flow Regime Rapid deposition

FIGURE 3.15   Ideal sequence of sedimentary structures (Bouma sequence) in a turbidite bed. (After Blatt, H., G. V. Middleton, and R. Murray, Origin of sedimentary rocks, 2nd ed. © 1980, Fig. 5.7, p. 145. Reprinted by permission of Prentice-Hall, Englewood Cliffs, N. J.)

gions. This concept has been challenged recently as greatly oversimplified (Nilsen, 1980). For example, thin-bedded, fine-grained turbidites can be deposited in proximal areas where turbidity currents overflow the banks of a channel and become more dilute as they spread out over the seafloor. Furthermore, coarse-grained turbidites may be deposited within the main channel at considerable distances from the source. Therefore, it may be best to abandon the terms "proximal" and "distal" when referring to different types of turbidites.

Some geologists have suggested that the deposits of a single turbidity current flow display horizontal size grading as well as vertical size grading. Thickest and coarsest deposits presumably occur in proximal areas or within the main channel in more distal areas. These coarser deposits are believed to grade laterally to thinner and finer grained sediments. Turbidites laid down near the source, or within the main transport channel where suspended sediment concentrations are high, typically are of the coarse-grained, massive or poorly laminated type. Sediment deposition in more distal regions of a flow or in proximal areas where a turbidity current has overflowed its channel takes place from parts of the flow in which suspended sediment concentrations are low, thus producing the thin-bedded, fine-grained type of turbidite.

Fluidized- and liquefied-flow deposits are typically thick, poorly sorted sand units that are characterized particularly by fluid escape structures such as dish structures, pipes, and sand volcanoes (Fig. 3.14B). **Grain-flow deposits** are massively bedded with little or no internal lamination and grading except possible reverse grading in the base. Deposits of a single grain flow are commonly less than about 5 cm thick. **Debris-flow deposits** are thick, poorly sorted units that lack internal layering. They typically consist of chaotic mixtures of particles that may range in size from clay to boulders. The large particles commonly show no preferred orientation. They are generally poorly graded, but if grading is present it may be either normal or reverse.

As pointed out above, the four principal types of sediment gravity flows are end members of a spectrum of gravity-flow processes that may grade into each other. The major types of sediment gravity-flow deposits described above are likewise end members of a spectrum of deposits that may be gradational in character. Thus, for example, gradations may exist between typical grain-flow deposits and some turbidity current deposits or between mudflow deposits and turbidites.

ADDITIONAL READINGS

Allen, J. R. L., 1970, Physical processes of sedimentation: George Allen & Unwin, London, 248 p.

Garde, R. J., and K. G. Ranga Raju, 1978, Mechanics of sediment transport and alluvial stream problems: Halsted Press, New York, 483 p.

Komar, P. D., 1976, Beach processes and sedimentation: Prentice-Hall, Englewood Cliffs, N.J., 429 p.

Lowe, D. R., 1982, Sediment gravity flows: II. Depositional models with special reference to the deposits of high-density turbidity currents: Jour. Sed. Petrology, v. 52, p. 279–297.

Middleton, G. V., and M. A. Hampton, 1976, Subaqueous sediment transport and deposition by sediment gravity flows, in D. J. Stanley and D. J. P. Swift (eds.), Marine sediment transport and environmental management, John Wiley & Sons, New York, p. 197–218.

Middleton, G. V., and J. B. Southard, 1984, Mechanics of sediment movement: Eastern Section, Soc. Econ. Paleontologists and Mineralogists Short Course Notes No. 3, 2nd ed., 401 p.

Saxov, S., and J. K. Nieuwenhuis (eds.), 1982, Marine slides and other mass movements: NATO Conference Series IV: Marine Sciences, v. 6: Plenum Press, New York, 353 p.

Stanley, D. J., and D. J. P. Swift (eds.), 1976, Marine sediment transport and environmental management: John Wiley & Sons, New York, 602 p.

Yalin, M. S., 1977, Mechanics of sediment transport, 2nd ed.: Pergamon Press, New York, 298 p.

4

Deposition of Nonsiliciclastic Sediments

4.1 INTRODUCTION

Chapter 3 explores the physical processes involved in transport and deposition of particulate silicate detritus generated by subaerial weathering and explosive volcanism. The present chapter carries this theme a step farther by examining the largely chemical and biochemical processes responsible for transport and deposition of the soluble products of subaerial and submarine weathering. Chemical and biochemical processes operate during weathering to bring about decomposition of rock-forming minerals and the release of soluble chemical constituents from parent-rock minerals. As pointed out in Chapter 2, some of these constituents recombine at the weathering site to form secondary minerals such as clay minerals and iron oxides. The remaining constituents are removed in solution, either in water that makes up the surface runoff or by water that infiltrates the soil and becomes part of the groundwater system. Some of the chemical constituents dissolved in surface waters and groundwaters make their way into lakes and ultimately may be precipitated to form nonmarine sedimentary rocks such as freshwater limestones, cherts, and evaporites. Most dissolved constituents are transported to the ocean, where they remain dissolved in seawater for periods ranging from hundreds to millions of years. The average amount of time that a particular chemical element remains in solution in the ocean before precipitating is called its **residence time.** Residence times of some common elements in seawater are given in Table 4.1.

This chapter begins by examining the origin of the chemical constituents that occur in river waters and the ocean. This discussion is followed by a brief look at the processes by which chemical constituents are transported to the ocean by rivers. Finally, the chemical and biochemical processes that cause dissolved chemical constituents to precipitate from seawater are discussed and related to the origin of the major kinds of nonsiliciclastic sedimentary rocks.

TABLE 4.1 Residence times of selected elements in seawater

Element	Residence time (yrs)
Cations	
Sodium (Na^+)	260,000,000
Magnesium (Mg^{2+})	12,000,000
Potassium (K^+)	11,000,000
Calcium (Ca^{2+})	1,000,000
Silicon (Si^{4+})	8,000
Manganese (Mn^{2+})	7,000
Iron (Fe^{2+}, Fe^{3+})	140
Aluminum (Al^{3+})	100
Anions	
Chlorine (Cl^-)	∞
Sulfate (SO_4^{2-})	11,000,000
Carbonate (CO_3^{2-})	110,000

Source: Ross, D. A., 1982, Introduction to oceanography: Prentice-Hall, Englewood Cliffs, N. J., and Stowe, K. S., 1979, Ocean science: John Wiley & Sons, New York.

4.2   ORIGIN AND RELATIVE ABUNDANCE OF DISSOLVED CONSTITUENTS

Some of the soluble constituents released from parent-rock materials during chemical weathering are carried away by surface runoff directly into streams. Others make their way into streams indirectly through the groundwater system. As rainwaters carrying dissolved constituents percolate through soils and into subsurface rock formations, they may lose some of their dissolved constituents owing to precipitation of mineral cements in sediment pores. On the other hand, groundwaters containing high concentrations of dissolved carbon dioxide or organic acids, picked up as the water moved through the soil zone, may increase their load of dissolved chemical constituents by reaction with various minerals in subsurface sediments and rocks. For example, the calcium and bicarbonate content of groundwaters that come in contact with limestones can be significantly increased by solution of the limestones. Groundwaters and their dissolved constituents may ultimately be discharged into streams at lower elevations. Additional chemical constituents can be added to river water by solution of rock materials that make up the stream bed, particularly in regions where the streams flow over easily dissolved rocks such as limestones and dolomites. Thus, the total dissolved load of rivers originates by (1) subaerial chemical weathering, (2) reaction of groundwaters with subsurface sediments and rocks before the groundwaters are discharged into streams, and (3) solution of sediment and rock by rivers as they flow across their beds.

The dissolved constituents in river waters are ultimately delivered to the ocean. Chemical constituents can also be added to the ocean by discharge of groundwaters directly into the ocean in coastal areas and by chemical and biochemical processes operating on rocks and sediments along the shoreline. Reaction of seawater with hot volcanic rocks along mid-ocean ridges and with other rocks and minerals on the deep seafloor also adds significant quantities of dissolved substances to ocean water, as discussed in Chapter 2.

The major dissolved constituents in average river water are shown in Table 4.2. River waters on the average contain about 120 parts per million (ppm) dissolved solids compared with about 35,000 ppm in the oceans. Note from Table 4.2 that the dominant

TABLE 4.2 Dissolved ion species in mean world river water and ocean water

Ionic species	Mean river water (a)		Ocean water (b)	
	ppm	% of total dissolved solids	ppm	% of total dissolved solids
HCO_3^-, CO_3^{2-}	58.7	48.6	140	0.4
Ca^{2+}	15.0	12.4	400	1.2
H_4SiO_4	13.1	10.8	1	<0.01
SO_4^{2-}	11.2	9.3	2,649	7.7
Cl^-	7.8	6.5	18,980	55.0
Na^+	6.3	5.2	10,556	30.6
Mg^{2+}	4.1	3.4	1,272	3.7
K^+	2.3	1.9	380	1.1
NO_3^-	1.0	0.8	0.5	<0.01
Fe^{2+}, Fe^{3+}	0.67	0.6	0.01	<0.01
$Al(OH)_4^-$	0.24	0.2	0.01	<0.01
F^-	0.09	0.07	1.3	<0.01
Sr^{2+}	0.09	0.07	8	0.02
$B(OH)_4^-$	0.1–0.01	0.08–<0.01	26	0.07
Mn^{2+}	0.02	0.02	—	—
Br^-	—	—	65	0.02
Total	120.8		34,479	

Source: (a) Livingston, D. A., 1963, Data of geochemistry. Chap. G, Chemical composition of rivers and lakes: U. S. Geol. Survey Prof. Paper 440-G. (b) Mason, B., 1966, Principles of geochemistry: John Wiley & Sons, New York.

dissolved constituents in river water are bicarbonate (HCO_3^-), calcium (Ca^{2+}), and silica (expressed as H_4SiO_4). Bicarbonate is derived mainly from soil CO_2 and by solution of carbonate minerals in limestones and dolomites. Calcium is generated by solution of both calcium carbonate minerals and silicate minerals such as calcium rich feldspars. Silica is derived by chemical weathering of silicate minerals, particularly by hydrolysis. The chemical composition of individual rivers is controlled by the major rock types in the drainage areas of the rivers. Thus, rivers draining different terranes such as limestone, shale, schist, and granite show distinct differences in relative abundance of dissolved constituents. For example, rivers draining limestone regions commonly have bicarbonate and calcium concentrations two or three times higher than those of average river water (Table 4.2), and rivers draining volcanic regions such as the volcanic terranes of Japan have silica contents up to three times greater than that of average river water.

Most rivers contain dissolved organic substances such as humic and fulvic acids in addition to dissolved inorganic chemical constituents. The average concentration of dissolved organic carbon in world rivers has been reported to be about 5.8 ppm (Meybeck, 1981). Concentrations of dissolved humic organic carbon in individual rivers may range up to 100 ppm (Varshal et al., 1979); however, the concentration in most natural surface waters ranges from about 0.1 to 10 ppm (Stumm and Morgan, 1981).

4.3 TRANSPORT OF CHEMICAL CONSTITUENTS

Most dissolved constituents in rivers are transported to the oceans in true solution. Some ions such as calcium, magnesium, and sodium are adsorbed onto clay minerals and transported along with these particles. Others such as iron and various trace metals may become chemically complexed to dissolved organic substances and transported with these substances. It has been suggested that some chemical constituents can be transported as colloids. Substances in the colloidal state consist of very tiny dispersed particles, on the order of 10^{-3} to 10^{-6} mm (1 micron to 1 nanometer), that carry electrical charges. Chemical constituents that may possibly be transported by rivers as colloids include silica (negative charge), ferric hydroxide (usually a positive charge, but may also be negative), organic substances (negative charge), and carbonates (usually a positive charge). Colloids presumably are deposited when they encounter electrolytes in seawater. Electrolytes, which are dissolved cations and anions, neutralize the charges on the colloids. This process causes them to aggregate or flocculate into small lumps, which then settle out of the water column under the action of gravity. River transport of colloidal substances and their subsequent deposition in the ocean are very poorly understood processes, and the details and overall importance of colloidal transport remain to be established.

4.4   DEPOSITION OF DISSOLVED CONSTITUENTS

Chemical Controls on Precipitation

The fundamental controls on precipitation of dissolved constituents from water are pH and Eh. The symbol **pH** represents a measure of the acidity or alkalinity of a solution and is defined as the negative logarithm to the base 10 of the approximate hydrogen-ion concentration in moles per liter. The pH scale extends from 0 to 14, corresponding to H^+ concentrations ranging from 10^0 to 10^{-14}. For example, a solution containing an H^+ concentration of 10^{-1} has a pH of 1; an H^+ concentration of 10^{-7} yields a pH of 7, and so forth. Solutions with a pH of 7 are considered neutral. Acids have pH values lower than 7 and bases have values greater than 7.

Eh, or redox potential, is a measure of the oxidizing or reducing capacity of a solution. Eh is particularly important to the behavior of polyvalent ions such as iron, manganese, and sulfur. Oxidation involves the loss of orbital electrons from an atom, whereas reduction involves gain of orbital electrons. Under natural conditions in sedimentary environments, Eh is controlled by the relative abundance of gaseous oxygen and organic matter available to be decomposed by oxidation. Oxidizing conditions prevail where gaseous oxygen is abundant; reducing conditions occur where gaseous oxygen is scarce and organic matter abundant, as in swampy environments. Eh is expressed on a scale ranging from slightly greater than $+1$ to -1. Positive values of Eh represent oxidizing conditions; negative values represent reducing conditions.

The pH of a solution can be measured easily and precisely in the field or laboratory by use of a pH meter or narrow-range litmus paper; however, field measurement of Eh is difficult and commonly imprecise. The range of pH and Eh conditions found in natural waters is shown in Figure 4.1. The stability of some common sedimentary minerals such as calcite and aragonite (the calcium carbonate minerals) depends primarily on pH. Solution and precipitation of these minerals is controlled by pH and is largely independent of Eh. The stability of some other common minerals, such as the iron oxides hematite and limonite, is controlled by both pH and Eh.

Precipitation of dissolved constituents is affected also by water temperature, the amount of carbon dioxide dissolved in the water, and the salinity of the water. These parameters are interrelated in various ways with pH and Eh. For example, increase in

FIGURE 4.1 Graph showing Eh and pH of waters in some natural environments. (Modified from Blatt H., G. V. Middleton, and R. Murray, Origin of sedimentary rocks, 2nd ed., © 1980, Fig. 6–12, p. 241. Reprinted by permission of Prentice-Hall, Englewood Cliffs, N. J.; based on data from Baas Becking, L. G. M., I. R. Kaplan, and D. Moore, 1960, Limits of the natural environment in terms of pH and oxidation-reduction potentials: Jour. Geology, v. 68, p. 243–284.)

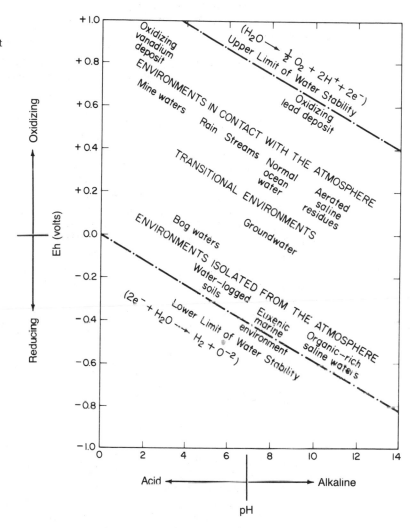

temperature or salinity causes a decrease in the solubility of carbon dioxide and oxygen in water; that is, an increase in temperature reduces the capacity of water to dissolve these gases. Because dissolved carbon dioxide increases the acidity of water by releasing H^+ ions, loss of carbon dioxide owing to increase in temperature or salinity causes an increase in pH, or decreased acidity. Loss of dissolved oxygen affects Eh by decreasing the oxidizing capacity of the system.

Increase in temperature also affects the solubility of dissolved mineral constituents. For example, the solubility of calcium carbonate minerals, which are the principal constituents of limestones, is decreased by increasing temperature. By contrast, the solubility of siliceous minerals such as quartz is increased. Decrease in solubility means that a mineral will be more likely to precipitate under a given set of conditions. Thus, calcium carbonate deposition is favored in the more tropical areas of the ocean, where surface water temperatures may reach almost 30°C, compared to about 0°C in the polar regions. It is not known that temperature change has any appreciable effect on precipitation of silica from ocean water; however, increase in temperature during sediment burial and diagenesis acts to cause solution of silicate minerals and prevent precipitation of silica cements in the pores of sediments.

The solubility of chemical constituents is affected also by salinity and the ionic strength of seawater. Ionic strength is a function of the concentration of ions in solution and the charges on these ions; thus, ionic strength increases as salinity increases. The solubility of calcium carbonate minerals, for example, is markedly enhanced at higher values of salinity because increase in ionic strength causes an increase in the concentration of foreign ions other than Ca^{2+} and CO_3^-. These foreign ions interfere with formation of the calcium carbonate crystal structure, making it more difficult for calcite or aragonite minerals to precipitate and grow. In general, the amount of solubility increase will depend upon the concentration of added salts (Krauskopf, 1979). Therefore, the solubility of calcium carbonate is several orders of magnitude higher in seawater than in freshwater (Degens, 1965). On the other hand, the influence of salinity on the solubility of calcium carbonate, as well as on the solubility of carbon dioxide and oxygen, in the surface waters of the open ocean may be slight because these waters range in salinities only from about 32 to 36 parts per thousand (‰).

Biochemical Controls on Precipitation

Although precipitation of minerals from water is fundamentally a chemical process, it has become increasingly clear in the last few decades that chemical processes are aided in a variety of ways by organisms. In fact, simple inorganic precipitation of minerals from normal-salinity seawater or freshwater may be relatively uncommon today. Furthermore, geologic evidence suggests that organisms may have played a very significant role in chemical sedimentation processes throughout at least most of Phanerozoic, or post-Precambrian, time.

The most important role of organisms in chemical sedimentation is probably the direct removal of dissolved constituents to build skeletal structures. The exact mechanisms by which organisms remove dissolved substances to build their shells is not well understood, but the process is very common. Numerous marine invertebrates build protective shells or other skeletal structures from calcium carbonate. These organisms range from freely drifting, planktonic species such as foraminifers and pteropods (winged snails) to bottom-dwelling benthonic organisms such as calcareous algae, corals, and molluscs. These organisms can remove $CaCO_3$ not only from calcium carbonate-saturated surface waters in tropical regions, but also from less saturated waters in temperate and colder regions. It is not definitely known that they can precipitate $CaCO_3$ from waters that are highly undersaturated in calcium carbonate (Krauskopf, 1979). The importance of biologic removal of calcium carbonate from the oceans is demonstrated by the fact that most Phanerozoic limestones contain some recognizable calcium carbonate fossils, and many are composed predominantly of such remains. Also, large areas of the modern ocean floor are covered by calcareous oozes composed predominantly of the shells of foraminifers and pteropods.

A few species of organisms, particularly planktonic diatoms and radiolarians, secrete siliceous shells or tests. These organisms can remove silica from ocean water that is greatly undersaturated in silica. In fact, biologic removal of silica from seawater appears to be the only important mechanism for precipitation of silica in the oceans today. The remains of silica-secreting organisms are important components of ancient chert beds and also form siliceous oozes in the deeper parts of the modern ocean. A few organisms such as certain primitive brachiopods can extract phosphorus from sea-water to build phosphatic shells, and many marine organisms extract phosphates from seawater as nutrients.

Another type of organic activity that is very important to some chemical depositional processes is removal of carbon dioxide from water by photosynthesizing plants. Aquatic plants, particularly blue-green algae (also referred to as cyanobacteria) and planktonic algae such as diatoms, remove carbon dioxide from water during the process of photosynthesis as shown by the following relationship:

$$6H_2O + 6CO_2 \rightarrow C_6H_{12}O_6 + 6O_2 \tag{4.1}$$

(water + carbon dioxide → carbohydrates + oxygen)

Blue-green algae and small phytoplankton such as diatoms and dinoflagellates are the most important users of carbon dioxide in the marine realm. The activities of photosynthesizing plants are at a peak in sunlight and at a minimum in the dark; therefore, the carbon dioxide content of water in which active photosynthesis is taking place can vary measurably from day to night. Removal of CO_2 by organisms increases pH, as mentioned, and may promote some depositional processes, particularly precipitation of calcium carbonate. Decay of dead organisms can also affect pH. Decay processes commonly cause pH to decrease owing to release of various organic acids and carbon dioxide to the water, but some decay products can be alkaline.

Bacterial activity may contribute to some depositional processes, although the influence and overall importance of bacterial activity is still poorly understood. Bacteria have been suggested to cause deposition, or at least to catalyze deposition, of ferric hydroxide deposits such as bog iron ores. They may also play some role in the formation of manganese nodules. Precipitation of iron sulfides such as pyrite may also be aided by sulfur-reducing bacteria. Some sulfur-reducing bacteria have the ability to break down calcium sulfates (gypsum and anhydrite) to form calcite (Friedman and Sanders, 1978). Breakdown of gypsum or anhydrite by sulfate-reducing bacteria can also lead to formation of native sulfur. This process may have been important in generating native sulfur deposits in salt domes.

4.5 DEPOSITION OF MAJOR TYPES OF NONSILICICLASTIC SEDIMENTARY ROCKS

Table 4.2 shows the principal ion species dissolved in both mean river water and ocean water. Comparison of the relative abundance of various constituents in river water and ocean water reveals some interesting differences. Note that the actual concentration (in ppm) of most constituents is greater in ocean water than in river water; however, the relative abundance (percent of total dissolved solids) of certain major constituents is much greater in river water than in ocean water. For example, bicarbonate ions (HCO_3^-) and carbonate ions (CO_3^{2-}) together make up almost 49 percent of the total dissolved solids in average river water but less than 1 percent of the dissolved solids in ocean water. The relative abundances of silica, expressed as H_4SiO_4, and calcium are also several orders of magnitude higher in river water than in the ocean. On the other hand, chlorine and sodium are relatively much more abundant in the ocean.

The explanation for these differences in relative abundance of ions in river water and ocean water is provided by examination of the relative abundance of various kinds of nonsilicilastic sedimentary rocks preserved in the geologic record. Measurements of the thickness and volume of different types of sedimentary rocks show that chemically and biochemically deposited sedimentary rocks make up about one-fifth to one-fourth of all sedimentary rocks. The majority of these rocks are limestones, composed of calcium carbonate minerals. Carbonate ions, bicarbonate ions, and calcium ions have been preferentially removed from the oceans throughout geologic time to form limestones, thus accounting for the low relative abundance of these ions in the modern ocean.

Evaporite deposits composed of salts of sodium, calcium, and potassium chlorides and sulfates are the second most abundant type of nonsiliciclastic sedimentary rocks. Siliceous deposits or cherts composed predominantly of the recrystallized remains of silica-secreting organisms are also common and moderately abundant in the rock record. Other less abundant, but economically important, nonsiliciclastic sedimentary rocks include sedimentary iron deposits and phosphorites. The selective removal from the oceans of ions such as calcium, silica, and iron to form these various kinds of sedimentary rocks throughout geologic time thus provides a rational explanation for the differences between the composition of average river water and the composition of the modern ocean.

The nonsiliciclastic sedimentary rocks are generated through chemical and biochemical processes. Most were deposited under marine conditions, although deposits of freshwater limestones, cherts, evaporites, and iron-rich sediments are known from the geologic record. We will now examine in greater detail the processes involved in deposition of each of these major groups of sedimentary rocks.

Calcium Carbonate Rocks

Limestones

Chemistry of Calcium Carbonate Deposition. The dissolution and precipitation of the calcium carbonate ($CaCO_3$) minerals calcite and aragonite are chiefly controlled by pH, as shown in Figure 4.2. Solution pH is linked in turn to the partial pressure of dissolved carbon dioxide in the water, as illustrated by the following reactions:

$$CO_2 + H_2O \rightleftharpoons H_2CO_3 \text{ (carbonic acid)} \qquad \textbf{(4.2)}$$

$$H_2CO_3 \rightleftharpoons H^+ + HCO_3^- \text{ (bicarbonate ion)} \qquad \textbf{(4.3)}$$

$$HCO_3^- \rightleftharpoons H^+ + CO_3^{2-} \text{ (carbonate ion)} \qquad \textbf{(4.4)}$$

These reactions show that the dissociation of carbonic acid to hydrogen ions and bicarbonate ions (equation 4.3) and the further dissociation of bicarbonate ions to hydrogen ions and carbonate ions (equation 4.4) release free hydrogen ions, thus lowering the pH of the solution.

If calcite or aragonite crystals are allowed to react with a carbonic acid solution, these minerals are readily dissolved. This reaction can be summarized as

$$H_2O + CO_2 + \underset{\substack{\text{(calcite or} \\ \text{aragonite)}}}{CaCO_3} \rightleftharpoons Ca^{2+} + 2HCO_3^- \qquad \textbf{(4.5)}$$

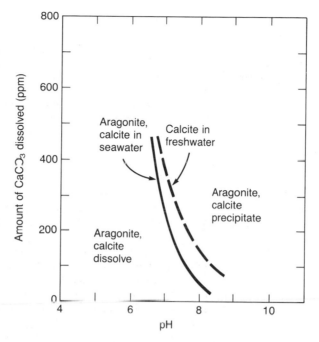

FIGURE 4.2 Effect of pH at approximately 25°C on the solubility of calcium carbonate. Solubility decreases with increasing temperature. (After Friedman, G. M., and J. E. Sanders, Principles of sedimentology: © 1978 by John Wiley & Sons, Inc. Fig. 5.23, p. 136, reprinted by permission of John Wiley & Sons, Inc., New York.)

Note by the presence of double arrows that this reaction is reversible. If equilibrium conditions are disturbed by loss of carbon dioxide, the concentration of hydrogen ions decreases and the pH increases. The reaction shifts toward the left, resulting in precipitation of solid $CaCO_3$. The partial pressure of carbon dioxide thus clearly exerts a major control on calcium carbonate precipitation. Anything that causes loss of carbon dioxide should theoretically trigger the onset of precipitation, although we shall see subsequently that inorganic precipitation of calcium carbonate owing to loss of CO_2 may not be as important under natural conditions in the open ocean, as suggested by equation 4.5.

Some of the mechanisms that can cause loss of carbon dioxide from water have already been discussed. Increase in temperature and salinity causes a decrease in the solubility of carbon dioxide in water, resulting in escape of carbon dioxide. Decrease in water pressure can also allow carbon dioxide to escape. Under natural conditions, pressure may be lowered by wave agitation caused by storm activity or breaking of waves in the surf zone or over shallow banks. Circulation of deep, pressurized waters to the surface can also release carbon dioxide, and even lowering of atmospheric pressure may cause slight loss of carbon dioxide from ocean water. The importance of carbon dioxide uptake by photosynthesizing plants has also been mentioned.

Inorganic Precipitation of Caclium Carbonate. According to the theoretical considerations illustrated in equation 4.5, significant loss of carbon dioxide by any of the mechanisms noted should lead to precipitation of calcium carbonate minerals. Considerable debate has been generated among geologists about the actual significance of inorganic precipitation of calcite or aragonite in the modern ocean. Does precipitation of calcium carbonate owing to loss of carbon dioxide occur on an important scale in the open-ocean environment today? One process considered by some investigators to constitute evidence of large-scale inorganic precipitation of $CaCO_3$ is the formation of whitings in such warm-water areas as the Bahamas, the Persian Gulf, and the Dead Sea. The sudden appearance of these **whitings,** which are milky patches of surface and near-surface water caused by dense concentrations of suspended aragonite crystals, is suggested to result from the spontaneous nucleation of aragonite crystals in waters supersaturated with calcium bicarbonate. This view has been challenged by workers who propose that mechanisms such as resuspension of aragonite from the shallow seafloor are responsible for whitings.

Unfortunately, the extent and overall importance of inorganic precipitation of calcium carbonate minerals under natural conditions in the ocean is not known from unequivocal evidence, but experimental work has demonstrated that neither calcite nor aragonite precipitate readily from seawater. There appear to be at least three reasons why inorganic precipitation of calcium carbonate may not be a widespread, quantitatively significant phenomenon in the modern ocean.

First, the magnitude of the pH changes that occur in the open ocean owing to loss of carbon dioxide is relatively small because seawater is a well-buffered solution. Buffering occurs because a considerable portion of the carbon dioxide dissolved in seawater forms undissociated H_2CO_3 rather than dissociating to H^+ ions, HCO_3^- ions, and CO_3^{2-} ions as predicted by equations 4.3 and 4.4. This buffering reaction is caused by the high alkalinity of ocean water; that is, the high concentrations of bicarbonate and carbonate ions already present in surface waters of the ocean inhibit breakdown of H_2CO_3 to form still more of these ions. Therefore, the actual change in pH in seawater owing to either gain or loss of carbon dioxide is small, and the pH values of seawater in the open ocean rarely fall outside the range of 7.8–8.3 (Bathurst, 1975).

Second, the presence of Mg^{2+} ions at the concentration levels found in seawater has been shown experimentally to have a strong inhibiting effect on the precipitation of calcite ($CaCO_3$). Experiments by Berner (1975) show that Mg^{2+} is readily absorbed onto the surface of calcite crystals and incorporated into their crystal structure. This nonequilibrium incorporation of Mg^{2+} into growing calcite crystals was interpreted by Berner to decrease their stability, resulting in an increase in calcite solubility. Thus, calcite crystals do not readily nucleate and grow in the presence of Mg^{2+} in seawater concentrations. If they form at all, the crystallization process takes place very slowly. Berner found no retardation of precipitation in similar experiments at low Mg^{2+} concentrations. Subsequent researchers have interpreted the inhibiting influence of Mg^{2+} in different ways. They have suggested difficulties in rapidly dehydrating the Mg^{2+} ion, which surrounds itself with water molecules (Mucci and Morse, 1983), and crystal poisoning by absorption of Mg^{2+} at reactive sites (Reddy and Wang, 1980).

Aragonite is also composed of $CaCO_3$ but has a different crystal structure from that of calcite. Mg^{2+} ions appear to be less prone to sorb to aragonite nuclei and disrupt crystal growth. Therefore, aragonite is less affected by Mg^{2+} and has a tendency in the presence of Mg^{2+} to precipitate in preference to calcite. Nonetheless, aragonite does not precipitate freely in ocean water, even when surface waters are supersaturated with respect to calcium carbonate. Experiments by Berner et al. (1978) show that the reluctance of aragonite to precipitate in ocean water may be due to the influence of organic compounds found in natural humic and fulvic acids or in phosphates. These compounds can apparently form thin organophosphatic coatings on aragonite seed nuclei, inhibiting nuclei growth and preventing or significantly delaying aragonite precipitation.

One process whereby calcium carbonate appears to precipitate largely by inorganic processes is the formation of oolites. Oolites are spherical, sand-size grains of calcium carbonate that, in modern environments at least, consist mainly of aragonite. These calcium carbonate spheres typically display an internal structure characterized by concentric growth rings around a nucleus of some kind, commonly a tiny shell fragment or other organic particle. Oolites form mainly under high-energy, agitated-water conditions in warm waters that are oversaturated with calcium carbonate. Warming and evaporation of cold ocean water driven onto shallow banks by tidal currents result in supersaturation of the water. Currents and waves keep the grains moving and intermittently suspended, allowing more or less even precipitation of calcium carbonate on all sides of the grains. Both supersaturation of the water and intermittent burial and resuspension of the oolites owing to agitation appear to be necessary for most oolites to form, although some oolites are known to form in quiet water. Further discussion of oolites and other carbonate grains is given in Chapter 8.

Biogenic Precipitation of Calcium Carbonate. The conclusion that emerges from the preceding discussion is that inorganic precipitation of calcium carbonate minerals may not be quantitatively important in the ocean today. How then can we account for the deposits of fine aragonite mud that occur on the modern seafloor in areas such as Florida Bay, the Bahama banks, and the Persian Gulf? The answer to this question appears to lie in the activities of various calcium carbonate-secreting organisms. As indicated in the discussion of organic controls on precipitation, many marine organisms can extract calcium carbonate from seawater to form skeletal hard parts. Because these organisms exist in large numbers in some parts of the ocean, disintegration of their skeletal remains to form fine particles has the potential to supply large quantities of fine calcium carbonate crystals to the ocean floor.

It is known, for example, that some green and red algae have calcareous skeletal elements composed of tiny, needlelike aragonite crystals that act as stiffeners for soft tissue (Fig. 4.3). When these organisms die, bacterial and chemical decomposition of the binding tissue releases the skeletal particles. This decay process yields a fine lime mud composed of aragonite crystals that are optically indistinguishable from inorgan-

FIGURE 4.3 Fresh specimens of erect calcareous green algae from the seafloor of the Great Bahama Banks. The upper part of the plant secretes an internal skeleton of aragonite and the basal rhizoids anchor the algae within the bottom sediment. From left to right, the specimens illustrated are *Halimeda incrassata, Penicillus capitatus, Rhipocephalus phoenix, Udotea flabellum.* (From Neuman, A. C., and L. S. Land, 1975, Lime mud deposition and calcareous algae in the Bight of Abaco, Bahamas: A budget: Jour. Sedimentary Petrology, v. 37. Fig. 6, p. 773, reprinted by permission of Society of Economic Paleontologists and Mineralogists, Tulsa, Okla.)

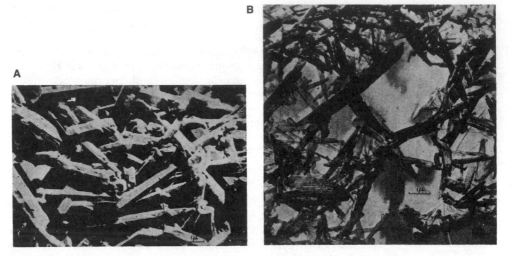

FIGURE 4.4   Electron photomicrographs of (A) aragonite crystals from the stem of *Penicillus* sp. and (B) aragonite crystals in fine lime mud taken from the seafloor west of Andros Island, Bahamas. (From Stockman, K. W., R. N. Ginsburg, and E. A. Shinn, 1967, The production of lime mud by algae in south Florida: Jour. Sedimentary Petrology, v. 37. Fig. 1 and 2b, p. 634–35, reprinted by permission of Society of Economic Paleontologists and Mineralogists, Tulsa, Okla.)

ically precipitated aragonite (Fig. 4.4). Quantitative studies of the rate of production of aragonite by disintegration of calcareous algae in the Florida Reef Tract (Stockman et al., 1967) and in the Bahamas (Neumann and Land, 1975) indicate that much or all of the aragonite mud deposited in these areas in the recent geologic past could have been supplied by skeletal disintegration of calcareous algae.

Depth Control of Calcium Carbonate Production. When precipitation of a mineral phase just equals dissolution, the solution is in equilibrium with the solid and is said to be **saturated** with this mineral phase. A solution that precipitates a mineral is **supersaturated,** and a solution that dissolves the mineral is **undersaturated.** Much of the surface water of the modern oceans is supersaturated with calcium carbonate. Little inorganic precipitation of $CaCO_3$ may actually be occurring in these waters, however, owing to Mg inhibition or other factors discussed. This condition of supersaturation changes rapidly with depth. The degree of calcium carbonate saturation drops off abruptly in waters below the surface layer; at depths greater than a few hundred meters seawater is undersaturated.

The undersaturated state of deeper waters is a function of several factors. Increase in carbon dioxide partial pressure is one of the most important variables. In shallower water, CO_2 production is increased closer to the ocean floor by the respiration of benthonic organisms. Oxidation of organic matter on the seafloor in both shallow and deeper water also increases CO_2 production. Furthermore, colder water found at depth can contain more dissolved CO_2 than warmer surface waters. Both decrease in temperature and increase in hydrostatic pressure with depth cause an increase in the solubility of calcium carbonate and thus the corrosiveness of seawater.

Because of decreasing calcium carbonate saturation of seawater with depth, calcium carbonate production is confined mainly to the very shallow-water areas of the ocean and the supersaturated surface waters of the deeper ocean. These are the waters in which most calcium carbonate-secreting organisms live. Calcium carbonate dissolution prevails in the deeper, undersaturated waters. The rate of dissolution does not, however, increase in a linear fashion with depth. Experiments in which calcite spheres were suspended on moorings at different depths in the ocean have demonstrated that little corrosion of the spheres occurred above a depth of about 3500 m, but solution of the spheres increased abruptly at that depth (Fig. 4.5). Effective solution of calcium carbonate thus occurs only at relatively great depths in the ocean. The particular depth at any locality at which the rate of dissolution of calcium carbonate equals the rate of supply of calcium carbonate to the seafloor, so that no net accumulation of carbonate takes place, is called the **calcium carbonate compensation depth** (CCD). The position of the CCD has been compared to the snowline of mountain ranges. Where biogenic oozes are accumulating in the modern ocean, white carbonate oozes cover elevated areas of the seafloor above the CCD, but give way to brown or gray pelagic clays or siliceous oozes below. The CCD ranges in depth in different parts of the modern ocean from about 3500 m to 5500 m owing to differences in rates of production of $CaCO_3$ in surface waters and variations in the factors that control carbonate saturation. The average depth of the CCD is about 4500 m.

Precambrian Limestone Deposition. The removal of calcium carbonate from seawater by organic activity has probably been the primary mechanism for precipitating calcium carbonate since at least early Paleozoic time. This conclusion is based on the fact that calcareous skeletal fragments and whole fossils are abundant constituents of most Phanerozoic limestones. We have a more difficult time explaining the formation of

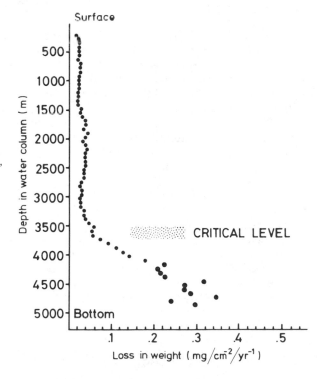

FIGURE 4.5 Weight loss of polished calcite spheres owing to solution at various depths in the ocean. Note the sharp increase in weight loss, indicating increased rate of solution, at about 3500 m. (After Peterson, M. N. A., 1966, Calcite: Rates of dissolution in a vertical profile in the Central Pacific: Science, v. 154. Fig. 2, p. 1543, reprinted by permission of American Association for the Advancement of Science, Washington, D.C.)

Precambrian limestones. The Precambrian record contains impressive thicknesses of carbonate rocks which, as far as we know, were deposited before the widespread appearance of calcium carbonate-secreting organisms. Thus, it does not seem likely on the basis of available evidence that organisms were directly responsible for deposition of large volumes of Precambrian limestone. Few, if any, Precambrian organisms could extract $CaCO_3$ to build shells. Blue-green algae (cyanobacteria) may have played an important indirect role in the precipitation of calcium carbonate through photosynthetic removal of carbon dioxide and by trapping and binding of fine carbonate sediment. These algae appear to have been particularly abundant in Precambrian time, probably because there were fewer numbers of grazing organisms that fed on the algal mats. Perhaps algae were responsible for deposition of most Precambrian limestones. Otherwise, we are left with little choice but to conclude that inorganic processes account for the formation of Precambrian carbonate rocks. Why large-scale inorganic precipitation of calcium carbonate may have been possible in Precambrian time but apparently not in post-Precambrian time remains an enigma, shrouded in our ignorance about differences in the chemical composition of the Precambrian ocean and the modern ocean.

Physical Processes in Carbonate Deposition. Calcium carbonate fossils, skeletal fragments, oolites, and other carbonate grains are subject to the same physical transport processes in the ocean as terrigenous grains. Thus, ultimate deposition of most limestones occurs through fluid-flow and sediment gravity-flow processes. Limestones may, therefore, display the same bedding characteristics and sedimentary structures as siliciclastic sedimentary rocks.

Dolomite

Dolomites are calcium carbonate rocks composed of more than 50 percent of the mineral dolomite [$CaMg(CO_3)_2$]. They are abundant and widely distributed in the geologic record, ranging in age from Precambrian to Holocene, although most dolomites are Paleozoic and older. Dolomites occur in close association with limestones and in many stratigraphic units as interbeds in the limestones; they are also commonly associated with evaporites.

Because dolomites recur so frequently in the stratigraphic record, they must have formed under environmental conditions that were relatively common and that were repeated over and over in various localities. Dolomites have been studied very extensively; therefore, in theory, we ought to understand their origin quite well. On the contrary, the origin of dolomite remains one of the most thoroughly researched but most poorly understood problems in sedimentary geology. Although it is clear from the presence of relict limestone textures and structures that many coarsely crystalline dolomites are secondary rocks formed by diagenetic replacement of older limestones, most fine-crystalline dolomites lack such textural evidence of replacement and cannot be proven to have originated by diagenetic alteration of limestones. It is these fine-crystalline dolomites that have created the so-called dolomite problem, which geologists have not been able to satisfactorily solve since dolomites were first recognized by the French naturalist Deodat de Dolomieu in 1791.

The dolomite problem arises from the fact that scientists have not yet been successful in the laboratory in precipitating perfectly ordered dolomite at the normal temperatures ($\sim25°C$) and pressure (~1 atm) that occur at Earth's surface. Perfectly ordered dolomite has 50 percent of the cation sites filled by Mg and 50 percent filled by

Ca. Elevated temperatures on the order of 200°C are required to produce a perfectly ordered dolomite in the laboratory. In laboratory experiments carried out at the normal temperatures found in natural environments, only a dolomitelike material called **protodolomite** forms. Protodolomite contains up to 20 mole percent excess $CaCO_3$ in its structure and is not a true dolomite. Thus, geochemists have been unable to determine from low-temperature experimental work what geochemical conditions, if any, favor the precipitation of dolomite in natural environments.

In spite of unsuccessful attempts to produce ordered dolomite in the laboratory, many early geologists nonetheless believed that some fine-grained ancient dolomites, particularly those associated with evaporites, are primary precipitates formed by direct precipitation of magnesium–calcium carbonate as dolomite. Prior to the middle 1940s, few occurrences of dolomite had been reported from modern depositional environments; consequently geologists had little direct support for the hypothesis of primary dolomite. Since about 1946, however, recent dolomite sediments have been reported from numerous localities, including some in Russia, South Australia, the Persian Gulf, the Bahamas, Bonaire Island off the Venezuela mainland, the Florida Keys, the Canary Islands, and the Netherlands Antilles (Table 4.3).

The ages of these dolomites have been estimated by radiocarbon methods to range from a few years to about 3000 years (Table 4.3). Most are not perfectly ordered dolomites. The mole percent $MgCO_3$ in these dolomites ranges from about 30 to 50 percent but falls mainly between 40 and 46 percent. Discovery of dolomite in modern environments was initially hailed as evidence that dolomite can be precipitated naturally as a primary deposit. Efforts were made to confirm primary precipitation through a variety of research efforts such as oxygen isotope studies of coexisting dolomite and calcite in a sample from a given environment. Experimental data extrapolated from high temperature to low temperature suggested that at 25°C primary precipitates of dolomite should be enriched, with respect to coexisting calcite, by detectable amounts

TABLE 4.3 Distribution and chemical properties of some modern marine dolomites

Location	Environment	Maximum (molar) Mg/Ca in brines	Dolomite crystal size (μ)	Mole % $MgCO_3$ in dolomite	Age (yrs)
Pekelmeer, Bonaire	Salt pan		2	44–46	1480 ± 140
Quatar, Persian Gulf	Supratidal flat		1–5	45–47	2450 ± 130
Andros Island, Bahamas	Supratidal flat	>40/1	1–2	44	0–160
Great Inagua Is., Bahamas	Salt pan	>600/1	<140 <5	40	<8(?) 2930—3420
Sugarloaf Key, Florida	Supratidal flat	>40/1	2–3	30–44	250–600
Coorong, South Australia	Supersaline lagoon and lakes	4–16/1	<20	45–50	300 ± 250
Abu Dhabi, Persian Gulf	Supratidal flats	>35/1	1–2		
Jarvis Atoll, Pacific	Brine pool	>35/1	?	calcium rich	2650 ± 200

Source: Milliman, J. D., 1974, Marine carbonates: Springer-Verlag, New York.

in heavy oxygen (^{18}O). Unfortunately, studies of dolomite–calcite pairs from natural environments failed to consistently produce the results predicted by these experimental data. Owing to these equivocal results, some investigators, such as Degens and Epstein (1964), suggested that most if not all modern dolomites probably form by rapid alteration of an initial precipitate of $CaCO_3$. That is, the $CaCO_3$ is replaced by dolomite, a process called **dolomitization.** Whether or not dolomite that formed by such rapid dolomitization should be called primary dolomite becomes a matter of semantics. One solution to the semantics problem is to label all early-formed dolomites as **penecontemporaneous** dolomites, regardless of their exact mode of formation. Early-formed dolomites thus include all those formed at or near the surface in the unconsolidated state as opposed to diagenetic dolomites that form during burial and uplift by replacement of older, consolidated, limestones. From a pragmatic point of view, it is not possible in ancient dolomites to distinguish between dolomite precipitated directly from solution and dolomite precipitated initially as $CaCO_3$ and then quickly dolomitized. In any case, this difference is probably not exceedingly important. The important factor is to understand the conditions that favor the early formation of dolomite in modern environments and thus by extension the formation of fine-grained, so-called primary, dolomite in ancient deposits.

Requirements for Dolomite Formation. The chemical reactions of interest with respect to formation of dolomite are:

$$Ca^{2+} \text{ (aq)} + Mg^{2+} \text{ (aq)} + 2CO_3^{2-} \text{ (aq)} = CaMg(CO_3)_2 \text{ (solid)} \qquad \textbf{(4.6)}$$

$$2CaCO_3 \text{ (solid)} + Mg^{2+} \text{ (aq)} = CaMg(CO_3)_2 \text{ (solid)} + Ca^{2+} \text{ (aq)} \qquad \textbf{(4.7)}$$

As suggested at the beginning of this discussion, the problem with the reaction shown in equation 4.6, which illustrates the direct precipitation of dolomite from aqueous solution, is that it will not take place at temperatures lower than about 200°C, at least in the laboratory. The reasons for this are far from well understood, but the problem is certainly related to kinetics, or reaction rates. It has been pointed out by a number of workers such as Gains (1980) that the Mg^{2+} ion is strongly bound by water (hydrated) in solution and must be desolvated before it can be incorporated into the solid dolomite crystal lattice. At low temperatures, Ca^{2+} ions, which are much less strongly bound by water, are more likely to enter the lattice and form $CaCO_3$ minerals. At elevated temperatures, Mg^{2+} ions are less strongly hydrated and thus more easily desolvated, allowing the naked Mg^{2+} ion to enter into the crystal lattice to form dolomite.

The highly ordered state of dolomite also creates a kinetics problem at low temperatures. Nucleation and growth of the highly ordered dolomite lattice in a solution saturated in calcium bicarbonate are so slow that in competition for calcium ions and carbonate ions well-ordered dolomite is prevented from forming, and minerals such as aragonite or cation-disordered, high-magnesium calcites form instead. Folk and Land (1975) draw an amusing analogy to "busing" to illustrate this kinetics problem. Imagine two buses, one a new, comfortable "dolomite bus" and the other an old, uncomfortable "calcite bus." Unfortunately, the "dolomite bus" has only one door, which is difficult to open, whereas the "calcite bus" has several easily opened doors. If a mob of people are waiting to get on a bus and it starts to rain, they will pile into the second-class "calcite bus" because they can get into it quickly. Few people will get on the dolomite bus because it is too hard to get into a hurry. The point here is that a high rate of precipitation or supersaturation means that calcite or aragonite will form because of its relative ease of crystallization, whereas the ordered dolomite structure is too difficult to develop. On the other hand, if only one person per minute comes by to get on a bus, that person will get on the more comfortable "dolomite bus," suggesting by analogy that when crystallization is very slow, as in freshwater, the difficult, ordered dolomite structure can form.

In spite of this suggestion that dolomite might form as a primary precipitate at very slow rates of crystallization, most geologists now appear to believe that the majority of ancient dolomites formed by replacement processes. Only a relatively insignificant amount of ancient dolomite is believed to be truly the product of primary precipitation at or above the sediment–water interface (Zenger and Dunham, 1980). Therefore, much of the recent and current research on dolomites has been focused on attempts to understand the mechanisms of dolomitization. With respect to formation of penecontemporaneous dolomites, two principal hypotheses are currently in vogue: dolomitization in (1) hypersaline environments and (2) a mixing zone of marine meteoric waters.

Dolomitization in Hypersaline Environments. According to the hypothesis of hypersaline environments, dolomites form principally under hypersaline conditions in environments such as sabkhas and the supratidal zones of arid climates, where rates of evaporation exceed rates of precipitation (Fig. 4.6). Under strongly evaporative conditions, Mg^{2+} ions become progressively enriched over Ca^{2+} ions owing to selective

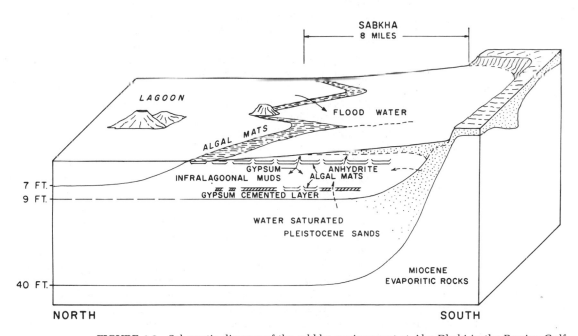

FIGURE 4.6 Schematic diagram of the sabkha environment at Abu Dhabi in the Persian Gulf. Penecontemporaneous dolomite (or protodolomite) is commonly associated with gypsum and anhydrite deposits in such environments. (From Butler, G. P., 1969, Modern evaporite deposition and geochemistry of coexisting brines, the sabkha, Trucial Coast, Arabian Gulf: Jour. Sedimentary Petrology, v. 39. Fig. 2, p. 72, reprinted by permission of Society of Economic Paleontologists and Mineralogists, Tulsa, Okla.)

removal of Ca^{2+} ions to form calcium sulfate minerals (gypsum and anhydrite). The Mg^{2+}/Ca^{2+} ratio is about $5:1$ in normal seawater. When this ratio increases to a sufficiently high level, perhaps in excess of $10:1$, dolomite forms, presumably mainly by replacement of aragonite or calcite.

Dolomitization in a Mixing Zone of Marine–Meteoric Waters. Several studies published since the early 1970s (e.g., Hanshaw et al., 1971; Folk and Land, 1975) have suggested that brackish groundwaters produced by mixing of seawater with meteoric water could be saturated with respect to dolomite at Mg^{2+}/Ca^{2+} ratios much lower than those required under hypersaline conditions. Mixing of freshwater and saline water in environments such as the subsurface zones of coastal areas where meteoric waters come in contact with seawater lowers salinities sufficiently so that dolomites can form at Mg^{2+}/Ca^{2+} ratios ranging from normal seawater values to as low as $1:1$ (Fig. 4.7). Evidently, dolomite can form at lower Mg^{2+}/Ca^{2+} ratios in these mixed waters than in seawater, owing to less competition by other ions in the less saline water.

Other Factors Influencing Dolomitization. Presumably, most early, near-surface, penecontemporaneous dolomitization can be explained by the hypersaline model, the marine–meteoric mixing model, or some variant or combination of these models. Other factors may also affect dolomitization, however, including the presence of dissolved and particulate organic matter, the activities of organisms such as bacteria and algae, and the presence of dissolved SO_4^{2+} ions. In all of the dolomite models discussed above, the Mg^{2+}/Ca^{2+} ratio of the environment is thought to be an important or controlling factor. By contrast, Baker and Kastner (1981) propose that dolomitization is not primarily controlled by the Mg^{2+}/Ca^{2+} ratio. Experimental work by these authors on formation of dolomite at 200°C demonstrated that the presence of dissolved SO_4^{2+} ions strongly inhibits the formation of dolomite. On the basis of their experiments, they suggest that the reason for the scarcity of open marine dolomite is the SO_4^{2+} ions dissolved in seawater. Dissolved SO_4^{2+} ions can apparently inhibit dolomitization of calcite even at concentrations less than 5 percent of their seawater value, although the mechanism of inhibition is still unknown. Dolomitization of aragonite can occur at somewhat higher concentrations of SO_4^{2+}.

Baker and Kastner (1981) propose that dolomite can form rapidly in nature only where the SO_4^{2+} concentration is low. They indicate that the most effective process for SO_4^{2+} removal in marine pore waters is microbial reduction of the ions in organic-rich sediments. The bacterial reduction of SO_4^{2+} promotes dolomitization in three ways: (1) by removal of the SO_4^{2+} inhibitor, (2) by production of alkalinity, and (3) by production of ammonia (NH_4^+). The NH_4^+ produced during reduction may exchange with magnesium held in exchange sites in marine silicate-rich sediments, freeing the magnesium for dolomitization. The concentration of SO_4^{2+} can also be reduced by precipitation of calcium sulfate in evaporative environments to form gypsum. Mixing of freshwater with seawater in groundwater environments likewise lowers the SO_4^{2+} concentration of seawater. Baker and Kastner believe that it is the reduction in SO_4^{2+} in these environments rather than the influence of the Mg/Ca ratio that allows the formation of dolomite.

Evaporite Deposits (Salts)

Sedimentary rocks composed of minerals formed by evaporation of saline waters are common in the geologic record. Evaporite deposits occur in rocks as old as the Pre-

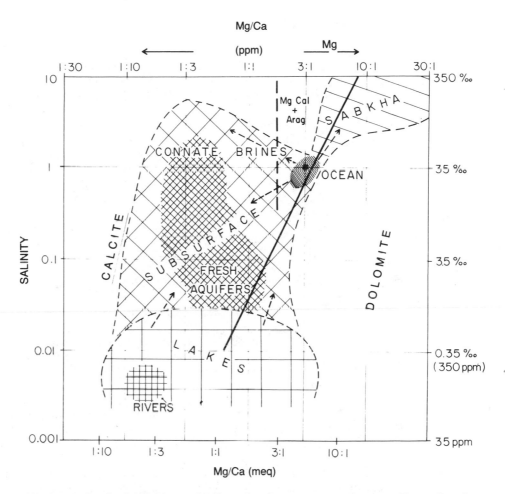

FIGURE 4.7 Graph of salinities vs. Mg/Ca ratios of common natural waters. The preferred fields of occurrence of dolomite, calcite, magnesian calcite, and aragonite are also shown. Note that the dolomite field lies to the right of the heavy, inclined line. Owing to slow crystallization rates and the scarcity of competing foreign ions at low salinities, dolomites can form at Mg/Ca ratios near 1:1. (From Folk, R. L., and L. S. Land, 1975, Mg/Ca ratio and salinity: Two controls over crystallization of dolomite: Am. Assoc. Petroleum Geologists Bull., v. 59. Fig. 1, p. 61, reprinted by permission of AAPG, Tulsa, Okla.)

cambrian, but are most abundant in Paleozoic and Mesozoic stratigraphic sequences. The minerals that make up evaporite deposits are dominantly chlorides; sulfates; and carbonates of calcium, sodium, potassium, and magnesium. Only about a dozen evaporite minerals (Table 8.3), including gypsum, anhydrite, and halite, are quantitatively important, however, more than seventy evaporite minerals are recognized (Stewart, 1963). Ocean water is undersaturated with respect to most evaporite minerals except the calcium carbonate minerals, but precipitation of these minerals can occur when water volume is sufficiently reduced by evaporation. Evaporites form under both marine and nonmarine conditions. Marine evaporites tend to be thicker and more laterally extensive than nonmarine evaporites and are of greatest geologic interest.

Evaporation Sequence

When ocean water is evaporated in the laboratory, evaporite minerals are precipitated in a definite sequence that was first demonstrated by Usiglio in 1848 (reported in Clark, 1924). Minor quantities of carbonate minerals begin to form when the original volume of seawater is reduced by evaporation to about one-half. Gypsum appears when the original volume has been reduced to about 20 percent, and halite forms when the water volume reaches approximately 10 percent of the original volume. Magnesium and potassium salts are deposited when less than about 5 percent of the original volume of water remains. The same general sequence of evaporite minerals occurs in natural evaporite deposits, although many discrepancies exist between the theoretical sequences predicted on the basis of laboratory experiments and the sequences actually observed in the rock record. In general, the proportion of $CaSO_4$ (gypsum and anhydrite) is greater and the proportion of Na–Mg sulfates is less in natural deposits than predicted from theoretical considerations (Borchert and Muir, 1964).

Depositional Models for Evaporites

Many evaporite deposits are quite thick, some exceeding two kilometers, yet it has long been recognized that evaporation of a column of seawater 1000 m thick will produce only about 15 m of evaporites. Evaporation of all the water of the Mediterranean Sea, for example, would yield a mean thickness of evaporites of only about 60 m. Obviously, special geologic conditions operating over a long period of time are required to deposit thick sequences of natural evaporites. The basic requirements for deposition of marine evaporites are a relatively arid climate, where rates of evaporation exceed rates of precipitation, and partial isolation of the depositional basin from the open ocean. Isolation is achieved by means of some type of barrier that restricts free circulation of ocean water into and out of the basin. Under these restricted conditions, the brines formed by evaporation are prevented from returning to the open ocean, allowing them to become concentrated to the point where evaporite minerals are precipitated.

Although most geologists agree on these general requirements for formation of evaporites, considerable controversy still exists regarding deep-water versus shallow-water depositional mechanisms. Figure 4.8 shows three possible models for deposition of thick sequences of marine evaporites. The deep-water, deep-basin model assumes existence of a deep basin separated from the open ocean by some type of topographic sill. The sill acts as a barrier to prevent free interchange of water in the basin with water in the open ocean, but allows enough water into the basin to replenish that lost by evaporation. Seaward escape of some brine allows a particular concentration of brine to be maintained for a long time, leading to thick deposits of certain evaporite minerals such as gypsum. The shallow-water, shallow-basin model assumes concentration of brines in a shallow, silled basin but allows for accumulation of great thicknesses of evaporites owing to continued subsidence of the floor of the basin. The shallow-water, deep-basin model requires that the brine level in the basin be reduced below the level of the sill, a process called **evaporative drawdown,** with recharge of water from the open ocean taking place only by seepage through the sill or by periodic overflow of the sill. Total desiccation of the floors of such basins could presumably occur periodically, allowing the evaporative process to go to completion and thereby deposit a complete evaporite sequence, including magnesium and potassium salts. One or another of these models can theoretically be applied to interpretation of all

FIGURE 4.8 Schematic diagram illustrating three models for deposition of marine evaporites in basins where water circulation is restricted by the presence of a topographic sill. (From Kendall, A. C., 1979, Subaqueous evaporites, *in* R. G. Walker (ed.), Facies models: Geoscience Canada Reprint. Ser. 1. Fig. 17, p. 170, reprinted by permission of Geological Survey of Canada, Ottawa.)

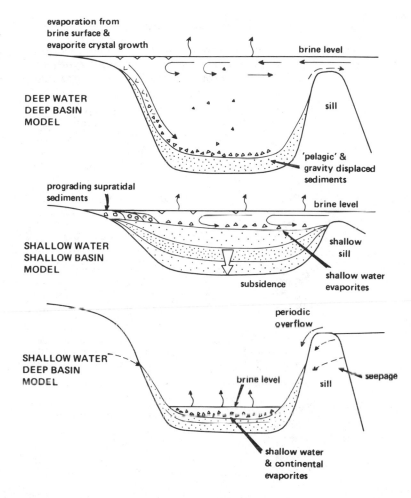

marine evaporite deposits. Nonetheless, geologists are far from unanimous in their interpretations of such deposits, and different models have been applied by different geologists to interpretation of the same deposit. Some of the criteria that are used to interpret deep-water versus shallow-water origin of evaporites are discussed in Chapter 8.

Physical Processes in Deposition of Evaporites

Although we tend to regard evaporite deposits as simply the products of chemical precipitation resulting from evaporation, many evaporite deposits are not just passive chemical precipitates. The evaporite minerals have, in fact, been transported and reworked in the same way as the constituents of siliciclastic and carbonate deposits (Kendall, 1979). Transport can occur by normal fluid-flow processes or by mass-transport processes such as slumps and turbidity currents. Therefore, evaporite deposits may display clastic textures, including both normal and reverse size grading, and various types of sedimentary structures such as cross-bedding and ripple marks.

Siliceous Deposits (Cherts)

In average river water the concentration of silica in transport as H_4SiO_4 is about 13 ppm (Table 4.2). In addition to silica transported to the oceans by rivers, silica is added to the oceans through reaction of seawater with hot volcanic rocks along mid-ocean ridges and by low-temperature alteration of oceanic basalts and detrital silicate particles on the seafloor, as described in Chapter 2. Some silica may also escape from silica-enriched pore waters of pelagic sediments on the seafloor. These silica sources are summarized in Figure 4.9. Despite contributions of silica from these various sources, the silica concentration in different parts of the ocean ranges from less than 0.01 ppm to a maximum of about 11 ppm; the average dissolved silica content of the ocean is only 1 ppm (Heath, 1974). Clearly, silica is constantly being removed by some process and has a relatively short residence time in the ocean.

Silica Solubility

Solubility studies show that the solubility of silica in seawater differs for different silicate minerals. The solubility of solid SiO_2 at 25°C ranges from ~6 to 10 ppm for quartz to ~60–130 ppm for amorphous or noncrystalline varieties of silica such as opal (Krauskopf, 1959; Morey et al., 1962, 1964; Iller, 1979). The solubility of amorphous silica appears to control the precipitation of silica. That is, silica concentrations must reach values on the order of 115 ppm or higher at 25°C in order for seawater to be saturated with silica and precipitation to occur. Therefore, the ocean, with an average dissolved silica content of only 1 ppm, is grossly undersaturated with respect to silica, in sharp contrast to the saturated state of the ocean with respect to calcium carbonate. This fact raises a very intriguing question. What mechanism (or mecha-

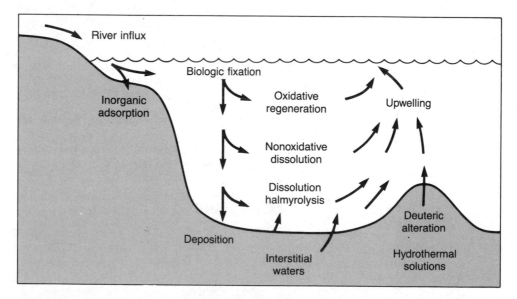

FIGURE 4.9   Sources of dissolved silica in ocean water. (After Heath, G. R., 1974, Dissolved silica and deep-sea sediments, *in* W. W. Hay (ed.), Studies in paleooceanography: Soc. Econ. Paleontologists and Mineralogists Spec. Pub. 20. Fig. 7, p. 81, reprinted by permission of SEPM, Tulsa, Okla.)

FIGURE 4.10 Solubility of silica at 25°C. The solid line shows the variation in solubility of amorphous silica with pH as determined experimentally. The upper dashed curve is a calculated curve for amorphous silica, based on an assumed constant solubility of 120 ppm SiO_2 at pH below 8. The lower dashed line is the calculated solubility of quartz based on the approximately known solubility of 6 ppm SiO_2 in neutral and acid solutions. (From Krauskopf, K. B., Introduction to geochemistry. © 1979. Fig. 6.3, p. 133, reprinted by permission of McGraw-Hill, New York.)

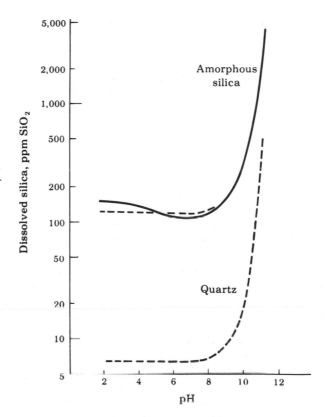

nisms) is capable of removing silica from highly undersaturated ocean water to form chert beds and maintain the low concentrations of dissolved silica in the ocean?

The solubility of silica is affected by both temperature and pH. Solubility increases with increasing temperature in essentially a linear fashion, and solubility at 100°C is approximately three to four times that at 25°C. Change in solubility of silica with pH is illustrated in Figure 4.10. Solubility changes only slightly with increase in pH up to about 9, but rises sharply at pH values above 9. Once silica is in solution under a given set of temperature and pH conditions, it does not readily crystallize to form quartz, even from solutions that have silica concentrations greatly exceeding the solubility of quartz (6–10 ppm). Therefore, it is unlikely that chert, which consists of microcrystalline quartz, can be precipitated by inorganic processes from highly undersaturated ocean water. Chert might be precipitated in some local basins where waters are silica saturated, owing perhaps to dissolution of volcanic ash. Some silica may be removed from seawater in the open ocean by adsorption onto clay minerals or other silicate particles (Bien et al., 1959; Heath, 1974); however, such processes cannot account for the many bedded sequences of nearly pure chert found in the geologic record.

Silica Extraction from Seawater

Removal of silica from ocean water by silica-secreting organisms appears to be the only mechanism capable of large-scale silica extraction from undersaturated seawater. This biologic process has operated since at least early Paleozoic time to regulate the balance of silica in the ocean. **Radiolarians** (Cambrian/Ordovician–Holocene), **diatoms** (Jurassic?–Holocene), and **silicoflagellates** (Cretaceous–Holocene) are microplankton that build skeletons of opaline silica. These siliceous microplankton have apparently been abundant enough in the ocean during Phanerozoic time to extract most of the silica delivered to the oceans by rock weathering and other processes. Diatoms are probably responsible for the bulk of silica extraction from ocean waters, at least in the modern ocean (Calvert, 1983). Heath (1974) calculates that the residence time in the ocean for dissolved silica ranges from 200 to 300 years for biologic utilization to 11,000–16,000 years for incorporation into the geologic record—a very short time from a geologic point of view.

While silica-secreting organisms are alive, their siliceous skeletons are protected by an organic coating that prevents them from dissolving in highly undersaturated and corrosive seawater. After death, this coating is destroyed by biochemical decomposition and the opaline skeletons began to undergo dissolution; however, in areas of the ocean where siliceous organisms flourish, the rate of production of siliceous skeletons may be so high that they cannot all be dissolved as rapidly as they are produced. Under such conditions, a sufficient number of the siliceous skeletons may survive total dissolution to accumulate on the seafloor as siliceous oozes (sediments containing >30 percent siliceous skeletal material). It is postulated that after burial by additional siliceous ooze or clayey sediment, these opaline skeletal materials continue to undergo solution; however, after burial the dissolving silica is trapped in the pore spaces of the sediment and cannot escape back to the open ocean. The pore waters thus become increasingly enriched in silica, perhaps to concentrations of 120 ppm or more. Cherts are thought to then slowly precipitate from these concentrated interstitial solutions. Thus, the formation of cherts is in part a sedimentation process involving the depositional concentration of biogenic opaline tests and in part a diagenetic process with crystallization and recrystallization of the chert taking place after sediment burial.

Transformation of Opal to Chert

Biogenic silica deposited in marine environments is principally in the form of opal, commonly called **opal-A** to indicate amorphous silica. During the process of transformation of biogenic silica to chert, opal-A may not convert directly to chert, which is microcrystalline quartz, but commonly goes through an intermediate, metastable phase called opal-CT. **Opal-CT** is low-temperature cristobalite disordered by interlayered tridymite lattices. Cristobalite and tridymite are varieties of quartz. Opal-CT may occur in open spaces in sediments as **lepispheres,** which are microcrystalline aggregates of blade-shaped crystals, but the most common form is massive. The rates of diagenetic evolution of silica from biogenic opal-A to opal-CT and finally chert are controlled by several physicochemical factors. Temperature is commonly considered to be a particularly important control, with increasing temperature promoting an increased rate of transformation. Kastner and Gieskes (1983) have demonstrated that the rate depends also upon the nature of the opal starting material and the presence of magnesium hydroxide compounds, which serve as a nucleus for the crystallization of opal-CT. Williams et al. (1985) conclude that increasing surface-to-volume ratio of siliceous particles—a ratio that increases with decreasing particle size—results in greater solubility of opal and an increase in the rate of transformation. These authors also suggest that under some conditions opal-A can transform directly to chert without going through the intermediate opal-CT stage. The transformation of opal to chert is discussed in greater detail in Chapter 9.

Replacement Chert

In addition to occurrence as bedded chert, chert can occur also in the form of small nodules, lenses, or thin, discontinuous beds, commonly in limestones. Relict textures in these nodular cherts suggest that most are formed by diagenetic replacement of limestones. Such replacement requires geochemical conditions where the diagenetic pore waters are simultaneously supersaturated with respect to crystalline silica and undersaturated with respect to calcite and where the pH is below about 9. Knauth (1979) suggests that the geochemical environment where these conditions are met most effectively is in areas where meteoric groundwaters mix with seawater in a coastal zone (Fig. 4.11). Silica is supplied in this environment by the dissolution of sponge spicules or other forms of biogenic opal-A within the sediment pile and is then transported into the zone of mixing, where replacement of $CaCO_3$ occurs. During the replacement process, opal-A is first transformed to opal-CT, which presumably then replaces the carbonate and is later diagenetically altered to chert. The carbonate could also be replaced directly by chert. Nodular cherts have also been reported to form by the silica replacement of anhydrite, with the silica again being derived by dissolution of biogenic opal (Chowns and Elkins, 1974).

Precambrian Cherts

Biogenic removal of silica from seawater is believed to account for deposition of most, if not all, Phanerozoic-age cherts, although siliceous fossil fragments are not preserved in all of these cherts. Bedded cherts are very common also in stratigraphic sequences of Precambrian age. Because the existence of large populations of silica-secreting or ganisms during Precambrian time has not yet been proven, the origin of Precambrian cherts, like the origin of Precambrian limestones, remains something of an enigma. In the absence of silica-secreting organisms, we must assume that deposition took place by inorganic processes. How these processes operated, given the geochemical con-

FIGURE 4.11 Formation of nodular cherts in a zone of mixing of seawater and meteoric water. The mixing zone is confined principally to a narrow interval above offshore sediments of low porosity and permeability, producing a relatively thin horizon of silicification. (From Knauth, L. P., 1979, A model for the origin of chert in limestone: Geology, v. 7, Fig. 3, p. 276.)

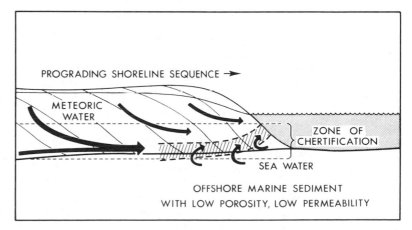

PROGRADING SHORELINE SEQUENCE →

METEORIC WATER

ZONE OF CHERTIFICATION

SEA WATER

OFFSHORE MARINE SEDIMENT WITH LOW POROSITY, LOW PERMEABILITY

straints discussed, is not understood, nor is the immediate source of the silica known. Perhaps the silica content of the Precambrian ocean was higher than that of the Phanerozoic ocean.

Iron-rich Sediments

Iron makes up about 6 percent of Earth's crust and is the fourth most abundant element in the crust, exceeded only by oxygen, silicon, and aluminum. Most sedimentary rocks contain at least some iron, and many siliciclastic sedimentary rocks contain moderate concentrations of iron chemically bound up in terrigenous silicate and oxide minerals and in secondary, diagenetic minerals. The average iron content of mudrocks, for example, is 4.8 percent; that of sandstones is 2.4 percent (Blatt, 1982). Sedimentary rocks greatly enriched in iron over these average values—that is, rocks containing more than about 15 percent iron—are referred to as iron rich. The iron-bearing minerals in these iron-rich rocks are commonly assumed to have originated by precipitation from seawater or freshwater; however, the origin of iron-rich sedimentary rocks is still not well understood, as we shall see.

Although iron-bearing minerals occur in siliciclastic sedimentary rocks of all ages, iron-rich sedimentary rocks are not particularly common in the geologic record. Whereas limestones, cherts, and evaporites occur in moderate abundance in stratigraphic sequences ranging in age from Precambrian to Cenozoic, iron-rich sedimentary rocks were deposited principally during three time periods: the Precambrian, particularly between 1.8 and 2.6 billion years B.P.; the early Paleozoic; and the Jurassic.

Types of Iron-rich Sediments

Iron-rich sedimentary rocks are customarily divided into two broad groups based on age and physical characteristics. The term **iron formation** is used for cherty, commonly well-banded, iron-rich rocks that are dominantly Precambrian in age. **Ironstones** are largely Phanerozoic-age, oolitic and generally fossiliferous, poorly banded or nonbanded, iron-rich sediments that are commonly associated with limestones and terrigenous sedimentary rocks. The major iron-bearing minerals in iron-rich sedimentary rocks are silicates, oxides, sulfides, and carbonates.

Depositional Environments of Modern Iron-bearing Minerals

There are no modern counterparts to ancient environments that presumably favored widespread deposition of iron-rich sediments to produce iron formations and ironstones, but iron-bearing minerals are being deposited on a small scale in a variety of modern environments. Iron sulfides, particularly pyrite (FeS_2), are forming in black muds that accumulate under reducing conditions in stagnant ocean basins, tidal flats, and organic-rich lakes. Iron sulfides are also accumulating around the vents of hot springs located on the crests of mid-ocean ridges, as discussed in Chapter 2. Chamosite, a complex Fe–Mg–Al silicate, has been reported in modern sediments at water depths as great as 150 m in the Orinoco and Niger deltas; on the ocean floor off Guinea, Babon, and Sarawak; and in the Malacca Straits. Glauconite is a K–Mg–Fe–Al silicate mineral that has been reported from Monterey Bay, California, and from various other parts of the ocean at water depths ranging down to about 2000 m. Iron oxides such as goethite (FeOOH) are accumulating in some modern lakes and bogs, as oolites on the floor of the North Sea, and in manganese nodules in both seawater and freshwater. Manganese nodules, which contain manganese, copper, cobalt, nickel, and other metals in addition to iron, are particularly widespread in the Pacific Ocean at water depths of 4–5 km.

Transport and Deposition of Iron

Transport and deposition of iron is governed by both Eh and pH of the environment. Eh–pH diagrams such as Figure 4.12 can be used to predict the stability of iron-bearing minerals and serve to illustrate that Eh is commonly more important than pH in determining which iron-bearing mineral will be deposited. For example, hematite (Fe_2O_3) is precipitated under oxidizing conditions at the pHs commonly found in the ocean and surface waters, siderite ($FeCO_3$) forms under moderately reducing conditions, and pyrite (FeS_2) forms under moderate to strong reducing conditions.

FIGURE 4.12 Eh–pH diagram showing the stability fields of the common iron minerals, sulfides, and carbonates in water at 25°C and 1 atmosphere total pressure. Total dissolved sulfur $= 10^{-6}$; total dissolved carbonate $= 10^0$. (From Garrels, R. M., and C. L. Christ, Solutions, minerals, and equilibria. © 1965. Fig. 7.21, p. 224, reprinted by permission of Harper & Row, New York.)

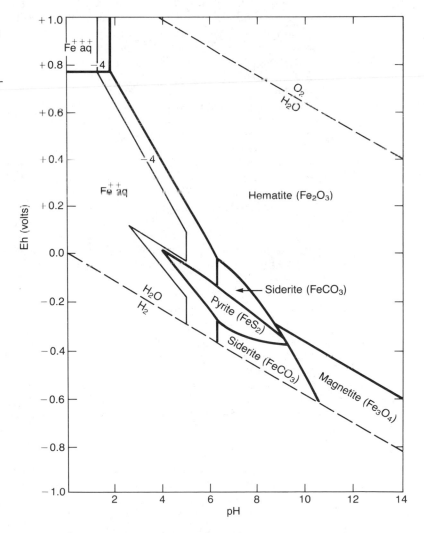

The iron geochemistry of natural systems is far more complex than the simplified conditions assumed in constructing Eh–pH diagrams. Such diagrams, therefore, are of only limited use in environmental interpretations. Many problems are associated with the formation of sedimentary iron deposits, and the mechanisms by which transport and deposition of iron occurred in the past to form iron formations and ironstones are still poorly understood and controversial. One of the principal problems stems from the fact that iron in the oxidized or ferric (Fe^{3+}) state is much less soluble than iron in the reduced or ferrous (Fe^{2+}) state (Fig. 4.12). Ferric iron is soluble only at pHs less than about 4; such values rarely occur under natural conditions. How then can large quantities of iron be taken into solution and transported from subaerial weathering sites under the oxidizing conditions that commonly prevail in streams and rivers?

This problem of solution and transport of iron under oxidizing conditions prompted some workers (Lepp and Goldich, 1964; Cloud, 1973) to postulate that low oxygen levels existed during the Precambrian, allowing great quantities of iron to be transported in the soluble, reduced state (Fe^{2+}) to marine basins. Presumably, owing to photosynthetically generated oxygen, local oxidizing conditions existed within some parts of broad, shallow-marine basins, where the iron was oxidized to the insoluble ferric state (Fe^{3+}) and precipitated. The concept of low oxygen levels during the late Precambrian has been questioned by other investigators (e.g., Dimroth and Kimberley, 1976). In any case, this argument cannot be used to explain the solution and transport of iron during the Phanerozoic when an oxidizing atmosphere clearly existed. Some workers have suggested that iron may have been transported as colloids by physical processes rather than in true solution or that it was sorbed to clay particles or organic materials and transported along with these substances. It appears unlikely, however, that mechanisms such as colloidal transport can account for transport of the large quantities of iron that occur in iron formations or ironstones (Ewers, 1983). Reducing conditions seem to be required for transport of large amounts of iron in solution; hence the dilemma.

Three problems have to be addressed to account for the formation of iron-rich rocks: (1) the source of the iron, (2) transport of iron to the depositional basin, presumably under reducing conditions, and (3) precipitation of iron within the basin, presumably under oxidizing conditions. Most workers originally assumed that the iron was derived by subaerial weathering of iron silicate minerals; however, the problem of transport of ferric iron to depositional basins remains an obstacle if iron is derived from the land. To get around this difficulty, some workers have suggested that the source of the iron was the depositional basin close to the depositional site. For example, Drever (1974) proposed an upwelling model in which iron and silica are precipitated from marine bottom waters as a result of upwelling of bottom waters into an oxidizing environment. That is, localization of iron deposition is related to areas of upwelling of deep anoxic water charged with ferrous iron as a result of reduction of ferric oxide in terrigenous siliciclastic bottom sediments. An intriguing recent hypothesis proposes that iron formations are primarily exhalative or hydrothermal in origin (Gross, 1980; Simonson, 1985). According to this hypothesis, the source of the iron is located basinward from the depositional site, not landward. Hydrothermal activity from hot springs situated along mid-ocean spreading ridges, such as those described in Chapter 2, provides a source for iron and silica. The effluents from these hot springs are postulated to be supersaturated with iron and silica, which would commonly be dispersed from the deeper basins upward and outward to shallower water, where precipitation could take place. A major objection to this model is the great distance over which the hydrothermal solutes would have to be dispersed.

Diagenetic Origin of Iron-rich Rocks

The difficulties in explaining large-scale solution transport and precipitation of iron have stimulated some investigators to propose that both iron-formations and ironstones may be secondary deposits, created by iron replacement of original carbonate minerals or other minerals during burial and diagenesis. It has long been recognized that some ironstone deposits are at least partly secondary in origin because of the presence in these deposits of relict calcium carbonate constituents, such as oolites and fossils, that have clearly been replaced by iron. Dimroth (1979) suggests that iron formations as well as ironstones were most likely precipitated initially as $CaCO_3$. The calcium carbonate deposits were subsequently replaced during diagenesis by silica and iron to form the cherty iron deposits. The sources of the iron needed to bring about such massive replacement and the mechanisms of selective iron and silica replacement that would be required to produce the alternating bands of silica-rich and iron-rich sediments are still highly speculative. This postulated replacement mechanism would probably require that ferrous iron be introduced uniformly into the carbonate deposits over a wide area and under reducing conditions, with subsequent intermittent change to oxidizing conditions to allow precipitation of the iron to occur. Ewers (1983) considers this process unlikely, at least for many iron formations.

Many puzzling aspects of the formation of sedimentary iron deposits remain. These puzzles include not only the origin of the iron itself but also the origin of the banding in iron formations and the manner in which the chert in cherty iron formations was deposited. Why, for example, were chert and iron not deposited together after the Precambrian? Also, why was deposition of iron-rich sediments particularly prevalent only during the Precambrian, early Paleozoic, and Jurassic? What role, if any, did organisms play in the deposition of iron formations and ironstones? Was the local production of oxygen by photosynthesizing organisms such as algae important? Did low forms of life such as bacteria and algae catalyze or initiate precipitation in some manner? If so, how did they cause precipitation and how important was such biologic activity? Obviously, much additional research will be required before the mystery surrounding the origin of iron-rich sediments is solved.

Phosphorites

Phosphorites are sedimentary deposits containing more than 15–20 percent P_2O_5. Phosphate-rich layers commonly occur interbedded with carbonate rocks, mudrocks, or chert. A characteristic feature of many major phosphorite accumulations is the triple association of phosphate, chert, and sediments containing abundant organic carbon. Phosphorites are predominantly of marine origin. They occur in rocks of all ages from Precambrian to Holocene and on all continents. One of the best studied examples of an ancient phosphorite deposit is the Permian-age Phosphoria Formation of western Wyoming and Idaho. Phosphatic members of this formation reach 30–35 m in thickness and cover an area of thousands of square kilometers. Much younger deposits of phosphorite nodules and phosphatic sediments occur on the present ocean floor, mainly at depths less than about 400 m and mostly in the vicinity of coastlines. They are concentrated particularly off the coasts of southern California and Baja California, eastern United States, Peru and Chili, and southwest Africa. They occur also on the floors of the Pacific, Indian, and Atlantic oceans and on some seamounts (Fig. 4.13). Most of these phosphate nodule accumulations are older than the Holocene and formed on the seafloor more than 700,000 years ago (Kolodny and Kaplan, 1970). Some

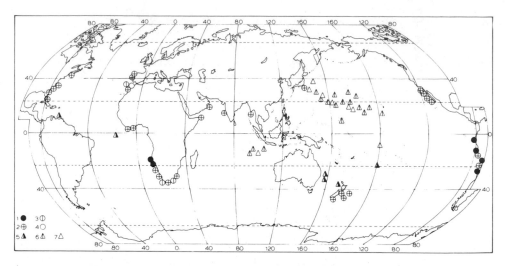

FIGURE 4.13   Distribution of phosphorite nodules and phosphatic sediments on the ocean floor: 1–4 show the locations of phosphorites on continental shelves; 5–7, phosphorites on sea- mounts. Ages of the phosphorites are 1, Holocene; 2 and 5, Neogene; 3 and 6, Paleogene; 4 and 7, Cretaceous. (From Baturin, G. N., 1982, Phosphorites on the sea floor: Developments in Sedi- mentology 33. Fig. 2.1, p. 56, reprinted by permission of Elsevier Science Publishers, Amster- dam.)

phosphorite nodules are now forming on the ocean floor in a few places such as the Peru–Chile continental slope and the Nambian shelf of southwest Africa.

Deposition of Phosphorites

The principal phosphate minerals found in sedimentary rocks are various varieties of apatites, of which carbonate apatite $[Ca_{10}CO_3(PO_4)_6]$ is most important. The conditions that favor precipitation of calcium carbonate also favor formation of carbonate apatite; carbonate apatite can precipitate at pH values possibly as low as 7.0, whereas calcium carbonates generally do not precipitate below a pH of about 7.5 (Bentor, 1980a). Car- bonate apatite precipitation appears to be favored also by slightly reducing conditions. The factors affecting phosphate solubility are not thoroughly understood, however, and the solubility of carbonate apatite has not been definitely established. It is not definitely known if the ocean is saturated with carbonate apatite, although it is be- lieved that it is very near saturation (Kolodny, 1981). The average concentration of phosphorus in the ocean is 70 parts per billion [ppb] (Gulbrandsen and Roberson, 1973) compared to 20 ppb in average river water. The concentration of phosphorus in ocean water ranges from only a few ppb in surface waters, which are strongly depleted by biologic uptake, to values of 50–100 ppb at depths greater than 200–400 m.

Phosphorus is removed from seawater in several ways. Some phosphorus is pre- cipitated along with calcium carbonate minerals during deposition of limestones; however, the average limestone contains only about 0.04 percent P_2O_5. Phosphorus is removed also by concentration in the tissues of organisms, but this phosphorus is returned to the ocean when organisms die unless they are quickly buried by sediment before decay of the organic tissue is complete. Some phosphorus may be removed from seawater by incorporation into metalliferous sediments as a result of adsorption onto metallic minerals such as iron hydroxides. Finally, phosphorus is removed from ocean water in some manner to form marine apatite deposits. We are particularly interested in this last process because the major problem associated with phosphorite deposition is identification of a mechanism that can explain how trace amounts of phosphorus in seawater can be concentrated to form phosphorite deposits. Assuming approximately the same average content of phosphorus in the ancient and modern ocean, how was the 70–100 ppb phosphorus content of ancient oceans upgraded to form widespread deposits of carbonate apatite containing as much as 40 percent P_2O_5, an enrichment of up to two millionfold?

A secondary, or replacement, origin has been suggested for some phosphorite deposits to account for this enrichment. The preservation in many other phosphorite deposits of clastic textures and primary sedimentary structures such as cross-bedding and laminations indicates that these sediments are primary deposits. Several geologists who made early studies of phosphorite deposits suggested an association between phosphorite deposition and areas of upwelling in the oceans. Studies of the distribution of ancient phosphorites show that most occur in lower latitudes in the trade wind belts along one side of a basin where deeper water could have upwelled adjacent to a continent. Most phosphate nodule deposits on the modern ocean floor also occur in areas of upwelling.

Early ideas on upwelling and phosphorite deposition assumed that inorganic precipitation of apatite occurred as cold, deep, phosphate-rich waters upwelled onto a shallow shelf. Under these postulated conditions, carbon dioxide would be lost from the upwelling waters because of pressure decrease, warming, or photosynthesis, causing pH to increase and carbonate apatite precipitation to occur. It has now been established, however, that Mg^{2+} ions in seawater have an inhibiting effect on the growth of carbonate apatite crystals in much the same way that they inhibit the precipitation of calcite (Martens and Harris, 1970). Also, most or all of the phosphorus brought to surface waters by upwelling currents is quickly used up by organisms that utilize phosphate as one of the essential nutrients needed for organic growth. Rapid biologic utilization prevents phosphate levels in the ocean from rising to the point of saturation. These two factors taken together thus appear to rule out inorganic precipitation of apatite from the open ocean.

Nonetheless, biologic utilization of phosphate to build soft body tissue appears to provide the answer to the problem of phosphate concentration in sediments. Modern phosphate nodules are forming in areas of oceanic upwelling where a steady supply of phosphate brought from the large deep-ocean reservoir allows continuous growth of organisms in large numbers. After death, organisms and organic debris not consumed by scavengers pile up on the ocean floor under reducing conditions, where decay is inhibited. These organic materials include the remains of phytoplankton and zooplankton, coprolites (feces), and the bones and scales of fish. Under the toxic, reducing conditions of the seafloor, some of the soft body tissue is thus preserved long enough to be buried and incorporated into accumulating sediment. Perhaps 1–2 percent of the total phosphorus involved in primary productivity in upwelling zones is ultimately incorporated into the sediments in this way (Baturin, 1982).

Slow decay of body tissue after burial releases phosphorus to the interstitial waters of the sediment. Studies of the chemistry of interstitial waters in sediments where modern phosphate nodules are forming and in other areas of the seafloor where organic-rich sediments are accumulating under reducing conditions have turned up phosphorus concentrations ranging from 1400 ppb to as much as 7500 ppb (Bentor, 1980a). At such high phosphorus concentrations, the waters are supersaturated with respect to calcium phosphate. The phosphate thus begins to precipitate on the surfaces of siliceous organisms, carbonate grains, particles of organic matter, fish scales and bones, siliciclastic mineral grains, or older phosphate particles (Baturin, 1982). Phosphorite nodules thus form within the sediments by diagenetic reactions between organic-rich sediments and their phosphate-enriched interstitial waters. Mg^{2+}/Ca^{2+} ratios are apparently lowered below the threshold values that inhibit apatite formation owing to magnesium–iron replacements in clay minerals in the anoxic marine sediments (Drever, 1971).

Physical Processes in Deposition

The presence of clastic textures and primary depositional sedimentary structures in some phosphorite deposits seems inconsistent with this proposed diagenetic concentration mechanism. Therefore, Kolodny (1980) suggested a two-stage process for the origin of ancient phosphorite deposits. In the first stage, apatite forms diagenetically in stagnant, reducing basins by phosphorus mobilization in interstitial waters in the manner postulated for formation of modern phosphorite nodules. The final stage involves reworking and enrichment of these diagenetically formed nodules by mechanical concentration processes. This stage is characterized by oxidizing conditions, and concentration presumably takes place in a high-energy environment during lower

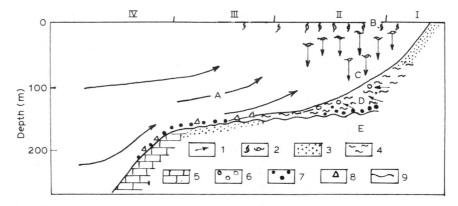

FIGURE 4.14 Schematic diagram illustrating formation of phosphorites in areas of upwelling on ocean shelves. The letters and numbers refer to: A, supply of phosphorus to the shelf by upwelling waters; B, consumption of phosphorus by organisms; C, deposition of phosphorus on the bottom in biogenic detritus and burial by accumulating sediment; D, formation of phosphate concretions in the biogenic sediment by diagenetic processes; E, mechanical reworking of sediments and concentration of phosphate concretions when sea level is low. Zone I is the zone of shallow-water clastic deposits; Zone II is the zone where high contents of phosphate-rich biogenic detritus accumulate in the sediment; Zone III is the zone of reworking of phosphate-rich sediments during lowered sea level; and Zone IV is a deeper water zone where carbonate sediments with local phosphate concretions occur. The arrows refer to 1, paths of movement of phosphorus in the ocean and interstitial waters; 2, plankton; 3, clastic sediments; 4, biogenic siliceous, siliceous-clastic, and siliceous-carbonate sediments; 5, carbonate sediments; 6, unconsolidated phosphate concretions; 7, dense phosphate concretions; 8, glauconite; 9, erosional surface. (From Baturin, G. N., 1982, Phosphorites on the sea floor: Developments in Sedimentology 33. Fig. 5.4, p. 227, reprinted by permission of Elsevier Science Publishers, Amsterdam.)

stands of sea level. This final stage, during which the original diagenetically formed phosphorite sediments are mechanically reworked under shallow-water conditions, accounts for the clastic textures and primary sedimentary structures found in many ancient phosphorites.

Summary of Phosphorite Deposition

In summary, upwelling of phosphate-rich waters from deeper parts of the ocean and biologic utilization of the phosphate in soft body tissue appear to be important factors in the origin of phosphorite deposits. Phosphorus is deposited on the seafloor in organic detritus and buried with accumulating sediment. Phosphate becomes concentrated in the pore waters of sediment during slow decay of phosphate-bearing, soft-bodied organisms and other organic detritus. Carbonate apatite precipitates diagenetically from these phosphate-enriched pore waters by some process not yet fully understood to form gel-like, gradually hardening, phosphate concretions. Subsequently, these diagenetic deposits are reworked mechanically owing to lowered sea levels, allowing final concentration and deposition of phosphatic sediments by waves and currents. These processes are summarized diagrammatically in Figure 4.14.

This postulated multistage process for formation of phosphorite deposits is a plausible hypothesis, but it has some flaws. It does not, for example, explain why phosphorites accumulated on a much vaster scale at some times in the geologic past than at present. Bentor (1980a) suggests that some additional mechanism, not realized or understood at the present time, may have been important in the past. Research on the origin of phosphorites is still very much in a state of flux.

ADDITIONAL READINGS

General

Berner, R. A., 1971, Principles of chemical sedimentology: McGraw-Hill, New York, 240 p.

Garrels, R. M., and C. L. Christ, 1965, Solutions, minerals, and equilibrium: Harper & Row, New York, 450 p.

Garrels, R. M., and F. T. Mackenzie, 1971, Evolution of sedimentary rocks: W. W. Norton, New York, 397 p.

Krauskopf, K. B., 1979, Introduction to geochemistry, 2nd ed.: McGraw-Hill, New York, 617 p.

Carbonate Rocks

Bathurst, R. G. C., 1975, Carbonate sediments and their diagenesis: Elsevier, New York, 658 p.

Milliman, J. D., 1974, Marine carbonates: Springer-Verlag, Berlin–New York, 357 p.

Zenger, D. H., J. B. Dunham, and R. L. Ethington (eds.), Concepts and models of dolomitization, Soc. Econ. Paleontologists and Mineralogists Spec. Pub. 28, 320 p.

Zenger, D. H., and S. J. Mazzullo (eds.), 1982, Dolomitization: Benchmark Papers in Geology, v. 65, Hutchinson Ross, Stroudsburg, Pa., 426 p.

Evaporites

Borchert, H., and R. O. Muir, 1964, Salt deposits: D. Van Nostrand, London–New York, 338 p.

Braitsch, O., 1971, Salt deposits, their origins and compositions: Springer-Verlag, Berlin–New York, 297 p.

Dean, W. E., and B. C. Schreiber, 1978, Marine evaporites: Soc. Econ. Paleontologists and Mineralogists Short Course Notes No. 4, 193 p.

Kirkland, D. W., and R. Evans (eds.), 1973, Marine evaporites: Origin, diagenesis and geochemistry: Dowden, Hutchinson and Ross, Stroudsburg, Pa., 426 p.

Sonnenfeld, P., 1984, Brines and evaporites: Academic Press, London, 624 p.

Siliceous Deposits

Aston, S. R. (ed.), 1983, Silicon geochemistry and biochemistry: Academic Press, London, 248 p.

Calvert, S. E., 1974, Deposition and diagenesis of silica in marine sediments, in K. J. Hsü and H. C. Jenkyns (eds.), Pelagic sediments: On land and under the sea: Internat. Assoc. Sedimentologists Spec. Pub. 1, p. 273–300.

Garrison, R. E., R. G. Douglas, K. E. Pisciotta, C. M. Isaacs, and J. C. Ingle (eds.), 1981, The Monterey Formation and related siliceous rocks of California: Pacific Sec., Soc. Econ. Paleontologists and Mineralogists, Tulsa, Okla., 327 p.

Heath, G. R., 1974, Dissolved silica and deep-sea sediments, in W. W. Hay (ed.), Studies in paleo-oceanography: Soc. Econ. Paleontologists and Mineralogists Spec. Paper 20, p. 77–94.

Iijima, A., J. R. Hein, and R. Siever (eds.), 1983, Siliceous deposits in the Pacific region: Elsevier, Amsterdam, 472 p.

Iller, R. K., 1979, Chemistry of silica: John Wiley & Sons, New York, 866 p.

McBride, E. F. (ed.), 1979, Silica in sediments: Nodular and bedded cherts: Soc. Econ. Paleontologists and Mineralogists Reprint Series 8, 184 p.

van der Linder, G. J. (ed.), 1977, Diagenesis of deep-sea biogenic sediments: Benchmark Papers in Geology, v. 40, Dowden, Hutchinson and Ross, Stroudsburg, Pa., 385 p.

Iron-rich Sediments

Lepp, H. (ed.), 1975, Geochemistry of iron: Benchmark Papers in Geology, v. 18, Dowden, Hutchinson and Ross, Stroudsburg, Pa., 464 p.

Melnik, Y. P., 1982, Precambrian banded iron-formations: Developments in Precambrian Geology 5, Elsevier, Amsterdam, 310 p. Translated from the Russian by Dorothy B. Vitaliano.

Trendall, A. F., and R. C. Morris (eds.), 1983, Iron-formation facts and problems: Developments in Precambrian Geology 6, Elsevier, Amsterdam, 558 p.

Phosphorites

Baturin, G. N., and C. W. Finkl, Jr., 1982, Phosphorites on the sea floor, origin, composition and distribution: Developments in Sedimentology 33, Elsevier, Amsterdam, 343 p. Translated from the Russian by Dorothy B. Vitaliano.

Bentor, Y. K. (ed.), 1980, Marine phosphorites—Geochemistry, occurrence, genesis: Soc. Econ. Paleontologists and Mineralogists Spec. Pub. 29, 249 p.

PART THREE
Physical Properties
of Sedimentary Rocks

5

Sedimentary Textures

5.1 INTRODUCTION

The transport and depositional processes described in Chapters 3 and 4 generate a wide variety of siliciclastic and nonsiliciclastic sedimentary rocks, each characterized by distinctive physical properties. The most important of these properties are sedimentary textures and structures. **Sedimentary texture** refers to the small-scale features that arise from the size, shape, and orientation of individual sediment grains. **Sedimentary structures** are larger scale features such as bedding and lamination, cross-bedding, ripple marks, and the tracks, trails and other markings made by organisms. These structures are discussed in Chapter 6. Geologists have long assumed that the texture of sedimentary rocks reflects the nature of transport and depositional processes and that characterization of texture can aid in interpreting ancient environmental settings and boundary conditions. An extensive literature has thus been published dealing with various aspects of sediment texture, particularly methods of measuring and expressing grain size and shape and interpretation of grain-size and shape data.

The texture of siliciclastic sedimentary rocks is produced primarily by physical processes of sedimentation and is considered to encompass grain **size, shape** (form, roundness, surface texture), and **fabric** (grain orientation and grain-to-grain relations). The interrelationship of these primary textural properties controls other, derived, textural properties such as **bulk density, porosity,** and **permeability.** The texture of some nonsiliciclastic sedimentary rocks such as certain limestones and evaporites is also generated partly or wholly by physical transport processes. The texture of others is due principally to chemical or biochemical sedimentation processes. Extensive recrystallization or other diagenetic changes may destroy the original textures of nonsiliciclastic sedimentary rocks and produce crystalline textural fabrics that are largely of secondary origin. Obviously the textural features of chemically or biochemically formed sedimentary rocks and of rocks with strong diagenetic fabrics have quite different genetic significance than those of unaltered siliciclastic sedimentary rocks.

The discussion that follows is focused primarily on the physically produced textures of siliciclastic sedimentary rocks. Some of the special textural features important to understanding the classification and genesis of limestones and other nonsiliciclastic sedimentary rocks are discussed in Chapter 8. In the present chapter, we examine the characteristic textural properties of grain size and shape, particle surface texture, and grain fabric and discuss the possible genetic significance of these properties. Although the study of sedimentary textures is probably not the most exciting aspect of sedimentology, it is nonetheless an important field of study. A thorough understanding of the nature and significance of sedimentary textures is fundamental to interpretation of ancient depositional environments and transport conditions, although much uncertainty still attends the genetic interpretation of textural data. Some long-standing ideas about the genetic significance of sediment textural data are now being challenged, while new ideas and techniques for studying and interpreting sediment texture continue to emerge. No textbook on sedimentology would be complete without a discussion of sediment texture and its genetic significance. This chapter provides a basic introduction to this important subject.

5.2 GRAIN SIZE

Grain size is a fundamental attribute of siliciclastic sedimentary rocks and thus one of the important descriptive properties of such rocks. Sedimentologists are particularly concerned with three aspects of particle size: (1) techniques for measuring grain size and expressing it in terms of some type of grain-size or grade scale, (2) methods for summarizing large amounts of grain-size data and presenting them in graphical or statistical form so they can be more easily analyzed, and (3) the genetic significance of these data. We will examine each of these concerns.

Methods for Measuring and Expressing Grain Size

Grade Scales. Particles in sediments and sedimentary rocks range in size from a few microns to a few meters. The natural distribution of particle sizes between these extremes forms a virtual continuum; however, it is convenient when discussing and presenting grain-size data to divide this spectrum of sizes into a number of discrete units, each having a definite upper and lower size limit. Because of the wide range of particle sizes that occur in rocks and sediments, logarithmic or geometric scales are more useful for expressing size than linear scales. In a geometric scale there is a succession of numbers such that a fixed ratio exists between successive elements of the series. Several grain-size or grade scales have been developed, but the scale that is used almost universally by sedimentologists is the **Udden–Wentworth scale.** This scale, first proposed by Udden in 1898 and modified and extended by Wentworth in 1922, is a geometric scale in which each value in the scale is two times larger than the preceding value, or one-half as large, depending upon the direction (Table 5.1). The Udden–Wentworth scale extends from <1/256 mm (0.0039 mm) to >256 mm and is divided into four major size categories (clay, silt, sand, and gravel), which can be further subdivided as illustrated in Table 5.1.

TABLE 5.1 Grain-size scale for sediments, showing Wentworth size classes, equivalent phi (ϕ) units, and sieve numbers of U.S. Standard Sieves corresponding to various millimeter and ϕ sizes

	U.S. Standard sieve mesh	Millimeters		Phi (ϕ) units	Wentworth size class
GRAVEL		4096		−12	
		1024		−10	Boulder
		256	256	− 8	
		64	64	− 6	Cobble
		16		− 4	Pebble
	5	4	4	− 2	
	6	3.36		− 1.75	
	7	2.83		− 1.5	Granule
	8	2.38		− 1.25	
SAND	10	2.00	2	− 1.0	
	12	1.68		− 0.75	
	14	1.41		− 0.5	Very coarse sand
	16	1.19		− 0.25	
	18	1.00	1	0.0	
	20	0.84		0.25	
	25	0.71		0.5	Coarse sand
	30	0.59		0.75	
	35	0.50	½	1.0	
	40	0.42		1.25	
	45	0.35		1.5	Medium sand
	50	0.30		1.75	
	60	0.25	¼	2.0	
	70	0.210		2.25	
	80	0.177		2.5	Fine sand
	100	0.149		2.75	
	120	0.125	⅛	3.0	
	140	0.105		3.25	
	170	0.088		3.5	Very fine sand
	200	0.074		3.75	
MUD — SILT	230	0.0625	¹⁄₁₆	4.0	
	270	0.053		4.25	
	325	0.044		4.5	Coarse silt
		0.037		4.75	
		0.031	¹⁄₃₂	5.0	
		0.0156	¹⁄₆₄	6.0	Medium silt
		0.0078	¹⁄₁₂₈	7.0	Fine silt
		0.0039	¹⁄₂₅₆	8.0	Very fine silt
MUD — CLAY		0.0020		9.0	
		0.00098		10.0	Clay
		0.00049		11.0	
		0.00024		12.0	
		0.00012		13.0	
		0.00006		14.0	

A useful modification of the Udden–Wentworth scale is the logarithmic **phi scale,** which allows grain-size data to be expressed in units of equal value for the purpose of graphical plotting and statistical calculations. This scale, proposed by Krumbein in 1934, is based on the following relationship

$$\phi = -\log_2 S \qquad (5.1)$$

where ϕ is phi size and S is the grain size in millimeters. Equivalent phi and millimeter sizes are shown in Table 5.1. Note that the phi scale yields both positive and negative numbers. The real size of particles, expressed in millimeters, decreases with increasing positive phi values and increases with increasing numerical negative values. Because sand-size and smaller grains are the most abundant grains in sedimentary rocks, Krumbein chose the negative logarithm of the grain size in millimeters so that grains of this size will have positive phi values, avoiding the bother of constantly working with negative numbers.

Methods of Grain-Size Measurement. The size of siliciclastic grains can be measured by several techniques (Table 5.2). The choice of methods is dictated by the purpose of the study, the range of grain sizes to be measured, and the degree of consolidation of sediment or sedimentary rock. Large particles (pebbles, cobbles, boulders) in either unconsolidated sediment or lithified sedimentary rock can be measured manually with a caliper. Grain size is commonly expressed in terms of either the long dimension or intermediate dimension of the particles. Granule- to silt-size particles in unconsolidated sediments or sedimentary rocks that can be disaggregated are commonly measured by sieving through a set of nested, wire-mesh screens. The sieve numbers of U.S. Standard Sieves that correspond to various millimeter and phi sizes are shown in Table 5.1. Sieving techniques measure the intermediate dimension of particles because the intermediate particle size determines whether or not a particle can go through a particular mesh.

Granule- to silt-size particles can be measured also by sedimentation techniques based on the settling velocity of particles. In these techniques, grains are allowed to settle through a column of water at a specified temperature in a settling tube, and the time required for the grains to settle is measured. The settling time of the particles is related empirically to a standard size-distribution curve (calibration curve) to obtain the equivalent millimeter or phi size. As mentioned in Chapter 3, settling velocity of particles is affected by particle shape. Spherical particles settle faster than nonspheri-

TABLE 5.2   Methods of measuring sediment grain size

Type of sample	Sample grade	Method of analysis
Unconsolidated sediment	Boulders Cobbles Pebbles	Manual measurement of individual clasts
	Granules Sand Silt	Sieving or settling tube analysis
	Clay	Pipette analysis, photohydrometer, Coulter counter
Lithified sedimentary rock	Boulders Cobbles Pebbles	Manual measurement of individual clasts
	Granules Sand Silt	Thin-section measurement
	Clay	Electron microscope

cal particles of the same mass. Therefore, determining the grain sizes of natural, non-spherical particles by sedimentation techniques may not yield exactly the same values as those determined by sieving.

The grain size of fine silt and clay particles can be determined by sedimentation methods based on Stokes Law (equations 3.16 and 3.17). If the settling velocity of small particles can be measured at a particular temperature, the diameter of the particles can be calculated by a simple mathematical rearrangement of equation 3.17 to yield

$$D = \frac{\sqrt{V}}{\sqrt{C}} \qquad\qquad (5.2)$$

where D is the diameter of the particles in centimeters, V is settling velocity of the particles, and C is a constant that depends upon the density of the particles and the density and viscosity of the fluid (usually water). The standard sedimentation method for measuring the sizes of these small particles is **pipette analysis.** To do a pipette analysis, fine sediment is stirred into suspension in a measured volume of water in a settling tube. Uniform-size aliquots of this suspension are withdrawn with a pipette at specified times and at specified depths, evaporated to dryness in an oven, and weighed. The data obtained can then be used in equation 5.2 to calculate grain size. Tables of withdrawal times and depths for specific particle sizes are available in many books and articles dealing with sedimentation analysis (e.g., Galehouse, 1971). Note that Stokes Law is based on the assumption that the settling particles are spheres. Because most natural particles are not spheres, use of Stokes Law yields grain sizes that are commonly smaller than the actual particle sizes.

Pipette analyses as well as manual settling-tube analyses of coarser sediment are very laborious processes because of the many operations involved. To simplify these procedures, automatic-recording settling tubes and sedimentation balances have been developed that allow the sizes of both sand-size and clay-size sediment to be easily and rapidly determined. Most automatic-recording settling tubes, commonly called rapid sediment analyzers, function by measuring changes with time in the weight of sediment that collects on a pan suspended in a column of water in the settling tube or by measuring changes in the pressure of the water column as sediment settles out of the column. Grain size can then be determined by comparing these weight- or pressure-vs.-time curves to a calibration curve. A special type of automatic settling tube for fine sediment is a **photohydrometer.** This instrument measures the intensity of a beam of light passed through a column of suspended sediment. As sediment settles out of suspension, less light is reflected by the fine particles and the light intensity increases. Changes in light intensity with time are automatically recorded on a chart recorder. The light intensity measured at predetermined times can be related empirically to settling velocity of the particles and thus to particle size (Jordan et al., 1971).

The grain size of small particles can also be measured by an electrical particle-counting device called a **Coulter counter.** The Coulter counter was originally developed for counting blood cells, but can also be applied to measuring particle sizes in the range of 0.5 microns to 1.0 mm. Size analysis with the Coulter counter is based on the principle that a particle passing through an electrical field maintained in an electrolyte will displace its own volume of the electrolyte and thus cause a change in the field. In practice, particles are dispersed in a suitable electrolyte and forced to flow through an aperture one at a time. As each particle passes through the aperture, the electrical properties of the field change. These changes are scaled and counted as voltage pulses. The magnitude of each pulse is proportional to the particle volume, and the number of pulses is a function of particle concentration. Thus by counting the number of pulses of various magnitudes, the volume percentage of particles of different sizes can be determined (Swift et al., 1972).

The grain size of particles in consolidated sedimentary rocks that cannot be disaggregated must be measured by techniques other than sieving or sedimentation analysis. The size and sorting of sand- and silt-size particles can be estimated by using a reflected-light binocular microscope and a standard size-comparison set, which consists of grains of specific sizes mounted on a card. More accurate size determination can be made by measuring grains in thin sections of rock by use of a transmitted-light petrographic microscope fitted with an ocular micrometer. Fine silt- and clay-size grains in consolidated rocks may be studied by use of an electron microscope.

The techniques for measuring grain size are described in detail in numerous books and papers, including several of the books listed at the end of this chapter under additional readings. Standard techniques for sieving, sedimentation analysis, and thin-section analysis are described particularly well in Carver (1971).

Graphical and Mathematical Treatment of Grain-size Data

Measurement of grain size by the techniques described generates large amounts of data that must be reduced to more condensed form before they can be used. Tables of data showing the weights of grains in various size classes must be simplified to yield such average properties of grain populations as mean grain size and sorting. Both graphical and mathematical data-reduction methods are in common use. Graphical plots are simple to construct and provide a readily understandable visual representation of grain-size distributions. On the other hand, mathematical methods, some of which are based on initial graphical treatment of data, yield statistical grain-size parameters that may be useful in environmental studies.

Figure 5.1 illustrates three common graphical methods for presenting grain-size data. Figure 5.1A shows typical grain-size data obtained by sieve analysis. Raw sieve weights are first converted to individual weight percents by dividing the weight in each size class by the total weight. Cumulative weight percent may be calculated by adding the weight of each succeeding size class to the total of the preceding classes. Figure 5.1B shows how individual weight percent can be plotted as a function of grain size to yield a grain-size **histogram**—a bar diagram in which grain size is plotted along the abscissa of the graph and individual weight percent along the ordinate. Histograms provide a quick, easy pictorial method for representing grain-size distributions because the approximate average grain size and the sorting—the spread of grain-size values around the average size—can be seen at a glance. Histograms have limited application, however, because the shape of the histogram is affected by the sieve interval used; such graphs cannot be used to obtain mathematical values for statistical calculations.

A **frequency curve** (Fig. 5.1B) is essentially a histogram in which a smooth curve takes the place of a discontinuous bar graph. Connecting the midpoints of each size class in a histogram with a smooth curve gives the approximate shape of the frequency curve. A frequency curve constructed in this manner does not, however, accurately fix the position of the highest point on the curve; this point is important for determining the modal size, to be described. A grain-size histogram plotted from data obtained by sieving at exceedingly small sieve intervals would yield the approximate shape of a

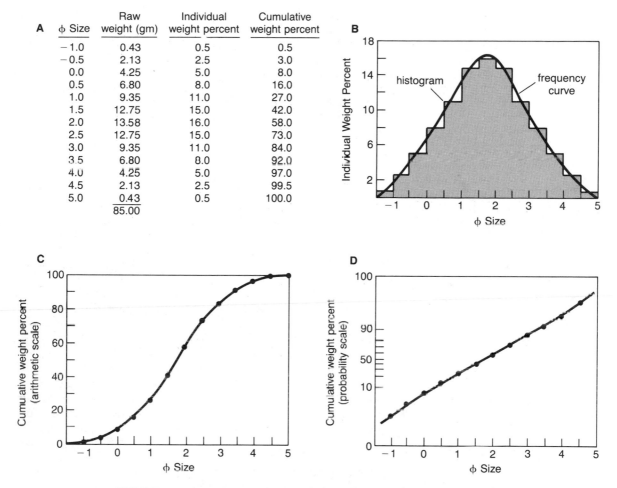

A ϕ Size	Raw weight (gm)	Individual weight percent	Cumulative weight percent
−1.0	0.43	0.5	0.5
−0.5	2.13	2.5	3.0
0.0	4.25	5.0	8.0
0.5	6.80	8.0	16.0
1.0	9.35	11.0	27.0
1.5	12.75	15.0	42.0
2.0	13.58	16.0	58.0
2.5	12.75	15.0	73.0
3.0	9.35	11.0	84.0
3.5	6.80	8.0	92.0
4.0	4.25	5.0	97.0
4.5	2.13	2.5	99.5
5.0	0.43	0.5	100.0
	85.00		

FIGURE 5.1 Common visual methods of displaying grain-size data. A. Grain-size data table. B. Histogram and frequency curve plotted from data in A. C. Cumulative curve with an arithmetic ordinate scale. D. Cumulative curve with a probability ordinate scale.

frequency curve, but such small sieve intervals are not practical. Accurate frequency curves can by derived from cumulative curves by special graphical methods described in detail by Folk (1974).

A grain-size **cumulative curve** is generated by plotting grain size against cumulative weight-percent frequency. The cumulative curve is the most useful of the grain-size plots. Although it does not give as good a pictorial representation of the grain-size distribution as a histogram or frequency curve, its shape is virtually independent of the sieve interval used. Also, data that can be derived from the cumulative curve allow calculation of several important grain-size statistical parameters. A cumulative curve can be plotted on an arithmetic ordinate scale (Fig. 5.1C) or on a log probability scale in which the arithmetic ordinate is replaced by a log probability ordinate (Fig. 5.1D). When phi-size data are plotted on an arithmetic ordinate, the cumulative curve typically has the S shape shown in Figure 5.1C. The slope of the central part of this curve reflects the sorting of the sample. A very steep slope indicates good sorting and a very gentle slope poor sorting. If the cumulative curve is plotted on log probability paper, the shape of the curve will tend toward a straight line if the population of grains has a normal distribution (Fig. 5.1D). In a normal distribution, the values show an even distribution, or spread, about the average value. In conventional statistics, a normally distributed population of values yields a perfect bell-shaped curve when plotted as a frequency curve. Deviations from normality of a grain-size distribution can thus be easily detected on log probability plots by deviation of the cumulative curve from a straight line. Most natural populations of grains in siliciclastic sediments or sedimentary rocks do not have a normal (or log-normal) distribution; the nearly normal distribution shown in Figure 5.1 is not typical of natural sediments. Some investigators believe that the shape of the log probability curve reflects conditions of the sediment-transport process and thus can be used as a tool in environmental interpretation. We shall return to this point subsequently.

Graphical plots permit quick, visual inspection of the grain-size characteristics of a given sample; however, comparison of graphical plots becomes cumbersome and inconvenient when large numbers of samples are involved. Also, average grain-size and sorting characteristics cannot be determined very accurately by visual inspection of grain-size curves. To overcome these disadvantages, mathematical methods that permit statistical treatment of grain-size data can be used to derive parameters that describe grain-size distributions in mathematical language. These statistical measures allow both the average size and the sorting characteristics of grain populations to be expressed mathematically. Mathematical values of size and sorting can be used to prepare a variety of graphs and charts that facilitate evaluation of grain-size data.

Average Grain Size. Three mathematical measures of average grain size are in common use. The **mode** is the most frequently occurring particle size in a population of grains. The diameter of the modal size corresponds to the diameter of grains represented by the steepest point (inflection point) on a cumulative curve. Siliciclastic sediments and sedimentary rocks tend to have a single modal size, but some sediments are bimodal, with one mode in the coarse end of the size distribution and one in the fine end. Some are even polymodal. The **median** size is the midpoint of the grain-size distribution. Half of the grains by weight are larger than the median size and half are smaller. The median size corresponds to the 50th percentile diameter on the cumulative curve (Fig. 5.2). The **mean** size is the arithmetic average of all the particle sizes in a sample. The true arithmetic mean of most sediment samples cannot be determined because we cannot count the total number of grains in a sample or measure each small grain. An approximation of the arithmetic mean can be arrived at by picking selected percentile values from the cumulative curve and averaging these values. As shown in Figure 5.2, the 16th, 50th, and 84th percentile values are commonly used for this calculation.

Sorting. The sorting of a grain population is a measure of the range of grain sizes present and the magnitude of the spread, or scatter, of these sizes around the mean size. Sorting can be estimated in the field or laboratory by use of a hand lens or microscope and reference to a visual estimation chart (Fig. 5.3). More accurate determination of sorting requires mathematical treatment of grain-size data. The mathematical expression of sorting is **standard deviation.** In conventional statistics, one standard

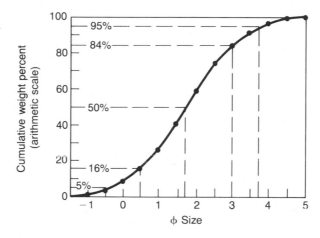

FIGURE 5.2 Method of calculating percentile values from the cumulative curve.

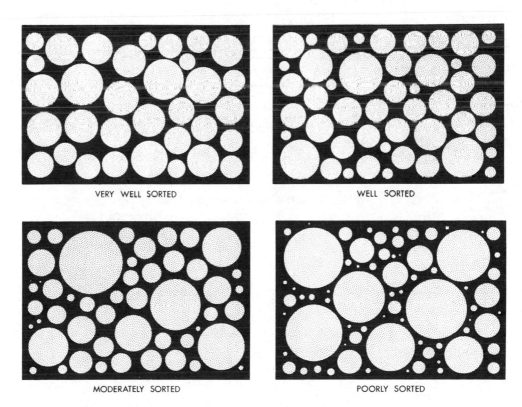

VERY WELL SORTED

WELL SORTED

MODERATELY SORTED

POORLY SORTED

FIGURE 5.3 Grain-sorting images for sediments with different degrees of sorting. (From Anstey, R. L., and T. L. Chase, 1974, Environments through time: Burgess, Minneapolis, Minn. Fig. 1.2, p. 2, reprinted by permission of Burgess Publishing Co.)

FIGURE 5.4 Frequency curve for a normal distribution of values showing the relationship of standard deviation to the mean. One standard deviation (1σ) on either side of the mean accounts for 68 percent of the area under the frequency curve. (After Friedman, G. M., and J. E. Sanders, Principles of sedimentology. © 1978 by John Wiley & Sons, Inc. Fig. 3.12, p. 70, reprinted by permission of John Wiley & Sons, Inc., New York.)

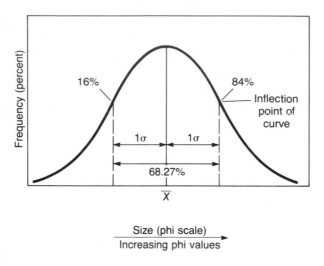

deviation encompasses the central 68 percent of the area under the frequency curve (Fig. 5.4). That is, 68 percent of the grain-size values lie within plus or minus one standard deviation of the mean size. A formula for calculating standard deviation by graphical-statistical methods is shown in Table 5.3. Note that the standard deviation calculated by this formula is expressed in phi (φ) values and is called phi standard deviation. The symbol φ must always be attached to the standard deviation value. Verbal terms for sorting corresponding to various values of standard deviation are given below, after Folk (1974).

Standard deviation

<0.35φ	very well sorted
0.35 to 0.50φ	well sorted
0.50 to 0.71φ	moderately well sorted
0.71 to 1.00φ	moderately sorted
1.00 to 2.00φ	poorly sorted
2.00 to 4.00φ	very poorly sorted
>4.00φ	extremely poorly sorted

TABLE 5.3   Formulas for calculating grain-size statistical parameters by graphical methods

Graphic mean	$M_z = \dfrac{\phi_{16} + \phi_{50} + \phi_{84}}{3}$	(1)
Inclusive graphic standard deviation	$\sigma_i = \dfrac{\phi_{84} - \phi_{16}}{4} + \dfrac{\phi_{95} - \phi_5}{6.6}$	(2)
Inclusive graphic skewness	$SK_i = \dfrac{(\phi_{84} + \phi_{16} - 2\phi_{50})}{2(\phi_{84} - \phi_{16})} + \dfrac{(\phi_{95} + \phi_5 - 2\phi_{50})}{2(\phi_{95} - \phi_5)}$	(3)
Graphic kurtosis	$K_G = \dfrac{(\phi_{95} - \phi_5)}{2.44(\phi_{75} - \phi_{25})}$	(4)

Source: Folk, R. L., and W. C. Ward, 1957, Brazos River bar: A study in the significance of grain-size parameters: Jour. Sedimentary Petrology, v. 27, p. 3–26.

FIGURE 5.5 Frequency curves illustrating the mode, median, mean, and the difference between normal frequency curves and asymmetrical (skewed) curves. (After Friedman, G. M., and J. E. Sanders, Principles of sedimentology. © 1978 by John Wiley & Sons, Inc. Fig. 3.18, p. 75. reprinted by permission of John Wiley & Sons, Inc., New York.)

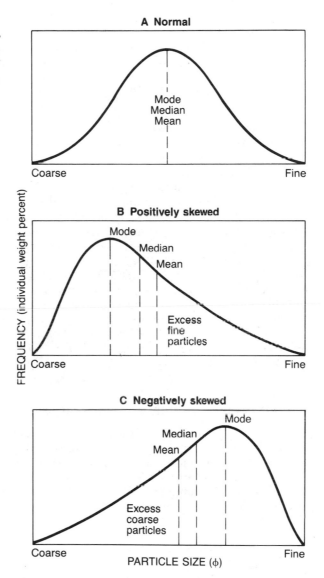

As mentioned, most natural sediment grain-size populations do not exhibit a normal or log-normal grain-size distribution. The frequency curves of such nonnormal populations are not perfect bell-shaped curves such as the example shown in Figure 5.5A. Instead, they show some degree of asymmetry, or **skewness.** The mode, mean, and median in a skewed population of grains are all different, as illustrated in Figures 5.5B and 5.5C. Skewness reflects sorting in the "tails" of a grain-size population. Populations that have a tail of excess fine particles (Fig. 5.5B) are said to be positively skewed or fine skewed, that is, skewed toward positive phi values. Populations with a tail of excess coarse particles (Fig. 5.5C) are negatively skewed or coarse skewed. Graphic skewness can be calculated by equation 3 in Table 5.3. Verbal skewness is related to calculated values of skewness, as shown below (Folk, 1974):

Skewness

> +0.30	strongly fine skewed
+0.30 to +0.10	fine skewed
+0.10 to −0.10	near symmetrical
−0.10 to −0.30	coarse skewed
< −0.30	strongly coarse skewed

Grain-size frequency curves can show various degrees of sharpness or peakedness. The degree of peakedness is called **kurtosis.** Sharp-peaked curves are said to be leptokurtic; flat-peaked curves are platykurtic. A formula for calculating kurtosis is shown in Table 5.3. Although kurtosis is commonly calculated along with other grain-size parameters, the geological significance of kurtosis is unknown; it appears to have little value in interpretative grain-size studies.

Grain-size statistical parameters can be calculated directly, without reference to graphical plots, by the mathematical **method of moments.** The procedure for calculating grain-size statistics by the method of moments has been known for many years (Krumbein and Pettijohn, 1938). The method has not been used extensively until comparatively recently because of the laborious calculations involved and because it has not been definitely proven that moment statistics are of greater value than graphical statistics in application to geologic problems. With the advent of modern computers, lengthy calculations no longer pose a problem, and moment statistics are now in common use. The computations in moment statistics involve multiplying a weight (weight frequency in percent) by a distance (from the midpoint of each size grade to the arbitrary origin of the abscissa). Equations for computing moment statistics are given in Table 5.4 and a sample computation form using ½φ size classes is given in Table 5.5.

Use and Interpretation of Grain-size Data

Grain size is a fundamental physical property of sedimentary rocks and, as such, is a useful descriptive property. Also, the related derived properties of porosity and permeability are of considerable interest to petroleum geologists and hydrologists. Because the size and sorting of sediment grains may reflect sedimentation mechanisms and depositional conditions, study of grain-size data has commonly been assumed to be a useful tool for interpreting the depositional environments of ancient sedimentary rocks.

TABLE 5.4 Formulas for calculating grain-size parameters by the moment method

Mean (1st moment)	$\bar{x}_\phi = \dfrac{\Sigma fm}{n}$	(1)
Standard deviation (2nd moment)	$\sigma_\phi = \sqrt{\dfrac{\Sigma f(m - \bar{x}_\phi)^2}{100}}$	(2)
Skewness (3rd moment)	$Sk_\phi = \dfrac{\Sigma f(m - \bar{x}_\phi)^3}{100\,\sigma_\phi{}^3}$	(3)
Kurtosis (4th moment)	$K_\phi = \dfrac{\Sigma f(m - \bar{x}_\phi)^4}{100\,\sigma_\phi{}^4}$	(4)

where f = weight percent (frequency) in each grain-size grade present
m = midpoint of each grain-size grade in phi values
n = total number in sample; 100 when f is in percent

TABLE 5.5 Form for computing moment statistics using ½φ classes

Class interval (φ)	m Midpoint (φ)	f Weight %	fm Product	m − x̄ Deviation	(m − x̄)² Deviation squared	f(m − x̄)² Product	(m − x̄)³ Deviation cubed	f(m − x̄)³ Product	(m − x̄)⁴ Deviation quadrupled	f(m − x̄)⁴ Product
0–0.5	0.25	0.9	0.2	−2.13	4.54	4.09	−9.67	−8.70	20.60	18.54
0.5–1.0	0.75	2.9	2.2	−1.63	2.66	7.71	−4.34	−12.59	7.07	20.50
1.0–1.5	1.25	12.2	15.3	−1.13	1.28	15.62	−1.45	−17.69	1.63	19.89
1.5–2.0	1.75	13.7	24.0	−0.63	0.40	5.48	−0.25	−3.43	0.16	2.19
2.0–2.5	2.25	23.7	53.3	−.13	0.02	0.47	0.00	0.00	0.00	0.00
2.5–3.0	2.75	26.8	73.7	0.37	0.13	3.48	0.05	1.34	0.02	0.54
3.0–3.5	3.25	12.2	39.7	0.87	0.76	9.27	0.66	8.05	0.57	6.95
3.5–4.0	3.75	5.6	21.0	1.37	1.88	10.53	2.57	14.39	3.52	19.71
>4.0	4.25	2.0	8.5	1.87	3.50	7.00	6.55	13.10	12.25	24.50
Total		100.0	237.9			63.65		−5.53		112.82

Source: McBride, E. F., Mathematical treatment of size distribution data, in R. E. Carver (ed.), Procedures in sedimentary petrology. © 1971 by John Wiley & Sons, Inc. Table 2, p. 119, reprinted by permission of John Wiley & Sons, Inc., New York.

Basis for Environmental Interpretation. It is this assumption that grain-size characteristics reflect conditions of the depositional environment that has sparked most of the interest in grain-size analyses. Geologists have studied the grain-size properties of sediments and sedimentary rocks for more than a century, and research efforts since the 1950s have focused particularly on statistical treatment of grain-size data. This prolonged period of grain-size research has generated hundreds of learned papers in geological journals and swelled the bibliographies of numerous sedimentologists. So, it would be logical to assume that by now the relationship between grain-size characteristics and depositional environments has been firmly established. Unfortunately, such is not the case! Let us briefly examine the actual progress that has been made in the application of sediment grain-size characteristics to environmental interpretation.

The ultimate characteristics of a sediment grain-size distribution are a function of several interacting processes, beginning with weathering in the source area. Grains of various sizes are released from parent rocks by weathering and fed into the sediment-transport mill. During transport by wind and water, changes occur in grain-size populations owing to breakdown of soft or brittle grains and selective sorting arising from differences in size, shape, and density of grains. The grain population that eventually reaches a given depositional site may be further modified by local depositional processes such as reworking by the swash on a beach. Finally, diagenetic processes operating during burial can increase the size of grains by adding cement as an overgrowth or decrease the size by solution or other alteration of grains. Thus, the ultimate size distribution of grains in a sedimentary rock is a function of: (1) availability of grains of different sizes at the source, (2) transport and depositional processes, and (3) postdepositional diagenetic changes.

Geologists commonly assume that transport and depositional processes are the most important of these factors in determining particle size and sorting of grain populations and that each major depositional environment leaves its impress on the sediments deposited in that environment in the form of specific grain-size and sorting characteristics. Since about the 1950s, the research approach to environmental analysis by use of grain-size data has consisted mainly of analyzing sediments from known modern environments in order to "fingerprint" the grain-size characteristics of sediments in these environments. It has generally been assumed that once the grain-size characteristics of siliciclastic sediments in various modern environments have been established, knowledge of these characteristics can then be used to interpret the depositional environments of ancient siliciclastic sedimentary rocks. Let's see how successful this approach has actually been.

To illustrate the possible relationship of grain-size characteristics to depositional environments, let's consider as an example the differences in characteristics of fluvial (river-deposited) sediments, beach sediments, and dune sediments. For one thing, fluvial sediments generally have coarser grains than sediments deposited in beach or dune environments. Mean grain size and maximum grain size commonly reflect the average energy and maximum energy of the depositional medium, although in some cases grain size of deposits may be controlled simply by the availability of particles of a particular size. For example, eolian sediments consist mainly of sand and finer sediment because wind does not have the capacity to move larger particles. By contrast, the flow velocity of some streams is so high that the streams can easily transport particles of boulder size. On the other hand, if only particles of sand size are available for transport, then streams will transport and eventually deposit sand regardless of the velocity of stream flow. Many beach deposits consist mainly of sand for this reason. That is to say, even though the energy of the surf may be high enough to move pebbles as bedload, beach processes result mainly in transport and deposition of sand because few pebbles are available for transport on many beaches.

Another difference between fluvial sediments and beach or dune sediments is that the fluvial sediments are likely to be more poorly sorted. Poor sorting is typical of fluvial deposits because a wide range of particle sizes is commonly available for transport by rivers, and transport by traction, saltation, and suspension can occur at the same time during flood stage. Different-size particles undergoing simultaneous transport may all get deposited together when flood stage wanes. Furthermore, these deposits generally undergo little or no reworking after deposition. Unlike grain size, sorting tends to reflect the persistence of the depositional process to a greater degree than it reflects the actual energy of the environment. Thus, sediment reworking has a very important influence on sorting, and the best sorted sediments occur in environments such as beaches and dunes, where the sediments undergo extensive reworking under moderate energy conditions before final deposition. Extremely high energy conditions such as those that occur during severe storms on the shallow continental shelf may actually produce very poorly sorted sediments. This is because high-energy events stir up and mix together sediment of different sizes; the sediment is then quickly deposited without reworking as the energy of the event diminishes. Sediments deposited by sediment gravity flows or by glaciers also tend to be poorly sorted.

In addition to variable sorting, the grain-size populations of sediments are typically skewed to varying degrees. Fluvial sediments, for example, may be positively skewed. Owing to the large amounts of clay and silt commonly carried by rivers, excess fine sediment becomes trapped among coarser particles during deposition, resulting in positive skewness. Dune sands also tend to be positively skewed because wind velocity is not strong enough to move coarser grains and transport them along with the finer grains undergoing traction and saltation transport. Thus, fine, positively skewed sands end up in the dunes, and coarser grains are left behind as a negatively skewed lag deposit. Beach sands, on the other hand, tend to be negatively skewed because fine sediment is winnowed out of beach sediment and removed owing to bidirectional transport by currents of unequal velocity. Currents moving up the front of a beach commonly have a greater velocity than the ebb flow back down the beach. Therefore, fine particles may be removed by the return flow, leaving a negatively skewed population of well-sorted, coarser grains.

Methods of Environmental Analysis. Because sediments from different environments tend to have distinctive grain-size characteristics, early research efforts in environmental analysis focused on attempts to characterize environments in terms of specific values of grain-size parameters. In other words, investigators tried to fingerprint each environment in terms of specific values (or narrow ranges of values) of mean grain size, grain-size standard deviation, or skewness. As more grain-size data accumulated, however, it soon became evident that a wide range of values for these grain-size parameters can be present in sediments from the same environment, and much overlap of the parameters occurs from environment to environment. It became apparent, for example, that not all fluvial sediments are more poorly sorted than beach sediments, nor are all fluvial deposits positively skewed. In an attempt to overcome these problems of overlap, Friedman (1961, 1967, 1979) popularized the use of two-component grain-size variation diagrams in which one statistical parameter is plotted against another. For example, mean size can be plotted against standard deviation, standard deviation against skewness, and so forth. By plotting large numbers of grain-size analyses of modern sediments on such diagrams, lines can be drawn separating the diagrams into major environmental fields. The idea behind these plots is that a single grain-size parameter such as mean size may not be adequate to distinguish sediments from different environments, but plots involving two grain-size parameters should allow more positive separation into environmental fields. Figure 5.6 is an example of a diagram in which modern beach sands are differentiated from river-deposited sands by plotting standard deviation against skewness. Plots of many other combinations of statistical

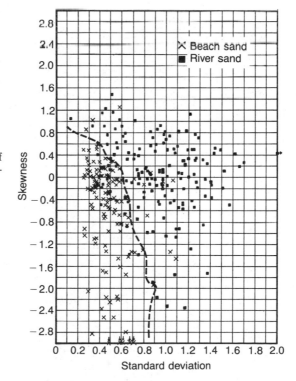

FIGURE 5.6 Bivariate plot of moment skewness versus moment standard deviation. Beach sands and dune sands are separated by this plot into two moderately well-defined fields, but some overlap occurs. (After Friedman, G. M., 1967, Dynamic processes and statistical parameters for size frequency distribution of beach and river sands: Jour. Sedimentary Petrology, v. 37. Fig. 5, p. 334, reprinted by permission of Society of Economic Paleontologists and Mineralogists, Tulsa, Okla.)

parameters have been constructed; all are claimed to be effective to varying degrees in differentiating sediments deposited in different major environments. This graphical technique has been used mainly to characterize and differentiate modern sediments of beach, river, and inland dune environments. Presumably, it can be used also to identify and differentiate ancient beach, river, and dune sediments.

Passega (1957, 1964, 1977) developed a different graphical approach to environmental analysis. This approach does not require calculation of statistical grain-size parameters, but makes use of what Passega calls C–M and L–M diagrams. Three properties of grain-size distributions are used in these diagrams: (1) the grain diameter (C) corresponding to the first percentile (1 percent) on the cumulative curve, to measure the coarsest grains in a deposit; (2) the median grain diameter (M); and (3) the percentile finer than 0.031 mm (L), to measure the fine tail of the size distribution. To use Passega's method, the results of size analyses of several samples from the same deposit or sedimentary unit are plotted on a C–M or an L–M diagram. Most of the samples from a given environment will theoretically fall within a specific environmental field of the diagram. Discrimination of different environmental fields on this basis is theoretically possible because water energy levels are sufficiently different in various major environments to produce distinctive differences in maximum particle size, mean grain size, and the finest particle sizes. Figure 5.7 shows an example of a C–M plot.

A third approach to environmental discrimination involves comparison of cumulative curves of grain-size data plotted on log probability paper (Visher, 1969; Sagoe

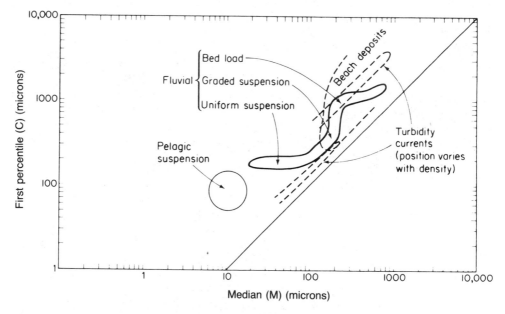

FIGURE 5.7   C–M diagram showing patterns of sediments deposited in different depositional environments. (Diagram from Blatt, H., G. V. Middleton, and R. Murray, Origin of sedimentary rocks, 2nd ed., © 1980, Fig. 3.17, p. 74. Reprinted by permission of Prentice-Hall, Englewood Cliffs, N.J. Modified from Passega, R., 1964, Grain size represented by CM patterns as a geological tool: Jour. Sedimentary Petrology. Fig. 1, p. 831, reprinted by permission of Society of Economic Paleontologists and Mineralogists, Tulsa, Okla.)

and Visher, 1977; Glaister and Nelson, 1974; Middleton, 1976). Cumulative log probability curves commonly do not plot as a single straight line but instead display two or three straight-line segments (Fig. 5.7). Each segment of the curve is interpreted to represent different subpopulations of grains that were transported simultaneously but by different transport modes; that is, by suspension, saltation, or bedload transport. Visher (1969) compared log probability plots of samples from several modern environments and concluded that sediments from different environments could be differentiated on the basis of the general shape of the curves, the slope of the curve segments, and the position of the truncation points (breaks in slope) between the straight-line segments (Fig. 5.8). Distinctive differences in the log probability grain-size curves of sediments deposited in different depositional environments are apparently related to variations in the relative volumes of sediment of particular sizes transported by suspension, saltation, or traction in these different environments. Visher applied this method of grain-size analysis to characterization of several types of sediments, including dune sands; fluvial deposits; beach, tidal, and nearshore sediments; and turbidites.

FIGURE 5.8 Relation of sediment transport dynamics to populations and truncation points in a grain-size distribution as revealed by plotting grain-size data as a cumulative curve on log probability paper. (After Visher, G. S., Grain size distributions and depositional processes: Jour. Sedimentary Petrology, v. 39. Fig. 4, p. 1079, reprinted by permission of Society of Economic Paleontologists and Mineralogists, Tulsa, Okla.)

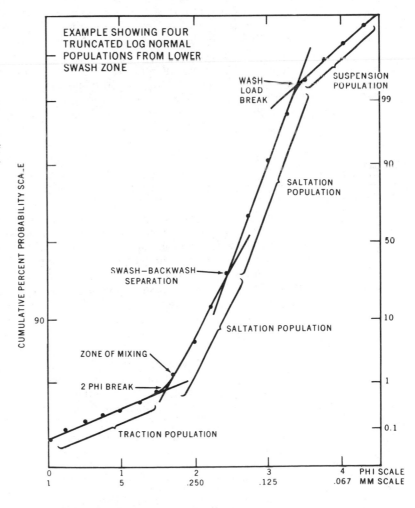

Reliability of Environmental Interpretation. How successful are these various approaches to characterizing the grain-size distributions of modern depositional environments and, more importantly, how reliable are they as tools for interpreting ancient sedimentary environments? Friedman (1979) and a small number of other investigators have claimed very positive results in using bivariate statistical plots for discriminating modern beach, river, and dune sands; however, much less satisfactory results have been reported by several other workers. For example, Tucker and Vacher (1980) found in a study of 970 samples from 40 different localities that the probability of error in assigning samples to the proper environmental field on the basis of grain-size data ranged from 25 to 35 percent. Reed et al. (1975), using settling velocity data analyzed with a statistical discriminant function, reported positive results in environmental discrimination, but found that the curve shape of sieved grain-size distributions plotted on log probability paper (Visher's method) was not a reliable guide for interpreting the depositional environment of the samples studied. The Sedimentation Seminar (1981) cast even more serious doubt on the validity of both log probability plots and bivariate plots for discriminating modern sediments. The seminar geologists found these techniques to be highly unreliable in a study of Amazon and Solimoes river sediments, most of which did not plot as river sediments when either of these techniques was used. Vandenberghe (1975) points out problems with use of C–M patterns for interpreting depositional environments of fine-grained sediments, arguing that a "uniform" suspension pattern can be due simply to the clay-rich nature of the sediment and does not necessarily indicate transport by a specific mechanism.

The reasons these grain-size techniques for identifying depositional environments of modern sediments fail to work consistently are probably related to variability in depositional conditions within major environmental settings. The energy conditions and sediment supply within river systems, for example, can differ considerably from one river to another and even within different parts of the same river system. Thus, in some cases, the grain-size characteristics of sediments may show as much variability within different parts of the same environmental setting as between different environments.

The reliability of grain-size data for identifying depositional environments becomes even more questionable when applied to ancient sedimentary rocks. Many ancient sedimentary rocks cannot be sufficiently disaggregated to permit accurate grain-size analysis by sieving or settling-tube methods. The results of size analyses of ancient sedimentary rocks are thus often more a reflection of the degree of success of disaggregation methods than of the actual grain-size distribution. The grain size of rocks that cannot be completely disaggregated must be analyzed by thin-section methods, which do not work well if large amounts of very fine-size grains are present. An even more serious problem may arise from alteration of grains during diagenesis. Alteration of feldspars or other minerals to clay minerals is a particularly common diagenetic process. There is growing evidence that much of the very fine-grained matrix material (grains <~0.03 mm) in ancient sandstones may have originated by diagenetic processes. If this is true, consistent, reliable results in interpreting paleoenvironments on the basis of grain size of texturally immature sandstones that contain abundant matrix are very unlikely. Thus, it appears that after several decades of intensive research into the techniques and significance of grain-size analysis, there is still need for more research on this problem. Such is science! In any case, grain-size data should be considered as only one of the available tools for environmental interpretation and should never be used alone for this purpose.

5.3 PARTICLE SHAPE

Particle shape is defined by three related but different aspects of grains. **Form** refers to the gross, overall configuration of particles and reflects variations in their proportions (lengths of major axes). Form is often confused with **roundness,** which is a measure of the sharpness of grain corners, that is, smoothness. **Surface texture** refers to small-scale, microrelief markings such as pits, scratches, and ridges that occur on the surfaces of grains. Form, roundness, and surface texture are independent properties, as illustrated in Figure 5.9, and each can theoretically vary without affecting the other. Actually, form and roundness tend to be positively correlated in sedimentary deposits—particles that are highly spherical in shape also tend to be well rounded. Surface texture can change without significantly changing form or roundness, but a change in form or roundness will affect surface texture because new surfaces are exposed. The three aspects of shape can be thought of as constituting a hierarchy, where form is a first-order property, roundness a second-order property superimposed on form, and surface texture a third-order property superimposed on both the corners of a grain and the surfaces between corners (Barrett, 1980).

Particle Form

Sphericity and Form Indices. Form reflects variations in the proportions of particles. Therefore, most parameters for estimating form require that the relative lengths of the three major axes (long, intermediate, short) of a particle be known. Numerous mathematical measures for expressing form have been proposed, but **sphericity** is probably the most widely used. The concept of sphericity was introduced by Wadell (1932), who defined sphericity mathematically as the ratio of the diameter of a sphere with the same volume as a particle to the diameter of the smallest circle that would just

FIGURE 5.9 Simplified representation of form, roundness, and surface texture to demonstrate their independence. Note that each of the three aspects can be represented by more than one dimension. (After Barrett, P. J., 1980, The shape of rock particles, a critical review: Sedimentology, v. 27. Fig. 1, p. 293, reprinted by permission of Elsevier Science Publishers, Amsterdam.)

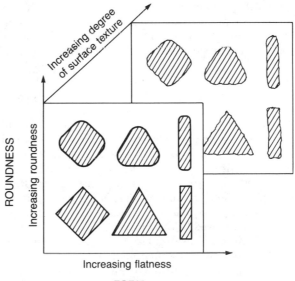

enclose or circumscribe the outline of the particle. Wadell determined the volumes of large particles by immersing single particles in water and measuring the volume change of the water. Krumbein (1941) modified Wadell's sphericity concept slightly to express sphericity (ψ) by the relationship

$$\psi = \sqrt[3]{\frac{\text{volume of the particle}}{\text{volume of the circumscribing sphere}}} \qquad (5.2)$$

The volume of a sphere is given by $\pi/6D^3$, where D is the diameter of the sphere. The approximate volume of natural particles can be calculated by assuming that the particles are triaxial ellipsoids having three diameters D_L, D_I, and D_S, where L, I, and S refer to the lengths of the long, intermediate, and short axes of the ellipsoid. After determining volume by substituting appropriate values into equation 5.2, sphericity is expressed by

$$\psi_I = \sqrt[3]{\frac{\pi 6 D_L D_I D_S}{\pi 6 D_L{}^3}} = \sqrt[3]{\frac{D_S D_I}{D_S{}^2}} \qquad (5.3)$$

The sphericity of a particle determined by this relationship is called **intercept sphericity** and can be calculated by measuring the long, intermediate, and short axes of a particle and substituting these values into equation 5.3.

Sneed and Folk (1958) suggest that Krumbein's intercept sphericity does not correctly express the behavior of particles as they settle in a fluid or are acted on by fluid flow. A rod-shaped particle, for example, settles faster than a disc, although the intercept sphericity equation suggests the opposite. Particles falling through water tend to settle with maximum projection areas (the plane of the long and intermediate axes) perpendicular to the direction of motion, and small particles resting on the bottom orient themselves perpendicular to current flow direction. Sneed and Folk thus proposed a different sphericity measure, called **maximum projection sphericity** (ψ_P), which they claim better expresses particle behavior. Maximum projection sphericity is defined mathematically as the ratio between the maximum projection area of a sphere with the same volume as the particle and the maximum projection area of the particle and is shown by the relationship

$$\psi_P = \sqrt[3]{\frac{D_S{}^2}{D_L D_I}} \qquad (5.4)$$

Maximum projection sphericity has gained favor for expressing the shape of particles deposited by water. Conceptually, it is not necessarily more valid than intercept sphericity when applied to other modes of particle transport and deposition—transport by ice and sediment gravity flows, for example.

Regardless of the sphericity measure used, experience has shown that particles having the same mathematical sphericity can differ considerably in their overall shape. Some additional measure or index is needed to more specifically define the form of particles. Two additional form indices that permit a more graphic representation of form are in wide use. Zingg (1935) proposed the use of two shape indices, D_I/D_L and D_S/D_I, to define four shape fields on a bivariate plot: oblate (disc), equant (spheres), bladed, and prolate (rollers) (Fig. 5.10A). Lines of equal intercept sphericity can be drawn on the Zingg shape fields (Fig. 5.10B), illustrating that particles of quite different form can have the same mathematical sphericity.

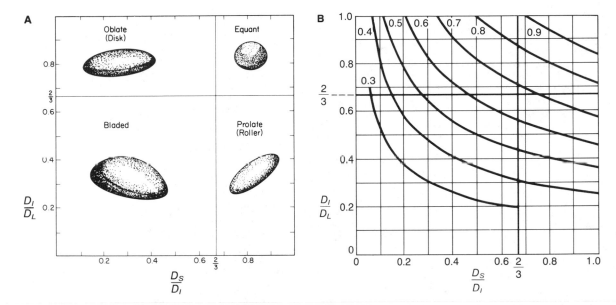

FIGURE 5.10 A. Classification of shapes of pebbles after Zingg (1935). B. Relationship between mathematical sphericity and Zingg shape fields. The curves represent lines of equal sphericity. (A, from Blatt, H., G. V. Middleton, and R. Murray, Origin of sedimentary rocks, 2nd ed. © 1980, Fig. 3.20, p. 80. Reprinted by permission of Prentice Hall, Englewood Cliffs, N.J. B, after Pettijohn, F. J., Sedimentary rocks © 1975. Fig. 3.19, p. 54, reprinted by permission of Harper & Row, New York.)

Sneed and Folk (1958) used two somewhat different shape indices to construct a triangular form diagram (Fig. 5.11). D_S/D_L is plotted against $D_L - D_I/D_L - D_S$ to create ten form fields (compact, platy, bladed, elongate, and so on). Lines of maximum projection sphericity drawn across the field (Fig. 5.11) again illustrate the disparity between mathematical sphericity and actual form.

Significance of Form. The form of sand-size and smaller mineral grains in sedimentary deposits is a function mainly of the original shapes of the minerals. Because of its superior hardness and durability, as well as its high average abundance in siliciclastic sedimentary rocks, quartz is commonly the only mineral examined in shape studies. Numerous studies have shown that the form of small quartz grains is not significantly modified during transport, although very slight changes may occur in the early stages of transport. The form of pebbles and larger fragments is also a function of the shape inherited from source rocks, although, owing to abrasion and breakage, pebbles are modified during transport to a greater extent than are sand-size grains.

The form of particles has a very pronounced effect on their settling velocity; in general the more nonspherical the particle the lower the settling velocity. Particle form thus affects relative transportability of particles traveling in suspension. That is, nonspherical particles tend to stay in suspension longer than spherical particles. Form also affects the transportability of larger particles that move by traction along the bed. In general, spheres and rollers are transported more readily than blades and discs having the same mass. Thus, owing to preferential transport of spheres and rollers,

FIGURE 5.11 Classification of pebble shapes after Sneed and Folk. The symbol V refers to the adjective very (e.g., very platy, very bladed, very elongated). (After Sneed, E. D., and R. L. Folk, 1958, Pebbles in the Lower Colorado River, Texas, a study in particle morphogenesis: Jour. Geology, v. 66. Fig. 2, p. 119, reprinted by permission of University of Chicago Press.)

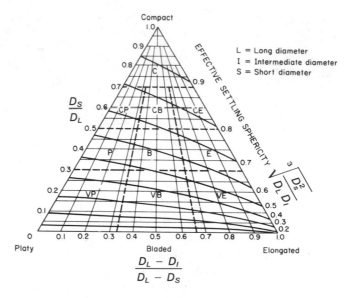

downstream changes in pebble form may occur in rivers. Such changes can be difficult or impossible to differentiate from changes caused by pebble abrasion. A long-standing and as yet unresolved controversy has to do with the flattened, disc shapes of many pebbles found on beaches. The prevalence of disc-shaped beach pebbles has been attributed alternatively to selective transport that leaves flattened pebbles behind on the beach as a lag deposit (Kuenen, 1964) and to flattening of pebbles on the beach by abrasion in the surf (Dobkins and Folk, 1970).

It has not yet been demonstrated that the form of particles alone can be used as a reliable tool for interpreting depositional environments. Although empirical studies show some relative differences in sphericity or form indices of grains from different environments, these differences have not proven to be sufficiently distinctive to permit environmental discrimination. Furthermore, it is very difficult to accurately measure the form of sand-size grains, particularly in consolidated sediments. In a thin section, for example, we can see only two dimensions of a particle, and neither dimension may represent the true length of the particle axis.

Roundness

Definition and Measurement. Wadell (1932) defined mathematical roundness as the arithmetic mean of the roundness of the individual corners of a grain in the plane of measurement. Roundness of individual corners is given by the ratio of the radius of curvature of the corners to the radius of the maximum-size circle that can be inscribed within the outline of the grain in the plane of measurement. The degree of Wadell roundness (R_W) is thus expressed as

$$R_W = \frac{\Sigma(r/R)}{N} = \frac{\Sigma(r)}{RN} \tag{5.6}$$

where r is the radius of curvature of individual corners, R is the radius of the maximum inscribed circle, and N is the number of corners. The relationship of r to R is illustrated in Figure 5.12.

FIGURE 5.12 Diagram of enlarged grain image illustrating the method of measuring the radius (R) of the maximum inscribed circle and the radii of curvature (r) of the corners of the grain. (After Boggs, S., Jr., 1967, Measurement of roundness and sphericity parameters using an electronic particle size analyzer: Jour. Sedimentary Petrology, v. 37. Fig. 3, p. 912, reprinted by permission of Society of Economic Paleontologists and Mineralogists, Tulsa, Okla.)

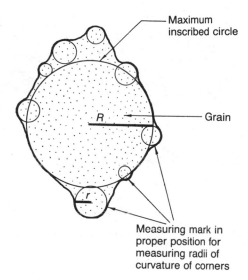

Maximum inscribed circle

Grain

Measuring mark in proper position for measuring radii of curvature of corners

Owing to the numerous radius measurements that must be made, it is very time consuming to determine the Wadell roundness of large numbers of grains. Simpler roundness measures have been proposed that require only that the radius of the sharpest corner be divided by the radius of the inscribed circle (Wentworth, 1919; Dobkins and Folk, 1970); however, Wadell's roundness measure is still used by most workers. Even if the simpler roundness formula is used, measuring the radii of large numbers of small grains is a very laborious process, requiring use of either a circular protractor or an electronic particle-size analyzer to measure enlarged images of grains (Boggs, 1967a). Consequently, visual comparison scales or charts consisting of sets of grain images of known roundness are often used to make rapid visual estimates of grain roundness. The visual charts of Krumbein (1941) and Powers (1953) are the most widely used of these comparison scales. The Powers visual and verbal roundness scale is shown in Figure 5.13, and the mathematical limits of each Powers verbal roundness class are given in Table 5.6. The interval between roundness classes is approximately $\sqrt{2}$; that is, the interval of each class is 1.41 times greater than the interval of the preceding class. Folk (1955) developed a logarithmic transformation of this scale; he called it the rho (ρ) scale, also shown in Table 5.6. Operator error in estimating grain roundness from visual charts is very high, and reproducibility of results even by the same operator is very low. Visual estimation methods yield only a rough approximation of the true roundness distribution in a population of grains and should not be used for serious research studies.

Significance of Roundness. The roundness of grains in a sedimentary deposit is a function of grain composition, grain size, type of transport process, and distance of transport. Hard, resistant grains such as quartz and zircon are rounded less readily during transport than weakly durable grains such as feldspars and pyroxenes. Pebble-to cobble-size grains commonly are more easily rounded than sand-size grains. Resistant mineral grains smaller than 0.05–0.1 mm do not appear to become rounded by any transport process. Because of these factors, it is always necessary to work with particles of the same size and composition when doing roundness studies.

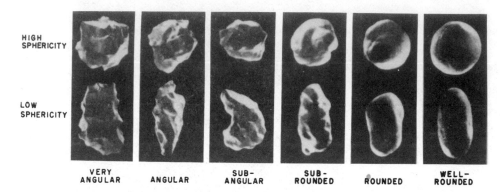

FIGURE 5.13 Powers grain images for estimating the roundness of sedimentary particles. (After Powers, M. C., 1953, A new roundness scale for sedimentary particles: Jour. Sedimentary Petrology, v. 23. Fig. 1, p. 118, reprinted by permission of Society of Economic Paleontologists and Mineralogists, Tulsa, Okla.)

Experimental studies in flumes and wind tunnels of the effects of abrasion on transport of sand-size quartz grains show that transport by wind is 100 to 1000 times more effective in rounding these grains than transport by water (Kuenen, 1959, 1960). In fact, almost no rounding occurs in as much as 100 km of transport by water. Most roundness studies of small quartz grains in rivers have corroborated these experimental results. For example, Russell and Taylor (1937) observed no increase in rounding of quartz grains over a distance of 1100 mi (1775 km) in the Mississippi River between Cairo, Illinois, and the Gulf of Mexico. The effectiveness of surf action on beaches in rounding sand-size quartz grains is not well understood. In general, surf processes appear to be less effective in rounding grains than wind transport but more effective than river transport.

Once acquired, the roundness of quartz grains is not easily lost and may be preserved through several sedimentation cycles. The presence of well-rounded quartz grains in an ancient sandstone may well indicate an episode of wind transport in its history, but it may be difficult or impossible to determine whether rounding took place during the last episode of transport or during some previous cycle.

TABLE 5.6 Relation of Powers' verbal rounding classes to Wadell roundness and Folk's rho (ρ) scale

Powers verbal class	Corresponding Wadell class interval	Folk's rho (ρ) scale
Very angular	0.12–0.17	0.00–1.00
Angular	0.17–0.25	1.00–2.00
Subangular	0.25–0.35	2.00–3.00
Subrounded	0.35–0.49	3.00–4.00
Rounded	0.49–0.70	4.00–5.00
Well rounded	0.70–1.0	5.00–6.00

Source: Powers, M. C., 1953, A new roundness scale for sedimentary particles: Jour. Sedimentary Petrology, v. 23, p. 117–119. Folk, R. L., 1955, Student operator error in determination of roundness, sphericity, and grain size: Jour. Sedimentary Petrology, v. 25, p. 297–301.

The roundness of transported pebbles is strongly related to pebble composition and size (Boggs, 1969). Soft pebbles such as shale and limestone become rounded much more readily than quartzite or chert pebbles, and large pebbles and cobbles are commonly better rounded than smaller pebbles. Although stream transport is relatively ineffective in rounding small quartz grains, pebble-size grains can become well rounded by stream transport. Depending upon composition and size, pebbles can become well rounded (roundness about 0.6) by stream transport in distances ranging from 11 km (7 mi) for limestones to 300 km (186 mi) for quartz (Pettijohn, 1975).

The presence of well-rounded pebbles in ancient sedimentary rocks is generally indicative of fluvial transport. The degree of rounding cannot, however, be depended upon to give reliable estimates of the distance of transport. The greatest amount of rounding takes place in the early stages of transport, generally within the first few kilometers. Also, the roundness of pebbles is not an unequivocal indicator of fluvial environments because pebbles can become rounded in beach environments and possibly on lakeshores. Furthermore, rounded fluvial pebbles may eventually be transported into nearshore marine environments where they may be reentrained by turbidity currents and resedimented in deeper parts of the ocean.

Surface Texture

The surfaces of pebbles and mineral grains may be polished, frosted (dull, matte texture like frosted glass), or marked by a variety of small-scale, low-relief features such as pits, scratches, fractures, and ridges. These surface textures originate in diverse ways, including mechanical abrasion during sediment transport; tectonic polishing during deformation; and chemical corrosion, etching, and precipitation of authigenic growths on grain surfaces during diagenesis and weathering. Gross surface textural features such as polishing and frosting can be observed with an ordinary binocular or petrographic microscope, however, detailed study of surface texture requires high magnifications and the use of an electron microscope. Krinsley (1962) pioneered use of the transmission electron microscope for studying grain-surface texture; the technique has now been superseded by use of the scanning electron microscope.

Most investigators who study the surface texture of sediment grains carry out their studies on quartz grains because the physical hardness and chemical stability of quartz grains allows these particles to retain surface markings for geologically long periods of time. Through study of thousands of quartz grains, investigators have now been able to fingerprint the markings on grains from various modern depositional environments. More than twenty different surface textural features have been identified, including conchoidal fractures, straight and curved scratches and striations, upturned plates, meander ridges, chemically etched Vs, mechanically formed Vs, and dish-shaped concavities. Examples of some of these markings are shown in Figure 5.14. Many other excellent electron micrographs of quartz surfaces may be found in Krinsley and Doornkamp's (1973) *Atlas of Quartz Sand Surface Textures.*

Surface texture appears to be more susceptible to change during sediment transport and deposition than do sphericity and roundness. Removal of old surface textural features and generation of new features is more likely to occur than marked changes in sphericity and roundness, and surface texture is more likely to record the last cycle of sediment transport or the last depositional environment. Therefore, geologists are interested in surface textural features as possible indicators of ancient transport conditions and depositional environment. The usefulness of surface texture in environmental analysis is limited, however, because similar types of surface markings can be

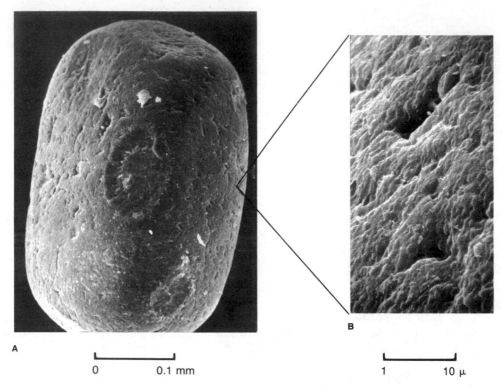

0   0.1 mm   1   10 μ

FIGURE 5.14 Electron micrograph of surface markings on a quartz grain, St. Peters Sandstone (Ordovician), United States midcontinent region. A. Quartz grain magnified 260×. B. Section of the grain surface magnified 2400×. (Photographs by M. B. Shaffer.)

produced in different environments. Also, the markings produced on grains in one environment may be retained on the grains when they are transported into another environment. Although less abrasion is required to remove surface markings from grains than is required to change roundness or sphericity, inherited markings from a previous environment may remain on grains for a long time before they are removed or replaced by different markings produced in the new environment. For example, grains on an arctic marine shelf may still retain surface microrelief features acquired during glacial transport of the grains to the shelf.

With care and the use of statistical methods, it has proven possible on the basis of surface textures to distinguish quartz grains from at least three major modern environmental settings: littoral (beach and nearshore), eolian (desert), and glacial. Quartz grains from littoral environments are characterized especially by V-shaped percussion marks and conchoidal breakage patterns. Grains deposited in eolian environments show surface smoothness and rounding, irregular upturned plates, and silica solution and precipitation features. Grains from glacial deposits have conchoidal fracture patterns and parallel to semiparallel striations. Techniques for studying surface texture of quartz grains in modern environments have been extended to study of ancient sedimentary deposits. Interpretation of paleoenvironments on the basis of surface texture is complicated by the fact that surface microtextures can be changed during diagenesis by addition of cementing overgrowths or by chemical etching and solution.

Fourier Shape Analysis

The two-dimensional shape of particles can be described with a high degree of precision by use of a method based on Fourier analysis, which is a method of representing periodic mathematical functions as an infinite series of summed sine and cosine terms (Ehrlich and Weinberg, 1970). Grain shape is represented by a series of terms called harmonics. Lower order harmonics summarize gross form (e.g., elongation), whereas roundness and surface texture are measured by progressively higher order harmonics (Fig. 5.15). The Fourier series is estimated for each grain projection from analysis of polar coordinates of peripheral points with their origin at the center of gravity of the maximum projection profile. The shape information contained in the first twenty harmonics is considered to be of fine enough resolution to represent a particle's shape, including surface texture (Dowdeswell, 1982). The method involves digitizing the periphery of a grain by projecting the grain onto a grid and recording, either manually or with an automatic digitizer, intercepts of the grain outline with the grid. Digitized data are reduced by computer to obtain the harmonics and produce computer-generated grain outlines.

      Fourier analysis has been used to study the source of sediment grains and to characterize grains from particular depositional environments. It is a fairly new technique, having been introduced in the 1970s, but preliminary results suggest that it has the potential to differentiate quartz grains derived from sedimentary sources from those derived from granitic or volcanic rocks and to differentiate between beach, dune, fluvial, and glaciofluvial sediments. It has been used also to map the source and transport paths of abyssal silts in the ocean. Because the technique is fairly new, additional work will be needed to confirm and extend these early results.

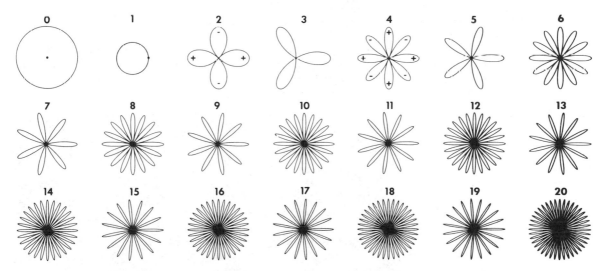

FIGURE 5.15 Graphic representation of selected Fourier harmonics. These first twenty harmonics are considered adequate to uniquely represent the two-dimensional shape of particles. (From Eppler, D. T., R. Ehrlich, D. Nummedahl, and P. H. Schulz, 1983, Sources of shape variations in lunar impact craters: Fourier shape analysis: Geol. Soc. America Bull., v. 94, Fig. 1, p. 275.)

5.4   FABRIC

The fabric of sedimentary rocks is a function of grain orientation and packing and is thus a property of grain aggregates. Orientation and grain packing in turn control such physical properties of sedimentary rocks as bulk density, porosity, and permeability.

Grain Orientation

Sedimentary rock particles that have a platy (blade or disc) shape or an elongated (rod or roller) shape commonly show some degree of preferred orientation. Platy particles tend to be aligned in planes that are roughly parallel to the bedding surfaces of the deposits. Elongated particles show a further tendency to be oriented with their long axes pointing roughly in the same direction. The preferred orientation of these particles is caused by transport and depositional processes and is related particularly to flow velocities and other hydraulic conditions at the depositional site. Most orientation studies have shown that sand-size particles deposited by fluid flows tend to become aligned parallel to the current direction (Parkash and Middleton, 1970), although a secondary mode of grains oriented normal to current flow may be present. If the grains have a streamlined or teardrop shape, the blunt ends of the grains commonly point upstream. Sand grains can also show well-developed imbrication, with long axes generally dipping upcurrent at angles less than about 20 degrees. Imbrication refers to an overlapping arrangement like that of shingles on a roof. Particles of sandy sediment deposited by turbidity currents or grain-flow or sandy debris-flow processes also tend to be aligned parallel to flow direction and display upstream imbrication at angles exceeding 20 degrees (Hiscott and Middleton, 1980); however, in some gravity-flow deposits, orientation and imbrication directions can be variable or polymodal. Fabric inconsistencies or bimodality appear to be related mainly to very rapid deposition from suspension or from sandy debris flow.

Pebbles in many gravel deposits and in ancient conglomerates also display preferred orientation and imbrication (Fig. 5.16). River-deposited pebbles are commonly oriented with their long axes normal to flow direction and display upstream imbrication of up to 15 degrees. Orientation can also be parallel to flow, or even bimodal, and increasing flow intensity appears to favor orientation with long axes parallel to current

FIGURE 5.16   Well-developed imbrication in river cobbles, Kiso River, Japan. The imbrication was produced by river currents flowing from right to left.

flow (Johansson, 1976). Pebbles deposited by turbidity currents or other gravity-flow processes also take up orientations with their long axes mainly parallel to flow direction, although orientation in some deposits can be random. Pebbles in glacial tills show preferred orientation parallel to flow, with a minor mode oriented normal to flow.

Grain Packing and Grain-to-Grain Relations

Grain packing refers to the spacing or density patterns of grains in a sedimentary rock and is a function mainly of grain size, shape, and degree of compaction of the sediment. Packing strongly affects the bulk density of the rocks as well as their porosity and permeability. The effects of packing on porosity can be illustrated by considering the change in porosity that takes place when even-size spheres are rearranged from loosest packing (cubic packing) to tightest packing (rhombohedral packing) as shown in Figure 5.17. Cubic packing yields porosity of 47.6 percent, whereas the porosity of rhombohedrally packed spheres is only 26.0 percent. The packing of natural particles is much more complex because of variations in size, shape, and sorting and is further complicated in lithified sedimentary rocks by the effects of compaction.

Poorly sorted sediments tend to have lower porosities and permeabilities than well-sorted sediments because, owing to the filling of pore spaces among larger grains by finer sediment, grains are packed more tightly in these sediments. Porosity is decreased also by compaction during diagenesis. Compaction forces grains into closer contact and causes changes in the types of grain-to-grain contacts. Taylor (1950) identified four types of grain contacts that can be observed in thin sections: **tangential contact,** or point contact; **long contact,** appearing as a straight line in the plane of a thin section; **concavoconvex contact,** appearing as a curved line in the plane of a thin section; and **sutured contact,** caused by mutual stylolitic interpenetration of two or more grains (Fig. 5.18). In very loosely packed fabrics, some grains may not make

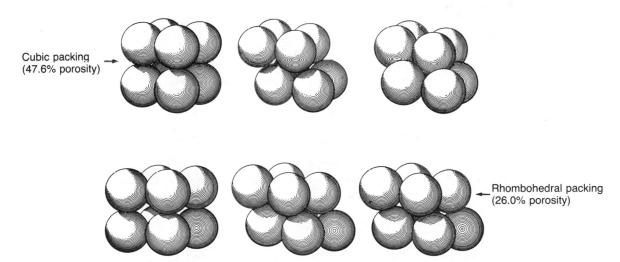

FIGURE 5.17 Progressive decrease in porosity of spheres owing to increasingly tight packing. (After Graton, L. C., and H. J. Fraser, 1935, Systematic packing of spheres with particular relation to porosity and permeability: Jour. Geology, v. 43. Fig. 3, p. 796, reprinted by permission of University of Chicago Press.)

FIGURE 5.18 Diagrammatic illustration of principal types of grain contacts. A. Tangential. B. Long. C. Concavoconvex. D. Sutured. (Based on Taylor, J. M., 1950, Pore-space reduction in sandstones: Am. Assoc. Petroleum Geologists Bull., v. 34, p. 701–716.)

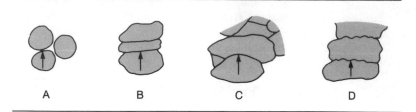

A   B   C   D

contact with other grains in the plane of the thin section and are referred to as "floating grains." Contact types are related to both particle shape and packing. Tangential contacts occur only in loosely packed sediments or sedimentary rocks, whereas concavoconvex contacts and sutured contacts occur in rocks that have undergone considerable compaction during burial. The relative abundance of these various types of contacts can be used as a measure of the degree of compaction and thus the depth of burial of sandstones. The effects of compaction are discussed further in Chapter 9.

The sand-size grains in sandstones are commonly in continuous grain-to-grain contact when considered in three dimensions and form a **grain-supported** fabric. Conglomerates deposited by fluid flows also generally have a grain-supported fabric. On the other hand, conglomerates in glacial deposits, mudflow deposits, and debris-flow deposits commonly have a **matrix-supported** fabric. In this type of fabric, the pebbles are not in grain-to-grain contact but "float" in a matrix of sand or mud. Matrix-supported conglomerates indicate deposition under conditions where fine sediment is abundant and deposition occurs by mass-transport processes or by processes that cause little reworking at the depositional site.

ADDITIONAL READINGS

Blatt, H., G. V. Middleton, and R. Murray, 1980, Origin of sedimentary rocks, 2nd ed., Chap. 3: Prentice-Hall, Englewood Cliffs, N.J., 782 p.

Carver, R. E. (ed.), Procedures in sedimentary petrology: John Wiley & Sons, New York, 653 p.

Folk, R. L., 1974, Petrology of sedimentary rocks: Hemphill, Austin, Tex., 182 p.

Griffith, J. C., 1967, Scientific methods in analysis of sediments: McGraw-Hill, New York, 508 p.

Krinsley, D., and J. Doornkamp, 1973, Atlas of quartz sand surface textures: Cambridge University Press, Cambridge, 91 p.

Lewis, D. W., 1984, Practical sedimentology, p. 58–108: Hutchinson Ross, Stroudsburg, Pa., 229 p.

Pettijohn, F. J., 1975, Sedimentary rocks, 3rd ed., Chap. 3: Harper & Row, New York, 628 p.

Pettijohn, F. J., P. E. Potter, and R. Siever, 1972, Sand and sandstone, Chap. 3, New York, Springer-Verlag, 618 p.

6

Primary Sedimentary Structures

6.1 INTRODUCTION

Primary sedimentary structures are large-scale features of sedimentary rocks—bedding units, ripple marks, and mudcracks, for example. Unlike sedimentary textures, which require laboratory study, most sedimentary structures must be studied in the field. The structures are generated by a variety of sedimentary processes, including fluid flow, sediment gravity flow, soft-sediment deformation, and biogenic activity. Because the structures reflect environmental conditions that prevailed at or very shortly after the time of deposition, they are of special interest to geologists as a tool for interpreting ancient depositional environments. We now know enough about the origin of sedimentary structures from experimental investigations and field studies to use them for evaluating such aspects of ancient sedimentary environments as sediment-transport mechanisms, paleocurrent flow directions, relative water depths, and relative current velocities. Some sedimentary structures can be used also to identify the tops and bottoms of beds and thus to determine if sedimentary sequences are in depositional stratigraphic order or have been overturned by tectonic forces. Sedimentary structures are particularly abundant in coarse siliciclastic sedimentary rocks that originate through traction transport or turbidity current transport. They occur also in nonsiliciclastic sedimentary rocks such as limestones and evaporites.

Because of the potential usefulness of sedimentary structures in environmental interpretation and paleocurrent analysis, a very large body of literature has developed since the 1950s. These publications include several important monographs that contain excellent photographs and drawings illustrating a large variety of primary sedimentary structures. Monographs that cover all types of sedimentary structure include those of J. R. L. Allen (1982), Collinson and Thompson (1982), Conybeare and Crook (1968), Pettijohn and Potter (1964), Potter and Pettijohn (1977), and Reineck and Singh

(1980). Allen (1968) gives a more specialized treatment of current ripples and associated structures. Dzulynski and Walton (1965) discuss sole markings on the bases of sandstone beds, particularly turbidite sandstones, and Picard and High (1973) cover the special sedimentary structures of ephemeral streams. Basan (1978), Crimes and Harper (1970), Curran (1985), Ekdale et al. (1984), and Frey (1975) discuss and illustrate biogenic sedimentary structures. Bouma (1969) deals mainly with methods of studying sedimentary structures.

Only a fraction of the large amount of available information on sedimentary structures can be included in a short summary chapter such as this, but within the limits of available space, I have tried to describe and discuss all the major sedimentary structures. The discussions are brief, but they include a summary of current ideas on mechanisms of formation and, where appropriate, an analysis of the usefulness and limitations of the structures in environmental interpretation. We begin by examining classification of the major primary sedimentary structures.

6.2 CLASSIFICATION OF PRIMARY SEDIMENTARY STRUCTURES

Several types of classifications of sedimentary structures have been proposed (Conybeare and Crook, 1968). Some classifications arrange structures into categories solely on the basis of morphology or descriptive physical characteristics of the structures. Other classifications are based on genetic factors or on both descriptive and genetic elements. Descriptive characteristics form the most objective basis for classifying sedimentary structures. As a rule, genetic classifications are less satisfactory because of the subjectivity involved in their definition and use. Nonetheless, sedimentary structures are often grouped into such genetic categories as depositional structures, erosional structures, deformation structures, and biogenic structures. In any case, more than 400 names have been applied to primary sedimentary structures of physical origin, and many other names have been used for structures of biogenic origin (Pettijohn and Potter, 1964). Fortunately, a much smaller number of more or less standardized names are now in common use for the more important structures.

The classification shown in Table 6.1 is fundamentally descriptive. Sedimentary structures are classified broadly as stratification structures and bedforms, bedding-plane markings, and other structures. Stratification structures and bedforms are further subdivided into four descriptive categories: (1) bedding and lamination, (2) bedforms, (3) cross-lamination, and (4) irregular stratification. Table 6.1 includes also a genetic classification, which categorizes structures according to their probable origin. Sedimentary structures fall into four broad groups on the basis of origin: (1) structures formed by sedimentation processes, (2) structures formed by erosion, (3) structures formed by soft-sediment deformation (penecontemporaneous deformation), and (4) structures of biogenic origin. In the following discussion, sedimentary structures are listed and described under the descriptive headings shown in Table 6.1, although the discussion does not in all cases follow the exact order shown in the table. In some parts of the discussion, structures listed under a particular descriptive heading in Table 6.1 are further subdivided into genetic categories. In the discussion of bedding-plane structures, for example, these structures are divided into (1) bedding-plane markings generated by erosion, (2) bedding-plane markings generated by deformation, (3) biogenic markings, and (4) markings of miscellaneous origin.

TABLE 6.1 Classification of common primary sedimentary structures

GENETIC CLASSIFICATION / MORPHOLOGICAL CLASSIFICATION	Depositional structures			Erosional structures		Deformation structures						Biogenic structures	
	Suspension-settling and current- and wave-formed structures	Wind-formed structures	Chemically and biochemically precipitated structures	Scour marks	Tool marks	Slump structures	Load and founder structures	Injection (fluidization) structures	Fluid-escape structures	Desiccation structures	Impact structures (rain, hail, spray)	Bioturbation structures	Biostratification structures
STRATIFICATION AND BED-FORMS													
Bedding and lamination													
Laminated bedding	X	X	X										
Graded bedding	X											X	
Massive (structureless) bedding	X											X	
Bedforms													
Ripples	X	X											
Sand waves	X												
Dunes	X	X											
Antidunes	X												
Cross-lamination													
Cross-bedding	X	X											
Ripple cross-lamination	X	X											
Flaser and lenticular bedding	X												
Hummocky cross-bedding	X												
Irregular stratification													
Convolute bedding and lamination							X						
Flame structures							X						
Ball and pillow structures							X						
Synsedimentary folds and faults						X							
Dish and pillar* structures									X				
Channels				X									
Scour-and-fill structures				X									
Mottled bedding												X	
Stromatolites													X
BEDDING-PLANE MARKINGS													
Groove casts; striations; bounce, brush, prod, and roll marks					X								
Flute casts				X									
Parting lineation	X												
Load casts							X						
Tracks, trails, burrows†												X	
Mudcracks and synersis cracks										X			
Pits and small impressions											X		
Rill and swash marks	X												
OTHER STRUCTURES													
Sedimentary sills and dikes								X					

*Not wholly stratification structures   †Not wholly bedding-plane markings

137

6.3 STRATIFICATION AND BEDFORMS

Bedding and Lamination

Concept of Bedding

Bedding is a fundamental characteristic of sedimentary rocks. **Beds,** or **strata,** are tabular or lenticular layers of sedimentary rock that have lithologic, textural, or structural unity that clearly distinguishes them from layers above and below. The upper and lower surfaces of beds are known as **bedding planes** or **bounding planes.** Otto (1938) regarded beds as **sedimentation units;** that is, the thickness of sediment deposited under essentially constant physical conditions. It is not always possible, however, to identify individual sedimentation units. Many beds defined by the criteria above may contain several true sedimentation units.

Beds are generally considered to be strata thicker than 1 cm (McKee and Weir, 1953), although Campbell (1967) suggests that beds have no limiting thickness and can range in thickness from a few millimeters to a few tens of meters. According to McKee and Weir's definition, layers less than 1 cm thick are **laminae.** On the other hand, Campbell (1967) defines laminae as the smallest megascopic units visible in a sedimentary sequence and suggests that although laminae are commonly measured in millimeters they can range in thickness to as much as 25 cm. According to Campbell, laminae are differentiated from beds because they are contained within beds and are not themselves internally laminated. In practice, Campbell's definition of beds and laminae is somewhat more difficult to apply than McKee and Weir's and is less commonly used by geologists. Terms used for describing the thickness of beds and laminae are shown in Figure 6.1.

FIGURE 6.1 Terms used for describing the thickness of beds and laminae. (Modified from McKee, E. D., and G. W. Weir, 1953, Terminology for stratification and cross-stratification is sedimentary rocks: Geol. Soc. America Bull., v. 64, Table 2, p. 383; and Ingram, R. L., 1954, Terminology for the thickness of stratification and parting units in sedimentary rocks: Geol. Soc. America Bull., v. 65, Fig. 1, p. 937.)

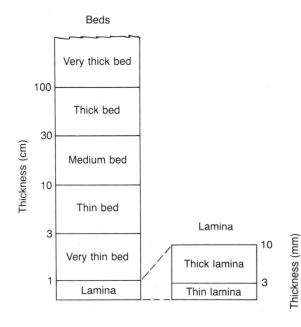

FIGURE 6.2 Informal subdivisions of beds based on internal structures. (From Blatt, H., G. V. Middleton, and R. Murray, Origin of sedimentary rocks, 2nd ed., © 1980, Fig. 5.1, p. 130. Reprinted by permission of Prentice-Hall, Englewood Cliffs, N.J.)

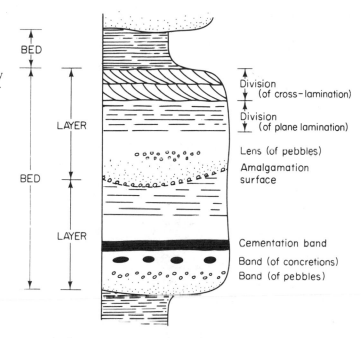

Beds can be differentiated internally into a number of informal units (Fig. 6.2). Blatt et al. (1980) suggest use of the term **layers** for parts of a bed thicker than laminae and separated by minor but distinct discontinuities in texture or composition. Note, however, that layer is used also in a much looser and more informal sense for any bed or stratum of rock. Marked discontinuities within beds are called **amalgamation surfaces**. **Divisions** are subunits that do not have distinct discontinuities but are characterized by a particular association of sedimentary structures. **Bands** and **lenses** are subdivisions based on color, composition, texture, or cementation. The term lens is also used less formally for any body of rock that is thick in the middle and thin at the edges.

Beds are separated by bedding planes, or bedding surfaces, most of which represent a plane of nondeposition, an abrupt change in depositional conditions, or an erosion surface (Campbell, 1967). Some bedding surfaces may be postdepositional features formed by diagenetic processes or weathering. The gross geometry of a bed depends upon the relationship between bedding-plane surfaces, a relationship that may be either parallel or nonparallel. The bedding surfaces themselves may be even, wavy, or curved (Fig. 6.3). Depending upon the combination of these characteristics, beds can have a variety of geometric forms; for example, uniform-tabular, tabular-lenticular, curved-tabular, wedge shaped, or irregular. Internal layers and laminae that are essentially parallel to the bedding planes constitute laminated bedding, or **planar stratification**. Layers and laminae that make up the internal structure of some beds are deposited at an angle to the bounding surfaces of the bed and are therefore called **cross-strata** or **cross-laminae**. Beds composed of cross-laminated or cross-stratified units are called **cross-beds**.

Groups of similar beds or cross-beds are called **bedsets** or **cosets**. A **simple bedset** consists of two or more superimposed beds characterized by similar composition, tex-

FIGURE 6.3 Descriptive terms used for the configuration of bedding surfaces. (From Campbell, C. V., 1967, Lamina, lamina set, bed and bedset: Sedimentology, v. 8. Fig. 2, p. 18, reprinted by permission of Elsevier Science Publishers, Amsterdam.)

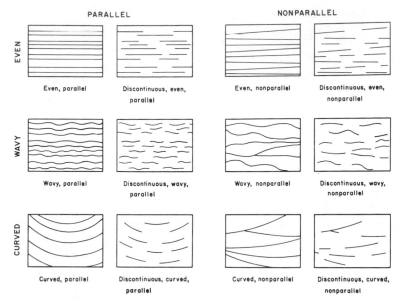

ture, and internal structures. A bedset is bounded above and below by bedset (bedding) surfaces. A **composite bedset** refers to a group of beds differing in composition, texture, and internal structures but associated genetically, representing a common type of deposited sequence (Reineck and Singh, 1980). The terminology of bedsets is illustrated in Figure 6.4.

Beds are characterized by lateral continuity, and some beds can be traced for many kilometers. Others may terminate within a single outcrop. Beds terminate laterally by (1) convergence and merging of upper and lower bounding surfaces (pinchout); (2) lateral gradation of a bed of one composition into another bed of different composition so that the bounding bed surfaces die out; or (3) meeting a cross-cutting feature such as a channel, fault, or unconformity.

Origin of Bedding

Individual beds are produced under essentially constant physical, chemical, or biological conditions. Many beds must have been produced very rapidly by a single event, such as a flood, that lasted only a few hours or days. Even more rapid deposition, lasting perhaps only seconds or minutes, occurs in some environments; for example, deposition of a sand layer by grain flow down the slip face of a sand dune. On the other hand, deposition from suspension of beds of very fine clay could take months or years. Thus, an individual bed may be produced rapidly by an event such as a single flood or a single sediment gravity flow or more slowly by a single episode of deposition of fine sediment from suspension. The true bedding planes or bounding surfaces between beds represent periods of nondeposition, erosion, or changes to completely different depositional conditions. Many beds are not preserved to become part of the geologic record but are destroyed by succeeding erosional episodes. The preservation potential for beds appears to be greater for beds deposited by a depositional event of

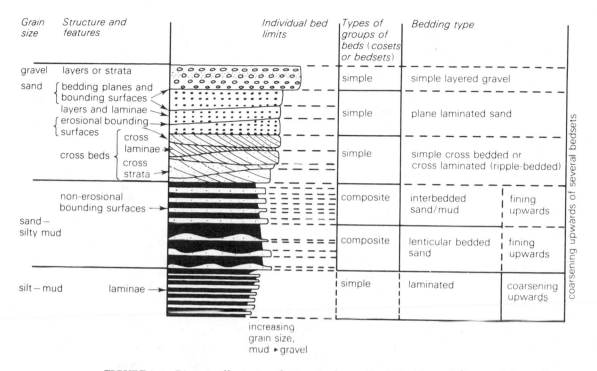

FIGURE 6.4 Diagram illustrating the terminology of bedsets. (From Collinson, J. D., and D. B. Thompson, 1982, Sedimentary structures. Fig. 2.2, p. 8, reprinted by permission of George Allen & Unwin, London.)

great magnitude—a very large flood, for example—than for beds formed by very small-scale events.

Origin of Lamination

Laminae are produced by less severe, or shorter lived, fluctuations in sedimentation conditions than the fluctuations that generate beds. The laminae result from changing depositional conditions that cause variations in (1) grain size, (2) content of clay and organic material, (3) mineral composition, or (4) microfossil content of sediments. Laminae produced by alternating layers of finer and coarser grained sediment are probably the most common kind. The size of grains within individual laminae may be uniform or may show either normal or reverse vertical grain-size grading. Boundaries between laminae due to grain-size differences can be either sharp or gradational. Changes in clay content of layers that otherwise have even-sized coarser grains can also create laminae. Laminations may be produced by such differences in mineral composition as those that occur in alternating mica-enriched and mica-poor laminae; alternating heavy-mineral (black sand) laminae and light-mineral laminae, as in some beach deposits; and alternating laminae of anhydrite and dolomite in evaporite deposits. Alternations of detrital minerals and the tests, or shells, of pelagic organisms are also known to produce laminae. Color changes may accentuate the presence of laminae. Color changes are commonly the result of variations in the content of distinctively colored minerals such as black, heavy minerals; the content of fine, dark-colored organic matter; iron content; or the oxidation state of iron. For example, reduced iron yields green colors; oxidized iron gives red or brown colors.

Parallel laminae, as opposed to cross-laminae, are produced by deposition both from suspension and by traction currents and form in a variety of sedimentary environments. Because the burrowing and feeding activities of organisms in many environments quickly destroy laminations, laminae have the greatest potential for preservation in reducing or toxic environments, where organic activity is minimal, or in environments where deposition is so rapid that sediment is buried below the depth of active organic reworking before organisms can destroy stratification.

Deposition of Laminae by Suspension Mechanisms. Parallel laminae composed of clay or fine silt can be generated by deposition of sediment from suspension in a number of different environmental settings. The most important depositional mechanisms and settings include (1) slow suspension settling in lakes, where levels of organic reworking are commonly low; (2) sedimentation on some parts of deltas, where abundant fine sediment periodically supplied by distributaries leads to rapid deposition; (3) deposition on tidal flats in response to fluctuations in energy levels and sediment supply during tidal cycles; (4) deposition in subtidal shelf areas, where thin sand layers that accumulate owing to storm activity may alternate with very thin mud laminae formed during periods of slower accumulation; (5) slow sedimentation in deep-sea environments, where deposition takes place from nepheloid layers; and (6) chemical sedimentation in evaporite basins—deposition of laminated anhydrites, for example.

Deposition of Laminae by Traction Mechanisms. The formation of parallel laminae in sand-size sediment during traction transport has been attributed to a variety of mechanisms, most of which are based on deductive reasoning rather than on actual observation. **Swash and backwash on beaches** is probably the most common mechanism responsible for formation of evenly laminated sands. This process leads to generation of laminae that may show reverse size grading and concentration of fine, heavy minerals in the base of the laminae (Clifton, 1969). **Steady flow of currents** may also produce laminae under three different types of conditions: (1) during the plane-bed phase of upper-flow regime transport (Harms and Fahnestock, 1965; J. R. L. Allen, 1984); (2) under shallow-flow conditions in the lower-flow regime by migration of low-relief ripples in which lack of avalanche faces prevents cross-laminae from forming (McBride et al., 1975); and (3) at low velocities below the critical velocity of ripple formation, at least for particles >0.7 mm (Guy et al., 1966). Laminated sands may also develop owing to **wind transport** (McKee et al, 1971, Hunter, 1977). Hunter observed that parallel laminae formed by (1) traction transport and deposition at very high wind velocities; (2) grainfall deposition in zones of flow separation that occur leeward of dune crests; and (3) deposition accompanying the migration of wind ripples, an eolian analog of the subaqueous process described by McBride et al. (1975). **Phases of upper-flow-regime transport during turbidity current flow** can generate laminated sands to form Bouma B subdivisions of turbidites (Fig. 3.15). Finally, **sheet flow** (the oscillatory equivalent of plane-bed transport in the upper-flow regime; Clifton, 1976) in shallow-marine environments or **migration of ripple forms accompanied by a very slow rate of deposition** can produce laminations in sandy deposits (Newton, 1968).

(Some terminology in the preceding paragraph—plane-bed, flow regime, and sheet flow, for example—may be unfamiliar to some readers. The meaning of these terms will become apparent in succeeding sections of this chapter.)

Graded Bedding

Graded beds are sedimentation units characterized by distinct vertical gradations in grain size and may range in thickness from a few centimeters to a few meters or more. They are commonly devoid of internal laminations, although the upper part of graded turbidite sequences (Bouma B, C, D divisions, Fig. 3.15) may show parallel or wavy laminae. Beds that show gradation from coarser particles at the base to finer particles at the top have **normal grading** (Fig. 6.5). More rarely, beds display **reverse grading,** with coarser particles at the top grading downward to finer particles. Graded beds commonly have sharp basal contacts.

Normal graded bedding can form by several processes (Klein, 1965); however, most graded beds in the geologic record have been attributed to turbidity currents. Differences in the rate at which particles of different sizes settle from suspension during the waning stages of turbidity current flow appear to account for the grading, but

FIGURE 6.5 Normally graded turbidites in the Cretaceous Hudspeth Formation, north-central Oregon. Four graded units are visible.

the exact manner in which the grading process operates is not well understood. The graded materials may be silt, sand, or, more rarely, gravel. Some graded turbidite units display an ideal sequence—a complete Bouma sequence (Fig. 3.15)—of sedimentary structures, but more commonly the sequence is truncated at the top or bottom. The basal A division may be present and some or all of the overlying divisions absent, or the A division itself may be missing. Graded beds occur also in shallower water environments than those in which turbidites form. Suggested mechanisms of formation of shallow-water graded beds include sedimentation from suspension clouds generated by storm activity on the shelf, periodic silting of delta distributaries, deposition in the last phases of a heavy flood, settling of volcanic ash after an eruption, deposition by waning currents on intertidal flats, and mixing of an underlying coarser sediment layer with an overlying mud layer owing to bioturbation activities of burrowing and feeding organisms.

Reverse size grading is much less common than normal grading in sediments and sedimentary rocks. It is known to occur in the individual laminae of beach sediments owing to segregation of fine-size heavy minerals and coarser grained light minerals (Clifton, 1969), in some pyroclastic flows or volcanic base-surge deposits, in some grain-flow deposits, and in laminae formed by the migration of wind ripples. It is also alleged to occur in some turbidite deposits that resulted from rapid deceleration of high-concentration flows. Reverse grading has been attributed to two types of mechanisms: (1) dispersive pressures and (2) kinetic sieving. Dispersive pressures (Chapter 3) are believed to be proportional to grain size: In a sediment of mixed grain size, the higher dispersive pressures acting on the larger particles tend to force them up into the zone of least shear. Alternatively, reverse grading may be explained by a kinetic sieve mechanism. In a mixture of grains undergoing agitation, the smaller grains presumably fall down through the larger grains as grain motion opens up spaces between the larger particles. Overall, reverse grading is a relatively rare phenomenon and its origin is still poorly understood.

Massive (Structureless) Bedding

The term **massive bedding** is used to describe beds that appear to be homogeneous and lacking in internal structures. Use of X-radiography techniques (Hamblin, 1965) or of etching and staining methods often reveals that such beds are not truly massive but rather that they contain very faintly developed structures. Nonetheless, one occasionally finds beds, particularly thick sandstone beds, in which internal structures cannot be recognized even with the aid of X-radiographic or staining techniques. Such beds are rare, which is fortunate for us because they are very difficult to explain. Reported occurrences of massive beds include both graded bed units in turbidites, which may lack internal structures other than size grading, and certain thick, nongraded sandstones.

Some massive bedding may be a secondary feature produced by extensive bioturbation by organisms, although bioturbation commonly produces recognizable mottled structures. Liquefaction of sediment by sudden shocking or other mechanisms shortly after deposition has also been suggested as a means of destroying original stratification. Otherwise, it is assumed that lack of stratification is a primary feature that results when traction transport is absent and sediments are deposited very rapidly from suspension or from very highly concentrated sediment dispersions during sediment gravity flows. The sediment is presumably dumped very rapidly without subsequent reworking and forms a more or less homogeneous mass.

Bedforms and Cross-stratification

Ripples, Sand Waves, and Dunes

Bedforms Generated by Unidirectional Currents. Anyone who has examined the sandy bed of a clear, shallow stream has certainly noticed that the bed is rarely perfectly flat and even. Instead, it is commonly marked by ripples and similar bedforms of various sizes. Such bedforms occur also in eolian and submarine environments, where they range in size from small ripples a few centimeters in length and a fraction of a centimeter in height to gigantic sand dunes and undersea sand waves tens to hundreds of meters in length and several meters to several tens of meters in height. If we carefully dissect a ripple exposed on the dry bed of a stream to reveal its internal structure, we almost invariably find internal cross-laminae that dip in a downcurrent direction. Clearly, there is a close genetic relationship between bedforms and cross-lamination and cross-bedding. We will explore that relationship further in this section.

The preservation potential of ripples is relatively low; they are therefore not extremely common features on the bedding planes of ancient sedimentary rocks. On the other hand, cross-beds are exceedingly common in many ancient sandstone sequences. In an attempt to better understand the origin of bedforms and cross-stratification, many investigators have turned to flumes for the study of sediment transport. Flumes are long, slightly sloping troughs with glass sides to allow observation. Sand or other sediment is placed on the floor of the flume and water is made to flow over the floor at various depths and velocities.

Numerous flume experiments have established that, under unidirectional flow, bedforms begin to develop in sandy sediment as soon as the critical entrainment velocity for the sediment is reached (Chapter 3). The exact sequence of bedforms that develop with increasing velocity depends upon the grain size of the material. If flow is over a bed of sediment ranging in size from about 0.25 to 0.7 mm (medium to coarse sand), the succession of bedforms illustrated in Figure 6.6 is generated, beginning with **ripples**.

The shape of ripples and the terminology used to describe ripples and larger bedforms are illustrated in Figure 6.7. Ripples are the smallest bedform, ranging in length from about 0.05 to 0.2 m and in height from about 0.005 to 0.03 m. Thus, they have a ripple index (ratio of ripple length to ripple height) ranging from about 8 for coarse sand to 20 for fine sand. They form in sediment ranging in size from silt (0.0625 mm) to sand as coarse as 0.7 mm. **Sand waves,** called **bars** by some authors, are low, straight- to sinuous-crested bedforms that have wave lengths ranging from 5 to 100 + m (under natural conditions) and a ripple index of about 50. They form at higher flow velocities than ripples and in sediment coarser than about 0.25 mm. **Dunes** form at still higher velocities and except for size are similar in appearance to ripples. Dunes range in length from 0.5 to 10 m, and dune ripple index ranges from 5 in finer sands to 50 in coarser sediment. They form in sediment coarser than about 0.1 mm. In the lower part of the dune stability field, ripples may be superimposed on the backs of dunes. The characteristics of bedforms that develop under unidirectional flow are summarized in Table 6.2.

Readers should be aware that some workers have cast doubt on the reality of distinctions made between sand waves and dunes (Harms et al., 1982). To avoid or sidestep the problem of distinction, these workers refer to sand waves and dunes collectively as large ripples. They call ripples, as defined above, small ripples. I have retained the terms sand waves and dunes in this book, but I also occasionally refer to these bedforms as large ripples or large-scale ripples.

FIGURE 6.6 The succession of bedforms that develop during unidirectional flow of sandy sediment (0.25–0.7 mm) in shallow water as flow velocity increases. (From Blatt, H., G. V. Middleton, and R. Murray, Origin of sedimentary rocks, 2nd ed., © 1980, Fig. 5.3, p. 137. Reprinted by permission of Prentice-Hall, Englewood Cliffs, N.J.)

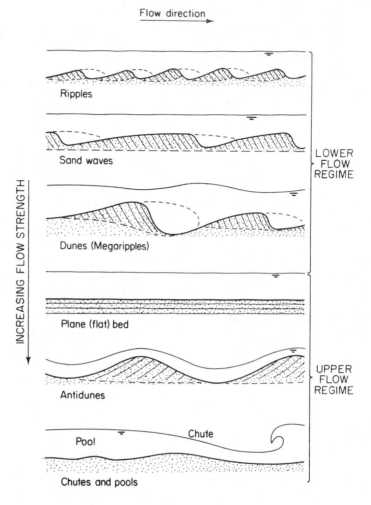

During the formation of ripples, sand waves, and dunes, either the water surface shows little disturbance or water waves are **out of phase** with bedforms (Fig. 6.6). Out-of-phase waves may show slight disturbance of the water surface over large-scale ripples and large swirls, or "boils," that rise to the surface. The hydraulic conditions that generate these bedforms and out-of-phase surface waves distinguish what is called the **lower-flow regime** (Simons and Richardson, 1961). Ripples, sand waves, and dunes generated in the lower-flow regime migrate downstream because sediment is eroded from the stoss side of these bedforms and carried up to the crest, where it avalanches down the lee slope (Fig. 6.6). Avalanching leads to formation of cross-laminations that dip downstream at angles up to about 30 degrees.

With further increase in flow velocity, dunes are destroyed and give way to an **upper-flow regime** stage of flow. Sheetlike, rapid flow of water generates symmetrical surface water waves that are **in phase** with the bedforms and distinguish the upper-flow regime. Owing to very rapid water flow, intense sediment transport takes place

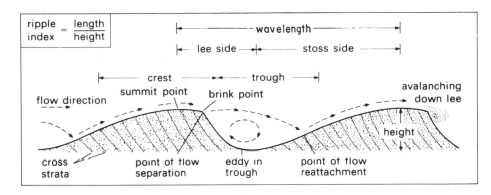

$$\frac{\text{ripple}}{\text{index}} = \frac{\text{length}}{\text{height}}$$

FIGURE 6.7 The terminology used to describe asymmetric ripples. (From Tucker, M. E., Sedimentary petrology, an introduction. © 1981 by John Wiley & Sons, Inc. Fig. 2.18, p. 28, reproduced by permission of Open University Educational Enterprises Ltd., Stony Stratford, England.)

TABLE 6.2 Characteristics of bedforms developed under unidirectional flow

	Ripples	Sand waves	Dunes	Lower plane bed	Upper plane bed	Antidunes
Length (spacing)	0.1–0.2 m	a few dm to 100s of m	a few dm to 10s of m (or more?)	—	—	dm–m
Height	a few cm	cm to a few 10s of m	dm to a few m (or more?)	—	—	cm–dm
Ripple index (length/height)	relatively low	relatively high	relatively low	—	—	relatively high
Plan geometry	strongly irregular/short crested	straight/sinuous and long crested	strongly irregular/short crested	—	—	long crested and short crested
Characteristic flow velocity	low	low/moderate	moderate/high	low	high	high
Characteristic flow depth	> a few cm	> a few dm	> a few dm	all	all	shallow flows
Characteristic sediment size	0.03–0.6 mm	>0.3 mm?	>0.2 mm	>0.6 mm	all	all

Source: Modified from Harms, J. C., J. B. Southard, and R. G. Walker, 1982, Structures and sequences in clastic rocks: Soc. Econ. Paleontologists and Mineralogists Short Course No. 9. Table 2–1, p. 2–11, reprinted by permission of SEPM, Tulsa, Okla.

over an initially relatively flat bed during what is referred to as the plane-bed stage of flow. Plane-bed flow gives rise to internal planar laminations ranging in thickness from a few millimeters to a few tens of millimeters. The preservation potential of these plane bed laminae appears to be low; nonetheless, upper-flow regime plane-bed laminae have been reported in ancient sandstone deposits. The most important environments where plane beds are likely to be preserved are in stream channels, on beaches and in other nearshore areas where strong shoaling waves prevail, and under high velocity turbidity currents (Harms et al., 1982)

At still higher velocities of flow, plane beds give way to **antidunes,** which are low, undulating bedforms up to 5 m in length with a ripple index ranging from about 7 to 100+. Antidunes form in very fast, shallow flows at Froude numbers (Chapter 3) greater than about 0.8. They migrate upstream during flow, giving rise to low-angle (<10 degrees) cross-bedding directed upstream. Antidunes have very low preservation potential and antidune cross-bedding is probably rarely preserved, but antidune cross-bedding has been reported at the base of some turbidite flow units.

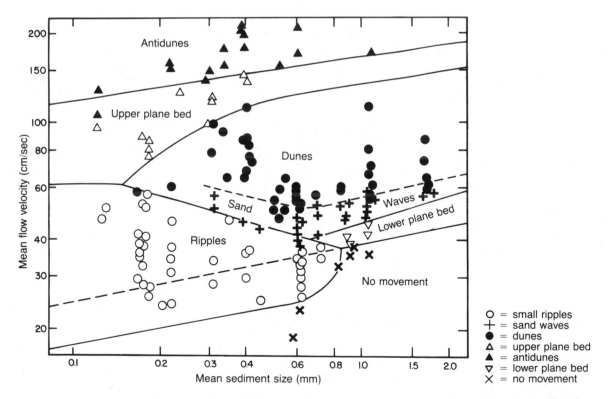

FIGURE 6.8 Size–velocity diagram for flow depths of 18–22 cm at a water temperature of 10°C, showing the hydraulic relationship of bedforms to sediment size and flow velocity. Ripples below the dashed line will not form from a plane bed unless the bed is perturbed, and will continue to migrate once formed. (Modified slightly from Harms, J. C., J. B. Southard, and R. G. Walker, 1982, Structures and sequences in clastic rocks: Soc. Econ. Paleontologists and Mineralogists Short Course No. 9. Fig. 2–5, p. 2–14, reprinted by permission of SEPM, Tulsa, Okla.)

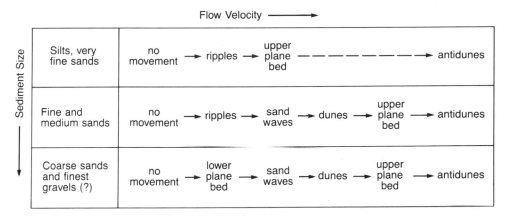

FIGURE 6.9 Sequence of bedforms that develop with increasing flow velocity, for sediment of various sizes. (Modified from Harms, J. C., J. B. Southard, and R. G. Walker, 1982, Structures and sequences in clastic rocks: Soc. Econ. Paleontologists and Mineralogists Short Course No. 9. Fig. 2–6, p. 2–15, reproduced by permission of SEPM, Tulsa, Okla.)

When flow velocity increases above the antidune stage, one final type of bedform, **chute and pool structure** (Fig. 6.6), develops at very high flow velocities. These bedforms develop where shallow, rapid (supercritical) flow forms chutes that end abruptly in a deeper pool where flow is tranquil. Sediment accumulations occur in the tranquil pool region where steeply dipping backset laminations develop (Leeder, 1982). The preservation potential of chute and pool structures is exceedingly poor, and this bedform is seldom found in nature.

Effects of Grain Size and Water Depth. Experimental studies show that the succession of bedforms that develops at a given water depth during flow depends not only upon flow velocity but also on grain size; therefore, the succession of bedforms shown in Figure 6.6 does not occur in sediment of all particle sizes. Figure 6.8 shows the relationship of bedforms to flow velocity and grain size at a water depth of 18–20 cm. If flow takes place over sediment coarser than about 0.7 mm, for example, the ripple phase does not develop. A lower plane-bed phase forms instead. Note also from Figure 6.8 that the sand-wave phase does not develop in sediment with grain size smaller than 0.25 mm and that below a grain size of about 0.15–0.2 mm dunes do not form. The relationships shown in Figure 6.8 are summarized in tabular form in Figure 6.9. For flow of grains of a given size in shallow water, increase in water depth has the general effect of increasing the velocity at which change from one bedform phase to another takes place. Figure 6.10 shows this relationship for fine, medium, and coarse sand.

Most studies of bedforms have been carried out in laboratory flumes or under shallow-water conditions in natural environments. Therefore, most available sediment-size/velocity data pertain to the formation of bedforms under shallow-water conditions (commonly less than about 1 m). Much less is known about the development of bedforms under deeper water conditions. On the basis of limited available information, Harms et al. (1982) suggest that the nature of small ripples is approximately the same in deep-water flows as in shallow-water flows; however, the larger bedforms

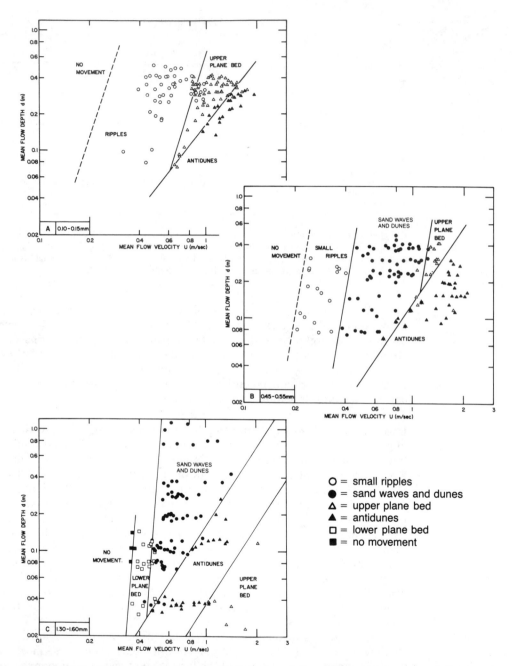

FIGURE 6.10 Depth–velocity diagram for A, fine sand, B, medium-coarse sand, and C, very coarse sand showing the hydraulic relationship of bedforms to flow depth and flow velocity at a water temperature of 10°C. (After Harms, J. C., J. B. Southard, and R. G. Walker, 1982, Structures and sequences in clastic rocks: Soc. Econ. Paleontologists and Mineralogists Short Course No. 9. Fig. 2–4, p. 2–13, reproduced by permission of SEPM, Tulsa, Okla.)

(sand waves and dunes) can grow much larger in deep-water flows. The hydraulic relationships in deep water are the same as for shallow water; that is, sand waves and dunes form at higher velocities than do ripples and at lower velocities than do plane beds and antidunes. The exact relationship between grain size, flow velocity, and bedform phase is not well documented for deeper water, but a generalized relationship is shown in Figure 6.11. Note from the figure that exceedingly high velocities are required to produce antidunes at a water depth greater than a few meters; therefore, it appears that antidunes are unlikely to occur under natural conditions in deep water.

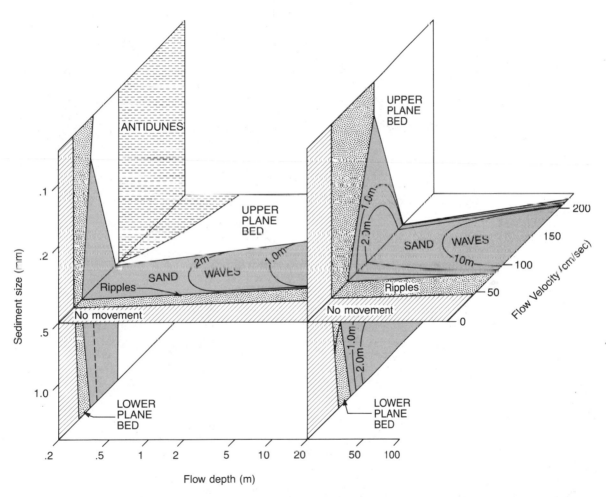

FIGURE 6.11 Generalized three-dimensional depth–velocity–grain-size diagram showing the relationship among bed phases and grain size for a wide variety of flow velocities and flow depths. The field labeled sand waves includes both sand waves and dunes. Diagram based on both flume data and observations on natural flows. (Modified slightly from Rubin, D. M., and D. S. McCulloch, 1980, Single and superimposed bedforms: A synthesis of San Francisco Bay and fluvial observations: Sedimentary Geology, v. 26. Fig. 11, p. 224, reprinted by permission of Elsevier Science Publishers, Amsterdam.)

The mechanisms of sediment transport responsible for formation of the different bedforms are very complex. In general, the formation of transverse bedforms is related to a phenomenon called **flow separation.** Sediment is transported in suspension or by traction up the stoss side of the bedform to the brink, or crest. At the brink, the flow separates from the bed to form a zone of reverse circulation, or backflow, producing a separation eddy (Fig. 6.7). Owing to turbulent mixing with the main flow, a zone of diffusion is present between the zone of backflow and the main flow above. Downstream from the point of separation a distance several times the height of the bedform, the flow becomes reattached to the bottom. Flow separation causes separation of the transported sediment into bedload and suspended-load fractions. The bedload fraction accumulates at the ripple crest until the lee slope exceeds the angle of repose and avalanching takes place. The suspended-load fraction is transported downcurrent, where the coarser particles in the suspended load settle through the zone of diffusion into the zone of backflow and are deposited in the lee of the ripple. It is these processes that cause development and movement of the bedforms.

The bedforms described above develop in response to unidirectional flow of water currents. They are asymmetrical in shape, with the steep or lee side facing downstream in the direction of current flow. Asymmetrical ripples formed in this fashion

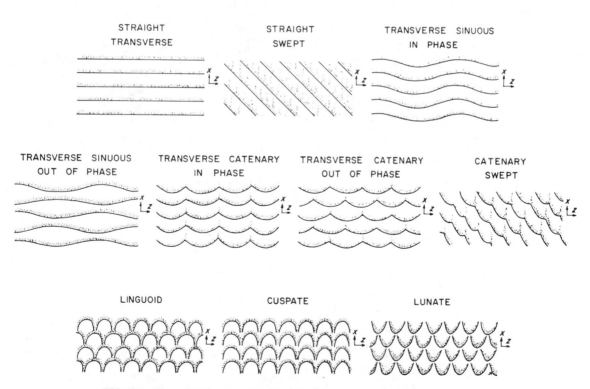

FIGURE 6.12 Idealized classification of current ripples and dunes based on plan-view shape. Flow is from the bottom to the top in each case. (After Allen, J. R. L., 1968, Current ripples: Their relation to patterns of water motion: North Holland Pub., Amsterdam. Fig. 4.6, p. 65, reprinted by permission of Elsevier Science Publishers, Amsterdam.)

are called **current ripples.** Under natural conditions they form by river and stream flow, backwash on beaches, longshore currents, tidal currents, and deep-ocean bottom currents. In plan view, the crests of current ripples and dunes have a variety of shapes: **straight, sinuous, catenary, linguoid,** and **lunate** (Fig. 6.12). The plan-view shape of ripples and dunes is apparently related to water depth and velocity (J. R. L. Allen, 1968); however, the factors that control the shape are not well understood. It has been observed under natural conditions that the more complex forms tend to develop in shallower water and at higher velocities than do the less complex forms and that the order in which the succession of bedforms develops with decreasing water depth and velocity is straight to sinuous to symmetric linguoid to asymmetric linguoid for ripples and straight to sinuous to catenary to lunate for dunes.

Bedforms Generated by Wind Transport. The bedforms that develop during wind transport range from ripples as small as 0.01 m long and a few millimeters in height to dunes 500–600 m long and 100 m high. Less commonly, gigantic bedforms called **draas,** which may have wave lengths measured in kilometers (up to 5.5 km) and heights up to 400 m, may also form by wind transport (I. G. Wilson, 1972). The wave length of wind-transported bedforms increases with increasing wind velocity, and wave height tends to increase with increasing grain size. Wilson suggests, however, that ripples, dunes, and draas can coexist under a given set of conditions of grain size and wind velocity. Thus, dunes exist on the backs of draas, and ripples are created on the backs of dunes.

The plan-view shape of eolian ripples is predominantly straight (Fig. 6.13), although sinuous forms also occur. More complex, three-dimensional forms are uncommon, although Leeder (1982) suggests that linguoid ripples analogous to those formed from water flows may occur in faster wind flows that are blowing very fine sand. The preservation potential of complete eolian bedforms is exceedingly low; they are rarely, if ever, found in ancient sediments. The former existence of dunes and other eolian bedforms is revealed in ancient sedimentary rocks only by the presence of preserved cross-beds; however, considerable controversy has been generated regarding differentiation of eolian cross-bedding from subaqueous cross-bedding in ancient sedimentary rocks.

FIGURE 6.13 Straight-crested eolian ripples, northern Padre Island, Texas. Wind direction is from left to right. (Photograph by R. E. Hunter.)

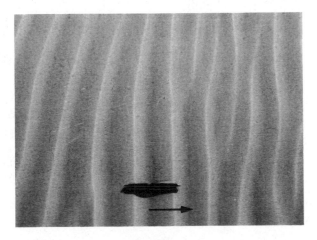

Bedforms Generated by Waves. The preceding discussion deals with development of bedforms by unidirectional movement of water currents or wind. Ripples and larger bedforms form also in lakes and oceans under the influence of orbital wave motion (Chapter 3). Oscillatory motion close to the bed creates **oscillation ripples** when near-bottom velocities become great enough to move the particles. Eddies created by these small-scale, wave-generated "orbital currents" throw sediment into suspension; the sediment alternately moves landward as the wave crest passes and seaward as the trough passes (Fig. 6.14).

There is a common misconception that oscillation ripples are always symmetrical in cross-sectional shape; however, Komar (1976) has shown that they can be either symmetrical or asymmetrical (Fig. 6.14). If the forward and backward orbital wave velocities are equal, symmetrical ripples develop. If these orbital velocities are unequal, or if a unidirectional current is superimposed on the orbital motion, asymmetrical ripples develop. Clifton (1976) suggests that if the velocity asymmetry of the maximum bottom orbital velocity (difference between forward and backward orbital velocity) is less than 1 cm/sec, symmetrical bedforms develop. If the orbital velocity asymmetry exceeds 5 cm/sec, asymmetrical bedforms develop, with the lee side of the bedform facing in the direction of wave movement. Intermediate velocities—between 1 and 5 cm/sec—presumably produce transitional forms. The issue of asymmetry of wave-formed bedforms is complex and poorly understood.

The size, spacing, and symmetry of bedforms produced by shoaling waves depend upon four parameters: maximum bottom orbital velocity, velocity asymmetry, median grain size, and wave period (the time required for passage of a single wave) (Clifton, 1976). Clifton defined three kinds of ripples on the basis of the relations of ripple wave length to grain size and maximum bottom orbital velocity. **Orbital ripples** form under short-period waves; ripple spacing is independent of grain size and depends directly on the length of the orbital diameter of the oscillatory current. **Suborbital ripples** form under longer period waves. Their wave lengths increase with increasing grain size but decrease with increasing orbital diameter. **Anorbital ripples** are produced by relatively long-period waves. Ripple spacing depends directly on grain size and is independent of orbital diameter.

Oscillation ripples are destroyed by increasing bottom orbital velocity, and a plane-bed phase of transport, called **sheet flow,** occurs. This phase is analogous to the plane-bed phase of upper-flow regime transport in unidirectional flow (Fig. 6.15). Under natural conditions in shallow offshore areas where waves are shoaling, a specific sequence of bedforms develops in a shoreward direction with increasing bottom orbital velocity and velocity asymmetry. This sequence grades from symmetrical waves in deeper water to asymmetrical ripples, lunate megaripples, and planar or flat beds (Fig. 6.16). The abrupt increase in bedform size from ripples to lunate megaripples resembles the change from ripples to dunes in unidirectional flow (Clifton, 1976).

The crests of oscillation ripples as seen in plan view are commonly straight to sinuous and tend to bifurcate, or fork, in a characteristic manner (Fig. 6.17). The wave lengths of oscillation ripples generally range from about 10 cm to 2 m and ripple heights range from about 3 to 25 cm; however, lunate megaripples can have a spacing ranging from 1 to 4 m and heights between 30 and 100 cm. The internal structure of oscillation ripples varies from shoreward-dipping cross-laminations in asymmetric and many symmetric ripples to both shoreward- and seaward-dipping cross-bedding in some symmetrical ripples. Shoreward-dipping cross-laminations are most common, even in symmetrical ripples.

FIGURE 6.14 The relationship between orbital motion owing to wave action and sand motion on a rippled sea bed. Note that wave motion with a superimposed unidirectional current produces asymmetrical ripples, whereas wave motion alone tends to produce symmetrical ripples. (From Komar, P. D., Beach processes and sedimentation, © 1976, Fig. 11.15, p. 315. Reprinted by permission of Prentice-Hall, Englewood Cliffs, N.J. Based on D. L. Inman and A. J. Bowen, 1963, Flume experiments on sand transport by waves and currents: Proc. 8th Conf. on Coastal Engineering, p. 137–150.)

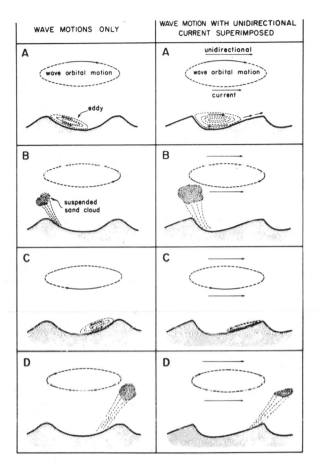

FIGURE 6.15 Threshold velocity for initiation of grain movement in uniform quartz sand and conversion of rippled bed to flat bed (sheet flow) under orbital flow. (After Clifton, H. E., 1976, Wave-formed sedimentary structures: A conceptual model, in Davis, R. A., Jr., and R. L. Ethington (eds.), Beach and nearshore sedimentation: Soc. Econ. Paleontologists and Mineralogists, Spec. Pub. 24. Fig. 4, p. 130, reprinted by permission of SEPM, Tulsa, Okla.)

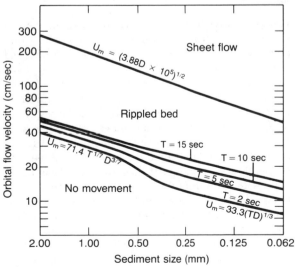

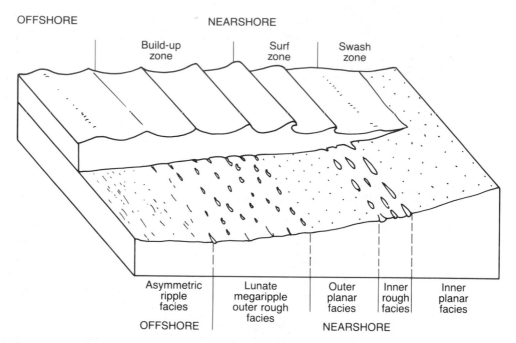

FIGURE 6.16   The sequence of bedforms that develop in the offshore and nearshore zone with increasing bottom orbital velocity and velocity asymmetry in the shoreward direction. (After Clifton, H. E., R. E. Hunter, and R. L. Phillips, 1971, Depositional structures and processes in the non-barred high-energy nearshore: Jour. Sedimentary Petrology, v. 41. Fig. 7, p. 656, reprinted by permission of Society of Economic Paleontologists and Mineralogists, Tulsa, Okla.)

Cross-Bedding

Cross-bedding forms primarily by sediment deposition resulting from avalanching or suspension settling in the zone of separation on the lee sides of ripples and other bedforms, as described. If most of the sediment is too coarse to be transported in suspension, avalanching of the bedload sediment down the lee side of the ripple will cause formation of laminae that are steep and straight. These inclined **foreset** laminae make contact with the nearly horizontal, thin **bottomset** laminae (deposited from suspension) at a distinct angle, which is approximately the same as the angle of repose. Roughly the same effect is achieved if the height of the lee slope is large compared to total flow depth, so that the suspended load falls mainly on the lee slope. If the suspended load is large, or if the height of the lee slope is small compared to flow depth, suspended sediment will pile up at the base of the lee slope rapidly enough to keep pace with growth of the avalanche deposits. This process causes the lower part of the foreset laminae to curve outward and approach the bottomset laminae asymptotically (Blatt et al., 1980). Thus, the cross-laminae are said to be **tangential** rather than angular.

The preservation potential of cross-laminae is much higher than that of the bedforms themselves; therefore, cross-bedding is a very common type of sedimentary

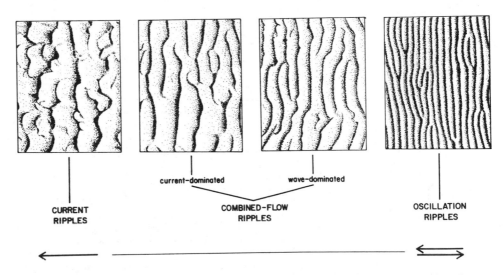

CURRENT
RIPPLES

current-dominated   wove-dominated

COMBINED-FLOW
RIPPLES

OSCILLATION
RIPPLES

FIGURE 6.17   The crest shape of oscillation ripples, as seen in plan view, compared to the shape of current ripples and current-dominated ripples. (From Harms, J. C., J. B. Southard, and R. G. Walker, 1982, Structures and sequences in clastic rocks: Soc. Econ. Paleontologists and Mineralogists Short Course No. 9. Fig. 2–19, p. 2–48, reprinted by permission of SEPM, Tulsa, Okla.)

structure in ancient sedimentary rocks. Cross-stratification can be formed also by filling of scour pits and channels, deposition on the point bars of meandering streams, and deposition on the inclined surfaces of beaches and marine bars. Cross bedding formed under different environmental conditions can be very similar in appearance; it is often difficult in field studies of ancient sedimentary rocks to differentiate cross-bedding formed in fluvial, eolian, and marine environments.

Cross-beds commonly occur in sets or cosets (Fig. 6.4). Cross-bedding in sets less than about 5 cm thick is called small-scale cross bedding; that in sets thicker than 5 cm is large-scale cross-bedding. Because of their diverse origins, many types of cross-beds occur. J. R. L. Allen (1963) proposed a very elaborate classification of cross-bedding based upon such properties as grouping of cross-bed sets, scale, nature of the bounding surfaces of the beds, angular relation of cross-strata in a set or coset to the bounding surfaces, and degree of grain size uniformity in different laminae. The much simpler scheme used by Potter and Pettijohn (1977), based on the classification of McKee and Weir (1953), is adopted herein. Cross-beds are divided into two principal types on the basis of overall geometry and the nature of the bounding surfaces of the cross-bedded units (Fig. 6.18). **Tabular cross-bedding** consists of cross-bedded units that are broad in lateral dimensions with respect to set thickness and have bounding surfaces that are essentially planar (Fig. 6.19). The laminae of tabular cross-beds are also commonly planar, but curved laminae that have a tangential relationship to the basal surface also occur. **Trough cross-bedding** consists of cross-bedded units whose bounding surfaces are curved (Fig. 6.20). The units are trough-shaped sets consisting of an elongate scour filled with curved laminae that commonly have a tangential relationship to the base of the set.

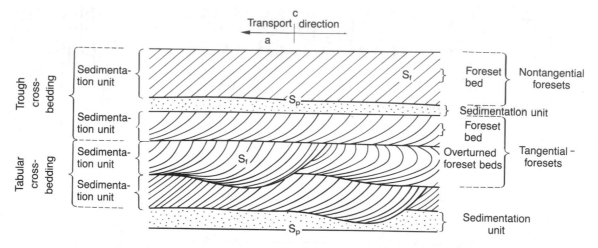

FIGURE 6.18 The terminology and defining characteristics of cross-bedding. Symbols: a, direction parallel to the average sediment transport direction; c, direction perpendicular to (a) and the transport plane (bed) in which (a) lies, (S_p) the principal bedding surface or bedding plane, (S_f) the foreset surface of cross-bedding. (After Potter, P. E., and F. J. Pettijohn, 1977, Paleocurrents and basin analysis, 2nd ed. Fig. 4.1, p. 91, reprinted by permission of Springer-Verlag, Heidelberg.)

FIGURE 6.19 Tabular cross-bedding in pebbley sands of the Coquille Formation (Pleistocene), southern Oregon coast. Note the opposing dip directions in the two cross-bedded units, suggesting possible deposition by reversing tidal currents (current direction from right to left in the lower unit and from left to right in the upper unit).

FIGURE 6.20 Trough cross-bedding in pebbly sands of the Coquille Formation (Pleistocene), southern Oregon coast. Note erosion of basal, parallel-bedded sands to produce the first trough cross-bedded unit (on the right), which was truncated in turn by current scour to form the second trough cross-bedded unit.

Tabular cross-bedding is formed mainly by the migration of large-scale ripples (Fig. 6.21). Individual beds range in thickness from a few tens of centimeters to a meter or more, but bed thicknesses up to 10 m have been observed (Harms et al., 1975). Trough cross-bedding can originate both by migration of small current ripples, producing small-scale cross-bed sets, or by migration of large-scale ripples (Fig. 6.22). Trough cross-bedding formed by migration of large-scale ripples commonly ranges in thickness up to a few tens of centimeters and in width from less than 1 m to more than 4 m.

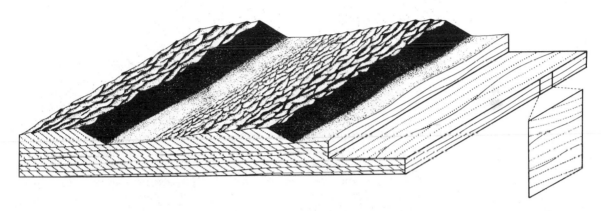

FIGURE 6.21   Diagram illustrating large-scale tabular cross-bedding formed by migrating sand waves. Flow is from left to right. (From Harms, J. C., J. B. Southard, and R. G. Walker, 1982, Structures and sequences in clastic rocks: Soc. Econ. Paleontologists and Mineralogists Short Course No. 9. Fig. 3–11, p. 3–21, reprinted by permission of SEPM, Tulsa, Okla.)

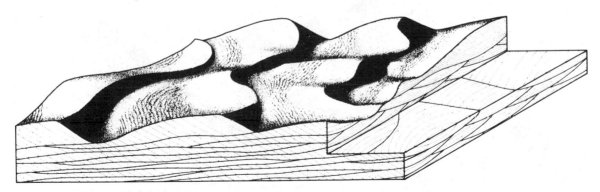

FIGURE 6.22   Diagram illustrating large-scale trough cross-bedding formed by migrating dunes. Flow is from left to right. (From Harms, J. C., J. B. Southard, and R. G. Walker, 1982, Structures and sequences in clastic rocks: Soc. Econ. Paleontologists and Mineralogists Short Course No. 9. Fig. 3–10, p. 3–19, reprinted by permission of SEPM, Tulsa, Okla.)

Ripple Cross-Lamination

Ripple cross-lamination, also called climbing-ripple lamination and ripple-drift cross-stratification, forms when deposition takes place very rapidly during migration of current or wave ripples (McKee, 1965; Jopling and Walker, 1968). A series of cross-laminae are produced by superimposing migrating ripples (Fig. 6.23). The ripples climb one on another in such a manner that the crests of vertically succeeding laminae are out of phase and appear to be advancing upslope. This process results in cross-bedded units that have the general appearance of waves (Fig. 6.24) in outcrop sections cut normal to the wave crests. In sections with other orientations, the laminae may appear horizontal or trough-shaped depending upon the orientation and the shape of the ripples. Ripple cross-lamination can be separated into two types: one in which both the lee side and stoss side of the ripples are preserved and one in which only the lee side is preserved.

The formation of climbing ripples appears to require an abundance of sediment, especially sediment in suspension, which quickly buries and preserves original rippled layers. Abundant suspended sediment supply must be combined with just enough traction transport to produce rippling of the bed but not enough to cause complete erosion of laminae from the stoss side of ripples. Some ripple laminae may be in phase—one ripple crest lies directly above the other—indicating that the ripples did not migrate. In-phase ripple laminae form under conditions where a balance is achieved between traction transport and sediment supply so that the ripples do not migrate despite a growing sediment surface. Ripple cross-lamination occurs in sediments deposited in environments characterized by rapid sedimentation from suspension—fluvial flood plains, point bars, river deltas subject to periodic flooding—and in environments of turbidite sedimentation.

Flaser and Lenticular Bedding.

Flaser bedding is a type of ripple bedding in which thin streaks of mud occur between sets of cross-laminations (Fig. 6.25). Mud is concentrated mainly in the ripple troughs

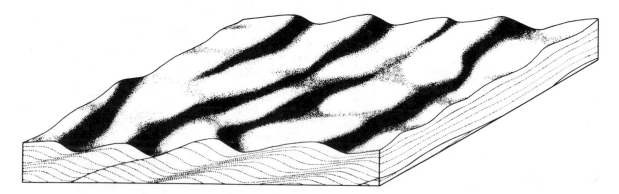

FIGURE 6.23 Diagram showing ripple cross-lamination produced by small current ripples climbing at a large angle. (From Harms, J. C., J. B. Southard, and R. G. Walker, 1982, Structures and sequences in clastic rocks: Soc. Econ. Paleontologists and Mineralogists Short Course No. 9. Fig. 3–7, p. 3–15, reprinted by permission of SEPM, Tulsa, Okla.)

but may also partly cover the crests. Flaser bedding suggests deposition under fluctuating hydraulic conditions. Periods of current activity, when traction transport and deposition of rippled sand take place, alternate with periods of quiescence, when mud is deposited. Repeated episodes of current activity result in erosion of previously deposited ripple crests, allowing new rippled sand to bury and preserve rippled beds with mud flasers in the troughs (Reineck and Singh, 1980). **Lenticular bedding** is a structure formed by interbedded mud and ripple cross-laminated sand in which the ripples or sand lenses are discontinuous and isolated in both vertical and horizontal directions (Fig. 6.26). Reineck and Singh (1980) suggest that flaser bedding is produced in environments in which conditions for deposition and preservation of sand are more favorable than for mud, but that lenticular bedding is produced in environments in which conditions favor deposition and preservation of mud over sand. Flaser and lenticular bedding appear to form particularly on tidal flats and in subtidal environments

FIGURE 6.24 (above left) Ripple cross-lamination (below ballpoint pen) in flood deposits of the Illinois River, southwestern Oregon.

FIGURE 6.25 (above right) Flaser bedding in the Elkton Siltstone (Eocene), near Cape Arago, southern Oregon coast.

FIGURE 6.26 (right) Lenticular bedding. Lenses of light-colored fine sandstone are interbedded with dark mudstone. Elkton Siltstone (Eocene), near Cape Arago, southern Oregon coast.

where conditions of current flow or wave action that cause sand deposition alternate with slack-water conditions when mud is deposited. Such bedding forms also in marine delta-front environments, where fluctuations in sediment supply and current velocity are common; in lake environments in front of small deltas; and possibly on shallow-marine shelves owing to storm-related transport of sand into deeper water.

Hummocky Cross-bedding

The name for **hummocky cross-bedding** was introduced by Harms et al. in 1975, although the structure has been recognized and described under different names by earlier workers. Hummocky cross-bedding is characterized by undulating sets of cross-laminae that are both concave-up (swales) and convex-up (hummocks). The cross-bed sets cut gently into each other with curved erosion surfaces (Fig. 6.27). Hummocky cross-bedding commonly occurs in sets 15–20 cm thick with wavy erosional bases and rippled, bioturbated tops (Harms et al., 1975). Spacing of hummocks and swales is from 50 cm to several meters. The lower bounding surface of a hummocky unit is sharp and is commonly an erosional surface. Current-formed sole marks may be present on the base. Hummocky cross-bedding occurs most typically in fine sandstone to coarse siltstone that commonly contain abundant mica and fine carbonaceous plant debris (Dott and Bourgeois, 1982).

Hummocky cross-bedding has not yet been produced in flumes or reported from modern environments, but it has been reported in ancient strata from numerous localities. Harms et al. (1975, 1982) suggest that this structure is formed by strong surges of varied direction (oscillatory flow) that are generated by relatively large storm waves. Strong storm-wave action first erodes the seabed into low hummocks and swales that lack any significant orientation. This topography is then mantled by laminae of material swept over the hummocks and swales. Although hummocky cross-bedding is commonly confined to marine sedimentary rocks, Duke (1985) reports the occurrence of this structure in some lacustrine sedimentary rocks. The occurrence of hummocky cross-bedding in lake deposits suggests that storm effects may strongly influence the sedimentary record of some lakes.

FIGURE 6.27 Hummocky cross-stratification in fine-grained sandstone of the Lower Member of the Coaledo Formation (Eocene), near Sunset Bay, southern Oregon coast. Arrows point to the hummocky erosional surface.

Irregular Stratification

Deformation Structures

Convolute Bedding and Lamination. **Convolute bedding** is a structure formed by complex folding or intricate crumpling of beds or laminations into irregular, generally small-scale anticlines and synclines. It is commonly, but not necessarily, confined to a single sedimentation unit or bed and the strata above and below this bed may show little evidence of deformation (Fig. 6.28). Convolute bedding is most common in fine sands or silty sands, and the laminae can typically be traced through the folds. Faulting generally does not occur, but the convolutions may be truncated by erosional surfaces that may also be convoluted. The convolutions increase in complexity and amplitude upward from undisturbed laminae in the lower part of the unit. They may either die out in the top part of the unit or be truncated by the upper bedding surface. Beds containing convolute laminations commonly range in thickness from about 3 to 25 cm (Potter and Pettijohn, 1977), but convoluted units up to several meters thick have been reported in both eolian and subaqueous deposits.

Convolute bedding is most common in turbidite sequences. It also occurs in intertidal-flat sediments, river floodplain sediments, and point-bar deposits. The origin of convolute bedding is still not thoroughly understood, but it appears to be caused by plastic deformation of partially liquefied sediment soon after deposition. The axes of some convoluted folds have a preferred orientation that commonly coincides with the paleocurrent direction, suggesting that the process that produces convolutions occurs during deposition, at least in these cases. Liquefaction of sediment can be caused by such processes as differential overloading, earthquake shocks, and breaking waves.

Flame Structures. **Flame structures** are wavy or flame-shaped tongues of mud that project upward into an overlying layer, which is commonly sandstone (Fig. 6.29). The crests of some flames are bent over or overturned and tend to all point in the same direction. Flame structures are commonly associated with other structures caused by

FIGURE 6.28 Convolute lamination in fine-grained sandstone. Note parallel laminated sandstone below. Coaledo Formation (Eocene), near Sunset Bay, southern Oregon coast.

FIGURE 6.29 Flame structures in base of fine-grained sandstone unit overlying laminated mudstones and siltstones. Some flames are overturned toward the left, suggesting slight downslope movement of the sand during loading. Coaledo Formation (Eocene), near Sunset Bay, southern Oregon coast. (Photograph by E. M. Baldwin.)

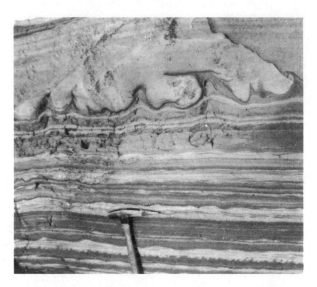

sediment loading. They are probably caused mainly by loading of water-saturated mud layers less dense than overlying sands and are consequently squeezed upward into the sand layers. The orientation of overturned crests suggests that loading may be accompanied by some horizontal drag or movement between the mud and sand bed.

Ball and Pillow Structures. Ball and **pillow structures** are found in the lower part of sandstone beds—and less commonly in limestone beds—that overlie mudrocks (Fig. 6.30). They consist of hemispherical or kidney-shaped sandstone or limestone masses that show internal laminations. In some hemispheres, the laminae may be gently curved or deformed, particularly next to the outside edge of the hemispheres, where they tend to conform to the shape of the edge. The balls and pillows may remain connected to the overlying bed, or they may be completely isolated from the bed and enclosed in the underlying mud. Such isolated masses are also called **pseudonodules.** Ball and pillow structures are believed to form as a result of foundering and breakup of semiconsolidated sand, or limy sediment, owing to partial liquefaction of underlying mud, possibly caused by shocking. Liquefaction of the mud causes the overlying sand beds or limy sediment to deform into hemispherical masses, which may subsequently break apart from the bed and sink into the mud. Kuenen (1958) experimentally produced structures that closely resemble natural ball and pillow structures by applying a shock to a layer of sand deposited over a thixotropic clay.

Synsedimentary Folds and Faults. The general term **slump structures** has been applied to structures produced by penecontemporaneous deformation resulting from movement and displacement of unconsolidated or semiconsolidated sediment, mainly under the influence of gravity. Potter and Pettijohn (1977) describe slump structures as being the products of either (1) pervasive movement involving the interior of the transported mass, producing a chaotic mixture of different types of sediments, such as broken mud layers embedded in sandy sediment or (2) a décollement type of movement in which the lateral displacement is concentrated along a sole, producing beds

FIGURE 6.30 Ball and pillow structures (arrows) on the base of a thin sandstone bed. Lookingglass Formation (Eocene), near Illahe, southwest Oregon.

that are tightly folded and piled into nappelike structures (Fig. 6.31). Slump structures may involve many sedimentation units; they are commonly faulted. Thicknesses of slump units have been reported to range from less than 1 m to as much as 55 m. Slump units may be bounded above and below by strata that show no evidence of deformation. It may be difficult in some stratigraphic sequences, however, to differentiate between slump units and incompetent beds such as shale that were deformed between competent sandstone or limestone beds during tectonic folding.

Slump structures typically occur in mudstones and sandy shales, and less commonly in sandstones, limestones, and evaporites. The structures are generally found in units that were deposited rapidly, and they have been reported from a variety of environments where rapid sedimentation and oversteepened slopes lead to instability. They occur in glacial sediments, varved silts and clays of lacustrine origin, eolian

FIGURE 6.31 Décollement-type synsedimentary folds (arrows) in thin, fine-grained sandstone layers interbedded with mudstone. Elkton Siltstone (Eocene), near Cape Arago, southern Oregon coast.

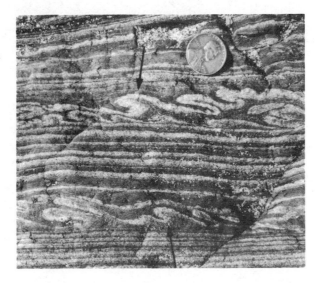

dune sands, turbidites, delta and reef-front sediments, and subaqueous dune sediments, and in sediments from the heads of submarine canyons, continental shelves, and the walls of deep-sea trenches.

Dish and Pillar Structures. Dish structures are thin, dark-colored, subhorizontal, flat to concave-upward clayey laminations (Fig. 6.32) that occur principally in sandstone and siltstone units (Lowe and LoPiccolo, 1974; Rautman and Dott, 1977). The laminations are commonly only a few millimeters thick, but individual dishes may range from 1 cm to more than 50 cm in width. They typically occur in thick beds where dish and pillar structures may be the only structures visible. They occur also in beds less than about 0.5 m thick, where they commonly cut across primary flat laminations and other laminations. **Pillar structures** generally occur in association with dish structures (Fig. 6.32). Pillars are vertical to near-vertical, cross-cutting columns and sheets of structureless or swirled sand that cut through either massive or laminated sands that commonly also contain dish structures and convolute laminations. Pillars range in size from tubes a few millimeters in diameter to large structures greater than 1 m in diameter and several meters in length. Pillars are not actually stratification structures. They are discussed here with dish structures because of their close association with these structures and because they form by a similar mechanism.

Dish and pillar structures were first observed in turbidite sequences and are most abundant in such sequences; however, they have now been reported in sediments from deltaic, alluvial, lacustrine, and shallow-marine deposits, as well as from volcanic ash layers. They are indicative of rapid deposition, and form by escape of water during

FIGURE 6.32 Strongly curved to nearly flat dish structures (large arrow) and pillars (small arrow) formed by dewatering of siliciclastic sediment. Jackfork Group (Pennsylvanian), southeast Oklahoma. (From Lowe, D. R., 1975, Water escape structures in coarse-grained sediment: Sedimentology, v. 22. Fig. 8, p. 175, reprinted by permission of Elsevier Science Publishers, Amsterdam.)

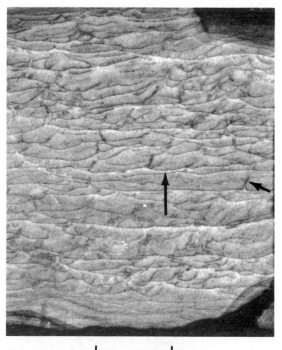

0   4 cm

consolidation of sediment. During gradual compaction and dewatering, semipermeable laminations act as partial barriers to upward-moving water carrying fine sediment. The fine particles are retarded by the laminations and added to them, forming the dishes. Some of the water is forced horizontally beneath the laminations until it finds an easier escape route upward. This forceful upward escape of water forms the pillars. Therefore, both dish structures and pillars are dewatering structures.

Erosion Structures

Channels. Channels are structures that show a U- or V-shape in cross section and which cut across earlier formed bedding and laminations (Fig. 6.33). They are formed by erosion, principally by currents but in some cases by mass movements. Channels are commonly filled with sediment that is texturally different from the beds they truncate. Channels visible in outcrop range in width and depth from a few centimeters to many meters. Even larger channels may be definable by mapping or drilling. It is seldom possible to trace their length in outcrop, but they can presumably extend for distances many times their width. They are very common in fluvial and tidal sediments. They occur also in turbidite sediments, where the long dimensions of the channels tend to be oriented parallel to current direction, as are other directional structures.

Scour-and-Fill Structures. Scour-and-fill structures, which are also called **cut-and-fill structures,** can be confused with channels. They consist of small, filled asymmetrical troughs a few centimeters to a few meters in size with long axes that point downcurrent and which commonly have a steep upcurrent slope and a more gentle downcurrent slope. They may be filled either with coarser grained or finer grained material than the substrate. These structures are most common in sandy sediment and are thought to form as a result of scour by currents and subsequent backfilling as current velocity decreases. In contrast to channels, several scour-and-fill structures may occur together closely spaced in a row. They are primarily structures of fluvial origin and can occur in river, alluvial-fan, or glacial outwash-plain environments.

FIGURE 6.33 Large channel (upper left) in gravelly, cross-bedded sands of the Coquille Formation (Pleistocene) near Bandon, southern Oregon coast. The channel is filled at the base with gravels, overlain by cross-bedded sand.

Biogenic Structures: Stromatolitic Bedding

Stromatolites are laminated structures composed of fine silt- or clay-size sediment or, more rarely, sand-size sediment. Most ancient stromatolites occur in limestones; however, stromatolites have been reported in siliciclastic sediments. Stromatolitic bedding ranges from nearly flat laminations that may be difficult to differentiate from sedimentary laminations of other origins to hemispherical forms in which the laminae are crinkled or deformed to various degrees (Fig. 6.34). The hemispherical forms range in shape from biscuit- and cabbagelike forms to columns. Logan et al. (1964) classified these hemispherical stromatolites into three basic types: (1) laterally linked hemispheroids; (2) discrete, vertically stacked hemispheroids; and (3) discrete spheroids, or spheroidal structures (Fig. 6.35). Laterally linked hemispheroids and discrete, vertically stacked hemispheroids can combine in various ways to create several different kinds of compound stromatolites. The term **thrombolite** was proposed by Aitken (1967) for structures that resemble stromatolites in external form and size but lack laminations. The laminations of stromatolites are generally less than 1 mm in thickness and are caused by concentrations of fine calcium carbonate minerals, fine organic matter, and detrital clay and silt. Stromatolites composed of quartz grains have also been reported (R. A. Davis, 1968).

Stromatolites were considered to be true body fossils by early workers, but are now known to be organosedimentary structures formed largely by the trapping and binding activities of blue-green algae (cyanobacteria). Although some geologists consider stromatolites to be a trace fossil (discussed subsequently), they are included here, because of their distinctive lamination, as a type of irregular stratification structure. They are forming today in many localities, where they occur mainly in the shallow subtidal, intertidal, and supratidal zones of the ocean. They have been found also in lacustrine environments. Because they are related to the activities of blue-green algae, which carry out photosynthesis, they are restricted to water depths and environments where enough light is available for photosynthesis. The laminated structure forms as a result of trapping of fine sediment in the very fine filaments of algal mats. Once a

FIGURE 6.34 Algal stromatolites in the Snowslip Formation (Precambrian), west of Logan Pass, Glacier National Park, Montana. (Photograph by G. J. Retallack.)

Types	Description	Vertical section of stromatolite structure
Laterally linked hemispheroids	Space-linked hemispheroids with close-linked hemispheroids as a microstructure in the constituent laminae	
Discrete, vertically stacked hemispheroids	Discrete, vertically stacked hemispheroids composed of close-linked hemispheroidal laminae on a microscale	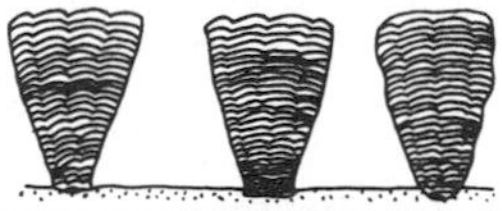
Discrete spheroids	Spheroidal structures consisting of inverted, stacked hemispheroids	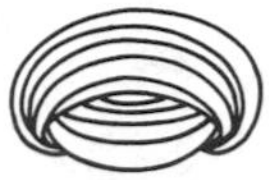
	Spheroidal structures consisting of concentrically stacked hemispheroids	
	Spheroidal structures consisting of randomly stacked hemispheroids	
Combination forms	Initial space-linked hemispheroids passing into discrete, vertically stacked hemispheroids with upward growth of structures	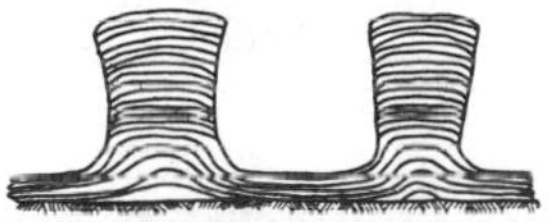
	Initial discrete, vertically stacked hemispheroids passing into close-linked hemispheroids by upward growth	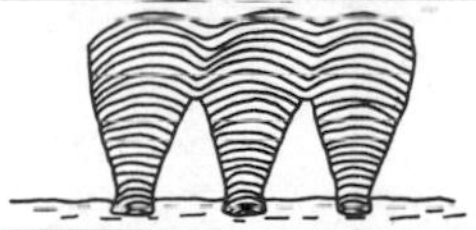
	Alternation of discrete, vertically stacked hemispheroids and space-linked hemispheroids due to periodic sediment infilling of interstructure spaces	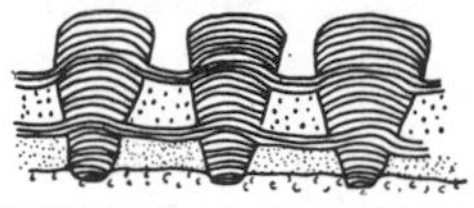
	Initial space-linked hemispheroids passing into discrete, vertically stacked hemispheroids; both with laminae of close-linked hemispheroids	
	Initial discrete, vertically stacked hemispheroids passing into close-linked hemispheroids; both with laminae of close-linked hemispheroids	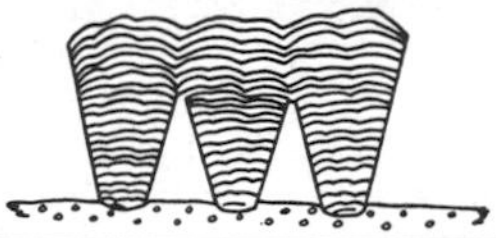

FIGURE 6.35 Structure of hemispherical stromatolites showing examples of laterally linked hemispheroids, vertically stacked hemispheroids, and discrete spheroids. (After Logan, B. W., R. Rezak, and R. N. Ginsburg, 1964, Classification and environmental significance of algal stromatolites: Jour. Geology, v. 72. Fig. 4, p. 76, and Fig. 5, p. 78, reprinted by permission of University of Chicago Press.)

thin layer of sediment covers the mat, the algal filaments grow up and around sediment grains to form a new mat, which traps another thin layer of sediment. This successive growth of mats produces the laminated structure. The shapes of the hemispheres are related to water energy and scouring effects in the depositional environment. Laterally linked hemispheroids tend to form in low-energy environments, where scouring effects are minimal. In higher energy environments, scouring by currents prevents linking of the stromatolite heads; thus, vertically stacked or discrete hemispheroids form. Further details on the origin and significance of stromatolites are discussed in Walter (1976).

6.4   BEDDING-PLANE MARKINGS

Markings Generated by Erosion and Deposition

Many bedding-plane markings occur on the undersides of beds as positive-relief casts and irregular markings. Owing to their location on the bases or soles of beds, they are often referred to as **sole markings.** Sole markings are preserved particularly well on the undersides of sandstones and other coarser grained sedimentary rocks that overlie mudstone or shale beds. Many sole markings are formed by erosional processes; consequently, they commonly show directional features that make them very useful for interpreting the flow directions of ancient currents.

These so-called erosional sole markings are actually formed by a two-stage process involving both erosion and deposition. First, a cohesive fine-sediment bottom is eroded by some mechanism to produce grooves, or depressions. Because of the cohesiveness of the sediment, the depressions may be preserved long enough to be filled in and buried during subsequent deposition, typically by sediment coarser grained than the bottom mud. This coarser sediment is probably deposited very shortly after erosion of the depression, possibly in some cases by the same current that formed the depression. After burial and lithification, the coarser-grained fill sediment remains attached to the base of the overlying bed. If the beds subsequently undergo tectonic uplift, these structures may be exposed by weathering and subaerial erosion as positive

FIGURE 6.36 Postulated stages of development of sole markings owing to erosion of a mud bottom followed by deposition of coarser sediment. The diagram also illustrates how the sole markings appear as positive-relief features on the base of the infilling bed after tectonic uplift and subaerial weathering and suggests how sole markings can be used to tell top and bottom of overturned beds. (From Collinson, J. D., and D. B. Thompson, 1982, Sedimentary structures. Fig. 4.1, p. 37, reprinted by permission of George Allen & Unwin, London. After Ricci-Lucchi, F., 1970, Sedimentografia: Zanchelli, Bologna, Italy.)

Erosion of bed   **Deposition**   **Burial and lithification**

Tectonic tilting   **Subaerial erosion**   **Tectonic overturning**   **Subaerial erosion**

relief features (Fig. 6.36). The initial erosional event that creates the depressions in a mud bottom can take the form of current scour, or it can result from the action of objects called **tools** that are carried by the current and that intermittently or continuously make contact with the bottom. These tools can be pieces of wood, the shells of organisms, or any similar object that can be rolled or dragged along the bottom. Erosional structures may thus be classified genetically as either **current-formed structures** or **tool-formed structures.**

Erosional sole markings are most common on the soles of turbidite sandstones, but they are found also in sedimentary rocks deposited in other environments. They can form in any environment where the requisite conditions of an erosive event followed reasonably quickly by a depositional event are met. They have been reported in both fluvial and shelf deposits in addition to turbidites.

Groove Casts. Groove casts are elongate, nearly straight ridges that result from infilling of erosional relief produced as a result of a pebble, shell, piece of wood, or other object being dragged or rolled across the surface of cohesive sediment (Fig. 6.37). They typically range in width from a few millimeters to tens of centimeters and have a relief of a few millimeters to a centimeter or two; however, much larger groove casts also occur. Groove casts are greatly elongated in comparison to their widths. Thus, they are directional features that are oriented parallel to the flow direction of the ancient currents that produced them. Unfortunately, many groove casts do not have features that show the unique flow direction; we cannot tell from them which direction was downcurrent and which upcurrent. Groove casts on the same bed commonly have the same general orientation, although they may diverge at slight angles and even cross. **Chevrons** are a variety of groove casts made up of continuous V-shaped crenulations that close in a downstream direction; thus, this type of groove cast can be used to determine the true direction of flow. Dzulynski and Walton (1965) suggest that chevrons are formed by tools moving just above the sediment surface, but not touching the surface, causing rucking-up of the sides of the groove. Groove casts are especially common on the soles of turbidite beds. The grooves are caused by shell fragments, pieces of wood, or other tools that are carried in the bases of turbidity current flows and dragged across a mud bottom. They occur also on the soles of beds deposited in shallow-water environments such as tidal flats and floodplains, where floating tools may touch bottom and leave grooves.

FIGURE 6.37 Groove casts on the base of a graded sandstone unit, Tyee Formation (Eocene), near Reedsport, southwest Oregon. Small drag features along the groove casts suggest that the paleo-current moved from right to left. (Photograph by P. D. Snavely, Jr.)

Bounce, Brush, Prod, Roll, and Skip Marks. Small gouge marks are produced by tools that make intermittent contact with the bottom. In the case of **brush** and **prod marks,** these gouge marks are asymmetrical in cross section, with the deeper, broad part of the mark oriented downcurrent. **Bounce marks** are roughly symmetrical. **Roll** and **skip marks** are formed either by a saltating tool or by rolling of a tool over the surface, producing a continuous track. The genesis of these structures is illustrated in Figure 6.38.

Flute Casts. Flute casts are elongated welts or ridges that have at one end a bulbous nose that flares out in the other direction and merges gradually with the surface of the bed (Fig. 6.39). The casts occur singly or in swarms in which all of the flutes are oriented in the same general direction. On a given sole, the flutes tend to be about the same size; however, flute casts on different beds can range in width from a centimeter

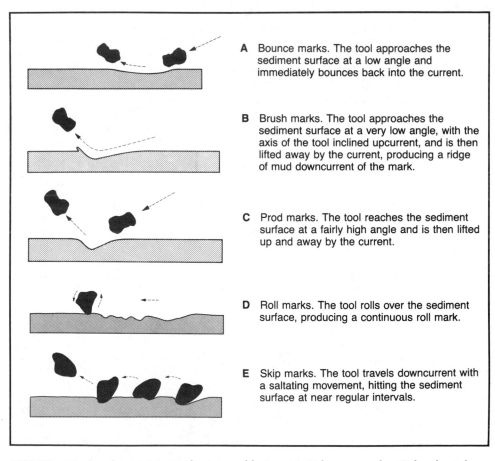

A   Bounce marks. The tool approaches the sediment surface at a low angle and immediately bounces back into the current.

B   Brush marks. The tool approaches the sediment surface at a very low angle, with the axis of the tool inclined upcurrent, and is then lifted away by the current, producing a ridge of mud downcurrent of the mark.

C   Prod marks. The tool reaches the sediment surface at a fairly high angle and is then lifted up and away by the current.

D   Roll marks. The tool rolls over the sediment surface, producing a continuous roll mark.

E   Skip marks. The tool travels downcurrent with a saltating movement, hitting the sediment surface at near regular intervals.

FIGURE 6.38   Development in a cohesive mud bottom of (A) bounce marks, (B) brush marks, (C) prod marks, (D) roll marks, and (E) skip marks by the action of tools making contact with the bottom in various ways. These tool-formed depressions are subsequently filled with coarser sediment to produce positive-relief casts. (After Reineck H. E., and I. B. Singh, 1980, Depositional sedimentary environments, 2nd ed. Figs. 127, 129, 125, 132, 131, p. 82, 83, reprinted by permission of Springer-Verlag, Heidelberg.)

or two to 20 cm or more, in height (relief) from a few centimeters to 10 cm or more, and in length from a few centimeters to a meter or more. The plan-view shape of flutes varies from nearly streamlined, bilaterally symmetrical forms to more elongate and irregular forms, some of which are highly twisted.

Flute casts are formed by filling of a depression scoured in cohesive sediment by current eddies created behind some obstacle or by chance eddy scour. This type of current scour produces asymmetrical depressions in which the steepest and deepest part of the depression is oriented upstream. Therefore, when such depressions are filled, the filling forms a positive relief structure with a bulbous nose oriented upstream. Flute casts thus make excellent paleocurrent indicators because they show the unique direction of current flow. Flutes are particularly prevalent on the soles of turbidite sequences, but they occur also in sediments deposited in shallow-marine and nonmarine environments. They have been reported on the soles of limestone beds as well as sandstone beds.

Current Crescents. Current crescents, also called **obstacle scours,** occur in modern environments as narrow semicircular or horseshoe-shaped troughs that form around small obstacles such as pebbles or shells as currents scour a mud or sand bottom (Fig. 6.40). In sandy sediment, these structures form on the downcurrent side of an obstacle

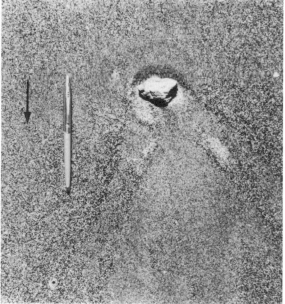

FIGURE 6.39 Flute casts on the base of a turbidite unit, Tyee Formation (Eocene), near Valsetz, northwestern Oregon. The bulbous terminations of the flute casts indicate that paleocurrents moved from the bottom toward the top. (Photograph by P. D. Snavely, Jr.)

FIGURE 6.40 Current crescent formed around a pebble on a modern beach, southern Oregon coast. Current flow was from top to bottom.

as small ridges. In ancient sedimentary rocks, they commonly occur as casts on the undersides of sandstone beds. Although they are very common in modern beach environments, in ancient sedimentary rocks they are most characteristic of fluvial sandstones with shale interbeds. They have also been reported from turbidite sequences. Somewhat similar structures are formed by sand blown around obstacles by wind, forming a ridge or tail of sand downwind from the object. Such wind-produced structures are rarely preserved in ancient sedimentary rocks.

Markings Generated by Deformation: Load Casts

Load casts are described by Potter and Pettijohn (1977, p. 198) as "swellings ranging from slight bulges, deep or shallow rounded sacks, knobby excrescences, or highly irregular protuberances." They commonly occur on the soles of sandstone beds that overlie mudstones or shales, and they tend to cover the entire bedding surface (Fig. 6.41). They range in diameter and relief from a few centimeters to a few tens of centimeters. Load casts may superficially resemble flute casts; however, they can be distinguished from flutes by their greater irregularity in shape and their lack of definite upcurrent and downcurrent ends. Also, load casts do not display a preferred orientation with respect to current direction.

Although they are called casts, load casts are not true casts because they are not fillings of a preexisting cavity or mold. They are formed by deformation of uncompacted, hydroplastic mud beds as the beds are subjected to unequal loading by overlying sand layers. Uncompacted muds with excess fluid pore pressures or muds liquefied by an externally generated shock can be deformed by the weight of overlying sand, which may sink unequally into the incompetent mud. Such loading forces protrusions of sand down into the mud, creating on the bases of the sandstone beds positive relief features that may resemble some erosional structures. Load casts are closely related genetically to ball and pillow structures and flame structures. Flute and groove casts may be modified by loading, a process that tends to exaggerate their relief and destroy original shapes.

FIGURE 6.41   Large load casts on the underside of a thin-bedded sandstone. Tar Springs Sandstone (Mississippian). (From Pettijohn, F. J., and P. E. Potter, 1964, Atlas and glossary of primary sedimentary structures. Pl. 52A. Reprinted by permission of Springer-Verlag, New York. Photograph courtesy of P. E. Potter.)

Load casts can form in any environment where water-saturated muds are quickly buried by sand before dewatering can take place. The casts are not indicative of any particular environment, although they tend to be most common in turbidite sequences. Their presence on the bases of some beds and not on others seems to reflect the hydroplastic state of the underlying mud. They apparently will not form on the bases of sand beds deposited on muds that have already been compacted or dewatered prior to deposition of the sand.

Markings Generated by Organisms: Trace Fossils

Definition. The burrowing, boring, feeding, and locomotion activities of organisms can produce a variety of trails, depressions, and open burrows and borings in mud or semiconsolidated sediment bottoms. Filling of these depressions and burrows with sediment of a different type or with different packing creates structures that may be either positive relief features such as trails on the bases of overlying beds or features that show up as burrow or bore fillings on the tops of underlying mud beds. Burrows and borings commonly extend down into beds; therefore, these structures are not exclusively bedding-plane structures.

Tracks, trails, burrows, borings, and other structures made by organisms on bedding surfaces or within beds are known collectively as **trace fossils.** They are also called **ichnofossils** and **lebensspuren.** Although geologists have long been aware of the presence of biogenic structures in sedimentary rocks, the recognition and naming of the many varieties of trace fossils now known, as well as a fuller understanding of the environmental significance of these structures, has come about to a large extent since the middle 1950s. Numerous research papers dealing with trace fossils have been published since that time, in addition to several full-length monographs. Only a very brief summary of the classification, occurrence, and significance of trace fossils is presented here. Additional details may be found in the books of Basan (1970), Crimes and Harper (1970, 1977), Curran (1985), Ekdale et al. (1984), Frey (1975), Frey and Pemberton (1984), Hantzschel (1975), and Seilacher (1964).

Classification of Trace Fossils. Trace fossils are not true bodily preserved fossils, but are simply biogenic structures that originated through the locomotion, feeding, burrowing, or resting activities of organisms. Interpreted broadly, biogenic structures can be considered to include (1) bioturbation structures (burrows, tracks, trails, root penetration structures), (2) biostratification structures (algal stromatolites, graded bedding of biogenic origin), (3) bioerosion structures (borings, scrapings, bitings), and (4) excrement (coprolites, such as fecal pellets or fecal castings). Not all geologists regard biostratification structures as trace fossils, and the structures are not commonly included in published discussions of trace fossils.

Trace fossils can be classified in several ways on the basis of morphology (taxonomy), presumed behavior of the organism that produced the structures, and preservational process (Simpson, 1975; Frey, 1978). On the basis of morphology, they can be grouped into such categories as tracks, trails, burrows, borings, and bioturbate texture, as shown in Table 6.3. Tracks, trails, burrows, and bioturbate texture are features formed in soft sediments. Borings are formed in hard substrates. Figures 6.42, 6.43, and 6.44 illustrate some of these features. Classification of trace fossils on the basis of behavior of the generating organism is referred to as ethological classification. Classified this way, trace fossils are divided into resting traces, crawling traces, grazing traces, feeding traces or structures, and dwelling structures (Fig. 6.45). Further description of these behavioral structures and the processes by which they are assumed to

TABLE 6.3 Descriptive-genetic classification of trace fossils

A   Tracks and Trails

Track—impression left in underlying sediment by an individual foot or podium

Trackway—succession of tracks reflecting directed locomotion by an animal

Trail—trace produced during directed locomotion and consisting either of a surficial groove made by an animal having part of its body in continuous contact with the substrate surface or of a continuous subsurface structure made by a mobile endobenthic organism

B   Burrows and Borings

Boring—excavation made in consolidated or otherwise firm substrates, such as rock, shell, bone, or wood

Burrow—excavation made in loose, unconsolidated sediments

Burrow or *boring system*—highly ramified and/or interconnected burrows or borings, typically involving shafts and tunnels

Shaft—dominantly vertical burrow or boring or a dominantly vertical component of a burrow or boring system having prominent vertical and horizontal parts

Tunnel (= gallery)—dominantly horizontal burrow or boring or a dominantly horizontal component of a burrow or boring system having prominent vertical and horizontal parts

Burrow lining—thickened burrow wall constructed by organisms as a structural reinforcement; may consist of (1) host sediments retained essentially by mucus impregnation, (2) pelletoidal aggregates of sediment shoved into the wall, like mud-daubed chimneys, (3) detrital particles selected and cemented like masonry, or (4) leathery or felted tubes consisting mostly of chitinophosphatic secretions by organisms; burrow linings of types (3) and (4) commonly called "dwelling tubes"

Burrow cast—sediments infilling a burrow (= burrow fill); may be either "active," if done by animals, or "passive," if done by gravity or physical processes; active fill termed "back fill" wherever U-in-U laminae, etc., show that the animal packed sediment behind itself as it moved through the substrate

C   Bioturbation

Bioturbate texture—gross texture or fabric imparted to sediments by extensive bioturbation; typically consists of dense, contorted, truncated, or interpenetrating burrows or other traces, few of which remain distinct morphologically. Where burrows are somewhat less crowded and thus are more distinct individually, the sediment is said to be "burrow mottled"

D   Miscellaneous

Configuration—in ichnology, the spatial relationships of traces, including the disposition of component parts and their orientation with respect to bedding and (or) azimuth

Spreite—bladelike to sinuous, U-shaped, or spiraled structure consisting of sets or cosets of closely juxtaposed, repetitive parallel or concentric feeding or dwelling burrows or grazing traces. Individual burrows or grooves comprising the spreite commonly anastomose into a single trunk or stem (as in *Daedalus*) or are strung between peripheral "support" stems (as in *Rhizocorallium*). "Retrusive" spreiten extended upward or promimal to the initial point of entry by the animal, and "protrusive" spreiten extended downward, or distal to the point of entry

Source: After Frey, R. W., 1978, Behavioral and ecological implications of trace fossils, *in* P. B. Basan (ed.), Trace fossil concepts: Soc. Econ. Paleontologists and Mineralogists Short Course No. 5. Table 2, p. 49, reprinted by permission of SEPM, Tulsa, Okla.

form is given in Table 6.4. Trace fossils can be classified on the basis of type of preservation by use of such terms as full relief, semirelief, concave, and convex (Fig. 6.46).
Traces formed at the sediment surface are called exogenic traces and those formed
within strata are called endogenic traces.

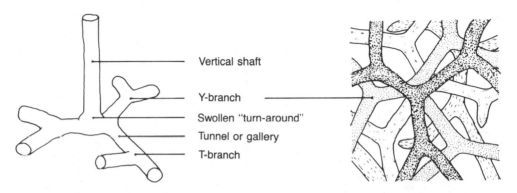

FIGURE 6.42 Diagrams illustrating standard ichnologic terminology. The diagram on the left
illustrates an open burrow; the one on the right represents a composite burrow composed of
successive cross-cutting mazes. (From Ekdale, A. A., R. G. Bromley, and S. B. Pemberton, 1984,
Ichnology: Trace fossils in sedimentology and stratigraphy: Soc. Econ. Paleontologists and Mineralogists Short Course No. 15. Fig. 2.1, p. 14, reprinted by permission of SEPM, Tulsa, Okla.)

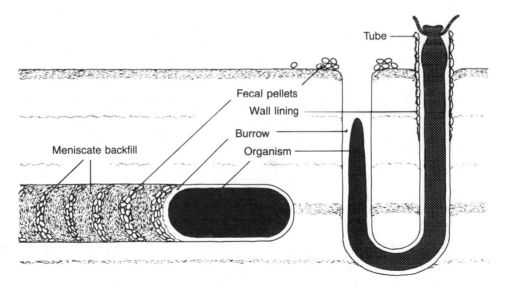

FIGURE 6.43 Two kinds of burrows: a mobile crawling trace or track of an irregular echinoid
(left) and a permanent, U-shaped dwelling burrow of an unidentified worm (right). Where the
wall lining consists of cemented grains, it is called a tube. (From Ekdale, A. A., R. G. Bromley,
and S. B. Pemberton, 1984, Ichnology: Trace fossils in sedimentology and stratigraphy: Soc.
Econ. Paleontologists and Mineralogists Short Course No. 15. Fig. 2.3, p. 15, reprinted by permission of SEPM, Tulsa, Okla.)

FIGURE 6.44 Examples of spreite structure, which is a type of structure produced as a burrow is shifted broadside by a burrower. An animal that moves its subhorizontal burrow up or down creates a wall-like, vertical spreite (e.g., *Teichichnus*). Lateral migration produces a bladelike, horizontal spreite (e.g., *Rhizocorallium*). Migration of a U-shaped burrow toward its aperture at the sediment surface causes the spreite to develop on the outside of the U; lengthening of the burrow downward produces a spreite on the inside of the U (e.g., the *Diplocraterion* specimens above). From Ekdale, A. A., R. G. Bromley, and S. B. Pemberton, 1984, Ichnology: Trace fossils in sedimentology and stratigraphy: Soc. Econ. Paleontologists and Mineralogists Short Course No. 15. Fig. 2.4, p. 16, reprinted by permission of SEPM, Tulsa, Okla.)

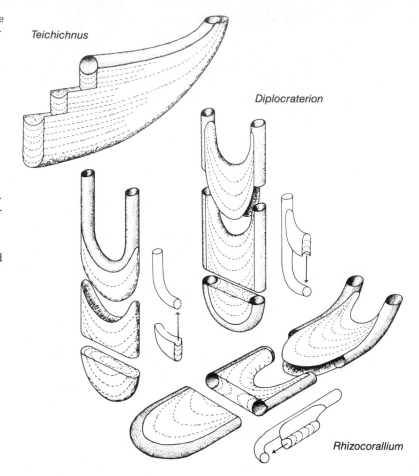

Teichichnus

Diplocraterion

Rhizocorallium

FIGURE 6.45 Classification of trace fossils on the basis of presumed behavior of the organisms producing the structures and the relationship of these traces to body fossils. Note the overlap of some categories of traces. Escape structures overlap several categories of behavioral traces and are not included here. (From Simpson, S., 1975, Classification of trace fossils, *in* R. W. Frey (ed.), The study of trace fossils. Fig. 3.2, p. 49, reprinted by permission of Springer-Verlag, Heidelberg. As translated from Seilacher, A., 1953: Neues Jahrb. Geologie u. Paläontologie, Abh. 96, p. 421–452.)

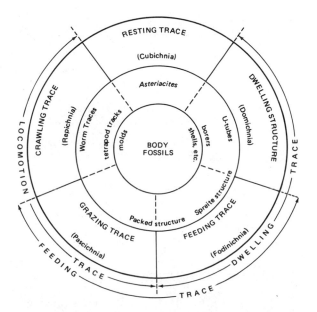

TABLE 6.4 Ethological classification of trace fossils

Categories of lebensspuren	Definition	Characteristic morphology
Resting traces (Cubichnia)	Shallow depressions made by animals that temporarily settle onto, or dig into, the substrate surface; emphasis on reclusion	Troughlike relief, recording to some extent the lateroventral morphology of the animal; structures isolated, ideally, but may intergrade with crawling traces or escape structures
Crawling traces (Repichnia)	Trackways, surficial trails, and shallow horizontal structures made by organisms traveling from one place to another; emphasis on locomotion	Linear or sinuous overall structures, some branched; footprints or continuous grooves, commonly annulated; complete form may be preserved or may appear as cleavage reliefs
Grazing traces (Pascichnia)	Grooves, pits, and furrows, many of them discontinuous, made by mobile deposit feeders at or near the substrate surface; emphasis on feeding behavior analogous to "strip mining"	Unbranched, nonoverlapping, curved to tightly coiled patterns or delicately constructed spreiten dominate; patterns reflect maximum utilization of surficial feeding area; complete form may be preserved
Feeding structures (Fodinichnia)	More or less temporary burrows constructed by deposit feeders; the structures may also provide shelter for the organisms; emphasis on feeding behavior analogous to "underground mining"	Single, branched or unbranched, cylindrical to sinuous shafts or U-shaped burrows, or complex, parallel to concentric burrow repetitions (spreiten structures); walls not commonly lined, unless by mucus; oriented at various angles with respect to bedding; complete form may be preserved
Dwelling structures (Domichnia)	Burrows or dwelling tubes providing more or less permanent domiciles, mostly for hemisessile suspension feeders or, in some cases, carnivores; emphasis on habitation	Simple, bifurcated, or U-shaped structures perpendicular or inclined at various angles to bedding, or branching burrow systems having vertical and horizontal components; walls typically lined; complete form may be preserved
Escape structures	Lebensspuren of various kinds modified or made anew by animals in direct response to substrate degradation or aggradation; emphasis on readjustment, or equilibrium between relative substrate position and the configuration of contained traces	Vertically repetitive resting traces; biogenic laminae either in echelon or as nested funnels or chevrons; U-in-U spreiten burrows; and other structures reflecting displacement of animals upward or downward with respect to the original substrate surface; complete form may be preserved, especially in aggraded substrates

Source: Frey, R. W., 1978, Behavioral and ecological implications of trace fossils, in P. B. Basan (ed.) Trace fossil concepts: Soc. Econ. Paleontologists and Mineralogists Short Course No. 5. Table 3, p. 51, reprinted by permission of SEPM, Tulsa, Okla.

179

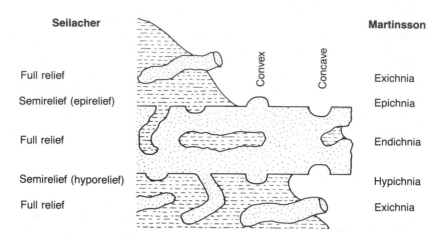

FIGURE 6.46 Terminology for preservational classification of trace fossils used by Seilacher and Martinsson. (From Ekdale, A. A., R. G. Bromley, and S. B. Pemberton, 1984, Ichnology: Trace fossils in sedimentology and stratigraphy: Soc. Econ. Paleontologists and Mineralogists Short Course No. 15. Fig. 2.6, p. 22, reprinted by permission of SEPM, Tulsa, Okla. From Seilacher, A., Sedimentological classification and nomenclature of trace fossils: Sedimentology, v. 3, p. 253–256; Martinsson, A., 1970, Toponomy of trace fossils, p. 109–130, in T. P. Crimes and J. C. Harper (eds.), The study of trace fossils: Springer-Verlag, New York.)

Environmental Significance. Trace fossils are produced by a variety or organisms such as crabs, flatfish, clams, molluscs, worms, shrimp, and eel. Because different organisms engage in similar types of behavior (crawling, grazing, feeding, and so forth), essentially identical traces may be produced by quite different organisms. Therefore, it is not always possible to identify the organism that produced a particular type of structure. It has been determined, however, that certain associations of biogenic structures tend to characterize particular sedimentary facies (Chapter 14). These facies, in turn, can be related to depositional environments. The term **ichnofacies** was introduced by Seilacher (1964) for sedimentary facies characterized by a particular association of trace fossils. Salinity, water depth, and consistency of the substrate (soft or hard bottom) appear to exert the primary controls on the distribution of trace fossils. Trace fossils occur in sediments deposited in environments ranging from subaerial continental to deep marine. In subaerial environments organisms such as insects, spiders, worms, millipeds, snails, and lizards can produce a variety of burrows and tunnels; vertebrate organisms leave tracks; and plants leave root traces. Freshwater fluvial and lacustrine environments are inhabited by organisms such as worms, crustaceans, insects, bivalves, gastropods, fish, birds, amphibians, mammals, and reptiles that can produce various kinds of traces. Trace fossils in freshwater, continental deposits are grouped into what is called the *Scoyenia* ichnofacies (Frey et al., 1984). This ichnofacies is rather nondistinctive, consisting of a low-diversity suite of invertebrate and vertebrate tracks, trails, and burrows (Ekdale et al., 1984).

Most interest in ichnofacies has focused on those facies that occur in marine sedimentary rocks. Marine trace fossils are produced by a wide variety of mainly invertebrate organisms such as worms, shrimp, lobsters, crabs, gastropods, and pelecypods. Some traces are also produced by fish. Seven marine ichnofacies are now recognized, each named for a representative trace fossil: *Terodolites, Trypanites, Glossifungites, Skolithos, Cruziana, Zoophycos,* and *Nereites* (Fig. 6.47). The *Teredo-*

lites ichnofacies (not shown in Fig. 6.47) occurs only in woody materials. The *Trypanites* ichnofacies is characteristic of hard, fully indurated substrates, and the *Glossifungites* ichnofacies typically occurs in firm, but uncemented, substrates. The remaining marine ichnofacies are all soft-sediment ichnofacies whose distribution appears to be controlled mainly by water depth.

The supratidal zone, intertidal zone, subtidal zone, and deeper zones of the marine realm are each distinguished by characteristic associations of trace fossils (Fig. 6.47). In general, the biogenic structures that characterize the *Trypanites* ichnofacies

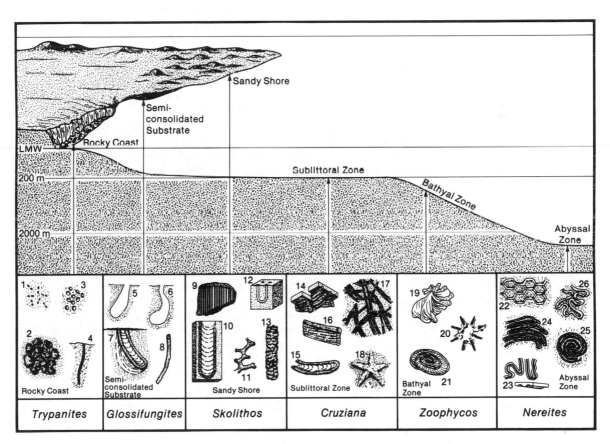

FIGURE 6.47 Schematic representation of the relationship of characteristic trace fossils to sedimentary facies and depth zones in the ocean. Borings of 1, *Polydora*; 2, *Entobia*; 3, echinoid borings; 4, *Trypanites*; 5, 6, pholadid burrows; 7, *Diplocraterion*; 8, unlined crab burrow; 9, *Skolithos*; 10, *Diplocraterion*, 11, *Thalassinoides*; 12, *Arenicolites*; 13, *Ophiomorpha*; 14, *Phycodes*; 15, *Rhizocorallium*; 16, *Teichichnus*; 17, *Crossopodia*; 18, *Asteriacites*; 19, *Zoophycos*; 20, *Lorenzinia*; 21, *Zoophycos*; 22, *Paleodictyon*; 23, *Taphrhelminthopsis*; 24, *Helminthoida*; 25, *Spirorhaphe*; 26, *Cosmorhaphe*. (From Ekdale, A. A., R. G. Bromley, and S. B. Pemberton, 1984, Ichnology: Trace fossils in sedimentology and stratigraphy: Soc. Econ. Paleontologists and Mineralogists Short Course No. 15. Fig. 15.2, p. 187, reprinted by permission of SEPM, Tulsa, Okla. Modified from Crimes, T. P., 1975, The stratigraphical significance of trace fossils, *in* T. P. Crimes and J. C. Harper (eds.), The study of trace fossils, Fig. 7.2, p. 118: Springer-Verlag, New York.)

of rocky coasts and pebbly shores are rock borings, most of which are dwelling structures for suspension-feeding organisms (Fig. 6.47, 1–4). Other structures in this ichnofacies include rasping and scraping traces made by feeding organisms, holes drilled by predatory gastropods, and microborings made by algae and fungi. The *Glossifungites* ichnofacies, redefined by Frey and Seilacher (1980), is now considered to be restricted to firm, uncemented surfaces that typically consist of dewatered, cohesive muds. The trace fossils produced in this environment are mainly vertical, U-shaped, and branched dwelling burrows of suspension feeders or carnivores such as shrimp, crabs, worms, and pholadid bivalves (Fig. 6.47, 5–8). The littoral zone or intertidal zone of sandy coasts is distinguished by harsh conditions resulting from high-energy waves and currents, desiccation, and large temperature and salinity fluctuations. Organisms adapt to these harsh conditions by burrowing into the sand to escape. Thus, vertical and U-shaped dwelling burrows, such as the *Skolithos, Diplocraterion, Arenicolites,* and *Ophiomorpha* burrows shown in Figure 6.47, 9–13, characterize the *Skolithos* ichnofacies of this zone; some of these burrows have protective linings. The sublittoral zone, or subtidal zone, extending from the low-tide zone to the edge of the continental shelf (at about 200-m water depth) is a less demanding environment, although erosive currents may be present. Vertical dwelling burrows and protected, U-shaped burrows are less common in this zone. Burrows tend to be shorter, and surface markings made by organisms such as crustaceans (or trilobites during early Paleozoic time) are more common. In the deeper part of the sublittoral zone, organic matter becomes abundant enough for sediment feeders to become established and produce feeding burrows. In these deeper waters, vertical escape burrows thus tend to give way to horizontal feeding burrows. This zone of the ocean is distinguished by the *Cruziana* ichnofacies, characterized by such traces as those shown in Figure 6.47, 14–18.

The deep bathyal and abyssal zones of the ocean exist below wave base, where low-energy conditions generally prevail, although erosion and deposition can occur in these zones owing to turbidity currents or deep-bottom currents. Complex feeding burrows such as those of *Zoophycos* and *Lorenzinia* (Fig. 6.47, 19–21) are particularly common in the bathyal zone. These traces make up the *Zoophycos* ichnofacies. In the even deeper waters of the abyssal zone, where bottom sediment is almost exclusively fine-grained clay, more complex spiral, winding, and meandering forms such as *Spirorhaphe* (Fig. 6.47, 25) or patterned trace fossils such as *Paleodictyon* (Fig. 6.47, 22) occur. This association of trace fossils constitutes the *Nereites* ichnofacies, named for *Nereites,* a type of horizontal grazing trail.

Although each of these marine ichnofacies tends to be characteristic of a particular bathymetric zone of the ocean, as shown in Figure 6.47, we now know that individual trace fossils can overlap depth zones. No single biogenic structure is an infallible indicator of depth and environment. Trace fossils should be studied as assemblages of structures in conjunction with other physical, chemical, and biological characteristics of the same substrates.

Trace fossils occur in rocks of all ages, including some Precambrian rocks. They have been reported in most types of sedimentary rocks except evaporites and rocks deposited in highly reducing (euxinic) environments. Highly saline environments or euxinic environments, where toxic conditions are caused by lack of oxygen and the presence of hydrogen sulfide gas, preclude or greatly reduce organic activity. Studies of bioturbation in modern open-ocean environments show that organisms may rework sediment so thoroughly that primary laminations and other physically produced structures are completely destroyed. Exceedingly intense bioturbation can produce bedding so homogenized that it has a mottled or stirred appearance or is completely devoid of

all structures. In order for bedding and other physically produced sedimentary structures to escape destruction by biogenic activity and become preserved in the geologic
record, they must be formed either in an environment where sedimentation rates are
so high that organisms do not have time to rework sediments and destroy original
structures or in euxinic or highly saline environments, as mentioned, where organic
activity is limited.

Other Applications of Trace Fossils. In addition to their usefulness as environmental
indicators, trace fossils are useful also in several other ways (Crimes, 1975). They may,
for example, serve as indicators of relative sedimentation rates on the basis of the
assumption that rapidly deposited sediments contain relatively fewer trace fossils than
slowly deposited sediments. They can help to show whether sedimentation was continuous or marked by erosional breaks, and they provide a record of the behavior
patterns of extinct organisms. They may even be useful in paleocurrent analysis; study
of the orientation of resting marks may indicate that some organisms preferred to face
into the current while resting. Trace fossils such as U-shaped burrows, which opened
upward when formed, can be used to tell the top and bottom orientation of beds. Trace
fossils also have biostratigraphic and chronostratigraphic significance for zoning and
correlation.

Bedding-plane Markings of Miscellaneous Origin

Mudcracks and Syneresis Cracks. Mudcracks in modern sediment are downward-
tapering, V-shaped fractures that display a crudely polygonal pattern in plan view. The
area between the cracks is commonly curved upward into a concave shape. Mudcracks
form in both siliciclastic and carbonate mud as desiccation takes place. Subsequent
sedimentation over a cracked surface fills the cracks. In ancient sedimentary rocks,
mudcracks are commonly preserved on the tops of bedding surfaces as positive-relief
fillings of the original cracks (Fig. 6.48). Mudcrack polygons range in diameter from a
few centimeters to a few meters. The cracks themselves commonly range in width up
to a few centimeters and in depth to a few tens of centimeters, but cracks up to a few
meters in depth have been reported. The presence of undoubted mudcracks indicates
intermittent subaerial exposure; however, mudcracks can be confused with syneresis
cracks, which form under water. Mudcracks occur in estuarine, lagoonal, tidal-flat,
river floodplain, playa lake, and other environments where muddy sediment is intermittently exposed and allowed to dry. They may be associated with raindrop or hailstone imprints, bubble imprints and foam impressions, flat-topped ripple marks, and
vertebrate tracks (Plummer and Gostin, 1981).

In contrast to the continuous, polygonal network of mudcracks that occurs on
bedding surfaces, **syneresis cracks** tend to be discontinuous and vary in shape from
polygonal to spindle shaped or sinuous (Plummer and Gostin, 1981). These cracks
commonly occur in thin mudstones interbedded with sandstones as either positive
relief features on the base of the sandstones or negative relief features on the top of
the mudstones. Syneresis cracks are subaqueous shrinkage cracks that form in clayey
sediment by loss of pore water from clays that have flocculated rapidly or that have
undergone shrinkage of swelling-clay mineral lattices owing to changes in salinity of
surrounding water (Burst, 1965). They are known in ancient sedimentary rocks from
both marine and nonmarine environments. They may be confused with mudcracks and
even some trace fossils. For example, the lenticular shape of the crack fill in plan view

FIGURE 6.48 Mudcracks on the upper surface of a Miocene mudstone bed, Bangladesh. (Photograph by E. M. Baldwin.)

may resemble burrow traces. Because some syneresis cracks do closely resemble mudcracks, it is important in trying to differentiate them to look for subaerial-exposure features associated with mudcracks—features such as raindrop imprints and vertebrate tracks.

Pits and Small Impressions. Small craterlike pits with slightly raised rims commonly occur with mudcracks and are thought to be impressions made by the impact of rain **(raindrop imprints)** or hail **(hailstone imprints).** They are commonly only a few millimeters deep and less than 1 cm in diameter and may occur as either widely scattered pits or very closely spaced impressions. When they can be unambiguously recognized, their presence indicates subaerial exposure; however, small circular depressions created by bubbles breaking on the surface of sediment **(bubble imprints),** escaping gas, and some types of organic markings can be confused with raindrop or hailstone imprints.

Rill and Swash Marks. **Rill marks** are small dendritic channels or grooves that form on beaches by the discharge of pore waters at low tide or by small streams debouching onto a sand or mud flat. They have very low preservation potential and are seldom found in ancient sedimentary rocks. **Swash marks** are very thin, arcuate lines or small ridges on a beach formed by concentrations of fine sediment and organic debris. They are caused by wave swash and mark the farthest advance of wave uprush. They likewise have low preservation potential, but when found and recognized in ancient sedimentary rocks they indicate either a beach or lakeshore environment.

Parting Lineation. **Parting lineation,** sometimes called **current lineation,** forms on the bedding surfaces of parallel-laminated sandstones. It consists of subparallel ridges and grooves a few millimeters wide and many centimeters long (Fig. 6.49). Relief on the ridges and hollows is commonly on the order of the diameter of the sandstone grains. The grains in the sandstone generally have a mean orientation of their long axes parallel to the lineation. The lineation is oriented parallel to current flow, and thus its

FIGURE 6.49 Current lineation on the surface of a tidal river bar near the mouth of the Rogue River, southern Oregon coast. The current that produced the lineation flowed roughly parallel to the handle of the small entrenching tool.

presence in ancient sandstones is useful in paleocurrent studies—although it shows only that the current flowed parallel to the parting lineations and does not show which of the two diametrically opposed directions was the flow direction. Parting lineation occurs in newly deposited sands on beaches and in fluvial environments. It is most common in ancient deposits in thin, evenly bedded sandstones. Its origin is obviously related to current flow and grain orientation, probably owing to flow over upper-flow regime plane beds, but the exact mechanism by which parting lineation forms is poorly understood.

6.5 OTHER STRUCTURES: SANDSTONE DIKES AND SILLS

Sandstone dikes and sills are tabular bodies of massive sandstone that fill fractures in any type of host rock. They range in thickness from a few centimeters to more than 10 m. They lack internal structures except for oriented mica flakes and other elongated particles that are commonly aligned parallel to the dike walls. Sandstone dikes are not common structures but they have been reported from numerous localities in rocks ranging in age from Precambrian to Pleistocene. They occur in a wide variety of depositional environments, ranging from deep marine to subaerial.

Sandstone dikes are formed by forceful injection of liquefied sand into fractures, commonly in overlying rock; however, injection appears to have been downward in some rocks. Sandstone sills are similar features that formed by injection parallel to bedding. These sills may be difficult or impossible to distinguish from normally deposited sandstone beds unless they can be traced into sandstone dikes or be traced far enough to show a cross-cutting relationship with other beds. Suggested causes of liquefaction of sand include shocks owing to earthquakes or triggering effects related to slumps, slides, or rapid emplacement of sediment by mass flow.

ADDITIONAL READINGS

Allen, J. R. L., 1982, Sedimentary structures: Their character and physical basis, v. 1–2: Elsevier, Amsterdam, 664 p.

Allen, J. R. L., 1968, Current ripples: Their relation to patterns of water and sediment motion: North Holland Pub., Amsterdam, 443 p.

Basan, P. B. (ed.), 1978, Trace fossil concepts: Soc. Econ. Paleontologists and Mineralogists Short Course No. 5, 181 p.

Bouma, A. H., 1969, Methods for the study of sedimentary structures: John Wiley & Sons, New York, 457 p.

Collinson, J. D., and D. B. Thompson, 1982, Sedimentary structures: George Allen & Unwin, London, 194 p.

Conybeare, C. E. B., and K. A. W. Crook, 1968, Manual of sedimentary structures: Dept. Nate. Development, Bur. Mineral Resources, Geology and Geophysics, Bull. 102, 327 p.

Crimes, T. P., and J. C. Harper (eds.), 1970, Trace fossils: Seel House Press, Liverpool, p. 547

Crimes, T. P., and J. C. Harper (eds.), 1977, Trace fossils 2: Seel House Press, Liverpool, p. 351

Curran, H. A. (ed.), 1985, Biogenic structures: Their use in interpreting depositional environments: Soc. Econ. Paleontologists and Mineralogists Spec. Pub. 35, 347 p.

Dzulnyski, S., and E. K. Walton, 1965, Sedimentary features of flysch and greywackes: Developments in Sedimentology, v. 7, Elsevier, Amsterdam, 274 p.

Ekdale, A. A., R. G. Bromley, and S. G. Pemberton, 1984, Ichnology, trace fossils in sedimentology and stratigraphy: Soc. Econ. Paleontologists and Mineralogists Short Course No. 15, 317 p.

Frey, R. W. (ed.), 1975, The study of trace fossils: Springer-Verlag, New York, 562 p.

Harms, J. C., J. B. Southard, D. R. Spearing, and R. G. Walker, 1975, Depositional environments as interpreted from primary sedimentary structures and stratification sequences: Soc. Econ. Paleontologists and Mineralogists Short Course No. 2, 161 p.

Harms, J. C., J. B. Southard, and R. G. Walker, 1982, Structures and sequences in clastic rocks: Soc. Econ. Paleontologists and Mineralogists Short Course No. 9,

Middleton, G. V. (ed.), 1965, Primary sedimentary structures and their hydrodynamic interpretation: Soc. Econ. Paleontologists and Mineralogists Spec. Pub. 12, 265 p.

Miller, M. F., A. A. Ekdale, and M. D. Picard (eds.), 1984, Trace fossils and paleoenvironments: Marine carbonate, marginal marine terrigenous and continental terrigenous settings: Jour. Paleontology, v. 58, p. 283–597.

Pettijohn, F. J., and P. E. Potter, 1964, Atlas and glossary of primary sedimentary structures: Springer-Verlag, New York, 370 p.

Picard, M. D., and L. R. High, Jr., 1973, Sedimentary structures of ephemeral streams: Elsevier, New York, 223 p.

Potter, P. E., and F. J. Pettijohn, 1977, Paleocurrents and basin analysis, 2nd ed.: Springer-Verlag, New York, 460 p.

Reineck, H. E., and I. B. Singh, 1980, Depositional sedimentary environments, 2nd ed.: Springer-Verlag, New York, 439 p.

Sarjeant, W. A. S. (ed.), 1983, Terrestrial trace fossils: Benchmark Papers in Geology, v. 76, Hutchinson Ross, 415 p.,

PART FOUR
Composition and
Classification of
Sedimentary Rocks

7
Siliciclastic Sedimentary Rocks

7.1 INTRODUCTION

Composition, like sedimentary textures and structures, is a fundamental property of sedimentary rocks. Furthermore, it is the most useful sediment property for the purpose of rock classification. We commonly use the general term **mineralogy** to refer to the identity of all the particles or grains in rocks. These siliciclastic grains may be weathering residues composed of resistant minerals or rock fragments, pyroclastic particles, or secondary minerals such as clay minerals (Chapter 2). Bulk chemistry is another aspect of the overall composition of sedimentary rocks. Chemical composition is directly related to the mineralogy of the rocks, but it has not proven as useful as mineralogy in classifying sedimentary rocks or in interpreting their geologic history. Consequently, geologists have generally been less interested in the chemical composition of siliciclastic sedimentary rocks than in particle composition.

The methods used for determining the mineralogy of siliciclastic sedimentary rocks depends upon the grain size of the rocks. The particle composition of very coarse-grained sedimentary rocks such as conglomerates can be determined by field examination of pebbles with the unaided eye or a hand lens. Rocks composed of sand- and silt-size grains are commonly analyzed by polarizing petrographic microscope study of thin sections of rock specimens. Very fine-grained rocks such as shales may be composed mainly of clay-size grains too small to be studied effectively with a petrographic microscope. These rocks must be studied by X-ray diffraction techniques or by use of a scanning electron microscope. The chemical composition of sedimentary rocks may be determined either by wet-chemical methods or by one of several rapid and accurate physical analytical techniques such as X-ray fluorescence or electron microprobe analysis.

Sedimentary rocks are classified to make written and oral communication easier. Classifications are simply arbitrary schemes for grouping rocks into "pigeonholes" on the basis of some characteristic property or properties of the rocks. Classifications based solely upon observable, measurable, rock properties such as mineral composition, without regard to the genetic significance of these properties, are called **descriptive classifications.** Classifications that provide some insight into depositional environment or sediment source are **genetic classifications.** Although genetic rock classifications are conceptually more appealing than descriptive classifications, it is often difficult or impossible to make the subjective, deductive interpretations necessary to effectively use such classifications. Therefore, descriptive classifications have gained favor in recent years.

Descriptive classification of siliciclastic sedimentary rocks is based fundamentally on mineralogy, although grain size plays a role in classifying some rocks, particularly fine-grained mudrocks. Chemical composition is not used in classifying siliciclastic sedimentary rocks. Classification of sandstones has received far more attention from geologists than classification of conglomerates and mudrocks, primarily because sandstones can be studied much more readily with a petrographic microscope than can either mudrocks or conglomerates. Although mineralogy is the principal basis for classifying sandstones, finding a classification that is suitable for all types of sandstones and acceptable to most geologists has proven to be an elusive goal. In fact, more than fifty different classifications for sandstones have been proposed; none, however, has received widespread acceptance. Classifications that are all-inclusive tend to be too complicated and unwieldly for general use, and classifications that are oversimplified may convey too little useful information.

7.2 PARTICLE COMPOSITION

Terrigenous Siliciclastic Particles

Siliciclastic sedimentary rocks are composed predominantly of terrigenous constituents derived by subaerial weathering processes and terrestrial volcanism. Local submarine volcanism may supply pyroclastic particles within a marine depositional basin. Minor amounts of chemically formed minerals can be added to assemblages of siliciclastic constituents during diagenesis (Chapter 9). These minerals are formed by chemical precipitation in the pore spaces of sediments or by alteration or replacement of existing minerals, a process called **authigenesis.** Authigenic siliciclastic minerals can generally be distinguished from terrigenous siliciclastic minerals by their small size and well-developed crystal faces. Although hundreds of silicate minerals are known, a very small number of mineral varieties, together with rock fragments, make up all siliciclastic sedimentary rocks. The most common siliciclastic minerals and rock fragments are shown in Table 7.1. These particulate constituents are discussed in greater detail in the following paragraphs.

Quartz. Quartz (SiO_2) is the dominant mineral in most siliciclastic sedimentary rocks. Quartz makes up about 65 percent of the average sandstone and 30 percent of the average mudrock (Blatt et al., 1980). Fortunately, it is a comparatively easy mineral to identify, both megascopically in hand specimens and by petrographic examination in

TABLE 7.1   Common minerals and rock fragments in siliciclastic sedimentary rocks

MAJOR MINERALS (abundance >~1%)
   Stable minerals (greatest resistance to chemical decomposition)
      Quartz—makes up approximately 65% of average sandstone, 30% of average shale, 5%
         of average carbonate rock
      Chert—makes up about 1–4% of siliciclastic sedimentary rocks
   Less stable minerals
      Feldspars—include K-feldspars: orthoclase, microcline, sanidine, anorthoclase, and pla-
         gioclase feldspars: albite, oligoclase, andesine, laboradorite, bytonite, anorthite; make
         up about 10–15% of average sandstone, 5% of average shale, <1% of average carbon-
         ate rock.
      Clay minerals and fine micas—clay minerals include the kaolinite group, illite group,
         smectite group (montmorillonite a principal variety), and chlorite group; fine micas
         are principally muscovite (sericite) and biotite; make up approximately 25–35% of to-
         tal siliciclastic minerals, but may comprise >60% of the minerals in shales

ACCESSORY MINERALS (abundance <~1%)
   Coarse micas—principally muscovite and biotite
   Heavy minerals (specific gravity >~2.9)
      Stable nonopaque minerals—zircon, tourmaline, rutile
      Metastable nonopaque minerals—amphiboles, pyroxenes, chlorite, garnet, apatite, stau-
         rolite, epidote, olivine, sphene, zoisite, clinozoisite, topaz, monazite, plus about 100
         others of minor importance volumetrically
      Stable opaque minerals—hematite, limonite
      Metastable opaque minerals—magnetite, ilmenite, leucoxene

ROCK FRAGMENTS (make up about 10–15% of the siliciclastic grains in average sandstone,
   and most of the nonmatrix grains in conglomerates; shales contain few rock fragments)
   Igneous rock fragments—may include clasts of any igneous rock, but fine-grained frag-
      ments of volcanic rocks and volcanic glass are most common in sandstones
   Metamorphic rock fragments—include metaquartzite, slate, phyllite, schist, and less com-
      monly gneiss clasts
   Sedimentary rock fragments—any type of sedimentary rock fragment possible in conglom-
      erates; clasts of fine sandstone, siltstone, shale, and chert are most common in sand-
      stones; limestone clasts occur rarely in sandstones

CHEMICAL CEMENTS (abundance variable)
   Silicate minerals—predominantly quartz; others may include chert, opal, feldspars, and
      zeolites
   Carbonate minerals—principally calcite; less commonly aragonite, dolomite, siderite
   Iron oxide minerals—hematite, limonite
   Sulfate minerals—anhydrite, gypsum, barite

thin sections. It can, however, be confused with feldspars. Because of its superior hardness and chemical stability, quartz can survive multiple recycling. Therefore, the quartz grains in many sandstones display some degree of rounding acquired by abrasion during one or more episodes of transport, particularly eolian transport.

Quartz can occur as single grains or as polycrystalline, composite grains (Fig. 7.1). Composite quartz grains that consist of exceedingly small crystals, referred to as microcrystalline quartz, are called **chert.** When examined under crossed polarizing prisms with a petrographic microscope, quartz grains commonly display sweeping patterns of extinction as the stage is rotated. This property is called **undulatory extinction.** Some authors (Folk, 1974; Basu et al., 1975) believe that the properties of polycrystallinity and undulatory extinction can be used to distinguish quartz derived from different sources—igneous vs. metamorphic, for example. Other workers (e.g., Blatt, 1967) have questioned the validity of such provenance determinations.

Feldspars. Feldspars are less abundant than quartz in most siliciclastic sedimentary rocks, but they are the second most abundant mineral in siltstones and sandstones. Feldspars make up about 10–15 percent of the constituents of average sandstones and about 5 percent of average mudrocks (Blatt et al., 1980). They belong to a group of minerals having similar atomic structures and properties, but several different varieties of feldspars are recognized on the basis of differences in chemical composition and optical properties (Fig. 7.2). They are divided into two broad groups: alkali feldspars and plagioclase feldspars.

Alkali feldspars constitute a group of minerals in which chemical composition can range through a complete solid solution series from $K(AlSi_3O_8)$ through $(K,Na)(AlSi_3O_8)$ to $Na(AlSi_3O_8)$. Because potassium-rich feldspars are such common members of this group, it has become widespread practice to call the alkali feldspars **potassium feldspars.** Common members of the potassium feldspar group include orthoclase, microcline, and sanidine. **Plagioclase feldspars** form a complex solid solution series ranging in composition from $NaAlSi_3O_8$ (albite) through $CaAl_2Si_2O_8$ (anorthite). A general formula for the series is $(Na,Ca)(Al,Si)Si_2O_8$. Potassium feldspars and plagioclase feldspars can usually be distinguished on the basis of optical properties by examination with a petrographic microscope, but may be difficult or impossible to differentiate megascopically. Potassium feldspars are generally considered to be somewhat more abundant overall in sedimentary rocks than plagioclase feldspars; however, plagioclase is more abundant in siliciclastic sedimentary rocks derived from volcanic sources.

Feldspars are chemically less stable than quartz and are more susceptible to chemical destruction during weathering and diagenesis. Because they are also softer than quartz, feldspars are more readily rounded during transport. They also appear to be somewhat more prone to mechanical shattering and breakup. Feldspars are less likely than quartz to survive several episodes of recycling, although they can survive

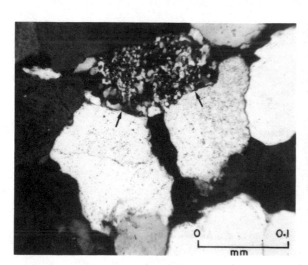

FIGURE 7.1   Photomicrograph of a polycrystalline quartz grain (arrow) surrounded by monocrystalline quartz grains. Polarized light.

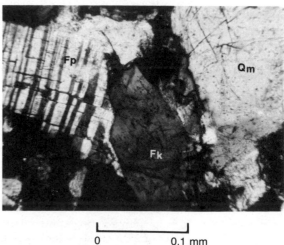

FIGURE 7.2   Plagioclase feldspar grain (Fp), potassium feldspar grain (Fk), and monocrystalline quartz (Qm). Polarized light.

more than one cycle if weathering occurs in a moderately arid climate. Owing to this possibility for recycling, the presence of a few feldspar grains in a sedimentary rock does not necessarily mean that the rock is composed of first-cycle sediments derived directly from crystalline igneous or metamorphic rocks. On the other hand, a high content of feldspars, particularly on the order of 25 percent or more, probably indicates derivation directly from crystalline source rocks.

Clay Minerals. Clay minerals make up about 25–35 percent of the constituents of siliciclastic sedimentary rocks as a whole, but they may compose more than 60 percent of the minerals in mudrocks. Because of their small size, clay minerals cannot be identified by routine petrographic microscopy. They must be identified by X-ray diffraction techniques, electron microscopy (Fig. 7.3), or other nonoptical methods. Clay minerals are compositionally diverse. They belong to the phyllosilicate mineral group, which is characterized by two-dimensional layer structures arranged in indefinitely extending sheets.

The most common clay mineral groups are **illite** $[K_2(Si_6Al_2)Al_4O_{20}(OH)_4]$, **smectite** (montmorillonite) $[(Al,Mg)_8(Si_4O_{10})_3(OH)_{10}\cdot12H_2O]$, **kaolinite** $[Al_2Si_2O_5(OH)_4]$, and **chlorite** $[(Mg,Fe)_5(Al,Fe^{3+})_2Si_3O_{10}(OH)_8]$. Kaolinite is a two-layer clay; the others are three-layer clays. Smectite is a clay mineral group, with montmorillonite a principal variety of smectite. Clay minerals form principally as secondary minerals during subaerial weathering and hydrolysis, although they can form also by subaqueous weathering in the marine environment and during burial diagenesis and metamorphism. The types of clay minerals generated during weathering are a function of weathering conditions and composition of the parent rock. For example, kaolinite tends to form under very severe chemical leaching conditions that cause removal of all soluble cations

FIGURE 7.3 Electronmicrograph of kaolinite clay minerals magnified 4700×. Note the arrangement of the pseudohexagonal kaolinite plates in distinct "books." (Photograph by M.B. Shaffer.)

1   10 μ

except silicon and aluminum. Illite forms under less severe leaching conditions in a potassium-rich environment, especially by weathering of potassium feldspar-rich rocks such as granites. Montmorillonite clays form in magnesium-rich environments under moderate leaching conditions and are common alteration or weathering products of volcanic rocks containing plagioclase feldspars. Because of these genetic affinities, many attempts have been made by geologists to use clay minerals in ancient sedimentary rocks as indicators of ancient weathering conditions (paleoclimates) and provenance (source). These efforts have been largely unsuccessful, owing to the fact that one type of clay mineral can alter to another after deposition, as a result of either submarine weathering on the seafloor or diagenetic alteration during burial.

Accessory Minerals. Minerals that have an average abundance in sedimentary rocks of less than about one percent are called accessory minerals. These minerals include the common **micas,** muscovite and biotite, and a large number of so-called **heavy minerals.**

The average abundance of coarse micas in siliciclastic sedimentary rocks is less than about 0.5 percent, although some sandstones may contain 2 to 3 percent coarse micas (Fig. 7.4). Muscovite is the most abundant coarse mica in sedimentary rocks. Fine-size muscovite and biotite are commonly present in mudrocks in considerably higher concentrations than accessory amounts, but few data are available on the actual concentrations of fine micas in siliciclastic sedimentary rocks.

Minerals that have a specific gravity greater than about 2.85–2.90 are called heavy minerals. These minerals include both chemically stable and less stable, or labile, varieties, as shown in Table 7.1. Stable heavy minerals such as zircon and rutile can survive multiple recycling and are commonly rounded, indicating a recycled, sedimentary source. Less stable minerals such as magnetite, pyroxenes, and amphiboles are less likely to survive recycling and thus are commonly first-cycle sediments that reflect the composition of proximate source rocks. Heavy minerals, therefore, are useful indicators of sediment source rocks because different types of source rocks yield different suites of heavy minerals. A few heavy minerals such as pyrite form postdepositionally during diagenesis and thus have no significance as source-rock indicators.

Because of their low abundance in siliciclastic rocks, heavy minerals are commonly concentrated for study by separating them from the light-mineral fraction by use of heavy liquids such as bromoform (specific gravity 2.89). In this separation process, disaggregated sediment is stirred into a heavy liquid contained in a funnel. The light minerals float on the surface of the heavy liquid, but the heavy minerals gradually sink into the stem of the funnel where they can be drawn off and separated from the light fraction. These heavy-mineral concentrates can then be mounted on a glass microscope slide (Fig. 7.5) and studied with a petrographic microscope.

Rock Fragments. Most of the framework grains in conglomerates are rock fragments consisting of rounded pebbles or cobbles. Rock fragments also make up about 10–15 percent of the grains in average sandstones, although the rock-fragment content of sandstones is highly variable, ranging from zero to more than 95 percent. On the other hand, most mudrocks contain few rock fragments other than occasional isolated pebbles.

The most common pebble-size fragments in ancient conglomerates are clasts of quartzite and chert. These clasts are both mechanically durable and chemically stable and are thus able to survive multiple recycling. Pebbles of easily erodable rocks such as shale or limestone cannot survive intensive fluvial transport; therefore, the presence of low-durability pebbles in ancient conglomerates suggests first-cycle sediments that have undergone only moderate fluvial transport.

Fragments of any kind of igneous, metamorphic, or sedimentary rock can occur in sandstones (Fig. 7.6); however, clasts of fine-grained source rocks are most likely to be preserved as sand-size fragments (Boggs, 1968). Coarse-grained source rocks such as granites do not yield fine sand-size clasts. The most common rock fragments in sandstones are clasts of volcanic rocks, volcanic glass (in younger rocks), and fine-grained metamorphic rocks such as slate, phyllite, and schist. Sand-size fragments of silica-cemented siltstone, very fine-grained sandstone, and shale are less common.

Rock fragments are particularly important in studies of sediment source rocks. They are commonly easily identified and are more reliable indicators of source rock types than individual minerals such as quartz or feldspar, which can be derived from different types of source rocks.

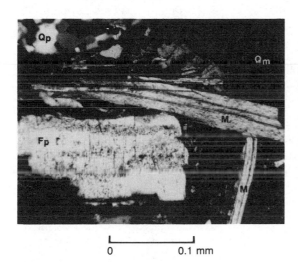

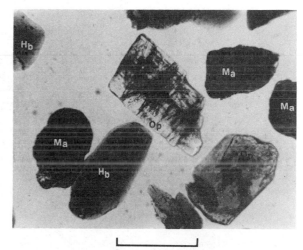

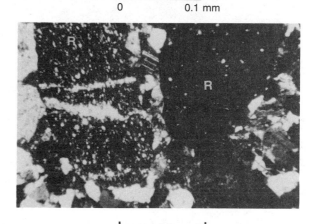

FIGURE 7.4 (above left) Photomicrograph showing a coarse muscovite mica grain (M). Polycrystalline quartz (Qp), monocrystalline quartz (Qm), and plagioclase feldspar (Fp) grains are also visible. Polarized light.

FIGURE 7.5 (above right) Heavy mineral grains from a modern shelf sand as viewed in a grain mount. The minerals shown here include orthopyroxene (Op), hornblende (Hb), and magnetite (Ma).

FIGURE 7.6 (right) Two large volcanic(?) rock fragments (R) in a fine-grained lithic sandstone. Polarized light.

Cements

The framework grains in most siliciclastic sedimentary rocks are bound together by some type of authigenic mineral cement. These cementing materials may be either silicate minerals such as quartz and chert or nonsilicate minerals such as calcite and dolomite. Quartz is the most common silicate mineral that acts as a cement. In most sandstones, the quartz cement is chemically attached to the crystal lattice of existing quartz grains, forming rims of cement called **overgrowths** (Fig. 7.7). Overgrowths that retain crystallographic continuity of a grain are said to be **syntaxial.** Because syntaxial overgrowths are optically continuous with the original grain, they go to extinction in the same position as the original grain when rotated on the stage of a polarizing microscope. Overgrowths can be recognized by a line of impurities or bubbles that mark the surface of the original grain. Quartz overgrowths are particularly common in quartz-rich sandstones. Less commonly, chert acts as a cement in sandstones. When silica cement is deposited as chert, it forms a mosaic of very tiny quartz crystals that fill the interstitial spaces among framework silicate grains. Not uncommonly, the crystals next to the framework grains are slightly elongated and are oriented normal to the surfaces of the framework grain (Fig. 7.8). More rarely, opal occurs as a cement in sandstones, particularly in sandstones rich in volcanogenic materials. Like quartz and chert, opal is composed of SiO_2, but unlike these minerals opal contains some water and lacks a definite crystal structure. Thus, it is said to be amorphous. Opal is metastable and crystallizes in time to quartz or chert.

Carbonate minerals are the most abundant nonsilicate mineral cements in siliciclastic sedimentary rocks. Calcite is a particularly common carbonate cement. It is precipitated in the pore spaces among framework grains, typically forming a mosaic of smaller crystals (Fig. 7.9). These crystals adhere to the larger framework grains and bind them together. Less common carbonate cements include dolomite, siderite (iron carbonate), and, in younger rocks, aragonite.

Other minerals that act as cements include the iron oxide minerals hematite and limonite, feldspars, anhydrite, gypsum, barite, clay minerals, and zeolite minerals. Zeolites are hydrous aluminosilicate minerals that occur as cements primarily in volcaniclastic sedimentary rocks (discussed in a succeeding section).

7.3 CHEMICAL COMPOSITION

Sedimentologists, in sharp contrast to geologists who study igneous and metamorphic rocks, have traditionally placed relatively little emphasis on study of the chemical composition of siliciclastic sedimentary rocks. This lack of interest is probably due mainly to the common belief that chemical composition is less useful than mineral composition for interpreting depositional history and provenance. Also, the present chemical composition of sedimentary rocks may not accurately reflect their composition at the time of deposition because crystallization of new minerals during sediment burial and diagenesis can change the original chemical composition. The formerly high cost of doing chemical analyses is an additional factor that has helped to discourage extensive chemical studies of sedimentary rocks. Several new tools such as the electron microprobe and X-ray fluorescence equipment now make it possible to perform rapid and comparatively inexpensive chemical analyses of rock composition.

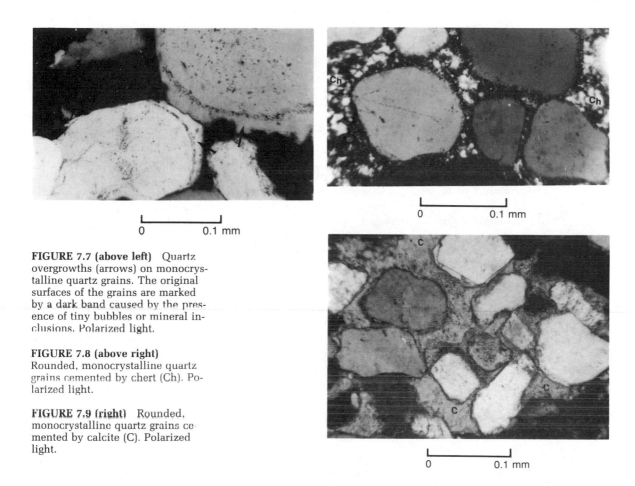

FIGURE 7.7 (above left) Quartz overgrowths (arrows) on monocrystalline quartz grains. The original surfaces of the grains are marked by a dark band caused by the presence of tiny bubbles or mineral inclusions. Polarized light.

FIGURE 7.8 (above right) Rounded, monocrystalline quartz grains cemented by chert (Ch). Polarized light.

FIGURE 7.9 (right) Rounded, monocrystalline quartz grains cemented by calcite (C). Polarized light.

These new tools, as well as changing attitudes regarding the significance of chemical composition, are causing sedimentologists to develop a deeper interest in the chemistry of sedimentary rocks. Published studies involving chemical composition of sedimentary rocks are steadily growing in number.

Because most grains in siliciclastic sedimentary rocks are derived from various types of igneous, metamorphic, and sedimentary rocks, the mineralogy and chemical composition of siliciclastic rocks is clearly a function of parent-rock composition. Nonetheless, sedimentary rocks display distinct chemical differences from parent source rocks owing to chemical changes that occur during weathering and diagenesis. For example, they tend to be enriched in silica and depleted in iron, magnesium, calcium, sodium, and potassium compared to the parent rocks. This enrichment in silica occurs because siliceous minerals are resistant to chemical weathering; thus, silica is concentrated in weathering residues with respect to more soluble cations. Also, the superior chemical stability of SiO_2 minerals such as quartz and chert causes sedimentary rocks to become progressively enriched in these minerals during multiple recycling. Thus, overall silica content increases at the expense of less stable iron- and magnesium-rich minerals.

The average chemical composition of principal classes of sandstones is shown in Table 7.2. I stress that this table shows *average composition*. Specimens of sandstone from different formations may have chemical compositions that deviate considerably from these average values. Note that chemical composition differs markedly in mineralogically different sandstones, reflecting the close relationship between mineral composition and chemical composition. Owing to the abundance of quartz in most sandstones, silicon, expressed as SiO_2, is the most abundant chemical constituent in all types of sandstones. Aluminum (as Al_2O_3) is moderately abundant in sandstones containing feldspars or in lithic arenites that contain a matrix of clay minerals. It is much less abundant in quartz-rich sandstones, which commonly do not have a clay matrix. On the average, iron, magnesium, calcium, sodium, and potassium are all less abundant in sandstones than is aluminum. Relative concentrations of these elements vary as a function of the mineralogy of the sand-size grains and the types of matrix clay minerals and diagenetic cements in the rock. For example, sandstones with abundant calcium carbonate cement may have anomalously high calcium content.

Table 7.3 shows the chemical composition of average shale or mudrock. The SiO_2 content of shales is much lower and the Al_2O_3 content much higher than those of sandstones. These differences reflect the higher content of aluminum-bearing clay minerals in shales. Iron, calcium, sodium, and potassium also tend to be more abundant in shales than in sandstones because of the greater abundance in shales of clay minerals and fine-size, iron-bearing minerals such as biotite.

Tables 7.2 and 7.3 suggest that both mineral and chemical composition of siliciclastic sedimentary rocks are functions of grain size of the sediment. Quartz, a primary source of silica, is particularly abundant in sandstones. By contrast, clay minerals, which are the principal sources of aluminum, potassium, and calcium, are much more abundant in mudrocks. Thus, chemical constituents tend to show either progres-

TABLE 7.2 Average chemical composition of principal classes of sandstones

Constituent	Quartz arenite	Lithic arenite	Graywacke	Feldspathic arenite
SiO_2	95.4	66.1	66.7	77.1
TiO_2	0.2	0.3	0.6	0.3
Al_2O_3	1.1	8.1	13.5	8.7
Fe_2O_3	0.4	3.8	1.6	1.5
FeO	0.2	1.4	3.5	0.7
MnO	—	0.1	0.1	0.2
MgO	0.1	2.4	2.1	0.5
CaO	1.6	6.2	2.5	2.7
Na_2O	0.1	0.9	2.9	1.5
K_2O	0.2	1.3	2.0	2.8
H_2O^+	0.3	3.6	2.4	0.9
H_2O^-	—	0.7	0.6	—
P_2O_5	—	0.1	0.2	0.1
CO_2	1.1	5.0	1.2	3.0
SO_3	—	—	0.3	—
S	—	—	0.1	—
C	—	—	0.1	—
Total	100.7	100.0	100.4	100.0

Source: Pettijohn, F. J., 1975, Sedimentary rocks, 3rd ed. Table 7.3, p. 210, reprinted by permission of Harper & Row, Publishers, New York.

TABLE 7.3 Chemical composition of average shales and related mudrocks

Constituent	A	B	C	D	E	F
SiO_2	58.10	55.43	60.15	60.64	56.30	69.96
TiO_2	0.65	0.46	0.76	0.73	0.77	0.59
Al_2O_3	15.40	13.84	16.45	17.32	17.24	10.52
Fe_2O_3	4.02	4.00	4.04	2.25	3.83	} 3.47
FeO	2.45	1.74	2.90	3.66	5.09	
MnO	—	trace	trace	—	0.10	0.06
MgO	2.44	2.67	2.32	2.60	2.54	1.41
CaO	3.11	5.96	1.41	1.54	1.00	2.17
Na_2O	1.30	1.80	1.01	1.19	1.23	1.51
K_2O	3.24	2.67	3.60	3.69	3.79	2.30
H_2O^+	} 5.00	{ 3.45	3.82	3.51	3.31	1.96
H_2O^-		{ 2.11	0.89	0.62	0.38	3.78
P_2O_5	0.17	0.20	0.15	—	0.14	0.18
CO_2	2.63	4.62	1.46	1.47	0.84	1.40
SO_3	0.64	0.78	0.58	—	0.28	0.03
Cl	—	—	—	—	—	0.30
Organic	0.80[a]	0.69[a]	0.88[a]	—	1.18[a]	0.66
Misc.	—	0.06[c]	0.04[c]	0.38[b]	1.98[b]	0.32
Total	99.95	100.48	100.46	99.60	100.00	100.62

[a]Carbon.

[b]FeS_2

[c]BaO

A. Average shale (Clarke, 1924, p. 24). Based on cols. B and C

B. Composite sample of 27 Mesozoic and Cenozoic shales. H. N. Stokes, analyst (Clarke, 1924, p. 552)

C. Composite sample of 51 Paleozoic shales. H. N. Stokes, analyst (Clarke, 1924, p. 552).

D. Unweighted average of 30 analyses of slate (29 Paleozoic, 1 Mesozoic, 0 early Paleozoic or Precambrian) (Eckel, 1904)

E. Unweighted average of 33 analyses of Precambrian slates (Nanz, 1953)

F. Composite analysis of 235 samples of Mississippi delta. G. Steiger, analyst (Clarke, 1924, p. 509)

Source: Pettijohn, F. J., 1975, Sedimentary rocks, 3rd ed. Table 8.7, p. 274, reprinted by permission of Harper & Row, New York.

sive enrichment or progressive depletion as grain size decreases from sand or silt through coarse clay and fine clay (Table 7.4).

7.4 CLASSIFICATION OF SILICICLASTIC SEDIMENTARY ROCKS

Textural Nomenclature of Mixed Sediments

Unconsolidated siliciclastic sediments are classified as gravel, sand, or mud, depending upon grain size. The lithified rock equivalents of these sediments are conglomerates, sandstones, and mudrocks. Because many siliciclastic sedimentary rocks are composed of grains of mixed sizes, it is not always easy to decide which textural name should be applied to a particular deposit. It might be difficult to decide, for example, whether a sedimentary rock composed of nearly equal portions of mud-size and sand-

TABLE 7.4 Relation of chemical composition of terrigenous sedimentary rocks to grain size of the sediment

Constituent	Fine sand	Silt	Coarse clay	Fine clay
SiO_2	71.15	61.29	48.07	40.61
TiO_2	0.50	0.85	0.89	0.79
Al_2O_3	10.16	13.30	18.83	10.97
Iron oxides	3.72	3.94	6.91	7.42
MgO	1.66	3.31	3.56	3.19
CaO	3.65	5.11	4.96	6.24
Na_2O	0.86	1.32	1.17	1.19
K_2O	2.20	2.33	2.57	2.62
Ignition loss	5.08	7.05	10.91	12.51

Note: Based on average of 12 clays: 1 residual clay, 1 Ordovician shale, 2 Cretaceous clays, and the remainder (8) of glacial or Recent origin. "Fine clay" is under 1 micron; "coarse clay" is 1 to 5 microns; and "silt" is 5 to 50 microns.

Source: Grout, F. F., 1925, Clays and shales of Minnesota: Geol. Soc. America Bull., v. 36, p. 393–416.

size particles should be called a mudrock or a sandstone. Various schemes of classification have been devised for naming texturally mixed sediments, most of which make use of triangular texture diagrams such as those shown in Figure 7.10.

The textural classification illustrated in Figure 7.10A includes particles ranging in size from mud (clay and fine silt) to gravel. Note that the textural boundaries in this classification scheme are not entirely symmetrical. Ideally, we might expect the boundary between gravel and mud–sand to be set at 50 percent; however, this is not always done, as Figure 7.10A shows. Because particles of gravel size are commonly less abundant than sand and mud particles, many geologists consider a sediment with as little as 30 percent gravel-size fragments to be a gravel. If sediments contain only particles of sand size and smaller, classification schemes such as those shown in Figure 7.10B or 7.10C—using sand, silt, and clay as end members of the classification—are more appropriate. Once the textural nomenclature of siliciclastic sedimentary rocks has been established, further classification within each textural group can be made on the basis of composition.

Classification of Sandstones

Sandstones make up about one-fourth of the total volume of sedimentary rocks in the geologic record. They are an extremely interesting group of rocks because the diverse textures and structures preserved in these rocks provide a wealth of information about their transport and depositional history. Their compositions and volumes also provide clues to the compositions and volumes of ancient, vanished source areas. Because the grains in sandstones are large enough to be viewed with a hand lens as well as a microscope, sandstones can be studied in both the field and the laboratory. Thus, they have received more study and research from geologists than any other type of sedimentary rock (except perhaps limestones). It is natural, therefore, that a great deal of attention has been directed toward classifying sandstones.

Most sandstones are made up of mixtures of a very small number of dominant components. Quartz plus chert, feldspars, rock fragments, and fine-grained matrix materials are the only constituents that are commonly abundant enough to be important in sandstone classification. **Matrix** consists of clay minerals and other minerals less

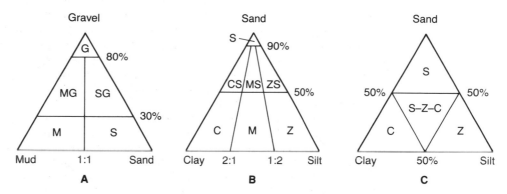

FIGURE 7.10 Nomenclature of mixed sediments. A, B. Simplified from Folk. C. After Robinson. G = gravel, S = sand, M = mud, C = clay, Z = silt, MG = muddy gravel, SG = sandy gravel, CS = clayey sand, MS = muddy sand, ZS = silty sand. (From Folk, R. L., 1954, The distinction between grain size and mineral composition in sedimentary rock nomenclature: Jour. Geology, v. 62. Fig. 1a, p. 346, and 1b, p. 349, reprinted by permission of University of Chicago Press. Robinson, G. W., 1949, Soils, their origin, constitution, and classification, 3rd ed.: Murby, London.)

than about 0.03 mm in size that fill interstitial space among sand-size grains. In spite of the very simple composition of sandstones, geologists have not been able to agree on a single acceptable sandstone classification. As mentioned in the introduction, more than fifty classifications for sandstones have been published since about 1950 (Friedman and Sanders, 1978).

Published classifications range from those that have a strong genetic orientation to those based strictly on observable, descriptive properties of sandstones. Most authors of sandstone classifications use a classification scheme that involves a QFR plot. These plots are triangular diagrams on which quartz (Q), feldspars, (F), and rock fragments (R) are plotted as end members at the poles of the classification triangle. There are numerous possible ways that such a QFR triangle can be subdivided into classification fields, and geologists have explored the full range of these possibilities. The result has been a proliferation of classifications and considerable confusion owing to overlapping and conflicting usage of names. Different classification schemes may use the same names for rocks of different composition and different names for rocks of the same composition.

Part of the problem of adopting a single sandstone classification has to do with the wide variation in relative abundance of QFR constituents in different suites of sandstones. A geologist who works with sandstones composed predominantly of quartz, minor feldspar, and essentially no rock fragments and matrix is likely to feel differently about sandstone classification than a geologist who works with sandstones containing abundant rock fragments and clay matrix, but little quartz. Also, certain types of grains create troublesome classification problems because they require head-scratching, subjective decisions about their appropriate end-member group. For example, should chert—a sedimentary rock fragment—and quartzite—a metamorphic rock fragment—be included with other rock fragments in the rock fragment pole or should they be included with quartz in the quartz pole because their superior hardness and chemical stability are similar to those of quartz? Different classifiers have handled this problem in different ways.

Matrix creates special problems in classification because matrix is a textural (grain-size) property, whereas quartz, feldspars, and rock fragments are compositional properties. The most common practice in classification has been to divide sandstones into two groups on the basis of matrix content: matrix-poor sandstones and matrix-rich sandstones. Little uniformity has prevailed, however, regarding what defines matrix poor and matrix rich. The suggested boundary between these two groups has ranged from as little as 5 percent to as much as 15 percent matrix. Some authors do not use matrix at all in their classification but instead use a separate **textural maturity** classification (Fig. 7.11). Matrix is troublesome also because there is now doubt about the primary depositional origin of some matrix. Early geologists assumed that matrix was deposited at the same time as sand-size grains and accepted its detrital origin without question. Recent work has suggested that much of the matrix in sandstones may be of secondary origin, produced during diagenesis largely by alteration of feldspars (Chapter 9). If this is so, the inclusion of matrix in sandstone classifications that are based on primary depositional properties is inappropriate.

Although geologists have not yet agreed on a single sandstone classification, there has been a tendency in recently published classifications to standardize the QFR members. Thus quartz, chert, and quartzite fragments are commonly included in the Q pole and total feldspars in the F pole. Unstable (labile) rock fragments such as shale,

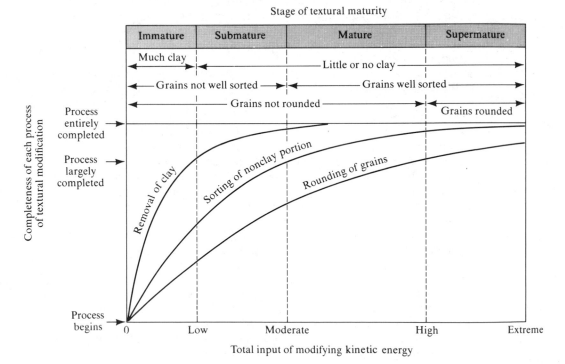

FIGURE 7.11 Textural maturity classification of Folk. Textural maturity of sands is shown as a function of input of kinetic energy. (From Folk, R. L., 1951, Stages of textural maturity in sedimentary rocks: Jour. Sedimentary Petrology, v. 21. Fig. 1, p. 128, reprinted by permission of Society of Economic Paleontologists and Mineralogists, Tulsa, Okla.)

phyllite, and basalt that do not easily withstand recycling are included in the R pole. The order of stability of grains during recycling is quartz>feldspars>unstable rock fragments, where quartz is much more stable than either feldspars or rock fragments. Use of these end members in classification thus constitutes a compositional **maturity index.** Sandstones with compositions that fall near the Q pole of a classification diagram have greater recycling potential, or have been recycled more times, than those with compositions that fall near the F or R poles.

One of the simplest of these maturity index classifications is the classification of Gilbert (H. Williams et al., 1982), shown in Figure 7.12. In this classification, sandstones that are effectively free of matrix are classified as either **quartz arenites, feldspathic arenites,** or **lithic arenites** depending upon the relative abundance of QFR constituents. If matrix can be recognized, the terms **quartz wacke, feldspathic wacke,** and **lithic wacke** are used instead. I recommend this classification for its simplicity and its utility for field study as well as for more detailed petrographic studies in the laboratory. This classification is adequate for naming most sandstones, but readers

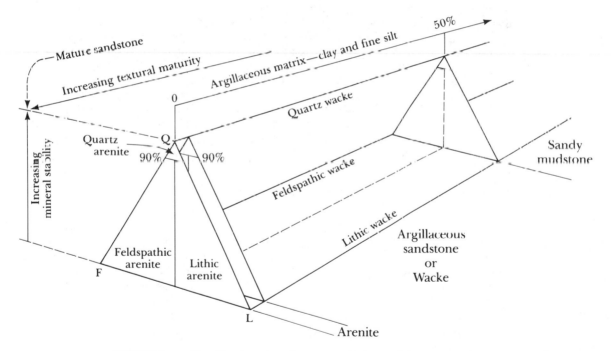

FIGURE 7.12 Classification of sandstones on the basis of three mineral components: Q = quartz, chert, quartzite fragments; F = feldspars; L = unstable, lithic grains (rock fragments). Points within the triangles represent relative proportions of Q, F, R end members. Percentage of argillaceous matrix is represented by a vector extending toward the rear of the diagram. The term arenite is restricted to sandstones essentially free of matrix; sandstones containing matrix are wackes. (From Williams, H., F. J. Turner, and C. M. Gilbert, Petrography, an introduction to the study of rocks in thin sections, 2nd ed. W. H. Freeman and Company, San Francisco. © 1982, Fig. 13.1, p. 327. Modified from Dott, R. H., Jr., 1964, Wacke, graywacke, and matrix—what approach to immature sandstone classification?: Jour. Sedimentary Petrology, v. 34. Fig. 3, p. 629, reprinted by permission of Society of Economic Paleontologists and Mineralogists, Tulsa, Okla.)

should be aware of two other names for sandstones that have been much used in the past. Feldspathic arenites particularly rich in potassium feldspars ($>\sim$25%) are commonly called **arkoses** or arkosic sandstones. Matrix-rich sandstones of any composition that have undergone deep burial, that commonly have a chloritic matrix, and that are dark gray to dark green, very hard, and dense have long been called **graywackes.** This term has been much misused and its continued use is controversial. Some geologists think that the term should be abandoned entirely and that we should substitute the word wacke for graywacke. In any case, the term is best restricted to field use.

Sandstone classifications somewhat more complex than Gilbert's and that have been rather widely used by American geologists include the classifications of McBride (1963) and Folk (1970), shown in Figure 7.13. Note the greater number of classification fields in these diagrams compared to Gilbert's and thus the increased number of sandstone names. Folk's classification is particularly interesting because it includes "daughter" triangles at the F and L poles. These daughter triangles allow arkoses to be further subdivided on the basis of relative abundance of different types of feldspars. Similarly, litharenites can be subdivided on the basis of relative abundance of igneous, metamorphic, and sedimentary rock fragments. Note also in Folk's classification that chert is included with rock fragments at the L pole (or R pole) rather than with quartz at the Q pole. These classification schemes are not very practical for classifying sandstones in the field. They are used particularly for classification of sandstones by detailed petrographic study in the laboratory.

Boggs (1967b) proposed the use of a numerical classification for sandstones to avoid the problems of conflicting nomenclature arising from the many sandstone classifications now in use; however, numerical classifications lack the appeal of verbal classifications for general communication. Reviews and discussions of other sandstone classifications and the problems of sandstone classification in general are given by Klein (1963), Boggs (1967b), Okada (1971), and Yanov (1978).

Characteristics of Major Classes of Sandstones

Quartz Arenites

Composition. Quartz arenites are white to cream, or occasionally red, sandstones that commonly contain more than 90–95 percent quartz, including minor chert and quartzite fragments. These sandstones have been referred to in the past as orthoquartzites. They should not be confused with quartzites, which are metamorphosed sandstones that have undergone extensive cementation by silica. Quartz arenites may contain also a small amount of the potassium feldspars microcline and orthoclase or sodium-rich plagioclase (albite), muscovite mica, and very minor amounts of ultrastable heavy minerals such as well-rounded zircon, tourmaline, and rutile. The quartz grains in quartz arenites tend to be monocrystalline, with low undulatory extinction, indicating a high proportion of unstrained quartz grains. The predominance of single-crystal, unstrained quartz grains is probably due to selective destruction of less stable polycrystalline quartz and highly strained monocrystalline quartz grains during transport and recycling.

Texture. Quartz arenites tend to be well sorted and have little or no matrix, owing both to depositional winnowing of fine mud and the fact that few minerals that can

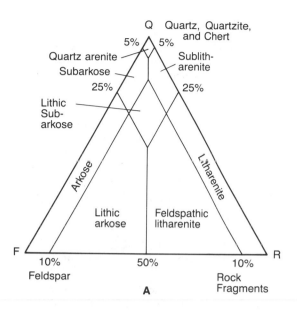

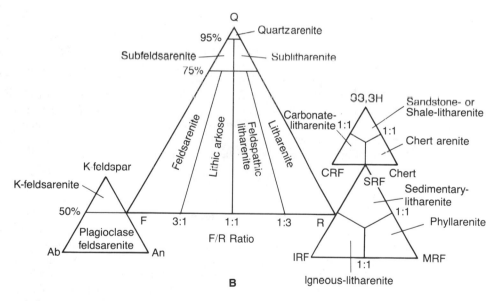

FIGURE 7.13 Classification of sandstones according to (A) McBride and (B) Folk. In Folk's classification, chert is included with rock fragments at the R pole, and granite and gneiss fragments are included with feldspars at the F pole. SS = sandstone, SH = shale, CRF = carbonate rock fragments, SRF = sedimentary rock fragments, IRF = igneous rock fragments, MRF = metamorphic rock fragments. (A, from McBride, E. F., 1963, A classification of common sandstones: Jour. Sedimentary Petrology, v. 34. Fig. 1, p. 667, reprinted by permission of Society of Economic Paleontologists and Mineralogists, Tulsa, Okla. B, from Folk, R. L., P. B. Andrews, and D. W. Lewis, 1970, Detrital sedimentary rock classification and nomenclature for use in New Zealand: New Zealand Jour. of Geology and Geophysics, v. 13. Fig. 8, p. 955, and Fig. 9, p. 959, British Crown copyright, reprinted by permission.)

alter diagenetically to matrix are present in these sandstones. Because quartz-rich sandstones rarely have matrix, quartz wackes are relatively rare rocks. The quartz grains in quartz arenites are commonly well rounded and may have a "frosted" or matte surface texture. Bimodal size distribution is common, particularly in sands of eolian origin. Selective removal of the fine-sand fraction (0.1–0.3 mm) by eolian processes leaves behind both the coarser and finer sizes.

Cements. The grains in quartz arenites are commonly cemented with silica (SiO_2) or calcium carbonate ($CaCO_3$). Silica cements are mainly quartz, which is added to existing quartz grains in the form of overgrowths, or rims. Less commonly, silica cement takes the form of chert or opal. Calcium carbonate cement is typically calcite rather than aragonite. More rarely dolomite forms a cement. The ferric oxide minerals hematite and goethite can also act as cements, either alone or in conjunction with other cements.

Chemical Composition. The chemical composition of quartz arenites is dominated by SiO_2 (Table 7.2), reflecting the high content of the SiO_2 minerals quartz and chert in these sandstones. Calcium derived from calcium carbonate cements can be present in amounts ranging up to about 10 percent. Minor aluminum and magnesium are contained in matrix clay minerals or feldspars, and iron may be present as hematite coatings or cements.

Origin. Quartz arenites can originate as first-cycle deposits derived from primary crystalline or metamorphic rocks, but they are more likely to be the product of multiple recycling of quartz grains from sedimentary source rocks. If they are first-cycle deposits, quartz arenites are probably of eolian origin. A single episode of fluvial transport of first-cycle sediment cannot account for the high degree of rounding and sorting typical of these sandstones. Reworking by surf processes on beaches may possibly produce first-cycle quartz arenites; however, origin by this mechanism has not yet been proven. Most quartz arenites are probably polycyclic; their history has likely included at least one episode of eolian transport, although not necessarily during the last depositional cycle.

Examples. Quartz arenites are very common rocks in the geologic record. They make up one-third or more of all sandstones and are most common in Mesozoic and Paleozoic stratigraphic sequences. Some well-known examples of quartz arenites in North America include the Ordovician St. Peter Sandstone in midcontinent United States, the Jurassic Navajo Sandstone of the Colorado Plateau, the Ordovician Eureka Quartzite in Nevada and California, parts of the Cretaceous Dakota Sandstone in the Colorado Plateau, and many Cambro-Ordovician sandstones in the Upper Mississippi Valley. Numerous examples of quartz arenites are known from other continents. Detailed, case-history discussion of quartz arenites and other sandstones, such as discussion of thickness, grain size, and facies relationships, is outside the scope of this book. Readers may find additional information of this type in Pettijohn et al. (1973), R. A. Davis (1983), and Miall (1984a).

Feldspathic Arenites and Wackes

Composition. Feldspathic arenites are typically pink or red owing to pink feldspar and/or secondary hematite cement; they may also be gray or white. They are characterized by abundant quartz, but less than 90 percent, and feldspar that commonly exceeds 10 percent. As noted, if feldspar content is greater than about 25 percent, feldspathic sandstones are often called arkoses. The feldspars in these sandstones generally include sodic plagioclase and potassium feldspars such as orthoclase, microcline, and perthite. Either potassium feldspars or sodic plagioclase may predominate, but calcic plagioclase is not common. Other constituents may include coarse muscovite and biotite, various types of rock fragments, and varied suites of unstable heavy minerals. Unstable heavy minerals may not be present in older feldspathic sandstones owing to postdepositional intrastratal solution. Rock fragments are very abundant in some feldspathic sandstones, and these sandstones may grade laterally or vertically into lithic arenites.

Texture. Feldspathic arenites tend to have little matrix, although some arkoses contain a small amount of clay matrix, commonly kaolinitic and iron stained. Some of the clay matrix in these sandstones may be diagenetic. The sorting of sand-size grains in feldspathic sandstones may range from very poor in residual arkoses developed directly upon granitoid bedrock or in arkoses transported only short distances to very good in more extensively transported and reworked feldspathic arenites. The feldspar grains are typically angular to subangular, but they are commonly better rounded than associated quartz grains. Some feldspar and quartz grains may be subrounded, depending upon the distance and intensity of transport.

Cement. Calcite is the most common and abundant cement in feldspathic arenite and arkoses. Quartz overgrowth may be present in some quartz grains, and hematite also may act as a cement. In some sandstones, hematite occurs as spots in calcite cement.

Chemical Composition. SiO_2 is the dominant chemical constituent in feldspathic arenites, although these rocks contain less SiO_2 than do quartz arenites. Owing to greater content of feldspars and clay minerals, Al, Mg, Ca, Na, and K are generally more abundant than in quartz arenites (Table 7.2). Ca and Mg can occur also in calcite or dolomite cements. Iron can occur both as Fe_2O_3 in hematite and as FeO in unstable iron-bearing minerals such as biotite.

Origin. Some arkoses originate essentially in situ by disintegration of granite and related rocks to produce a granular sediment called **grus.** These residual arkosic materials may be shifted a short distance downslope and deposited as fans or aprons of waste material, commonly referred to as **clastic wedges.** These fans may extend into basins and become intercalated, or interbedded, with more normal (better stratified and better sorted) sediments. Other feldspathic arenites undergo considerable transport and reworking by rivers or the sea before they are deposited. These reworked sandstones commonly contain smaller amounts of feldspars than do residual arkoses and are better sorted and better rounded. They also have lower feldspar contents, commonly less than 25 percent.

Most feldspathic sandstones are thought to be derived from granitic-type primary crystalline rocks, such as coarse granite or metasomatic rocks containing abundant potassium feldspar. Feldspathic arenites containing predominantly plagioclase feldspars, derived from igneous rocks such as quartz diorites, are also known. Such rocks are very uncommon. Although some feldspars may possibly survive recycling from a sedimentary source, it appears unlikely that sedimentary source rocks can furnish enough feldspar to produce a feldspathic arenite or arkose.

Examples. Feldspathic arenites make up about 15 percent of all sandstones. They occur in sedimentary sequences of all ages, although they appear to be particularly abundant in Mesozoic and Paleozoic strata. Some common examples include the Carboniferous Old Red Sandstone in Scotland, the Triassic Newark Group in the New Jersey area; the Pennsylvanian Fountain and Lyon formations of the Colorado Front Range, and the Paleocene Swauk Formation of Washington. The Swauk Formation is particularly interesting because it is a plagioclase arkose.

Lithic Arenites and Wackes

Composition. Lithic arenites are typically light-gray, salt-and-pepper colored rocks, although lithic arenites rich in volcanic rock fragments tend to be dark gray or dark greenish gray. Lithic arenites are characterized by an excess of rock fragments relative to feldspars and a quartz content of less than 90 percent. In some lithic sandstones, rock fragments may predominate over quartz; however, quartz is generally more abundant. Rock-fragment content commonly ranges from about 10 to more than 50 percent. The rock fragments in lithic sandstones fall into three principal rock groups: (1) low-grade metamorphic fragments such as slate, phyllite, and mica schist; (2) aphanitic, volcanic-flow rock fragments such as basalt, andesite, and felsite (less common); and (3) fine-grained sedimentary fragments such as shale, argillite, siltstone, chert, micritic limestone, and dolomite. Coarse flakes of either muscovite or biotite mica may be common. Lithic sandstones may also contain fine carbonaceous plant fragments, shale pebbles, and a variety of accessory heavy minerals, depending upon the type of source rocks. Lithic arenites composed of abundant rock fragments of a particular type are commonly given special names to indicate the predominant clast type. Thus, we speak of phyllarenites, schist arenites, volcanic arenites, chert arenites, and so forth.

Texture. Lithic arenites may range from well sorted to poorly sorted. Most petrology textbooks suggest that primary detrital matrix is absent in lithic sandstones and that the matrix consists of squashed rock fragments, sometimes called pseudomatrix, and/ or authigenically precipitated clays. Lithic arenites rich in volcanic rock fragments commonly contain substantial material that appears to be matrix. Careful petrographic study usually reveals the presence of many squashed, soft rock fragments that have been squeezed around harder framework grains and superficially resemble matrix. Much true clay matrix may also exist, but it is commonly very difficult to determine unequivocally whether this matrix is detrital or authigenic. Many geologists now believe that most matrix in sandstones may have been produced during diagenesis by alteration of feldspars or other minerals, as discussed under sandstone classification. The Eocene volcanic wackes of western Oregon are examples of lithic sandstones that are very poorly sorted and that contain abundant matrix commonly altered to chlorite. Much of this matrix may indeed be authigenic.

The roundness of grains in lithic arenites is quite variable. Although quartz grains tend to be less rounded than in quartz arenites, they are typically better rounded than the quartz grains in feldspathic arenites because many of the quartz grains in lithic sandstones are derived from sedimentary source rocks in which the quartz has been recycled. On the other hand, lithic sandstones derived from metamorphic or primary igneous sources may contain many angular or subangular quartz grains, as well as angular to subangular feldspars and rock fragments.

Cement. The most common cements in lithic sandstones are calcite and quartz. Smectite clay minerals, chlorite, and corrensite (interlayered montmorillonite–chlorite) also can occur as fibrous or microgranular cements, particularly in volcanic arenites. Lithic arenites that are particularly enriched in chert may be cemented with chert rather than quartz. Many lithic arenites are uncemented or poorly cemented, but have been firmly lithified by compaction and close packing of the sand-size grains.

Chemical Composition. Many lithic arenites contain less quartz and thus lower levels of SiO_2 than do feldspathic arenites. The concentrations of other major elements in lithic arenites are quite variable, as shown in Table 7.2, depending upon the relative abundance of various types of rock fragments, feldspars, and clay matrix.

Origin. Lithic arenites, with the exclusion of the so-called graywackes, make up about 20–25 percent of all sandstones. They are compositionally immature sandstones that originate under conditions favoring the production and deposition of large volumes of relatively unstable materials. The mechanically weak character of many of the lithic fragments in lithic arenites suggests that the fragments are probably derived from nearby source areas without significant transport. Some lithic arenites may have been derived from within uplifted parts of the same sedimentary basin in which they are deposited. On the basis of probable depositional environment and provenance, Pettijohn et al. (1973) recognize three types of lithic sandstones: (1) alluvial sandstones deposited on the flanks of marked uplifts or thick accumulations in closely associated molasse basins, (2) alluvial sandstones deposited in cratons by large rivers that derived much of their detritus from marginal and distant uplifts, and (3) marine turbidite deposits in geosynclinal depressions. Molasse basins are basins filled with sedimentary deposits believed to be eroded from adjacent, elevated ranges during and immediately after a period of major mountain building, or diastrophism.

Examples. Common examples of lithic sandstones include the Paleozoic sandstone sequences of the central Appalachians in the eastern United States, many sandstones associated with the Coal Measures throughout the world, many Jurassic and Cretaceous sandstones of the Rocky Mountains and the West Coast, and Tertiary sandstones of the Gulf Coast, West Coast, and the Alps. There are numerous other examples from stratigraphic sequences of all ages.

Graywackes

Graywackes are either lithic wackes or feldspathic wackes that are tough, well-indurated, dark-gray to black rocks. Matrix is generally abundant—some investigators say

>15 percent—and commonly consists of a chloritic "paste." As mentioned under sandstone classification, the term graywacke has been extremely controversial, with various meanings attached to it. More recent practice has been to restrict its use to a field term that can be applied to rocks having the general characteristics just described. Graywackes are abundant in the geologic record, making up about 20 percent of all sandstones.

Other Sandstones

The sandstones discussed above are composed of constituents derived primarily by weathering of preexisting rocks. A few less abundant types of sandstones, whose constituents formed largely within the depositional basin by chemical or biochemical processes or by explosive volcanism, are known. These sandstones are called **hybrid** sandstones by some authors. They include such uncommon varieties of sandstones as greensands (glauconitic sands); phosphatic sands; and calcarenaceous sands, which are sands composed predominantly of calcium carbonate grains. A particularly important group of hybrid sandstones is that of the tuffaceous sandstones, or **volcaniclastic** sandstones. Volcaniclastic sandstones are composed largely of pyroclastic materials that have been transported and reworked. They are characterized especially by the presence of euhedral feldspars, pumice fragments, glass shards, and low quartz content. Volcaniclastic sandstones and other hybrid sandstones make up about 5 percent of all sandstones.

Classification of Conglomerates

In contrast to their preoccupation with sandstone classifications, geologists have proposed few formal classification schemes for conglomerates. Pettijohn (1975) includes a formal classification for conglomerates in his textbook on sedimentary rocks (Table 7.5). He divided conglomerates into two basic types: (1) **orthoconglomerates,** containing less than 15 percent matrix, and (2) **paraconglomerates,** containing more than 15 percent matrix. Orthoconglomerates composed of clasts of a single type, commonly quartzose pebbles, are **oligomict** conglomerates. Those composed of clasts of mixed lithology are **petromict** conglomerates. The matrix-rich paraconglomerates are also called **diamictites,** which is a general term used for rocks composed of nonsorted or poorly sorted siliciclastic sediment that contains larger particles in a muddy or sandy matrix. Paraconglomerates that are known to be of glacial origin are called **tillites.** Many workers use informal names for conglomerates, based either on clast size or on dominant composition. Thus, on the basis of clast size, a conglomerate may be called a pebble conglomerate, cobble conglomerate, or boulder conglomerate. If based on clast composition, names such as quartzite conglomerate, limestone conglomerate, and basalt conglomerate are applied. Textural and compositional terms may be combined to yield names such as quartzite cobble conglomerate or petromict pebble conglomerate.

Most conglomerates have a **clast-supported** fabric in which the pebbles or cobbles touch each other. This fabric is called the gravel framework. Interstitial spaces in the framework are voids, which are commonly filled with sand or mud matrix. Note that the term matrix when applied to conglomerates has a different meaning than when applied to sandstones. The matrix in sandstones is defined as material less than about 0.03 mm in size. Matrix in conglomerates can be any size material that is fine enough to fill the interstitial spaces among the pebbles or cobbles that make up the

TABLE 7.5   Classification of sedimentary conglomerates and breccias

EPICLASTIC	Extraformational (clasts derived from outside depositional basin)	Orthoconglomerates (matrix <15%)	Metastable clasts <10%	Orthoquartzitic (oligomict) conglomerate (composed of pebbles of one type)
			Metastable clasts >10%	Petromict conglomerate (composed of pebbles of more than one type) (specify dominant type—e.g., limestone conglomerate)
		Paraconglomerates (matrix >15%); also called diamictites	Laminated matrix	Laminated conglomeratic mudstone or argillite
			Nonlaminated matrix	Tillite (glacial) Tilloid (nonglacial)—e.g., pebbly mudstones, olistostromes
	Intraformational (clasts derived from erosion of sediments within the depositional basin)	Intraformational conglomerates and breccias		
PYROCLASTIC	Volcanic breccias and agglomerates			

Source: Pettijohn, F. J., 1975, Sedimentary rocks, 3rd ed., Table 6.2, p. 165, reprinted by permission of Harper & Row, New York.

gravel framework. Rare conglomerates with unfilled voids are called **openwork** conglomerates. Because some conglomerates may contain as much as 70 percent mud or sand matrix, not all conglomerates are clast supported. **Matrix-supported** conglomerates are those in which clasts are scattered or floating in the matrix and do not touch.

Origin and Occurrence of Conglomerates

Clast-supported Conglomerates. Overall, conglomerates are much less abundant than sandstones, probably making up no more than 1 or 2 percent of the total rock record. They are formed by a variety of fluid-flow and sediment gravity-flow processes. Most clast-supported conglomerates, or orthoconglomerates, originate in shallow water by transport and deposition in high-energy streams or in the surf zone of beaches. A few seem to have been deposited in deep water by turbidity currents. Conglomerates deposited by turbidity currents are sometimes referred to as **resedimented** conglomerates. Oligomict conglomerates are composed of highly resistant clasts of quartzite, quartz, or chert, suggesting transport and deposition under high-energy conditions that destroyed less resistant clasts. Owing to their resistance to chemical and mechanical destruction, quartzose clasts may be reworked and redeposited through several cycles of sedimentation. Other clast-supported conglomerates are made up of less stable clasts of rocks such as granite, basalt, phyllite, shale, or limestone that must have accumulated under much lower energy conditions and that have much lower recycling potential. These conglomerates are commonly composed of mixed clast types and thus are petromict conglomerates.

Conglomerates occur in rocks of all ages. For example, quartzose conglomerates occur in some Precambrian rocks of the Canadian Shield, such as the Baraboo Conglomerate in the Great Lakes region—and are common in many Paleozoic formations of the Appalachians, including the Silurian Tuscarora Formation, the Mississippian Pocono Formation, and the Pennsylvanian Pottsville Formation. They occur also in Mesozoic rocks of the Colorado Plateau—for example, the Cretaceous Dakota Formation. In northern California, the Cretaceous Hornbrook Formation contains clasts composed mainly of quartzite pebbles and cobbles. Quartzose gravels of Tertiary age have been reported from Maryland and western Kentucky. Petromict conglomerates are also widespread. Examples include the mixed-pebble Ogishke Conglomerate and other conglomerates of Precambrian age in the Lake Superior region; Triassic limestone conglomerates of Maryland and Pennsylvania; Pennsylvanian-age granite-pebble conglomerates of the Fountain and Weber formations in Colorado; the Cretaceous Humbug Mountain Formation of southwestern Oregon, probably a deep-water, turbidite unit; and the Tertiary Bushnell Rock Formation of southwestern Oregon.

Mud-supported Conglomerates. Mud-supported conglomerates are formed by various processes, both in subaerial and subaqueous environments. Some are deposited by mudflows or debris flows on land; others are glacial deposits. Still others may be deposited by underwater mudflows or by turbidity currents. Thus, they form in environments where total energy is high enough to transport gravel-size sediment, but where there is insufficient reworking by currents to remove sand and mud. Mud-supported conglomerates are characterized by more matrix than clasts; therefore, they are sometimes called **conglomerate mudstones** or **pebbly mudstones.**

Many examples of mud-supported conglomerates are known from rocks of widely different ages. Glacial tillites occur mainly in rocks of middle and late Precambrian, Permocarboniferous, and Pleistocene ages. Examples include the Precambrian Gowganda Formation of Ontario and Quebec, the Permocarboniferous Dwyka tillites of South Africa, and the widespread Pleistocene glacial deposits of North America and Europe. Many paraconglomerates originally thought to be tillites have been reinterpreted as either mudflow deposits or turbidites. For example, the so-called Squantum "Tillite" of the Boston Bay area has been reinterpreted as an underwater mudflow. The Pennsylvanian Haymond Formation of the Marathon region in Texas, originally interpreted as a tillite, is now thought to have been formed by submarine slumping, mudflows, and turbidity currents. Other deposits, such as the Tertiary Gunnison "Tillite" of Colorado and the Jurassic pebbly mudstones at Cape Blanco, Oregon, are also probably mudflow deposits. Upper Jurassic and Lower Cretaceous pebbly mudstones in many parts of California have been interpreted as turbidity current deposits.

Intraformational Conglomerates. Intraformational conglomerates are composed of clasts of sediments believed to have been formed within the depositional basin, in contrast to extraformational conglomerates, in which the clasts are derived from outside the depositional basin. Intraformational conglomerates originate by penecontemporaneous deformation of semiconsolidated sediment and redeposition of the fragments fairly close to the site of deformation. Penecontemporaneous breakup of sediment to form clasts may take place subaerially, such as by drying out of mud on a tidal flat, or under water. Subaqueous ripups of semiconsolidated muds by tidal currents, storm waves, or sediment gravity flows are possible causes. In any case, sedimentation is interrupted only a short time during this process. The most common types of fragments found in intraformational conglomerates are mud clasts and lime clasts. The clasts are commonly angular or only slightly rounded, suggesting little transport. In some beds, flattened clasts are stacked virtually on edge, apparently owing to unusually strong wave or current agitation, to form what is called **edgewise conglomerates**.

Intraformational conglomerates commonly form thin beds, a few centimeters to a meter in thickness, that may be laterally extensive. Although much less abundant than extraformational conglomerates, they nonetheless occur in rocks of many ages. So-called flat-pebble conglomerates composed of carbonate or limy siltstone clasts are particularly common in Cambrian-age rocks in various parts of North America. Examples include conglomerate beds in the Conococheague Formation of Maryland, the Gros Ventre and Gallitan formations of Wyoming, and the Muav Limestone of the Grand Canyon area in Arizona. They occur also in many early Paleozoic limestones of the Appalachian region. Intraformational conglomerates composed of shale ripup clasts embedded in the basal parts of sandstone units are very common in sedimentary sequences deposited by sediment gravity-flow processes.

Classification of Mudrocks (Shales)

Mudrocks are siliciclastic sedimentary rocks composed of particles of silt and clay. **Shale** is a historically accepted class name for this group of rocks (Tourtelot, 1960) and is equivalent to the class name sandstone, although some authors restrict the usage of shale to mudrocks that show lamination. The fine particle size of most mudrocks— commonly less than about 0.06 mm—makes them difficult to study with a hand lens or an ordinary petrographic microscope. Quantitative X-ray analysis or other quantitative identification techniques are generally required for their study. Such techniques are time consuming and expensive. Because of these difficulties, geologists do not routinely determine the mineral composition of mudrocks. Therefore, most classifications that have been proposed for mudrocks have not been based on mineral composition, or at least not entirely on mineral composition. These classifications, none of which has been widely accepted, commonly emphasize the relative amounts of silt and clay, the hardness or degree of induration of the mudrocks, and the presence or absence of laminations. Shales that exhibit lamination are said to possess **fissility**. Exceptions to this general practice of classification are the Picard (1971) classification, which emphasizes mineral composition of the silt-size grains in mudrocks, and the Lewan (1978) classification, which requires semiquantitative X-ray diffraction analysis to determine mineralogy.

The classification of Potter et al. (1980), shown in Table 7.6, is based on grain size, stratification, and degree of induration. It is similar to the field classification of shales proposed by Lundegard and Samuels (1980). This classification emphasizes the importance of clay-size constituents and bedding thickness (bedded or laminated). For example, a mudrock containing more than two-thirds clay-size particles is called a claystone if bedded (layers thicker than 10 mm) or a clayshale if laminated (layers thinner than 10 mm). Additional informal terms can be used with this classification to provide further information about the properties of mudrocks. These may include terms that express color; type of cementation (calcareous, or limy; ferruginous, or iron-rich; siliceous); degree of induration (hard, soft); mineralogy if known (quartzose, feldspathic, micaceous, or other); fossil content (fossiliferous, foraminifer-rich, or other); organic matter content (carbonaceous, kerogen-rich, coaly, or other); type of fracturing (conchoidal, hackly, blocky); or nature of bedding (wavy, lenticular, parallel, or other).

TABLE 7.6 Classification of shales

Percentage clay-size constituents			0–32	33–65	66–100
Field adjective			Gritty	Loamy	Fat or slick
NONINDURATED	Beds	Greater than 10 mm	Bedded silt	Bedded mud	Bedded claymud
NONINDURATED	Laminae	Less than 10 mm	Laminated silt	Laminated mud	Laminated claymud
INDURATED	Beds	Greater than 10 mm	Bedded siltstone	Mudstone	Claystone
INDURATED	Laminae	Less than 10 mm	Laminated siltstone	Mudshale	Clayshale
METAMORPHOSED	Degree of metamorphism	Low	Quartz argillite	Argillite	
METAMORPHOSED	Degree of metamorphism		Quartz slate	Slate	
METAMORPHOSED	Degree of metamorphism	High	Phyllite and/or mica schist		

Source: Potter, P. E., J. B. Maynard, and W. A. Pryor, 1980, Sedimentology of shales. Table 1.2, p. 14, reprinted by permission of Springer-Verlag, New York.

Origin and Occurrence of Shales

Shales form under any environmental conditions in which fine sediment is abundant and water energy is sufficiently low to allow suspension settling of fine silt and clay. Shales are particularly characteristic of marine environments adjacent to major continents at places where the seafloor lies below storm wave base, but they can form also in lakes and quiet-water parts of rivers and in lagoonal, tidal-flat, and deltaic environments. The fine-grained siliciclastic products of weathering greatly exceed coarser particles; thus fine sediment is abundant in many sedimentary systems. Because fine sediment is so abundant and can be deposited in a variety of quiet-water environments, shales are by far the most abundant type of sedimentary rock. They make up 50 60 percent of the total sedimentary rock record. Shales commonly occur interbedded with sandstones or limestones in units ranging in thickness from a few millimeters to several meters or tens of meters. Nearly pure shale units hundreds of meters thick also occur. Shale units in marine sequences tend to be laterally extensive.

Shales are so common, abundant, and difficult to study petrographically that they tend to hold little interest for many geologists. A few shales that are particularly well known owing to their thickness, widespread areal extent, stratigraphic position, or fossil content include the Cambrian Burgess Shale of western Canada, which is famous for its well-preserved imprints of soft-bodied animals; the Eocene Green River Oil Shale of Colorado; the Cretaceous Mancos Shale of western North America, which forms a thick eastward-thinning wedge stretching from New Mexico to Saskatchewan and Alberta; the Devonian–Mississippian Chattanooga Shale and equivalent formations that cover much of North America; the Silurian Gothlandian shales of western Europe, northern Africa, and the Persian Gulf region that contain a pelecypod and graptolite faunal association; and the Precambrian Figtree Formation of South Africa, well known for studies of its early fossils. The origin and occurrence of shales are discussed in detail by Potter et al. (1980).

7.5 PROVENANCE SIGNIFICANCE OF MINERAL COMPOSITION

The silicate mineralogy and rock-fragment composition of siliciclastic sedimentary rocks are fundamental properties of these rocks that set them apart from other sedimentary rocks. Mineralogy is a particularly important property for studying the origin of siliciclastic sedimentary rocks because it provides almost the only available clue to the nature of vanished source areas. The kinds of siliciclastic minerals and rock fragments preserved in sedimentary rocks furnish important evidence of the lithology of the source rocks. Rock fragments provide the most direct lithologic evidence, but feldspars and other minerals are also important source-rock indicators. For example, potassium feldspars suggest derivation mainly from alkaline plutonic igneous or metamorphic rocks, whereas sodic plagioclases are derived principally from alkaline volcanic rocks and calcic plagioclases from basic volcanic rocks. Particular suites of heavy minerals are also used for source-rock determination. Thus, a suite of heavy minerals consisting of apatite, biotite, hornblende, monazite, rutile, titanite, pink tourmaline, and zircon is indicative of alkaline igneous source rocks. A suite consisting of augite, chromite, diopside, hypersthene, ilmenite, magnetite, and olivine suggests derivation from basic igneous rocks. Andalusite, garnet, staurolite, topaz, kyanite, sillimanite, and staurolite constitute a mineral suite diagnostic of metamorphic rocks, whereas a suite of heavy minerals consisting of barite, iron ores, leucoxene, rounded tourmaline, and rounded zircon suggests a recycled sediment source. Even quartz may have some value as a provenance indicator, although its use in source-rock studies has been controversial, as suggested in the discussion of quartz. As an example of its provenance significance, some workers believe that a high percentage of quartz grains with undulose extinction greater than 5 degrees, combined with a high percentage of polycrystalline grains containing more than three crystal units per grain, distinguish low-rank metamorphic source rocks from high-rank metamorphic or plutonic igneous source rocks.

In addition to providing information about source-rock lithology, the relative chemical stabilities and the degree of weathering and alteration or certain minerals can be used as tools for interpreting the climate and relief of source areas. For example, the presence of large, fresh, angular feldspars in a sandstone suggests derivation from a high-relief source area where grains were eroded rapidly before extensive weathering occurred. Alternatively, such feldspars may have been derived from a source area having a very arid or extremely cold climate that retarded chemical weathering. Small, rounded, highly weathered feldspar grains indicate a source area of low relief and/or a warm, humid climate where chemical weathering was moderately intense. Absence of feldspars may indicate either that weathering was so intense that all feldspars were destroyed or that no feldspars were present in the source rocks. Such analyses of mineral constituents provide only very tentative conclusions about climate and relief. Also, they are subject to misinterpretations if there has been diagenetic alteration or destruction of source-rock minerals.

Geologists are interested also in the tectonic setting of source areas and associated depositional sites. With development of the theory of seafloor spreading and plate tectonics, this interest has focused sharply on interpreting the tectonic setting in terms of plate tectonic provinces (Dickinson and Suczek, 1979; Dickinson, 1982, Dickinson et al., 1983). In other words, geologists want to know if a particular deposit was derived from source rocks located within a continent, in a volcanic arc associated with a subduction zone, or in other tectonic settings. Three principal types of tectonic settings, or provenances, have been identified: (1) continental-block provenances, (2) magmatic-arc provenances, and (3) recycled-orogen provenances (Dickinson and Suczek, 1979).

Continental-block provenances are located within continental masses, which may be bordered on one side by a passive continental margin and on the other by an orogenic belt or zone of plate convergence. Source rocks consist of plutonic igneous, metamorphic, and sedimentary rocks, but include few volcanic rocks. Sediment eroded from these sources typically consists of quartzose sand, feldspars with high potassium feldspar to plagioclase feldspar ratios, and metamorphic and sedimentary rock fragments. Sediment eroded from continental sources may be transported off the continent into adjacent marginal ocean basins or it may be deposited in local basins within the continent.

Magmatic-arc provenances are located in zones of plate convergence, where sediment is eroded mainly from volcanic-arc sources consisting of volcanogenic highlands. Volcaniclastic debris shed from these highlands consists largely of volcanic lithic fragments and plagioclase feldspars. Quartz and potassium feldspars are commonly very sparse except where the volcanic cover is dissected by erosion to expose

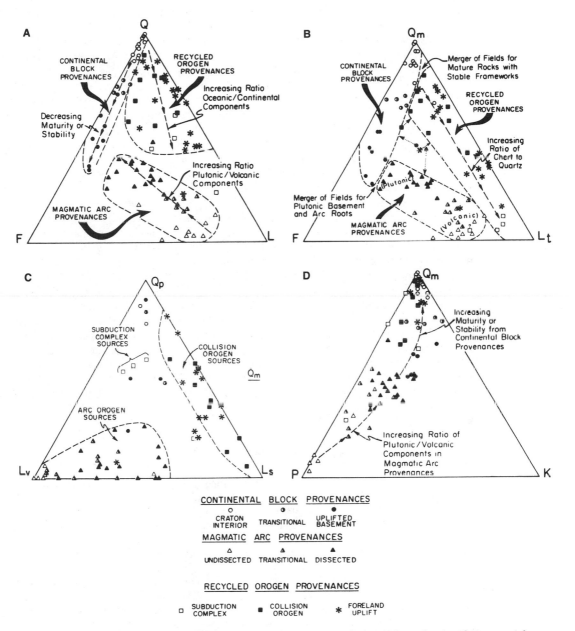

FIGURE 7.14 Four triangular plots showing mean framework modes (sand-size particle composition) for selected sandstone suites derived from different types of provenances: (A) QFL plot, (B) Q_mFL_t plot, (C) $Q_pL_vL_s$ plot, (D) Q_mPK plot. Q is total quartz, including monocrystalline (Q_m) and polycrystalline (Q_p) varieties, F is total feldspar grains, P is plagioclase feldspar grains, K is K-feldspar grains, L_t is total rock fragments, including stable quartzose (Q_p) and unstable (L) varieties, L_v is volcanic-metavolcanic rock fragments, and L_s is sedimentary-metasedimentary rock fragments. (After Dickinson, W. R., and C. A. Suczek, 1979, Plate tectonics and sandstone composition: Am. Assoc. Petroleum Geologists Bull., v. 63. Figs. 1–4, pp. 2171, 2172, reprinted by permission of AAPG, Tulsa, Okla.)

underlying plutonic rocks. Sediment shed from volcanic highlands may be transported to an adjacent trench or deposited in fore-arc and back-arc basins.

Recycled-orogen provenances are zones of plate convergence, where collision of major plates creates uplifted source areas along the collision suture belt. Where two continental masses collide, source rocks in the collision uplifts are typically sedimentary and metamorphic rocks that were present along the continental margins prior to collision. Detritus stripped from these source rocks commonly consists of abundant sedimentary-metasedimentary rock fragments, intermediate quartz content, and a high ratio of quartz to feldspars. Where a continental mass collides with a magmatic-arc complex, uplifted source rocks may include deformed ultramafic rocks, basalts, other oceanic rocks, and a variety of other rock types such as greenstone, chert, argillite, graywacke sandstones, and limestones. Sediment derived from these sources may include many types of rock fragments, quartz, feldspars, and chert. Chert is a particularly abundant constituent of sediments derived from this provenance.

To differentiate sediment derived from these three major tectonic provenances, Dickinson and Suczek (1979) suggest the use of triangular composition diagrams showing framework proportions of monocrystalline quartz, polycrystalline quartz, potassium and plagioclase feldspars, and volcanic and sedimentary-metasedimentary rock fragments. On the basis of study of sandstone compositions from many parts of the world, they generated the provenance diagrams shown in Figure 7.14. To use these diagrams for provenance determination of other sandstones, the composition of the sand-size grains in a sandstone is determined and plotted as QFL, Q_mFL_t, $Q_pL_vL_s$, or Q_mPK diagrams (Fig. 7.14). The explanation of these symbols is given in the caption for Figure 7.14. Comparison of plotted diagrams with Dickinson and Suczek's models (Fig. 7.14) allows the tectonic provenance of the sediments to be identified. The reason for constructing different framework composition diagrams is that plotting the composition data on different compositional diagrams allows a more reliable interpretation than one based on a single diagram.

ADDITIONAL READINGS

Adams, A. E., W. S. Mackenzie, and C. Guilford, 1984, Atlas of sedimentary rocks under the microscope: John Wiley & Sons, New York, 104 p.

Folk, R. L., 1974, Petrology of sedimentary rocks: Hemphill, Austin, Tex., 182 p.

Koster, E. H., and R. J. Steel (eds.), 1984, Sedimentology of gravels and conglomerates: Canadian Soc. Petroleum Geologists Mem. 10, 441 p.

Milner, H. B., 1962, Sedimentary petrography. V. 2, Principles and applications: MacMillan, New York, 715 p.

Pettijohn, F. J., P. E. Potter, and R. Siever, 1972, Sand and sandstone: Springer-Verlag, New York, 618 p.

Potter, P. E., J. B. Maynard, and W. A. Pryor, 1980, Sedimentology of shale: Springer-Verlag, New York, 306 p.

Scholle, P. A., 1979, A color illustrated guide to constituents, textures, cements, and porosities of sandstones and associated rocks: Am. Assoc. Petroleum Geologists Mem. 28, 201 p.

Tickell, F. G., 1965, The techniques of sedimentary mineralogy: Elsevier, New York, 220 p.

Williams, H., F. J. Turner, and C. M. Gilbert, 1982, Petrography, 2nd ed. Part 2, Sedimentary rocks, p. 277–427: W. H. Freeman, San Francisco.

Zuffa, G. G. (ed.), 1985, Provenance of arenites: D. Reidel, Dordrecht, Netherlands, 408 p.

8

Carbonates and Other Nonsiliciclastic Sedimentary Rocks

8.1 INTRODUCTION

The so-called "chemical" sedimentary rocks originate by precipitation of minerals from water through various chemical or biochemical processes (Chapter 4). These rocks are distinguished from siliciclastic sedimentary rocks by their chemistry, mineralogy, and texture. They can be divided on the basis of mineralogy and chemistry into five fundamental types: (1) carbonates, (2) evaporites, (3) siliceous sedimentary rocks (cherts), (4) iron-rich sedimentary rocks, and (5) phosphorites. A sixth group of nonsiliciclastic rocks is that of the carbonaceous sedimentary rocks. These are sedimentary rocks such as coals and oil shales that contain at least 10–20 percent nonskeletal organic matter. The remaining constituents of carbonaceous rocks may be either siliciclastic minerals or nonsiliciclastic constituents, commonly carbonate minerals.

The carbonate rocks are by far the most abundant type of nonsiliciclastic sedimentary rock and are the only group of these rocks for which formal classification schemes are in common use. They can be divided on the basis of mineralogy into limestones and dolomites (dolostones). Limestones are commonly further divided into several subtypes based on relative abundance of various kinds of carbonate grains, such as fossils and oolites, and the ratio of total carbonate grains to carbonate-mud matrix. Evaporites are the second most abundant group of nonsiliciclastic rocks. They can be subdivided by mineralogy into carbonates, sulfates, and chlorides. Cherts, iron-rich rocks, phosphorites, and carbonaceous rocks are volumetrically less important groups of nonsiliciclastic sedimentary rocks; however, many of these rocks have considerable economic significance and are therefore of great interest to geologists.

In this chapter, I describe the mineral composition, chemistry, and, where appropriate, the special textures that characterize the nonsiliciclastic sedimentary rocks. The classification of carbonate rocks is discussed in detail. Major types of evaporites, siliceous sedimentary rocks, iron-rich sedimentary rocks, phosphorites, and carbonaceous sedimentary rocks are identified and described.

8.2   CARBONATE SEDIMENTARY ROCKS

Carbonate sedimentary rocks include all sedimentary rocks composed predominantly of carbonate minerals (Table 8.1). The two principal types of carbonate rocks are **limestones,** which are composed mainly of calcite, and **dolomites,** or dolostones, composed mainly of dolomite. These carbonate sedimentary rocks make up 10–20 percent of the known sedimentary record. They are the only volumetrically significant group of non-siliciclastic sedimentary rocks. Limestones contain richly varied textures, structures, and fossils that yield important information about ancient marine environments, paleoecological conditions, and the evolution of life forms, particularly marine organisms, through time. Carbonate sedimentary rocks are also an economically important group of rocks because limestones and dolomites are useful for agricultural and industrial purposes; they make good building stones; and, most importantly, they act as reservoir rocks for more than one-third of the world's petroleum reserves. Because of their environmental and economic significance, limestones have been extensively studied and their mineralogy, chemistry, and textural characteristics described in hundreds of research papers. The characteristic properties of carbonate rocks have been summarized in several books, such as those of Bathurst (1975), Chilingarian et al. (1967), Lippman (1973), MacQueen (1983), Milliman (1974), Reeder (1983), and Scholle (1978).

Mineralogy

The chemistry and structure of the common carbonate minerals are shown in Table 8.1. Much more detailed analyses of the crystal chemistry of the carbonates are given by Lippman (1973) and Reeder (1983). Modern carbonate sediments are composed mainly of aragonite but they also include calcite and dolomite. Much of the aragonite in modern sediments appears to be of biogenic origin (Chapter 4). Calcite ($CaCO_3$) can contain several percent magnesium in its formula; magnesium can readily substitute for calcium in the lattice of calcite crystals because magnesium ions and calcium ions are similar in size and charge. Thus, we recognize both low-magnesian calcite containing less than about 4 percent $MgCO_3$ and magnesian calcite, or high-magnesian calcite, containing more than 4 percent $MgCO_3$. Magnesian calcite retains the crystal structure of calcite in spite of the presence of Mg ions, which randomly substitute for Ca ions in the calcite crystal lattice. By contrast, true dolomite is a totally different mineral: Mg ions occupy half of the cation sites in the crystal lattice and are arranged in well-ordered planes that alternate with planes of CO_3 ions and Ca ions. Dolomite occurs in a few restricted modern environments, particularly in certain supratidal environments

TABLE 8.1   Chemical composition and crystal structure of the common carbonate minerals

Mineral	Chemical formula	Crystal system
Aragonite	$CaCO_3$	Orthorhombic
Calcite	$CaCO_3$	Hexagonal (rhombohedral)
Magnesite	$MgCO_3$	Hexagonal (rhombohedral)
Dolomite	$CaMg(CO_3)_2$	Hexagonal (rhombohedral)
Ankerite (ferroan dolomite)	$Ca(Fe,Mg)(CO_3)_2$	Hexagonal (rhombohedral)
Siderite	$FeCO_3$	Hexagonal (rhombohedral)

and freshwater lakes, but it is of minor importance in modern carbonate environments in comparison to aragonite and calcite. Other carbonate minerals, such as magnesite, ankerite, and siderite, are even less common in modern sediments.

In contrast to modern carbonate sediments, in which aragonite dominates, ancient carbonate rocks are composed mainly of calcite and dolomite. Calcite is the principal mineral in ancient limestones; dolomite is the dominant mineral in dolomites. Aragonite is rarely found in ancient carbonate rocks, particularly in those older than about Cretaceous age. Aragonite is the metastable polymorph of $CaCO_3$ and inverts fairly rapidly under aqueous conditions to calcite. The ratio of dolomite to calcite is much greater in ancient carbonate rocks than in modern carbonate sediments because $CaCO_3$ minerals exposed to magnesium-rich interstitial waters during burial and diagenesis are converted to dolomite by replacement (Chapter 4).

Chemistry

The elemental chemistry of carbonate rocks is dominated by calcium, magnesium, carbon, and oxygen. The relative abundance of these elements (expressed as oxides) in average limestones and nearly pure dolomite is shown in Figure 8.1. Numerous other elements are present in carbonate rocks in minor or trace amounts. Many of the elements that occur in minor concentrations are contained in noncarbonate impurities. For example, Si, Al, K, Na, and Fe occur mainly in siliciclastic silicate minerals such as quartz, feldspars, and clay minerals that are present in minor amounts in most carbonate rocks. Trace elements that are common in carbonate rocks include B, Be, Br, Cl, Co, Cr, Cu, Ga, Ge, and Li. The concentrations of these trace elements are controlled not only by the mineralogy of the rocks but also by the type and relative abundance of fossil skeletal grains in the rock. Many organisms concentrate and incorporate trace elements into their skeletal structures. A good discussion of the distribution of trace elements in various kinds of skeletal and nonskeletal constituents in carbonate rocks is given by Wolf et al. (1967).

Carbonate Textures

Ancient limestones are essentially monomineralic rocks composed of calcite. Calcite can be present in at least three distinct textural forms: (1) **carbonate grains,** such as oolites and skeletal grains, which are silt-size or larger aggregates of calcite crystals; (2) **microcrystalline calcite,** or carbonate mud, which is texturally analogous to the mud in siliciclastic sedimentary rocks but which is composed of extremely fine-size calcite crystals; and (3) **sparry calcite,** consisting of much coarser grained calcite crystals that appear clear to translucent in plane light. Before classification of limestones can be considered, a fuller understanding of these contrasting carbonate textural elements must be developed.

Carbonate Grains

Early geologists tended to regard limestones as simply crystalline rocks that commonly contained fossils and that presumably formed largely by passive precipitation from seawater. We now know that many, and perhaps most, carbonate rocks are not simple crystalline precipitates. Instead, they are composed in part of aggregate particles or grains that may have undergone mechanical transport before deposition (Chapter 4). Folk (1959) suggested use of the general term **allochems** for these carbonate grains to emphasize that they are not normal chemical precipitates. Carbonate grains typically range in size from coarse silt (0.02 mm) to sand (up to 2 mm), but larger particles also occur. The grains can be divided into five basic types, each characterized by distinct shape, internal structure, and mode of origin: carbonate clasts; skeletal particles; ooliths; peloids; and lumps, or grapestones.

Carbonate Clasts. Carbonate clasts are rock fragments that were derived either by erosion of ancient limestones exposed on land or by erosion of partially or completely lithified carbonate sediments within a depositional basin. If carbonate clasts are derived from land sources located outside the depositional basin, they are called **lithoclasts.** If they are derived from within the basin by erosion of semiconsolidated carbonate sediments from the seafloor, adjacent tidal flats, or a carbonate beach (beach

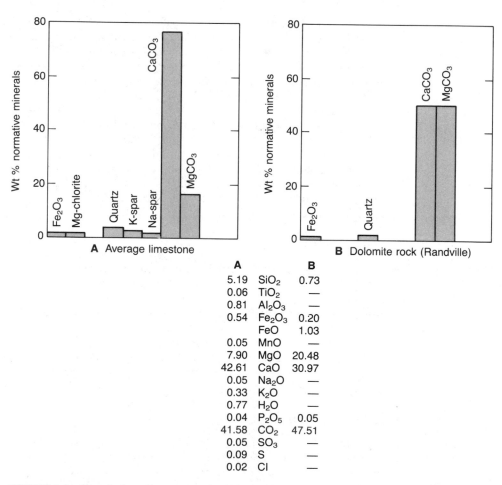

A		B
5.19	SiO$_2$	0.73
0.06	TiO$_2$	—
0.81	Al$_2$O$_3$	—
0.54	Fe$_2$O$_3$	0.20
	FeO	1.03
0.05	MnO	—
7.90	MgO	20.48
42.61	CaO	30.97
0.05	Na$_2$O	—
0.33	K$_2$O	—
0.77	H$_2$O	—
0.04	P$_2$O$_5$	0.05
41.58	CO$_2$	47.51
0.05	SO$_3$	—
0.09	S	—
0.02	Cl	—

FIGURE 8.1 Chemical composition and normative mineral composition. A. An average limestone. B. A nearly pure dolomite (dolostone). (After Garrels, R. M., and F. T. Mackenzie, 1971, Evolution of sedimentary rocks. Fig. 8.3, p. 211, reprinted by permission of W. W. Norton & Company, Inc., New York.)

rock), they are called **intraclasts.** The distinction between lithoclasts and intraclasts has important implications for interpreting the transport and depositional history of limestones; however, this distinction is commonly very difficult to make. Carbonate clasts generally lack distinctive internal features, such as iron staining resulting from weathering, that would allow differentiation of fragments of ancient limestones from penecontemporaneously produced intraclasts. **Limeclast** is a nonspecific term that can be used for carbonate clasts when this distinction cannot be made. If limeclasts can be recognized as intraclasts, the existence in the depositional basin of a high-energy event capable of ripping up the substrate is indicated.

Limeclasts range in size from very fine sand to gravel, although sand-size fragments are most common. They generally show some degree of rounding, indicative of transport, but subangular or even angular clasts are not unusual. Some clasts display internal textures or structures such as lamination, older clasts, siliciclastic grains, fossils, oolites, or pellets, but others are internally homogeneous. A limestone composed of large limeclasts is a type of intraformational conglomerate. Clasts are not the most abundant type of carbonate grain in ancient limestones, but they occur with sufficient frequency in the geologic record to show that the clast-forming mechanism was a common process. Examples of clasts are shown in Figure 8.2.

Skeletal Particles. Skeletal fragments occur in limestones as whole microfossils, whole larger fossils, or broken fragments of larger fossils. They are by far the most common allochem in carbonate rocks and are so abundant in some limestones that they make up most of the rock. Fossils representing all the major phyla of calcareous marine invertebrates are found in limestones. The specific kinds of skeletal particles that occur depend upon both the age of the rocks and the paleoenvironmental conditions under which they were deposited. Owing to evolutionary changes in fossil assemblages through time, different kinds of fossil remains dominate rocks of different ages. For example, trilobite skeletal remains characterize early Paleozoic-age rocks but do not occur in Cenozoic rocks, which instead commonly contain abundant foraminifers. Likewise, certain kinds of skeletal particles characterize limestones formed in different environments. To illustrate, the remains of colonial corals, which build rigid, wave-resistant skeletal structures, are commonly restricted to limestones deposited in

FIGURE 8.2 Photomicrograph of angular limeclasts in a matrix of dark, organic-rich micrite. The large clast in the middle of the photograph is partially replaced by chert. Polarized light. Figure 8.7 shows another example of lime-clasts.

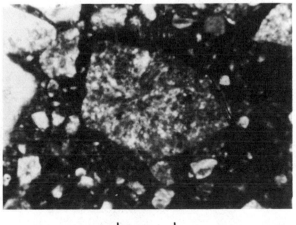

0   0.3 mm

shallow-water, high-energy environments where the water was well agitated and oxygen levels high. By contrast, branching types of bryozoans are fragile organisms that cannot withstand the rigors of high wave-energy environments. Thus, their remains are found mainly in limestones deposited under quiet-water conditions.

Depending upon paleoenvironmental conditions, skeletal remains in a given specimen of limestone may consist entirely or almost entirely of one species of organism; commonly, however, they include several species. An example of a mixed assemblage of skeletal particles is shown in Figure 8.3. The serious student of carbonate rocks must learn to identify the many kinds of fossils and fossil fragments that occur in limestones because fossils have special significance for paleoenvironmental and paleoecological interpretation (Chapter 10).

Ooliths. The term oolith, or coated grain, is applied as a general name to carbonate grains that contain a nucleus of some kind—a shell fragment, pellet, or quartz grain—surrounded by one or more thin layers or coatings consisting of fine calcite or aragonite crystals. Spherical to subspherical coated grains that exhibit several internal concentric layers with a total thickness greater than that of the nucleus are called **oolites** (Fig. 8.4). Oolites form where strong bottom currents and agitated-water conditions exist and where saturation levels of calcium bicarbonate are high (Chapter 4). Modern oolites are composed mainly of aragonite, whereas ancient oolites are composed principally of calcite. Whether or not all ancient oolites were composed originally of aragonite that later inverted to calcite is an unsolved problem. Some authors cite petrographic evidence suggesting that some ancient oolites may have originated as calcite. If true, this may indicate that magnesium levels in the ancient ocean were lower than in the modern ocean, thus allowing calcite to precipitate in preference to aragonite.

Although most oolites display an internal structure consisting of concentric layers (Fig. 8.4), some oolites show a radial internal structure. Radial oolites probably form, in many cases, as a result of recrystallization. The coating on some ooliths consists only of one or two very thin layers with a total thickness less than that of the nucleus. Such ooids have been called **superficial oolites** or **pseudo-oolites.** Ooliths that have an internal structure similar to oolites but are much larger—that is, greater than 2 mm—are called **pisolites.** Pisolites are generally less spherical than oolites and are commonly crenulated. Some pisolites are of algal origin, formed by the trapping and

FIGURE 8.3 Skeletal grains cemented with sparry calcite cement. The skeletal materials include foraminifers (F), echinoderms (E), and molluscs (M). Polarized light.

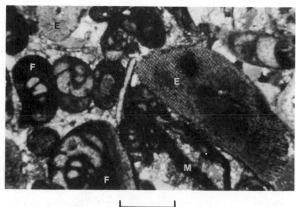

0   0.3 mm

binding activities of blue-green algae (cyanobacteria) in the same way stromatolites are formed (Chapter 4). Spheroidal stromatolites that reach a size exceeding 1–2 cm are called **oncolites.**

Peloids. Peloid is a nongenetic term for carbonate grains that are composed of micro-crystalline or cryptocrystalline calcite or aragonite but do not display distinctive internal structures (Fig. 8.5). Peloids are smaller than oolites and are generally silt to fine sand size (0.03–0.1 mm). The most common peloids are fecal pellets, produced by organisms that ingest calcium carbonate muds. Fecal pellets tend to be small, oval to rounded in shape, and uniform in size. They commonly have a high content of organic matter, which causes them to appear opaque or dark colored. Pellets can be differentiated from oolites by their lack of concentric or radial internal structure and from rounded intraclasts by their uniformity of shape, good sorting, and small size. Because they are produced by organisms, their sizes and shapes are not related to current transport, although pellets may be transported by currents and redeposited after initial deposition by organisms.

Peloids may be produced also by other processes, such as **micritization** of small oolites or rounded skeletal fragments by boring activities of certain organisms, particularly endolithic (boring) algae. These boring activities convert the original grains into a nearly uniform, homogeneous mass of microcrystalline calcite. Some peloids may simply be very small, well-rounded intraclasts of fine carbonate mud.

Lumps and Grapestones. Lump is a rather inelegant name that is applied to irregularly shaped composite or aggregate carbonate grains. Lumps appear to be peloids, or possibly coated grains, that are bound together at points of contact with dark-colored,

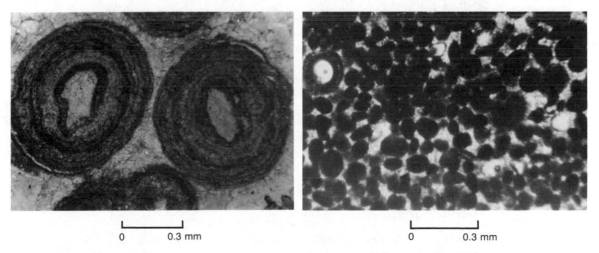

0   0.3 mm   0   0.3 mm

FIGURE 8.4 Large oolites cemented with sparry calcite cement. Note that each oolite has a distinct nucleus, probably a shell fragment, surrounded by well-developed concentric layers of calcite formed by precipitation around the nucleus. Polarized light.

FIGURE 8.5 Small, even-sized pellets cemented with sparry calcite cement. A few oolites (larger grains) are also visible. Note the lack of internal structure in the pellets compared to that of the oolites shown in Figure 8.4. Plane light.

organic-rich, very fine-crystalline calcium carbonate. Lumps in modern carbonate environments are composed mainly of aragonite; lumps in ancient limestones are dominantly calcite. The lumps in some modern environments such as the Bahama Banks resemble a bunch of grapes and are referred to as **grapestones** (Illing, 1954). Such lumps may be aggregates of fecal pellets. Lumps in modern environments can commonly be recognized by their botryoidal shapes and lack of internal structures; however, they can be confused with intraclasts. In fact, they are considered a type of intraclast by some geologists (Scholle, 1978). Lumps are only rarely reported in ancient limestones, possibly because their shapes become distorted beyond recognition during diagenesis.

Microcrystalline Calcite

Carbonate mud composed of very fine-size calcite crystals is present in many ancient limestones in addition to sand-size carbonate grains. Carbonate mud or lime mud occurs also in modern environments, where it is composed predominantly of needle-shaped crystals of aragonite about 1–5 microns (0.001–0.005 mm) in length. The carbonate mud in ancient limestones is composed of similar-sized crystals of calcite. Lime muds may also contain small amounts of fine-grained detrital minerals such as clay minerals, quartz, feldspar, and fine-size organic matter. They have a grayish to brownish, subtranslucent appearance under the microscope and are easily distinguished from carbonate grains and sparry calcite crystals (defined in the next section) by their extremely small crystal size (Fig. 8.6). Folk (1959) proposed the contraction **micrite** for microcrystalline calcite and the term has been universally adopted to signify very fine-grained carbonate sediments.

Micrite may be present as matrix among carbonate grains or it may make up most or all of a limestone. A limestone composed mostly of micrite is analogous texturally to a siliciclastic mudrock or shale. The presence of micrite in an ancient limestone is commonly interpreted to indicate deposition under quiet-water conditions, where little winnowing of fine mud takes place. By contrast, carbonate sediments deposited in environments where bottom currents or wave energy are strong are commonly mud free because carbonate mud is selectively removed in these environments. This winnowing process produces well-washed carbonate sands with unfilled intergranular voids that may be filled later, during diagenesis, with carbonate cement. On purely chemical considerations, carbonate mud, or micrite, can theoretically form by inorganic precipitation of aragonite, later inverted to calcite, from surface waters supersaturated with calcium bicarbonate. As discussed in Chapter 4, however, relatively little aragonite seems to be generated by inorganic processes in the modern ocean. Most carbonate muds appear to originate through organic processes.

Sparry Calcite

Some limestones contain large crystals of calcite, on the order of 0.02–0.1 mm, that appear clear or white when viewed with a hand lens or in plane light under a polarizing microscope. Such crystals are called sparry calcite. They are distinguished from micrite by their larger size and clarity and from allochems by their crystal shapes and lack of internal texture. Some sparry calcite can be seen under the microscope to fill interstitial pore spaces among grains or to fill solution cavities as a cement (Fig. 8.7). The presence of sparry calcite cement in intergranular pore spaces indicates that grain

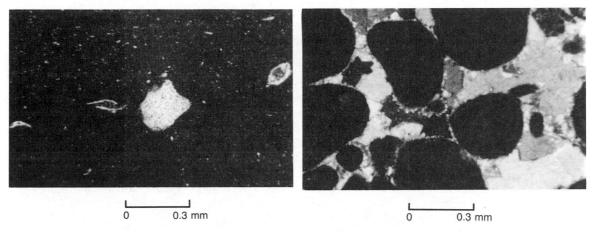

0   0.3 mm

0   0.3 mm

FIGURE 8.6 Dark, organic-rich micrite with a few small fossil fragments. Plane light.

FIGURE 8.7 Sparry calcite cementing rounded, dark-colored limeclasts. Note that the cement displays drusy texture. The calcite crystals around the margins of the clasts are small and display an orientation of their long dimensions perpendicular to the clast surfaces. These small, oriented crystals grade toward the center of the pores to larger, randomly oriented calcite crystals. Polarized light.

framework voids were empty of lime mud at the time of deposition, suggesting deposition under agitated-water conditions that removed fine mud, as mentioned.

Sparry calcite can form also in ancient limestones by recrystallization of primary depositional grains and micrite during diagenesis. Sparry calcite formed by recrystallization (Chapter 9) can be very difficult in some cases to differentiate from sparry calcite cement. It is important to make a distinction between the two types of sparry calcite because incorrect identification of recrystallized spar as sparry calcite cement can cause errors in both environmental interpretation and limestone classification.

Classification of Carbonate Rocks

Attempts to classify carbonate rocks date back to at least 1904 with the publication of Grabau's classic textbook on the classification of sedimentary rocks. Additional classifications were proposed by other authors in the thirties, forties, and fifties. Most of these early classifications were basically genetic schemes in which names such as "fore-reef talus limestone" or "low-energy limestone" were used to identify limestones according to their presumed environment of deposition (Ham and Pray, 1962). These classifications failed to recognize the clear distinction between carbonate grains and carbonate mud or to exploit difference in identity of the various kinds of carbonate grains. Publication in 1959 of Folk's largely descriptive "Practical Petrographic Classification of Limestones" marked the beginning of the modern period of limestone classification. In 1962 several additional classifications appeared (Ham, 1962); with one exception, these classifications are mainly descriptive. Unlike the confusion attending the proliferation of sandstone classifications, the appearance of several modern limestone classifications seems to have had a largely positive effect. It has forced geologists to become more keenly aware of the varied constituents that make up limestones, as well as the environmental significance of these constituents.

Mineralogy plays only a small role in classification of carbonate rocks because most carbonate rocks are essentially monomineralic. Mineralogy is used primarily to differentiate dolomite from limestone or carbonate rocks from noncarbonate rocks, as shown in Figure 8.8. The principal constituents or parameters used in modern carbonate classification are (1) the types of carbonate grains or allochems and (2) the grain/micrite ratio. The nature of the grain packing, or fabric, is also used in some classifications in which the fabric is referred to as either grain supported or mud supported. A grain-supported fabric is one in which grains are in contact, creating an intact grain

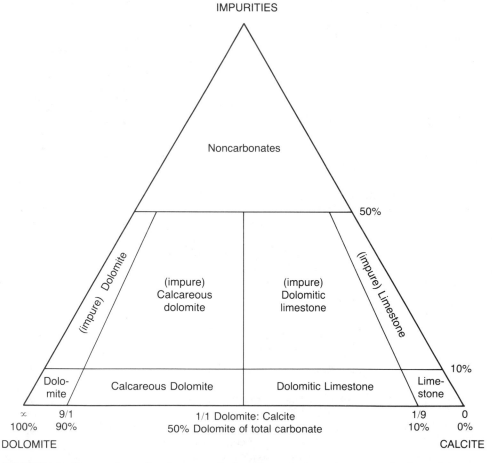

FIGURE 8.8 Terminology of carbonate rocks based on relative percentages of calcite, dolomite, and noncarbonate impurities. (After Leighton, M. W., and C. Pendexter, 1962, Carbonate rock types, *in* W. E. Ham (ed.), Classification of carbonate rocks: Am. Assoc. Petroleum Geologists Mem. 1. Fig. 2, p. 51, reprinted by permission of AAPG, Tulsa, Okla.)

TABLE 8.2 Classification of carbonate rocks

	Limestones, partly dolomitized limestones, and primary dolomites (see Notes 1 to 6)						Replacement dolomites[7] (V)	
	>10% Allochems — Allochemical rocks (I and II)		<10% Allochems — Microcrystalline rocks (III)		Undisturbed bioherm rocks (IV)		Allochem ghosts	No allochem ghosts
Volumetric allochem composition	Sparry calcite cement > microcrystalline ooze matrix (I) — Sparry allochemical rocks	Microcrystalline ooze matrix > sparry calcite cement (II) — Microcrystalline allochemical rocks	1–10% Allochems — Most abundant allochem	<1% Allochems			Evident allochem	
>25% Intraclasts (i)	Intrasparrudite (Ii:Lr) / Intrasparite (Ii:La)	Intramicrudite (IIi:Lr) / Intramicrite (IIi:La)	Intraclasts: intraclast-bearing micrite* (IIIi:Lr or La)	Micrite (IIIm:L); if disturbed, dismicrite (IIIm:X, L); if primary dolomite, dolomicrite (IIIm:D)	Biolithite (IV:L)		Finely crystalline intraclastic dolomite (Vi:D3) etc.	Medium crystalline dolomite (V:D4)
<25% Intraclasts, >25% Oölites (o)	Oosparrudite (Io:Lr) / Oosparite (Io:La)	Oomicrudite (IIo:Lr) / Oomicrite (IIo:La)	Oölites: oölite-bearing micrite* (IIIo:Lr or La)				Coarsely crystalline oölitic dolomite (Vo:D5) etc.	Finely crystalline dolomite (V:D3)
<25% Oölites, Volume ratio of fossils to pellets >3:1 (b)	Biosparrudite (Ib:Lr) / Biosparite (Ib:La)	Biomicrudite (IIb:Lr) / Biomicrite (IIb:La)	Fossils: fossiliferous micrite (IIIb:L, La or L1)				Aphanocrystalline biogenic dolomite (Vb:D1) etc.	etc.
3:1–1:3 (bp)	Biopelsparite (Ibp:La)	Biopelmicrite (IIbp:La)						
<1:3 (p)	Pelsparite (Ip:La)	Pelmicrite (IIp:La)	Pellets: pelletiferous micrite (IIIp:La)				Very finely crystalline pellet dolomite (Vp:D2) etc.	

*Designates rare rock types.

[1] Names and symbols in the body of the table refer to limestones. If the rock contains more than 10 percent replacement dolomite, prefix the term "dolomitized" to the rock name, and use DLr or DLa for the symbol (e.g., dolomitized intrasparite, Li:DLa). If the rock contains more than 10 percent dolomite of uncertain origin, prefix the term "dolomitic" to the rock name, and use dLr or dLa for the symbol (e.g., dolomitic pelsparite, Ip:dLa). If the rock consists of primary (directly deposited) dolomite, prefix the term "primary dolomite" to the rock name, and use Dr or Da for the symbol (e.g., primary dolomite intramicrite, IIi:Da). Instead of "primary dolomite micrite" (IIIm:D) the term "dolomicrite" may be used.

[2] Upper name in each box refers to calcirudites (median allochem size larger than 1.0 mm); lower name refers to all rocks with median allochem size smaller than 1.0 mm. Grain size and quantity of ooze matrix, cements, or terrigenous grains are ignored

[3] If the rock contains more than 10 percent terrigenous material, prefix "sandy," "silty," or "clayey" to the rock name and "Ts," "Tz," or "Tc" to the symbol, depending on which is dominant (e.g., sandy biosparite, Tslb:La; or silty dolomitized pelmicrite Tzllp:DLa). Glauconite, collophane, chert, pyrite, or other modifiers may also be prefixed.

[4] If the rock contains other allochems in significant quantities that are not mentioned in the main rock name, these should be prefixed as qualifiers proceeding the main rock name (e.g., fossiliferous intrasparite, oölitic pelmicrite, pelletiferous oösparite, or intraclastic biomicrudite). This can be shown symbolically as Ii(b), Io(p), IIb(i), respectively.

[5] If the fossils are of rather uniform type or one type is dominant, this fact should be shown in the rock name (e.g., pelecypod biosparrudite, crinoid biomicrite).

[6] If the rock was originally microcrystalline and can be shown to have recrystallized to microspar (5–15 microns, clear calcite) the terms "microsparite," "biomicrosparite," etc., can be used instead of "micrite" or "biomicrite."

[7] Specify crystal size as shown in the examples.

Source: Folk, R. L., 1962, Spectral subdivision of limestone types, in W. E. Ham (ed.), Classification of carbonate rocks: Am. Assoc. Petroleum Geologists Mem. 1. Table 1, p. 70, reprinted by permission of AAPG, Tulsa, Okla.

framework in which voids may or may not be filled with mud (matrix). In a mud-supported fabric, most grains do not touch—they appear to float in the carbonate mud.

Folk's (1959, 1962) classification has probably been the most widely accepted modern limestone classification because of its applicability to a wide range of carbonate rock types and the ease with which its terms can be utilized and understood. The classification is based on the relative abundance of three major types of constituents: (1) carbonate grains, or allochems; (2) microcrystalline carbonate mud (micrite); and (3) sparry calcite cement. Folk refers to micrite and sparry calcite cement as orthochemical particles, or **orthochems.** As illustrated in Table 8.2, classification is made by first determining the relative abundance of total allochems to micrite plus sparry calcite cement. Further subdivision is then made on the basis of relative abundance of the various types of carbonate grains (Fig. 8.9) and the abundance of micrite relative to sparry calcite cement. This classification approach yields a bipartite name that re-

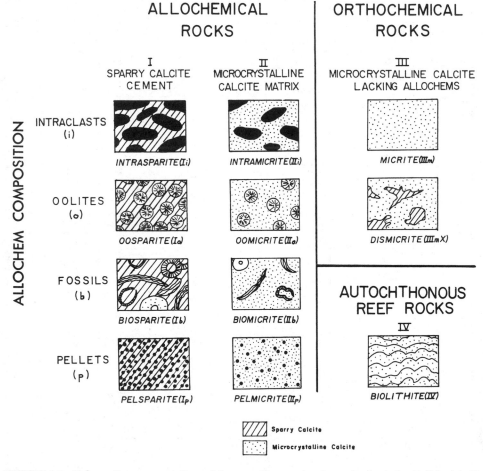

FIGURE 8.9 Schematic representation of the constituents that form the basis for Folk's classification of carbonate rocks (Table 8.2). (After Folk, R. L., 1962, Spectral subdivision of limestone types, in W. E. Ham (ed.), Classification of carbonate rocks: Am. Assoc. Petroleum Geologists Mem. 1. Fig. 3, p. 71, reprinted by permission of AAPG, Tulsa, Okla.)

flects both the major type of carbonate grain in the limestone and the relative abundance of micrite and sparry calcite cement. Thus, an **oosparite** is an oolite-rich rock cemented with sparry calcite and containing little or no micrite, whereas an **oomicrite** is an oolite-rich limestone in which micrite is abundant and sparry calcite is subordinate. Additional textural information can be added by use of the textural maturity terms shown in Figure 8.10. Thus, a **packed oomicrite** indicates a grain-supported oolitic limestone, and a **sparse oomicrite** is an oolitic rock with a mud-supported fabric. Note that Folk's classification can be used also to classify dolomite rock if "ghosts" of the original allochems are still identifiable in the dolomite.

The terms used by Folk (1959, 1962) to differentiate depositional textures are purely descriptive; however, they also have a genetic connotation. The term **biomicrite,** for example, conveys an interpretation of deposition under quiet-water conditions where micrite is abundant and winnowing of the lime mud is minimal. Thus micrite accumulates along with skeletal particles. On the other hand, the term **biosparite** suggests deposition in a wave-agitated environment where micrite is removed by winnowing currents, allowing mud-free carbonate grains to accumulate. These grains are subsequently cemented with sparry calcite during diagenesis.

Dunham (1962) has a somewhat different type of classification that stresses the relative abundance of allochems and micrite, but does not consider the different kinds of carbonate grains. Dunham's classification is based solely upon depositional texture and considers two aspects of texture: (1) grain packing and the relative abundance of grains to micrite and (2) depositional binding of grains. By depositional binding, I mean whether or not carbonate grains show evidence of having been bound together at the time of deposition, as in a colonial reef complex or algal stromatolite bed. Dunham's classification (Fig. 8.11) separates components that were not bound together at the time of deposition into those that lack lime mud and those that contain lime mud. Rocks that contain no mud are obviously grain supported. Rocks that contain mud

Percent Allochems	OVER 2/3 LIME MUD MATRIX				SUBEQUAL SPAR & LIME MUD	OVER 2/3 SPAR CEMENT		
	0-1 %	1-10 %	10-50%	OVER 50%		SORTING POOR	SORTING GOOD	ROUNDED & ABRADED
Representative Rock Terms	MICRITE & DISMICRITE	FOSSILI- FEROUS MICRITE	SPARSE BIOMICRITE	PACKED BIOMICRITE	POORLY WASHED BIOSPARITE	UNSORTED BIOSPARITE	SORTED BIOSPARITE	ROUNDED BIOSPARITE
Terminology	Micrite & Dismicrite	Fossiliferous Micrite	B i o m i c r i t e			B i o s p a r i t e		
Terrigenous Analogues	C l a y s t o n e		Sandy Claystone	Clayey or Immature Sandstone		Submature Sandstone	Mature Sandstone	Supermature Sandstone

■ LIME MUD MATRIX   ▨ SPARRY CALCITE CEMENT

FIGURE 8.10 Textural classification of carbonate sediments based on relative abundance of lime mud matrix and sparry calcite cement, and on the abundance and sorting of carbonate grains (allochems). (After Folk, R. L., 1962, Spectral subdivision of limestone types, *in* W. E. Ham (ed.), Classification of carbonate rocks: Am. Assoc. Petroleum Geologists Mem. 1. Fig. 4, p. 76, reprinted by permission of AAPG, Tulsa, Okla.)

DEPOSITIONAL TEXTURE RECOGNIZABLE					DEPOSITIONAL TEXTURE NOT RECOGNIZABLE
Original components not bound together during deposition				Original components were bound together during deposition . . . as shown by intergrown skeletal matter, lamination contrary to gravity, or sediment-floored cavities are roofed over by organic or questionably organic matter and are too large to be interstices.	CRYSTALLINE CARBONATE
Contains mud (particles of clay and fine silt size)			Lacks mud and is grain-supported		
Mud-supported		Grain-supported			
Less than 10% grains	More than 10% grains				(Subdivide according to classifications designed to bear on physical texture or diagenesis.)
MUDSTONE	WACKESTONE	PACKSTONE	GRAINSTONE	BOUNDSTONE	

FIGURE 8.11 Classification of carbonate rocks on the basis of depositional texture. (After Dunham, R. J., 1962, Classification of carbonate rocks according to depositional texture, *in* W. E. Ham (ed.), Classification of carbonate rocks: Am. Assoc. Petroleum Geologists Mem. 1. Table 1, p. 117, reprinted by permission of AAPG, Tulsa, Okla.)

may be either grain supported or mud supported. Note, however, that grain support does not depend upon the absolute grain-to-mud ratio because grain support is a function also of shapes of the carbonate grains. Platy or elongate grains such as bivalve shells may form a grain-supported fabric at much lower grain abundances than more spherical particles such as oolites. Therefore, Dunham's boundary between grain-supported and mud-supported limestones is not based on a fixed grain/micrite ratio. Because Dunham's classification does not consider the identity of the carbonate grains, it is commonly used in conjunction with another classification such as Folk's. Thus, a limestone identified as a packed oomicrite in Folk's classification could alternatively be called an **oomicrite packstone** in a combination of Folk's and Dunham's classifications. Additional limestone classifications are discussed in the symposium volume edited by Ham (1962).

Classification of Mixed Carbonate and Siliciclastic Rocks

The classifications of carbonate rocks presented are designed primarily for classification of relatively pure carbonate rocks containing few siliciclastic constituents. Likewise the sandstone classifications discussed in Chapter 7 are applicable mainly to classification of sedimentary rocks composed predominantly of siliciclastic constituents. None of these classifications is particularly suitable for classifying mixed carbonate and siliciclastic sediments. Although such mixed sediments are not abundant in the geologic record, they are common enough to merit consideration in classification schemes for sedimentary rocks. Furthermore, the fact that carbonate and siliciclastic constituents are mixed together has some important environmental implications.

Only a few attempts have been made to classify mixed carbonate and siliciclastic sedimentary rocks (e.g., Mount, 1985; Zuffa, 1980). Most workers have been content to use such informal terms as calcareous sandstone or sandy limestone for these mixed rocks. Perhaps the most useful of the formal classification schemes proposed for mixed sediments is that of Mount (1985), who attacked the classification problem by grouping siliciclastic and carbonate constituents into four components: siliciclastic sand, siliciclastic mud, carbonate allochems, and carbonate mud. These components are used as end members of a tetrahedral diagram, yielding four triangular classification diagrams—one on each face of the tetrahedron. When the tetrahedron is unfolded, this tetrahedral classification system appears as shown in Figure 8.12. Thus, depending upon the mix of carbonate and siliciclastic constituents, a mixed sedimentary rock might be called by such names as an **allochemic sandstone, sandy allochem limestone,** or **micritic mudrock.** Note that these names do not reflect the mineral composition of the siliciclastic rocks or the kinds of carbonate particles in the limestones. To provide additional information of this type, adjectives can be added as prefixes to yield names such as **feldspathic sandy allochem limestone** and **fossil-bearing allochemic mudstone.** Unfortunately, complex classifications schemes of this type tend to get a bit unwieldy. Many geologists are likely to continue using informal terms such as fossiliferous mudstone instead of fossil-bearing allochemic mudstone.

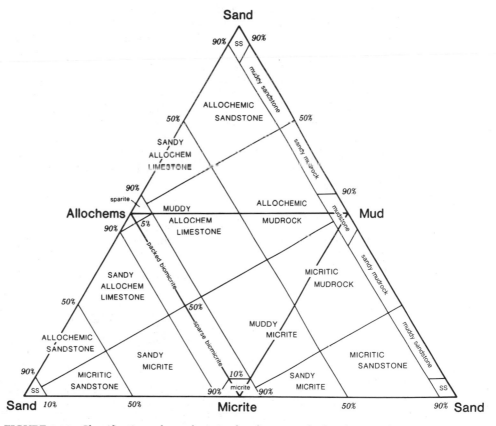

FIGURE 8.12 Classification scheme for mixed carbonate and siliciclastic sediments. The diagram represents an unfolded tetrahedron. The central triangle is the base of the tetrahedron. The three triangles that radiate from the central triangle are the upper three faces of the tetrahedron. (From Mount, J., Mixed siliciclastic and carbonate sediments: A proposed first-order textural and compositional classification: Sedimentology, v. 32. Fig. 2, p. 438, reprinted by permission of Elsevier Science Publishers, Amsterdam.)

8.3 EVAPORITES

Composition

Evaporites include all deposits that are precipitated from solutions concentrated by evaporation (Chapter 4). Evaporites occur in rocks of all ages but are particularly common in Cambrian, Permian, Triassic, and some Jurassic and Cretaceous sequences (Ronov et al., 1980). Although the total volume of evaporites in the geologic record is much less than that of carbonate rocks, some individual evaporite deposits reach thicknesses of several hundred meters. Evaporite deposits are composed predominantly of varying proportions of halite (rock salt), anhydrite, and gypsum. Deposits range from those that are composed almost entirely of anhydrite and gypsum to those that are mainly halite.

Although approximately eighty minerals have been reported from evaporite deposits (Stewart, 1963), only about a dozen of these minerals are common enough to be considered important evaporite rock formers. Evaporites can be classified basically into those of marine origin and those of nonmarine origin. If carbonate minerals, most of which are not evaporites, are excluded, the most common marine evaporite minerals are the calcium sulfate minerals gypsum and anhydrite. Halite is next in abundance, followed by the potash salts sylvite, carnallite, langbeinite, polyhalite, and kainite, and the magnesium sulfate kieserite. The marine evaporite minerals can be grouped for the purpose of classification into chlorides, sulfates, and carbonates (Table 8.3).

Marine evaporites commonly contain mixtures of minerals, although gypsum or anhydrite and halite predominate in most horizons. Evaporite deposits may also contain various amounts of impurities such as clay minerals, quartz, feldspar, or sulfur. Only a few rock names have been applied to evaporite deposits. Rocks composed predominantly of the mineral halite are called halite or **rock salt.** Rocks made up predom-

TABLE 8.3 Classification of marine evaporites on the basis of mineral composition

Mineral class	Mineral name	Chemical composition	Rock name
Chlorides	Halite	NaCl	Halite; rock salt
	Sylvite Carnallite	KCl $KMgCl_3 \cdot 6H_2O$	Potash salts
Sulfates	Langbeinite Polyhalite Kainite	$K_2Mg_2(SO_4)_3$ $K_2Ca_2Mg(SO_4)_6 \cdot H_2O$ $KMg(SO_4)Cl \cdot 3H_2O$	
	Anhydrite Gypsum Kieserite	$CaSO_4$ $CaSO_4 \cdot 2H_2O$ $MgSO_4 \cdot H_2O$	Anhydrite; anhydrock Gypsum; gyprock —
Carbonates	Calcite Magnesite Dolomite	$CaCO_3$ $MgCO_3$ $CaMg(CO_3)_2$	Limestone — Dolomite; dolostone

Source: Data from Stewart, F. H., 1963, Marine evaporites, in M. Fleischer (ed.), Data of geochemistry: U.S. Geol. Survey Prof. Paper 440-Y; Borchert, H., and R. O. Muir, 1964, Salt deposits: The origin, metamorphism, and deformation of evaporites: Van Nostrand, London.

inantly of gypsum or anhydrite are simply called gypsum or anhydrite, although some geologists use the names **rock gypsum** or **rock anhydrite.** Less commonly they are called gyprock and anhydrock. Few evaporite beds are composed predominantly of minerals other than the calcium sulfates and halite. No formal names have been proposed for rocks enriched in other evaporite minerals, although the term potash salts is used informally for potassium-rich evaporites.

Nonmarine evaporites are characterized by evaporite minerals that are not common in marine evaporites. These minerals may include trona [$Na_3H(CO_3)_2 \cdot 2H_2O$], mirabilite [$Na_2SO_4 \cdot 10H_2O$], glauberite [$Na_2Ca(SO_4)$], borax [$Na_2B_4O_5(OH)_4 \cdot 8H_2O$], epsomite [$MgSO_4 \cdot 7H_2O$], thenardite [$NaSO_4$], gaylussite [$Na_2CO_3 \cdot CaCo_3 \cdot 5H_2O$], and bloedite [$Na_2SO_4 \cdot MgSO_4 \cdot 4H_2O$]. Nonmarine deposits may also contain anhydrite, gypsum, and halite.

Structural Classification of Evaporites

Calcium sulfates are deposited predominantly as gypsum. Gypsum can be altered into, and pseudomorphosed by, anhydrite while the sediments are still in their general depositional environment (Shearman, 1978). Gypsum is also dehydrated to anhydrite with burial to a few hundred meters; this loss of water is accompanied by a 38 percent decrease in solid volume of the gypsum. Because of this rapid dehydration with burial, most ancient calcium sulfate deposits are composed of anhydrite. Anhydrite can be hydrated back to gypsum after uplift and exposure to low-salinity surface waters, with an accompanying increase in volume. These volume changes can cause distortion of original depositional structures and textures; many calcium sulfate deposits are characterized by distorted fabrics. Maiklem et al. (1969) propose a structural classification for anhydrites based on fabric, bedding, and the presence or absence of distortion. This classification divides anhydrites into about two dozen structural types. These structural types can be lumped into three fundamental structural groups: nodular anhydrites, laminated anhydrites, and massive anhydrites.

Nodular anhydrites are irregularly shaped lumps of anhydrite that are partly or completely separated from each other by a salt or carbonate matrix (Fig. 8.13). Maiklem et al. (1969) make a distinction between nodular anhydrite and **mosaic anhydrite,** in which the anhydrite masses or lumps are approximately equidimensional and are separated by very thin stringers of dark carbonate mud or clay. Many other authors do not make this distinction and use the term nodular anhydrite in a general sense to include mosaic anhydrite. The term **chickenwire** structure is used for a particular type of mosaic or nodular anhydrite that consists of slightly elongated, irregular polygonal masses of anhydrite separated by thin dark stringers of other minerals such as carbonate or clay minerals (Fig. 8.14).

The formation of nodular anhydrite is initiated by displacive growth of gypsum in carbonate or clayey sediments. Gypsum crystals subsequently alter to anhydrite pseudomorphs, which continue to enlarge by addition of Ca^{2+} and SO_4^{2-} from an external source, ultimately growing displacively into anhydrite nodules. Chickenwire anhydrite forms when, with increasing size, the nodules ultimately coalesce and interfere. Most of the enclosing sediment is pushed aside and what remains forms thin stringers between the nodules (Shearman, 1978).

Nodular anhydrites have been observed in many modern coastal **sabkha** environments. Coastal sabkhas are flat, salt-encrusted supratidal zones, in areas such as the Trucial Coast of the Persian Gulf, where climates are arid and evaporation rates high (Fig. 4.6). Because of the occurrence of nodular anhydrites in sabkha environments,

FIGURE 8.13 Nodular anhydrite in a core sample of the Buckner Anhydrite (Jurassic), Texas. Dark-colored carbonate separates and surrounds the lighter colored anhydrite nodules.

many geologists have assumed that nodular anhydrites found in ancient evaporite deposits also signify deposition under subaerial, sabkha conditions. Dean et al. (1975) point out, however, that both nodular anhydrites and laminated anhydrites, which form in standing water, occur together in some environments. This association is taken to mean that nodular anhydrites can form also in deeper water environments. In fact, all that is needed for formation of nodular anhydrite is growth of crystals in mud in contact with highly saline brines, which can occur in deep or shallow standing water as well as in a sabkha environment. Therefore, caution must be used in interpreting all ancient nodular anhydrites as sabkha evaporites.

   Laminated anhydrites, sometimes called laminites, consist of thin, nearly white anhydrite or gypsum laminations that alternate with dark-gray or black laminae rich in dolomite or organic matter (Fig. 8.15). These laminae are commonly only a few millimeters thick and rarely reach one centimeter. Many of these thin laminae are remarkably uniform, with sharp planar contacts that can be traced for long distances laterally. Also, they may make up vertical sequences hundreds of meters thick in

FIGURE 8.14 Chickenwire structure in anhydrite. Evaporite series of the Lower Lias (Jurassic), Aquitaine Basin, southwest France. (From Bouroullec, J., 1981, Sequential study of the top of the evaporitic series of the Lower Lias in a well in the Aquitaine Basin (Auch 1), southwestern France, *in* Chambre Syndical de la Recherche et de la Production due Pétrole et du Gaz Naturel (eds.), Evaporite deposits: Illustration and interpretation of some environmental sequences. Pl. 36, p. 157, reprinted by permission of Editions Technip, Paris, and Gulf Publishing Co., Houston, Tex. Photograph courtesy of J. Bouroullec.)

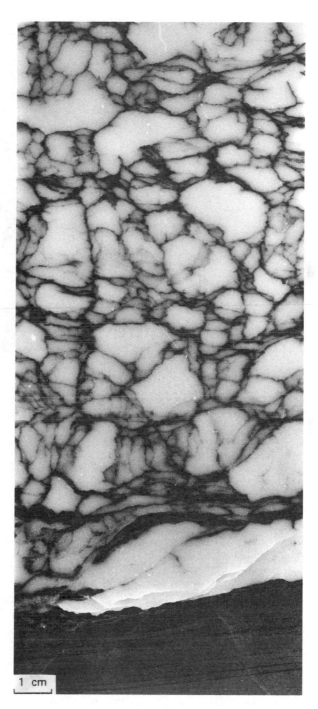

1 cm

FIGURE 8.15 Laminated anhydrite from the Prairie Evaporite
(Devonian), Canada.

which hundreds of thousands of laminae may be present. Alternating light and dark
pairs of bands have been suggested to be annual varves resulting from seasonal
changes in water chemistry and temperature; however, they might equally well represent cyclic changes or disturbances of longer duration. Also, laminae of anhydrite
can alternate with thicker layers of halite, producing laminated halite.

Individual anhydrite or gypsum laminae have been reported to be traceable for
distances up to 290 km (Anderson and Kirkland, 1966). Because of this lateral persistence, which indicates uniform depositional conditions over a wide area, laminites are
commonly interpreted to form by precipitation of evaporites in quiet water below
wave base. They could presumably form either in a shallow-water area protected in
some manner from strong bottom currents and wave agitation or in a deeper water
environment. Some ancient laminated sulfate deposits have been interpreted as algal
stromatolites. Such an origin is unlikely, however, because algal laminations are either
wavy and discontinuous or, if regular and parallel, can be traced only for short distances laterally in contrast to laterally persistent laminites. It has been proposed that
laminated anhydrite can form also by coalescing of anhydrite nodules, which by continued growth in a lateral direction merge into one another to produce a layer. Layers
formed by this mechanism are thought to be thicker and less distinct and continuous
than laminae formed by precipitation. A special type of contorted layering that has
resulted from coalescing nodules has been observed in some modern sabkha deposits

where continued growth of nodules creates a demand for space. The lateral pressures that result from this demand cause the layers to become contorted, forming ropy bedding or **enterolithic structures** (Fig. 8.16).

   Massive anhydrite is anhydrite that lacks perceptible internal structures. True massive anhydrite appears to be less common than nodular and laminated anhydrite, and little information is available regarding its formation. Presumably, it represents sustained, uniform conditions of deposition. Haney and Briggs (1964) suggest that massive anhydrite forms by evaporation at brine salinities of approximately 200–275 parts per thousand (‰), just below the salinities at which halite precipitation begins.

8.4 SILICEOUS SEDIMENTARY ROCKS

Siliceous sedimentary rocks are fine grained, dense, very hard rocks composed predominantly of the SiO_2 minerals quartz, chalcedony, and opal, with minor impurities such as siliciclastic grains and diagenetic minerals. **Chert** is the general term used for siliceous rocks as a group, but several other names are also applied to specific types of chert. **Flint** is used both as a synonym for chert and for a variety of chert, particularly chert nodules that occur in Cretaceous chalks. **Jasper** is a variety of chert colored red by impurities of disseminated hematite. Jasper interbedded with hematite in Precambrian iron formations is called **jaspilite**. **Novaculite** is a very dense, fine-grained, even-textured chert that occurs mainly in mid-Paleozoic rocks of the Arkansas, Oklahoma, and Texas region of south-central United States. **Porcellanite** is a term used for fine-grained siliceous rocks with a texture and fracture resembling that of unglazed porcelain. **Siliceous sinter** is porous, low-density, light-colored siliceous rock deposited by waters of hot springs and geysers. Although most siliceous rocks consist predominantly of chert, some have high contents of detrital clays or micrite. These impure cherts grade into siliceous shales or siliceous limestones.

FIGURE 8.16 Enterolithic structure in nodular gypsum of the Grenada Basin, southern Spain. (Photograph courtesy of J. M. Rouchy.)

Classification and Occurrence

Siliceous sedimentary rocks can be divided on the basis of gross morphology into two principal types: (1) bedded cherts and (2) nodular cherts. Bedded cherts are further distinguished by their content of siliceous organisms of various kinds. Mineralogy cannot be used as a basis for classifying siliceous sedimentary rocks because these rocks are all composed mainly of fine-size quartz (chert). The principal distinguishing characteristics of bedded and nodular cherts are described in the following paragraphs.

Bedded Chert

Bedded chert consists of layers of nearly pure chert ranging up to several centimeters in thickness and commonly interbedded with millimeter-thick partings, or laminae, of siliceous shale (Fig. 8.17). Bedding may be even and uniform or show pinching and swelling. Most chert beds lack internal sedimentary structures; however, graded bedding, cross-bedding, ripple marks, and sole markings have been reported in some cherts. The presence of these structures indicates that mechanical transport was involved in deposition of these rocks. Bedded cherts are commonly associated with submarine volcanic rocks, pelagic limestones, and siliciclastic or carbonate turbidites (Garrison, 1974; Nisbet and Price, 1974). Many bedded cherts are composed predominantly of the remains of siliceous organisms, which are commonly altered to some degree by solution and recrystallization (Chapter 9). Bedded cherts can be subdivided on the basis of type and abundance of siliceous organic constituents into four principal types: (1) diatomaceous deposits, (2) radiolarian deposits, (3) siliceous spicule deposits, and (4) bedded cherts containing few or no siliceous skeletal remains.

Diatomaceous Deposits. Diatomaceous deposits include both diatomites and diatomaceous cherts. **Diatomites** are light-colored, soft, friable siliceous rocks composed

FIGURE 8.17 Thin, well-bedded cherts in the Mino Belt Group (Triassic), near Inuyama, Honshu, Japan.

chiefly of the opaline frustules of diatoms, a unicellular aquatic plant related to the algae. Thus, they are fossil diatomaceous oozes. Diatomites of both marine and lacustrine origin are recognized. Marine diatomites are commonly associated with sandstones, volcanic tuffs, mudstones or clay shales, impure limestones (marls), and, less commonly, gypsum. Lacustrine diatomites are almost invariably associated with volcanic rocks. **Diatomaceous chert** consists of beds and lenses of diatomite that have well-developed silica cement or groundmass that has converted the diatomite into dense, hard chert. Beds of diatomaceous chert making up strata several hundred meters in thickness have been reported from some sedimentary sequences such as the Miocene Monterey Formation of California (Garrison et al., 1981).

Radiolarian Deposits. Radiolarian deposits consist predominantly of the remains of radiolarians, which are marine planktonic protozoans with a latticelike skeletal framework. Radiolarian deposits can be divided into radiolarite and radiolarian chert. **Radiolarite** is the comparatively hard, fine-grained, chertlike equivalent of radiolarian ooze, that is, indurated radiolarian ooze. **Radiolarian chert** is well-bedded, microcrystalline radiolarite that has a well-developed siliceous cement or groundmass. Radiolarian cherts are commonly associated with tuffs, mafic volcanic rocks such as pillow basalts, pelagic limestones, and turbidite sandstones that are believed to indicate a deep-water origin. On the other hand, some radiolarian cherts are associated with micritic limestones and other rocks that suggest deposition at shallower depths of perhaps 200 1000 m (Iijima et al., 1979).

Siliceous Spicule Deposits. Spicularite (spiculite) is a siliceous rock composed principally of the siliceous spicules of invertebrate organisms, particularly sponges. Spicularite is loosely cemented in contrast to **spicular chert,** which is hard and dense. Spicular cherts are mainly marine in origin and occur associated with glauconitic sandstones, black shales, dolomite, argillaceous limestones, and phosphorites. They are not generally associated with volcanic rocks and are probably deposited mainly in relatively shallow water a few hundred meters deep.

Bedded Cherts Containing Few or No Siliceous Skeletal Remains. Many bedded chert deposits that have been described contain few or no recognizable remains of siliceous organisms. Some of these reported occurrences of barren cherts may simply be the result of inadequate microscopic examination of the cherts, which might be found upon closer examination to contain siliceous organisms. Other samples have been examined closely and clearly contain few siliceous organisms. Cherts in this latter group include most cherts associated with the Precambrian iron formations, as well as many Phanerozoic-age cherts such as the Mississippian–Devonian-age Arkansas Novaculite of Arkansas and Oklahoma and the Caballos Novaculite of Texas. Except for the absence of skeletal remains, these cherts resemble radiolarian cherts both megascopically and in their lithologic associations (Cressman, 1962).

As discussed in Chapter 4, the origin of cherts that do not contain siliceous organic remains is poorly understood. Direct, inorganic precipitation of amorphous silica has been reported in some ephemeral Australian lakes (Peterson and von der Borch, 1965). In the open marine environment, however, there have been no reports of similar occurrences that could help explain the presence of widespread nonfossiliferous chert deposits such as the Arkansas Novaculite. The scarcity of radiolarians and sponge spicules in the Arkansas Novaculite and similar Phanerozoic-age cherts does not preclude the possibility that these cherts were formed by organisms. They could have been derived from siliceous oozes that were subsequently almost completely dissolved and recrystallized, leaving few recognizable siliceous organic remains (Weaver and Wise, 1974). We simply don't know how Precambrian bedded cherts were formed during a time when siliceous organisms are not known to have existed.

Nodular Cherts

Nodular cherts are subspheroidal masses, lenses, or irregular layers or bodies that range in size from a few centimeters to several tens of centimeters (Fig. 8.18). They commonly lack internal structures, but some nodular cherts contain silicified fossils or relict structures such as bedding. Colors of these cherts vary from green to tan and black. They typically occur in shelf-type carbonate rocks, and they tend to be concentrated along certain horizons parallel to bedding. More rarely they occur in sandstones and mudrocks, lacustrine sediments, and evaporites. Nodular cherts originate mainly by diagenetic replacement of carbonate minerals and fossils; they occur in both limestones and dolomites. Diagenetic origin is clearly demonstrated in many nodules by the presence of partly or wholly silicified remains of calcareous fossils or ooliths. Nodular cherts have also been reported to form by replacement of anhydrite.

Mineralogy and Texture

Chert that makes up siliceous sedimentary rocks is composed mainly of fine-size quartz; depending upon age, minor opal may also be present. Cherts can be divided into three main textural types: (1) **microquartz,** consisting of nearly equidimensional grains of quartz 1–5 microns in size; (2) **chalcedonic quartz,** forming sheaflike bundles of radiating, extremely thin crystals about 0.1 mm long; and (3) **megaquartz,** composed of equant to elongated grains greater than 20 microns in length (Folk, 1974). The silica in siliceous organisms is amorphous silica or opal, commonly called opal-A. Therefore, opal-A is present in some cherts, particularly those of Tertiary age. Opal-A is metastable and crystallizes in time to opal-CT and finally to chert (Chapter 4). Depending upon the age of the chert deposits and the conditions of burial, all gradations, from pure opal to pure chert, may be present. Opal is found mainly in rocks of Mesozoic and younger ages.

FIGURE 8.18 Nodular chert (arrows) in limestones of the Onondaga Formation (Devonian), New York. (Photograph by E. M. Baldwin.)

Chemical Composition

Cherts are composed predominantly of SiO_2 but may also include minor amounts of Al, Fe, Mn, Ca, Na, K, Mg, and a few other elements. These additional elements are contained mainly in impurities such as authigenic hematite and pyrite, siliciclastic minerals, and pyroclastic particles. The amount of SiO_2 differs markedly in different types of cherts, ranging from more than 99 percent in very pure cherts such as the Arkansas Novaculite to less than 65 percent in some nodular cherts (Cressman, 1962). Aluminum is commonly the second most abundant element in cherts, followed by Fe, Mg or K, Ca, and Na.

8.5 IRON-BEARING SEDIMENTARY ROCKS

Some iron is present in almost all sedimentary rocks. For example, the average iron content of siliciclastic mudrocks is 4.8 percent. Sandstones contain 2.4 percent iron on the average, and limestones contain 0.4 percent (Blatt, 1982). Iron-rich sedimentary rocks have much higher concentrations of iron than these average sedimentary rocks. The term "iron-rich" is usually reserved for sedimentary rocks that contain at least 15 percent iron, corresponding to 21.3 percent Fe_2O_3 or 19.4 percent FeO. Iron-rich rocks can be classified into three broad groups: detrital chemical iron-rich sediments, iron-rich shales, and miscellaneous iron-rich deposits (Table 8.4). Of these, iron formations

TABLE 8.4 Principal classes of iron-rich sedimentary rocks

I. Detrital chemical iron-rich sediments
  A. Cherty iron formation
     Texture: analogous to limestone texture
     Composition: iron-rich chert containing hematite, magnetite, siderite, ankerite, or (predominantly alumina-poor) silicates as predominating iron minerals; relatively poor in Al and P
  B. Minette-type ironstone
     Texture: analogous to limestone texture
     Composition: aluminous iron silicates (chamosite, chlorite, stilpnomelane), iron oxides, and carbonates; relatively rich in Al and P

II. Iron-rich shales
  C. Pyritic shales
     Bituminous shales containing nodules or laminae of pyrite; grade into massive pyrite bodies by coalescence of pyrite laminae and nodules
  D. Siderite-rich shales
     Bituminous shales with siderite concretions; grade into massive siderite bodies by coalescence of concretions

III. Miscellaneous iron-rich deposits
  E. Iron-rich laterites
  F. Bog iron ores
  G. Manganese nodules and oceanic iron crusts
  H. Iron-rich muds precipitated from hydrothermal brines, Lahn-Dill type iron oxide ores, and stratiform, volcanogenic sulfide deposits
  I. Placers of magnetite, hematite, or ilmenite sand

Source: Modified slightly from Dimroth, E., 1979, Models of physical sedimentation of iron formations, in R. G. Walker (ed.), Facies models: Geoscience Canada Reprint Ser. 1. Table I, p. 175, reprinted by permission of Geological Association of Canada.

and ironstones, which are detrital chemical iron-rich sediments, are the only volumetrically important iron-rich sedimentary rocks. Iron formations and ironstones are differentiated on the basis of age, physical characteristics, and mineralogy, as shown in Table 8.5.

Iron Formations and Ironstones

Iron Formations

Iron formations are iron-rich deposits that range in age from early Precambrian to Cambrian, although they are primarily of Precambrian age. They consist of distinctively banded sequences (Fig. 8.19) 50–600 m thick composed of layers enriched in iron alternating with layers rich in chert. Cherty iron formations can grade into slightly cherty iron-rich sandstones, siltstones, and shales. The textures of iron formations resemble those of limestones. Dimroth (1979) recognizes textural types in iron formations equivalent to micritic, pelleted, intraclastic, oolitic, pisolitic, and stromatolitic limestone textures (Table 8.6). Sedimentary structures reported from banded iron formations include cross-bedding, graded bedding, load casts, ripple marks, erosion channels, shrinkage cracks, and slump structures. These structures show that many of the constituents of iron formations have undergone mechanical transport and deposition.

TABLE 8.5   Differences between ironstone and iron formation

	Ironstone	Iron formation
Age	Pliocene to Middle Precambrian; principal beds from Lower Paleozoic and Jurassic	Cambrian to Early Precambrian; principal formations approximately 2000 million years old
Thickness	Major units a few meters to a few tens of meters	Major units 50–600 m
Original areal extent	Individual depositional basins rarely more than 150 km in maximum dimension	Difficult to determine; some deposits with continuity over many hundreds of kilometers
Physical character	Massive to poorly banded; silicate and oxide facies oolitic	Thinly bedded; layers of dominantly hematite, magnetite, siderite, or silicate alternating with chert, which makes up approximately half the rock; oolites rare
Mineralogy	Dominant oxide goethite; hematite very common; magnetite relatively rare; chamosite primary silicate; calcite and dolomite common constituents	No goethite; magnetite and hematite about equally abundant; primary silicate greenalite; chert a major constituent; dolomite present in some units, but calcite rare or absent
Chemistry	Except for high iron content, no distinctive aspects	Remarkably low content of Na, K, and Al; low P
Associated rocks	Both typically interbedded with shale, sandstone, or graywacke; yet the iron formation has few or no clastics compared to the ironstone.	
Relative abundance of facies	No gross differences apparent; probable order of abundance for ironstone: oxide, silicate, siderite, sulfide; for iron formation the order is similar, but siderite facies may be more abundant than silicate facies	

Source: James, H. L., 1966, Chemistry of the iron-rich sedimentary rocks, *in* M. Fleischer (ed.), Data of geochemistry: U.S. Geol. Survey Prof. Paper 440-W.

FIGURE 8.19 Banded iron formation from the Negaunee Iron Formation (Precambrian), Michigan. The light-colored bands are chert; dark layers are the iron-rich units. Length 8 cm. (Specimen furnished by M. H. Reed.)

Ironstones

Ironstones are predominantly Phanerozoic-age sedimentary deposits. They occur mainly in Early Paleozoic and Jurassic rocks, but they range in age from Pliocene to Middle Precambrian. They form thin, massive or poorly banded sequences a few meters to a few tens of meters thick, in sharp contrast to the much thicker, well-banded iron formations. Ironstones are commonly interbedded with carbonates, mudrocks, and fine-grained sandstones of shelf to shallow-marine origin. They generally have an oolitic texture, and they may contain fossils that have been partly or completely replaced by iron minerals. Sedimentary structures in ironstones include cross-bedding, ripple marks, scour-and-fill structures, clasts, and burrows.

Mineral Facies of Iron Formations and Ironstones

On the basis of relative abundance of major kinds of iron-bearing minerals, James (1966) defines four main mineral facies in iron-rich sedimentary rocks: (1) oxides, (2) silicates, (3) carbonates, and (4) sulfides (Table 8.7).

TABLE 8.6 Textural types of cherty iron formation equivalent to limestone textural types

Type	Description
Micrite type	Deposited as a mud whose particles are too fine grained to survive diagenesis; only lamination and stratification visible as depositional structures; small-scale cross-beds here and there prove deposition as particulate, noncohesive matter
Pelleted	Fine pellet texture of silt or very fine sand
Intraclastic	Containing gravel-size fragments (intraclasts) whose internal textures prove derivation from penecontemporaneous sediment; fragments embedded in a micrite-type matrix or bound by a cement introduced during diagenesis
Peloidal	Containing sand-size fragments (peloids) without internal textures; peloids embedded in a micrite-type matrix or bound by a clear chert cement introduced during diagenesis
Oolitic	Containing concentrically laminated ooids, either set in a micrite-type matrix or, more commonly, bound by a clear chert cement introduced during diagenesis
Pisolitic	Containing pisolites set either in a micrite-type matrix or cemented by clear chert
Stromatolitic	Wavy, columnar, or digitating stromatolites

Source: After Dimroth, E., 1979, Models of physical sedimentation of iron formations, in R. G. Walker (ed.), Facies models: Geoscience Canada Reprint Ser. 1. Table II, p. 176, reprinted by permission of Geological Association of Canada.

TABLE 8.7 Principal iron-bearing minerals in iron-rich sedimentary rocks

Mineral class	Mineral	Chemical formula
Oxides	Goethite*	$FeOOH$
	Hematite	Fe_2O_3
	Magnetite	Fe_3O_4
Silicates	Chamosite	$3(Fe,Mg)O \cdot (Al,Fe)_2O_3 \cdot 2SiO_2 \cdot nH_2O$
	Greenalite	$FeSiO_3 \cdot nH_2O$
	Glauconite	$KMg(Fe,Al)(SiO_3)_6 \cdot 3H_2O$
	Stilpnomelane	$2(Fe,Mg)O \cdot (Fe,Al)_2O_3 \cdot 5SiO_2 \cdot 3H_2O$
	Minnesotaite (iron talc)	$(OH)_2(Fe,Mg)_3Si_4O_{10}$
Sulfides	Pyrite	FeS_2
	Marcasite	FeS_2
Carbonates	Siderite	$FeCO_3$
	Ankerite	$Ca(Mg,Fe)(CO_3)_2$
	Dolomite	$CaMg(CO_3)_2$
	Calcite	$CaCO_3$

*Not found in Precambrian iron formations

Oxides. Hematite, goethite, and magnetite are the common iron oxides in iron-rich sediments. Hematite is present in both iron formations and ironstones; goethite occurs in ironstones but is absent in Precambrian iron deposits. Magnetite is most abundant in Precambrian deposits, but occurs also in Phanerozoic deposits.

Silicates. The principal iron silicate minerals are chamosite and greenalite. Chamosite is the primary silicate mineral in ironstones, whereas greenalite is predominant in Precambrian iron formations.

Carbonates. Carbonate minerals in iron-rich facies include siderite, dolomite, calcite, and ankerite. Siderite is an important constituent in both Precambrian and Phanerozoic iron-rich sediments, where it commonly consists of flattened nodules or more or less continuous beds. Dolomite is common also in both iron formations and ironstones. Calcite is common in ironstones but is rare in iron formations, and ankerite is most common in iron formations.

Sulfides. Pyrite is the predominant sulfide mineral in iron-rich rocks, but marcasite is common also. Sulfide minerals rarely form the major iron mineral in these rocks, but locally they can predominate in some thin beds.

Iron-rich Shales

Pyritic black shales occur in association with both Precambrian iron formations and Phanerozoic ironstones. They commonly occur in thin beds in which sulfide content may range as high as 75 percent. Pyrite occurs disseminated in these black carbonaceous shales and in some limestones. It occurs also as nodules and laminae and as a replacement of fossil fragments and other iron minerals. Pyrite-rich layers have also been reported in some limestones. **Siderite-rich shales** (clay ironstones) occur primarily in association with other iron-rich deposits. They are found also in the coal measures of both Great Britain and the United States. Siderite occurs disseminated in the mudrocks or as flattened nodules and more or less continuous beds.

Bog Iron Ores, Iron-rich Laterites, Manganese Crusts and Nodules

Bog iron ores are minor accumulations of iron-rich sediments that occur particularly in small freshwater lakes of high altitude. These ores range from hard, oolitic, pisolitic, and concretionary forms to soft, earthy types. **Iron-rich laterites** are residual iron-rich deposits that form as a product of intense chemical weathering. They are basically highly weathered soils in which iron is enriched. **Manganese crusts and nodules** are widely distributed on the modern seafloor in parts of the Pacific, Atlantic, and Indian oceans in areas where sedimentation rates are low. They have been reported also from ancient sedimentary deposits in association with such oceanic sediments as red shales, cherts, and pelagic limestones. Both iron-rich (15–20 percent Fe) and iron-poor (less than about 6 percent Fe) varieties of manganese nodules are found. These nodules contain various amounts of Cu, Co, Ni, Cr, and V in addition to manganese and iron. Owing to the presence of these valuable metals in manganese nodules, considerable interest has developed in the possibility of mining them from the seafloor. Recovery vessels are already being planned and designed, and political negotiations are under way among the major nations of the world regarding mining rights to these potentially valuable deposits.

Iron-rich Muds (Metalliferous Sediments)

Metalliferous sediments occur in oceanic settings, particularly near active mid-ocean spreading ridges. They are believed to form by precipitation from metal-rich hydro-thermal fluids that have become enriched through contact and interaction with hot basaltic rocks. These sediments are enriched in Fe, Mn, Cu, Pb, Zn, Co, Ni, Cr, and V. Metal-enriched sediments have also been reported from some ancient sedimentary deposits in association with submarine pillow basalts and ophiolite sequences of ocean crustal rocks.

Heavy Mineral Placers

Placers are sedimentary deposits that form by mechanical concentration of mineral particles of high specific gravity, commonly in beach or alluvial environments. Magnetite, ilmenite, and hematite sands are common constituents of placers, particularly beach and marine placers. Placers are local accumulations that occur mainly in Pleistocene- to Holocene-age sediments and that commonly do not exceed 1–2 m in thickness. Marine placer deposits containing about 5 percent iron ore have been mined off the southern tip of Kyushu, Japan, for many years (Mero, 1965), and offshore placers containing up to 10 percent magnetite and ilmenite have been reported off the southeastern coast of Taiwan (Boggs, 1975). Beach placers containing ilmenite have been exploited commercially in Australia since about 1965 (Hails, 1976). "Fossil" placer deposits are comparatively rare, although thin, heavy-mineral laminae are common in some ancient beach deposits. Hails (1976) reports that outcrops of ilmenite- and magnetite-bearing placers of Cretaceous age are exposed discontinuously, subparallel to the Rocky Mountains, through New Mexico, Colorado, Wyoming, and Montana.

Chemical Composition of Iron-rich Sedimentary Rocks

Depending upon the type of deposit, the chemical composition of iron-rich sedimentary rocks varies over a wide range; it is impossible to establish a representative average composition. Iron expressed as Fe_2O_3, FeO, or FeS is the dominant chemical constituent in some iron-rich sediments; however, the iron content of iron-rich rocks is frequently exceeded by the content of silica (Table 8.8). Other common chemical constituents of iron-rich sedimentary rocks include Al, Mn, Mg, and Ca. Manganese concentration, in particular, may reach considerable percentages in some iron formations.

8.6   SEDIMENTARY PHOSPHORITES

The phosphorus or phosphate content of rocks is generally expressed as percentage P_2O_5. Sedimentary phosphorites are rocks that contain more than 20 percent P_2O_5 and are thus significantly enriched in phosphorus over other types of rocks. Phosphorus-rich sedimentary rocks are called **phosphate rock, rock phosphate, phosphates,** or **phosphorites.** Mudrocks or limestones that contain less than 20 percent P_2O_5 but which are enriched in phosphorus over that found in average sediments are called

TABLE 8.8 Chemical composition of sedimentary facies of iron formations

	Oxide facies	Silicate facies	Carbonate facies	Sulfide facies
Fe	37.80	26.5	21.23	20.0
FeO	2.10	28.9	22.22	2.35
Fe_2O_3	51.69	5.6	5.74	—
FeS_2	—	—	—	38.70
SiO_2	42.89	50.7	48.72	36.67
Al_2O_3	0.42	0.4	0.15	6.90
Mn	0.3	0.4	0.50	0.001
P	0.03	—	0.07	0.09
CaO	0.1	0.1	4.60	0.13
MgO	+	4.2	0.84	0.65
K_2O	+	—	+	1.81
Na_2O	+	—	0.01	0.26
TiO_2	+	+	+	0.39
CO_2	—	5.1	14.10	—
S	—	+	2.76	—
SO_3	—	—	—	2.60
C	—	+	+ +	7.60
H_2O^+	0.43	5.2	2.67	1.25

+ = trace; + + = larger trace; — = not present

Source: Eichler, J., 1970, Origin of Precambrian banded iron formations, in K. H. Wolf (ed.), Handbook of strata-bound and stratiform ore deposits, v. 7. Table VI, p. 187, reprinted by permission of Elsevier Science Publishers, Amsterdam.

phosphatic shales and phosphatic limestones. The phosphorus content of average mud-rocks is 0.17 percent P_2O_5 and the content of average limestones is 0.04 percent.

The total volume of sedimentary phosphates in the geologic record is quite small; however, phosphates are of special economic interest because they contribute more than 80 percent of the world's production of phosphate rock. Sedimentary phosphates occur in rocks of all ages, from Precambrian to Holocene, but phosphorite deposition appears to have been particularly prevalent during the Precambrian and Cambrian in central and southeast Asia, the Permian in North America, the Jurassic and Early Cretaceous in eastern Europe, the Late Cretaceous to Eocene in the Tethyan Province of the Middle East and North Africa, and the Miocene of southeastern North America (P. J. Cook, 1976). Modern phosphorites occur also as nodules on some parts of the ocean floor (Chapter 4).

Principal Types of Phosphate Deposits

Phosphate-rich sedimentary rocks may occur in layers ranging from thin laminae a few millimeters thick to beds a few meters thick. Some phosphate sequences such as the Phosphoria Formation of northwestern United States may reach several hundred meters in thickness, although such sequences are not composed entirely of phosphate-rich rocks. Phosphorites are generally interbedded with mudstones or shales; cherts; limestones and dolomites; or, more rarely, sandstones. Phosphatic rocks commonly grade regionally into nonphosphatic sedimentary rocks of the same age.

No widely accepted scheme for classifying phosphorites appears to be in general use. P. J. Cook (1976) mentions use of a threefold classification based essentially on analogy with the characteristics of phosphorite deposits in three well-studied areas of the United States: (1) **geosynclinal** or **west-coast type** (e.g., the western phosphate fields of the United States), (2) **platform** or **east-coast type** (e.g., the North Carolina deposits), and (3) **weathered** or **residual type** (e.g., the brown rock deposits of Tennessee). The principal characteristics of the west-coast and east-coast types are shown in Table 8.9. Residual phosphorites form as a result of concentration of phosphate by chemical or mechanical weathering and are thus not primary sedimentary phosphorites. Cook points out that phosphorite deposits may have characteristics of more than one of these types of deposits and that there are also undesirable genetic implications involved in using terms such as geosynclinal and platform that detract from the usefulness of this classification scheme.

A second approach to classification of phosphorites is to divide them into groups based on bedding characteristics and the principal types of phosphate materials that make up the deposits. Five principal groups can be identified: (1) bedded phosphorites, (2) bioclastic phosphorites, (3) nodular phosphorites, (4) pebble-bed phosphorites, and (5) guano deposits.

Bedded phosphorites form distinct beds of varied thickness, commonly interbedded and interfingering with carbonaceous mudrocks, cherts, and carbonate rocks. The phosphorite occurs as pellets, oolites, pisolites, phosphatized brachiopods and other skeletal fragments, and cements. Perhaps the best studied example of a bedded phosphorite deposit is the Permian Phosphoria Formation of the northwestern United States. This formation has a total thickness of 420 m and extends over an area of about 350,000 km^2 (McKelvey et al., 1959). Bedded phosphorites are believed to form on shelf areas associated with zones of upwelling in the ocean (Chapter 4).

Bioclastic phosphorites are a special type of bedded phosphate deposit composed largely of vertebrate skeletal fragments such as fish bones, shark teeth, fish scales, coprolites, and similar remains. Deposits composed mainly of invertebrate fossil remains such as phosphatized brachiopod shells are also known. These phosphate-bearing organic materials commonly become further enriched in P_2O_5 during diagenesis and may be cemented by phosphate minerals.

Nodular phosphorites are brownish to black spherical to irregular-shaped nodules ranging in size from few centimeters to a meter or more. Internal structure of phosphate nodules ranges from homogeneous (structureless) to layered or concentri-

TABLE 8.9   Differences in west-coast and east-coast types of phosphorite deposits

Feature	Geosynclinal or west-coast type	Platform or east-coast type
Phosphorite type	Generally pelletal; minor oolitic	Pelletal, nodular, and non-pelletal
Matrix to pellets	Argillaceous or siliceous	Quartzose (sandy) or calcareous
Grade of the phosphorite	High grade	Low grade
Nature of the deposit	Thick extensive continuous beds	Discontinuous beds
Fauna	Pelagic	Shallow water
Inferred water depth	Hundreds of meters	Tens of meters
Sediment association	Black shale-chert	Carbonates, sands
Influence of synsedimentary structures	Generally lacking	Deposits commonly in synclines or the flanks of anticlines
Tectonics	Strongly folded	Gently folded

Source: Cook, P. J., 1976, Sedimentary phosphate deposits, *in* K. H. Wolf (ed.), Handbook of strata-bound and stratiform ore deposits, v. 7. Table I, p. 511, reprinted by permission of Elsevier Science Publishers, Amsterdam.

cally banded. Phosphatic grains, pellets, shark teeth, and other fossils may be present within the nodules. Phosphate nodules are forming today in zones of upwelling in the ocean, and many ancient nodular phosphorites may have a similar origin. Some ancient phosphorite nodules, however, may be of diagenetic origin. Some are relict features of disconformity and unconformity surfaces.

Pebble-bed phosphorites are composed of phosphatic nodules, phosphatized limestone fragments, or phosphatic fossils that have been mechanically concentrated by reworking of earlier formed phosphate deposits.

Guano deposits are composed of bird and possibly bat excrement that has been leached to form an insoluble residue of calcium phosphate. Guano is being deposited today on small oceanic islands in the Eastern Pacific and the West Indies. Guano deposits are not important in the geologic record.

Composition of Phosphorites

Sedimentary phosphates are composed of phosphate minerals, all of which are varieties of apatite. The principal varieties include: fluorapatite [$Ca_5(PO_4)_3F$], chlorapatite [$Ca_5(PO_4)_3Cl$], and hydroxyapatite [$Ca_5(PO_4)_3OH$]. Most sedimentary phosphates are carbonate hydroxyl fluorapatites in which up to 10 percent carbonate ions can be substituted for phosphate ions to yield the general formula $Ca_{10}(PO_4,CO_3)6F_{2-3}$. These carbonate hydroxyl fluorapatites are commonly called **francolite**. The wastebasket term **collophane** is often used for sedimentary apatites for which the exact chemical composition has not been determined.

Additional details of the origin, distribution, and composition of sedimentary phosphorites are given in Bentor (1980b) and Volume 136 of the *Journal of the Geological Society (London)* (1980).

8.7 CARBONACEOUS SEDIMENTARY ROCKS

Introduction

Most sedimentary rocks contain at least a small amount of organic matter that consists of the preserved residue of plant or animal tissue. When the tissue of organisms decays in an oxygen-deficient environment, organic degradation is incomplete and more decay-resistant fractions of organic substances such as cellulose, fats, resins, and waxes are not immediately decomposed. If a depositional basin happens to be in a highly reducing environment—such as a restricted basin, stagnant swamp, or bog—decay-resistant organic matter may be preserved long enough to become incorporated into accumulating sediment, where it may persist for hundreds of millions of years after burial.

Some organic matter occurs in most sedimentary rocks, including those of Precambrian age. The average content of organic matter in sedimentary rocks is 2.1 weight percent in mudrocks, 0.29 percent in limestones, and 0.05 percent in sandstones (Degens, 1965). The average in all sedimentary rocks is about 1.5 percent. Organic matter contains 50–60 percent carbon; therefore, the average sedimentary rock contains about 1 percent organic carbon.

A few special types of sedimentary rocks contain significantly more organic material than these average rocks. Black shales and other carbonaceous and bituminous mudrocks typically contain 3–10 percent organic matter. Oil shale or kerogen shale contains even higher percentages, ranging up to 25 percent or more, and coals may be composed of more than 70 percent organic matter. Certain solid hydrocarbon accumulations, such as asphalt and bitumen deposits formed from petroleum by oxidation and loss of volatiles, are another example of a sedimentary deposit greatly enriched in organic carbon.

Kinds of Organic Matter in Sedimentary Rocks

Three basic kinds of organic matter are accumulating in subaerial and subaqueous environments under present conditions: humus, peat, and sapropel. Soil **humus** is plant organic matter that accumulates in soils to form a number of decay products such as humic and fulvic acids. Most soil humus is eventually oxidized and destroyed; little is preserved in sedimentary rocks. **Peat** is also humic organic matter, but peat accumulates in freshwater or brackish-water swamps and bogs where stagnant, anaerobic conditions prevent total oxidation and bacterial decay. Therefore, some of the humus that accumulates under reducing conditions can be preserved in sediments. **Sapropel** refers to fine organic matter that accumulates subaqueously in lakes, lagoons, or marine basins where oxygen levels are low. It consists of the remains of phytoplankton, zooplankton, and spores and fragments of higher plants. Phytoplankton are tiny plants such as algae that drift about in currents in the upper water column, and zooplankton are small drifting animals such as foraminifers.

It is often difficult to differentiate accurately between the types of organic matter found in ancient sediments; however, both humic and sapropelic types are recognized. Humic organic matter is the chief constituent of most coals, although a few are formed of sapropel. The organic matter in oil shales and other carbonaceous mudrocks and limestones is also sapropelic, but it is so finely disseminated and altered that it is difficult to identify. This type of organic matter is called **kerogen.** It resembles coal dust in appearance but contains a wide range of organic structures that have been identified in living organisms.

Classification of Carbonaceous Sedimentary Rocks

The predominant organic constituents of carbonaceous sediments are thus humic and sapropelic organic matter. The nonorganic constituents are either siliciclastic grains or carbonate materials. Carbonaceous sediments can be classified on the basis of relative abundance of these constituents and the kind of organic matter that composes the constituents (humic vs. sapropelic) into three basic types of organic-rich rocks: coal, oil shale, and asphaltic substances (Fig. 8.20). Each of these types of rocks contains at least 10–20 percent organic constituents.

Coals

Coals are the most abundant type of carbonaceous sediment. They are composed predominantly of combustible organic matter, but contain various amounts of impurities (ash) that are largely siliciclastic materials. The amount of ash that coals can contain and still retain the name of coal is not precisely fixed. Some very impure coals (bone coals) may contain 70–80 percent ash, but most coals have less than 50 percent ash by weight. Most coals are humic coals, although a few are sapropelic coals that are made up mostly of spores, algae, and fine plant debris. Cannel coals and boghead coals are

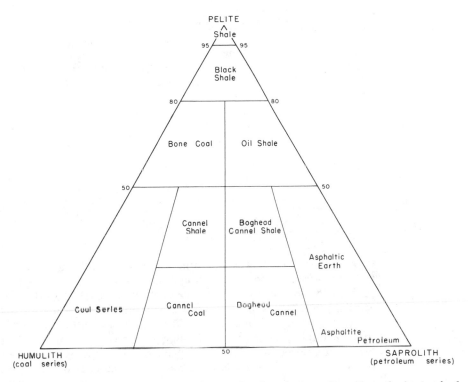

FIGURE 8.20 Classification and nomenclature of carbonaceous sediments on the basis of relative abundance of humic organic constituents (humulith), sapropelic organic constituents (saprolith), and fine-grained terrigenous constituents (pelite). (From Pettijohn, F. J., 1975, Sedimentary rocks, 3rd ed. Fig. 11.37, p. 445, reprinted by permission of Harper & Row, New York.)

sapropelic coals. Coals are defined in various ways, but a commonly accepted definition is that of J. M. Schopf (1956, p. 527):

> Coal is a readily combustible rock containing more than 50 percent by weight and more than 70 percent by volume of carbonaceous material, formed from compaction or induration of variously altered plant remains similar to those of peaty deposits. Differences in the kinds of plant materials (type), in degree of metamorphism (rank), and range of impurities (grade), are characteristic of the varieties of coal.

A common method of classifying coals is by **rank,** which is based on the degree of coalification or carbonification (increase in organic carbon) attained by a given coal owing to burial and metamorphism (Table 8.10). Peat is included in Table 8.10 but is actually not a true coal. **Peat** consists of unconsolidated, semicarbonized plant remains with high moisture content. **Lignite,** or brown coal, is the lowest rank coal. Lignites are brown to brownish-black coals that have high moisture content and commonly retain many of the structures of the original woody plant fragments. They are dominantly Cretaceous or Tertiary in age. **Bituminous coals** are hard, black coals that contain lesser amounts of volatiles and moisture than lignite and have a higher carbon content. They commonly display thin layers consisting of alternating bright and dull bands (Fig. 8.21). **Subbituminous coal** has properties intermediate between those of

TABLE 8.10 Classification of coals based on rank

Rank stages		Carbon content (%) (dry ash free)	Volatile content (approx. %)	Calorific value (Kcal/kg)	Vitrinite reflectance
Peat		60			
Soft brown coal			53	4000	0.3
Hard brown coal	Lignite	71	49	5500	0.5
	Subbitum.	77	42	7000	
Bituminous hard coal	Low rank	87	29	8650	1.1
	High rank	91	8	8650	2.5
Anthracite		100	0		

Source: Tucker, M. E., 1981, Sedimentary petrology, an introduction: John Wiley & Sons (Halsted Press), New York. Table 8.1, p. 199, reprinted by permission of Open University Educational Enterprises Ltd., Stony Stratford, England.

lignite and bituminous coal. **Anthracite** is a hard, dense, black coal containing more than 90 percent carbon. It is a bright, shiny rock that breaks with conchoidal fracture, such as the fractures in broken glass. Bituminous coals and anthracite are largely of Mississipian and Pennsylvanian (Carboniferous) ages. **Cannel coal** and **boghead coal** are nonbanded, dull, black coals that also break with conchoidal fracture; however, they have bituminous rank and much higher volatile content than anthracite. Cannel coal is composed of conspicuous percentages of spores. Boghead coals are composed predominantly of nonspore algal remains. **Bone coal** is very impure coal containing high ash content.

Coals are classified also on the basis of megascopic textural appearance and recognizable petrographic or microscopic constituents. Stopes (1919) recognized four types of coal, now called lithotypes, on the basis of megascopic appearance. These lithotypes—vitrain, clarain, durain, and fusain—occur as millimeter-thick bands or layers of humic coal.

Vitrain is a brilliant, glossy, vitreous, black coal that occurs in thin horizontal bands that are commonly 3–5 mm thick. It breaks with a conchoidal fracture and is clean to the touch.

Clarain has a smooth fracture that displays a pronounced gloss or shine and is distinguished from vitrain by the presence of dull intercalations or striations. Small-scale sublaminations occur within the layers or bands and give the surface a silky luster. Clarain is the most widely distributed and common macroscopic constituent of humic coals.

Durain occurs in bands up to a few centimeters in thickness and has a close, firm texture that appears somewhat granular; broken surfaces are not smooth but have a fine lumpy or matte texture. Durain is characterized by its lack of luster, gray or brownish-black color, and earthy appearance.

Fusain is soft, friable, and black. It resembles common charcoal and has been described as "mineral charcoal." It occurs chiefly as irregular wedges and is friable and porous if not mineralized.

Under the microscope, coal can be seen to consist of several kinds of organic units that are single fragments of plant debris or in some cases are fragments consisting of more than one type of plant tissue. Stopes (1935) suggested the name **maceral** for these organic units as a parallel word for the term mineral used for the constituents of

FIGURE 8.21 Layered and banded bituminous coal, Cedar Grove Seam (Pennsylvanian), Logan County, West Virginia. Thickness of the coal seam is about 2.3 m. (Photograph courtesy of Island Creek Coal Company.)

inorganic rocks. The starting materials for macerals are woody tissues, bark, fungi, spores, and similar items; however, these materials are not always recognizable in coals. Macerals are divided into three major groups: vitrinite, inertinite, and liptinite.

Vitrinites are macerals that originated as wood or bark. They are a major humic constituent of bright coals. Two types of vitrinites are known. **Collinite** is a structureless or nearly structureless maceral that commonly occurs as a matrix or impregnating material for fragments of other macerals. **Telinite** is derived from cell-wall material of bark and wood and preserves some of the cellular texture.

Inertinites are macerals composed of woody tissues, fungal remains, or fine organic debris of uncertain origin and are characterized by relatively high carbon content. They can be divided into five subtypes:

Fusinite—displays cell structures composed of carbonized or oxidized cell walls and hollow lumens (the spaces bounded by the walls of organs) that are commonly mineral filled; characteristic of fusain
Semifusinite—a transition state between fusinite and vitrinite
Sclerotinite—composed of the remains of fungal schlerotia (a hardened mass of tubular filaments or threads) or altered resins; characterized by oval shape and varying size
Micrinite and **macrinite**—structureless, granular macerals derived from fine-grained organic detritus; opaque and generally less than 10 microns in size (micrinite) but ranges up to 100 microns (macrinite)

Inertodetrinite—finely divided, structureless clastic form of inertinite in which fragments of various kinds of inertinite maceral occur as dispersed particles

Liptinites (exinites) originate from spores, cuticles, resins, and algae. They can be recognized by their shapes and structures, although original constituents may be compacted and squashed. The major types of liptinites are

Sporinite—composed of the remains of yellow, translucent bodies (spore exines) that are commonly flattened parallel to the bedding
Cutinite—formed from macerated fragments of cuticles (layers covering the outer walls of a plant's epidermal cells)
Resinite—the remains of plant resins and waxes; occurs as isolated rounded to oval or spindle-shaped, translucent, reddish bodies; occurs also as diffuse impregnations or as fillings in cell cavities
Alginite—macerals composed of the remains of algal bodies; serrated, oval shape; characteristic maceral of boghead coal

Oil Shale (Kerogen Shale)

The term "oil shale" is applied to fine-grained sedimentary rocks from which substantial quantities of oil can be derived by heating. The term is actually a misnomer; relatively little free oil occurs in these rocks, although small blebs, pockets, or veins of asphaltic bitumens may be present. More than 80 percent of the organic matter in oil shales is present in the form of kerogen, which yields oil when heated to a temperature of about 350°C. The principal constituents of oil shales are shown in Figure 8.22. Organic constituents commonly do not exceed about 25 percent of the rock. Kerogen is disseminated organic matter that is insoluble in nonoxidizing acids, bases, or organic solvents. It consists of masses of almost completely macerated organic debris, chiefly plant remains such as algae, spores, spore coats, pollen, resins, and waxes. On the basis of the type of organic remains from which it was derived, kerogen has been classified into five principal types: (1) **algal**—composed dominantly of the remains of algae, (2) **amorphous**—composed largely of sapropelic organic matter from plankton

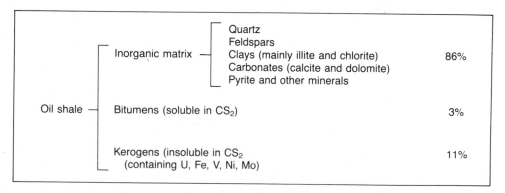

FIGURE 8.22 Principal constituents in oil shales. (After Yen, T. F., and G. V. Chilingarian, 1976, Introduction to oil shales, *in* T. F. Yen and G. V. Chilingarian (eds.), Oil shales: Developments in petroleum science 5. Fig. 1.2, p. 3, reprinted by permission of Elsevier Science Publishers, Amsterdam.)

and other low forms of life, (3) **herbaceous**—composed of pollen, spores, cuticles, and similar materials, (4) **woody,** and (5) **coaly** (inertinite) (Hunt, 1979).

Not all oil shales are actually shales. Some are organic-rich siltstones, mudrocks, limestones, and impure coals. Three basic types are recognized: (1) **Carbonate-rich oil shales** are those in which the principal nonkerogen constituents are calcite, dolomite, ankerite, and various amounts of siliciclastic silt. They are generally hard, tough, and resistant to weathering (Duncan, 1976). (2) **Silica-rich oil shales** are shales in which the main constituents apart from kerogen are fine-grained quartz, feldspar, and clay minerals. These shales also contain chert, opal, and phosphatic nodules. Siliceous oil shales are generally dark brown or black and are less resistant to weathering than the carbonate-rich shales. (3) **Cannel shale** is an oil shale that consists predominantly of organic matter that completely encloses other mineral grains. The organic matter is composed largely of algal remains. Cannel shales are sometimes classified as impure cannel coals and are referred to as **torbanites.** Many oil shales are characterized by distinct lamination caused by alternations of millimeter-thick organic laminae with either siliciclastic or carbonate laminae. The amount of oil that can be extracted from oil shales ranges from about 4 percent to more than 50 percent of the weight of the rock; that is, between 10 and 150 gallons of oil per ton of rock (Duncan, 1976).

Oil shales form in environments where organic matter is abundant and where anaerobic, or reducing, conditions prevent oxidation and total bacterial decomposition. The shales are deposited in both lacustrine and marine environments where the above conditions are met. The principal environments are (1) large lakes; (2) shallow seas or continental platforms and continental shelves in areas where water circulation was restricted and reducing or weakly oxidizing conditions existed; and (3) small lakes, bogs, and lagoons associated with coal-producing swamps. Oil shales formed in lakes or swamps may be associated with impure cannel or boghead type coal, tuffs and other volcanic rocks, or even evaporites. Many oil shales deposited in large lakes are carbonate-rich types and tend to have high oil yields. Oil shales deposited in marine environments are characteristically the silica-rich type and have lower oil yields, although some Tertiary- and Mesozoic-age siliceous oil shales have rich oil yields. Oil shales extend over wide geographic areas and are commonly associated with limestones, cherts, sandstones, and phosphatic deposits. Additional details of the origin, distribution, and composition of oil shales are given in Yen and Chilingarian (1976b).

Petroleum and Solid Bitumens

Petroleum. Petroleums are not sedimentary rocks but they are carbon-rich organic substances that occur as liquid and gas accumulations predominantly in sandstones and carbonate rocks. For this reason, they are included here with the carbonaceous sedimentary rocks. Petroleum forms from plant and animal organic matter by a complex maturation process during burial; the process involves initial microbial alteration and subsequent thermal alteration and cracking. The source materials for petroleum are contained primarily in organic-rich shales and carbonate rocks. After petroleum has formed from organic source materials at substantial burial depths, it migrates out of the fine-grained source rocks into coarser grained, porous and permeable sandstone or carbonate reservoir rocks, where it eventually accumulates in traps such as anticlines.

Petroleum is composed predominantly of carbon (about 84 percent) and hydrogen (about 13 percent). It also contains an average of about 1.5 percent sulfur, 0.5

percent nitrogen, and 0.5 percent oxygen (Hunt, 1979). Despite its simple elemental chemical composition, the molecular structure of petroleum can be exceedingly complex. The molecules in petroleum range from the simple methane gas molecule (CH_4), with a molecular weight of 16, to molecules with molecular weights in the thousands. Several hundred different hydrocarbons have been recorded in natural crude oils; however, all hydrocarbons can be grouped into a few basic classes, or series, having common molecular structural form. These structural forms are complex and are not explained in detail here, but the main hydrocarbon series are (1) **paraffins (alkanes)**—open chain molecules with single covalent bonds between carbon atoms (Fig. 8.23); (2) **naphthenes (cycloparaffins)**—closed-ring molecules with single covalent bonds between carbon atoms (Fig. 8.24); and (3) **aromatics (arenes)**—one or more benzene ring structures with double covalent bonds between some carbon atoms (Fig. 8.25).

Most natural gases as well as many liquid petroleums belong to the paraffin series of hydrocarbons. Most naphthene hydrocarbons are liquid petroleums, although two occur as gases at normal temperatures. The aromatics, which are named for their strong aromatic odor, are liquid petroleums. They commonly make up only a small percentage of the petroleums in natural crude oils.

FIGURE 8.23 Schematic structure of paraffin hydrocarbons having the general formula C_nH_{2n+2}, where *n* refers to the number of carbon or hydrogen atoms. A. Butane. B. Pentane.

A Butane, C_4H_{10} $CH_3(CH_2)_2CH_3$

B Pentane, C_5H_{12} $CH_3(CH_2)_3CH_3$

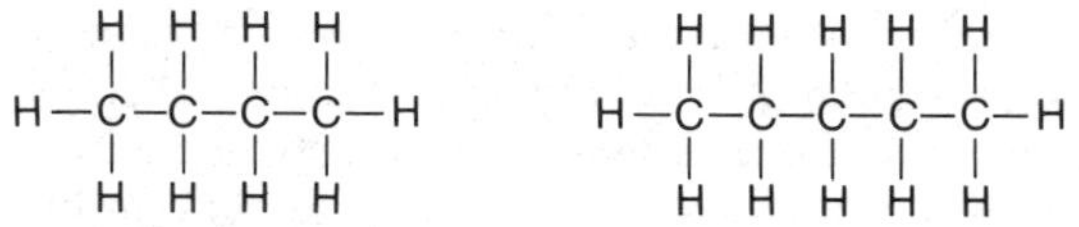

FIGURE 8.24 Schematic structure of naphthene (cycloparaffin) hydrocarbons having the general formula C_nH_{2n}. A. Cyclopentane. B. Cyclohexane.

A Cyclopentane, C_5H_{10} $CH_2CH_2CH_2CH_2CH_2$

B Cyclohexane, C_6H_{12} $CH_2CH_2CH_2CH_2CH_2CH_2$

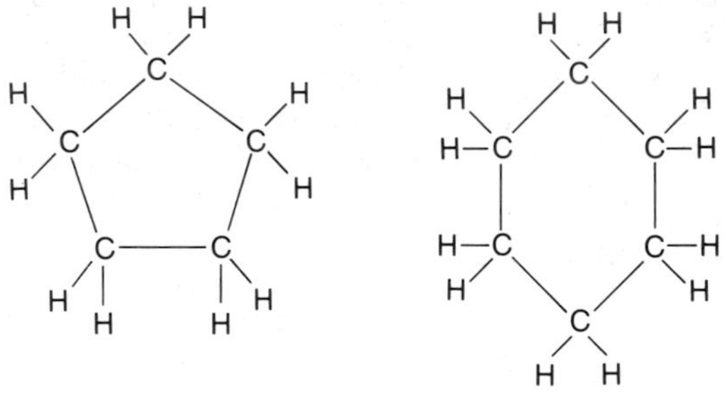

FIGURE 8.25 Schematic structure of aromatic hydrocarbons having the general formula C_nH_{2n-6}. A. Benzene. B. Toluene.

A Benzene, C_6H_6

B Toulene, $C_6H_5CH_3$

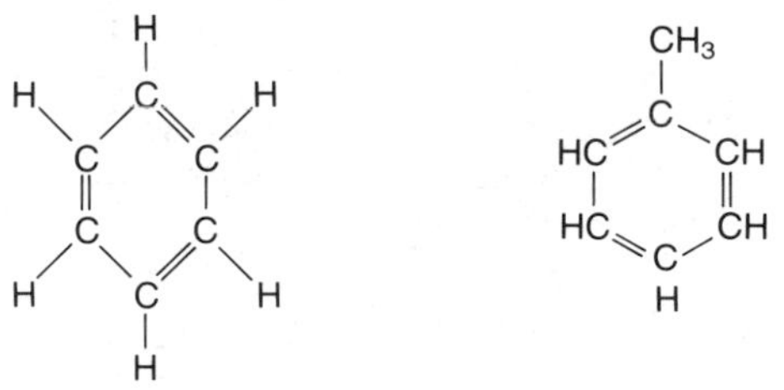

Solid Bitumens. These substances are hydrocarbons such as natural asphalts and mineral waxes that occur in a semisolid or solid state. Most solid hydrocarbons probably formed from liquid petroleums that were subjected to loss of volatiles, oxidation, and biologic degradation after seepage to the surface. Others may never have existed as light oils. Solid bitumens occur as seepages, surface accumulations, and impregnations occupying the pore spaces of sandstones or other sedimentary rock and in veins and dikes. They are black or dark brown and have a characteristic odor of pitch or paraffin.

Solid hydrocarbons have roughly the same elemental chemical composition as liquid petroleum, but the percentages of carbon and hydrogen tend to be somewhat lower and the content of sulfur, nitrogen, and oxygen somewhat higher. These hydrocarbons are divided into four main varieties, or series, on the basis of fusibility (melting temperature) and solubility in carbon disulfide (CS_2), an organic solvent (Fig. 8.26): (1) asphalts, (2) asphaltites, (3) pyrobitumens, and (4) native mineral waxes.

Asphalts are soft, semisolid bitumens that occur as seeps, surface pools, or viscous impregnations in sediments (tar sands). They are dark colored, plastic to fairly hard, easily fusible, and soluble in carbon disulfide. Varietal names for asphalts from different areas are shown in Figure 8.26. Asphalts are commonly associated with active oil seeps.

Asphaltites occur primarily in dikes and veins that cut sediment beds. They are harder and denser than asphalts and melt at higher temperatures. They are largely

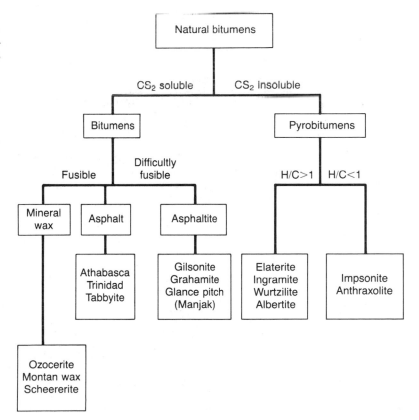

FIGURE 8.26 Terminology of principal kinds of naturally occurring solid hydrocarbons. (From Petroleum geochemistry and geology, by J. M. Hunt, W. H. Freeman and Company, © 1979, Fig. 8.28, p. 400).

soluble in carbon disulfide. Names applied to varieties of asphaltites that differ slightly in density, fusibility, and solubility are **gilsonite, glance pitch,** and **grahamite.**

Pyrobitumens occur in dikes and veins as do asphaltites but are infusible and largely insoluble in carbon disulfide. Several varieties of pyrobitumens are recognized. Softer forms include **elaterite,** a soft elastic substance rather like India rubber, and **wurtzlite.** More indurated forms are **albertite,** a black, solid bitumen with a brilliant jetlike luster and conchoidal fracture; **ingramite;** and the metamorphosed pyrobitumens **impsonite** and **anthraxolite.**

Native mineral waxes are solid, waxy, light-colored substances that consist largely of paraffinic hydrocarbons of high molecular weight. They represent the residuum of high-wax oils exposed at the surface. The most important native mineral wax is **ozocerite,** which consists of veinlike deposits of greenish or brown wax. **Montan wax** is an extract obtained from some kinds of brown coals or lignites.

The solid hydrocarbons are of interest to geologists because their presence at the surface is an indication of petroleum at depth in a region and because study of their occurrence may help to solve problems related to the origin and alteration of petroleum. Also, many of the solid hydrocarbons are of commercial value themselves. Details of the geochemistry, origin, distribution, and exploitation of bitumen and other solid hydrocarbons are discussed in Chilingarian and Yen (1978).

ADDITIONAL READINGS

Carbonate Rocks

Bathurst, R. G. C., 1975, Carbonate sediments and their diagenesis, 2nd ed.: Elsevier, New York, 658 p.

Chilingarian, G. V., H. J. Bissell, and R. W. Fairbridge (eds.), 1967, Carbonate rocks. Part A: Origin, occurrence, and classification, 471 p.; Part B: Physical and chemical aspects, 413 p.: Elsevier, New York.

Ham, W. E. (ed.), 1962, Classification of carbonate rocks: Am. Assoc. Petroleum Geologists Mem. 1, 279 p.

Lippman, F., 1973, Sedimentary carbonate minerals: Springer-Verlag, New York, 228 p.

Milliman, J. D., 1974, Marine carbonates: Springer-Verlag, New York, 375 p.

Reeder, R. J. (ed.), 1983, Carbonates: Mineralogy and chemistry: Reviews in Mineralogy, v. 11, Min. Soc. America, 394 p.

Scholle, P. A., 1978, A color illustrated guide to carbonate rock constituents, textures, cements, and porosities: Am. Assoc. Petroleum Geologists Mem. 27, 241 p.

Wilson, J. L., 1975, Carbonate facies in geologic history: Springer-Verlag, New York, 471 p.

Evaporites

Borchert, H., and R. O. Muir, 1964, Salt deposits: The origin, metamorphism and deformation of evaporites: Van Nostrand, New York, 338 p.

Braitsch, O., 1971, Salt deposits: Their origin and composition: Springer-Verlag, New York, 279 p.

Chambre Sydnical de la Recherche et de la Production due Pétrole et du Gaz Naturel (eds.), 1980, Evaporite deposits: Illustration and interpretation of some environmental sequences: Editions Technip, Paris, and Gulf Publishing, Houston, 284 p.

Dean, W. E., and B. C. Schreiber, 1978, Marine evaporites: Soc. Econ. Paleontologists and Mineralogists Short Course Notes No. 4, Tulsa, Okla., 193 p.

Kirkland, D. W., and R. Evans (eds.), 1973, Marine evaporites: Origin, diagenesis, and geochemistry: Dowden, Hutchinson and Ross, Stroudsburg, Pa., 444 p.

Sonnenfeld, P., 1984, Brines and evaporites: Academic Press, London, 624 p.

Siliceous Sedimentary Rocks

Cressman, E. R., 1962, Nondetrital siliceous sediments: U. S. Geol. Survey Prof. Paper 440-T, 22 p.

Ireland, H. A., 1959, Silica in sediments: Spec. Pub. 7, Soc. Econ. Paleontologists and Mineralogists, 185 p.

McBride, E. F. (ed.), 1979, Silica in sediments: Nodular and bedded chert: Soc. Econ. Paleontologists and Mineralogists Reprint Ser. 8, 184 p.

Iron-rich Sedimentary Rocks

Dimroth, E., 1976, Aspects of the sedimentary petrology of cherty iron-formation, in K. H. Wolf (ed.), Handbook of strata-bound and stratiform ore deposits: Elsevier, New York, p. 203–254.

Eichler, J., 1976, Origin of the Precambrian iron-formation, in K. H. Wolf (ed.), Handbook of strata-bound and stratiform ore deposits: Elsevier, New York, v. 7, p. 157–201.

James, H. L., 1966, Chemistry of the iron-rich sedimentary rocks: Data of geochemistry, 6th ed., U.S. Geol. Survey Prof. Paper 440-W, 61 p.

James, H. L., and P. K. Sims (eds.), 1973, Precambrian iron formations of the world: Econ. Geology, v. 68, p. 913–1179.

Van Houton, F. B., and D. P. Bhattacharyya, 1982, Phanerozoic oolitic ironstone: Geologic record and facies models: Rev. Earth and Planetary Sci., v. 10, p. 441–457.

Phosphorites

Bentor, Y. K. (ed.), 1980, Marine phosphorites: Soc. Econ. Paleontologists and Mineralogists Spec. Pub. 29, 249 p.

Cook, P. J., 1976, Sedimentary phosphate deposits, in K. H. Wolf (ed.), Handbook of strata-bound and stratiform ore deposits: Elsevier, New York, vol. 7, p. 505–536.

Journal Geological Society (London), 1980, v. 136, pt. 6; An issue devoted to phosphatic and glauconitic sediments, p. 657–805.

Nriagu, J. O., and P. B. Moore (eds.), 1984, Phosphate minerals: Springer-Verlag, New York, 434 p.

Carbonaceous Sedimentary Rocks

Crelling, J. C., and R. R. Dutcher, 1980, Principles and applications of coal petrology: Soc. Econ. Paleontologists and Mineralogists Short Course Notes No. 8, 127 p.

Chilingarian, G. V., and T. F. Yen, 1978, Bitumens, asphalts and tar sands: Elsevier, New York, 331 p.

Hunt, J. M., 1979, Petroleum geochemistry and geology: W. H. Freeman, San Francisco, 617 p.

International Committee for Coal Petrology, 1963, International hand-book of coal petrology, 2nd ed., Centre National de la Recherche Scientifique, Paris. Supplements published in 1971, 1975.

Petrakis, L., and D. W. Grandy, 1980, Coal analysis, characterization and petrography: Jour. Chem. Education, v. 57, p. 689–694.

Rahmani, R. A., and R. M. Flores (eds.), 1985, Sedimentology of coal and coal-bearing sequences: Intl. Assoc. Sedimentologists Spec. Pub. 7, Blackwell Scientific Publications, Oxford, 412 p.

Ross, C. A., and J. R. P. Ross (eds.), 1984, Geology of coal: Benchmark Papers in Geology, v. 77, Hutchinson Ross, Stroudsburg, Pa., 349 p.

Stach, E., 1975, Handbook of coal petrology, 2nd ed.: Gebrüder Borntraeger, Berlin, 428 p.

Ward, C. R. (ed.), 1984, Coal geology and coal technology: Blackwell Scientific Publications, Oxford, 345 p.

Yen, T. F., and G. V. Chilingarian (eds.), 1976, Oil shale: Elsevier, New York, 292 p.

PART FIVE
Diagenesis

9

Sediment to Rock: Lithification and Diagenesis

9.1 INTRODUCTION

The sedimentation processes described in Chapters 3 and 4 lead to deposition of various kinds of unconsolidated sediments in continental and marine settings. Newly deposited sediments are characterized by loosely packed, uncemented fabrics, high porosities, and high interstitial water content. As sedimentation continues in subsiding basins, older sediments are progressively buried by younger sediment to depths that may reach tens of kilometers. Burial is accompanied by physical and chemical changes that take place in the sediments in response to increase in pressure from the weight of overlying sediment and increase in temperature downward in the subsurface. These changes act in concert to bring about compaction and **lithification** of the sediment, ultimately converting it into consolidated sedimentary rock. Thus, unconsolidated sand is eventually lithified to sandstone, siliciclastic mud is hardened into mudrock or shale, carbonate sediments change to limestones or dolomites, evaporite deposits become evaporite rocks such as gypsum, and so forth. The collective process that brings about change in sediments during burial and lithification is called **diagenesis.** Diagenesis occurs at temperatures and pressures higher than those of the weathering environment but below those that produce metamorphism. There is no clear-cut boundary between the realms of diagenesis and metamorphism (Fig. 9.1)—the effects of diagenesis grade into those of burial metamorphism.

Diagenetic alteration of sediment can begin, owing to organic activity or early cementation, while sediment is still on the ocean floor (or other basin floor). Alteration may continue through deep burial and eventual uplift, although most diagenetic change takes place during shallow to intermediate burial. The mechanisms of diagenesis lead to lithification of sedimentary rocks through physical processes associated with sediment compaction and chemical processes that bring about cementation and recrystallization. Original mineral composition may be altered by chemical reactions

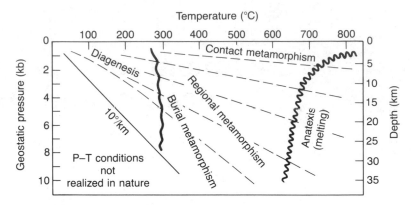

FIGURE 9.1 Generalized pressure–temperature conditions that control diagenesis and metamorphism. The pressure–temperature region below a geothermal gradient of 10°C/km is not realized in nature (After Winkler, H. G. F., 1967, Petrogenesis of metamorphic rocks, 2nd ed. Fig. 1, p. 4, reprinted by permission of Springer-Verlag, New York.)

involving mineral replacement, formation of new minerals, and solution. These processes affect the ultimate porosity of sedimentary rocks as well as their final mineral composition, texture, and structure. Porosity is of particular interest to petroleum geologists and hydrologists because it controls the amount of oil, gas, or water that can be contained within a rock. Porosity is decreased during diagenesis by compaction and the addition of mineral cements to pore spaces, but it is increased by solution processes that can selectively remove unstable grains or cements. Through burrowing, boring and sediment-ingesting activities, commonly referred to as bioturbation, organisms may play a role in early diagenesis by reworking sediments on the ocean floor before deep burial. Bioturbation can cause obliteration of primary sedimentary structures such as laminations and can cause mixing together of coarse sediment and fine sediment from different layers.

Diagenetic processes thus play an important role in shaping the ultimate characteristics of sedimentary rocks. To properly interpret the depositional history of these rocks, it is necessary to recognize and distinguish between the primary features caused by depositional processes and the secondary features resulting from diagenetic alteration. Diagenesis is the final stage in the process of forming sedimentary rocks, a process that begins with weathering and erosion of parent rocks and continues through sediment transport, deposition, and burial. Because diagenesis is a complex and highly technical subject, a comprehensive, detailed discussion of diagenetic processes and the compositional and textural changes produced by diagenesis is outside the scope of this book. This chapter is intended only to introduce students to the fundamental principles of diagenesis. Additional details, as well as case-history examples of diagenetic textures and fabrics, may be found in several monographs devoted to diagenesis; for example, those by Bathurst (1975), Berner (1980), Larsen and Chilingar (1979, 1983), D. A. McDonald and Surdam (1984), and Scholle and Schluger (1979). We begin study of diagenesis by examining the major processes that cause postdepositional changes in sediment and the nature of these changes. Subsequently, we will extend this examination to subsurface environmental conditions to see how the temperature, pressure, and chemistry of the subsurface environment control the diagenetic process.

9.2 MAJOR DIAGENETIC PROCESSES

The principal physical, chemical, and biologic processes that bring about diagenesis are summarized in very general terms in Table 9.1. Not all of these processes cause

TABLE 9.1 Major diagenetic processes that operate to change depositional assemblages of minerals and convert sediment into sedimentary rock

Diagenetic process		Process mechanism	Examples
Mainly physical	Compaction	Reorganization of grains into a tighter packed fabric, causing decrease in water content and porosity and thinning of beds	Siliciclastic mud → mudstone; porosity decreases from 60–80% to 10–20% Siliciclastic sand → sandstone; porosity decreases from 35–40% to 20±%
Mainly chemical	Cementation	Precipitation of new minerals in sediment pore space or onto the surface of existing minerals of the same kind to form overgrowths	Precipitation of calcite crystals in pores of terrigenous sands or carbonate sediments Precipitation of silica onto rounded quartz grains, creating new crystal faces
Mainly chemical	Authigenesis	Alteration of one mineral to form a new mineral, which may or may not act as a cement (in a broad sense, all processes that cause new minerals to form in sediment or sedimentary rock)	Fe-bearing minerals → pyrite (reduction) Fe-bearing minerals → hematite (oxidation) Feldspars → clay minerals
Mainly chemical	Recrystallization	Change in size (commonly increase) or shape of mineral crystals without significant change in composition; original textures and structures commonly destroyed	Lime mud → coarse crystalline limestone Oolites → coarse mosaic of crystals; concentric structure obscured
Mainly chemical	Inversion	Replacement of a mineral by its polymorph (a mineral having the same chemical composition but a different crystal form); commonly accompanied by recrystallization	Aragonite (orthorhombic $CaCO_3$) → calcite (rhombohedral $CaCO_3$) Fibrous aragonite shells → coarse, mosaic calcite shells
Mainly chemical	Replacement	Crystallization of a new mineral in the body of an old mineral or mineral aggregate of different composition by practically simultaneous capillary solution and deposition; original textures and structures commonly well preserved	Fossils (calcite) → fossils (chert) Clay minerals → calcite Chert grains → calcite Calcite shells → glauconite
Mainly chemical	Dissolution	Solution of a less stable mineral in an assemblage of minerals, leaving a cavity	Calcite shells → cavity Silica shells → cavity Calcite/aragonite crystals → cavity
Mainly biological	Bioturbation	Boring, burrowing, and sediment ingestion activities of organisms such as molluscs, shrimp, and sea cucumbers, causing sediment degradation and mixing and alteration of primary sedimentary textures and structures	Chemical/mechanical erosion of carbonate substrate by boring molluscs, causing sediment degradation and formation of cavities (borings) in the substrate Burrowing and sediment ingestion by marine worms and sea cucumbers that alter textures and structures and aggregate carbonate sediment into fecal pellets

hardening, or lithification, of sediment, but they all contribute in some way to the final characteristics of sedimentary rocks. Alteration of sediments through organic activity is commonly the earliest diagenetic process. Cementation may also occur quite early, but can continue throughout burial and uplift. Compaction begins as soon as sediments are buried and continues through deep burial. The other diagenetic processes occur at various stages of sediment burial and uplift.

Organic Activity

Bioturbation. Bioturbation refers to the reworking of sediments through the various burrowing, boring, and sediment-ingesting activities of organisms. Most bioturbation occurs at or just below the depositional interface very shortly after the time of sediment deposition and at sediment burial depths of only a few tens of centimeters. Because it occurs so quickly after deposition, it is moot whether or not bioturbation should be included as a diagenetic process or considered a syndepositional process. In fact, in Chapter 6, I characterize the tracks, trails, and other traces produced in sediment by organisms as primary sedimentary structures. Nevertheless, even though it may seem slightly inconsistent to do so, I am including bioturbation here as a post-depositional process because it is a process that brings about modification of sediment after deposition. Indeed, it is one of the most important processes that acts to modify original depositional textures and structures.

Bioturbation of sediments is accomplished largely through the activities of burrowing, boring, and browsing organisms. **Borers** are organisms such as various types of algae, molluscs, and echinoids that are capable of penetrating a hard substrate. **Burrowers** such as shrimps, anemones, and polychaete worms are capable of excavating unconsolidated particles. Borers and burrowers act to break down skeletal material, bore shells and other carbonate grains, and bore and burrow through unconsolidated or partially lithified sediment. They thus degrade and destroy carbonate components and in the process create openings, or voids, within carbonate grains and sediments—voids that may facilitate later diagenetic processes such as recrystallization. **Browsers** such as sea cucumbers (holothurians) and gastropods ingest sediment in a nonselective manner and extract nutrients from the sediment. Large quantities of sediment pass through the intestinal tracts of these organisms. In the process, carbonate particles undergo mechanical and chemical degradation. Browsing organisms also alter primary sedimentary textures by mixing together sediment of different sizes, commonly obliterating primary sedimentary structures such as laminations.

The combined activities of burrowing, boring, and browsing organisms thus result in size reduction of carbonate particles and extensive alteration of primary sedimentary textures and structures. Under some conditions, organisms may rework sediment so thoroughly that they leave a nearly featureless, homogeneous deposit. More commonly, organic activity leaves the sediment marked by distinctive bioturbation structures such as burrows, tracks, and trails. These markings are the trace fossils, or ichnofossils, described and discussed in Chapter 6. In additional to the physical effects of bioturbation, sediment reworking by organisms may also affect the chemistry of sediments. Because the chemical properties of sediment are related to grain size (Table 7.4), changes in grain size owing to bioturbation can change the chemistry of the sediments. Bioturbation also affects the movement of fluids through sediment and may, for example, promote the interchange of sediment waters with overlying seawater.

Microbial Activity. Many important chemical reactions that take place in sediment are associated with the decomposition of organic substances. Bacteria are involved in decomposition processes through a variety of activities that can influence the pH and Eh of the sediment pore waters as well as the buildup of certain chemical reactants. These activities may include respiration, fermentation, nitrate reduction, manganese reduction, iron reduction, sulfate reduction, and methane gas production (Aller, 1982). They can result in formation of products such as carbon dioxide, nitrogen, phosphates, ammonia, bicarbonates, hydrogen sulfide, methane gas, and alcohols, as well as releasing metal ions such as iron and manganese into solution. Aerobic bacteria (requiring the presence of oxygen) are commonly confined to the upper few meters of sediment. Anaerobic bacteria (capable of existing in a reducing atmosphere) may survive in sediments to much greater depths. It is not known definitely to what burial depths anaerobic bacteria can survive, but they have been reported at depths of at least 140 m (Hunt, 1979). Bacterial alteration of organic matter is one of the steps involved in the formation of petroleum.

Compaction

Compaction is the squeezing together of sediment grains during burial owing to the weight of overlying sediment. This process causes reorganization of grain packing to produce a more tightly packed fabric. Closer packing results in reduced porosity and permeability of sediments, expulsion of pore fluids, and thinning of beds. The effects of compaction on siliciclastic and carbonate sediments are of particular interest because sandstones, limestones, and dolomites are the primary reservoir rocks for petroleum.

Compaction of Siliciclastic Sediment

In Chapter 5, I discuss the theoretical effect on porosity of closer packing of even-size spheres and point out that simple rearrangement of loosely packed spheres can reduce porosity from about 48 percent to about 26 percent (Fig. 5.17). The particles in natural terrigenous sands and muds are not uniform-size particles and they are not spheres; therefore, changes in porosity with burial depth cannot be predicted on purely theoretical grounds. The compressibility, or compactibility, of sands depends upon several factors, including the initial packing and porosity; particle shape, size, and sorting; and particle composition. Thus, different types of sands tend to undergo different rates of compaction depending upon the nature of the sand grains. For example, platy or elongated particles can cause a bridging effect that creates a strong skeletal framework that resists deformation and compaction (D. R. Allen and Chilingarian, 1975). On the other hand, a high content of easily deformed, soft particles such as shale or phyllite fragments makes sand more susceptible to compaction. Owing to these variables, it is difficult to generalize about porosity loss in sands relative to depth of burial, but it is believed that compaction of unconsolidated sand owing to overburden pressures alone can cause reduction of porosity from initial values of 40–50 percent to less than 20 percent at burial depths of about 6000 m (Blatt, 1979). Under natural conditions, however, porosity loss owing to compaction is commonly accompanied by porosity loss owing to cementation. Therefore, it is difficult to accurately evaluate the importance of porosity reduction by compaction alone. Figure 9.2 illustrates diagrammatically the

FIGURE 9.2 Schematic diagram illustrating theoretical reduction in porosity of a sand having an initial porosity of 50 percent (A). Owing to rearrangement of grains and tighter packing resulting from compaction, porosity is progressively reduced to 40 percent (B), 30 percent (C), and 20 percent (D). (From Bissell, H. J., and G. V. Chilingarian, 1975, in G. V. Chilingarian and K. H. Wolf (eds.), Compaction of coarse-grained sediments, I. Fig. 4.17, p. 235, reprinted by permission of Elsevier Science Publishers, Amsterdam.)

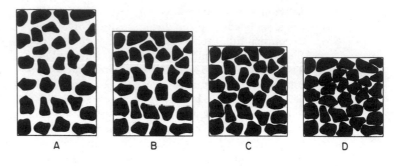

theoretical reduction of porosity in a natural sand from about 50 percent to 20 percent by a simple rearrangement of grains and tighter packing resulting from compaction.

Porosity reduction and bed thinning are even more pronounced during compaction of terrigenous muds. Initial porosities and water contents of muds may be as high as 60–80 percent. These porosities can be reduced during compaction to 10 or 15 percent or less at depths of a few thousand meters (Riecke and Chilingarian, 1974; Burst, 1976). Compaction of muds is accompanied by expulsion of pore waters on a massive scale. Waters squeezed from muds are believed to enter associated, more permeable interbedded sands and then flow laterally and upward toward the surface. This migrating water may play an important role in redistributing mineral matter through subsurface sandstones by chemical dissolution and precipitation.

After initial rearrangement of grains to yield tighter packing, continued compaction can result in physical changes to the grains as they undergo increasing pressure at grain contacts. Hard, brittle grains such as quartz may be fractured or shattered at shallow burial depths. After burial to greater depths and higher temperatures, they may yield plastically by bending or solid flow. At even greater burial depths, quartz grains undergo solution at grain contacts owing to increased solubility of the grains at points of increased pressure along the grain contact. This process of solution at grain contacts owing to overburden pressure is called **pressure solution.** Pressure solution is believed to cause significant solution of both silicate grains in sandstones and carbonate materials in limestones and further acts to reduce porosity by bringing grains in closer contact. The types of grain contacts produced by plastic flow and pressure solution of silicate grains are discussed in Chapter 5 and illustrated in Figure 5.18. Squeezing and plastic flow of micas (Fig. 9.3), clay minerals, and other relatively soft grains can also contribute to porosity reduction.

Compaction of Carbonate Sediments

Carbonate sediments also undergo porosity loss and pressure solution of grains as a result of compaction, although compaction may be less severe in carbonate sediments than in siliciclastic sediments. Unconsolidated carbonate muds or lime muds in modern environments have porosities ranging from 50 to 70 percent, yet many ancient limestones have porosities less than 5 percent. Sedimentologists differ in their interpretation of the cause of porosity loss in limestones. Some authors suggest that comparatively little compaction occurs in carbonate sediments because most ancient limestones show little evidence of shell breakage or grain breakage. They reason, therefore,

FIGURE 9.3 Photomicrograph showing a coarse muscovite mica grain (M) bent around a quartz grain by compaction. Polarized light.

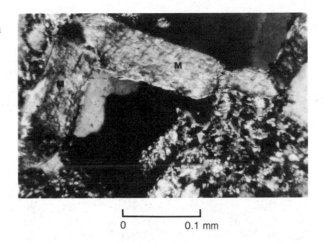

that porosity must have been lost primarily because of early cementation, which filled pore space with cements and prevented extensive compaction. On the other hand, some experimental studies of compaction in carbonate sediments have shown that considerable loss of porosity can occur by compaction and, further, that compaction may not necessarily result in extensive grain breakage. Experimental compaction of cores of fine-grained carbonate sediments by Shinn and Robbin (1983) resulted in reduction of sediment thickness by 50 percent and reduction of initial porosities of 65–75 percent to 35–45 percent at pressures equivalent to burial depth of about 1000 ft (305 m). In an earlier paper, Shinn et al. (1977) reported experimental compaction of unconsolidated carbonate sediment of approximately 75 percent under a pressure equivalent to a burial depth of 1–1.5 km. Furthermore, they found that foraminifers and other fossils were not noticeably crushed or broken.

Pressure solution owing to compaction is a very important phenomenon in carbonate rocks. It leads to significant thinning of carbonate beds and generates an important source of carbonate cement. Pressure solution is discussed in greater detail under the heading of solution.

Cementation

Cementation is the process whereby new minerals are precipitated from pore fluids onto the surface of grains, or chemical substances are added as syntaxial overgrowths on existing mineral grains. Cements can eventually fill almost all existing pore space and thus reduce porosity to virtually zero. Cements bind sedimentary particles together, and cementation is the principal chemical process that produces lithification of sediment. Cementation occurs mainly during the early to middle stages of diagenesis, but it can occur quite late, even after uplift into the zone of meteoric water (groundwater). When cementation occurs early, it becomes an important factor in limiting later mechanical compaction of sediment unless the cement is subsequently removed by solution processes.

Cementation of Sandstones

Cements in sandstones can be examined quite easily with an ordinary petrographic microscope; therefore, cementation of sandstones has been studied extensively by sedimentary petrologists. As discussed in Chapter 7, the most common cementing minerals in sandstones and other siliciclastic sedimentary rocks are calcite and quartz. Aragonite, dolomite, siderite, chert, opal, hematite, limonite, feldspars, anhydrite, gypsum, barite, zeolite minerals, and clay minerals are less common cementing materials. Cements such as calcite are precipitated from pore waters within sands and commonly form a mosaic of smaller crystals within the pore spaces among larger sand grains (Fig. 7.9). They adhere to the larger grains and thus act as a kind of glue to bind them together. A somewhat different type of cementation involves the addition of cementing material to minerals of like composition to form a syntaxial overgrowth. As described in Chapter 7, an existing mineral acts as a seed for nucleation of material of like composition; this material is added to the crystal lattice of the existing mineral, increasing its size. A very common example of this process in sandstones is the addition of quartz overgrowths to quartz grains, commonly changing rounded detrital quartz grains to grains having crystal outlines (Fig. 7.7). Syntaxial overgrowths commonly form on feldspar grains also. Some sandstones are cemented or bonded together by clay minerals precipitated authigenically as cements after deposition. Such sandstones are lithified by a simple bond by which large grains are aggregated through surface cohesion of clay-size particles (Dapples, 1979).

The addition of cements to sandstones during diagenesis is caused by changes in diagenetic conditions such as pH and temperature. Such changes result in pore waters becoming oversaturated with respect to a given cementing substance. The principal questions directed toward understanding sandstone cementation focus on timing and depths of cementation and on the source of the cement. Because quartz and calcite constitute the bulk of cementing materials in sandstones, most research has centered on the mechanisms of quartz and calcite cementation. Cementation by these minerals can occur at any time during diagenesis when the correct geochemical conditions arise, although cementation by quartz tends to precede cementation by calcite. The geochemical conditions that cause precipitation of quartz on the one hand and calcite or aragonite on the other are discussed in Chapter 4 and under the heading of replacement. Briefly, precipitation is controlled mainly by pH and temperature. Increase in pH or temperature favors precipitation of calcium carbonate minerals; decrease in pH or temperature favors precipitation of quartz or chert.

On the basis of kinetic arguments, Blatt (1979) suggests that it takes too long for pore fluids to migrate laterally through deeply buried sandstones to allow pore spaces to be filled with quartz cement in a geologically reasonable period of time. Therefore, he concludes that most silica cementation probably occurs at burial depths less than a few hundred meters and owing largely to vertical migration of pore fluids over relatively short distances. On the other hand, quartz cement in some depositional basins is much more abundant at depths exceeding 1500–2000 m than at shallower depths, suggesting possible derivation of silica from deeper sources. Quartz cementation can apparently take place either during basin subsidence or subsequently when tectonic forces cause uplift of sediments to shallow depths. Silica is assumed to migrate downward from surface waters or to be derived from deeper sources by pressure solution of quartz, chemical diagenesis of clay minerals accompanying the dewatering of shales, and dissolution of other silicate minerals such as feldspars.

Calcite cementation in sandstones tends to occur at greater depths than does quartz cementation, because calcite solubility decreases with increasing depth and temperature and also because of relatively great abundance of calcium carbonate-rich pore waters at depth. The calcium carbonate that forms cements in sandstone may be derived by pressure solution of associated limestones, dissolution of previously formed calcite cement in sandstones, and solution of calcium carbonate fossils.

Cementation of Carbonate Rocks

Cementation is also a widespread phenomenon in carbonate rocks. Together with recrystallization, it is responsible for lithification of limestones and dolomites. The cement in carbonate rocks may be aragonite, calcite, or dolomite, although calcite is the most common cement in ancient limestones. Aragonite appears to precipitate in preference to calcite in marine waters or other pore waters containing magnesium ions (Chapter 4), whereas calcite is the preferred species of carbonate cement precipitated under freshwater conditions. Aragonite cements alter diagenetically to calcite with time, so that most carbonate cements in Paleozoic and Mesozoic limestones are calcite.

Cementation in carbonate sediments can apparently occur at any stage during diagenesis—from very early, while carbonate sediments are still on the seafloor prior to burial, to very late, after uplift of limestones or dolomites into the zone of groundwater recharge. Carbonate cementation can take place both in saline pore waters in deeply buried formations and in meteoric pore waters in the shallow groundwater zone. Thus carbonate cementation can apparently occur at depths ranging from deep burial zones to the very shallow vadose zone (the zone of suspended water above the water table), depending upon the physicochemical conditions of the diagenetic environment. Good evidence is now available that some cementation can take place very early. For example, carbonate sediments can become lithified in the intertidal and sea-spray zone before burial to form **beachrock,** which results from cementation by aragonite and high-magnesian calcite (Tietz and Müller, 1971). Early cementation of carbonate grains on the seafloor also has been reported by several investigators (e.g., Roberts and Moore, 1971; Shinn, 1971). On the other hand, evidence based on oxygen isotope studies and petrographic study of cement fabrics indicates that some cementation occurs very late in the zone of meteoric waters. Cements precipitated in meteoric waters consist mainly of low-magnesian calcite. Calcite cements in limestones typically occur as a mosaic of clear, sparry crystals filling pore spaces among carbonate grains or in solution cavities (Figure 8.7). Calcite cement can occur also as syntaxial overgrowths on fossil echinoderm fragments, which consist of large single calcite crystals.

Authigenesis

In the broadest sense, the term authigenesis encompasses all processes, including cementation and replacement, that cause postdepositional formation of new minerals in sediments or sedimentary rocks. I use authigenesis here in a somewhat more restricted sense to refer to diagenetic processes that mainly involve formation of new minerals other than cements and replacement minerals. These new minerals may form by crystallization out of solution or by alteration of existing minerals and rock fragments through chemical reorganization and recombination of elements. Formation of new minerals by replacement is treated separately. There are six principal processes of authigenesis:

1. Formation of pyrite by alteration of iron-bearing minerals under reducing conditions. This process commonly occurs in the presence of organic carbon compounds. These compounds contribute electrons to drive iron into the ferrous state and permit

its fixation in the presence of sulfur as pyrite; hence the common association of pyrite with dark-colored, organic-rich sandstones and mudrocks.

2. Oxidation of iron-bearing minerals under conditions of largely positive Eh to yield the ferric oxide minerals limonite (goethite) or hematite. Hematite is the mineral that gives the red color to terrigenous redbeds; it is now believed that much, if not most, hematite in red sandstones and mudrocks is of authigenic origin (Walker, 1967). Authigenic hematite and hematite cement are very common in siliciclastic sedimentary rocks and occur also in carbonate rocks. Although most hematite probably forms at or near the sediment–atmosphere or sediment–water interface, hematite is known to be stable under reducing conditions at moderately high pHs (Figure 4.12).

3. Alteration of clay minerals to form muscovite, biotite, illite, chlorite, and glauconite. The formation of authigenic micas and clay minerals occurs mainly at temperatures and pressures higher than those that exist at the surface, although glauconite commonly forms also under low-pressure and low-temperature conditions on the seafloor. Authigenic micas are generated principally in clay-rich sandstones and mudrocks. Muscovite tends to be favored over other authigenic micas except under reducing conditions, which are indicated by the presence of associated pyrite or abundant preserved organic matter. Under these conditions, authigenic biotite or chlorite formed by alteration of montmorillonite may be abundant. Secondary glauconite derived from illite may occur in such organic-rich rocks. Montmorillonite or other smectite clays contain interlayer water that can be lost during diagenesis. Loss of water combined with replacement of sodium by potassium and some silicon by aluminum converts montmorillonite to illite. The dehydration of montmorillonite is temperature dependent, and the threshold temperature required for alteration to illite occurs at temperatures ranging from about 70°C to more than 150°C (Bruce, 1984). Temperatures in the range of 100° to 150°C are most common. Kaolinite also breaks down at about these same temperatures, although kaolinite and illite can coexist. Thus, illite and chlorite tend to increase during deep burial at the expense of montmorillonite and kaolinite. The dehydration of montmorillonite releases pore waters that may differ in chemical composition from the pore waters of adjacent sediments. The escape of these shale waters during diagenesis is believed to be an important mechanism for flushing petroleum hydrocarbons from the shales into more permeable sandstones or limestones.

4. Formation of clay minerals and fine-size muscovite (sericite) by alteration of feldspars (Fig. 9.4) and volcanic rock fragments. Potassium feldspars are commonly altered to sericite and kaolinite, and plagioclase feldspars may be altered to montmorillonite. This process is favored by increased temperature and occurs mainly under diagenetic conditions. Kaolinite is known to form also by alteration of feldspars under the low-temperature, low-pressure conditions of weathering. Kaolinite forms mainly in the temperature range below 100°C and may be destroyed with increasing burial and higher temperatures, as mentioned. Basic volcanic lithic fragments commonly break down to form montmorillonite clay.

5. Alteration of volcanic ash, particularly volcanic glass, to form smectite clays and zeolite minerals. The formation of zeolites is strongly temperature dependent and takes place at temperatures only slightly below those of metamorphism.

6. Formation of authigenic quartz and feldspars. Although authigenic quartz and feldspar are commonly precipitated as overgrowths on existing grains, new quartz and feldspar grains can develop authigenically. Authigenic quartz and feldspars can commonly be distinguished from terrigenous quartz and feldspar by smaller size and well-developed crystal faces. Of particular interest is the formation of authigenic albite (so-

FIGURE 9.4 Plagioclase feldspar grain showing extensive alteration to fine sericite (S). Polarized light.

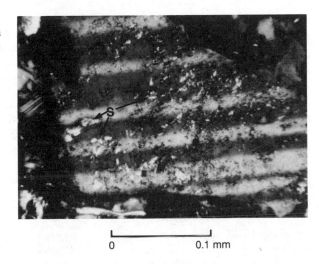

0   0.1 mm

dium plagioclase). Albite can apparently form by alteration of potassium feldspar or calcium plagioclase in a sodium-rich environment in sediments subjected to deep burial and high temperatures. This process is referred to as **albitization.** The conversion to albite can take place directly, or it may involve prior replacement of the feldspars by other minerals, such as anhydrite, calcite, or dolomite (T. R. Walker, 1984).

Replacement

Diagenetic replacement involves the slow dissolution of one mineral, or a part of a mineral, and the essentially simultaneous deposition of a new mineral in its place. The best known example of replacement is probably the formation of petrified wood, a process that involves the solution of organic woody tissue and its replacement by chert. Replacement during diagenesis is an extremely common process in both siliciclastic sedimentary rocks and carbonate rocks. Minerals, rock fragments, fossils and other carbonate grains, clay matrix in sandstones, lime mud in limestones, and cements may all be replaced to various degrees.

The factors that control replacement of minerals during diagenesis are basically the same factors that control solubility of the minerals. Thus pH, Eh, temperature, pressure, the presence of other ions in solution, and the partial pressure of CO_2 in solution may all play important roles in the replacement process. The two most common replacing minerals are calcite and chert. As discussed in Chapter 4, the solubility of these minerals is controlled largely by pH and temperature, and pH in turn is controlled mainly by dissolved CO_2. Because pH and temperature affect the solubility of calcium carbonate and silica in opposite ways, an inverse relationship exists between chert replacement and calcite replacement. Calcium carbonate solubility is decreased and silica solubility is increased by increasing pH and temperature. Chert is a common replacing mineral for calcite, and vice versa, and multiple episodes of carbonate–silica replacement have been reported (T. R. Walker, 1962) in response to reversals in pH, temperature, or other conditions during diagenesis.

In addition to calcite and chert, many other minerals—including dolomite, hematite, limonite, glauconite, anhydrite, and siderite—are known to be replacement minerals. For example, the clay matrix in many muddy sandstones is commonly replaced in part by calcite or chert. The matrix in some sandstones may be replaced by dolomite or siderite. Silicate minerals such as feldspar and quartz are most likely to be replaced by calcite, but the replacing calcite, as well as calcite cement, may in turn be replaced by chert or glauconite. Carbonate rocks are particularly susceptible to replacement. Silica is a common replacement material in limestones. Silicification tends to be selective; chert typically replaces fossils and other carbonate grains in preference to fine lime mud (Fig. 9.5). Replacement of calcite or aragonite by dolomite—a process called dolomitization—is also extremely common. As discussed in Chapter 4, dolomitization is a widespread and volumetrically important phenomenon. Although dolomitization of limestones tends to be a selective process, guided perhaps by zones of higher porosity and permeability in the limestones, it can also be pervasive. Pervasive dolomitization led to massive alteration of ancient limestones to dolomites.

Inversion

Inversion is the replacement of a mineral by its polymorph; that is, by a mineral having the same chemical composition but a different crystal structure. Inversion can occur by solid-state transformation or by a solution–reprecipitation process (Carlson, 1983). A very common example of inversion that occurs during diagenesis is the alteration of aragonite (orthorhombic $CaCO_3$) to low-magnesian calcite (rhombohedral $CaCO_3$). As discussed in Chapter 4, aragonite appears to be precipitated preferentially over calcite in the presence of Mg in seawater. Most modern lime muds consist of aragonite, but most ancient limestones are composed of calcite, indicating that aragonite is metastable and transforms to calcite with time. The factors that govern the rate of transformation of aragonite to calcite are complex and poorly understood. Water appears to catalyze the process of transformation, which may proceed rapidly under some aqueous conditions. Inversion has been reported to occur in as short a time as one year. On the other hand, aragonite has been found in rocks as old as the Paleozoic in some special types of deposits such as asphalts and oil shales. Nonetheless, aragonite occurs only rarely in rocks older than the Tertiary.

The conversion of opal-A to opal-CT and finally to quartz during the process of forming chert (Chapter 4) is another example of a mineral that can change during

FIGURE 9.5 A fusulinid foraminifer replaced along one edge by chert (Ch). Note the ragged boundary between the foraminifer and the chert caused by the replacement process.

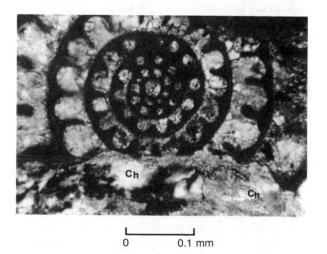

0   0.1 mm

diagenesis to a different mineral of the same composition. Opal-A is amorphous SiO_2; opal-CT consists of cristobalite (orthorhombic SiO_2) and tridymite (tetragonal SiO_2); and quartz is rhombohedral SiO_2. The rate of conversion of opal-A to quartz has generally been considered to be strongly temperature controlled. It depends also upon the nature of the opal-A starting material and the solution chemistry, particularly the availability of Mg^{2+} and OH^- ions, which form magnesium hydroxide compounds that serve as a nucleus for crystallization of opal-CT (Kastner and Gieskes, 1983). The conversion of opal-A to chert (quartz) is a solution–precipitation process (L. A. Williams and Crerar, 1985). Opal-A, such as in the skeletons of siliceous organisms, dissolves to yield solutions high in silica. The silica then flocculates to yield opal-CT, which becomes increasingly ordered before dissolving to yield pore waters of low silica concentration. Slow growth of quartz takes place from these waters. L. A. Williams et al. (1985) indicate that silica solubility increases with increasing surface area of siliceous particles. They suggest that this relationship between solubility and surface area can explain simple opal-A to quartz transformation and that temperature may thus be a less significant factor in the transformation than previously thought.

As mentioned, most previous investigators have indeed placed considerable emphasis on temperature as a major factor controlling the rate at which biogenic opal-A transforms to opal-CT and chert. The time required for this transformation in natural systems is thus believed to be a function of sedimentation and burial rates and the geothermal gradient (discussed in a succeeding section) in the region of burial (Siever, 1983). Depending upon these factors, the transformation from opal-A to chert can take from a few million years to a few tens of million, or even hundreds of million, years. Faster rates of transformation occur where rates of sedimentation and burial are high because sediment is quickly buried to depths where high temperatures bring about rapid transformation. High geothermal gradients in a region mean that siliceous sediments reach the high temperature required for transformation at shallower burial depths. The effects of burial on the diagenesis of opal-A is illustrated in the time–temperature diagrams of Figures 9.6 and 9.7. These diagrams show the time–temperature regions for conversion of opal-A to opal-CT and quartz for two geographic areas with different sedimentation and burial rates and geothermal gradients. The time–temperature curve in Figure 9.6 represents deposition of siliceous lacustrine sediments in a rift valley where the geothermal gradient is high and sedimentation rates are variable but high. Figure 9.7 is a time–temperature curve for slow sedimentation of pelagic oozes on a spreading seafloor in the open ocean, where the geothermal gradient is lower than in the rift valley. Note the longer time required for transformation of opal-A to quartz under the conditions assumed in Figure 9.6.

Recrystallization

Some confusion still exists regarding the exact meaning of the word recrystallization, a term that is sometimes used in a broad sense to include replacement and inversion. In its narrowest sense, recrystallization refers only to change in crystal size and/or shape, without accompanying change in chemical composition or mineralogy, of a given mineral species. Folk (1965) proposed the inclusive term **neomorphism** to cover all aspects of recrystallization, including inversion (Fig. 9.8). The term recrystallization is used in this book in its more restricted sense to imply only changes in grain size and shape of minerals.

Changes in grain size of minerals during diagenesis are primarily in the direction of increasing grain size, although grain size can decrease by recrystallization. Increase

FIGURE 9.6 Time–temperature diagram for the diagenetic alteration of siliceous sediments under conditions of variable, but moderately rapid, sedimentation in a rift zone with a high geothermal gradient. The diagram shows a time–temperature curve that is based on estimates of sedimentation rates and the geothermal gradient. Time–temperature regions for the conversion of opal-A to opal-CT and quartz (chert) are also shown. (From Siever, R., 1983, Evolution of chert at active and passive continental margins, *in* A. Iijima, J. R. Hein, and R. Siever (eds.), Siliceous deposits in the Pacific region: Developments in Sedimentology 36. Fig. 6, p. 17, reprinted by permission of Elsevier Science Publishers, Amsterdam.)

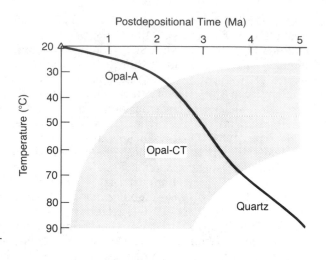

FIGURE 9.7 Time–temperature diagram for the diagenetic alteration of siliceous sediments under conditions of slow sedimentation on a spreading seafloor where the geothermal gradient is low. (From Siever, R., 1983. Fig. 8, p. 20, reprinted by permission of Elsevier Science Publishers, Amsterdam.)

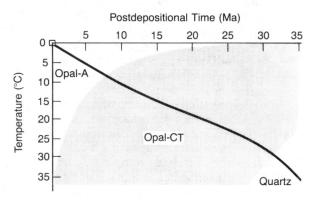

in crystal size owing to recrystallization is called **aggrading neomorphism** by Folk (1965), whereas the less common process of decrease in crystal size is called **degrading neomorphism.** The driving force behind recrystallization is not well understood. It has been proposed that the recrystallization force may come from the energy stored in strained crystals owing to deformation and, alternatively, to the force arising from the surface tension of curved crystal boundaries. Experimental work by Baker et al. (1980) indicates that the mechanisms of recrystallization involve both relief of strain energy at grain-to-grain contacts and decrease in surface free energy by decrease in the surface areas of the grains. These authors found that increase in effective pressure, temperature, and ionic strength of the experimental solutions also caused an increase in the extent of calcite recrystallization, whereas the presence of clay minerals and other noncarbonate minerals retarded recrystallization. By extrapolation, these experimental results can be taken to suggest that, under natural conditions, increases in temperature and pressure and the changing nature of interstitial fluids are important controls on recrystallization. Also, porosity and permeability of the sediments probably play some role in recrystallization by controlling the rate of flow of fluids through rocks. It is known that the energetics behind any recrystallization of a substance from one form

Process	Example	Metamorphic or Igneous Term	Metallurgical Term	Folk (1965)	
				Loose	Strict Usage
One mineral replaces another of a different composition	Calcite ⟶ Dolomite, Pyrite, etc.	Replacement or Metasomatism	———	Replacement	
A mineral is replaced by its polymorph	Aragonite mud or skeleton ⟶ Calcite mosaic	Inversion or Transformation	Allotropic Recrystallization	Replacement (loosely) / Recrystallization (loosely) / Neomorphism	Inversion
A deformed mineral changes to a mosaic of undeformed crystals of the same mineral	Strained calcite ⟶ Unstrained calcite	Recrystallization	(Strain) Recrystallization		Strain Recrystallization
An undeformed mineral changes its form, grain size or orientation	Calcite mud or fibers ⟶ Calcite mosaic, etc.	Recrystallization	Grain growth		Recrystallization (and degrading recrystallization)
A mineral dissolves leaving a cavity; cavity is filled later	Skeleton ⟶ Cavity ⟶ Calcite	Rare No special term	———	Solution-cavity Fill	

FIGURE 9.8 Terminology used to describe various types of mineral alteration as observed with a petrographic microscope. (After Folk, R. L., 1965, Some aspects of recrystallization in ancient limestones, in L. C. Pray and R. C. Murray (eds.), Dolomitization and limestone diagenesis: Soc. Econ. Paleontologists and Mineralogists Spec. Pub. 13. Fig. 1, p. 15, reprinted by permission of SEPM, Tulsa, Okla.)

to another more stable form is the tendency toward a minimum in the Gibbs free energy of the chemical system (Pettijohn et al., 1973). Yet, we know relatively little about the specific causes of recrystallization under natural conditions or why decrease in grain size, rather than increase, sometimes occurs during recrystallization.

Although recrystallization can occur in any type of sedimentary rock, it is particularly important in the nonsiliciclastic sedimentary rocks, especially carbonate rocks. Limestones are markedly susceptible to recrystallization, which may be either selective or pervasive. Some limestones become so thoroughly recrystallized during diagenesis that they are converted into coarsely crystalline rocks resembling dolomites, which may preserve no vestige of original depositional textures and structures. More commonly, limestones are selectively recrystallized. When viewed with a petrographic microscope, selective recrystallization shows up as patches of coarse, recrystallized sparry calcite or recrystallized fossil fragments or other carbonate grains in otherwise fine-textured, unrecrystallized limestones (Fig. 9.9).

FIGURE 9.9 Photomicrograph of dark-colored micrite partially recrystallized to fine crystals of sparry calcite (light-colored area) along a slightly embayed boundary. Note the dark patches of relict clay or organic material in the recrystallized sparry calcite. Polarized light.

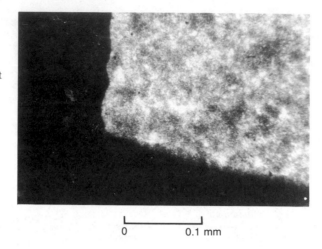

```
0          0.1 mm
```

Dissolution

Solution, or dissolution, is a very important diagenetic process that leads to an increase in porosity and thinning of beds, especially in easily dissolved rocks such as carbonates and evaporites. Because of its effect on porosity development, petroleum geologists, hydrologists, and other scientists interested in subsurface flow of fluids are particularly interested in solution processes. Therefore, considerable research has been done on the development of secondary porosity in both limestones and sandstones. Diagenetic dissolution is a highly selective process that tends to remove the most soluble constituents under a given set of environmental parameters. The factors that control solution are mainly pH, Eh, temperature, pressure, CO_2 partial pressure, and ionic strength and chemical composition of the pore fluids (Chapter 4). Solution during diagenesis is affected also by such other factors as existing porosity and permeability, mineralogy, and grain size of the sediment. Increase in pH and temperature favors dissolution of silicate minerals, whereas decrease in pH and temperature favors solution of carbonate minerals. Therefore, solution of silicate minerals is most likely to occur with deep burial and increased temperature. Carbonate rocks are more susceptible to solution at shallow burial depths, particularly if tectonic uplift brings them into the zone of circulating carbon dioxide-rich meteoric waters. Solution of carbonate minerals can occur also during deep burial if increased hydrostatic pressure causes carbon dioxide to dissolve in pore waters to a great enough extent to offset the effects of decreased carbon dioxide solution brought about by increased temperature.

The most soluble constituents in sandstones are calcite cements and rare fossils; therefore, the major effect of solution in sandstones is removal of cements, a process commonly referred to as **decementation.** Metastable framework grains in sandstones—for example, feldspars, rock fragments, and mafic heavy minerals—can also be selectively removed by solution. Selective removal of such unstable siliciclastic minerals has been referred to as **intrastratal solution.** Intrastratal solution is inferred to be responsible for the observed statistical tendency of unstable heavy minerals such as pyroxenes and amphiboles to become less abundant with increasing age of rocks (Pettijohn, 1941). The selective dissolution of minerals in a siliciclastic mineral assemblage thus leaves the assemblage biased toward the more chemically stable minerals, thereby complicating interpretation of source rocks on the basis of heavy mineral suites or other metastable minerals.

Solution of cements and selective solution of some framework minerals play very important roles in creating secondary porosity in sandstones. We have only recently come to realize that most of the porosity in sandstones at burial depths exceeding about 3000 m is secondary porosity (Schmidt and McDonald, 1979; McDonald and Surdam, 1984). It is now believed that most primary porosity in sandstones is reduced to virtually zero by compaction and cementation at burial depths exceeding about 3000 m, whereas secondary porosity generated by dissolution of carbonate cements and to a lesser extent of silicate and sulfate minerals may persist to depths exceeding 7000 m (Fig. 9.10). At burial depths of 2–3 km or more, solution of quartz grains owing to enhanced solubility at points of contact—that is, pressure solution—becomes volumetrically important. Such solution causes chemical compaction of framework grains and thus acts to reduce primary or secondary porosity. Pressure solution is also an important mechanism for releasing silica into solution and may account for as much as one-third of the silica cement present in some formations (Sibley and Blatt, 1976).

Selective dissolution is a particularly important mechanism for producing secondary porosity in carbonate rocks. Carbonate rocks are much more susceptible to solution than sandstones and other siliciclastic sedimentary rocks. Much, if not most, of the porosity in these rocks is either secondary porosity or primary porosity that has been enhanced by solution processes. Mineralogy and magnesium content of carbonates strongly influence their relative solubilities. High-Mg calcite is least stable, or most easily dissolved, followed by aragonite and calcite (low-Mg calcite). Because carbonate sediments commonly are composed initially of mixtures of aragonite, high-Mg calcite, and low-Mg calcite, selective solution of the more soluble forms of $CaCO_3$

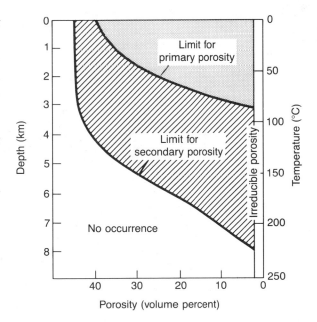

FIGURE 9.10 Changes in primary and secondary porosity in sandstones with depth. The example shows porosity change in a quartz-rich sandstone (quartz arenite) deposited at a sedimentation rate of 30.5 m/1,000,000 years in a basin having a geothermal gradient of 2.70°C/100 m. (After Schmidt, V., and D. A. McDonald, 1979, The role of secondary porosity in the course of sandstone diagenesis, *in* P. A. Scholle and P. R. Schluger (eds.), Aspects of diagenesis: Soc. Econ. Paleontologists and Mineralogists Spec. Pub. 26. Fig. 12, p. 189, reprinted by permission of SEPM, Tulsa, Okla.)

leads to formation of solution pores and thus enhanced porosity. The grain size of carbonate sediment also affects solubility, which decreases with increasing grain size. Solubility increases with increase in the rate of fluid circulation through the sediment, which in turn is governed by existing porosity and permeability. The relationship between environmental parameters and carbonate solubility is summarized diagrammatically in Figure 9.11.

Most solution porosity in carbonate rocks appears to take place at shallow depths, particularly in the zone of meteoric water (Longman, 1982). In meteoric waters, high-Mg calcite either dissolves or alters to low-Mg calcite by expulsion of Mg. Alternatively, in Mg-rich waters, it may alter to dolomite by addition of Mg. Aragonite may dissolve or invert to calcite, and calcite may dissolve or persist depending upon the geochemical conditions. Dolomite if present may also dissolve, alter to calcite by loss of Mg, or remain unchanged. Solution porosity in carbonate rocks is not thought to form to any great extent below a burial depth of a few hundred meters because pressure solution becomes the dominant solution process in carbonate rocks at greater depths. Pressure solution is a widespread diagenetic solution process that does not produce solution pores. On the contrary, it leads to chemical compaction and reduction in thickness of carbonate beds by as much as 35–40 percent (Park and Schot, 1968). The effects of pressure solution are manifested in carbonate rocks by the presence of **stylolites** and microstylolites, which are suturelike seams marked by the presence of insoluble clay minerals and other silicate minerals left behind after solution of more soluble carbonate constituents (Fig. 9.12). Pressure solution is an important mechanism for supplying the $CaCO_3$ that forms secondary pore-filling cements.

FIGURE 9.11   Effect on carbonate solubility of mineralogy, grain size, and diagenetic environmental factors. The arrows indicate the direction of increasing values. (After Longman, M. W., 1982, Carbonate diagenesis as a control on stratigraphic traps: Am. Assoc. Petroleum Geologists Educ. Course Notes Ser. 21. Fig. 46, p. 82, reprinted by permission of AAPG, Tulsa, Okla.)

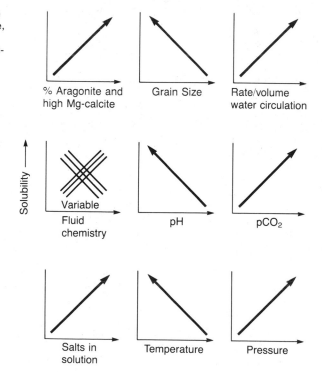

FIGURE 9.12 A stylolite seam in a slab of polished limestone. Note the dark-colored, insoluble residue of clay or organic matter concentrated along the stylolite seam.

9.3 THE DIAGENETIC ENVIRONMENT

The diagenetic processes discussed are controlled by the physical and chemical conditions of the subsurface environment. A brief description of these environmental conditions is given here to promote understanding of the underlying causes of diagenesis. Sediments accumulate in depositional basins at moderately low temperatures ($<\sim25°C$) and at atmospheric pressure (14.66 lb/in^2, or 1.03 kg/cm^2). The pH and Eh of depositional environments are variable, as shown in Figure 4.1. Under most depositional conditions, pH falls within the range of about 4 to 9. Oxidizing conditions also prevail in most depositional environments, but moderately reducing conditions may exist in some basins where water circulation is restricted and organic matter is abundant. The pore waters of freshly deposited sediments are either freshwaters, if deposited in continental settings, or waters of mainly normal marine salinity, if deposited on the seafloor. Temperature and pressure increase significantly with burial depth, and pH, Eh, and chemical composition and salinity of pore fluids in the sediment also change. These various controls on diagenesis are summarized in Figure 9.13. Assemblages of terrigenous minerals from mixed igneous and metamorphic sources and nonsiliciclastic minerals deposited under low-temperature, low-pressure conditions are thus, during burial, brought into an environment in which they may be out of equilibrium with the temperatures, pressures, and geochemical conditions that exist in the subsurface. Therefore, there is a general tendency during diagenesis for nonequilibrium assemblages of minerals to change chemically and mineralogically to approach chemical equilibrium.

Pressure in the Diagenetic Environment

Burial of sediment is accompanied by increase in pressure owing to the weight of overlying sediment, as well as increase in fluid pressure within the sediment pores. The pressure owing to the weight of the sediment is called **geostatic pressure** or **lithostatic pressure.** Change in geostatic pressure with depth is the **geostatic gradient,** which varies according to the density of the overlying sediment. The approximate geostatic gradient in sedimentary basins is 1.08 pounds per square inch (psi) per foot of depth (or about 0.249 kg/cm^2/m, or 244 bars/km, where 1 bar = 0.987 atmospheres of pressure). The geostatic gradient is illustrated in Figure 9.14. If we take as an example a burial depth of 10 km (6.2 mi), the geostatic pressure is about 2.4 kilobars (kb)

FIGURE 9.13 Major environmental controls on diagenesis. (After Longman, M. W., 1982, Carbonate diagenesis as a control on stratigraphic traps: Am. Assoc. Petroleum Geologists Educ. Course Notes Ser. 21. Fig. 45, p. 80, reprinted by permission of AAPG, Tulsa, Okla.)

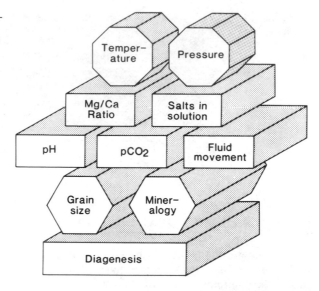

(1 kb = 1000 bars), about 2400 times greater than atmospheric pressure. The most pronounced effects of increased lithostatic pressure during diagenesis are compaction and porosity reduction in sediments. Pressure also influences the stability of mineral phases and thus helps to control the destruction of existing minerals and the crystallization of authigenic minerals.

Fluid pressure within sediments results from the weight of the column of interconnected fluid within the sediment and increases with depth at a rate that depends upon density of the pore fluids. Fluid pressure is referred to as **hydrostatic pressure.** The rate of change of hydrostatic pressure with depth is the **hydrostatic gradient.** The hydrostatic gradient for freshwaters is 0.433 psi per foot of depth (or ~0.099 kg/cm^2/m) and ranges up to about 0.50 psi/ft (0.115 kg/cm^2/m) for very dense brines. The average hydrostatic gradient in sedimentary basins is about 0.46 psi/ft (10.41 kg/cm^2, or 104 bars/km) (Fig. 9.14). At 10 km below the surface, the average fluid pressure is thus approximately 1.04 kb (or 15,092 psi). Fluid pressures determine the direction of fluid movement in the subsurface and the mixing of different types of subsurface fluids during diagenesis. Increase in hydrostatic pressure with depth also enhances the solution of gases such as CO_2 and natural gas in pore waters.

In some basins, such as parts of the Gulf Coast region of the United States, the fluid pressures encountered at depth are considerably higher than those predicted by the average hydrostatic gradient. Subsurface formations containing such abnormally high fluid pressures are said to be **geopressured.** Overpressured fluid conditions arise from several causes, the most important of which is related to tectonism and flow of soft sediments under lithostatic pressure loading. Thrust faulting and squeezing of soft, compressible sediments allows compressive tectonic stresses to be transmitted to the contained fluids. Consequently, if fluids are trapped within impermeable layers and cannot escape, fluid pressures may rise to abnormally high values. Some geopressured formations in the Gulf Coast region contain important quantities of natural gas dissolved in the sediment pore waters under these overpressured conditions.

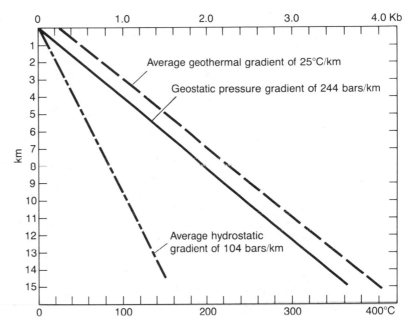

FIGURE 9.14 Average geothermal gradient, geostatic pressure gradient, and hydrostatic pressure gradient in sedimentary basins.

Diagenetic Temperatures

Temperature Gradients. Burial of sediment is accompanied by increase in temperature of approximately 20–30°C/km of depth. Change in temperature with depth is referred to as the **geothermal gradient.** Values for the geothermal gradient are controlled by the heat flux within Earth, and geothermal gradients are highest where heat flux is highest. Thus, they vary greatly in different parts of Earth, from extreme low values of about 6°C/km to extreme highs of about 65°C/km. Heat flux tends to be very high in areas of intense volcanism, in rift systems, and along mid-ocean spreading ridges. It is very low in most deep-sea trenches and has intermediate values in other parts of the ocean and in nonvolcanic areas of the continents. The geothermal gradient illustrated in Figure 9.2 was constructed assuming an average, uniform value of temperature change of 25°C/km in sedimentary basins. In fact, the rate of temperature increase with depth is probably not uniform and certainly must decrease significantly at great depth. In any case, if we again take a 10-km burial depth as an example, the temperature at that depth is approximately 280°C, or about eleven times the average surface temperature.

Estimating Paleotemperatures. Temperature has a particularly significant effect on many diagenetic processes. It exerts an important control on the solubility of minerals and thus on solution and precipitation reactions; it also affects recrystallization reactions and authigenic processes such as the alteration of feldspars to clay minerals and micas and the alteration of one type of clay mineral to another. Therefore, geologists are greatly interested in determining the temperatures at which particular diagenetic reactions take place; considerable research has been carried out to develop reliable techniques for paleotemperature analysis. Tools used for determining paleotemperatures are commonly referred to as **geothermometers.** The principal techniques now in use for determining diagenetic paleotemperatures include methods based on (1) conodont color alteration, (2) vitrinite reflectance, (3) graphitization levels in kerogen, (4) clay mineral assemblages, (5) zeolite mineral assemblages, (6) fluid inclusions, and (7) oxygen isotope ratios.

These methods are described briefly in Table 9.2, which also gives the useful temperature range of each method and some explanatory remarks about the materials that constitute the geothermometers. The methods have various degrees of reliability and none can be considered an infallible estimator of paleotemperatures; conodont color alteration and vitrinite reflectance are generally regarded to be the most useful methods. Methods based on analyses of mineral assemblages tend to be less sensitive and more equivocal. The formation of zeolite minerals, for example, depends upon pressure and the salinity and chemical composition of sediment pore waters, as well as temperature. Two or three different methods, which generally include examination of conodont color alteration and vitrinite reflectance, are commonly used together as a cross-check on reliability.

pH, Eh, and Chemical Composition of Subsurface Waters

Owing to the difficulty in making direct measurements of pH and Eh in subsurface formations, few reliable data are available on changes of these parameters with burial depth. Both pH and Eh measurements can be made on waters recovered from drill holes; however, these waters may not yield reliable values for formation pH and Eh. In addition to problems of contamination by drilling fluids and other substances, pH may change because of oxidation of reduced species in the water; loss of dissolved gases, particularly CO_2; or reaction between carbonate species:

$$H_2CO_3 \leftrightharpoons HCO_3^- + H^+ \leftrightharpoons CO_3^{2-} + H^+$$

Such a reaction may either tie up or release H^+ ions. Also, pH is temperature sensitive. Measured at wellhead or at laboratory temperatures that are much lower than formation temperatures, pH values are higher than those of actual formation waters unless a correction for temperature is made. Most published values for formation pH have probably not been corrected for temperature. The measurement of Eh in formation waters is even more difficult and imprecise; therefore, very little direct information is available on the oxidation state of formation waters.

pH. Published values for pH of formation waters are reported to vary rather widely. For example, Blatt et al. (1980) report ranges from 3 to 11 and Hanor (1979) gives the range as 5–9. The pH of most formation waters appears to fall between about 7 and 9. There are simply not enough reliable data on the pH of subsurface waters to draw definite conclusions about patterns of pH change with depth, although there appears to be a general trend of pH increase. This increase is apparently due mainly to loss of hydrogen from pore waters as hydrogen enters minerals during diagenetic alteration. The pH may also rise as a result of increasing temperature of pore waters, resulting in dissociation of carbonic acid to carbon dioxide and water and subsequent escape of CO_2.

TABLE 9.2 Methods for estimating diagenetic paleotemperatures

Method	Basis for technique	Useful temperature range	Remarks	References
Conodont color alteration	Color changes in conodonts from pale yellow (thermally unaltered) to brown to black, owing to carbon fixation within trace amounts of organic material in the conodonts as a result of increase in temperature; followed by change in color to white (extreme thermal alteration) caused by carbon loss and loss of water by crystallization and recrystallization.	<50°−>400°C Visually recognizable levels in the conodont color alteration index (CAI): 1 <50°−80°C (pale yellow) 1.5 50°−90° 2 60°−140° (bn/dk bn) 3 110°−200° 4 190°−300° 5 300°−400° (black)	Conodonts are marine microfossils composed of the mineral apatite, but also containing trace amounts of organic material. They range in age from Cambrian to Triassic and occur principally in carbonate rocks and shales.	Harris (1979) Epstein et al. (1977)
Vitrinite reflectance	Light reflection from vitrinite (thermally altered organic grains) is measured quantitatively with a reflectance microscope. Reflection increases with increasing degree of thermal alteration (metamorphism).	Up to about 240°C % vitrinite reflectance is related to minimum temperature by the scale: <0.48% <100°C 0.53 125 0.72 145 0.83 165 1.00 180 1.13 195 1.42 210 1.50 220 1.70 230 1.92 235 2.14 240	Vitrinite is structured or unstructured woody tissue plus tissue impregnations that occurs as disseminated grains in sediment and is a major constituent in coals (Chapter 8).	Bostock (1979) Diessel and Offler (1975)

TABLE 9.2 continued

Method	Basis for technique	Useful temperature range	Remarks	References
Analysis of graphitization levels in kerogen	The carbon atoms in kerogen become increasingly well ordered (structured) with increasing levels of thermal diagenesis—a process called graphitization. Increase in the degree of ordering, or level of graphitization, as determined by X-ray diffraction methods can thus be related to increase in diagenetic temperature.	Up to about 600°C	Kerogen is the disseminated organic matter of sedimentary rocks; insoluble in nonoxidizing acids, bases, and organic solvents. The organic matter initially deposited in sediments is converted to kerogen during diagenesis by thermocatalytic processes. Kerogen occurs principally in shales.	Harrison (1979) Landis (1971)
Analysis of clay mineral assemblages	With increasing temperature and metamorphic grade, smectite clay minerals convert to illite through a mixed layer illite/smectite series; chlorite appears, and kaolinite as well as K-feldspar disappears. Thus, the relative abundance of these clay minerals in rocks determines the metamorphic grade and can be related to paleotemperatures (e.g., smectites occur at temperatures below about 100°C; mixed-layer clays are stable up to about 200°C; illite forms at temperatures above 200°C).	Up to about 300°C	Clay minerals are phyllosilicate minerals made up of two-dimensional layer structures. They belong to four major clay mineral groups distinguished by differences in layer structure and composition: kaolinite, smectite, illite, and chlorite. Clay minerals are most abundant in shales, but also occur as a matrix in sandstones and in minor amounts in limestones.	Hoffman and Hower (1979) Dunoyer de Segonzac (1970)
Analysis of zeolite facies mineral assemblages	Temperature theoretically exerts a strong control on the types of zeolite minerals that occur together. For example, heulandite and analcite tend to occur at temperatures below about 100–125°C, whereas laumontite and pumpellyite occur at temperatures between about 100–125°C and 175–200°C. Thus paleotemperature can be established roughly on the basis of the assemblages of zeolite-facies minerals present in a rock.	Up to about 200–250°C	Zeolite facies minerals include a large group of minerals that develop authigenically in volcaniclastic sediments through alteration of chemically reactive volcanic materials. They form in the overlapping temperature range of diagenesis and metamorphism and show a progression of mineral facies that is clearly a reflection of temperature (and pressure) of burial.	Ghent (1979) Coombs (1971) Merino (1975a, b)

Analysis of fluid inclusions	Recrystallization of minerals or formation of overgrowths on minerals during diagenesis may trap fluid as minute inclusions in the crystals. Fluid inclusions commonly consist of a liquid plus a bubble of gas. Presumably at the time of formation, the inclusion consisted of a single fluid phase which separated into two upon cooling. By reheating the mineral until the phase boundary between the liquid and gas can be seen to just disappear, the approximate temperature at which the inclusion formed can be established—taking into account an estimate of the pressure of formation.	25°–150/200°C	Fluid inclusions are found in geodes, vugs, and veins in sediments; sedimentary ore deposits; carbonate and quartz cements in terrigenous sedimentary rocks; salt and sulfur deposits; petroleum reservoir rocks; and in sphalerite (zinc-bearing ore mineral) in bituminous coal beds	Roedder (1979) Roedder (1976)
Oxygen isotope ratio	The ratio of $^{18}O/^{16}O$ in two coexisting oxygen-bearing minerals (from the same specimen) such as quartz and illite is commonly different. The amount of this difference has been shown to be a function of the maximum temperature to which the rock containing the minerals has been heated during diagenesis. Therefore, the isotopic fractionation, or difference between the $^{18}O/^{16}O$ ratios of two minerals which have reached equilibrium with each other, can be used to calculate the maximum diagenetic temperature to which the rock was heated. For example, isotopic fractionation between quartz and illite pairs is greatest at low temperatures and decreases with increasing temperature.	Up to about 400°C	The fractionation factor α between two coexisting minerals is defined as $$\alpha_{A-B} = \frac{(^{18}O/^{16}O)A}{(^{18}O/^{16}O)B}$$ where A and B refer to two oxygen-containing minerals.	Eslinger et al. (1979) Yeh and Savin (1977)

It is not clear, however, if pH increases at all burial depths and at all temperatures. For example, the thermal maturation of organic matter in sediments—a process that ultimately leads to the formation of petroleum—is believed to release large quantities of carbon dioxide to formation waters at certain stages in the maturation process. Carbon dioxide is released from organic matter through a process called **decarboxylation,** by which the COOH functional group in organic matter is decomposed to furnish CO_2 plus water (Hunt, 1979). This process, which occurs at depths on the order of 2–4 km, may increase the partial pressure of CO_2 in formation waters sufficiently to overcome the opposite effect of CO_2 loss resulting from increased temperature and thus may bring about a decrease in the pH of formation waters. Such a process has been postulated to account for decrease in pH and consequent dissolution of carbonate cements in sandstones at depths down to about 4000 m (Schmidt and McDonald, 1979).

Eh. We seem to know even less about Eh of the diagenetic environment than about pH. Because most surface waters are oxidizing and because these waters may circulate to considerable depths in the subsurface, subsurface waters recharged from the surface also are assumed to be oxidizing. On the other hand, the water in many formations is connate water (defined in a succeeding section) that has been out of contact with the atmosphere for an appreciable period of geologic time. Some connate waters, particularly those associated with petroleum, are known to have negative values of Eh, ranging as low as -0.3 V (Collins, 1975). Negative Eh values probably occur also in organic-rich shales and coalbeds, as suggested by the common occurrence of pyrite, which forms under reducing conditions in these rocks.

Chemical Composition. The composition of subsurface waters in the diagenetic environment differs from those of both normal seawater and fresh surface waters. Composition depends upon several factors that can change in different diagenetic settings and may, therefore, differ in different sedimentary basins as well as differing with depth. Factors that control the composition of subsurface waters include (1) composition of the water at the depositional site; that is, seawater or freshwater, which determines the initial pore-water composition of the sediment; (2) changes in water composition that result from rock–water chemical interactions, and (3) changes in composition that result from mixing with other types of waters during migration of pore fluids at depth (Collins, 1980). On the basis of origin, three principal kinds of subsurface waters are recognized: (1) meteoric water, (2) connate water, and (3) juvenile water.

Meteoric water is water that was recently involved in atmospheric circulation and has a young age compared to that of enclosing rock. Meteoric waters enter sedimentary strata in elevated outcrop areas owing to rainfall or snowfall and then circulate downward into sedimentary basins toward discharge points at lower elevations.

Connate water is water that has been out of contact with the atmosphere for at least an appreciable part of a geologic period. It consists of either seawater or freshwater that has been modified through time by water–rock interactions and/or mixing with other water types. The term **diagenetic water** is used to refer to water that has changed chemically and physically owing to diagenetic processes such as bacterial alteration, dehydration of smectite clays, replacement reactions such as dolomitization, and membrane filtration. Membrane filtration, for example, is brought about by movement of fluids through clay-rich rocks and may cause concentration of certain ions and increase in salinity as discussed in a succeeding section.

Juvenile water is water that is derived from primary magmas and is therefore new water.

All of these waters may be encountered in the diagenetic environment, but connate waters and meteoric waters are volumetrically far more important than juvenile water. The general, nongenetic term **formation water** is commonly used to denote water that occurs naturally in subsurface formations before they are penetrated by drilling, regardless of the origin of the water.

The anion composition of formation waters is commonly used as a basis for chemical classification. Based on predominant anion composition, three principal types of subsurface waters can be differentiated: (1) bicarbonate-rich waters, (2) sulfate-rich waters, and (3) chloride-rich waters (Table 9.3). In general, bicarbonate and sulfate waters are characteristic of shallow and intermediate burial depths, and chloride waters characterize deeper formations (Table 9.4). Significant differences exist between composition and total salinity of formation waters and seawater. Formation waters range in salinity from much less than the average seawater value of about 35,000 ppm to almost ten times greater. As shown in Table 9.4, salinities of subsurface waters may range from less than 200 ppm to 300,000 ppm or more. In most sedimentary basins, meteoric waters of low salinity overlie connate waters at shallow depths. Below the zone of meteoric waters, the connate waters characteristically display a trend of increasing salinity with increasing depth and formation age (White, 1965; Dickey, 1969) (Fig. 9.15). In addition to increase in total salinity with depth, many cations and anions show changes in relative abundance compared to their abundance in seawater (Table 9.5). Chlorine is by far the predominant anion in deep waters and the relative abundance of other ions can usefully be compared to chlorine content. Note from Table 9.5 that compared to seawater the Na/Cl, Mg/Cl, SO_4/Cl, K/Cl, and HCO_3/Cl ratios decrease, whereas Ca/Cl, CO_3/Cl, Sr/Cl, and H_4SiO_4/Cl ratios increase. Also the Ca/Mg ratio increases; although not shown in Table 9.5, the Ca/Na ratio is also known to generally increase with increasing salinity.

It is beyond the scope of this chapter to attempt detailed explanation of the causes of salinity increase and changes in relative abundance of cations and anions in formation waters. Many chemical mechanisms have been suggested to explain these changes, and general agreement is still lacking. One of the most interesting ideas advanced to explain increase in salinity with depth and formation age is the concept of a **salt sieving,** or reverse osmosis, mechanism, which operates owing to passage of waters through fine-grained sediment containing clay minerals such as montmorillonite and illite (White, 1965). These fine sediments are believed to act as semipermeable membranes, which allow uncharged water molecules to pass unimpeded through the shales in the direction of decreasing pressure, while filtering out and retaining the dissolved ions. The water on the upflow sides of the membranes becomes increasingly saline with time, whereas the water on the downflow sides contains relatively few ions. Thus, most shales have lower salinity than associated sandstones. This salt-sieving mechanism has been invoked also as a possible explanation for changes in relative abundance of some ions under the assumption that the shale membranes are not ideal semipermeable membranes and will allow partial fractionation and passage of some ions. Presumably, the ions that get through are most likely to be those with the smallest ionic size and electrical charge. Therefore, ions with larger size and charge are preferentially retained in the increasingly saline water behind the membrane.

TABLE 9.3 Major geochemical groups of subsurface waters based on dominant anion content

Major group of water	Class	Genetic types of water	Reacting value in percent						Important cations
			$HCO_3^- + CO_3^-$	Cl^-	SO_4^{2-}	$Cl^- + SO_4^{2-}$	$HCO_3^- + Cl^-$	$HCO_3^- + SO_4^{2-}$	
Bicarbonate	I	Bicarbonate	>40	—	—	<10	—	—	$Na^+ + K^+$ prevail in all types of waters
	II	Bicarbonate-chloride	40–30	—	—	10–20	—	—	Ca^+ and Mg^{2+} less than 2.5 in the water of high saline facies and less than 19.0 in low saline facies
	III	Chloride-bicarbonate	30–15	—	—	20–35	—	—	
Sulfate	IV	Sulfate-chloride	15–5	<25	>25	—	—	—	$Na^+ + K^+$ prevail in all types of waters
		Sulfate	—	—	>40	—	<10	—	Ca^{2+} less than 4.5 in all waters Mg^{2+} less than 4.0 in all waters
Chloride	III	Chloride-bicarbonate	30–15	>20	—	—	—	—	$Na^+ + K^+$ prevail in all types of waters
	IV	Chloride-sulfate	15–5	>20	—	—	—	<25	Ca^{2+} less than 12.5 in the water of high saline facies
	V	Chloride	<5	>40	—	—	—	<10	Mg^{2+} less than 6.0 in all types of waters

Source: Chebotarev, I. I., 1955, Metamorphism of natural waters in the crust weathering: Geochim. et Cosmochim. Acta, v. 8. Table 10, p. 44, reprinted by permission.

TABLE 9.4 Characteristics and distribution of formation waters in the subsurface. Class refers to the major classes of water shown in Table 9.5

Hydrodynamic zone			Geochemistry of water				Geological environment		
Recharge–discharge cycle	Water exchange	Class	Hydro-chemical facies	Approximate salinity (ppm)	Common terms for water	Structures	Relation to water	Depth (ft)	Examples
Zone of recharge	Active exchange	I and II (sometimes III)	Low saline facies	180–2,400	Fresh	Different	Intensive flush	Usually less than 500	Everywhere
			Transitional (typical) facies	2,400–11,400	Brackish	Deep portions of structures with peculiar geochemical environment	Delayed flush	Sometimes 5,000–7,000	Great Artesian Basin, Rocky Mountain oil field, and others
			High saline facies	1,400–37,800	Saline		Hampered flush		
Zone of pressure	Delayed exchange	III and IV	Low saline facies	400–2,500	Fresh	Different	Inadequate flush	Usually less than 1,000	Everywhere
			Transitional (typical) facies	2,500–7,400	Brackish	Deeper portions of structures, folded zones	Circulation and drainage limited	Sometimes 3,000–4,000	Pre-Caucasian Basin, South Dakota Basin, some oil-field areas
			High saline facies	7,400–19,300	Saline				
Zone of accumulation	Stagnant condition	V	Low saline facies	1,500–20,000	Fresh and saline	Different	Salt accumulation prevails upon leaching	Different	Chiefly in arid regions; deeper portions of many artesian basins: some oilfields
			Transitional (typical) facies	20,000–90,000	Saline and brines	Deeper portions of structures, highly folded zones		Sometimes 8,000–13,000	
			High saline facies	90,000–300,000	Brines		Water exchange manifests on geological scale time		Many oilfield areas (Louisiana, Alberta, etc.)

Source: Chebotarev, I. I., 1955, Metamorphism of natural waters in the crust weathering: Geochim. et Cosmochim. Acta, v. 8. Table 58, p. 199, reprinted by permission.

Not all geologists fully accept the theory of salt-sieving, and many other explanations for the observed trends in salinity and ionic abundances in deep waters have been offered. These include solution of halite (salt) deposits, dolomitization, bacterial reduction of sulfate, cation exchange on clay minerals, solubility relationships that differ for different substances with increasing temperature, molecular diffusion from salt beds, gravitational segregation of ions, and thermal diffusion.

FIGURE 9.15 Changes in total salinity of subsurface waters with increasing depth in selected depositional basins. (After Petroleum geochemistry and geology by J. M. Hunt. W. H. Freeman and Company. Fig. 6.3, p. 194. Copyright © 1979. Based on data from P. A. Dickey, 1969, Increasing concentrations of subsurface brines with depth: Chem. Geology, v. 4.)

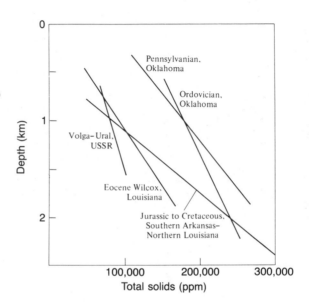

TABLE 9.5 Comparison between the chemical composition of formation waters and modern seawater

Characteristic	Modern seawater	Range in formation waters	Median in formation waters	Direction of change
pH	7.9–8.2	3–11	~9	Increase
Eh	−0.2 to +0.4 V	−0.4 to +0.7 v	?	?
Chlorinity	0.535 m	up to 7.61 m	2.48 m	Increase
mNa/mCl	0.86	0.15–0.95	0.72	Decrease
mMg/mCl	0.097	0.003–0.117	0.032	Decrease
mSO$_4$/mCl	0.052	7×10^{-6}–0.013	0.002	Decrease
mK/mCl	0.019	1×10^{-3}–0.064	0.008	Decrease
mCa/mCl	0.019	0.008–0.34	0.113	Increase
mHCO$_3$/mCl	0.004	2×10^{-5}–0.024	3.2×10^{-3}	Decrease
mCO$_3$/mCl	0.7×10^{-4}	?	4.6×10^{-4}	Increase
mSr/mCl	1.7×10^{-4}	1×10^{-3}–0.013	1.9×10^{-3}	Increase
mH$_4$SiO$_4$/mCl	1.9×10^{-5}	3×10^{-5}–5×10^{-4}	~10^{-4}	Increase
mCa/mMg	0.19	0.56–26.92	3.53	Increase
Percentage "total CO$_2$" as CO$_3$ ion	0.7%	?	10%	Increase

Source: Blatt, H., G. V. Middleton, and R. Murray, Origin of sedimentary rocks, 2nd ed., © 1980, Table 6.9, p. 231. Reprinted by permission of Prentice-Hall, Englewood Cliffs, N.J.

9.4   MAJOR CONTROLS ON DIAGENESIS

Influence of Original Mineral Composition

In addition to the specific physical and chemical conditions of the subsurface environment that control diagenesis, the nature of diagenetic changes in the rocks may be influenced to some extent by the original mineralogy of the sediments. It has been pointed out, for example, that siliciclastic sediments enriched in volcanic rock fragments and feldspars contain authigenic minerals such as chlorite and montmorillonite that require abundant magnesium, iron, calcium, and potassium. By contrast, the authigenic minerals in quartzose sandstones are mainly kaolinite and quartz. Likewise, the authigenic minerals in limestones are mainly other carbonate minerals and chert, which may form, in some limestones, from silica derived by solution of siliceous organisms in the limestones. The mineral assemblages in the rocks thus influence the formation of authigenic minerals by controlling the ions available for authigenic growth.

Influence of Water Chemistry in the Depositional Environment

The depositional environment is also believed to influence authigenic mineral assemblages owing to the chemistry of the waters at the depositional site. These waters fill the sediment pores at the time of deposition and thus influence the precipitation of authigenic minerals, at least to burial depths of several hundred meters. In a study of Triassic and Carboniferous sediments in Germany, for example, Füchtbauer (1974) determined that kaolinite, dioctahedral chlorite, and potassium feldspar cements are characteristic of sediments deposited in freshwater. Early chlorite, analcime (a zeolite mineral), and albite occur mainly in marine and evaporitic sandstones; quartz and micas can apparently form cements in sediments deposited in all environments. These results cannot be generalized to all sediments deposited in similar environments, but they do indicate that the original chemistry of sediment pore waters, which is determined by the depositional environment, influences the course of diagenesis. At the least, it may influence diagenesis until the pore waters themselves are changed by interaction with the rock.

Influence of Tectonic Setting

The physical and chemical conditions of the subsurface environment that control diagenesis are, in turn, controlled at least in part by the tectonic setting of the depositional basin. Siever (1979) has suggested that the plate tectonic setting exerts an important control on the diagenetic environment and can be used as a framework for relating diagenesis to patterns of regional tectonics (including rates of subsidence, uplift, and deformation), volcanism and plutonism, heat flow, and transport of formation waters. The principal plate tectonic settings and associated diagenetic environments include mid-ocean ridges, trailing-edge continental margins, subduction zones, continent-to-continent collision zones, rift valleys, and intraplate settings.

   Mid-ocean ridges are sites of thin pelagic sediment accumulation that are affected by initial high heat flow, hydrothermal activity, and volcanic activity followed by decreasing heat flow and slow burial as spreading continues away from a ridge.

Geothermal gradients are higher than on the continents. In this environment, muddy and sandy sediment remain largely unconsolidated but chert beds and limestones lithify.

Trailing-edge (passive) continental margins are characterized by lower heat-flow rates and lower geothermal gradients. Subsidence and burial rates of terrigenous and nonterrigenous sediments are high after the initial opening phase of the ocean, during which evaporites and heavy metal deposits are formed. Sedimentary rocks in this diagenetic environment tend to be moderately lithified and cemented with quartz and carbonate cements. Chlorite–illite mineral pairs are common in deeper zones. Volcaniclastic sediments are relatively uncommon deposits on trailing margins; therefore, zeolite minerals are rare.

Subduction zones are areas of low heat flow, rapid subsidence and sediment burial, and compressional crumpling of sediments. Sediments in this environment include turbidites and fine pelagic deposits and also include large amounts of volcaniclastic debris derived from adjacent island arcs. Both zeolites and smectite clays are produced diagenetically.

Continent-to-continent collision zones are the sites of deposition of large quantities of terrigenous sediments in both marine and continental environments ranging from intermontane basins with subaerial exposure and meteoric water infiltration to submarine fans and plains. This diagenetic environment is characterized by low geothermal gradients and rapid sedimentation and burial rates. The effects of diagenesis are generally mild except in stratigraphic units that have been deeply buried for hundreds of millions of years.

Rift valleys are characterized by high heat flow, hydrothermal activity, and rapid burial. Volcaniclastic sediments are abundant, mixed with alluvial and lacustrine sediments.

Intraplate (cratonic) settings are areas of generally low subsidence and low sedimentation rates in regimes of low continental heat flow; however, vertical and horizontal oscillatory tectonic movements are common and frequent, producing multiple changes in formation-water composition.

The variables that are affected by these different diagenetic environments are illustrated in Figure 9.16.

9.5   MAJOR EFFECTS OF DIAGENESIS

Diagenetic processes bring about a variety of textural, mineralogical, physical, and chemical changes in sedimentary rocks. Many of these changes have already been pointed out. The major diagenetic effects can be grouped under the general headings of (1) physical changes, (2) mineralogic changes, and (3) chemical changes.

Physical Changes

The principal physical changes that occur in sedimentary rocks during diagenesis are textural changes. Several diagenetic processes produce textural changes, including bioturbation, compaction, cementation, and solution. Bioturbation brings about changes in grain size and sorting owing to organisms mixing together sediment of different sizes from different layers. In the process of mixing, primary sedimentary structures such as stratification may be destroyed. Compaction produces changes in the arrangement of grains, resulting in the grains being packed into a tighter fabric.

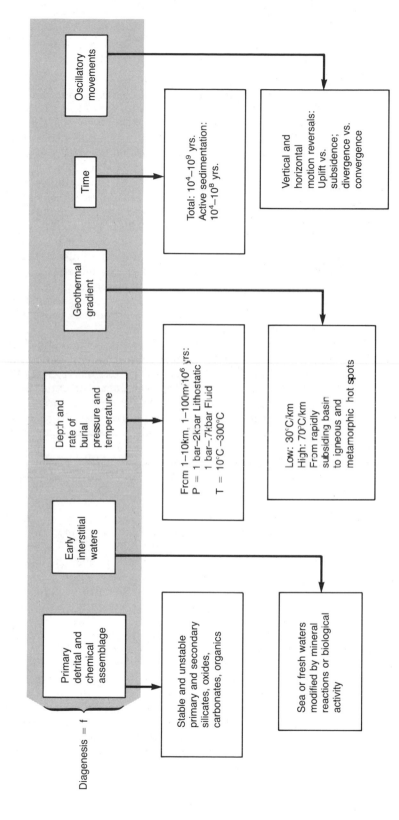

FIGURE 9.16 Diagenetic variables controlled by plate tectonic setting. (From Siever, R., 1979, Plate-tectonics controls on diagenesis: Jour. Geology, v. 87, Fig. 1, p. 129, reprinted by permission of University of Chicago Press.)

Compaction thus causes an increase in the numbers of contacts between grains. Increased compaction during deeper burial also causes plastic deformation of the grains, resulting in development of long contacts (grain elongation) and concavoconvex contacts in hard grains such as quartz. Soft minerals, easily deformed rock fragments such as shale clasts, or weakly cemented carbonate grains such as pellets can be squashed beyond recognition and squeezed among other grains, creating pseudomatrix. Severe compaction accompanied by pressure solution can produce sutured contacts in siliciclastic or carbonate grains and, on a large scale, generate stylolites and considerable thinning of beds. Cementation plugs pore spaces with mineral cements and adds syntaxial overgrowths to siliciclastic grains and some fossil fragments, resulting in increase in size and alteration of shape. Rounded quartz grains, for example, may take on crystalline shapes owing to the development of overgrowths. Cementation on a larger scale can create nodules and concretions in sediments. Authigenic processes involving alteration of feldspars and other grains to clay minerals and fine micas generate authigenic matrix that complicates the problems of rock classification and environmental interpretation. The processes of recrystallization cause coarsening in grain size, particularly in carbonate rocks and cherts. Solution may cause reduction in size of grains or complete destruction of the grains.

Textural modifications of these types can bring about marked changes in the porosity of sedimentary rocks. Closer packing owing to compaction causes significant decrease in porosity of sediments, accompanied by expulsion of pore fluids. On the other hand, solution processes commonly result in enhancement of porosity. Changes in porosity affect the bulk properties of sedimentary rocks in various ways. In particular, decrease in porosity causes a decrease in the permeability of rocks so that they transmit fluids less readily. Decreased porosity also causes an increase in bulk density of the rocks. In turn, bulk density affects other properties of the rocks, such as their ability to transmit sound waves or seismic waves; this ability increases with increasing bulk density.

Mineralogical Changes

Mineralogic changes can occur during diagenesis as a result of cementation, authigenesis, replacement, inversion, and solution. Depending upon the process, original minerals may be destroyed or new minerals created. Mineralogic changes may not always be easy to detect. Cements are commonly among the easiest diagenetic minerals to recognize because they fill interstitial pore spaces or form overgrowths. Other authigenic minerals may be recognizable owing to their generally small size in comparison to framework grains and to their well-developed crystal faces. Replacement minerals can often be recognized by cross-cutting relationships or incompletely replaced grains. Where original grains are completely replaced, recognition of the authigenic origin of grains may be difficult unless a pseudomorph that preserves the form or outline of the original grain remains. Examples of pseudomorphic replacement are chert replacement of a pelecypod shell, preserving the form and perhaps internal structural features of the shell; replacement of a feldspar grain by fine calcite crystals, preserving the form of the original cleavage faces of the feldspar; or replacement of a glass shard by montmorillonite clay, preserving the shape or outline of the shard. Solution may completely destroy labile or metastable particles such as heavy mineral grains or fossils. There is no way to recognize the former presence of dissolved grains. Therefore, intrastratal solution during diagenesis can bias a mineral assemblage toward more chemically resistant minerals, thereby confusing provenance determinations.

A variety of authigenic minerals can form during diagenesis. These are mentioned in the discussions of the various diagenetic processes and need not all be repeated here. Calcite and quartz cements probably account for the largest volume of authigenic minerals in most sedimentary rocks, but other cements, authigenic clay minerals and micas, and some replacement minerals may likewise be present in significant volumes. For example, dolomite is a mineral that can essentially replace all original minerals in limestones. The overall effect of mineralogic changes during diagenesis is to move depositional assemblages of minerals toward a greater degree of chemical equilibrium with the conditions that exist in the diagenetic environment. The processes of weathering, sediment transport, and deposition bring together in the depositional environment minerals derived from different igneous, metamorphic, and sedimentary sources. Such mixed groups of minerals could not have formed together under most temperature–pressure conditions of metamorphism or plutonism, and thus constitute nonequilibrium assemblages. Diagenetic alteration of siliciclastic minerals to form authigenic minerals tends to bring these incompatible assemblages of minerals into a greater degree of chemical equilibrium with their surroundings.

Chemical Changes

Changes in mineralogy during diagenesis are accompanied by changes in bulk chemical composition of the rocks. Depending upon the diagenetic processes involved, chemical substances may be either added to or removed from a group of sediments. Thus, for example, calcium and silica may be added owing to cementation, or they may be removed owing to dissolution. On the other hand, chemical constituents taken into solution locally may be transported only short distances by pore waters before they are reprecipitated within the same pile of sediments, resulting in little net change in chemical composition. Although it is reasonable to assume that fairly significant changes in chemical composition can occur during diagenesis, in general it is exceedingly difficult to make a quantitative determination of these changes because we don't know the initial composition of the rocks. If we consider just the chemistry of pore waters in sediments, we can make some estimate of the chemical change that has occurred in these waters owing to interaction with the rock during diagenesis. Such an estimate is possible because we can assume that either freshwater or seawater must have been present in the sediment pores at the time of deposition, depending upon the depositional environment. Because we know the approximate chemical composition of freshwaters and seawaters, this gives us a starting point for comparison. On the other hand, we have no exact way of knowing the original chemical composition of the rock itself prior to diagenesis. We can only infer what the original composition might have been on the basis of the present chemical and mineral composition and our interpretation of mineralogic changes during diagenesis. We may be able to determine, for example, that silica has been added to a formation during diagenesis because we can see silica cement in the rocks; however, we probably can't tell if a net change in silica composition has occurred because we don't know the original silica composition. Uncertainties of this type are one of the reasons why sedimentologists have not been more enthusiastic about using the chemistry of sedimentary rocks as a tool for studying provenance and depositional environments.

9.6 SUMMARY

Sediment burial brings unconsolidated terrigenous and nonterrigenous sediments into a postdepositional environment in which temperatures and pressures are higher than those of depositional environments and in which pH, Eh, and chemical composition of associated pore waters may differ significantly from the waters of depositional sites. The specific conditions of the diagenetic environment differ depending upon the tectonic setting. The combined effects of the physical and chemical diagenetic processes that take place in the burial environment act to convert or lithify unconsolidated sediment to consolidated sedimentary rock, largely through compaction and cementation. During the process of lithification, important changes can occur in depositional mineral assemblages owing to destruction or alteration of some minerals and creation of new authigenic minerals. These changes act to move genetically incompatible assemblages of siliciclastic minerals toward conditions of chemical equilibrium that are in adjustment with the diagenetic environment. Diagenetic processes also alter the porosity and permeability of sediments and thus affect their capacity to act as reservoir rocks for petroleum and groundwater. Porosity is reduced by the processes of compaction and cementation and enhanced by dissolution. Diagenetic processes can also change and obscure original depositional textures and structures. In order to correctly interpret ancient depositional environments and sediment provenance and extract other genetic information from ancient sedimentary rocks, geologists must learn to identify the diagenetic characteristics of these rocks and differentiate them from their original depositional characteristics.

ADDITIONAL READINGS

Bathurst, R. G. C., 1975, Carbonate sediments and their diagenesis, 2nd ed., Developments in Sedimentology 12: Elsevier, Amsterdam, 658 p.

Berner, R. A., 1980, Early diagenesis: A theoretical approach: Princeton University Press, Princeton, N.J., 224 p.

Bricker, O.P., 1971, Carbonate cements: John Hopkins University Press, Baltimore, Md., 376 p.

Chilingarian, G. V., and K. H. Wolf (eds.), 1976, Compaction of coarse-grained sediments, I: Elsevier, New York, 552 p.

Chilingarian, G. V., and K. H. Wolf (eds.), 1976, Compaction of coarse-grained sediments, II: Elsevier, New York, 808 p.

Larsen, G., and G. V. Chilingarian (eds.), 1967, Diagenesis in sediments: Developments in Sedimentology 8, Elsevier, Amsterdam, 551 p.

Larsen, G., and G. V. Chilingarian (eds.), 1979, Diagenesis in sediments and sedimentary rocks: Elsevier North Holland, New York, 579 p.

Larsen, G., and G. V. Chilingarian (eds.), 1983, Diagenesis in sediments and sedimentary rocks, 2: Elsevier, New York, 572 p.

McCall, P. L., and M. J. S. Tevesz, 1982, Animal–sediment relations. The biogenic alteration of sediments: Plenum, New York, 336 p.

McDonald, D. A., and R. C. Surdam (eds.), 1984, Clastic diagenesis: Am. Assoc. Petroleum Geologists Mem. 37, 434 p.

Parker, A., and B. W. Sellwood (eds.), 1983, Sediment diagenesis: D. Reidel, Boston, 427 p.

Pettijohn, F. J., P. E. Potter, and R. Siever, 1973, Sand and sandstone, Chap. 10, Diagenesis, p. 383–437: Springer-Verlag, New York, 618 p.

Pray, L. C., and R. C. Murray (eds.), 1965, Dolomitization and limestone diagenesis: A symposium: Soc. Econ. Paleontologists and Mineralogists, Tulsa, Okla., 180 p.

Riecke, H. H., III, and G. V. Chilingarian (eds.), 1974, Compaction of argillaceous sediments: Elsevier, New York, 424 p.

Scholle, P. A., and P. R. Schluger (eds.), 1979, Aspects of diagenesis: Soc. Econ. Paleontologists and Mineralogists Spec. Pub. 26, 400 p.

PART SIX
Sedimentary
Environments: The Link
Between Sedimentology
and Stratigraphy

The characteristic properties of sedimentary rocks are generated through the combined action of the various physical, chemical, and biological processes that make up the sedimentary cycle. Weathering, erosion, sediment transport, deposition, and diagenesis all leave their impress in some way on the final sedimentary rock product. It is the sedimentary processes and conditions that collectively constitute the depositional environment and play the primary role in determining textures, structures, bedding features, and stratigraphic characteristics of sedimentary rocks. This close genetic relationship between depositional processes and rock properties provides a potentially powerful tool for interpreting ancient depositional environments. If geologists can find a way to relate specific rock properties to particular depositional processes and conditions, they can work backward to infer the ancient depositional processes and environmental conditions that created these particular rock properties.

Unfortunately, we can never know the exact nature of depositional processes and conditions that operated in the past. We are forced to look to the rocks themselves for clues to these conditions long after the processes that produced the rocks ceased to operate. Therefore, relating depositional process to depositional product is not a simple procedure. We must turn to study of sediments and sedimentary processes in modern environments for help in understanding the link between sedimentary processes and sedimentary rock properties. Knowledge acquired through such study can in turn be applied to environmental interpretation of ancient sedimentary rocks.

This procedure of applying knowledge gained through study of modern sedimentary environments to interpretation of ancient depositional processes and settings is the essence of environmental analysis. It does, however, have limitations. For example, the distribution of lands and seas that we see today is not typical of much of the geologic past. Furthermore, the intensity of geologic processes has very likely varied at different times in the past as well as differing from the intensity at present. Also, some geologic events of the past were probably unique. Many environments may have existed in the past that are not available for study today (Reineck and Singh, 1980). Therefore, geologists must be careful in the interpretation of ancient depositional environments not to be guided too rigidly by environmental models based on modern conditions.

Study of ancient depositional environments is important because the insight gained through environmental analysis allows us to reconstruct the paleogeography of previous geologic periods—that is, the relationship of ancient lands and seas. It helps us also to develop a proper understanding of sedimentary processes in the history of Earth and an improved ability to interpret complex stratigraphic relationships such as lateral and vertical variations in lithology and texture. Furthermore, thorough understanding of depositional environments is an essential factor in evaluating the economic significance of sedimentary rocks—their potential as reservoir rocks and source rocks for petroleum, for example.

Chapter 10 focuses on discussion of the fundamental tools of environmental analysis and the methods geologists use to recognize and identify ancient depositional environments. Chapters 11 through 13 deal respectively with continental, marginal-marine, and marine depositional environments and the sedimentary rocks deposited in these environments.

10

Principles of Environmental Interpretation and Classification

10.1 DEFINITION AND SCOPE OF SEDIMENTARY ENVIRONMENTS

Concept of Environment

Although geologists agree on the general meaning of sedimentary environment, they have found it difficult to frame a precise definition of such an environment. To illustrate, a sedimentary environment has been variously described as "the place of deposition and the physical, chemical, and biological conditions that characterize the depositional setting" (Gould, 1972, p. 1), "the complex of physical, chemical, and biological conditions under which a sediment accumulates" (Krumbein and Sloss, 1963, p. 234), "a part of the Earth's surface which is physically, chemically, and biologically distinct from adjacent terrains" (Selley, 1978, p. 1), and "a spatial unit in which external physical, chemical, and biological conditions and influences affecting the development of a sediment are sufficiently constant to form a characteristic deposit" (Shepard and Moore, 1955, p. 1488).

These definitions differ somewhat, but they all have in common an emphasis on the physical, chemical, and biological conditions of the environment. Other workers have pointed out the desirability of considering depositional environments from a geomorphic point of view (Twenhofel, 1950; Potter, 1967). In this context, a sedimentary environment is considered to be a geomorphic unit in which deposition takes place. Such an environment is defined by a particular set of physical, chemical, and biological parameters that correspond to a geomorphic unit of a particular size and geometry. These processes operate at a rate and an intensity that generate characteristic textures, structures, or other properties, so that a distinctive deposit is produced. For example, a beach can be considered a geomorphic unit of specified size and shape on which specific physical processes (wave and current activity), chemical processes (solution and precipitation), and biologic processes (burrowing, sediment ingestion, and similar activities) take place to produce a body of beach sand characterized by particular geometry, sedimentary textures and structures, and mineralogy.

Environmental Parameters

The physical parameters referred to above encompass both **static** and **dynamic** elements of the environment. Static physical elements include basin geometry; depositional materials such as siliciclastic gravel, sand, and mud; water depth; temperature; and humidity. Dynamic physical elements are factors such as energy and flow direction of wind, water, and ice; rainfall; and snowfall. Chemical parameters include salinity, pH, Eh, and carbon dioxide and oxygen content of waters in the environment. The biological elements of the environment can be considered to encompass both the activities of organisms—plant growth, burrowing, boring, sediment ingestion, and extraction of silica and calcium carbonate to form skeletal materials—and the presence of organic remains as depositional materials.

10.2 SEDIMENTARY PROCESSES AND PRODUCTS

It is important in the study of depositional environments to make a clear distinction between sedimentary environments and sedimentary facies. Each sedimentary environment is characterized by a particular suite of physical, chemical, and biologic parameters that operate to produce a body of sediment characterized by specific textural, structural, and compositional properties. We refer to such distinctive bodies of sediment or sedimentary rock as **facies.** The term facies refers to stratigraphic units distinguished by lithologic, structural, and organic characteristics detectable in the field. A sedimentary facies is thus a unit of rock that, owing to deposition in a particular environment, has a characteristic set of properties. **Lithofacies** are distinguished by physical characteristics such as color, lithology, texture, and sedimentary structures. **Biofacies** are defined on the basis of paleontologic characteristics. The point emphasized here is that depositional environments generate sedimentary facies. The characteristic properties of the sedimentary facies are in turn a reflection of the conditions of the depositional environment. We will return to the definition of facies in Chapter 14.

Process and Response

We can take a very simplistic and optimistic point of view and assume that a particular set of environmental conditions operating at a particular intensity will produce a sedimentary deposit with a unique set of properties that identify it as the product of that particular environment. Although it is probably not true that each environment produces a unique sedimentary deposit, the basis of environmental interpretation rests on the assumption that particular environments generate deposits that bear the impress of environmental processes and conditions to a degree sufficient to allow discrimination of the environment. This linked set of reactions between environments and facies is commonly referred to as **process** and **response** (Fig. 10.1).

As Figure 10.1 illustrates, the term process is used rather loosely to include both dynamic and static elements of the environment. Together, these process elements are

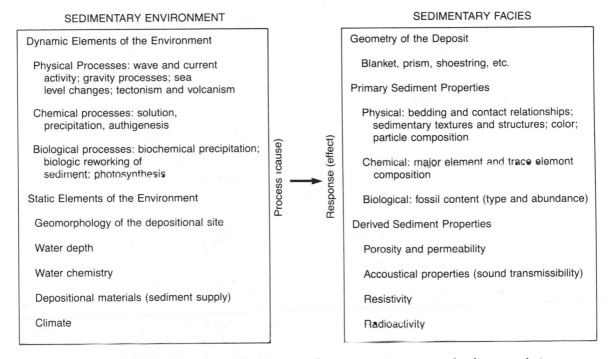

FIGURE 10.1 Relationship between sedimentary environments and sedimentary facies.

responsible for generating a particular response in the form of specific facies. When dealing with the depositional environments of ancient sedimentary rocks, geologists cannot, of course, observe the process elements of the environment. They have only the response element with which to work. Thus, the first step in environmental interpretation is always to characterize the facies in terms of specific physical, chemical, and biological properties. Geologists then attempt to work the process–response model backward and infer the conditions of the ancient depositional environment. In other words, as Middleton (1978, p. 324) puts it, "it is understood that (facies) will ultimately be given an environmental interpretation." Thus, environmental analysis must always begin with study of sedimentary facies. Only after the facies have been carefully and painstakingly analyzed and characterized can we make a reasonable interpretation of the depositional environment in which these facies were formed.

Facies Associations

Environmental interpretation is commonly hampered by the fact that very similar facies can be produced in different environmental settings. It is often impossible to make a unique environmental interpretation on the basis of a single depositional facies. For example, cross-bedded sandstones can be formed by either wind or water transport. If deposited in water, they can originate on a beach, in a river or tidal channel, on a shallow-marine shelf, or in any other environment where traction transport occurs. Environmental interpretation is improved if we study facies associations and sequences rather than individual facies. Facies associations can be thought of as groups of facies that occur together and are genetically or environmentally related (Reading, 1978a). For example, if cross-bedded sandstones are closely associated with overlying or underlying peat, coal, or silty shale containing roots, leaves, and stems, we could make an interpretation of deposition in a river system with some confidence. Such an interpretation might be very difficult to make on the basis of the cross-bedded sandstones alone. R. G. Walker (1979a, p. 1) stresses that "the key to (environmental) interpretation is to analyze all of the facies communally, in context. The sequence in which they occur thus contributes as much information as the facies themselves."

In the study of facies associations and sequences, careful attention must be given both to the nature of the contacts between facies and to the degree of randomness or nonrandomness of the sequences themselves. By application of stratigraphic principles discussed in Chapter 14, we can infer that two facies separated by a gradational contact or boundary represent environments that were once laterally adjacent. On the other hand, facies separated by sharp or erosive boundaries may or may not represent environments that were laterally adjacent. In fact, facies overlying erosive contacts commonly indicate a significant change in depositional conditions and the beginning of a new cycle of sedimentation (R. G. Walker, 1979a). The facies within a particular association of facies may be distributed vertically in an apparently random manner or they may show a definite or preferred pattern of vertical change. Two common types of vertical facies changes are **coarsening-upward sequences** and **fining-upward sequences.** Coarsening-upward sequences display an increase in grain size upward from a sharp or erosive base; fining-upward sequences are those in which the sequence becomes finer grained upward to a sharp or erosive top. In general, fining-upward sequences indicate a decrease in transporting power of currents during deposition, and coarsening-upward sequences indicate an increase. Fining- and coarsening-upward sequences should not be confused with graded bedding. Although such sequences can be on the scale of a single graded bed, they commonly involve many different beds that individually may not be graded. Each bed in the vertical succession is simply finer, or coarser, than the underlying bed. The significance of fining- and coarsening-upward sequences is discussed subsequently.

Facies Relationship Diagrams

In the study of facies associations, it may be possible to determine by visual inspection if facies are randomly or nonrandomly distributed; however, it is often necessary to resort to statistical techniques to detect whether or not one facies passes into another more often than would be predicted purely on a random basis. Statistical methods are also necessary to handle the large amounts of data involving different types of facies contacts and preferred relationships. A popular way of documenting the relationship between facies has been the use of a **facies relationship diagram** (Selley, 1970; Miall, 1973; Cant and Walker, 1976; R. G. Walker, 1979a). These diagrams show by means of symbols and arrows the transitions from one facies to another. The method involves converting the numbers of facies transitions to observed probabilities of transitions. The next step is to compare the observed probabilities with the probabilities that would apply if all the transitions between facies were random. Supposedly, transitions that occur a lot more commonly than random must have some geological significance. The problem has been to derive a statistical matrix of random probabilities. It now appears that the methods described by the cited authors for deriving a random matrix based on absolute facies abundances are statistically incorrect (R. G. Walker, 1984a), and should no longer be used. Much more complex statistical methods involving Markov chain analysis are required to evaluate facies transitions. Space limitations do not permit discussion of these methods here. Readers are referred to Carr (1982) and Powers and Easterling (1982) for details. Harper (1984) also describes improved statistical methods for analyzing facies sequences.

10.3 BASIC TOOLS OF ENVIRONMENTAL ANALYSIS

Interpretation and reconstruction of ancient depositional environments depends upon study of sedimentary rocks to identify physical, chemical, and biological characteristics that can be related to environmental parameters. The most important criteria for environmental recognition are listed in Table 10.1. No single criterion can generally

TABLE 10.1 Criteria for recognition of ancient sedimentary environments

Criteria based on primary depositional properties
Mainly physical properties
   Geometry of facies units
   Gross lithology and mineralogy of strata
   Facies associations (stratigraphic sequences)
   Sedimentary structures
      Nondirectional structures
      Directional structures (paleocurrent indicators)
   Sedimentary textures

Mainly chemical properties
   Major element composition
      Bulk composition
      Ratios of major elements
   Trace element composition
   Isotope ratios
   Type and relative abundance of organic matter

Mainly biologic properties
   Total faunal and floral assemblages
   Relative abundance and ratios of specific fossil types
   Endemic (indigenous) vs. displaced biota
   Ecologic characteristics of fossils
   Types of trace fossils
   Types and abundance of fecal pellets

Criteria based on derived sediment properties
Properties measured from mechanical well logs
   Resistivity
   Sonic velocity
   Radioactivity

Properties interpreted from well-log measurements
   Density/porosity
   Grain size
   Lithology
   Dip of bedding

Characteristics interpreted from seismic reflection records
   Major contact relationships (unconformities, conformities)
   Continuity of strata (pinchouts, truncations, etc.)
   Dip of strata
   Identification of seismic facies units

be relied upon to provide unequivocal environmental interpretation; geologists commonly must make use of all available properties of the sedimentary rocks. Only when several independent criteria yield the same interpretation can the environment be confidently assigned. Furthermore, because environmental interpretation can be severely hampered by diagenetic changes in sediments, geologists must be particularly careful to separate primary depositional features from postdepositional features caused by diagenesis.

Physical Criteria

Geometry. Geometry refers to the three-dimensional shape of sedimentary bodies. Most sedimentary units have irregular, complex shapes; nonetheless, two fundamental types of sediment bodies can be recognized on the basis of shape: (1) equidimensional bodies such as sheets, or blankets, and prisms and (2) elongate bodies such as pods, ribbons, or shoestrings, and dendroids (Potter, 1962). **Sheets,** or **blankets,** are sediment bodies that are more or less equidimensional in plan view, with length–width ratios of approximately 1:1. They have relatively uniform thickness, but thickness is quite small compared to lateral dimensions, which may be thousands of square miles. Sheet deposits can be generated in several ways, including deposition of fine sediment from suspension, transgression of shallow seas, and eolian transport and deposition. Thus, sheet sands may be indicative of shelf, beach, deep-water turbidite, desert, or even lake environments. **Prisms** are roughly equidimensional bodies as seen in plan view, but are wedge shaped in longitudinal cross section. The deposits of alluvial fans and deltas are examples of deposits having roughly prismatic geometry. Deltas are lobate in plan view, wedge shaped in longitudinal cross section, and lens shaped in transverse cross section (Fig. 10.2)

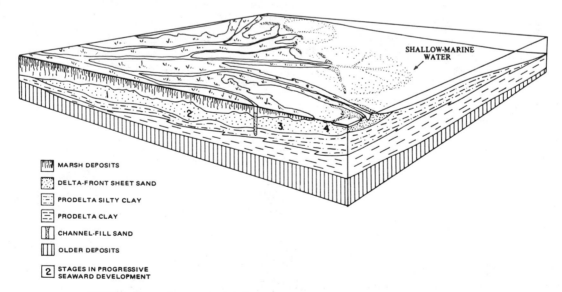

SHALLOW-MARINE WATER

⬚ MARSH DEPOSITS

⬚ DELTA-FRONT SHEET SAND

⬚ PRODELTA SILTY CLAY

⬚ PRODELTA CLAY

⬚ CHANNEL-FILL SAND

⬚ OLDER DEPOSITS

⬚ 2 STAGES IN PROGRESSIVE SEAWARD DEVELOPMENT

FIGURE 10.2   Geometry of deltaic deposits. (From Gould, H. R., 1970, The Mississippi delta complex, *in* J. P. Morgan and R. H. Shaver (eds.), Deltaic sedimentation, modern and ancient: Soc. Econ. Geologists and Paleontologists Spec. Pub. 15. Fig. 11, p. 17, reprinted by permission of SEPM, Tulsa, Okla.)

Pods are generally small bodies with length–width ratios of less than 3:1 (Fig. 10.3). **Ribbons, or shoestrings,** are narrow, elongate bodies with length–width ratios exceeding 3:1; length commonly greatly exceeds thickness. **Dendroids** are more sinuous in two-dimensional shape than ribbons and may have either tributary or distributary branches (Fig. 10.3). Pods, ribbons, and dendroids may run together laterally and coalesce to form **belts.** Elongate bodies of sediment such as pods, ribbons, and dendroids form either by filling channels or other elongated depressions, or by buildup on essentially flat sediment surfaces to form bars. Thus, they occur in a variety of environments, including fluvial, littoral, shallow-marine shelf, submarine canyons, and tidal flats.

The geometry of sedimentary bodies can be determined in outcrops by measuring thicknesses in good exposures and tracing beds laterally to determine changes in thickness and lateral extent. The geometry of subsurface formations is determined either by mapping the units from borehole data such as core and well-log information (Fig. 10.4), or by seismic-stratigraphic methods explained in Chapter 16. Geometry in itself is commonly not indicative of a unique sedimentary environment because similar-shaped bodies of sediment can form in different environments. However, distinc-

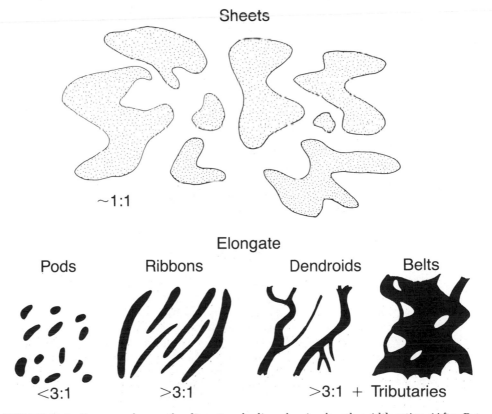

FIGURE 10.3 Common shapes of sedimentary bodies, showing length–width ratios. (After Pettijohn, F. J., P. E. Potter, and R. Siever, 1973, Sand and sandstone. Fig. 11.1, p. 441, reprinted by permission of Springer-Verlag, Heidelberg.)

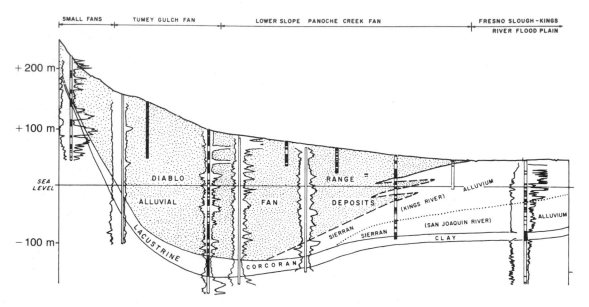

FIGURE 10.4 Mapping the geometry of subsurface bodies from well-log data. (From Bull, W. E., 1972, Recognition of alluvial fan deposits in the stratigraphic record, *in* J. K. Rigby and W. K. Hamblin (eds.), Recognition of ancient sedimentary environments: Soc. Econ. Paleontologists and Mineralogists Spec. Pub. 16. Fig. 13, p. 76, reprinted by permission of SEPM, Tulsa, Okla.)

tive shapes such as the ribbon geometry of bars and channels or the lobate shape of deltas can be very useful environmental indicators when combined with other environmental parameters.

Lithology. The gross lithology of sedimentary units may be a very general indicator of depositional environment. For example, limestones tend to be deposited on warm, shallow-marine shelves; mature, quartzose sandstones are especially common deposits in eolian environments or high-energy marine environments where extensive reworking takes place; evaporite deposits suggest deposition under conditions of high evaporation, restricted water circulation, and high salinity; conglomerates occur most commonly in fluvial environments; and coals occur in fluvial, swampy environments. Because there are many exceptions to these generalities, gross lithology is only the roughest kind of guide to depositional environments.

The particle composition of sedimentary rocks may yield more useful environmental information than gross lithology. In particular, the composition of limestones is controlled to a high degree by environmental conditions. For example, the presence of sparry calcite-cemented oolites suggests deposition under high-energy, agitated-water conditions, where lime mud is winnowed and removed. By contrast, micritic limestones such as pelmicrites or biomicrites are deposited mainly under quiet-water conditions, where water energy is too low to winnow and remove fine lime mud. J. L. Wilson (1975) utilized this principle to identify 24 different carbonate **microfacies** on the basis of specific types of carbonate grains, kinds of fossils, relative abundance of micrite, and other features. Microfacies are the small-scale characteristics of rocks that can be recognized in thin sections or hand specimens. Wilson was able on the basis of these microfacies to discriminate nine major carbonate depositional environments, ranging from shallow open-marine shelf to deep-water basin environments. Clearly, petrographic study of limestones can be a powerful tool for interpreting depositional environment.

The particle composition of siliciclastic sedimentary rocks is determined primarily by the nature of the source rocks and the conditions of weathering and transportation, rather than the environment. Still, some aspects of composition may have environmental significance. For example, Davies and Ethridge (1975) suggest that within a given depositional basin where sediments are derived from a single source, the imprint of different depositional environments is recorded in siliciclastic sediments as the relative abundance and size of individual siliciclastic particles. They maintain that statistical treatment of composition data can thus be useful in determining depositional environments. For example, sediments deposited in high-energy environments of intense winnowing are significantly enriched in quartz, which is highly resistant to mechanical abrasion, compared to low-energy environments, where less quartz but more fine-size matrix materials are deposited. By contrast, sediments subjected to the least transport and reworking tend to be enriched in rock fragments, which are commonly less durable than quartz.

The relative abundance of quartz has been used also as a measure of distance from shore in marine mudrocks. In a study of outcrop samples of marine shale, Blatt and Totten (1981) reported a decrease in quartz content from 47 percent at a distance of 60 km from a known shoreline (determined by other criteria) to 11 percent at a distance of 270 km. They suggest that the percentage of quartz can be a useful indicator of the position of the shoreline in ancient marine mudrocks. Coarse micas have been used also to indicate position within a basin. In modern environments, coarse mica flakes appear to be deposited preferentially in the narrow nearshore, tidal-flat environment and on the upper continental slope, whereas the central and outer continental shelves are areas of winnowing of coarse micas (Doyle et al., 1968). Abundant coarse mica in ancient marine sedimentary rocks may thus indicate either tidal-flat or continental-slope environments.

The simple presence of certain authigenic minerals in siliciclastic sediments may also have environmental significance. For example, the iron silicate minerals **glauconite** and **chamosite** are restricted principally to marine continental shelf environments (Porrenga, 1967). These minerals form in marine areas of low sedimentation, and in the presence of some organic matter, primarily by alteration of other minerals at the sediment–water interface or at very shallow burial depths. Therefore, these authigenic minerals are fairly reliable indicators of marine conditions, although glauconite has been reported rarely from nonmarine environments (Selley, 1978).

Lateral and Vertical Facies Associations. Lateral and vertical facies relationships observed in outcrops or determined from subsurface data are among the most valuable criteria for environmental discrimination. The concept involved here is an application of the stratigraphic principle known as Walther's Law (Chapter 14), which predicts that facies developed laterally may occur also in vertical sequences. This principle is illustrated in Figure 10.5. In this example, fine-grained shales containing marine fossils grade laterally to clean, well-sorted sands, which in turn grade to muddy deposits containing plant remains. The same sequence can be observed in vertical profile, where marine shales grade upward through mature sandstones to muddy, silty deposits. By using modern environments as a model, these facies relationships, plus the

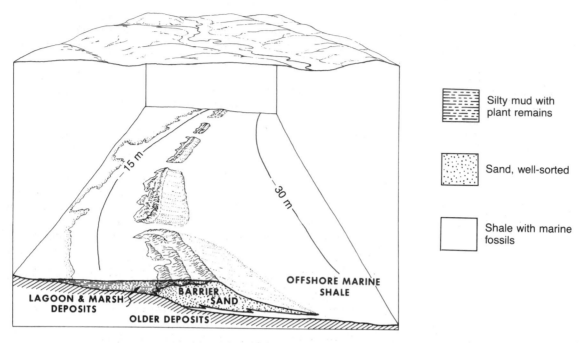

FIGURE 10.5 Diagram illustrating the use of vertical and lateral facies as clues to depositional environment. (Modified from Gould, H. R., 1972, Environmental indicators—a key to the stratigraphic record, *in* J. K. Rigby and W. K. Hamblin (eds.), Recognition of ancient sedimentary environments: Soc. Econ. Paleontologists and Mineralogists Spec. Pub. 16. Fig. 1, p. 2, reprinted by permission of SEPM, Tulsa, Okla.)

three-dimensional geometry of the deposits, suggest the interpretation shown in Figure 10.5 of offshore marine shales grading landward into barrier-bar sands and finally to lagoon-marsh deposits. A second example is illustrated in Figure 10.6. A fine-grained, well-bedded limestone containing pelagic fossils grades laterally and vertically to massive, poorly bedded limestones composed of reef-building organisms, which in turn grade to carbonate sands and eventually fine-grained, micritic limestones containing abundant shallow-water organisms. This sequence of facies is interpreted as a result of progressive seaward migration of a lagoonal lime mud environment over a back-barrier carbonate sand and barrier-reef environment to a deeper water environment (Fig. 10.6).

Fining- and Coarsening-Upward Sequences. Vertical trends of grain-size change are important also in environmental interpretation. Many environments are known to generate either fining-upward or coarsening-upward vertical sequences, as discussed in Section 10.3. The lateral shifting of meandering streams in a fluvial environment, for example, produces fining-upward vertical sequences that result from migration of fine-grained overbank deposits over sandy point-bar deposits and gravelly channel deposits (Fig. 10.7). By contrast, the seaward outbuilding of deltas commonly generates coarsening-upward sequences as coarse-grained distributary-channel deposits prograde seaward over delta-front sands, which in turn build seaward over prodelta muds. Marine

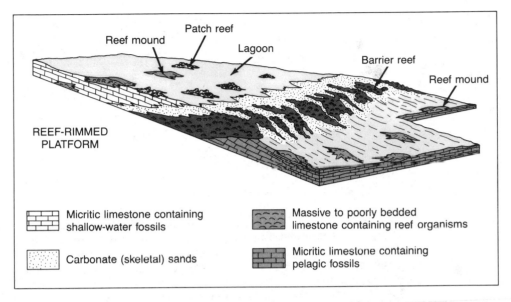

Reef mound

Patch reef

Lagoon

Barrier reef

Reef mound

REEF-RIMMED
PLATFORM

Micritic limestone containing
shallow-water fossils

Massive to poorly bedded
limestone containing reef organisms

Carbonate (skeletal) sands

Micritic limestone containing
pelagic fossils

FIGURES 10.6 Environmental interpretation based on vertical and lateral carbonate facies relationships. (Modified from James, N. P., 1983, Reef environment, in P. A. Scholle, D. G. Bebout, and C. H. Moore (eds.), Carbonate depositional environments: Am. Assoc. Petroleum Geologists Mem. 33. Fig. 202, p. 439, reprinted by permission of AAPG, Tulsa, Okla.)

deposits formed by shoreward transgression of the sea commonly display fining-upward sequences owing to landward migration of finer grained, deeper water deposits over shallower water deposits. Regression, or seaward migration, of the shoreline generates deposits that are characterized by coarsening-upward sequences. Owing to size grading within individual flow units, turbidite deposits are distinguished by fining-upward sequences on a small scale. Figure 10.8 illustrates highly idealized vertical grain-size profiles of sediments from some major environments. Because fining-upward or coarsening-upward sequences may form in several different environments, such sequences cannot be used alone for environmental interpretation. They are extremely valuable environmental criteria when used in conjunction with other environmental indicators.

Cyclic Facies. The processes that produce vertical successions of facies are commonly repeated in time, leading to generation of cycles of vertical successions that follow each other in a predictable pattern in which a particular vertical sequence (or sequences) is repeated. For example, repeated migration of a meandering stream over the floor of a subsiding alluvial basin could produce the succession of fining-upward sequences illustrated in the meandering-stream profile shown in Figure 10.8, or repeated turbidity current flows into a basin could generate the turbidite profile shown in the same figure. Successive episodes of marine transgression and regression may lead to generation of cycles of fining-upward sequences overlain by coarsening-upward sequences, and so on. Recognizing these cyclic repetitions of vertical sequences is important in recognizing and interpreting changes in environmental conditions. Some

FIGURE 10.7 Fining-upward sequences of fluvial deposits from the south Sasketchewan River, Canada. The sequences grade upward from basal gravelly in-channel sandy deposits through sandy-bar and channel deposits to muddy overbank (vertical accretion, VA) deposits. (From Walker, R. G., and D. J. Cant, 1984, Sandy fluvial systems, *in* R. G. Walker (ed.), Facies models, 2nd ed.: Geoscience Canada Reprint Ser. 1. Fig. 17, p. 81, reprinted by permission of Geological Association of Canada.)

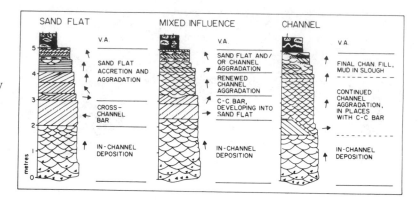

sedimentation cycles are so well developed and obvious that they can be detected by visual inspection and study of outcrops or subsurface data. Other cycles are much more subtle and their detection requires the use of statistical techniques aided by the computer (Schwarzacher, 1975). Computer analysis makes possible the evaluation of vertical profiles by rigorous comparison, with known limits of reliability, between model profiles and the vertical sequence being analyzed.

Sedimentary Structures. Sedimentary structures are commonly considered to be useful environmental indicators because they are generated almost exclusively by depositional processes; they are definitely formed in place. Unfortunately, the fact is that few, if any, individual sedimentary structures are generated by processes unique to a particular depositional environment. Similar structures can be produced in different environments and by different depositional agencies. For example, ripple marks and cross-bedding can be produced subaqueously by current flow and subaerially by wind transport. Although some individual sedimentary structures provide insight into environmental conditions such as relative water depth and the flow velocity of depositing currents, interpretation of environmental parameters on the basis of sedimentary structures must be made with caution. It is known, for example, that ripple marks and cross-bedding typically form by current and wave activity in very shallow water, but they occur also in deep water well below storm-wave base because of the presence of strong bottom currents. Nonetheless, in the use of sedimentary structures for environmental analysis, we must not lose sight of the necessity to first interpret from sedimentary structures the processes that created the structures. Only after we have interpreted process can we proceed to interpretation of depositional environment.

Interpretation based on preferred associations of sedimentary structures, rather than on individual sedimentary structures, has the greatest usefulness for environmental discrimination. The relative abundance of particular sedimentary structures may have environmental significance also. Figure 10.9 summarizes graphically the distribution of various kinds of sedimentary structures in major depositional environments. Although most sedimentary structures can occur in several environmental settings, certain associations of structures are more common in some environments than in others. For example, the association of graded bedding, flute casts, groove casts, and slump structures is more common and these structures generally more abundant in sediments from marine-basin to slope environments than in sediments from most other environments. By contrast, the association of high-angle and trough-festoon cross-bed-

FIGURE 10.8 Schematic illustration of highly idealized vertical grain-size profiles characteristic of certain environments. No scale. (From Selley, R. C., 1978, Ancient sedimentary environments, 2nd ed. Fig. 1.2, p. 18, reprinted by permission of Cornell University Press, Ithaca, N.Y.)

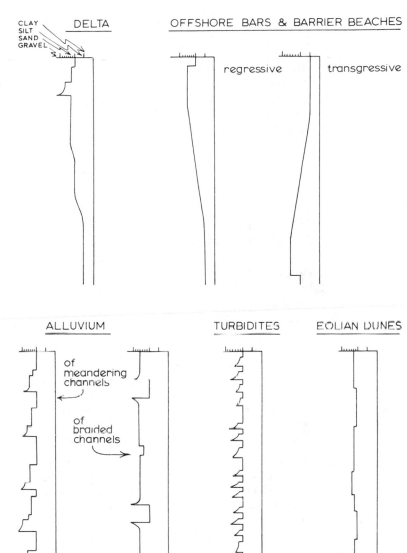

ding, ripple marks, mudcracks, and raindrop prints is most characteristic of sediments deposited in marginal and nonmarine environments. The more geologists know about the associations of structures that occur in different environmental settings and the processes that create these structures, the better able they are to use these associations in environmental analysis.

Paleocurrent Patterns. Some sedimentary structures yield directional data that show the direction ancient current flowed at the time of deposition. The dip direction of

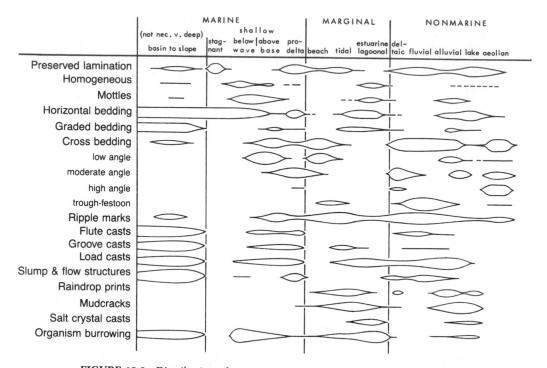

FIGURE 10.9 Distribution of various types of sedimentary structures in major depositional environments. (From Heckel, P. H., 1972, Recognition of ancient shallow marine environments, *in* J. K. Rigby and W. K. Hamblin (eds.), Recognition of ancient sedimentary environments: Soc. Econ. Paleontologists and Mineralogists Spec. Pub. 16. Fig. 6, p. 243, reprinted by permission of SEPM, Tulsa, Okla.)

cross-bed foresets; the asymmetry and orientation of the crests of current ripples; and the orientation of flute casts, groove casts, and current lineation are examples of directional data that can be obtained from sedimentary structures. Although directional data are used primarily to yield information about ancient slope orientations and current flow directions, they also have environmental significance.

The orientation of directional sedimentary structures is determined in the field with a Brunton compass by taking measurements from as many different outcrops and individual beds as possible and practical. The orientation of directional structures determined from a particular bed or stratigraphic unit commonly shows considerable scatter. Therefore, directional data must be treated statistically in some manner to reveal primary and secondary directional trends. For example, the dip direction of cross-bed foresets in the ancient deposits of a meandering river system may range from N 20° W to N 20° E owing to variations in flow direction of the stream in different parts of the meandering river system. By examining the orientation data statistically, we may be able to determine that the primary flow direction of the stream was approximately due north. Because all of the cross-bed foresets in this example indicate flow in the same general direction, in spite of some scatter, we say that flow was **unidirectional.** By contrast, the cross-bed foresets in sandy deposits of marine tidal channels may display two opposing dip directions owing to formation of cross-beds during both incoming and outgoing tides. This type of opposing flow is referred to as **bidirectional.** In some environments, such as the eolian environment, depositing currents may flow in several directions **(polydirectional)** at various times during deposition of a particular sedimentary unit.

The paleocurrent data collected from stratigraphic units that have undergone little or no tectonic deformation or tilting can be compiled and summarized directly. If the rocks have undergone considerable tilting, it is necessary to correct the measured orientation by restoring directions to their original attitude before tilting. A simple procedure using a stereogram can be used to reorient directional data collected from tilted stratigraphic units (Collinson and Thompson, 1982, p. 188). After any necessary reorientation of data has been done, the data are commonly plotted as a circular histogram, or "rose diagram" (Fig. 10.10). Such diagrams show the principal direction of paleocurrent flow and any secondary or tertiary modes of flow. If the paleocurrent flow as revealed by the rose diagram is predominantly in a single direction, the paleocurrent flow is said to be **unimodal.** If two principal directions of flow are indicated the flow is **bimodal,** and if three or more directions of flow are revealed by the directional data the paleocurrent flow is called **polymodal.**

Local paleocurrent directions may have environmental significance. For example, sediments from alluvial and deltaic environments tend to have unimodal paleocurrent vector patterns, whereas bimodal paleocurrent patterns are more common in shoreline and shelf sediments. Paleocurrent data have their greatest usefulness when plotted on a regional scale to reveal regional paleocurrent patterns. Table 10.2 shows some regional paleocurrent patterns from selected major depositional environments.

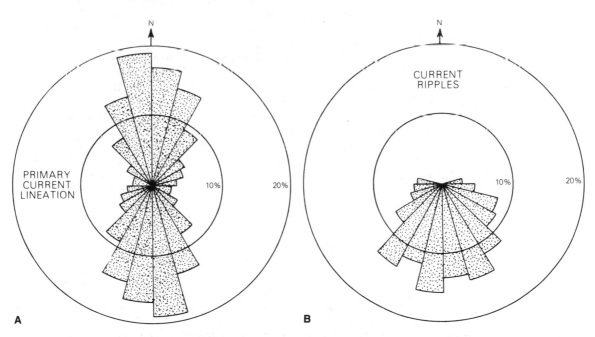

FIGURE 10.10 Paleocurrent data plotted as rose diagrams. A. Bimodal pattern. B. Unimodal pattern. (From Collinson, J. D., and D. B. Thompson, 1982, Sedimentary structures. Fig. 10.2, p. 176, reprinted by permission of George Allen & Unwin, London.)

TABLE 10.2   Environmental significance of paleocurrent patterns

Environment	Local current vector	Regional pattern
Alluvial { Braided { Meandering	Unimodal, low variability Unimodal, high variability	Often fan shaped Slope controlled; often centripetal basin fill
Eolian	Uni-, bi-, or polymodal	May swing round over hundreds of miles around high-pressure systems
Deltaic	Unimodal	Regionally radiating
Shorelines and shelves	Bimodal (due to tidal currents), sometimes unipolar or polymodal	Generally consistently oriented onshore, offshore, or longshore
Marine turbidite	Unimodal (some exceptions)	Fan shaped or, on a larger scale, trending into or along trough axes

Source: Selley, R. C., 1978, Ancient sedimentary environments, 2nd ed. Table 1.4, p. 14, reprinted by permission of Cornell University Press, Ithaca, N.Y.

Sedimentary Textures. Many attempts have been made to use particle size, sorting, shape, and surface texture as tools for environmental analysis. These approaches are discussed in considerable detail in Chapter 5 and need not be repeated here. On the whole, the use of grain-size data for environmental interpretation has proven to be disappointing. Techniques involving the use of bivariate plots of grain-size statistical parameters, such as skewness vs. standard deviation (Friedman, 1967, 1979), log probability plots (Visher, 1969, Sagoe and Visher, 1977), and C–M/L–M plots (Passega, 1964, 1977), have achieved only limited success in differentiating sediments from modern environments and even less success in reconstructing ancient sedimentary environments. Therefore, many geologists now have serious reservations about the reliability of environmental analysis based on grain-size data. Even the use of grain size as a general indicator of relative current energy must be viewed with caution. Sediment grain size provides a rough guide to the minimum energy of the environment, but it cannot be used reliably in many cases to estimate the maximum energy. It is obvious, for example, that boulders in an ancient sedimentary deposit could not have been moved and deposited by wind or weak traction currents; an estimate can be made of the minimum energy required to move such boulders. On the other hand, it is far from obvious that all fine-grained sediment must have been transported and deposited by low-energy currents. If only fine-size sediment is available for transport under a particular set of environmental conditions, then only fine sediment will be transported and eventually deposited, regardless of the energy of the transporting currents.

Measures of particle form—sphericity, for example—have generally not proven to be very useful in environmental discrimination, even though sphericity is known to affect particle transportability and settling velocity. This lack of usefulness may be due in part to our inability to accurately measure and express the form of particles. The relatively new technique involving computer-assisted Fourier analysis of particle shape (discussed in Chapter 5) appears to have the potential to describe particle form with a much higher degree of precision than previously possible. Some success has already been reported in using this technique to differentiate beach, dune, fluvial, and glaciofluvial sediments; future research on Fourier shape analysis may show that it has even broader applicability in environmental studies.

Particle roundness has significance in environmental studies, but it must be used with considerable caution. The rounding of sand-size quartz is known to take place primarily by wind transport, although some rounding may occur in the surf zone. Thus, the presence of abundant, well-rounded quartz grains in ancient sandstone deposits might reliably be taken to indicate deposition in an eolian, or possibly beach, environment if it were not for the problem of grain recycling. Because of the superior hardness and chemical stability of quartz, it can be recycled many times. Thus, roundness of grains in a particular deposit may have been acquired during a previous cycle, or cycles, of transport and does not necessarily reflect the most recent episode of transport and deposition. Pebbles and cobbles become rounded quite readily by stream transport. The presence of abundant well-rounded pebbles in ancient sedimentary deposits may indicate deposition in a fluvial environment; however, pebbles can become rounded in beach and lakeshore environments also. Furthermore, rounded fluvial pebbles can be transported by streams onto beaches or into nearshore environments, where they may become reentrained by turbidity currents and resedimented in deepwater environments.

Microrelief surface markings on quartz grains—striations, V-shaped pits, ridges, and conchoidal fractures, for example—are relatively more easily produced and destroyed by transport and depositional processes than changes in particle form and roundness. Therefore, these surface markings are more likely than form and roundness to reflect the conditions of the last environment of deposition. Although similar surface textures can be produced in different environments, certain markings are more common on quartz grains from particular environments. For example, striations are particularly common on grains from glacial environments, and quartz grains from littoral environments are characterized especially by V shaped percussion marks and conchoidal breakage patterns. By careful statistical analysis of surface texture data from large numbers of quartz grains, investigators have been able to characterize and distinguish at least three major modern depositional settings on the basis of quartz grain textures: (1) littoral (beach and nearshore), (2) eolian (desert), and (3) glacial. Extension of the results obtained by study of quartz surface textures in modern sediments to interpretation of ancient depositional environments is rendered more difficult because of diagenetic effects such as addition of quartz overgrowths or etching of grain surfaces by diagenetic solution processes. Also, investigators must be aware of the fact that even though surface markings are more easily changed than form and roundness, markings produced in one environment may persist for a considerable time after being moved into another environment, confusing environmental interpretation.

Geochemical Criteria

The major-element composition of siliciclastic sedimentary rocks is primarily a function of the chemical composition of the siliciclastic particles that make up the rocks. Elements such as Si, Al, K, and Na are so abundant in siliciclastic minerals that they cannot be overprinted by elements added in the depositional environment. Therefore, major-element composition is of relatively little value in environmental studies of siliciclastic sedimentary rocks. By contrast, the chemical composition of so-called "chemical" sedimentary rocks such as carbonates and evaporites is governed primarily by conditions of the depositional environment. Nevertheless, the major-element composition of these rocks is generally not a sufficiently sensitive indicator of environments to have significant diagnostic value in differentiating between major environments. The abundance of Ca and Co_3^{2-} in limestones, for example, does not vary markedly in different parts of the marine environment; it is not possible on the basis of concentrations of these elements to tell the difference between shelf limestones and deeper water limestones. Furthermore, diagenetic changes involving addition or removal of chemical elements may significantly alter initial concentrations of Ca, Fe, K, Mg, CO_3, and SO_4^{2-}, for example, making them of little use for geochemical environmental analysis (Ernst, 1970), even if they might otherwise be of value in environmental interpretation. Because of these problems, major-element geochemistry has not developed as a useful tool for differentiating between principal depositional environments.

On the other hand, certain trace elements and isotopes appear to have promise as tools for determining salinity and temperature of the water in which ancient sediments were deposited. Paleosalinity data from rocks that do not contain fossils or other environmental indicators are useful for differentiating marine and nonmarine environments, and paleotemperature data provide some insight into climatic conditions and possibly water depth.

Paleosalinity Determinations

Boron. Boron has been the most investigated element in paleosalinity studies (e.g., Bohor and Gluskoter, 1973; Couch, 1971; Harder, 1970; Potter et al. 1963; Shimp et al., 1969). The boron content of sedimentary rocks varies markedly with the type of sediment. Therefore, boron determinations must be made from rocks of a single type, or lithology. Shales contain relatively high concentrations of boron and are the rocks most commonly studied in paleosalinity determinations. Boron content increases with decreasing grain size of shales; it is affected also by the mineralogy of the shales and to some degree by diagenesis, although not to the extent that many major elements are affected. After eliminating these variables of size and mineralogy as nearly as possible by comparing strata of similar mineralogical composition, grain size, and diagenetic histories, several studies have shown that the concentration of boron in marine sediments is markedly higher than in nonmarine sediments. Examples of these differences are shown in Table 10.3. The reason for increased boron content in marine sediments has to do with the fact that the boron content of seawater is much higher than that of freshwater (4.8 mg/L vs. 0.01 mg/L; Ernst, 1970). Boron is adsorbed onto clay minerals and possibly incorporated into their crystal lattices when the clay minerals are brought into a marine environment by rivers. In the marine environment, boron content appears to increase seaward from the strandline because of decrease in grain size of clayey sediment in the seaward direction.

TABLE 10.3   Boron content in marine and freshwater carboniferous shales from various localities

Locality and reference	Boron content		
	Marine (ppm)	**Brackish (ppm)**	**Freshwater (ppm)**
Westfalen, Germany	100–200	45–110	15–80
England	300		150
Pennsylvania, U.S.A.	>80		<60

Source: Harder, H., 1970, Boron content of sediments as a tool in facies analysis: Sedimentary Geology, v. 4. Table I, p. 155, reprinted by permission of Elsevier Science Publishers, Amsterdam.

Considerable time is evidently required for boron to be adsorbed to capacity on clay particles; therefore, boron content is not sensitive to rapid salinity changes (Hallam, 1981). In fossiliferous sedimentary rocks, fossils are better indicators of salinity than boron. Boron may be a useful paleosalinity indicator in Precambrian rocks and other nonfossiliferous rocks if used with proper regard for the effects of grain size, mineralogy, and diagenetic history.

Other Trace Elements. Several other trace elements have been studied as potential paleosalinity indicators. Cr, Cu, Ga, Ni, and V have all been reported to be more abundant in marine than in freshwater clayey sediment; however, they are not as good as boron as salinity indicators (Shimp et al., 1969). Veizer and Demovic (1974) report on the basis of 1200 analyses of rocks from the central western Carpathians of Europe that strontium concentration is high in hypersaline, dark-colored, and deep-sea limestones and low in littoral, neritic, and shallow bathyal limestones. Authigenic iron sulfides have also been suggested as paleosalinity indicators. Berner et al. (1979) report that FeS_2/FeS ratios greater than 10.0 are characteristic of marine sediments, whereas ratios from brackish-water estuarine sediments or freshwater lake sediments are generally less than 1. The low ratios are believed to be the result of incomplete conversion of FeS to pyrite (FeS_2), owing to the low sulfate content of fresh- or brackish-water sediments.

Organic Matter. Even the characteristics of preserved organic matter in sediments have been used in attempts to differentiate marine sediments from continental sediments. For example, Tissot and Welte (1978) distinguish three types of organic matter in sediments on the basis of the type and molecular weight of organic carbon compounds in the sediments: (1) marine organic matter, derived mainly from phytoplankton and zooplankton; (2) continental organic matter, derived mainly from higher land plants; and (3) microbial organic matter, which occurs principally in lacustrine and paralic (marginal-marine or shallow-marine) environments and is derived mainly by microbial degradation of plant material. Didyk et al. (1978) used the high wax content of organic matter to indicate contribution from continental sources. A major problem with using organic matter to differentiate marine and nonmarine sediments is that organic matter derived from continents can be transported into marine environments.

Carbon and Oxygen Isotopes. Many investigators have attempted to use carbon and oxygen isotopes in fossil shells and limestones to differentiate marine and nonmarine sediments (e.g., Clayton and Degens, 1959; Keith and Weber, 1964; Keith et al., 1964; Dodd and Stanton, 1975; Rothe et al., 1974; Schidlowski, 1982). The use of carbon and oxygen isotopes for environmental determination is based on the premise that freshwaters are depleted in both heavy carbon (^{13}C) and heavy oxygen (^{18}O) relative to marine waters; therefore, the ^{13}C and ^{18}O values of freshwater carbonates and shells are measurably lower than those of marine carbonates and shells. The low $\delta^{13}C$ in freshwater environments is due mainly to contribution of plant-derived CO_2 (with a mean $\delta^{13}C$ of -25 per mil or less) to groundwaters and streams. The low $\delta^{18}O$ values in freshwaters is caused by fractionation of ^{18}O and ^{16}O during evaporation of ocean water, causing the heavier oxygen to remain in the ocean while lighter oxygen is carried off in water vapor and precipitated on land as rain or snow. The meaning of $\delta^{13}C$ and $\delta^{18}O$ is explained more fully in Section 18.7.

Figure 10.11 illustrates the range of $\delta^{13}C$ and $\delta^{18}O$ values in marine and fresh-water carbonates and in various kinds of calcareous marine organisms. Note that marine limestones are characterized by positive to slightly negative $\delta^{13}C$ and $\delta^{18}O$ values, whereas freshwater limestones commonly have strongly negative $\delta^{13}C$ values and strongly negative to slightly positive $\delta^{18}O$ values. The use of carbon and oxygen isotopes in environmental interpretation is complicated by the fact that diagenetic processes may alter isotope ratios after deposition; in the case of oxygen isotopes, temperature of the water may also affect the ratio. Carbon isotope ratios appear to be affected only slightly during diagenesis. Oxygen isotope ratios can be strongly affected owing to recrystallization and isotopic interchange with meteoric waters. Thus, in general, $\delta^{18}O$ values of marine carbonates decrease progressively with time; very old marine carbonate rocks may have $\delta^{18}O$ values close to those of freshwater limestones. The $\delta^{18}O$ values also decrease with increasing temperature of the water in which carbonates form, as discussed in a succeeding section; therefore, possible temperature effects and the age of the rocks must be taken into consideration when using oxygen isotopes for environmental reconstruction.

FIGURE 10.11 Distribution of $\delta^{18}O$ and $\delta^{13}C$ values in various types of marine carbonates. (From Milliman, J. D., 1974, Marine carbonates. Fig. 19, p. 33, reprinted by permission of Springer-Verlag, Heidelberg.)

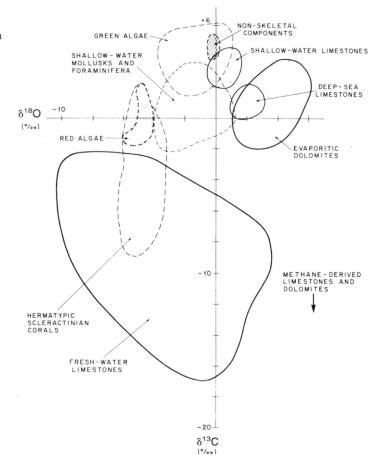

Paleotemperature Determinations

Both boron and bromine are temperature-dependent trace elements. If salinity remains constant, boron content increases with increasing temperature. Only relative changes in temperature can be read from boron distributions; absolute temperatures cannot (Ernst, 1970). The proportion of bromine incorporated into NaCl or KCl minerals is also a function of the temperature of evaporation of brine solutions and thus can be used to determine temperature of formation of evaporite deposits. With increasing temperature, magnesium and strontium increase preferentially over calcium in carbonate sediments and the shells of some calcareous organisms; therefore, Ca/Mg and Ca/Sr ratios have been used by some investigators as paleotemperature indicators. Broad, general indications of warm climates are provided by the simple presence of evaporite deposits—which tend to be deposited in warm climates where evaporation rates are high—and carbonate rocks—which are deposited primarily in warm waters.

The most widely used tool for determining paleotemperatures of sedimentary environments is oxygen isotopes: $^{18}O/^{16}O$ ratios are not only a function of salinity but also are strongly affected by water temperature. The fractionation of ^{18}O and ^{16}O during evaporation of surface ocean water decreases with increasing temperature; therefore, under conditions of constant salinity, $\delta^{18}O$ values decrease essentially linearly with increasing temperature (Fig. 10.12). Paleotemperature studies are commonly made by use of the shells of calcareous organisms such as foraminifers, belemnites, and oysters. Different types of organisms yield paleotemperatures for different parts of the oceanic regime. Thus, data from planktonic organisms give information about surface temperature of the ocean, whereas benthonic organisms provide information about the temperature of water on the ocean floor. Because oxygen isotope ratios are strongly affected by recrystallization and interchange with meteoric waters, only fossil shells that have escaped recrystallization can be used for paleotemperature determinations. This puts serious constraints on the use of isotope data obtained from many older rocks. The oxygen isotope method of paleotemperature analysis has been used most successfully in the study of foraminifer shells in cores of deep-sea sediments as old as late Cretaceous. It has been especially useful for determining paleoclimatic fluctuations during the Pleistocene.

Biologic Criteria

Of all the geologic tools available for paleoenvironmental reconstruction, fossils rank as one of the most useful. Of course, fossils are not found in all sedimentary rocks, particularly Precambrian rocks and nonmarine rocks of all ages. When they are present they have great environmental significance. They provide evidence that makes possible interpretations about salinity, water depth, water turbulence, oxygenation state of the water, temperature, and the nature of the substrate on or in which they lived. Observations of the ecological behavior of modern representatives of fossil organisms provide a basis for judging the response of ancient organisms to these variables of the environment. By knowing how modern organisms are distributed in the marine and nonmarine realms in relation to temperature, water depth, salinity, and turbidity conditions of the environment, paleontologists and paleoecologists can deduce the environmental conditions represented by most groups of ancient organisms. In the simplest and most broad-base application of fossil data to environmental reconstruction, marine strata can be differentiated from nonmarine strata by the simple presence of marine

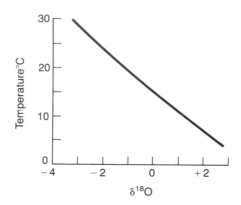

FIGURE 10.12 Variation of $\delta^{18}O$ with temperature at a salinity of 35‰. (From Milliman, J. D., 1974, Marine carbonates. Fig. 18, p. 32, reprinted by permission of Springer-Verlag, Heidelberg.)

fossils or nonmarine fossils. Much finer scale differentiation of marine paleoenvironments can be made on the basis of kind (diversity) and abundance of fossils, including trace fossils.

The use of fossils in environmental studies is not without problems. Some groups of fossil organisms have no modern representatives. Their ecological or environmental significance must be deduced in ways other than analogy, perhaps by relating the morphology of preserved hard parts to presumed environmental factors. Even interpretation of paleoecology on the basis of living descendants of ancient organisms has its limitations, because the modern descendants may not occupy all the environments or exactly the same environments as their ancestors. Perhaps an even more serious problem has to do with transport of fossils within the environment and reworking of fossils from older rocks. The assemblage of fossils that we see in an ancient sedimentary rock may not necessarily represent an assemblage of organisms that lived in the particular environment in which they died. They may have lived in one environment and have been transported after death into a different environment, or they may have been transported by inadvertence while still alive into a harsh or hostile environment in which they perished. Selley (1978) illustrates the former possibility by citing the example of the many drowned cats washed out to sea by the River Thames.

Geologists must be alert also to the possibility that fossils may have been eroded from older rocks during weathering and then transported into environments quite different from their original habitat. In spite of these limitations, careful workers have been able to use fossils with great effectiveness in environmental interpretation. Microfossils are particularly useful; because of their small size they can be found even in very small pieces of rock, such as subsurface well cuttings and cores. Trace fossils also have special advantage in environmental studies because they are in situ features and cannot be reworked and retransported. Almost all fossils are of some value in environmental studies if the investigator is knowledgable and skillful enough to know how to use them. An assemblage of organisms that live together in an interrelated community is called a **biocoenosis.** Because the assemblage of organisms brought together after death and buried at a particular depositional site may not be the same as the living assemblage at that site, the term **thanatocoenosis** has been used to distinguish such a death assemblage. The organic elements that make up a thanatocoenosis may be considered to include hard skeletal parts such as shells, teeth, and bones, excretory matter such as the fecal pellets found in limestones, fine-size organic matter, trace fossils of various types, and biostratification structures such as stromatolites. The relationship of these various kinds of fossil materials to specific environmental parameters is discussed in the following paragraphs.

Fossils as Salinity Indicators. The usefulness of fossils as paleosalinity indicators depends upon our ability to recognize fossil organisms that were restricted to waters of particular salinities. Among modern organisms, we recognize some groups, called **euryhaline organisms,** that are adapted to tolerate wide variations in salinity. Others that can tolerate only very slight variations in salinity are called **stenohaline organisms.** The distribution of modern-day, fossilizable invertebrate organisms with respect to salinity is shown in Figure 10.13. The majority of the modern ocean consists of water of normal salinity (~35‰) and is characterized by stenohaline organisms such as corals, radiolarians, brachiopods, echinoderms, most calcareous foraminifers, calcareous red and green algae, and scaphopods (Heckel, 1972). We have reasons to believe that ancient representatives of these modern marine stenohaline organisms were also sten-

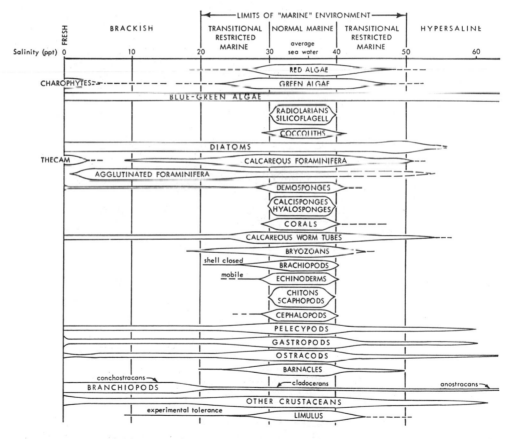

FIGURE 10.13 Distribution of modern invertebrate groups in relation to salinity. Thickness of bar roughly reflects diversity of taxa. (From Heckel, P. H., 1972, Recognition of ancient shallow marine environments, *in* J. K. Rigby and W. K. Hamblin (eds.), Recognition of ancient sedimentary environments: Soc. Econ. Paleontologists and Mineralogists Spec. Pub. 16. Fig. 3, p. 234, reprinted by permission of SEPM, Tulsa, Okla.)

ohaline. Some groups of organisms such as trilobites, archaeocyathids, tentaculitids, and graptolites that are known only as fossils are also believed to be marine stenohaline forms (Gall, 1983). Fossilizable freshwater stenohaline organisms that live in lakes and streams include charophytes (a type of calcified green algae) and some types of gastropods (snails), molluscs (bivalves), fish, and amphibians. Euryhaline organisms adapted to life in brackish water include such forms as oysters, certain gastropods, ostracods, foraminifers with agglutinated tests, the primitive *Lingula* brachiopods, fossil eurypterids, diatoms, and the blue-green algae (cyanobacteria) that form stromatolites. In general, brackish-water faunas are characterized by low diversity (few specific kinds), high faunal density (large numbers of individuals), smaller size compared to marine members of the same species, and shells that are commonly thinner than those of marine forms (Gall, 1983).

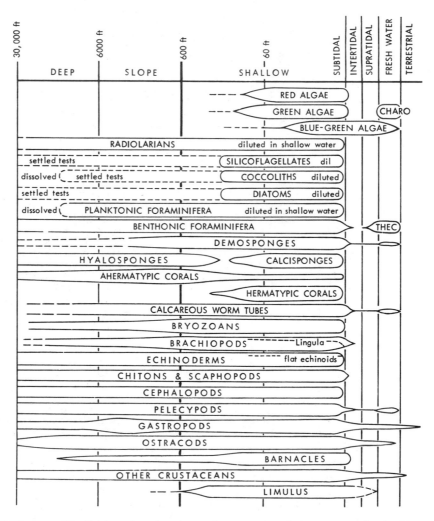

FIGURE 10.14 Distribution of modern invertebrate groups relative to water depth. (From Heckel, P. H., 1972, Fig. 4, p. 237, reprinted by permission of SEPM, Tulsa, Okla.)

Fossils as Indicators of Water Depth. Some fossil organisms are sufficiently restricted in their bathymetric range to be useful depth indicators. Figure 10.14 shows the distribution of major groups of modern invertebrate organisms. Chlorophyll-bearing plants, such as algae, are restricted by the need to carry on photosynthesis to the zone of light penetration in the ocean. This photic zone is generally less than about 200 m deep. Stromatolites, which are produced mainly by photosynthesizing blue-green algae, are thus indicators of shallow water. Hermatypic corals, which have a symbiotic relationship with unicellular algae, also are restricted to very shallow water. Many benthonic organisms tend to be more abundant at the shallow depths found on the continental shelves (<~200 m) than in deeper water, and certain groups such as flat echinoids or sand dollars, *Lingula* brachiopods, and calcisponges are almost exclusively shallow-water organisms. Planktonic organisms such as radiolarians and foraminifers can live in the near-surface waters of both the shallow and deep parts of the ocean and thus can theoretically be deposited in any part of the ocean. They are more abundant in deeper water, however, where there is less competition for nutrients by benthonic organisms. Also, after deposition they are less "diluted" in deeper water by the presence of siliciclastic or carbonate sediment and shallow-water benthonic organisms. Therefore, planktonic organisms commonly increase in numbers in bottom sediments in the offshore direction. Although earlier workers believed that diversity of organisms—expressed as number of species, for example—decreased in the direction of deeper water, later work suggests that this interpretation was in error owing to faulty methods of collecting deep-sea fauna (Heckel, 1972). Species diversity appears to be related more to stability of the environment than to depth; many stable deeper water environments contain more species than shallow-water environments.

Several problems complicate interpretation of water depths on the basis of nonvertebrate fossils. River currents may carry land plants into oceanic environments, and turbidity currents can transport shallow-marine benthonic organisms into deeper water. A more serious problem stems from the probability that some groups of organisms have changed their depth requirements during geologic time; the distribution of modern descendants of these groups may not accurately reflect past depth distribution. For example, hexactinellid sponges and some decapod crustaceans and attached crinoids seem to have migrated into deeper water in late Cretaceous time (Gall, 1983). Also, the depth distribution of organisms in the ocean is related to temperature and water turbidity and may not be a function entirely of depth per se.

Trace fossils have received considerable attention as depth indicators. Their usefulness as depth indicators is predicated on the assumption that organisms behave differently under the different environmental conditions that exist at different depths in the ocean; further, it is assumed that the ancestors of modern forms behaved in approximately the same way as their modern descendants. In the harsh, high-energy littoral environment of sandy coasts, organisms excavate vertical or U-shaped burrows into sand to escape the waves. On rocky or pebbly coasts, they bore into the rock. In the neritic, or subtidal, zone of continental shelves, where energy conditions are less harsh, vertical and U-shaped burrows are less common, burrows are shorter, and surface markings are more common. In deeper parts of the neritic zone, where organic matter is more abundant, vertical escape burrows give way to horizontal feeding burrows. In the deep bathyal to abyssal zones of the ocean, where low-energy conditions prevail below wave base, complex, spiral, winding, and meandering feeding traces develop. Because trace fossils cannot be reworked and redeposited like other fossils, they are considered to be one of the most useful depth indicators. Nevertheless, they are not infallible guides to depth. As better techniques for examining structures in deeper water develop, some trace fossils originally believed to be exclusively shallow-water forms are being found at greater depths. Also, study of trace fossils in some ancient sedimentary rocks has turned up supposedly deep-water forms such as *Zoophycos* in beds that are, on the basis of the presence of coal and algal limestones, clearly shallow-water deposits (Hallam, 1981). These inconsistencies suggest that factors other than depth and food supply may be involved in controlling distribution of trace fossils. Oxygen levels in the water may, for example, be an important control not related totally to water depth.

Finally, I stress that it is generally impossible to obtain reliable quantitative estimates of water depth on the basis of fossil evidence. Although we can sometimes say, for example, that water depth must have been less than 150–200 m because the organisms had to live in the photic zone, we can never pin the depth down to an exact figure. In most cases we have to be content with using terms that express relative water depth, such as littoral to tidal flat, shallow neritic, slope, and deep basin.

Fossils as Indicators of Water Temperature. Temperature exerts an important control on the distribution of organisms both on land and in the ocean; however, organisms differ greatly in their ability to tolerate temperature variations. Those fauna and flora that can withstand only minor variations in temperature are called **stenothermal.** Those that can withstand a much greater variation are called **eurythermal.** In the marine realm, eurothermal organisms are most characteristic of surface waters and shallow coastal waters where large fluctuations in temperature occur. Stenothermal organisms live mainly in the open ocean and at great depth, where temperature variations are small. Because most marine organisms are cold-blooded, temperature is probably more important in controlling distribution of organisms in the ocean than are either salinity or water depth. The modern ocean can be subdivided into five biogeographic temperature zones (Fig. 10.15), with different groups of organisms characterizing each zone. Thus, we can speak of polar, subpolar, temperate, subtropical, and tropical populations of organisms.

Marine organisms living in zones of high, constant temperature are characterized by high species diversity, with each species consisting of a relatively small number of individual organisms. The calcareous shells of warm-water species tend to be thicker and more highly ornamented than related cold-water species, presumably owing to the higher concentrations of dissolved calcium bicarbonate in warm waters. Reef-building corals are particularly partial to warm waters, and most modern coral reefs are restricted to water warmer than about 18°C. By inference, ancient coral reefs also probably formed in very warm water. Cold-water organisms tend to be less diversified, but large numbers of individual organisms of each species may exist. In cold waters, calcareous shells are smaller and thinner and true reefs are not found. Siliceous, planktonic diatoms and radiolarians are particularly abundant in cooler waters because of the generally greater availability of nutrients in these waters.

FIGURE 10.15 Temperature limits of biogeographic zones in the ocean. (After Hedgpeth, J. W., 1957, Marine biogeography, Fig. 5, p. 364, *in* J. W. Hedgpeth (ed.), Treatise on marine ecology and paleoecology; v. 1, Ecology: Geol. Soc. America.)

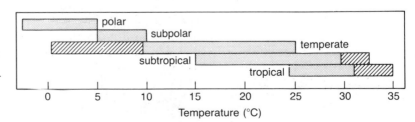

Land-dwelling animals include both warm-blooded and cold-blooded forms. Warm-blooded animals can withstand wide ranges of temperature variation and are thus less diagnostic of particular temperature zones than the cold-blooded reptiles, which can flourish only in areas of high annual temperature. Nonetheless, land animals tend to be distributed by climatic zones. Thus, elephants and rhinoceroses live in warm climates, reindeer and musk-oxen prefer very cold climates, and so on. Land plants are particularly sensitive to temperature change and show marked variation with latitude and temperature. Therefore, they make excellent paleoclimate indicators. For example, study of plant pollen and spores in Pleistocene sediments has revealed numerous fluctuations in temperature that can be related to glacial and interglacial stages.

To use fossils successfully as paleotemperature indicators requires first that we have adequate knowledge of the temperature distribution of modern organisms and the effect that temperature has on the morphology of shells and skeletal structures. I have already mentioned the relationship of water temperature to thickness of calcareous shells. Another illustration of how temperature affects morphology can be seen in the way the shells of some foraminifers coil in response to temperature changes. The foraminifer *Globorotalia truncatulinoides*, which is very abundant in the modern ocean, coils predominantly in a right-hand direction in warm waters and in a left-hand direction in cold water.

The second requirement for paleoclimatic interpretation is that we be able to extrapolate from the modern to the ancient. For example, knowledge about the coiling directions of modern formainifers has been applied successfully in the study of deep-sea cores to determine climatic variations during the Pleistocene. Also, we assume, because modern coral reefs develop mainly in warm water, that ancient coral reefs must likewise have formed in warm water and that similar tree pollen found in modern environments and Pleistocene sediments indicates similar climatic zones. Obviously, the validity of these kinds of assumptions decreases the farther we go back in time because we become less sure that ancient organisms exhibited the same response to temperature changes and patterns as their modern descendants. Also, many ancient organisms such as trilobites and ammonites are known only as fossils. Other problems in interpretation stem from the fact that it may be difficult to separate the behavioral effects caused by temperature from those caused by salinity and water depth. And finally, organisms adapted to one temperature zone may be transported accidentally into a different, hostile temperature zone, where they perish and become buried. Thus, they do not reflect the temperature of the depositional site.

Water Energy. Water turbulence is an important factor in controlling the distribution of organisms because it helps regulate the distribution of food, nutrients, oxygen, and carbon dioxide in the environment. In addition, current movements control the spreading and distribution of planktonic organisms and juvenile (larval) forms of benthonic organisms. Also, water energy affects the ability of some organisms to physically exist in a particular environment. Turbulence that is too vigorous may destroy fragile forms of organisms such as branching bryozoans or may uproot or dislodge organisms.

Water energy is related to water depth and tends to be greatest in shallow water. In the discussion of trace fossils as depth indicators, I pointed out that organisms living in the shallow-water, high-energy zone of sandy coasts construct vertical escape burrows in which they bury themselves in order to be protected from wave energy while suspension feeding. By contrast, detritus feeders live in quieter, and generally deeper, water, where organic matter settles out of the water. Thus, they do not need to build escape burrows. In general, organisms that live in high-energy environments tend to build thick, robust shells or to develop encrusting forms or other skeletal modifications that are wave resistant. Some types of corals and coralline algae, for example, adapt to a high-energy environment by building colonial structures in which the individual coral organisms are bound together to form a rigid, wave-resistant mass. High-energy environments are characterized particularly by organisms such as colonial corals; encrusting bryozoans; thick-shelled brachiopods, pelecypods, and gastropods; stromatoporoids (Silurian and Devonian predominantly); coralline algae; and rudistid pelecypods (Cretaceous predominantly). By contrast, organisms that live mainly in low-energy environments tend to build thin, fragile shells or delicate, branching skeletal structures. Thus, quiet-water environments are characterized by such organisms as fragile, branching types of bryozoans; solitary corals; ostracods; and thin-shelled brachiopods, gastropods, and pelecypods, as well as the fecal pellets of bottom-dwelling scavengers. Some organisms develop different skeletal modifications to allow them to live in water of different energy levels. For example, the stinging hydrozoan coral, *Millepora*, grows as an encrusting, hummocky colony in very shallow tropical water, but in water of increasing depth, where turbulence is reduced, the colonies assume a vertically bladed, labyrinthine shape (Laporte, 1968). Figure 10.16 illustrates other examples of shape modification in response to water turbulence. The rounding and sorting of fossil debris also provide clues to water energy. Shells in high-energy environments tend to become broken, rounded, and sorted by wave and current activity. Shells deposited in quiet-water environments either remain whole or, if broken by scavenging organisms, remain angular and unsorted.

Water Turbidity and Sedimentation Rates. Excessive turbidity of water owing to suspended clay and organic matter adversely affects organisms by limiting the amount of sunlight that filters through the water and thereby limiting photosynthesis by marine plants. Also, such particles may clog respiratory or food-gathering apparatus of organisms, particularly filter feeders. Figure 10.17 illustrates the distribution of modern fossilizable invertebrate benthonic organisms—and by inference their fossil ancestors—

FIGURE 10.16 Modification of skeletal morphology in response to water energy. The recent gastropod *Patella vulgata* develops a thick, conical shell (A) where exposed to tidal action; this shell allows it to stick more efficiently to the substrate. In quieter water areas, the shell is thinner and more flattened (B). The modern coral *Porites* has a "stubby," encrusting form (C) in turbulent water, but a much more fragile, branching form (D) in quieter water. (After Gall, J. C., 1983, Ancient sedimentary environments and the habitats of living organisms. Fig. 20, p. 25, reprinted by permission of Springer-Verlag, Heidelberg. Originally after Moore, Vaughan, and Wells, 1943.)

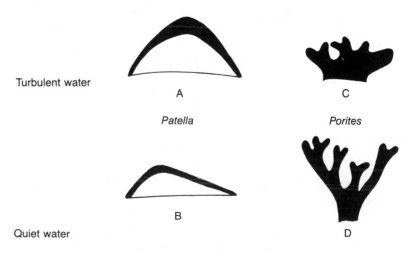

Turbulent water

A

Patella

C

Porites

B

Quiet water

D

with respect to water turbidity and sedimentation rates. Green and red algae, which require sunlight for photosynthesis, and suspension feeders such as sponges, corals, pelmatozoans, scaphopods, and barnacles are restricted in their distribution to relatively clear water. Deposit feeders such as certain clams, gastropods, and ophiuroids and carnivores and scavengers such as starfish, echinoids, and arthropods can commonly feed in water of much higher turbidity but they cannot live in water so highly charged with clay and silt that their breathing apparatus is clogged. Excessively high rates of suspension sedimentation are thus detrimental to most benthonic organisms;

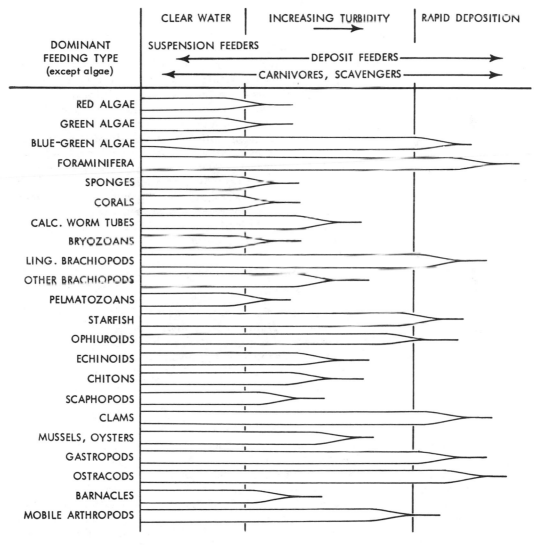

FIGURE 10.17 Distribution of modern, fossilizable invertebrate benthonic groups with respect to water turbidity and rapidity of deposition. (From Heckel, P. H., 1972, Fig. 9, p. 250, reprinted by permission of SEPM, Tulsa, Okla.)

only a very limited number of organisms—lingulid brachiopods, starfish, and certain burrowing clams, for example—can live in highly turbid environments of rapid sedimentation. Interpretation of water turbidity must be based on benthonic organisms only, because near-surface planktonic organisms are affected relatively little by turbidity. The usual problem of transport of organisms after death must be considered also.

Nature of the Substrate. In addition to the environmental parameters discussed, most marine benthonic organisms are influenced in their distribution by the nature of the substrate on or in which they live. Some of these organisms require hard, rocky bottoms to which they attach themselves or into which they bore. Others require soft muddy or sandy bottoms into which they burrow for escape or to forage for food. Benthonic organisms that live buried in soft bottom sediment are called **infauna.** Those that live attached to a hard substrate or that move over the bottom are **epifauna.** Epifauna that attach themselves to the bottom are said to be **sessile;** those that move about are **vagile.** In general, infauna and sessile epifauna are most useful for interpreting ancient substrates. Vagile epifauna can roam about over a variety of substrates and thus are not diagnostic of any particular type of bottom condition. Fossil assemblages consisting of colonial corals, calcareous worm tubes, barnacles, oysters, bryozoans, red algae, and chitons suggest an environment with a hard, well-cemented, or rocky substrate. Hard substrates may also be marked by borings, which are generally smooth, either vertical or inclined at various angles, and commonly filled with sand. Sponges, green algae, solitary corals, pelmatozoans, and mussels indicate a firm but uncemented sediment substrate. An assemblage consisting of lingulid brachiopods, scaphopods, and burrowing clams suggests a soft substrate. Vertical or U-shaped escape burrows also indicate a soft substrate, generally sand; horizontal burrows and complex feeding structures suggest a soft, probably muddy substrate.

Environmental Criteria Based on Instrumental Well Logs

Mechanical well logs are discussed in Chapter 14 as a tool for correlation and stratigraphic analysis. They also have limited use in environmental studies of subsurface formations. Well logs measure derived properties such as resistivity, sonic velocity, and radioactivity that are in turn a function of primary depositional properties (mainly grain size, mineralogy, and dip of beds), diagenetic properties (degree of compaction, cementation, and solution), and fluid content (water, oil, and gas) of the rocks.

The primary application of well-log data to environmental analysis comes in identifying vertical changes in grain size that may reveal patterns of change, that is, coarsening-upward or fining-upward cycles. Electric logs, which display both a formation resistivity curve and an S.P. (spontaneous potential) curve, are most commonly used in grain-size interpretation (Fig. 10.18). The S.P. curve is generated in response to voltage fluctuations caused by electrofiltration and electroosmosis of fluids in strata adjacent to the borehole. These phenomena are related to permeability of the rocks, which in turn is related to grain size. Shales and other mudrocks commonly have very poor permeability. Sandstones have much better permeability, which tends to increase with increasing grain size. In Figure 10.18, lithologic data obtained by coring are superimposed on the electric log to illustrate the relationship between the S.P. curve and grain size. Once this relationship between curve shape and grain size has been established empirically in a particular basin in a few cored wells, it can be extrapolated to nearby uncored wells. Grain-size information can be obtained also from gamma-ray

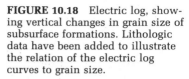

FIGURE 10.18 Electric log, showing vertical changes in grain size of subsurface formations. Lithologic data have been added to illustrate the relation of the electric log curves to grain size.

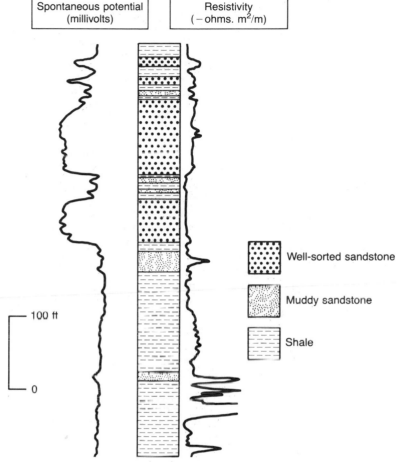

logs. Gamma-ray logs measure the natural radioactivity in formations. Clay minerals generally contain the highest natural radioactivity; thus, the highest levels of radioactivity on a gamma-ray log tend to correlate with the finer grained formations, which contain the greatest abundance of clay minerals. Grain-size profiles generated from well-log data are studied and interpreted in the manner described under "Lateral and Vertical Facies Associations."

Cross-section profiles based on well-log data can help to define the three-dimensional geometry of subsurface units on a regional scale, and well logs called **dipmeter logs** can be used to determine the local dip of subsurface formations. Dipmeter data can then be applied in subsurface environmental analysis to help determine the geometry and orientation of subsurface bodies such as channels, sand bars, and reefs. Dipmeter logs are a special type of electric log that measures resistivity at four opposing points on the wall of a borehole and simultaneously records the orientation of the dipmeter tool with respect to magnetic north. These data can then be utilized with the help of the computer to calculate the dip represented by the four opposing resistivity points, which are related to bedding surfaces.

10.4 CLASSIFICATION OF DEPOSITIONAL ENVIRONMENTS

Most textbooks that discuss sedimentary environments include some kind of environmental classification, either formally stated in tabular form or otherwise provided by the organization of chapters and subheadings. As mentioned in the introduction to this chapter, we commonly recognize three fundamental depositional environmental settings: (1) continental, (2) marginal marine, and (3) marine. Each of these primary environmental realms has been further divided by different workers into three to five or more major environments plus numerous subenvironments. The most comprehensive listing of depositional environments is probably that of Crosby (1972), who compiled a list of 18 major environments and more than 50 subenvironments. Although such a detailed listing of modern depositional environments is useful in illustrating the wide variety of depositional conditions under which sedimentary rocks can accumulate, it is impractical as a workable guide to ancient depositional environments because we cannot discriminate ancient sedimentary environments on such a fine scale. We are fortunate indeed in some cases just to be able to recognize whether an ancient sedimentary rock was deposited in a marine environment or a nonmarine environment. The most practical and useful kind of classification of ancient sedimentary environments is one that includes only a relatively small number of major environments and subenvironments, each of which is capable of being recognized and distinguished from other environments by the tools of environmental interpretation available to us.

Table 10.4 is a simplified classification that meets these general requirements; however, it is probably impossible to create a classification of depositional environments that is totally acceptable to all geologists. For example, the choice of major

TABLE 10.4 Simplified classification of ancient depositional environments

Primary depositional setting	Major environment	Subenvironment
Continental	*Fluvial	*Alluvial fan *Braided stream *Meandering stream
	*Desert Lacustrine *Glacial	
Marginal marine	*Deltaic	*Delta plain *Delta front *Prodelta
	*Beach/barrier bar *Estuarine/lagoonal Tidal flat	
Marine	Neritic	Shelf **Organic reef
	Oceanic	Slope Deep-ocean floor

*Dominantly siliciclastic deposition.

**Dominantly carbonate deposition.

Environments not marked by an asterisk(s) may be sites of siliciclastic carbonate, evaporite, or mixed sediment deposition depending upon depositional conditions.

environments and subenvironments shown in Table 10.4 may not be acceptable to all workers, some of whom may prefer to include more, fewer, or different environments. Also, the classification contains some inconsistencies and omissions. To illustrate, glacial environments may be the deposition sites of fluvial, eolian, and lake sediments as well as sediments transported and deposited directly by glaciers. Eolian deposits can form in back-beach, marginal-marine environments and on the tops of barrier islands, as well as in continental environments. Organic reefs may form in marginal-marine and possibly even freshwater environments, as well as in truly marine environments. Deltas can form in lakes in addition to shoreline or transitional environments, and turbidites can form in environments other than deep marine. In any case, it is probably best not to get too involved in the nomenclature used in classifications because no simple classification can adequately account for all possible environments. Classifications are not particularly important in themselves, but they provide a convenient framework to which we can relate facies models, which are necessary for environmental interpretation.

10.5 FACIES MODELS

Few, if any, of the tools for environmental analysis can be used alone for interpreting depositional environments. It is not possible in most cases to examine a single property of sedimentary rocks—geometry or sedimentary structures, for example—and, on the basis of that property, to confidently deduce the depositional environment of the rock. To interpret the depositional environment of an ancient sedimentary rock, we must examine many different properties of the rock and then compare these properties with some mental picture we have of the properties of rocks deposited in known depositional environments. This mental picture constitutes an environmental model. Few of us have had enough personal field experience, read enough books and papers, or have good enough memories to carry around a mental picture of every important depositional environment. Fortunately, we can draw on the experience of many geologists through their published data and ideas to construct facies models that will provide the reference framework we need for interpreting ancient depositional environments.

Definition of Models

R. G. Walker (1979a, p. 3) defines a facies model as a "general summary of a specific sedimentary environment, written in terms that make the summary usable." Facies models can be expressed as idealized sequences of facies, as block diagrams, or as graphs and equations. Such summary models act as norms for purposes of comparison and as frameworks and guides for future observations. They serve also as predictors of new geological situations and form bases for interpretation of environment in terms of hydrodynamic conditions. Facies models thus provide a method for simplifying, ordering, categorizing, and interpreting data that may otherwise seem random and confusing. They provide a means of distilling local details until only the "pure essence" of the environment remains (R. G. Walker, 1979a). This process of distilling away local variations to arrive at a model that can serve as a norm and a predictor is illustrated in Figure 10.19. Facies models can be developed for each depositional system or environment. Once such models have been developed, we can use them as frames of reference to which we can compare ancient sedimentary rocks. Each facies model can

FIGURE 10.19 How facies models
are constructed. (From Walker,
R. G., 1979, Facies and facies models, *in* R. G. Walker (ed.), Facies
models: Geoscience Canada Reprint Ser. 1. Fig. 4, p. 4, reprinted
by permission of Geological Association of Canada.)

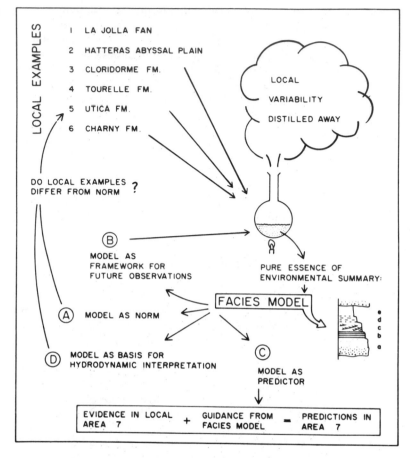

be described in terms of the properties that characterize the facies: geometry, sedimentary structures, sedimentary textures, particle and chemical composition, fossil content, vertical grain-size trends, and associated lithologic types. We can then use the
model to infer the depositional environment.

Types of Facies Models

Models take many different forms, including descriptive, geometric, and mathematical
or statistical types. Descriptive models are written summaries of the distinguishing
characteristics of particular environments. Geometric models may be topographic
maps, cross sections, three-dimensional block diagrams, and other forms that graphically illustrate basic depositional framework. Four-dimensional geometric models that
portray changes in erosion and deposition with time have also been utilized (R. A.
Davis and Fox, 1972). Statistical models employ techniques such as multiple linear
regression, trend-surface analysis, and factor analysis. Often, the objective of statistical
models is to examine several environmental parameters simultaneously in order to
predict the response of one element to another in a process–response model. The use
of statistical techniques and computer-simulation models in environmental studies has

the potential to yield results that in some cases may not be obtainable through less mathematically rigorous methods. Nonetheless, Hallam (1981, p. 12) warns against the danger of being "seduced away from significant geological problems by the beauty of the techniques" and of using "a sledgehammer to crack a nut" when much simpler approaches to modeling may produce equally good or better results and are generally better understood by the majority of geologists.

Construction and Use of Models

Facies models are commonly based on modern sedimentary environments. We assume that natural laws are constant in time and space and that we need not invoke hypothetical unknown processes to explain features of the rock record if presently observable processes and conditions can explain these features. Still, it is not possible in all cases to create facies models based on observation of sedimentation processes and products in modern environments. For example, no one has ever actually observed a modern turbidity current in the ocean, and few examples of historical turbidites are known. In developing a model for turbidite environments, we have to use general knowledge of sedimentation processes in modern environments supplemented with knowledge based on experimental laboratory investigations and studies of ancient examples of turbidites—examples identified as turbidites through the processes of reasoning rather than by comparison with modern deposits. Likewise, we cannot confidently use analogy with modern environments to develop models for carbonate sedimentation in shallow epicontinental seas that extended into the interior of continents—the Paleozoic epicontinental seas of the United States midcontinent region, for example—because we have no modern analogs of carbonate epicontinental seas on which to draw. Thus, it becomes necessary in some cases to turn to the ancient record for guidance in constructing facies models.

I interject a word of caution at this point about the use of facies models. Although such models are invaluable aids to the study of ancient depositional environments, too rigid adherence to a particular model can lead to problems in interpretation. As mentioned, many environments are complex and cannot be adequately represented by a simple model. Reading (1978a) points out that, although many environments can be described by one general model, it is only by concentrating on differences rather than similarities that the importance of the various processes that combine to make the facies in each environment can be evaluated. Common sense and good judgment must be exercised in the use of facies models, and we must be careful not to ignore or subordinate facts that do not fit in order to force interpretation of depositional environments on the basis of a particular model. The best sedimentologist is the one who has studied the most sediments, is familiar with the most facies models, and keeps an open mind when faced with data that do not fit preconceived concepts. "The conviction that one's own hypothesis is right is frequently the mark of one who is poorly informed about alternatives" (Reading, 1978a, p. 10).

ADDITIONAL READINGS

Anderton, R., 1985, Clastic facies models and facies analysis, *in* P. J. Brenchley and B. P. J. Williams (eds.), Sedimentology, recent developments and applied aspects: The Geological Society, Blackwell Scientific Publications, Oxford, p. 31–48.

Ernst, W., 1970, Geochemical facies analysis: Elsevier, Amsterdam, 152 p.

Gall, J. C., 1983, Ancient sedimentary environments and the habitats of living organisms: Springer-Verlag, Berlin, 219 p.

Hallam, A., 1981, Facies interpretation and the stratigraphic record: W. H. Freeman, San Francisco, 291 p.

Heckel, P. H., 1972, Recognition of ancient shallow marine environments, p. 226–286, *in* J. K. Rigby and W. K. Hamblin (eds.), Recognition of ancient sedimentary environments: Soc. Econ. Paleontologists and Mineralogists Spec. Pub. 16.

Krumbein, W. C., and F. A. Graybill, 1965, An introduction to statistical models in geology: McGraw-Hill, New York, 475 p.

Reading, H. G., 1978, Facies, p. 4–14, *in* H. G. Reading (ed.), Sedimentary environments and facies: Elsevier, New York.

Reading, H. G. (ed.), 1978, Sedimentary environments and facies: Elsevier, New York, 557 p.

Rigby, J. K., and W. K. Hamblin (eds.), 1972, Recognition of ancient sedimentary environments: Soc. Econ. Paleontologists and Mineralogists Spec. Pub. 16, 340 p.

Schwarzacher, W., 1975, Sedimentation models and quantitative stratigraphy: Developments in Sedimentology No. 19, Elsevier, Amsterdam, 382 p.

Selley, R. C., 1978, Ancient sedimentary environments, 2nd ed.: Cornell University Press, Ithaca, New York, 287 p.

Walker, R. G. (ed.), 1984, Facies models, 2nd ed.: Geoscience Canada Reprint Ser. 1, 317 p.

11
Continental Environments

11.1 INTRODUCTION

Following the discussion of environmental interpretation and facies analysis presented in Chapter 10, we turn to application of these principles in the study of depositional systems, beginning in this chapter with continental systems. Geologists recognize four major kinds of continental environments: fluvial, desert, lacustrine, and glacial (Table 10.4). Although treated as separate depositional systems, similar kinds of sediments can be generated in more than one of these environments. For example, eolian sediments can accumulate both in desert environments and in some parts of glacial environments. Lacustrine sediments form in lakes in any environment, including desert and glacial environments. Fluvial sediments are deposited mainly in river systems of humid regions, but they also form in rivers within desert areas and glacial environments.

Facies deposited in continental environments are predominantly siliciclastic sediments characterized by general scarcity of fossils and complete absence of marine fossils. Nonsiliciclastic sediments such as freshwater limestones and evaporites occur also in continental environments, but they are distinctly subordinate to siliciclastic deposits. Some continental deposits are distinguished by distinctive geometries; for example, fluvial channel sands and alluvial fan deposits. Continental sedimentary rocks are less abundant overall than marine and marginal marine sediments, but they nonetheless form an important part of the geologic record in some areas. Tertiary fluvial sediments of the Rocky Mountain–Great Plains region of the United States, Jurassic eolian sandstones of the Colorado Plateau, Tertiary lacustrine sediments (Green River Formation) of Wyoming and Colorado, and the late Paleozoic glacial deposits of South Africa and other parts of ancient Gondwanaland are all examples of continental deposits. Some continental sediments have economic significance also. They may contain important quantities of natural gas and petroleum, coal, oil shale, and uranium. We now examine in turn each of the major continental environments.

11.2 FLUVIAL SYSTEM

Fluvial deposits encompass a wide spectrum of sediments generated by the activities of rivers, streams, and associated sediment-gravity flow processes. Such deposits occur at the present time under a variety of climatic conditions and in various continental settings, ranging from desert areas to humid and glacial regions. Although many sub-environments of the fluvial system can be recognized, most ancient fluvial deposits can be assigned to one of three broad environmental settings: alluvial fan, braided river, or meandering river. These fluvial environments are interrelated and overlapping. To illustrate, braided rivers commonly occur on alluvial fans in humid regions and contribute to deposition of the fans; some humid alluvial fans merge downslope into braided-river systems; and some braided rivers may be transformed downslope into meandering rivers. Readers should be aware that in the analysis of ancient fluvial systems it may be extremely difficult to differentiate sediments from such overlapping environments.

Alluvial Fans

Depositional Setting

Alluvial fans are deposits with gross shape approximating a segment of a cone. Modern alluvial fans form in areas of high relief, commonly at the base of a mountain range, where an abundant supply of sediment is available. They are particularly common in sparsely vegetated arid or semiarid regions where sediment transport occurs infrequently but with great violence during sudden cloudbursts. The alluvial fans in Death Valley, California, are good examples of modern arid fans. Individual arid or semiarid fans are typically small, commonly less than about 30 km across. In arid or semiarid settings, alluvial fans may pass downslope into desert-floor environments with internal drainage, including playa lake environments. Alluvial fans occur also in humid regions, including proglacial settings, where they develop as outwash fans in front of melting glaciers. In these humid regions, they may merge downslope with alluvial or deltaic plains, beaches or tidal flats, or even build into lakes (Collinson, 1978a). Examples of modern humid fans include the giant Kosi Fan of Nepal and India, which is about 150 km from apex to toe, and the Scott Fan of Alaska, which is formed as an outwash fan from Scott Glacier. Along mountain fronts, alluvial fans developed in adjacent drainage systems may merge laterally to form an extensive piedmont, or bajada.

Depositional Processes

Streamflow, debris flow, mudflow, and landsliding are all important processes in transport and deposition of alluvial-fan deposits. Streamflow processes are particularly important on humid fans and tend to dominate over other transport processes. Streamflow leads to deposition of three types of fan deposits. **Stream-channel sediments** accumulate within the channels of streams that debouch onto fans. These deposits form long, narrow bodies consisting of the coarsest and most poorly sorted of the streamflow deposits (Nilsen, 1982). **Sheetflood deposits** are formed by surges of sediment-laden water that spread out from the end of a stream channel onto a fan. Deposition occurs because of widening of the flow into shallow bands or sheets, with concurrent decrease in water depth and velocity of flow (Bull, 1972). These processes result in sheetlike deposits of gravel, sand, or silt that tend to be well sorted and may be crossbedded, laminated, or nearly structureless. **Sieve deposits** consist of coarse gravel lobes. They occur on fans where the source supplies relatively little sand, silt, and clay to the fan. Under these conditions, highly permeable gravel deposits are generated; they allow water to pass through rather than over the gravels, holding back only the coarser material in transport. Streamflow in humid regions may be perennial; however, most sediment transport probably occurs as a result of periodic major flood events. As streams debouch onto areas of lower relief, they spread out, become more shallow, and lose their competence and capacity to carry sediment. Thus, sediment loads tend to be dumped quickly, causing stream channels to choke and the stream to shift sideways across the fan.

Debris flows and mudflows are more common on fans in arid or semiarid regions where rainfall is infrequent but violent, slopes are steep, and vegetation sparse. Such flows occur also where abundant unconsolidated volcaniclastic or glacial sediment is available. The fans of arid regions may also include streamflow deposits, which may even be dominant in some arid fans. Streamflow in arid regions is markedly periodic, with intense sediment transport taking place during occasional floods and little transport during intervening periods. **Debris-flow deposits** are characteristically poorly sorted and lacking in sedimentary structures except for possible reverse graded bedding in their basal parts. They may contain blocks of various sizes, including large boulders, and they are typically impermeable and nonporous owing to their high content of muddy matrix. Debris flows commonly "freeze up" and stop flowing after relatively short distances of transport over lower slopes on the fan; however, some flows have been reported to travel distances of up to 20 km (12 mi). **Mudflows** are similar to debris flows but consist mainly of sand-size and finer sediments. **Landslides** are commonly associated with debris flows, and in many cases landslide deposits form a source of sediment for debris flows. Landslides may include rock falls, slumps, slides, and snow avalanches. Material transported by landslide processes can include mud, sand, boulders, blocks of bedrock, soil, vegetation, and an occasional automobile or house. Owing to the similarity of much of the material transported by landslides and debris flows, landslide deposits may be difficult to differentiate from debris-flow deposits in ancient alluvial fans.

Characteristics of Alluvial-Fan Sediments

Geometry. Alluvial fans are cone shaped to arcuate in plan view, with a well-developed system of sinuous to anastomosing distributary channels that cross the fan (Fig. 11.1). The radial surface profile, from fanhead to fantoe, is commonly concave upward, but the cross-fan profile is generally convex upward. In cross section, alluvial fans are typically either wedge shaped or lens shaped. If wedge shaped, the wedge may be thick near the mountain front and thin out away from the mountain. Alternatively, the wedge may be thin near the mountain front and get thicker away from it (Bull, 1972). In longitudinal (radial) cross profile, alluvial fans can be divided into three parts. The **upper fan,** also called the proximal fan or fanhead, has the steepest slope and coarsest sediment. Streamflow in the upper fan tends to be confined to a single channel, which

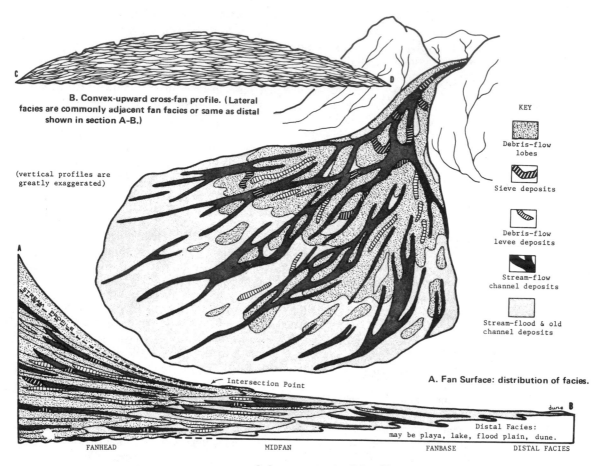

B. Convex-upward cross-fan profile. (Lateral facies are commonly adjacent fan facies or same as distal shown in section A-B.)

KEY

Debris-flow lobes

Sieve deposits

Debris-flow levee deposits

Stream-flow channel deposits

Stream-flood & old channel deposits

(vertical profiles are greatly exaggerated)

STREAM-PROFILE

Intersection Point

A. Fan Surface: distribution of facies.

Distal Facies:
may be playa, lake, flood plain, dune.

FANHEAD   MIDFAN   FANBASE   DISTAL FACIES

C. Concave-upward radial profile.

FIGURE 11.1 Typical surface features and cross-sectional profiles of alluvial fans. A. Fan surface. B. Profile across the fan. C. Longitudinal profile. (After Spearing, D. R., 1974, Summary sheets of sedimentary deposits, Mc-8, Sheet 1, Fig. 1C: Geol. Soc. America.)

may be entrenched as much as 20 to 30 m below the fan surface. Shifting of this channel can occur owing to clogging of the channel with streamflow or debris-flow deposits. The **midfan** is characterized by a more gentle slope and sediment of intermediate size. A branching network of shallower channels typically feeds different parts of the midfan. The **distal fan,** or fanbase, makes up the toe of the fan and is distinguished by the gentlest slopes, finest sediment, and lack of well-defined channels (Fig. 11.1).

Textures, Structures, and Vertical Sequences. The sediments of alluvial fans commonly show strong proximal-distal differences in grain size. The steep upper fan is characterized by coarse, extremely poorly sorted deposits with poorly developed sedimentary structures. These deposits consist largely of coarse-grained, matrix-rich conglomerates, mainly of debris-flow origin. Some clast-supported conglomerates may occur in stream channels. These proximal deposits grade downfan into somewhat finer grained, thinner midfan deposits. Midfan deposits include both extensive streamflow sediments and debris-flow units. Sheetlike deposits of sands and gravels predominate and may display both planar and trough cross-bedding. Coarse-grained conglomerates are commonly present, including both debris-flow and channeled, streamflow conglomerates. Gravels in the streamflow conglomerates commonly show well-developed imbrication with the clasts dipping upfan. Distal fan sediments are largely sand and silt deposits of sheetflood origin, although thin conglomerate layers may be present. Channeled deposits are rare. They tend to be better sorted than midfan and upper-fan deposits. They may show low-angle cross-stratification and, in the more distal part of the fan, trough stratification (Figs. 11.2 and 11.3).

Individual beds in alluvial fans may display no detectable vertical grain-size trends or may become finer or coarser upward. Overall, alluvial-fan deposits tend to

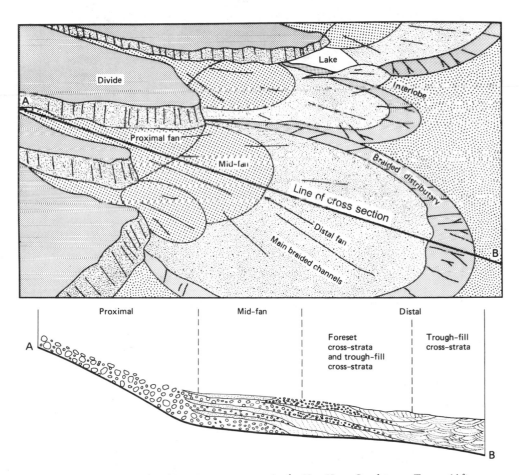

FIGURE 11.2 Facies and sedimentary structures in the Van Horn Sandstone, Texas. (After McGowen, J. H., and C. G. Groat, 1971, Van Horn Sandstone, West Texas: An alluvial fan model for mineral exploration: Texas Bur. Econ. Geology Rept. Inv. 72. Figs. 3, 31; p. 8, 39; reprinted by permission of University of Texas, Austin.)

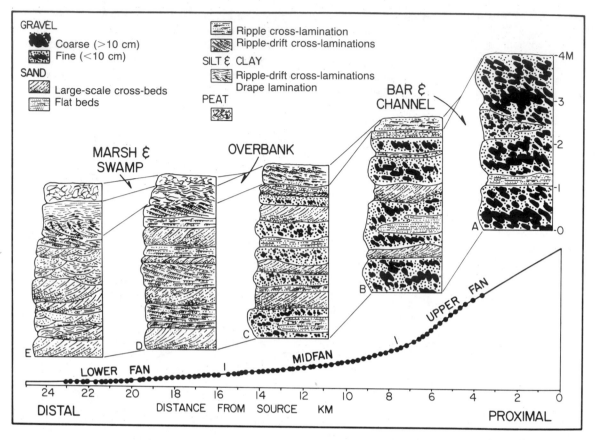

FIGURE 11.3 Downstream variations in facies and sedimentary structures in a glacial outwash fan. Bar and channel sequences do not fine upward, but are capped by a finer overbank facies that becomes more important in a downfan direction. (From Boothroyd, J. C., and G. M. Ashley, 1975, Processes, bar morphology and sedimentary structures on braided outwash fans, northwest Gulf of Alaska, *in* A. V. Joplin and B. C. McDonald (eds.), Glaciofluvial and glaciolacustrine sedimentation: Soc. Econ. Paleontologists and Mineralogists Spec. Pub. 23. Fig. 25, p. 218, reprinted by permission of SEPM, Tulsa, Okla.)

be characterized by strongly developed thickening- and coarsening-upward sequences, caused by active fan progradation or outbuilding. Nonetheless, some fans display thinning- and fining-upward sequences, which are indicative of relative inactivity of depositional processes or fan retrogradation (retreat) (Nilsen, 1982). The thickness of these fining- or coarsening-upward sequences may be hundreds or even thousand of meters. Alluvial-fan deposits grade laterally into nonfan deposits such as fluvial-plain sediments, windblown deposits, or playa-lake sediments.

Summary Characteristics. Alluvial-fan deposits are lobate in plan view and wedge shaped or lenticular in cross section. They are compositionally and texturally immature and show strong downfan decrease in grain size. Sedimentary structures consist mainly of cross-bedding and planar bedding, and paleocurrent patterns are radial from the fan apex. Abrupt facies changes, particularly in the downfan direction, are typical. The principal distinguishing characteristics of alluvial fan facies are summarized in Table 11.1. Alluvial-fan deposits have many characteristics in common with other types of fluvial deposits and ancient alluvial fan deposits may be difficult to differentiate from other fluvial deposits. Humid fans are deposited under very much the same conditions as braided-stream deposits; many, in fact, merge downslope with braided streams.

Ancient Examples. Alluvial fans appear to have been particularly important in Precambrian and early Paleozoic time before land plants were present to provide an adequate vegetation cover to inhibit erosion. Alluvial fans are known also from stratigraphic sequences of other ages, such as the Triassic sediments of the Newark Basin in northeastern North America and the Neogene Ogallala Formation of Texas. They are commonly associated with lake deposits or windblown deposits; the gravels and sands tend to be mineralogically and texturally immature. The Van Horn Sandstone of West Texas is a particularly well-studied example of a Precambrian wet-fan system. McGowen and Groat (1971) identify three facies in this fan system, as shown in Figure 11.2. The proximal fan or upper fan consists of massive conglomerate deposits, with

TABLE 11.1 Principal distinguishing characteristics of alluvial-fan deposits

Characteristic	Description
Texture	Sorting characteristically very poor; great range of grain sizes; clasts poorly rounded, reflecting short distance of transport; rapid downfan decrease in both average and maximum clast size
Composition	Compositionally immature; commonly composed of a wide variety of clast types; mineralogy and clast types dependent upon source rocks, thus composition may change laterally in coalescing fans from different drainage systems
Vertical sequences	Individual beds possibly showing no vertical change in grain size or coarsening or thinning upward; overall, alluvial-fan sequences displaying strong thickening- and coarsening-upward trend or thinning- and fining-upward trend, reflecting either progradation or retrogradation of the fan
Sedimentary structures	Limited suites of sedimentary structures; mainly medium- to large-scale trough or planar cross-bedding and planar stratification; overall, poorly stratified; may display many laterally discontinuous sediment-filled channels and cut-and-fill structures, particularly in upper-fan deposits; paleocurrent patterns radiating outward downfan; complex radiating paleocurrent patterns in coalescing fans
Geometry	Lobate in plan view; lenticular or wedge shaped in cross section; typically forming clastic wedges
Other	Deposits commonly oxidized (red, brown, yellow); contain very little fine organic matter and few fossils except rare vertebrate bones and plant remains; generally composed of a mixture of streamflow and debris-flow deposits—may also contain sieve deposits (gravel lobes) that are unique to alluvial-fan environments

Source: After Bull, W. B., 1972, Recognition of alluvial fan deposits in the stratigraphic record, in J. K. Rigby and W. K. Hamblin (eds.), Recognition of ancient sedimentary environments: Soc. Econ. Paleontologists and Mineralogists Spec. Pub. 16, p. 63–83; Nilsen, T. H., 1982, Alluvial fan deposits, in P. A. Scholle and D. Spearing (eds.), Sandstone depositional environments: Am. Assoc. Petroleum Geologists Mem. 31, p. 49–86.

boulders reaching meter size, representing channel streamflow deposition and debris flows. These coarse deposits grade downfan into alternating beds of conglomerates and thin, cross-stratified, pebbly sandstones, indicating a greater dominance of sheetflow deposition. The distal fan facies is characterized by thinner sedimentation units, an increasing proportion of trough to planar cross-beds, and increasing mudstone content. Relatively few examples of arid or semiarid fans have been reported in the geologic record. The New Red Sandstone of Scotland and the Permian Peranera Formation of Spain, which abuts the Hercynian core of the Pyrenees Mountains, are redbed formations that are cited examples of ancient arid fans.

River Systems

River systems have been studied extensively by geologists, geomorphologists, and engineers. As a result, the processes that form river deposits and the characteristics of these deposits are reasonably well understood. On the basis of stream morphology, four types of rivers are recognized: braided, anastomosing, straight or almost straight, and meandering (Fig. 11.4). Of these types, straight rivers are uncommon, and anastomosing rivers can be regarded as a special type of meandering river that has a relatively permanent and stable system of high-sinuosity channels, with cohesive banks and separated by large, stable, vegetation-covered islands (Miall, 1977). Therefore, braided rivers and meandering rivers are commonly regarded as the two principal types of rivers. Actually, a nearly continuous spectrum of river systems may exist between these distinct end-member types. Both predominantly gravelly and predominantly sandy rivers occur. Sandy systems are most common and best known.

Braided-River Systems

Depositional Setting and Depositional Processes. Braided rivers are distinguished from meandering rivers by their lower sinuosity, which is defined as the ratio of channel length to length of the valley containing the river. They are characterized also by

FIGURE 11.4 Principal types of rivers. (From Miall, A. D., 1977, A review of the braided-river depositional environment: Earth Science Reviews, v. 13. Fig. 1, p. 5, reprinted by permission of Elsevier Science Publishers, Amsterdam.)

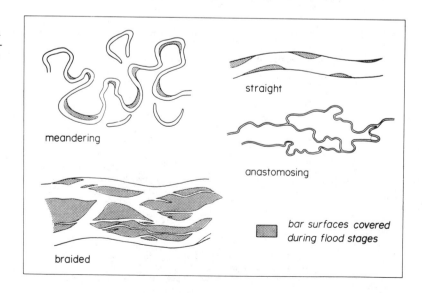

many channels separated by bars or small islands. Braided rivers are best developed in the distal parts of alluvial fans, on glacial outwash plains, and in the mountainous reaches of river systems. In these areas, sediment is abundant, water discharge is high and commonly sporadic, and little vegetation may be present to hinder runoff. Under these conditions, rivers are generally overloaded with sediment, leading to rapid deposition. Braiding apparently takes place because of rapid, large fluctuations in river discharge, abundance of coarse sediment, a high rate of supply of sediment, and easily erodable, noncohesive banks (Cant, 1982). According to Leopold and Wolman (1957), braiding is developed by sorting action as a stream leaves behind those sizes of particles that it is incompetent to handle. Deposition of the coarser bedload causes mid-channel bars to form. Thus, during periods of high discharge the stream channel is rapidly choked with coarse bedload detritus, creating bars around which the discharge is diverted. Repeated bar formation and channel branching generate a braided network of bars and channels over the entire stream bed. Although particles of all sizes can be transported during flood stage of the river, only very fine sediment or no sediment at all moves during periods of lower discharge. Braided rivers tend to have high slopes, or gradients, and are most characteristic of upstream reaches of river systems. In fact, many braided rivers grade or merge downslope into meandering rivers as both the river gradient and the grain size of the sediment load decrease. The processes of deposition in braided rivers are much the same as those that take place on humid fans, particularly in the lower, or distal, part of the fans. Consequently, it is often difficult to distinguish between the facies that form in these two environments.

Bedforms and Structures. Braided rivers are characterized particularly by large bedforms called bars, and several kinds of bars are recognized in modern braided rivers (Fig. 11.5). Although terminology used by different authors for braid bars is somewhat confusing, bars can be grouped into three basic types: (1) longitudinal bars, (2) linguoid and transverse bars, and (3) lateral bars, including point bars and side bars.

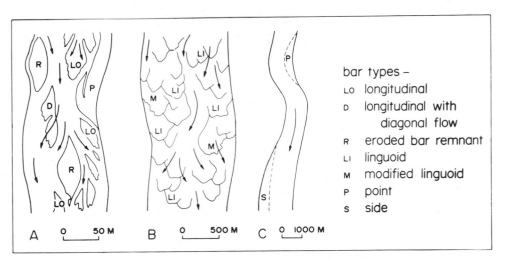

FIGURE 11.5 Principal types of bars in braided rivers. (From Miall, A. D., 1977, Fig. 3, p. 12, reprinted by permission of Elsevier Science Publishers, Amsterdam.)

Longitudinal bars are the midchannel bars that form when the coarsest part of the stream load is deposited as stream flow wanes or other factors cause loss of stream competency. They are oriented with their long axes roughly parallel to current flow. Bars are small when first formed, but continue to grow in length and height as fine particles are trapped in the interstices of the original deposit and as more bedload sediment is deposited downstream in the lee of the bar (Miall, 1977). Coarsest material is concentrated along the central axis and bottom of the bar, and grain size tends to decrease upward and downstream. The internal structure of longitudinal bars is characterized by massive or crude horizontal bedding that may indicate transport and deposition under upper-flow-regime conditions. **Linguoid** and **transverse bars** are oriented transverse to the direction of streamflow and are particularly characteristic of sandy

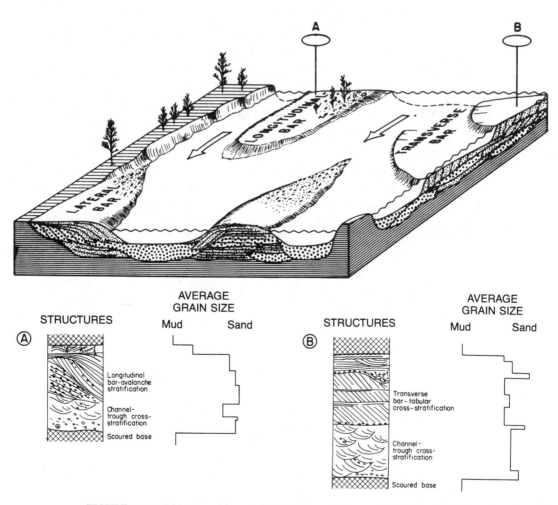

FIGURE 11.6 Structure of bars in braided rivers. Sequence A is dominated by migration of a gravelly longitudinal bar. Sequence B records deposition of successive transverse bar cross-bed sets upon a braid-channel fill. (After Galloway, W. E., and D. K. Hobday, 1983, Terrigenous clastic depositional systems. Fig. 4.4, p. 56, reprinted by permission of Springer-Verlag, Heidelberg.)

braided streams. Linguoid bars are lobate or rhombic in shape, with steep downstream avalanche faces. Transverse bars are similar except that they tend to have straight crests. Linguoid and transverse bars appear to be large ripple forms, that is, sand waves or dunes, that develop under high flood conditions. **Lateral bars** are typically very large bars that develop in areas of relatively lower energy along the sides of the stream channel. They are attached to the bank, as illustrated in Figure 11.6.

Owing to the generally high-flow conditions that prevail in braided rivers, most of the sediment deposited is gravel and sand. Mud is a distinctly subordinate constituent of braided-stream deposits. Longitudinal bars tend to be composed largely of gravels or mixtures of sand and gravels. Linguoid, transverse, and lateral bars are generally more sandy. Gravelly longitudinal bars show crude planar stratification to poorly developed cross-bedding. Cross-bedding is commonly much better developed in sandy units. Linguoid and transverse bars are extensively cross-bedded and both planar and trough cross-bed sets are common. The dip direction of cross-bed foresets is variable, but overall it is unidirectional downstream. Ripple marks are common on the surface of sandy bar deposits. The principal lithofacies and sedimentary structures that characterize braided-river deposits are summarized in Table 11.2. Both gravelly and sandy braided rivers migrate laterally, leaving sheetlike or wedge-shaped deposits

TABLE 11.2 Facies and characteristic structures of braided rivers

Facies identifier	Lithofacies	Sedimentary structures	Interpretation
Gm	Gravel, massive or crudely bedded; minor sand, silt, or clay lenses	Ripple marks, cross-strata in sand units, gravel imbrication	Longitudinal bars, channel-lag deposits
Gt	Gravel, stratified	Broad, shallow trough cross-strata, imbrication	Minor channel fills
Gp	Gravel, stratified	Planar cross-strata	Linguoid bars or deltaic growths from older bar remnants
St	Sand, medium to very coarse; may be pebbly	Solitary or grouped cross-strata	Dunes (lower-flow regime)
Sp	Sand, medium to very coarse; may be pebbly	Solitary or grouped planar cross-strata	Linguoid bars, sand waves (upper- and lower-flow regimes)
Sr	Sand, very fine to coarse	Ripple marks of all types, including climbing ripples	Ripples (lower-flow regime)
Sh	Sand, very fine to very coarse; may be pebbly	Horizontal lamination, parting or streaming lineation	Planar bed flow (lower- and upper-flow regimes)
Ss	Sand, fine to coarse; may be pebbly	Broad, shallow scours (including cross-stratification)	Minor channels or scour hollows
Fl	Sand (very fine), silt, mud, interbedded	Ripple marks, undulatory bedding, bioturbation, plant rootlets, caliche	Deposits of waning floods, overbank deposits
Fm	Mud, silt	Rootlets, desiccation cracks	Drape deposits formed in pools of standing water

Source: Miall, A. D., 1977, Table III, p. 20, reprinted by permission of Elsevier Science Publishers, Amsterdam.

of channel and bar complexes (Cant, 1982). Lateral migration combined with aggrada-tion leads to deposition of sheet sandstones or conglomerates with thin, nonpersistent shales enclosed within coarser sediments (Fig. 11.6B).

Vertical Sequence of Facies. Braided-river deposits are highly variable depending upon the sizes of bedload sediment transported, the depth of the stream channel, and the amount and variability of stream discharge. Some deposits are generated by lateral accretion, as side or point bars develop. Others form by vertical accretion on channel floors and bar tops. Channels may fill by aggradation during waning current flow, and flooding can cause beds formed under decreasing current velocity to be superimposed. These depositional processes generate vertical sequences of facies that either show no distinctive pattern or vertical grain-size change or display a fining-upward trend. Miall (1977) proposed four vertical profile models that can develop under different condi-tions of bedload and discharge (Fig. 11.7). The **Scott-type** model consists mainly of roughly horizontally bedded gravels and minor sand wedges; this model shows poorly developed cycles and reflects deposition in gravelly proximal streams during high river discharges. The **Donjek-type** model consists of fining-upward cycles of variable scale that reflect deposition in braided rivers with mixed bedloads of sand and gravel, where sedimentation may occur at different levels within channels or where channel aggradation is followed by channel shifting. Sandy braided rivers of more steady dis-

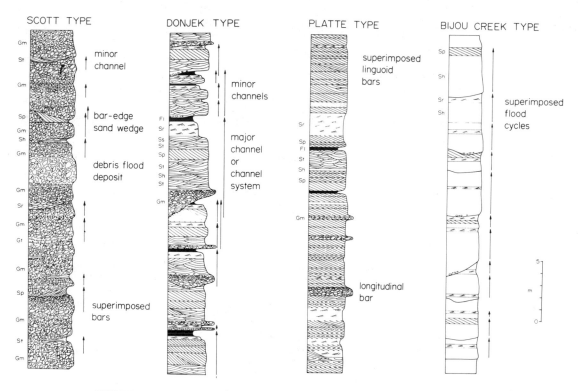

FIGURE 11.7 General stratigraphic models for sandy braided streams. See Table 11.2 for de-scription of individual facies. (From Miall, A. D., 1977, Figs. 12, 14, p. 46, 47, reprinted by permission of Elsevier Science Publishers, Amsterdam.)

charge are dominated by linguoid and transverse bars that generate largely cross-bedded, sandy deposits of the **Platte type** that do not have very distinct cycles, although some fining-upward sequences may be identified. Braided streams with markedly variable discharge, owing to periodic flooding and relatively little topographic differentiation between channels and bars, generate deposits of the **Bijou Creek type** (Cant, 1982), which are characterized by superimposed flood deposits that accumulate during waning current flow. Each flood event is represented by a fining-upward deposit.

Braided-River Deposits in the Geologic Record. Although modern braided rivers are known from many areas, braided-stream deposits appear to be rather poorly represented in the geologic record. Reported examples included the Westwater Canyon Member of the Jurassic Morrison Formation of New Mexico, which is known for its uranium deposits, and the Triassic Ivishak Formation on the Alaskan North Slope. The Ivishak Formation forms a reservoir rock for oil in the Prudhoe Bay Oil Field. A particularly well-studied example of an ancient braided-stream deposit is the Devonian Battery Point Sandstone of Quebec (Cant and Walker, 1976; R. G. Walker and Cant, 1984). A summary sequence for the Battery Point Sandstone is shown in Figure 11.8. This sequence begins with a channel-floor lag deposit lying on a scoured surface (SS) overlain by poorly defined trough cross-bedding (A). These deposits are succeeded upward by in-channel sands that display well-defined trough cross-bedding (B) and

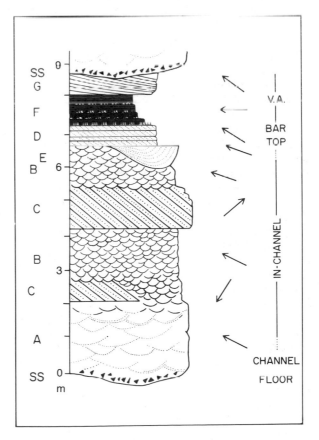

FIGURE 11.8 Summary vertical sequence of braided-stream facies for the Devonian Battery Point Sandstone, Quebec. The sequence was developed using statistical Markov analysis, and preferred facies relationships were drawn as a stratigraphic column using average thicknesses. Arrows show paleoflow directions. The letters indicate facies, which are explained in the text. (After Walker, R. G., and D. J. Cant, 1984, Sandy fluvial systems, in R. G. Walker (ed.), Facies models, 2nd ed.: Geoscience Canada Reprint Ser. 1. Fig. 20, p. 83, reprinted by permission of Geological Association of Canada.)

large sets of planar-tabular cross-bedding (C). Above this are bar-top deposits consisting of small sets of planar-tabular cross-bedding (D) with isolated scour fills (E), overlain by vertical accretion deposits (V.A.) of cross-laminated siltstones interbedded with mudstones (F), and finally low-angle cross-stratified sandstones (G). The sequence displays a generally fining-upward trend. Although the Battery Point Sandstone cannot be considered a general model for braided-river deposits, it illustrates many of the main features of such deposits.

Meandering-River Systems

In contrast to the network of channels that characterize braided rivers, meandering rivers tend to be confined within a single major channel, characterized by cohesive banks that are difficult to erode. Meandering rivers also differ from braided rivers by their much greater sinuosity, lower gradients, and finer sediment load. Many meandering rivers are simply downstream continuations of braided rivers, formed as stream slope and coarseness of bedload decrease and large-scale fluctuations in discharge become less marked. Others occur independently of braided rivers. Large meandering rivers commonly discharge into delta systems. Meandering rivers have been studied extensively, and the depositional model for these rivers is better established and understood than that for braided rivers. The fluvial sedimentology of meandering river systems has been reviewed and summarized by several workers, including J. R. L. Allen (1965a), Cant (1982), Collinson (1978a), Reineck and Singh (1980), and R. G. Walker and Cant (1984).

Depositional Setting and Sedimentation Processes. The morphological elements of the meandering river system are shown in Figure 11.9. These elements consist principally of the main meander channel, point bars that build outward on the inside bend of meander loops, natural levees, floodbasins alongside the levees, and oxbow lakes and abandoned cut-off meanders. Sediments accumulate in different parts of this system owing to channel flow and periodic overbank flooding.

Channel flow within the main meander channel in response to episodes of increased streamflow is responsible for much of the sediment erosion and deposition that takes place within the meandering river system. Periods of increased streamflow occur periodically, often seasonally. During low-flow conditions, the maximum velocity of channel flow swings back and forth across the channel, hugging the bank along the concave part of the meanders (Fig. 11.10A). During high-flow conditions, the current takes a straighter path. The lateral shifting of the current sets up a strong transverse spiral or helical flow that tends to deflect the water from the outer concave bank toward the inner convex bank of the meander (Fig. 11.10B). Because the deflecting force is larger near the surface, where stream velocity is higher than near the bed, the transverse spiral flow moves toward the outer bank near the stream surface and toward the inner bank near the bottom. This helical flow component carries sediment across the stream channel and up the sloping bank of adjacent point bars, where it is deposited under lower velocity conditions. Turbulence as well as current velocity is greatest in the deeper water along the outer bank of the meander, leading to further deepening of the channel and undermining of the bank. Only the coarsest sediment accumulates as a lag deposit in the deeper part of the channel. The remaining sediment eroded from the concave bank of the meander bend is transported laterally across the stream, as the zone of maximum current velocity shifts back and forth from one outside bend

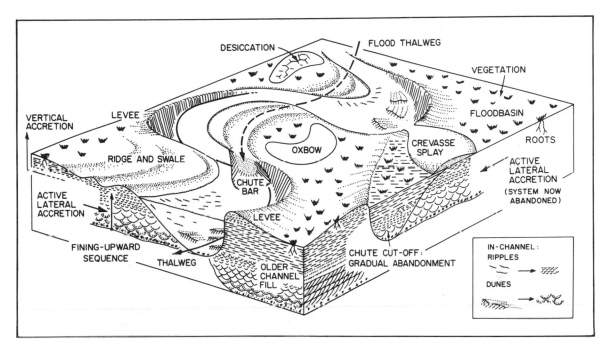

FIGURE 11.9 The morphological elements of a meandering river system. A thalweg is a line connecting the deepest points along a stream channel. It is commonly the line of maximum current velocity. (From Walker, R. G., and D. J. Cant, 1984, Fig. 1, p. 72, reprinted by permission of Geological Association of Canada.)

to the next, and coarser bedload sediment is deposited by lateral accretion on the next downstream point bar (Fig. 11.10B). This process causes lateral and downstream migration of the meanders.

During flood stage, the river becomes bank full, and overbank flooding takes place. Overbank flooding leads to deposition of fine silt and mud on natural levees, adjacent floodbasins, and in oxbow lakes largely by vertical accretion. Natural levees may occasionally become breached, allowing sediment-laden water containing both suspended-load and bedload materials to wash suddenly into flood basins. There the sediment load is rapidly deposited. This process is called **crevasse splay.** Further details of these sedimentation processes are given below in the discussion of meandering-river deposits.

Deposits. Fluvial depositional processes lead to accumulation of sediment in five different settings within the meandering steam system: (1) the main channel, (2) point bars, (3) natural levees, (4) the flood basin, and (5) oxbow lakes and meander (chute) cutoffs. Each of these subenvironments of the system generates deposits with characteristic grain sizes and sedimentary structures.

Channel sediments are primarily lag deposits composed of coarse material that the river can move only at maximum stream velocity during flood stage. These deposits include coarse bedload gravels, together with water-logged plant material and

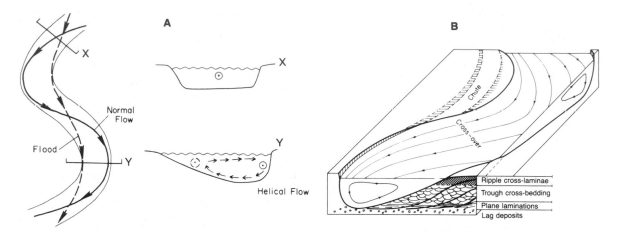

FIGURE 11.10   Flow patterns in a meandering stream. A. Pattern of the thread of maximum velocity as it shifts back and forth from one meander to the next. Cross profiles X and Y indicate differences in transverse current flow in different reaches of the stream. B. Details of helical flow and the nature of deposits formed as a result of such flow. (A after Galloway, W. E., and D. K. Hobday, 1983, Terrigenous clastic depositional systems. Fig. 4.2, p. 53, reprinted by permission of Springer-Verlag, Heidelberg, Germany. B from Blatt, H., G. V. Middleton, and R. Murray, Origin of sedimentary rocks, 2nd ed. © 1980, Fig. 19.5, p. 637. Reprinted by permission of Prentice-Hall, Englewood Cliffs, N.J.)

chunks of partly consolidated mud eroded from the channel wall (R. G. Walker and Cant, 1979). Bedding is indistinct in these coarse materials but imbrication of pebbles and cobbles is common. Channel-lag deposits are typically thin and discontinuous and may be absent altogether in sandy streams that transport little gravel.

At discharge rates lower than those required to move coarse lag gravels, sand is transported over the channel-lag deposits and accreted to the sloping surface of downstream **point bars.** When the stream is bank full of water, helical flow develops; as described above, this flow creates a vertical circulation cell normal to the stream bank and carries bottom water and sediment load up the sloping face of the point bar. Bed shear stress falls when velocities decrease with shallower depth as the upslope component of flow moves upward over the point-bar surface. Thus, coarsest grains tend to be deposited on the lower part of the point bar and finer grains higher up on the bar, leading to a fining-upward point-bar sequence (Fig. 11.9). Large dune bedforms are generated on the lower part of the bar, whereas ripples form on the higher parts of the bar. The preserved sedimentary structures of the point bar thus pass from large-scale trough cross-bedded coarse sands in the lower part of the bar to small-scale trough cross-beds higher on the bar. Cross-bed dip directions may be highly variable, but overall they show unidirectional dip in the downstream direction. Upper-flow-regime, plane-bed flow can be achieved at different heights on the bar depending upon flow velocity; thus, plane-bed parallel laminations may be preserved, interbedded with small-scale or large-scale trough cross-beds (R. G. Walker and Cant, 1979).

When the stream becomes overfull during flood stage and overtops its banks, deposition of fine sediment occurs on natural levees, adjacent flood basins, and in oxbow lakes. Deposition from overbank waters results in upbuilding of the sediment surface and is thus called vertical accretion, in contrast to the lateral accretion that takes place on point bars. **Natural-levee deposits** form primarily on the concave or steep-bank side of meander loops immediately adjacent to the channel (Fig. 11.9) as a result of sudden loss of competence of streams as they overtop their banks. These deposits are thickest and coarsest near the channel bank and become thinner and finer grained toward the flood basin. Sedimentary structures consist of rippled and horizontally stratified fine sands overlain by laminated mud (R. A. Davis, 1983). **Flood-basin deposits** are fine-grained sediments that settle out of suspension from floodwaters carried into the flood basin, which may be a broad, low-relief plain, a swamp, or even a shallow lake. These thin, fine-grained deposits commonly contain considerable plant debris and may be bioturbated by land-dwelling organisms. **Crevasse-splay deposits** may occur on floodplains where rising floodwaters breach natural levees (Fig. 11.9), as mentioned above. Sedimentation from traction and suspension occurs rapidly as water containing both coarse bedload sediment and suspended sediment debouches suddenly onto the plain, resulting in graded deposits that may resemble a Bouma turbidite sequence (R. G. Walker and Cant, 1979). **Oxbow-lake deposits** consist of fine silt and mud introduced into the lakes from the main stream during overbank flooding. They are commonly well laminated and may contain plant remains and the shells of ostracods and freshwater molluscs.

Vertical Sequence of Facies. At any particular time in a meandering stream system, sedimentation may take place essentially simultaneously in the lag channel, on point bars, and in the various overbank environments. As lateral shifting of these different environments takes place owing to stream meandering, sediments from laterally contiguous environments will become superimposed or vertically stacked (Fig. 11.11). As a result of meander migration, coarse lag deposits thus become overlain by sandy, fining-upward point-bar deposits, which themselves are overlain by silty and muddy overbank deposits, producing an overall fining-upward sequence. Thus, the classic model for meandering stream deposits, as proposed by J. R. L. Allen (1970b), consists of a fining-upward sequence that begins with a basal lag-gravel conglomerate lying above an erosion surface. This unit is replaced upward by cross-bedded point-bar sands, which in turn are overlain by fine overbank mud and silt containing root traces, desiccation cracks, and possibly bioturbation structures (Fig. 11.12). The sands that

FIGURE 11.11 Diagrammatic cross section of a point bar showing contiguous environments and the structure of the bar. (From Bernard, H. A., C. F. Major, Jr., B. S. Parrott, and J. J. LeBlanc, Sr., 1970, Recent sediments of southeast Texas—A field guide to the Brazos alluvial and deltaic plains and the Galveston barrier island complex: Texas Bur. Econ. Geology Guidebook 11; described by W. F. Fisher and L. F. Brown, 1972, Clastic depositional systems—A genetic approach to facies analysis, annotated outline and bibliography: Texas Bur. Econ. Geology. Reprinted by permission of The University of Texas, Austin.)

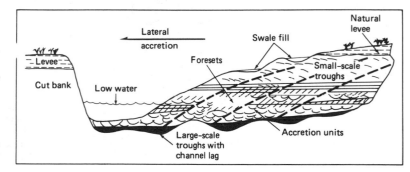

form the basal unit decrease in size upward and display trough cross-bedding with upward reduction in cross-bed set size. Parallel laminations may be interbedded with the cross-bedded units at various levels. Although Allen's fining-upward model can be utilized as a very useful norm to which fluvial deposits may be compared, Collinson (1978a) points out that variations from this norm can occur in some meandering stream deposits. Complications caused by deposition when discharge is less than bank full, flow patterns that diverge from the helicoidal modal, or deposition on coarse, gravelly point bars may result in point-bar deposits in which vertical grain-size changes may not be clear or in which grain size may even coarsen upward. Also, the distribution of bedforms and sedimentary structures may be much less ordered than in the ideal model, as seen, for example, in the development of ripples rather than dunes in the deeper part of the channel.

Ancient Examples. Meandering stream deposits are moderately common in the geologic record, particularly in rocks of Devonian and younger ages. Many examples from the United States are known, including Paleozoic sequences of the Appalachian region—the Devonian Catskill Formation, for example—and parts of the Jurassic Morrison Formation, the Cretaceous Mesaverde Group, and the Tertiary Wasatch and Fort Union formations of the Colorado Plateau. The lower part of the Devonian Old Red

FIGURE 11.12 Classic fining-upward sequence in a meandering river deposit. (After Walker, R. G., and D. J. Cant, 1984, Fig. 2, p. 73, reprinted by permission of Geological Association of Canada. Modified from J. R. L. Allen, 1970, Studies in fluviatile sedimentation: A comparison of fining-upwards cyclothems, with special reference to coarse-member composition and interpretation: Jour. Sedimentary Petrology, v. 40, Fig. 12, p. 331.)

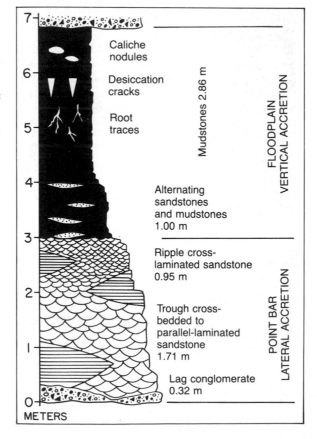

Sandstone of Wales and England is a particularly famous and well-studied European example of a meandering stream deposit. The characteristics of this sandstone sequence largely formed the basis for J. R. L. Allen's (1970b) model of fining-upward facies shown in Figure 11.12. This formation has been studied particularly carefully by Allen (1964, 1965b), who described several detailed sequences of the Lower Old Red Sandstone facies from different localities in southern Wales and southeastern England. A generalized stratigraphic section of the sequence measured at Abergavenny is shown in Figure 11.13. The sequence begins with a basal unit of conglomerates and sandstones, lying on a scoured surface, interpreted as channel-fill and channel-lag deposits. This unit is overlain by ripple cross-stratified and planar cross-bedded sandstones that probably represent mixed channel fill deposits and lateral accretion deposits on point bars. The upper part of the sequence consists of coarse siltstones containing invertebrate burrows and thin lenses of ripple-bedded fine sandstones. These deposits are interpreted as overbank flood deposits. The overall sequence is thought to represent the deposits of a shallow, small meandering stream subject to periodic, intense flooding.

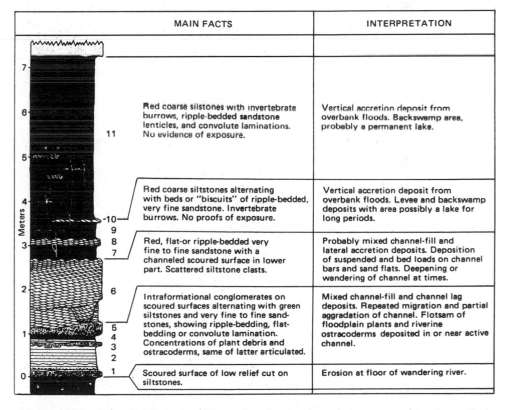

FIGURE 11.13 Generalized stratigraphic section showing meandering stream deposits at Abergavenny, Wales. Channel-lag, lateral-accretion, and vertical-accretion deposits form a fining-upward sequence. (From Allen, J. R. L., 1964, Studies in fluviatile sedimentation: Six cyclothems from the Lower Old Red Sandstone, Anglo-Welsh Basin: Sedimentology, v. 3. Fig. 11, p. 184, reprinted by permission of Elsevier Science Publishers, Amsterdam.)

Geometry of River Deposits

Owing to extensive lateral channel migration and channel aggradation, braided rivers produce sheet sandstones or conglomerates containing thin beds or lenses of shales enclosed within the thicker sediments, as pointed out (Fig. 11.14). By contrast, meandering streams, which are confined within the rather narrow, sandy meander belt of stream floodplains, generate linear "shoestring" sand bodies oriented parallel to the river course (Cant, 1982). These shoestring sands are surrounded by finer grained, overbank floodplain sediments (Fig. 11.14). As the river aggrades its channel and builds up above the floodplain, a natural levee may eventually break at some point during flood stage, causing abandonment of the old channel and creation of a new channel, a process called **avulsion.** Successive episodes of avulsion can lead to formation of several linear sand bodies within a major stream valley (Fig. 11.14).

Recognizing Ancient Fluvial Deposits

There are no unequivocal criteria for recognition of ancient fluvial deposits. In general, they are recognized by total absence of marine fossils, relatively poor sorting, red color (in some cases), unidirectional paleocurrent patterns, and a generally strong downstream decrease in particle size. Meandering-river deposits are further characterized by distinct fining-upward grain-size trends. Braided-stream deposits may display either no detectable vertical size trends or a fining-upward trend. Alluvial-fan deposits may also show no vertical grain-size trend or they may display either a fining-upward or coarsening-upward trend. Because of many similarities in the characteristics of alluvial-fan, braided-river, and meandering-river deposits, it may be difficult to differentiate these deposits in ancient sedimentary sequences. Also, fluvial deposits may be difficult to differentiate from glacial-outwash deposits and delta-plain deposits.

Ancient fluvial deposits are common in the geologic record, although not exceptionally abundant, and are found in many parts of the world in strata of virtually all ages from Precambrian to Quaternary. In the United States, they appear to be particularly common in late Paleozoic strata of the Appalachian region, where they have been reported to make up as much as 65 percent of the sandstones in the central Appalachian folded Paleozoic sequence (Pettijohn et al., 1973).

11.3 EOLIAN DESERT SYSTEMS

Introduction

Deserts cover broad areas of the world today, particularly within the latitudinal belts of about 10 to 30 degrees north and south of the equator, where dry, descending air masses create prevailing wind systems that sweep toward the equator. Deserts are areas in which potential rates of evaporation greatly exceed rates of precipitation. They cover about 20 to 25 percent of the present land surface. Owing to their generally low rainfall, commonly less than about 25 cm/yr, we tend to think of deserts as extremely dry areas dominated by wind activity and covered by sand. In reality, a variety of subenvironments exist within deserts, including alluvial fans; ephemeral streams that run intermittently in response to occasional rains; ephemeral saline lakes, also called playas or inland sabkhas; sand-dune fields; interdune areas covered by sediments, bare rocks, or deflation pavement; and areas around the fringes of deserts where wind-

FIGURE 11.14   Contrasting geometry of meandering (A) and braided (B) rivers. (From Walker, R. G., and D. J. Cant, 1984, Fig. 9, p. 77, reprinted by permission of Geological Association of Canada.)

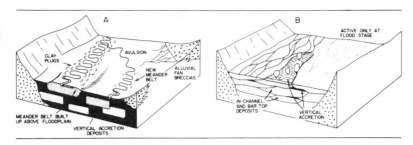

blown dust (loess) accumulates. Large areas of the desert environment may indeed be carpeted by windblown, or eolian, sand. Such areas, called **sand seas** or **ergs,** cover about 20 percent of modern deserts. The remaining areas are covered by eroding mountains, rocky areas, and desert flats. The largest desert in the world, the Sahara (7 million km²), contains several ergs arranged in belts. The larger belts cover areas as extensive as 500,000 km² (R. G. Walker and Middleton, 1979). The deposits of deserts have been studied extensively; several recent monographs and research papers describe these deposits in detail. Readers may find the publications of Ahlbrandt and Fryberger (1982), Bigarella (1972), Brookfield (1984), Brookfield and Ahlbrandt (1983), Collinson, (1978b), Glennie (1970), Hunter (1977), McKee (1979a), Reineck and Singh (1980), R. G. Walker and Middleton (1979), and I. G. Wilson (1972) particularly interesting.

Depositional Processes

Most deserts are characterized by extreme fluctuations in temperature and wind, on both a daily and a seasonal basis. Rainfall rates are low, as mentioned, and the rains are very sporadic. Vegetation is generally extremely sparse. When rains do come, they tend, owing to the lack of vegetative cover, to create flash floods. Rainwater typically drains toward the centers of desert basins, where playas or inland sabkhas may develop and become sites of deposition for carbonate and evaporite minerals. Because periodic rains create flash floods and ephemeral streams and mobilize debris flows and mudflows, they are extremely important agents in sediment transport in deserts. Nonetheless, much of the time water plays a relatively small role in sediment transport in deserts. Most of the time, wind is the dominant agent of sediment transport and deposition. Wind is much less effective than water as an agent of erosion, but is an extremely effective medium of transport for loose sand and finer sediment. It not only accounts for the transport of vast quantities of siliciclastic sand in deserts, but is also responsible for sediment transport in glacial environments, on river floodplains, and along many coastal areas, where both carbonate and siliciclastic sands may be transported inland. The windblown, or eolian, deposits of these environments are quite small compared to the sand seas of desert areas. Wind storms, or dust storms, may also carry silt and clay far from their sources and are responsible for transport of much of the pelagic sediment to deep ocean basins.

Wind transports sediment in much the same way as water, separating the sediment into three transport populations: traction, saltation, and suspension. Wind effectively separates sediment finer than about 0.05 mm from coarser sediment and transports this sediment long distances in suspension. Except at unusually high wind velocities, coarser sediment travels by traction and saltation close to the ground. Saltation is a particularly important mode of wind transport, aided by downslope creep of grains owing to the impact of saltating grains as they strike the bed. Wind appears to be especially effective in transport of medium to fine sand and finer sediment, but coarse particles (up to 2 mm or somewhat larger) may also undergo transport by rolling and surface creep under high-velocity winds. Eolian sediment transport has been investigated by many workers; however, Bagnold's (1954) study dealing with the physics of blown sand remains the classic piece of research in this field. The transporting and sorting action of wind tends to produce three kinds of deposits: (1) dust deposits, sometimes referred to as loess, that commonly accumulate far from the source; (2) sand deposits; and (3) lag deposits, consisting of gravel-size particles too large to be transported by wind and that form a deflation pavement. Wind transport and deposition generate many of the same kinds of bedforms and sedimentary structures—including ripples, dunes, and cross-beds—as those produced by water transport. The processes involved in the transport and deposition of sand to produce sand dunes are much the same as those described in the discussion of bedforms in Chapter 6.

Deposits of Modern Deserts

The various environments of deserts mentioned above can be grouped into three main desert subenvironments: dune, interdune, and sand sheet (Ahlbrandt and Fryberger, 1982). The dune environment is primarily the site of wind transport and deposition of sand, which accumulates in a variety of dune forms, many of which have steeply dipping slip faces or avalanche faces. Interdune areas can receive both windblown sediment and sediment transported and deposited by ephemeral streams in stream floodplains or playa lakes. The sand-sheet environment exists around the margins of dune fields. The deposits of this environment form a transitional facies between dune and interdune deposits and deposits of other environments.

Dunes

Many types of dunes occur in modern deserts, ranging from those with no slip faces to those with three or more slip faces (Fig. 11.15). Eolian bedforms range in scale from small ripples to transverse and longitudinal dunes 0.1–100 m high to complex pyramidal dunes, called **draas,** with heights of 20–450 m (I. G. Wilson, 1972). **Barchans, barchanoid ridges,** and **transverse dunes** form under the influence of unidirectional winds. They have single slip faces and appear to represent a gradational series corresponding to an increase in sand supply (McKee, 1979b). **Parabolic** and **blowout dunes** have one or more slip faces. They are related to the preceding group of dunes, but their development is controlled by vegetation cover. **Dome dunes** are circular in plan view and have no definite slip faces. They may originate by modification of barchanoid dunes by strong winds. **Linear dunes,** also called seif, or longitudinal, dunes, have nearly symmetrical ridges, and **reversing dunes** have asymmetrical ridges; both types have two slip faces (Fig. 11.15). Linear dunes form in areas of uniform sand accumulation under generally high-velocity winds of variable directions. Reversing dunes, in which slip faces form on opposite sides at different times, result from a close balance between opposed winds (McKee, 1979c). **Star dunes,** also called draas, are huge stellate dunes up to 450 m high, with a high central peak and radiating arms and commonly with superposed complex or compound dunes (Fig. 11.15). They have three or more slip faces and apparently form under intense, multidirectional wind systems in areas of high sand-drift potential (Fryberger and Dean, 1979).

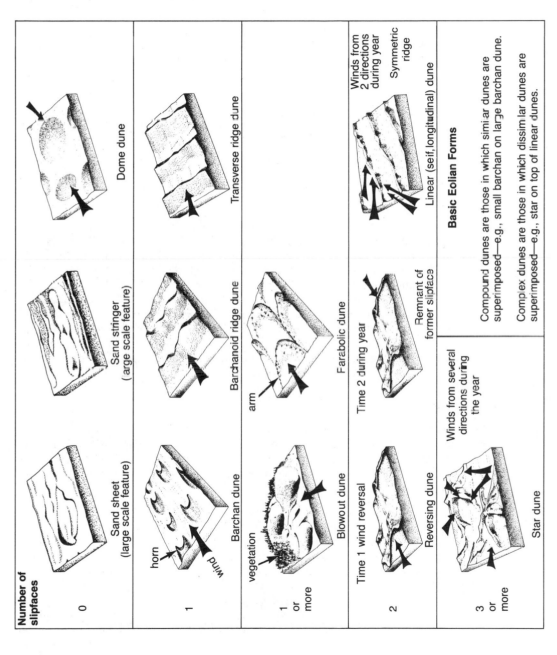

FIGURE 11.15 Basic eolian dune forms grouped by number of slipfaces. [After Ahlbrandt, T. S., and S. G. Fryberger, 1982, Introduction to eolian deposits, in P. A. Scholle and D. Spearing (eds.), Sandstone depositional environments: Am. Assoc. Petroleum Geologists Mem. 31. Fig. 3, p. 14, reprinted by permission of AAPG, Tulsa, Okla.)

Dune deposits commonly consist of texturally mature sands that are well sorted and well rounded; however, considerable textural variation can occur. They are also typically quartz rich, although many coastal dune deposits contain high concentrations of heavy minerals and unstable rock fragments. Coastal dunes in some tropical areas may consist largely of oolites, skeletal fragments, or other carbonate grains; dunes composed of gypsum occur in some desert areas, such as White Sands, New Mexico. Eolian dunes are characterized particularly by the presence of cross-bedding (Fig. 11.16). McKee (1979b) suggests that the following sedimentary structures are common to most types of eolian dunes:

1. Sets of medium- to large-scale cross-strata that typically consist of foresets dipping in the direction of wind transport (leeward) at angles of repose ranging as high as 30–34 degrees
2. Sets of tabular-planar cross-strata that in vertical section tend to be progressively thinned from the base upward
3. Bounding planes between individual sets of cross-strata that are mainly horizontal or dip leeward at low angles

In addition to these large-scale, gross structural features of dune deposits, several kinds of small-scale internal structures may be present. Hunter (1977) groups the sedimentary structures in thin strata of eolian dunes into six classes: plane-bed laminae, rippleform laminae, ripple-foreset cross-laminae, climbing translatent strata, grainfall laminae, and sandflow cross-strata (Table 11.3).

Owing to the variety of dune types that can form under different wind conditions, local paleocurrent vectors derived from eolian cross-bed data can range from unimodal to polymodal. Paleocurrent data may thus show a high degree of scatter that complicates calculation of ancient prevailing sediment-transport directions. On a regional scale, eolian paleocurrent patterns may swing around over hundreds of miles around high-pressure wind systems (Selley, 1978).

Interdunes

Interdune areas occur between dunes and are bounded by dunes or other eolian deposits such as sand sheets. Interdunes may be either deflationary (erosional) or depositional. Very little sediment accumulates in most deflationary interdunes except coarse, granule-size lag sediments that may show rippled surfaces and inverse grading.

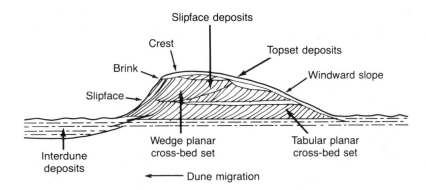

FIGURE 11.16 Typical geometry and internal structure of a barchanoid dune ridge. (From Ahlbrandt, T. S., and S. G. Fryberger, 1980, Eolian deposits in the Nebraska Sand Hills: U.S. Geol. Survey Prof. Paper 1120A, Fig. 7, p. 9.)

TABLE 11.3 Basic types of stratification in eolian deposits

Depositional process	Character of depositional surface	Type of stratification	Dip angle	Thickness of strata / Sharpness of contacts	Segregation of grain types / Size grading	Packing	Form of strata
Tractional deposition	Rippled	Subcritically climbing translatent stratification	Stratification: low (typically 0–20°, maximum ~30°). Depositional surfaces: similarly low	Thin (typically 1–10 mm, maximum ~5 cm). Sharp, erosional	Distinct. Inverse	Close	Tabular, planar
	Rippled	Supercritically climbing translatent stratification	Stratification: variable 0–30°. Depositional surface: intermed. (10–25°)	Intermediate (typically 5–15 mm). Gradational	Distinct. Inverse except in contact zones	Close	Tabular, commonly curved
	Rippled	Ripple-foreset cross-lamination	Relative to translatent stratification: intermed. (5–20°)	Individual laminae: thin (typically 1–3 mm). Sharp or gradational, nonerosional		Close	Tabular, concave-up or sigmoidal
	Rippled	Rippleform lamination	Generalized: intermediate (typically 10–25°)		Individual laminae and sets of laminae: indistinct. Normal and inverse, neither greatly predominating	Close	Very tabular, wavy
	Smooth	Plane-bed lamination	Low (typically 0–15° max.?)	Sets of laminae: intermediate (typically 1–10 cm). Sharp or gradational, nonerosional		Close	Very tabular, planar
Largely grainfall deposition	Smooth	Grainfall lamination	Intermediate (typically 20–30°, min. 0°, max. ~40°)			Intermediate	Very tabular, follows preexistent topography
Grainflow deposition	Marked by avalanches	Sandflow cross-stratification	High (angle of repose) (typically 28–34°)	Thick (typically 2–5 cm). Sharp, erosional or nonerosional	Distinct to indistinct. Inverse except near toe	Open	Cone-shaped, tongue-shaped, or roughly tabular

Source: Hunter, R.E., 1977, Basic types of stratification in small eolian dunes: Sedimentology. v. 24. Table 1, p. 364, reprinted by permission of Elsevier Science Publishers. Amsterdam.

Deflationary interdunes are preserved in the rock record as a disconformity overlain by thin, discontinuous winnowed lag deposits. Sediments deposited in depositional interdunes can include both subaqueous and subaerial deposits, depending upon whether they are deposited in wet, dry, or evaporite interdunes (Ahlbrandt and Fryberger, 1981). All interdune deposits are characterized by low-angle stratification (<~10 degrees), although many deposits may be almost structureless owing to secondary processes, largely bioturbation, that destroy stratification.

Dry interdunes or ephemerally wetted interdunes are most common. Deposits in dry interdunes are generated by ripple-related wind-transport processes, grainfall in the wind shadow in the lee of dunes, or sand flow (avalanching) from adjacent dunes. The deposits tend to be relatively coarse, bimodal, and poorly sorted, with gently dipping, poorly laminated layers. They are also commonly extensively bioturbated by both animals and plants. Wet interdune areas are the sites of lakes or ponds where silts and clays are trapped by semipermanent standing bodies of water rather than being deflated and removed. These sediments may contain freshwater organisms such as gastropods, pelecypods, diatoms, and ostracods. They are also commonly bioturbated and may contain vertebrate footprints. Some wet interdune sediments become contorted owing to loading by dune sediments. Evaporite interdunes, or inland sabkhas, occur where drying up of shallow ephemeral lakes or evaporation of damp surfaces causes precipitation of carbonate minerals, gypsum, or anhydrite. Growth of carbonate minerals or gypsum in sandy sediment tends to disrupt and modify primary depositional features. Desiccation cracks, raindrop imprints, evaporite layers, and pseudomorphs may characterize these sediments (Galloway and Hobday, 1983).

Sheet Sands

Sheet sands are flat to gently undulating bodies of sand that commonly surround dune fields (Fig. 11.17). They are typically characterized by low to moderately dipping (0–20 degrees) cross-stratification and may be interbedded in some parts with ephemeral stream deposits. Sheet-sand deposits may also contain gently dipping, curved, or irregular surfaces of erosion several meters in length; abundant bioturbation traces formed by insects and plants; small-scale cut-and-fill structures; gently dipping, poorly laminated layers resulting from adjacent grainfall deposition; discontinuous thin layers of coarse sand intercalated with fine sand; and occasional intercalations of high-angle eolian deposits (Ahlbrandt and Fryberger, 1982).

Ancient Dune Deposits

Stratification

The most striking features of modern sandy deserts are various bedforms ranging from ripples to dunes to draas; however, these bedforms are not commonly preserved in ancient eolian deposits. Instead, we see only cross-bedding and other internal features that remain as a record of bedform migration across ancient deserts. Furthermore, the preserved remains commonly represent only the lowest parts of the original eolian bedform. Brookfield (1984) suggests that migration of eolian bedforms leads to generation of three types of bounding surfaces within ancient dune deposits (Fig. 11.18). **First-order surfaces** are flat-lying bedding planes that cut across all other eolian structures. These surfaces are believed to be caused by the migration of very large bedforms

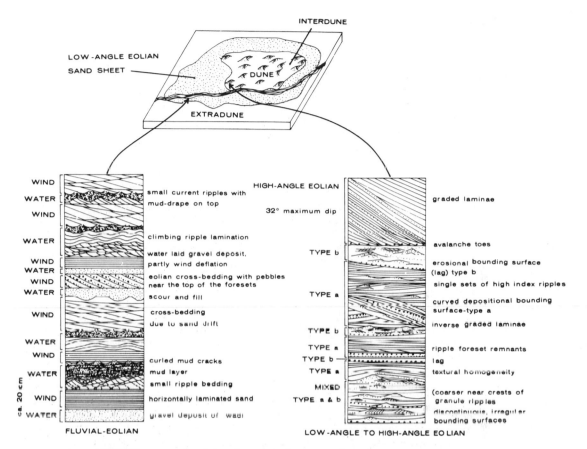

FIGURE 11.17 Areal distribution and stratigraphic relationships of sheet sands and eolian dune sands. Type a and type b beds are both low-angle cross-laminated units. Type a beds are generally finer grained and better sorted than type b beds, which commonly contain isolated horizons of coarse-grained sediment. (From Fryberger, S. G., T. S. Ahlbrandt, and S. Andrews, 1979, Origin, sedimentary features, and significance of low-angle eolian "sand sheet" deposits, Great Sand Dunes National Monument and vicinity, Colorado: Jour. Sedimentary Petrology, v. 49. Fig. 12, p. 745, reprinted by permission of Society of Economic Paleontologists and Mineralogists, Tulsa, Okla.)

(draas) across the desert. **Second-order surfaces** lie at an angle between first-order surfaces and dip downwind. They are apparently caused by migration of dunes down the lee slopes or slip faces of draas or by lateral migration of longitudinal dunes across the lee slope. **Third-order surfaces** form bounding surfaces for bundles of laminae within cosets of cross-laminae. They are analogous to the reactivation surfaces in tidal bedforms in that they are attributed to erosion followed by renewed deposition owing to fluctuations in wind direction and velocity.

The nature of the internal cross-stratification and bounding surfaces preserved in ancient eolian deposits thus depends upon the types of dune forms that migrated across the ancient desert floors and the migration patterns of these dunes. Figure 11.19

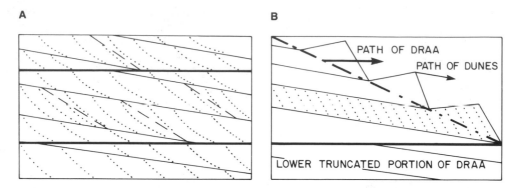

FIGURE 11.18   Relationship and origin of bounding surfaces in dune deposits. A. A synthetic section of ancient eolian sandstone parallel to bedform migration. Heavy lines = first-order bounding surfaces; thin lines = second-order bounding surfaces; dashed lines = third-order bounding surfaces; dotted lines = eolian lamination. B. Origin of bounding surfaces by dunes climbing down draa lee slopes. Bulk of draa is truncated by migration of succeeding draa. (From Brookfield, M. E., 1984, Eolian sands, *in* R. G. Walker (ed.), Facies models, 2nd ed.: Geoscience Canada Reprint Ser. 1. Fig. 8, p. 97, reprinted by permission of Geological Association of Canada.)

shows theoretical or hypothetical stratification models for different types of dunes developed by unidirectional climbing and migration of ideal bedforms. Depending upon local conditions and migration patterns, the actual stratification of ancient eolian deposits may deviate from these ideal forms.

Vertical Sequences

Owing to the unpredictable distribution of dune types in deserts, it has commonly been assumed that eolian deposits do not display a preferred vertical sequence of sedimentary structures or grain-size patterns or any consistent lateral change (R. G. Walker and Middleton, 1979). Brookfield (1984) suggests, however, that large-scale vertical sequences of desert deposits can develop that may reflect the growth and decay of the desert environment during slow environmental change. Figure 11.20 shows a vertical sequence of structures and facies in a Permian deposit that can be interpreted to have been generated during a desert cycle that began with the onset of arid conditions followed by gradual change to a semiarid climate. Thus, the sequence begins with small-scale cross-bed units, representing onset of desert conditions; passes upward through large-scale cross-bed units, generated during the main eolian phase, to small-scale eolian cross-beds and alluvial-fan deposits formed during waning desert conditions; and culminates in alluvial fans and desert-floor fluvial deposits that signal the end of desert conditions. This model may not, of course, be applicable to all ancient eolian deposits, but it illustrates the point that facies models for ancient eolian deposits are possible.

FIGURE 11.19 (opposite)   Stratification of simple and complex/compound dunes. Transverse sections are perpendicular to resultant wind direction; longitudinal sections are parallel to resultant wind direction. (From Brookfield, M. E., 1984, Fig. 7, p. 96, reprinted by permission of Geological Association of Canada.)

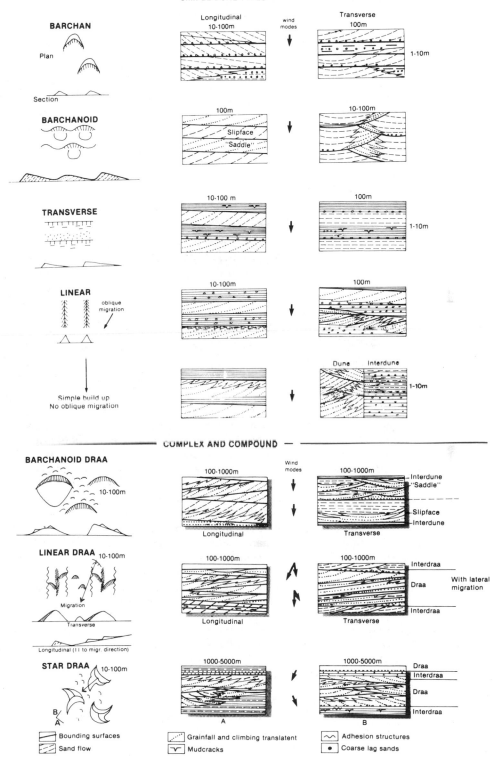

SIMPLE DUNE TYPES

BARCHAN

Plan

Section

Longitudinal
10-100m

wind
modes

Transverse
100m

1-10m

BARCHANOID

100m

Slipface

"Saddle"

10-100m

TRANSVERSE

10-100 m

100m

1-10m

LINEAR

oblique
migration

10-100m

100m

Simple build up
No oblique migration

Dune   Interdune

1-10m

COMPLEX AND COMPOUND

BARCHANOID DRAA

10-100m

Wind
modes

100-1000m

Longitudinal

100-1000m

Interdune
"Saddle"

Slipface
Interdune

Transverse

LINEAR DRAA   10-100m

Migration

Transverse

Longitudinal (11 to migr. direction)

100-1000m

Longitudinal

100-1000m

Interdraa

Draa

With lateral
migration

Interdraa

Transverse

STAR DRAA   10-100m

B

A

1000-5000m

A

1000-5000m

Draa

Interdraa

Draa

Interdraa

B

Bounding surfaces

Sand flow

Grainfall and climbing translatent

Mudcracks

Adhesion structures

Coarse lag sands

369

FIGURE 11.20 Example of vertical sequence of eolian deposits. Section through the Lower Permian deposits of the Thornhill and Dumfries intermontane basins. (From Brookfield, M. E., 1984, Fig. 15, p. 101, reprinted by permission of Geological Association of Canada.)

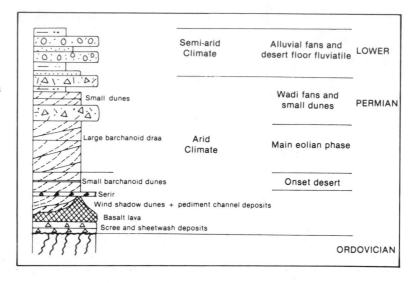

Summary Characteristics

Geologists tend to characterize ancient eolian deposits as sedimentary units composed predominantly of fine-size, well-sorted, well-rounded, commonly frosted, quartz sand grains. Eolian strata are believed to typically display large-scale, high-angle, planar-tabular or planar-wedge cross-bedding and may preserve tracks and trails of desert fauna. Although these characteristics probably are typical of many dune deposits, the preceding discussion shows that eolian deposits as a whole can be much more complex. They can include interdune and sheet-sand deposits, with different structural and textural properties than dune deposits, and may also include noneolian sediments such as ephemeral stream deposits. Therefore, the simple facies model that depicts eolian deposits as high-angle cross-bedded, lithologically homogeneous units must be expanded to include these more complex, lithologically heterogeneous sediments. Troublesome problems still exist in differentiating eolian deposits from some water-deposited sediments. Identification of eolian deposits can be difficult and equivocal, and the eolian origin of some presumed ancient sandstones is controversial. As a case in point, the Jurassic Navajo Sandstone of the Colorado Plateau has often been cited as a classic example of an ancient eolian sandstone. Yet, the Navajo has been reinterpreted by some recent workers to be a water-laid deposit, and its origin is still being debated. Future recognition of eolian deposits may have to rely more on detailed study of styles of stratification and cross-bedding, particularly study of smaller scale structures such as those proposed by Hunter (1977).

Examples of Ancient Eolian Deposits

Ancient sandstones interpreted to be windblown deposits have been described from sedimentary sequences as old as the Precambrian from many parts of the world. Particularly noteworthy are the Jurassic Navajo Sandstone, as mentioned, and the Jurassic

Entrada Sandstone, Triassic Wingate Sandstone, Permian White Rim Sandstone, Permian Coconino Sandstone, and Permian Lyons Sandstone of the Colorado Plateau. All of these sandstones are characterized by compositional and textural maturity and well-developed, generally high-angle cross-stratification. Whether or not all of these reported examples are actually eolian sandstones is an open question. The Lyons Sandstone in Colorado is a particularly well-studied example (T. R. Walker and Harms, 1972) that appears to be definitely of eolian origin. This formation consists of quartz-rich, well-sorted sandstone with well-developed cross-stratification. The cross-stratification sets range from about 3 to 13 m, with dip angles of the foresets ranging from about 25 to 28 degrees. Ripples located on the lee slopes of the steeply inclined cross-beds have high ripple indices. Sedimentary structures suggesting subaerial exposure include raindrop imprints and the tracks of reptiles. Avalanche bedding, which typically forms on the lee slopes of dunes, and thin lag deposits that may represent wind deflation pavements are also present.

Examples from other continents include the Permian Rotliegendes of northwestern Europe, the Jurassic–Cretaceous Botucatu Formation of the Parana Basin, Brazil, and the Permian Lower Bunter Sandstone of Great Britain. The characteristics of the upper Rotliegendes, summarized by Glennie (1972), illustrate well the various kinds of eolian and noneolian sediments that can occur in desert environments (Fig. 11.21). Note in the lower part of the section shown in Figure 11.21 the interbedding of cross-bedded sandstones, interpreted as eolian deposits, with pebbly conglomerates and conglomeratic sandstones, interpreted as wadi deposits. A wadi is an intermittent or ephemeral stream. In the upper part of the section, cross-bedded sandstones give way to mud-cracked clays, silts, and sands containing anhydrite nodules, indicating deposition in an inland sabkha. These deposits are overlain by evaporite and red clay deposits that formed in a desert lake. Marine evaporites overlie the lake-bed deposits, indicating the end of desert conditions and encroachment of a marine basin.

11.4 LACUSTRINE SYSTEMS

Introduction

Lakes form about 1 percent of the present continental surface of Earth. Because the world's continents are presently in a higher state of emergence than was typical of much of Phanerozoic time, lake sedimentation is more prevalent today than it was during much of the geologic past. In fact, ancient lake sediments appear to be of only minor importance volumetrically in the overall stratigraphic record, although they have been reported in stratigraphic sequences ranging in age from Precambrian to Holocene. Even though lake sediments are not abundant in the geologic record, they are nonetheless important because their chemistry is sensitive to climatic conditions, making them useful indicators of past climates (Collinson, 1978c). Also, some lake deposits contain economically significant quantities of oil shales, evaporite minerals, coal, uranium, or iron. Many lake sediments also contain abundant fine organic matter that may act after burial as a source material for petroleum. Lacustrine sedimentation has not been studied overall as thoroughly as fluvial and eolian sedimentation, but the better known lacustrine deposits have been examined in detail and their characteristics are reasonably well known. Useful descriptions and summaries of lake deposits

FIGURE 11.21 Generalized stratigraphic section from the Rotliegendes of northwestern Europe, south-central part of Rotliegendes basin, showing interbedded eolian and noneolian desert deposits. (After Glennie, K. W., 1972, Permian Rotliegendes of northwest Europe interpreted in light of modern desert sedimentation studies: Am. Assoc. Petroleum Geologists Bull., v. 56. Fig. 9, p. 1055, reprinted by permission of AAPG, Tulsa, Okla.)

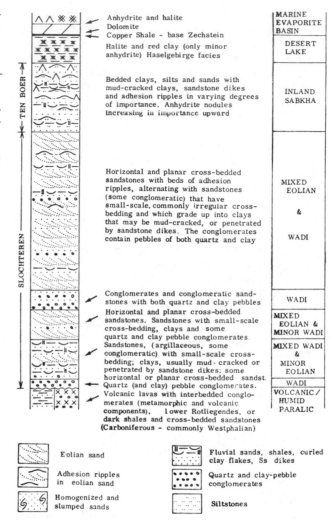

Anhydrite and halite
Dolomite
Copper Shale - base Zechstein
Halite and red clay (only minor anhydrite) Haselgebirge facies

Bedded clays, silts and sands with mud-cracked clays, sandstone dikes and adhesion ripples in varying degrees of importance. Anhydrite nodules increasing in importance upward

Horizontal and planar cross-bedded sandstones with beds of adhesion ripples, alternating with sandstones (some conglomeratic) that have small-scale, commonly irregular cross-bedding and which grade up into clays that may be mud-cracked, or penetrated by sandstone dikes. The conglomerates contain pebbles of both quartz and clay

Conglomerates and conglomeratic sandstones with both quartz and clay pebbles
Horizontal and planar cross-bedded sandstones. Sandstones with small-scale cross-bedding, clays and some quartz and clay pebble conglomerates.
Sandstones, (argillaceous, some conglomeratic) with small-scale cross-bedding; clays, usually mud-cracked or penetrated by sandstone dikes; some horizontal or planar cross-bedded sandst.
Quartz (and clay) pebble conglomerates.
Volcanic lavas with interbedded conglomerates (metamorphic and volcanic components), lower Rotliegendes, or dark shales and cross-bedded sandstones (**Carboniferous** - commonly Westphalian)

TEN BOER
SLOCHTEREN

MARINE EVAPORITE BASIN
DESERT LAKE
INLAND SABKHA
MIXED EOLIAN & WADI
WADI
MIXED EOLIAN & MINOR WADI
MIXED WADI & MINOR EOLIAN
WADI
VOLCANIC / HUMID PARALIC

Eolian sand

Adhesion ripples in eolian sand

Homogenized and slumped sands

Fluvial sands, shales, curled clay flakes, Ss dikes

Quartz and clay-pebble conglomerates

Siltstones

may be found in Collinson (1978c), Fouch and Dean (1982), Lerman (1978), Matter and Tucker (1978), Picard and High (1972), Picard and High (1981), and Reeves (1968).

Origin and Size of Lakes

The basins or depressions in which lakes form can be created by a variety of mechanisms, including tectonic movements such as faulting and rifting; glacial processes such as ice scouring, ice damming, and moraine damming; landslides or other mass movements; volcanic activity such as lava damming or crater explosion; deflation by wind scour or damming by windblown sand; and fluvial activity such as the formation of oxbow lakes and levee lakes. Most existing lakes appear to have originated directly or indirectly by glacial processes (Picard and High, 1981) and thus may not be typical of ancient lakes, which formed predominantly by tectonic processes.

Modern lakes range in areal dimensions from a few tens of square meters to tens of thousands of square kilometers. The largest modern lake is the saline, inland Caspian Sea, with a surface area of 436,000 km^2 (Van der Leeden, 1975). Other large lakes with surface areas ranging between 50,000 and 100,000 km^2 include Lake Superior, Lake Huron, and Lake Michigan in the United States, Lake Victoria in South Africa, and Lake Aral in the USSR. Water depths of modern lakes range from a few meters in small ponds to more than 1700 m in the world's deepest lake, Lake Baikal, USSR. Water depth and surface area are not necessarily related; thus, some of the largest lakes have very shallow depths and vice versa. For example, Lake Victoria has a surface area of 68,000 km^2 but a maximum depth of only 79 m, whereas Crater Lake, Oregon, with a surface area of about 52 km^2, has a maximum depth exceeding 580 m. Preserved lacustrine sediments show that ancient lakes also ranged in size from small ponds to large bodies of water exceeding 100,000 km^2. The largest ancient lake recognized is the Late Triassic Popo Agie lake of Wyoming and Utah, which had a minimum areal extent, based on the preserved sediment record, of 130,000 km^2 (Picard and High, 1981). Reported thickness of preserved ancient lake sediments ranges from less than 20 m to as much as 4000 m.

Sedimentation Processes in Lakes

Modern lakes occur in a variety of environmental settings, including glaciated inland plains and mountain valleys, nonglaciated inland plains and mountain regions, deserts, and coastal plains. The lakes exist under a spectrum of climatic conditions ranging from very hot to very cold and from highly arid to very humid. Most lakes are filled with freshwater, but others, such as the Caspian Sea and many lakes in arid regions—Great Salt Lake, Utah, for example—are highly saline. Many lakes are associated with other types of depositional systems, notably glacial, fluvial, eolian, and deltaic systems. The depositional processes that occur in lakes are influenced both by climatic conditions and by a variety of physical, chemical, and biological processes that take place in the lakes and their associated depositional systems.

Influence of Climate. Climatic factors affect lake sedimentation in numerous ways. First, water level in lakes is maintained by the balance between evaporation and precipitation. Also, chemical sedimentation in lakes strongly reflects climatic conditions. For example, chemical sedimentation in lakes of arid regions is dominated by precipitation of gypsum, halite, and various other salts. In humid climates, chemical sedimentation is dominated by carbonate deposition. Sediment input to lakes is influenced by the vegetation cover in the drainage area of the lakes and is greatest in arid regions with low vegetation cover. In cold climates, seasonal drops in temperature lead to freezing of lakes, causing decrease in sediment input and cessation of wave activity, allowing deposition of fine suspended sediment during these quiet-water conditions. Climate and the physiography of lake settings also determine the local weather conditions over lakes. Severe, localized storms with high winds can cause considerable shore erosion, coupled with sediment transport and deposition, during short periods of time.

Physical Processes. The physical processes that interact in lakes to bring about sediment transport and deposition include wind, river inflow, atmospheric heating, surface barometric pressure, and gravity (Fig. 11.22). Surface barometric pressure and

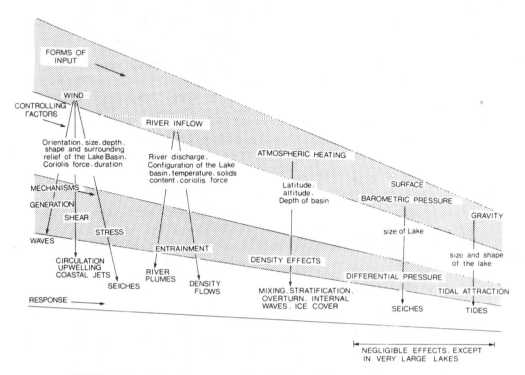

FIGURE 11.22 Lake response to various forms of physical input. (From Sly, P. G., 1978, Sedimentation processes in lakes, *in* A. Lerman (ed.), Lakes: Chemistry, geology, physics. Fig. 1, p. 68, reprinted by permission of Springer-Verlag, Heidelberg.)

gravity effects are of least importance (Sly, 1978). Except in very large lakes, tides commonly play only a relatively minor role in lake processes. Wind processes are of major importance because winds create waves and currents. River inflow may generate plumes of fine sediment that extend in surface waters far out into a lake, or it may generate density underflows, or turbidity currents, that carry sediment along the bottom toward the basin center. River inflow can also create currents that flow along the margins of lakes. Other currents may be generated by flowthrough of water along the lake bottom toward a point of lake discharge. A more extended discussion of interaction of river inflow into standing bodies of water is given in Chapter 12 under the heading of deltaic sedimentation. Atmospheric heating, which is a function of climate, is responsible for density differences in lake water. These differences can cause stratification or, under some conditions, generation of density currents that produce mixing and lake overturn. Also, temperature variations may cause alternate freezing and melting of lake surface waters, thereby affecting sediment transport within the lake.

Thus, a variety of sediment-transport and depositional mechanisms operate in lakes. Deposition of siliciclastic sediment in the calmer and deeper portions of lakes can take place by settling of fine particles that were suspended in the water column by wave and current activity, or deposition may occur from turbidity currents generated where sediment-laden streams discharge into lakes. Sedimentation can occur also along the shallow shorelines of lakes by the action of wind-generated traction currents or river-inflow currents deflected along the lake margins.

The physical depositional processes that take place in lakes are analogous in many ways to those that occur in marine environments and may include turbidite, deltaic, and beach sedimentation. One type of sedimentation process that appears to be particularly characteristic of cold-climate lakes is the formation of varves, which are very thin, alternating light- and dark-colored sediment layers. Varves are presumably generated in so-called glacial lakes owing to seasonal freezing and melting. These seasonal changes cause periodic fluctuations in sediment input to the lakes by rivers. The fluctuations are reflected in differences in the thickness and grain size of laminae deposited during winter and summer. Varves are reported to form also in nonglacial lakes (Picard and High, 1981) owing, at least in some lakes, to seasonal variations in carbonate production. Greater carbonate production during the summer tends to mask fine organic matter, which accumulates very slowly throughout the year. Thus, an annual couplet is formed consisting of a very thin organic layer produced during the winter and a thicker, lighter colored layer of carbonate sediment produced during the summer. Fine clastic sediments produced during spring floods may mix with the upper part of an organic layer, also creating varves. Sly (1978) points out that physical sedimentary processes in lakes differ from those in marine environments in three other major aspects: (1) The small size of most lakes greatly limits the generation of long-period wind waves; thus, the energy level of all but the very largest lakes is well below that of marine systems. This means that deposition of coarse sands and gravels is confined mainly to the shallow-water areas of lakes. (2) Lakes are nearly closed systems with respect to sediment transport. The ratio between the area of land drainage and lake area is commonly high, and sedimentation rates in lakes are much higher than in marine environments, generally at least ten times that of marine environments. (3) Lakes are almost tideless. Tidal currents are of negligible importance in sediment transport, and littoral zones are absent or much reduced compared to marine littoral zones.

Physicochemical Processes. Although siliciclastic sedimentation is dominant in lakes of humid regions, physicochemical precipitation of various minerals may also be important. Deposition of carbonates is most common; phosphates, sulfides, cherts, and iron and manganese oxides have also been reported in a number of lakes. In arid regions, deposition of evaporite minerals is a particularly important type of lake sedimentation. The evaporite deposits of lakes include many common marine evaporite minerals such as gypsum, anhydrite, halite, and sylvite, but also include several minerals such as trona, borax, epsomite, and bloedite that are not common in marine evaporites (Chapter 8). The general conditions and processes affecting the physicochemical precipitation of evaporite and nonevaporite minerals are discussed in Chapter 4. Readers should be aware, however, that differences may exist in chemical sedimentation processes in freshwater and in seawater. For example, owing to the low concentrations of magnesium in freshwater, calcite tends to be the preferred species of carbonate mineral precipitated in freshwater lakes. By contrast, aragonite precipitates preferentially in seawater. A useful discussion of carbonate deposition in lakes is given by Kelts and Hsü (1978), and a more general discussion of the mineralogy and related chemistry of lake sediments is covered by Jones and Bowser (1978). Depositional processes in saline lakes are reviewed by Eugster and Hardie (1978).

Organic Processes. Organisms play an important role in lake sedimentation through (1) extraction of chemical elements from lake water to build shells, or tests, and subsequent deposition of these shells, (2) CO_2 assimilation during photosynthesis, (3) bio-turbation, and (4) contribution of plant remains to form plant deposits. Many kinds of organisms live in lakes and contribute their skeletal and nonskeletal remains to lake sediments. Siliceous diatoms are particularly widespread and noteworthy. Diatoms carry out photosynthesis and are the only important type of lake organisms that produce siliceous tests. Their remains form important diatomite deposits in many Pleistocene lakes of North America. Pelecypods, gastropods, calcareous algae, and ostracods also abound in many lakes and are important contributors of calcium carbonate sediments to lakes. Blue-green algae (cyanobacteria) carry on photosynthesis and also trap fine sediment to form stromatolites. Many different types of higher plants live in lakes. Under the reducing conditions and high sedimentation rates that exist in some lakes, the remains of higher plants may be partially preserved to eventually form peat and coal. Considering the small size of many lakes and their lower alkalinity, or lower buffering capacity, compared to that of the open ocean, the assimilation of CO_2 by plants during photosynthesis is probably a much more important factor in controlling the pH of lakes than of the ocean. Finally, organisms such as pelecypods, freshwater shrimp, and worms may burrow and rework lake sediments, destroying laminations and other primary sedimentary structures.

Characteristics of Lacustrine Deposits

The preceding discussion shows that lakes vary greatly in size. Also, depositional processes in lakes are highly variable, depending upon climate, physical characteristics of the lake, and the influence of associated depositional systems such as deltas. Owing to these factors, the preserved deposits of lakes may display highly variable characteristics. Lacustrine deposits may thus include siliciclastic mudstones, sandstones, or conglomerates; chemically or biochemically deposited sediments such as evaporites, carbonates, and phosphorites; and biogenically formed sediments such as diatomites, skeletal carbonates, and peat deposits. All of these sediment types can be formed in other environments; therefore, lake sediments may resemble other kinds of nonmarine deposits or some kinds of shallow-marine sediments. For this reason, lacustrine deposits can be difficult to differentiate from other deposits, and no single criterion is generally adequate to make such a distinction. The principal distinguishing characteristics of lacustrine deposits are described in two excellent papers by Picard and High (1972, 1981), and much of the following discussion is based on these descriptions.

Size and Geometry. The areal extent of recognized ancient lacustrine deposits ranges from a few hundred square kilometers to a maximum of about 130,000 km^2. Thickness of lacustrine deposits ranges from a few meters to more than 1000 m (R. A. Davis, 1983); however, most deposits are less than 300 m thick. Lacustrine deposits are roughly circular in plan view and tend to be lens shaped in cross section. Wedge-shaped deposits also may form in some tectonic basins.

Sedimentary Structures. Lake sediments may contain numerous types of sedimentary structures, including laminated bedding, varves, stromatolites, cross-bedding, ripple marks, parting lineations, graded bedding, groove casts, load casts, soft-sediment deformation structures, burrows and worm trails, raindrop impressions, mudcracks, and vertebrate footprints. Many of these structures are illustrated by Fouch and Dean (1982) in a sequence of excellent color photographs. Most sedimentary structures that characterize lake deposits occur also in the deposits of other environments. Varves are believed to form mainly in lakes and are one of the more diagnostic characteristics of lake sediments, but light and dark laminae resembling varves have been reported in nonlacustrine sediments. Another distinguishing characteristic of lake sediments is that lake beds tend to be thin and laterally continuous compared to those of associated fluvial deposits. Otherwise, no uniquely diagnostic structures occur in lake sediments. The sedimentary structures of lacustrine deposits are particularly difficult to differentiate from those of shallow-marine sediments.

Fossils. Plant life is commonly abundant in shallow water around lake margins, and plant deposits may become important during the late stages of lake filling. Fossil plant spores and pollen are common constituents of ancient lacustrine sediments. Various types of invertebrate organisms also live in lakes, mostly at water depths less than about 10 m (Reineck and Singh, 1980). Neither the shells of these organisms nor their bioturbation structures are common in deeper water. The principal types of invertebrate remains in lacustrine sediments include bivalves, ostracods, gastropods, diatoms, and charophytes and other algae. Stromatolites produced by blue-green algae are also common in lake deposits, and aquatic and terrestrial vertebrate remains may occur.

Unfortunately for the purpose of environmental interpretation, many of these fossil organisms are not unique to lakes. Freshwater lake faunas may be difficult to distinguish from stream faunas (Picard and High, 1972), and freshwater faunas can differ from faunas of saline lakes. Saline-lake faunas, in turn, may be difficult to distinguish from some marine faunas.

Lateral Facies Patterns. The lateral facies patterns displayed by lake sediments differ for siliciclastic facies and "chemical" facies. The facies of siliciclastic sediments are ideally arranged in concentric belts that grade basinward from gravels at the lakeshore through sand, sandy marly mud, to mud (Fig. 11.23). This is the pattern that might be expected if numerous sediment sources existed around a lake. If only a few sediment sources, or a single source, existed at the time of deposition, or some special depositional conditions prevailed, the lateral facies may deviate greatly from this ideal pattern. They may, for example, be markedly irregular or asymmetric. The so-called chemical facies of carbonate lakes may display two contrasting patterns of deposition (Picard and High, 1981). One pattern shows decreasing carbonate content toward the center of the lake owing to nearshore mixing of siliciclastic sediments. The second

FIGURE 11.23 Ideal lacustrine facies pattern. Such ideal patterns are not found in most ancient lake sediments. (After Picard, M. D., and L. R. High, 1972, Criteria for recognizing lacustrine rocks, *in* J. K. Rigby and W. K. Hamblin (eds.), Recognition of ancient sedimentary environments: Soc. Econ. Paleontologists and Mineralogists Spec. Pub. 16. Fig. 8, p. 115, reprinted by permission of SEPM, Tulsa, Okla.)

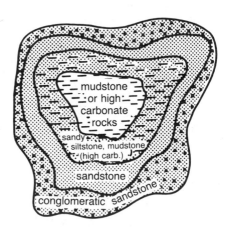

pattern shows high carbonate concentrations near the lake margins and decreasing content inward, presumably owing to greater carbonate productivity in shallow water. In saline lakes, shoreline facies are likely to be muddy, rather than sandy or gravelly, because low water-inflow rates in arid regions lead to reduced clastic input. As the lake dries up with increased evaporation, salts are deposited. The least soluble salts precipitate around the lake margin and the most soluble precipitate in the center of the lake basin.

Vertical Sequences. Owing to high sedimentation rates in lakes and the fact that they are closed systems with respect to sediment transport, all lakes are ephemeral features. Lake basins eventually fill with sediment, and most are converted into fluvial plains as they are overrun by fluvial systems. Therefore, lake filling is commonly regarded as a regressive process. That is, coarser nearshore sediments are believed to gradually encroach on finer lake basin sediments and in turn to be covered with fluvial sediments. This postulated process of filling theoretically generates coarsening-upward sequences of lake facies, as illustrated in the "ideal" lacustrine sequences shown in Figure 11.24. Detailed studies of ancient lake sediments show, however, that most sequences of lake sediments are far more complex than these ideal models. They commonly display evidence of multiple transgressive-regressive cycles, indicating more than one episode of lake development and filling, which may be related to climate or tectonism. Although the ultimate filling of lakes and encroachment by prograding fluvial or other coarser grained deposits may generate a gross coarsening-upward pattern of facies, the ideal sequences illustrated in Figure 11.24 probably rarely occur, except perhaps in some very small lakes.

FIGURE 11.24 Postulated ideal vertical sequence in lacustrine deposits. Such sequences are abstractions and are rarely found in ancient lake deposits. (From Picard, M. D., and L. E. High, 1981, Physical stratigraphy of ancient lacustrine deposits, *in* F. G. Ethridge and R. M. Flores (eds.), Recent and ancient nonmarine depositional environments: Soc. Econ. Paleontologists and Mineralogists Spec. Pub. 31. Fig. 14, p. 247, reprinted by permission of SEPM, Tulsa, Okla.)

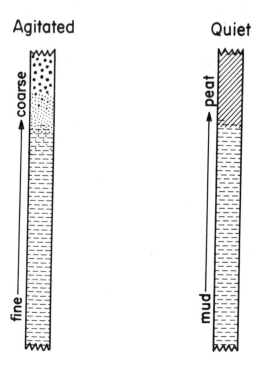

Ancient Lake Deposits

Ancient lake sediments may be difficult to differentiate from some other types of sediments, particularly shallow-marine deposits. Freshwater fossils, vertebrate tracks, and desiccation structures in lake-margin sediments can help establish the deposits as nonmarine, although faunas from saline lakes may be difficult to differentiate from marine faunas. The presence of varves appears to be diagnostic, but most sedimentary structures of lake sediments form by processes that can occur also in other environments. Lake sediments are commonly better sorted than fluvial sediments and may display a general tendency toward fining upward and inward toward the basin center. They occur in association with other continental sediments, particularly fluvial and periglacial sediments (described in Section 11.5). Lake deposits are predominantly fine-grained sediments, either siliciclastic muds or carbonate sediments and evaporites. Sandstones and conglomerates also may be present, but they are distinctly subordinate to fine-grained sediments.

Many ancient lake deposits are characterized by cyclic sequences of facies that result from climatic and tectonic changes. These cycles may include glacial varves, nonglacial varves, transgressive-regressive cycles owing to alternate expansion and shrinking of the lake, and longer cycles represented by bundles of varves or other cycles (Picard and High, 1981). The deposits of the Lockatong Formation of the Triassic Newark Group of New Jersey are a good example of cyclic development of lake sediments. The entire formation is characterized by small-scale cycles or sequences, a few meters thick, that can be traced for distances of a kilometer or so. Van Houten (1964) grouped these cycles into two major types: detrital cycles and chemical cycles (Fig. 11.25). The detrital cycles are generally coarsening-upward sequences, 4–6 m thick, that begin with pyritic black mudstone. The pyritic mudstones pass upward through interlaminated dolomitic mudstones to massive dolomitic mudstones, commonly mudcracked and bioturbated. Fine-grained, cross-stratified sandstones with convolutions may cap the cycles. The cycles are interpreted as small-scale regressive units that formed during lake shrinkage following deepening of the lake. The chemical cycles are more abundant. They grade upward from dark, laminated mudstone with dolomite and calcite laminae and lenses into more massive dolomite or analcime-rich (zeolite-rich) mudstone, characterized by mudcracks and small-scale syneresis cracks. The chemical cycles are interpreted to result from progressive reduction of the lake area and subsequent shallowing as siliciclastic input is reduced. Van Houten (1964) interpreted the entire Lockatong Formation as the deposits of a perennial lake that was thermally stratified most of the time. The climate was warm, with cyclic rainfall that brought about changing water levels, leading to cyclic sedimentation.

Many other ancient deposits interpreted as lacustrine sediments have been reported from rocks of various ages. Some of the better studied lacustrine deposits in North America include the Eocene Green River Formation of Utah, Colorado, and Wyoming, known for its oil-shale deposits; much of the Jurassic Morrison Formation of the Colorado Plateau; parts of the Triassic Chugwater Group of Wyoming; and the Devonian Escuminac Formation of southern Quebec. Well-known lake deposits from other parts of the world include Middle Devonian sediments from the Old Red Sandstone of the Orcadian Basin of northeast Scotland; parts of the Lower Permian Rotliegendes deposits of southwest and eastern Germany; clastics, evaporites and carbonates of the Triassic Keuper Marl of South Wales; and the Permo-Triassic Beaufort strata of the Eastern Karoo Basin, Natal, South Africa. Space does not allow discussion of these

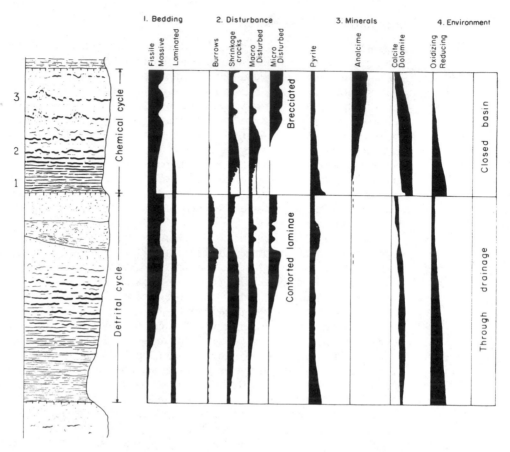

FIGURE 11.25 Depositional model for terrigenous (detrital) and chemical cycles in sediments of the Lockatong Formation, New Jersey, interpreted as lake deposits. (After Van Houten, F. B., 1964, Cyclic lacustrine sedimentation, Upper Triassic Lockatong Formation, central New Jersey and adjacent Pennsylvania, in D. F. Merriam (ed.), Symposium on cyclic sedimentation: Kansas Geol. Survey Bull. 169, v. 11. Fig. 6, p. 505, reprinted by permission of Kansas Geological Survey, Lawrence.)

various lake deposits. Readers are referred to Collinson (1978c), R. A. Davis (1983), and Picard and High (1972, 1981) for summaries of their characteristics.

11.5 GLACIAL SYSTEMS

Introduction

I have placed glacial systems last in this discussion of continental environments because the glacial environment, in a broad sense, is a composite environment that includes fluvial, eolian, and lacustrine environments. It may also include parts of the shallow-marine environment. Glacial deposits make up only a relatively minor part of the rock record as a whole, although glaciation was locally important at several times in the geologic past, particularly during the Late Precambrian, Late Ordovician, Permian, and Pleistocene. Because of their minor abundance and because they are generally regarded to have little economic potential, glacial deposits have not caught the attention of petroleum geologists or other economic geologists to the same extent as other continental deposits. Glaciers presently cover about 10 percent of Earth's surface, mainly at high latitudes. They exist primarily as large ice masses on Greenland and Antarctica and smaller masses on Iceland, Baffin Island, and Spitsbergen. Small mountain glaciers occur at high elevations in all latitudes. By contrast to their present distribution, ice sheets covered about 30 percent of Earth during maximum expansion of glaciers in the Pleistocene and extended into much lower latitudes and elevations than those currently affected by continental glaciation.

The glacial environment is confined specifically to those areas where more or less permanent accumulations of snow and ice exist above the snow line—the elevation above which snow does not melt in summer. Environments temporarily occupied by ice during colder periods of the year are not considered glacial environments. Glaciers form above the snow line by accumulation of snow. They move downslope below the snow line only if rates of accumulation of snow above the snow line exceed rates of melting of ice below. The factors affecting glacier movement and the mechanisms of ice flow are not of primary interest here, and readers are referred to Flint (1971), Sugden and John (1976), and J. Shaw (1985) for details of these processes. Our primary concerns are the sediment-transport and depositional processes associated with glacial movement and melting and the sediments deposited by glaciers. The symposium volume *Glaciofluvial and Glaciolacustrine Sedimentation* (Jopling and McDonald, 1975) is an important source of information on glacial sedimentation processes and sediments in proglacial environments. Other sources of information on glacial sedimentation include Ashley et al. (1985), Easterbrook (1982), Edwards (1978), Eyles and Miall (1984; also contains an extensive reference list of recent papers dealing with glacial facies), Goldthwait (1971, 1975), Leggett (1976), Molnia (1983), Sugden and John (1976), and A. E. Wright and Mosley (1975).

Environmental Setting

The glacial environment proper is defined as all those areas in direct contact with glacial ice. It is divided into (1) the **basal** or **subglacial zone,** influenced by contact with the bed; (2) the **supraglacial zone,** which is the upper surface of the glacier; (3) the **ice-contact zone** around the margin of the glacier; and (4) the **englacial zone** within the glacier interior (Edwards, 1978). Important depositional environments around the margins of the glacier are influenced by melting ice but are not in direct contact with the ice. These environments make up the **proglacial environment,** which includes glaciofluvial, glaciolacustrine, and glaciomarine (where glaciers extend into the ocean) settings (Fig. 11.26). The area extending beyond and overlapping the proglacial environment is the **periglacial environment.**

The basal zone of a glacier is characterized by erosion and plucking of the underlying bed. Debris removed by erosion is incorporated into the bed of the glacier. This debris causes increased friction with the bed as the glacier moves and thus aids in abrasion and erosion of the bed. The supraglacial and ice-contact zones are zones of melting or ablation where englacial debris carried by the glacier accumulates as the

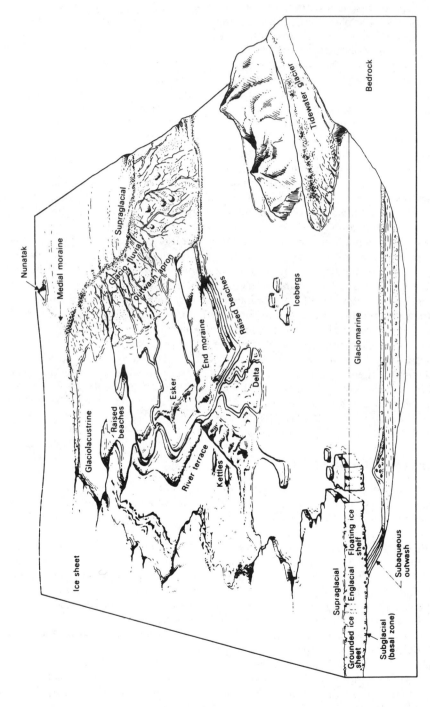

FIGURE 11.26 Glacial and associated proglacial environments. (From Edwards, M. B., 1978, Glacial environments, in H. G. Reading (ed.), Sedimentary environments and facies. Fig. 13.2, p. 418, reprinted by permission of Elsevier Science Publishers, Amsterdam.)

Labels in figure: Nunatak; Medial moraine; Supraglacial; Glacio fluvial; Outwash apron; Ice sheet; Glaciolacustrine; Raised beaches; Esker; River terrace; Kettles; End moraine; Raised beaches; Delta; Icebergs; Tidewater glacier; Bedrock; Glaciomarine; Supraglacial; Englacial; Floating ice shelf; Grounded ice sheet; Subglacial (basal zone); Subaqueous outwash

glacier melts. The glaciofluvial environment is situated downslope from the glacier front and is characterized by fluctuating meltwater flow and abundant coarse englacial debris that is available for fluvial transport. The glaciofluvial environment is one of the characteristic environments in which braided streams develop. Extensive outwash plains, or aprons, may be present along the margins of outwash glaciers (Flint, 1971). Lakes are very common proglacial features, created by ice damming or damming by glacially deposited sediments. Meltwater streams draining into these lakes may create large coarse-grained deltas along the lake edge, while finer sediment is carried outward in the lake by suspension or as a density underflow. Glaciers that extend out to sea create an important environment of glaciomarine sedimentation where sediments are deposited close to shore by melting of the glacier in contact with the ocean or farther out on the shelf or slope by melting of ice blocks, or icebergs.

The glacial environment may range in size from very small to very large. **Valley glaciers** are relatively small ice masses confined within valley walls of a mountain. **Piedmont glaciers** are larger masses or sheets of ice formed at the bases of mountain fronts where mountain glaciers have debouched from several valleys and coalesced. **Ice sheets,** or continental glaciers, are huge sheets of ice that spread over large continental areas or plateaus.

Transport and Deposition in Glacial Environments

Glaciers acquire their sediment load by abrasion and plucking of material from the bedrock and adjacent valley walls and by free fall of rock debris from the valley walls above. Much of this material is carried in the base of the glacier, but other rock debris is concentrated along the glacier margin or dispersed within the body of the glacier (see Fig. 11.27). Large and small blocks of rock can be quarried by moving ice and incorporated into the base or sides of the glacier. Extremely fine sediment, called **rock flour,** is produced by grinding of the rock-studded glacier base over bedrock. Thus, the glacier sediment load typically consists of an extremely heterogeneous assortment of particles ranging from clay-size to meter-size boulders.

As glaciers move downslope below the snow line, they eventually reach an elevation where the rate of melting at the front of the glacier equals or exceeds the rate of new snow accumulation above the snow line. If the rate of melting approximately equals the rate of accumulation, the glacier achieves a state of equilibrium in which it neither advances nor retreats. Within such an equilibrium glacier, internal movement of ice continues to carry along the rock load and supply rock debris to the melting snout of the glacier. This process causes a ridge of unsorted sediment, called an **end moraine,** or terminal moraine, to accumulate in front of the glacier. **Lateral moraines,** or marginal moraines, can accumulate from concentrations of debris carried along the edges of the glacier where ice is in contact with the valley wall. Medial moraines may form where the lateral moraines of two glaciers join (Fig. 11.27). When the rate of melting at the snout of a glacier exceeds the rate of new snow accumulation above the snow line, the glacier retreats back up the valley. If a glacier retreats steadily, it drops its load of rock debris as lateral moraines, medial moraines, and a more or less evenly distributed sheet of **ground moraine.** If it retreats in pulses, it leaves a succession of end moraines, called **recessional moraines.**

As glaciers melt on land, large quantities of water run along the margins, beneath, and out from the front of the glacier to create a meltwater stream. Such streams flow with high but variable discharge in response to seasonal and diurnal temperature variations. Near the glacier front, the meltwater quickly becomes choked with sus-

FIGURE 11.27   Modes of transport in glaciers and various types of sediment loads. (From Reineck, H. E., and I. B. Singh, 1980, Depositional sedimentary environments, 2nd ed. Fig. 274, p. 190, reprinted by permission of Springer-Verlag, Heidelberg. Modified from R. P. Sharp, 1960, Glaciers: The Condon Lecture Series, Oregon State System of Higher Education, Eugene, Ore.)

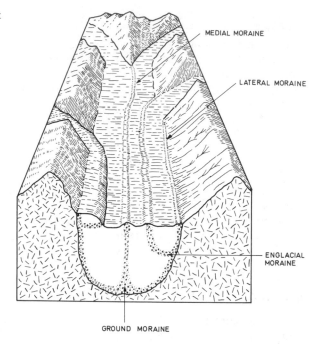

pended sediment and loose bedload sand and gravel, leading to formation of branching and anastomosing braided-stream channels. The characteristics of these braided streams are discussed under fluvial systems. Streams that discharge into glacial lakes tend to build prograding delta systems with steeply inclined foresets. Very fine sediment discharged into the lake from streams may be dispersed in suspension basinward by wind-driven waves or currents. If a large enough concentration of sediment is present in suspension to create a density difference in the water, a density underflow or turbidity current will result that can carry sediment along the lake bottom into the middle of the basin. Strong winds blowing over a glacier or ice sheet pick up fine sand from exposed, dry outwash plains and deposit it downwind in nearby areas as sand dunes. Fine dust picked up by wind can be kept in suspension and transported long distances before being deposited as loess in the periglacial environment.

   Where glaciers extend beyond the mouths of river valleys to enter the sea, their sediment load is dumped into the ocean to form **glacial-marine** sediments. Sedimentation under these circumstances may take place in four different ways (Edwards, 1978). (1) Melting beneath the terminus of the glacier allows large quantities of glacial debris to be released onto the seafloor with little reworking (Fig. 11.28). (2) Large blocks of ice calve from the front of the glacier and float away as icebergs. These icebergs gradually melt, allowing their sediment load to drop onto the seafloor, either on the shelf or in deeper water. (3) Fresh glacial meltwater charged with fine sediment can rise to the surface to form a low-density overflow above denser saline water. Silt and flocculated clays then gradually settle out of suspension from this freshwater plume. (4) Mixing of fresh meltwater and seawater may produce a high-density underflow that can carry sand-size sediment seaward.

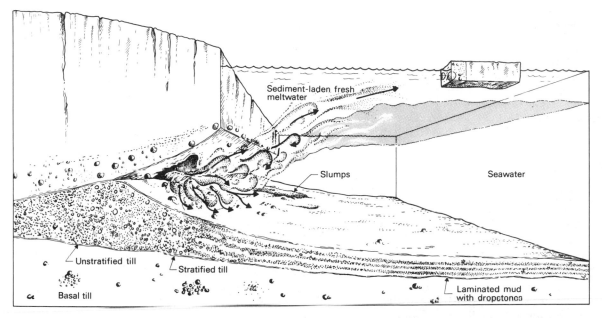

FIGURE 11.28 Hypothetical model for glaciomarine sedimentation. Most of the fresh glacial meltwater rises to the surface of the sea as a low-density overflow layer. This layer gradually mixes with seawater. Silt and flocculated clay are gradually deposited from suspension. Rapid mixing of freshwater and seawater adjacent to tunnel mouths may produce a high density undorflow capable of transporting sand-grade sediment and possibly coarser material. (From Edwards, M. B., 1978, Glacial environments, in H. G. Reading (ed.), Sedimentary environments and facies. Fig. 13.5, p. 423, reprinted by permission of Elsevier Science Publishers, Amsterdam.)

Glacial Facies

Because the broad glacial environment encompasses the proglacial and periglacial environments as well as the glacial environment proper, it is necessary to order to avoid confusion to distinguish between glacial facies deposited directly from the glacier and facies transported and reworked by processes operating beyond the margins of glaciers. Furthermore, it is desirable to distinguish between glacial facies deposited on land and those deposited on the seafloor. Table 11.4 illustrates the full range of sedimentary facies that are affected in some way by glacial processes. Many of these facies are subtypes of facies deposited in fluvial, lacustrine, eolian, shallow-marine, and deep-marine environments, which are treated elsewhere in this book. Therefore, we will focus our discussion here primarily on grounded ice facies and proximal marine-glacial facies. Only very minor emphasis will be given to other glacially related facies.

Poorly sorted glacial deposits are called **diamicts.** Their consolidated equivalents are **diamictites,** sometimes called diamictons. Glacial diamicts are deposited more or less directly from ice without additional winnowing or reworking by water. Easterbrook (1982) suggests that glacial diamicts can be divided into two types: (1) **till,** which is deposited on land, and (2) **glaciomarine drift,** which is glacial debris melted

TABLE 11.4   Facies of glacial environments

Facies of continental glacial environments
   Grounded ice facies
   Glaciofluvial facies
   Glaciolacustrine facies
      Facies of proglacial lakes
      Facies of periglacial lakes
   Cold-climate periglacial facies
Facies of marine glacial environments
   Proximal facies
   Continental shelf facies
   Deep-water facies

Source: Eyles, N., and A. D. Miall, 1984, Glacial facies, in R. G. Walker (ed.), Facies models, 2nd ed.: Geoscience Canada Reprint Ser. 1, p. 15–38.

out of floating ice in marine water. Owing to some disagreement among glacial geologists about the exact meaning of till, Eyles and Miall (1984) recommend that we use the nongenetic term diamict, or diamictite, as appropriate, for all poorly sorted gravel–sand–mud deposits.

Continental Glacial Facies

Grounded Ice Facies

Unstratified Diamicts. Diamicts deposited directly from ice on land consist of unstratified, unsorted pebbles, cobbles, and boulders with an interstitial matrix of sand, silt, and clay. They are thus characterized by a bimodal particle-size distribution in which pebbles predominate in the coarser fraction, with cobbles and boulders scattered throughout (Easterbrook, 1982). Some pebbles are rounded, indicating that they are probably stream pebbles entrained by the ice. Others may be faceted, striated, or polished owing to glacial abrasion. Elongated pebbles and cobbles tend to show some preferred orientation, commonly with their long dimensions parallel to the direction of glacial advance. They may also be crudely imbricated, with long axes dipping upstream. Pebble composition can be highly diverse and may include rock types derived from bedrock located hundreds of kilometers distant. Sands and silts are commonly angular or subangular. Much of the silt in glacial deposits is produced by glacial abrasion and grinding.

Stratified Diamicts. In addition to deposition directly from melting ice, deposition of glacial debris can occur also from meltwaters flowing upon (supraglacial), within (englacial), underneath (subglacial), or marginal to the glacier. The deposits of these meltwaters form on, against, or beneath the ice and thus are commonly known as **ice-contact** sediments. They are reworked to some degree by meltwater and thus exhibit some stratification. They are also better sorted than sediments deposited directly from ice, commonly lack the characteristic bimodal size distribution of direct deposits, and may contain pebbles rounded by meltwater transport. These stratified deposits can accumulate in channels or as mounds or ridges known as kames, kame terraces, or eskers (Fig. 11.29). **Kames** are small mound-shaped accumulations of sand or gravel that form in pockets or crevasses in the ice. **Kame terraces** are similar accumulations deposited as terraces along the margins of valley glaciers. **Eskers** are narrow, sinuous

FIGURE 11.29 Location and geometry of glacial bodies. (From Allen, J. R. L., 1970, Physical processes of sedimentation. Fig. 7.7, p. 229, reprinted by permission of George Allen & Unwin, London.)

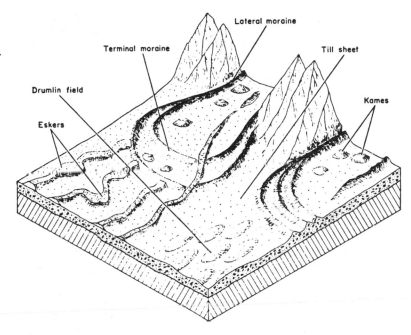

ridges of sediment oriented parallel to the direction of glacial advance. They are the deposits of meltwater streams that probably flowed through tunnels within the glacier. The deposits were then let down onto the subglacial surface after the ice melted. Stratified diamicts are commonly characterized by slump or ice collapse features, including contorted bedding and small gravity faults.

Facies of Proglacial and Periglacial Environments

As discussed, meltwaters issuing from glaciers transport large quantities of glacial debris downslope and deposit it as **glaciofluvial** sediment in braided streams or as **glaciolacustrine** sediment in glacial lakes formed by ice damming or moraine damming. These transported and reworked deposits take on the typical characteristics of the environment in which they are deposited; however, they may retain some characteristics that identify them as glacially derived materials. For example, the large daily to seasonal fluctuations in meltwater discharge may be reflected in abrupt changes in particle size of sediments deposited in meltwater streams or lacustrine deltas. Sediments deposited in streams or lakes very close to the glacier front may also display various slump deformation structures caused by melting of supporting ice. As mentioned in the discussion of lakes, one of the most characteristic properties of glacial lakes is the presence of varves, which form in response to seasonal variations in meltwater flow. Additional details on the characteristics of glaciofluvial and glaciolacustrine sediments are given by Jopling and McDonald (1975). Although sand dunes accumulate in periglacial areas adjacent to some glaciers as a result of wind transport of sand from outwash plains, the primary wind deposits in these environments are silts. Deflation of rock flour and other fine sediment from outwash plains and alluvial plains provides enormous quantities of silt-size sediment that is transported by wind and deposited as widespread sheets, or layers, of fine, well-sorted loess. Owing to the even size of its grains, loess typically lacks well-defined stratification. It is composed predominantly of angular grains of quartz, but may also contain some clays.

Glacial Marine Facies

Proximal Facies

In environments where water is in direct contact with the glacier margin, substantial quantities of sediment are deposited directly from meltwater conduits or tunnels into subaqueous fans, with additional sediment supplied by melting of rafted ice (Eyles and Miall, 1984) (Fig. 11.30). Coarse cobbles and gravels accumulate at the apices of fans, and sands and gravels accumulate within channels as a result of sediment-gravity underflows. Mud and sand are contributed by ice melting and "rain-out" of suspended

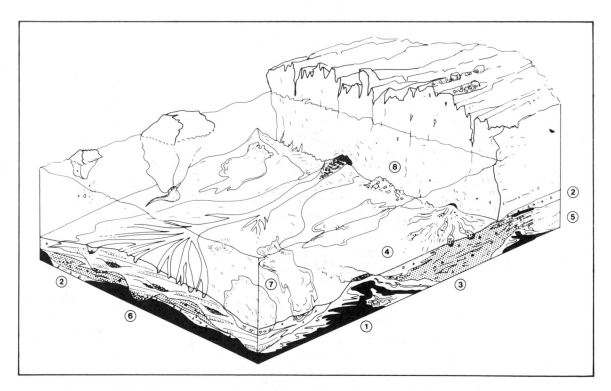

FIGURE 11.30 Proximal subaqueous sedimentation from glaciers: (1) glacially tectonized marine sediment, (2) lensate lodged diamict units, (3) coarse-grained stratified diamicts; (4) pelagic muds and diamicts, (5) coarse-grained proximal outwash, (6) interchannel cross-stratified sands with channel gravels, (7) resedimented facies (debris flows, slides, turbidites), and (8) supraglacial debris. Sediment deformation results from ice advances, melt of buried ice, and iceberg turbation. (From Eyles, N., and A. D. Miall, 1984, Glacial facies, in R. G. Walker (ed.), Facies models: Geoscience Canada Reprint Ser. 1. Fig. 8, p. 20, reprinted by permission of Geological Association of Canada.)

sediment. Some reworking of sediment occurs by downslope sediment-gravity flows and episodic traction-current activity. Proximal glacial-marine sediments may thus range from poorly sorted, poorly stratified diamicts that resemble those deposited on land to coarse-grained stratified diamicts, with a muddy sandy matrix, that may display current-produced structures. Melting of buried ice masses can cause surface subsidence and associated deformation and faulting of sediment.

Distal Facies

Away from the proximal environment in which glaciers are in direct contact with marine water, glacial sediment is supplied by floating ice masses and deposition is dominated by marine processes. Melting of icebergs supplies both fine sediment and coarse debris to the ocean floor by "rain-out," or fallout. Ice-rafted debris deposited on the continental shelves may be reworked to some degree by marine waves and currents (and possibly turbidity currents) and may be affected by iceberg grounding (where icebergs touch bottom). In deeper water on the continental shelf, the debris fallout from floating ice may or may not be retransported by turbidity currents to deeper water. Ice-rafted debris that settles to the deep ocean floor is probably little modified by further depositional processes, except for deposition of a mantle of hemipelagic or pelagic sediment. In general, glacial-marine sediments are distinguished from glacial diamicts by the presence of some stratification, and especially by the presence of marine fossils. Fossil evidence that particularly suggests a glacial-marine origin includes fossils preserved in growth position as whole shells—that is, fossils entombed by fallout sediment; marine molluscs or barnacles attached to glacially faceted pebbles; preservation of delicate ornamentation on shells; and the presence of foraminifers and diatoms in the matrix material (Easterbrook, 1982).

Vertical Facies Sequences

Successive advances and retreats of valley glaciers and ice sheets produce complex vertical sequences of facies as ice progressively overrides proglacial environments during glacial advance and, conversely, as direct ice deposits and ice-contact deposits are reworked in the proglacial environment as a glacier retreats. These facies are much too varied and complex to attempt description here; however, Figures 11.31 and 11.32 are highly generalized sketches that provide some insight into the nature of the vertical sequences that develop under these conditions. Figure 11.31 depicts hypothetical continental facies resulting from advance and retreat of an ice sheet. Figure 11.32 shows the facies that form as the result of an ice sheet moving into a marine basin.

Ancient Glacial Deposits

Glacial deposits range in size from small bodies deposited by valley glaciers to diamictite sheets, deposited by continental glaciers, that cover many thousands of square kilometers. The most characteristic feature of continental diamictites, or grounded ice facies, is their extremely poor sorting and lack of stratification. Facies of glaciofluvial, glaciolacustrine, and glaciomarine environments tend to be better stratified and better sorted. Because the characteristics of these proglacial sediments reflect the environment in which they are deposited, it may be very difficult in ancient stratigraphic sequences to distinguish proglacial sediments from other types of continental sediments. For example, glaciofluvial sediments may appear much the same as other flu-

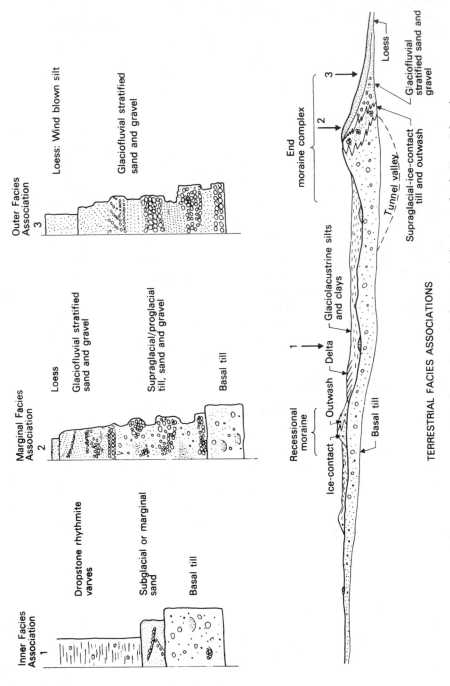

Inner Facies Association

1

Dropstone rhythmite varves

Subglacial or marginal sand

Basal till

Marginal Facies Association

2

Loess

Glaciofluvial stratified sand and gravel

Supraglacial/proglacial till, sand and gravel

Basal till

Outer Facies Association

3

Loess: Wind blown silt

Glaciofluvial stratified sand and gravel

Ice-contact

Recessional moraine

Outwash

Delta

1

Glaciolacustrine silts and clays

Basal till

End moraine complex

2

3

Tunnel valley

Supraglacial-ice-contact till and outwash

Glaciofluvial stratified sand and gravel

Loess

TERRESTRIAL FACIES ASSOCIATIONS

FIGURE 11.31 Vertical facies associations resulting from advance and retreat of glaciers on land. (From Edwards, M. B., 1978, Glacial environments, in H. G. Reading (ed.), Sedimentary environments and facies. Fig. 13.11A, p. 436, reprinted by permission of Elsevier Science Publishers, Amsterdam.)

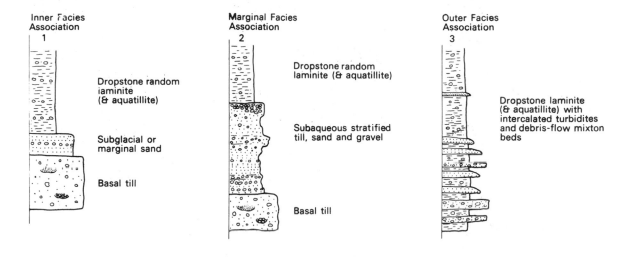

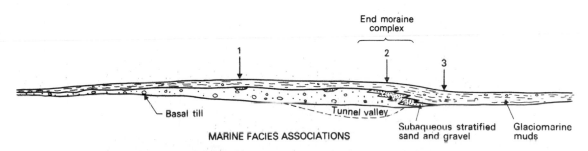

FIGURE 11.32 Vertical facies associations that result from advance and retreat of glaciers that flow into the ocean. Aquatillite is a glaciomarine till-like deposit, such as one deposited from a melting iceberg. (From Edwards, M. B., 1978, Fig. 13.11B, p. 437, reprinted by permission of Elsevier Science Publishers, Amsterdam.)

vial sediments. Nevertheless, a few characteristics of these deposits may reveal their relationship to glacial environments. As mentioned above, the presence of varves may be diagnostic of glacial lakes, and abrupt changes in sediment size related to variable meltwater discharge may be suggestive of proglacial deposition in general. Glacio-marine sediments are distinguished from other types of glacial deposits by the presence of marine fossils and possibly from other marine deposits by generally poorer sorting and stratification. Also the presence of scattered cobbles or boulders, called dropstones, in marine muds suggests deposition from rafted ice.

Ancient glacial deposits are best known from sedimentary units of Pleistocene age, which are widespread in many parts of the world. At least four major pulses of continental glaciation occurred during the Pleistocene, plus numerous smaller pulses. Extensive continental glaciation appears to have been important also during the late Paleozoic, late Ordovician, late Precambrian, and early Proterozoic. Carboniferous- to Permian-age glacial deposits are known from South America, southern Africa, Antarc-

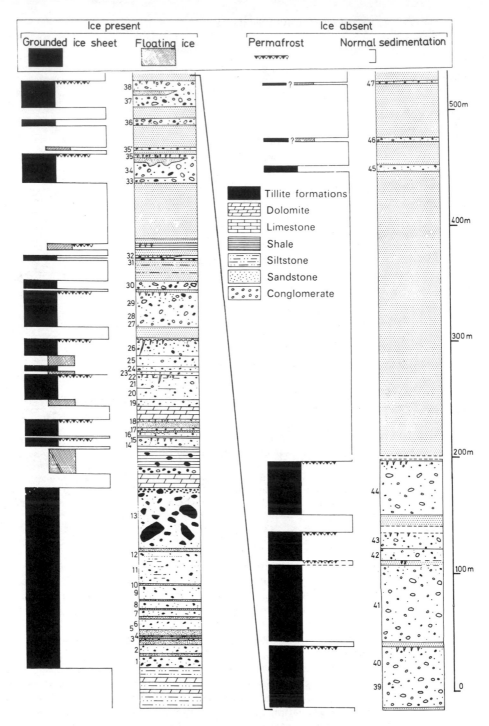

FIGURE 11.33 Composite stratigraphic section of the Port Askaig Tillite sequences for Greenland and Scotland. (After Spencer, A. M., 1975, Late Precambrian glaciation in the North Atlantic region, *in* A. E. Wright and F. Moseley (eds.), Ice ages: Ancient and modern: Seel House Press, Liverpool. Fig. 5, p. 226, reprinted by permission of John Wiley & Sons, Ltd.)

tica, India, and Australia. Late Ordovician diamictites have been reported from South America, several parts of Africa, and possibly Ethiopia. Late Precambrian deposits are known on all continents except Antarctica, and early Proterozoic glacial sediments have been reported in North America in a belt extending from Wyoming to Quebec.

The Port Askaig Tillite in Scotland and Ireland is a good example of a Late Precambrian glacial sequence that illustrates some of the characteristics of ancient tillites and related periglacial deposits. This sequence is more than 700 m thick and is composed of numerous thick tillite units interbedded with glacial-marine, glaciofluvial, and nonglacial sediments (Fig. 11.33). More than half of the sequence shown in Figure 11.33 was deposited by grounded ice sheets (Spencer, 1975). These deposits are mainly conglomerates, but also include pebbly sandstones, siltstones, and shales. Spencer interpreted them as grounded ice deposits because of the sharp contacts at the bases of the units and the lenticularity of the beds. A few thinner beds of glacial-marine shale or siltstone, which are characterized by gradational contacts with underlying units and the presence of dropstone pebbles, are also present. The remaining deposits in the sequence are the product of normal marine sedimentation processes. Summaries of other ancient glacial deposits are given by Edwards (1978) and R. A. Davis (1983).

ADDITIONAL READINGS

Fluvial Systems

Collinson, J. D., and J. Lewin (eds.), 1983, Modern and ancient fluvial systems: Internat. Assoc. Sedimentologists Spec. Pub. 6: Blackwell, Oxford, 575 p.

Ethridge, F. G., and R. M. Flores (eds.), 1981, Recent and ancient nonmarine depositional environments. Part II: Alluvial fan and fluvial deposits: Soc. Econ. Paleontologists and Mineralogists Spec. Pub. 31, p. 49–212.

Miall, A. D. (ed.), 1978, Fluvial sedimentology: Can. Soc. Petroleum Geologists Mem. 5, 589 p.

Miall, A. D., 1982, Analysis of fluvial depositional systems: Am. Assoc. Petroleum Geologists Education Course Note Ser. 20, 75 p.

Nilsen, T. H. (ed.), 1985, Modern and ancient alluvial fan deposits: Benchmark Papers in Geology 87, Van Nostrand Reinhold, New York.

Rachocki, A., 1981, Alluvial fans: John Wiley & Sons, New York, 161 p.

Schumm, S. A., 1977, The fluvial system: John Wiley & Sons, New York, 338 p.

Van Houten, F. B. (ed.), 1977, Ancient continental deposits: Benchmark Papers in Geology 43, Dowden, Hutchinson and Ross, Stroudsburg, Pa., 367 p.

Eolian Systems

Bigarella, J. J., 1972, Eolian environments: Their characteristics, recognition, and importance, in J. K. Rigby and W. K. Hamblin (eds.), Recognition of ancient sedimentary environments: Soc. Econ. Paleontologists and Mineralogists Spec. Pub. 16, p. 12–62.

Brookfield, M. E., and T. S. Ahlbrandt (eds.), 1983, Eolian sediments and processes: Elsevier, Amsterdam, 660 p.

Ethridge, F. G., and R. M. Flores (eds.), 1981, Recent and ancient nonmarine depositional environments. Part IV: Eolian deposits: Soc. Econ. Paleontologists and Mineralogists Spec. Pub. 31, p. 279–349.

Glennie, K. W., 1970, Desert sedimentary environments: Developments in Sedimentology 14, Elsevier, Amsterdam, 222 p.

Mckee, E. D. (ed.), 1979, A study of global sand seas: U.S. Geol. Survey Prof. Paper 1052, 429 p.

Lacustrine Systems

Cole, G. A., 1975, Textbook of limnology: C. V. Mosby, St. Louis, Mo., 283 p.

Ethridge, F. G., and R. M. Flores (eds.), 1981, Recent and ancient nonmarine depositional environments. Part III: Lacustrine deposits: Soc. Econ. Paleontologists and Mineralogists Spec. Pub. 31, p. 213–278.

Hákanson, L., and M. Jansson, 1983, Lake sedimentation: Springer-Verlag, Berlin, 320 p.

Haworth, E. Y., and J. W. G. Lund (eds.), 1984, Lake sediments and environmental history: University of Minnesota Press, Minneapolis, 411 p.

Lerman, A. (ed.), 1978, Lakes: Chemistry, geology, physics: Springer-Verlag, New York, 363 p.

Matter, A., and M. E. Tucker (eds.), 1978, Modern and ancient lake sediments: Internat. Assoc. Sedimentologists Spec. Pub. 2, 290 p.

Picard, M. D., and L. R. High, Jr., 1972, Criteria for recognizing lacustrine rocks, in J. K. Rigby and W. K. Hamblin (eds.), Recognition of ancient sedimentary environments: Soc. Econ. Paleontologists and Mineralogists Spec. Pub. 16, p. 108–145.

Glacial Systems

Ashley, G. M., N. D. Smith, and J. D. Shaw, 1985, Glacial sedimentary environments: Soc. Econ. Paleontologists and Mineralogists Short Course Notes No. 16, 246 p.

Eyles, N. (ed.), 1983, Glacial geology: An introduction for engineers and earth scientists: Pergamon, Oxford, 409 p.

Evenson, E. B., Ch. Schluchter, and J. Rabassa, 1983, Till and related deposits: A. A. Balkema, Rotterdam, 454 p.

Flint, R. F., 1971, Glacial and Quaternary geology: John Wiley & Sons, New York, 892 p.

Goldthwait, R. P. (ed.), 1971, Till: A symposium: Ohio State University Press, Columbus, Oh., 402 p.

Goldthwait, R. P. (ed.), 1975, Glacial deposits: Dowden, Hutchinson and Ross, Stroudsburg, Pa., 464 p.

Jopling, A. V., and B. C. McDonald, 1975, Glaciofluvial and glaciolacustrine sedimentation: Soc. Econ. Paleontologists and Mineralogists Spec. Pub. 23, 320 p.

Legget, R. F. (ed.), 1976, Glacial till: An inter-disciplinary study: Royal Soc. Canada Spec. Pub. 12, 412 p.

Molnia, B. F., 1983, Glacial-marine sedimentation: Plenum, New York, 844 p.

Wright, A. E., and F. Moseley, 1975, Ice ages: Ancient and modern: Geol. Journal Spec. Issue 6, Seel House Press, Liverpool, 320 p.

12
Marginal-Marine Environments

12.1 INTRODUCTION

The marginal-marine setting lies along the boundary between the continental and the marine depositional realms. It is a narrow zone dominated by riverine, wave, and tidal processes. Salinities may range in different parts of the system from freshwater through brackish water to supersaline, depending upon river discharge and climatic conditions. Intermittent to nearly constant subaerial exposure characterizes some environments of the marginal-marine setting. Others are continuously covered by shallow water. Many marginal-marine environments are further characterized by high-energy waves and currents, although some lagoonal and estuarine environments are dominated by quiet-water conditions. Owing to intermittent exposure, high-energy conditions, or marked variations in salinity or temperature, much of the marginal-marine realm is a high-stress environment for organisms. Therefore, organisms that live in marginal-marine environments tend to be species with high tolerance for salinity or temperature changes. They adapt to high-energy conditions by burrowing into sandy bottoms or boring into rocky substrates.

A wide variety of sediment types, including conglomerates, sandstones, shales, carbonates, and evaporites, can accumulate in marginal-marine environments. Owing to the large quantities of siliciclastic sediment delivered by rivers to the coastal zone throughout geologic time, the volume of marginal-marine deposits preserved in the geologic record is significant. The principal depositional settings for marginal-marine sediments are deltas, beaches and barrier bars, estuaries, lagoons, and tidal flats.

12.2 DELTAIC SYSTEMS

Introduction

The word **delta** was used by the Greek philosopher Herodotus about 490 B.C. to describe the triangular-shaped alluvial plain formed at the mouth of the Nile River by deposits of the Nile distributaries. Most other modern deltas are less triangular and more irregular in shape than the Nile delta. Nevertheless, the term delta is still applied to any deposit, subaerial or subaqueous, formed by fluvial sediments that build into a standing body of water. Deltas can form in any body of water where rivers discharge greater quantities of clastic sediment than can be removed by wave and current activity. Thus, they can occur in lakes and inland seas as well as the ocean. They are most important in the open ocean, and much of the siliciclastic sediment transported to coastal zones throughout geologic time has been deposited in deltas.

Ancient deltaic deposits have been identified in stratigraphic sequences of many ages, and deltaic sediments are known to be important hosts for petroleum and natural gas, coal, and some minerals such as uranium. Although ancient deltaic sediments are common in the rock record, much of what we know about delta systems comes from study of modern deltas. Deltas are particularly common in the modern ocean owing to post-Pleistocene sea-level rise coupled with high sediment loads carried by many rivers. Rising sea level causes increased sedimentation rates on deltas because sediment is trapped by the rising water, inhibiting sediment removal by currents. The locations of major modern deltas are shown in Figure 12.1, and the dimensions and discharge characteristics of some representative deltas are given in Table 12.1.

Deltas occur on all continents (except Antarctica and Greenland, which are covered by ice sheets) where large, active drainage systems with heavy sediment loads exist. These conditions appear to be met particularly well on trailing-edge or passive coasts such as the east coasts of Asia and the Americas, where tectonic activity is low.

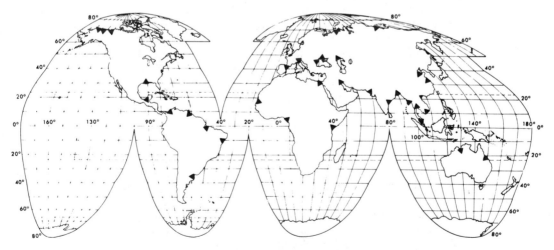

FIGURE 12.1 Locations of major modern deltas (black triangles). (From Coleman, J. M., and L. D. Wright, 1975, Modern river deltas: variability of process and sand bodies, *in* M. L. Broussard (ed.), Deltas, models for exploration, 2nd ed. Fig. 2, p. 101, reprinted by permission of Houston Geological Society, Houston, Tex.)

TABLE 12.1 Characteristics of some modern deltas

Delta	Subaerial area (km² · 10³)	Average water discharge (m³/sec · 10³)	Annual sediment discharge (t · 10⁶)	Annual rate of growth (m)
Chao Phraya	25	1	5	—
Danube	4	6	91	—
Ganges-Brahmaputra	91	39	635	—
Hwang Ho and N. Yellow Plain	127	4	2	268
Irrawaddy	31	28	272	46–61
Lena	28	9	—	—
Mekong	52	11	—	61
Mississippi	29	17	469	91
Niger	19	6	23	—
Nile	16	3	54	—
Orinoco	57	17	—	—
Po	14	1	61	26–61
Red	8	2	118	—
Rhine	22	2	—	—
Rhône	3	2	41	—
Rio Grande	8	0.1	17	—
Ural	9	—	—	—
Volga	11		—	—
Yangtze and S. Yellow Plain	124	22	544	23

Source: Reineck, H. E., and I. B. Singh, 1980, Depositional environments, 2nd ed.: Table 25, p. 321, reprinted by permission of Springer-Verlag, Heidelberg.

Less than ten percent of major modern deltas occur on collision coasts, where tectonic activity is high and drainage divides are close to the sea (Inman and Nordstrom, 1971; L. D. Wright, 1978). Under such conditions, the large drainage systems necessary to supply heavy sediment loads are not developed. Owing to their potential importance as oil and gas reservoirs, considerable interest has been generated in deltaic deposits since the 1950s. Consequently, the literature on deltas and deltaic deposits is extensive. Particularly useful discussions of deltas are given in Broussard (1975), Coleman (1981), Coleman and Prior (1982, 1983), Eliott (1978a), Galloway (1975), Miall (1984b), Morgan (1970), Shirley and Ragsdale (1966), and L. D. Wright (1977, 1978). Many of the important papers on deltas published prior to 1975 are summarized by LeBlanc (1975).

Sedimentation Processes on Deltas

The distribution and characteristics of deltas are controlled by a complex set of inter-related fluvial and marine processes and environmental conditions. These factors include climate, water and sediment discharge, river-mouth processes, nearshore wave power, tides, nearshore currents, and winds (Coleman, 1981). Other factors that influence the formation of deltas are the slope of the shelf, rates of subsidence and other tectonic activity at the depositional site, and the geometry of the depositional basin. Among these variables, **sediment input, wave-energy flux,** and **tidal flux** are the most important processes that control the geometry, trend, and internal features of the progradational framework sand bodies of deltas (Galloway, 1975; Galloway and Hobday,

1983). These three processes constitute the basis for recognition of three fundamental types of deltas: (1) fluvial-dominated, (2) tide-dominated, and (3) wave-dominated (Fig. 12.2).

Fluvial-dominated Deltas. In an early but important paper dealing with deltaic processes, C. C. Bates (1953) contrasted the behavior of sediment-laden river water as it enters equally dense, more dense, and less dense basin water. River water entering basin water of almost equal density, referred to by Bates as **homopycnal flow,** leads to

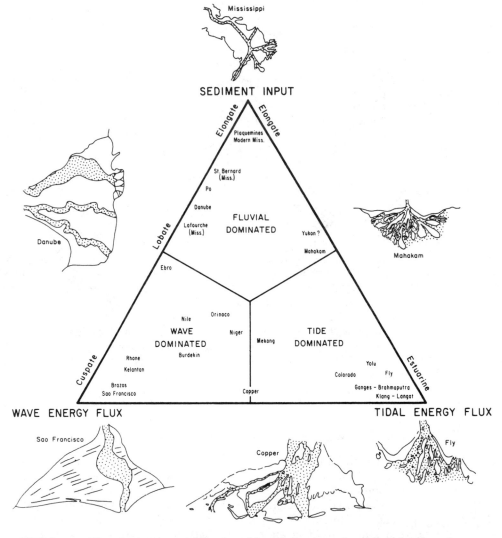

FIGURE 12.2 Three basic types of deltas based on relative intensity of fluvial and marine processes. (From Galloway, W. E., and D. K. Hobday, 1983, Terrigenous clastic depositional systems. Fig. 5.8, p. 91, reprinted by permission of Springer-Verlag, Heidelberg. Modified from W. E. Galloway, 1975, Process framework for describing the morphologic and stratigraphic evolution of deltaic depositional systems, in M. L. Broussard (ed.), Deltas, models for exploration, 2nd ed. Fig. 3, p. 92, reprinted by permission of Houston Geological Society, Houston, Tex.)

rapid, thorough mixing and deposition of much of the sediment load at this point (Fig. 12.3A). This type of jet outflow presumably causes the formation of Gilbert-type deltas that display a topset, foreset, and bottomset arrangement of beds (Fig. 12.4), created as sediment deposition progrades basinward. River water that has higher density than basin water flows beneath the basin water, generating a vertically oriented plane-jet flow called **hyperpycnal flow** (Fig. 12.3B). This type of jet flow moves along the bottom as a density current that may be erosive in its initial stages but eventually deposits its load along the more gentle slopes of the delta front to form turbidites. Such turbidity currents have been observed in lakes and may occur also under marine conditions, although they have not yet been observed in the ocean. If river outflow is less dense than basin water, as when rivers flow into denser seawater, it flows outward on top of the basin water as a horizontally oriented plane jet called **hypopycnal flow** (Fig. 12.3C). Fine sediment may thus be carried in suspension some distance outward from the river mouth before it flocculates and settles from suspension. Flocculation involves aggregation of fine sediment into small lumps owing to the presence in seawater of electrolytes that neutralize negative charges on clay particles. Hypopycnal flow tends to generate a large, active delta-front area, typically dipping at 1 degree or less, as contrasted with the 10–20 degree dip of most Gilbert-type dunes (Miall, 1984b). Hypopycnal flow is probably the most important type of river outflow in marine basins (Eliott, 1978a).

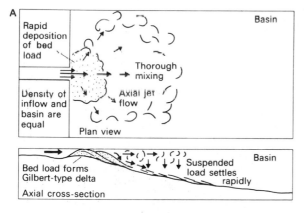

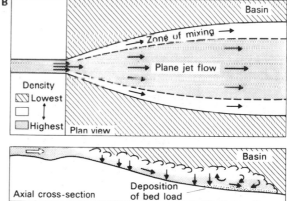

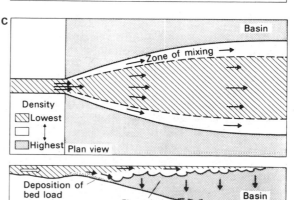

FIGURE 12.3 Differences in interaction of sediment-laden river water and basin water owing to differences in relative densities of the water bodies. A. Homopycnal flow. B. Hyperpycnal flow. C. Hypopycnal flow. (From Fisher, W. L., L. F. Brown, A. J. Scott, and J. H. McGowen (eds.), 1969, Delta systems in exploration for oil and gas: Texas Bur. Econ. Geology. Fig. 3, p. 92, reprinted by permission of University of Texas, Austin. Originally after C. C. Bates, 1953, Rational theory of delta formation: Am. Assoc. Petroleum Geologists Bull., v. 37. Fig. 3, p. 2124, reprinted by permission of AAPG, Tulsa, Okla.)

FIGURE 12.4 Gilbert-type delta. A. Section through a Pleistocene delta in Lake Bonneville. B. Vertical facies sequence produced by delta progradation. (From Elliot, T., 1978, Deltas, *in* H. G. Reading (ed.), Sedimentary environments and facies. Fig. 6.1, p. 97, reprinted by permission of Elsevier Science Publishers, Amsterdam. Originally after G. K. Gilbert, 1885, The topographic features of lake shores: U.S. Geol. Survey Ann. Rept.; J. Barrell, 1912, Criteria for recognition of ancient delta deposits: Geol. Soc. America Bull., v. 23.)

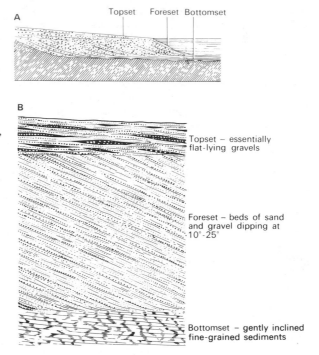

L. D. Wright (1977) suggests that the depositional patterns of sediments deposited by river-mouth processes in coastal areas with low tidal range and low wave energy depend upon the relative dominance of (1) outflow inertia, (2) turbulent bed friction seaward of the river mouth, and (3) outflow buoyancy. Outflows dominated by inertial forces—that is, characterized by large Reynolds numbers and high flow velocities—generate fully turbulent, homopycnal jet flows with negligible interference from the bottom. They exhibit low lateral spreading angles and progressive lateral and longitudinal deceleration and produce narrow river-mouth bars of the Gilbert type (Fig. 12.5). These types of outflows are considered to be rare. More commonly, water depths seaward of river mouths are shallow, and turbulent bed friction becomes dominant owing to high outflow velocities and bed shear stresses. This friction causes rapid deceleration and lateral expansion of the outflow and leads to formation of subaqueous levees, triangular-shaped "middle-ground" bars that become finer grained basinward, and channel bifurcation (Fig. 12.6). Where river mouths are relatively deep and tidal range is low, fine-grained sediment loads prevail. Mixing is minimal, and strong density stratification can develop with freshwater flowing out over an underlying saltwater wedge. The outflow then spreads as a buoyant plume above the underlying saltwater, the hypopycnal inflow of Bates. Buoyant outflows lead to formation of elongate distributaries with parallel banks, called subaqueous levees; few channel bifurcations; and narrow distributary-mouth bars that grade seaward to fine-grained distal-bar deposits and prodelta clays (Fig. 12.7). Bar sands or **bar-finger** sands are typical components of such deltaic assemblages.

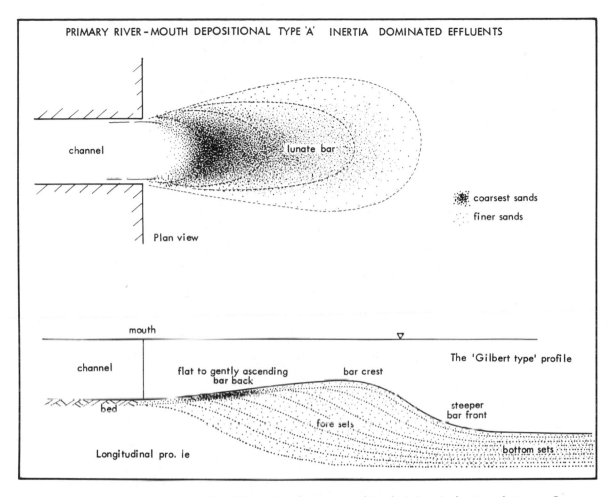

FIGURE 12.5 Idealized depositional pattern resulting from inertia-dominated river outflow. (From Wright, L. D., 1977, Sediment transport and deposition at river mouths: A synthesis: Geol. Soc. America Bull., v. 88, Fig. 4, p. 860.)

The modern Mississippi River delta is a classic example of a birdsfoot-type, river-dominated delta. It is among the largest deltas in the world outside of Asia (Table 12.1). The Mississippi delta consists of seven distinct sedimentary lobes that have been active during the last 5000–6000 years, indicating that periodic channel or distributary abandonment, to be discussed, is a common process. The generalized characteristics of the Mississippi delta system are shown in Figure 12.8. This figure illustrates the well-developed birdsfoot distributary system, typical of the delta, with barfinger sands developed at the mouths of the distributaries. Common sediment facies on the Mississippi delta include marsh and natural-levee deposits, delta-front silts and sands, and prodelta clays. These types of deltaic deposits are described in greater detail in the discussion of sediment characteristics.

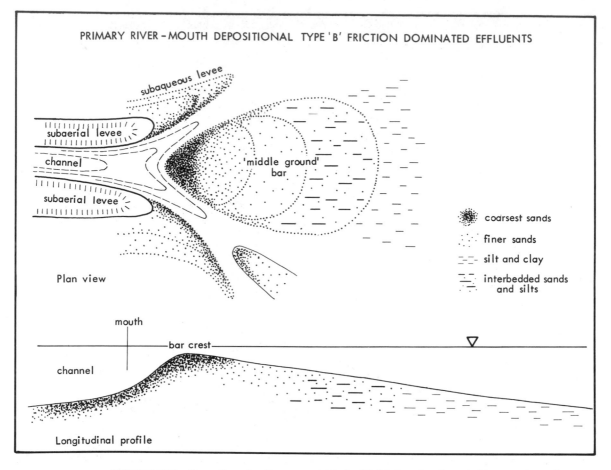

FIGURE 12.6 Depositional patterns associated with friction-dominated river-mouth outflow. (From Wright, L. D., 1977, Fig. 5, p. 861.)

Tide-dominated Deltas. The processes and deposits described above may be significantly modified under conditions of high tidal range or high wave energy. If tidal currents are stronger than river outflow, these bidirectional currents can redistribute river-mouth sediments, producing sand-filled, funnel-shaped distributaries. The distributary-mouth bar may be reworked into a series of linear tidal ridges that replace the bar and extend from within the channel mouth out onto the subaqueous delta-front platform (L. D. Wright, 1977, 1978).

The modern Ganges–Brahmaputra delta is a well-known example of a tide-dominated delta (Fig. 12.9). The areal size of this delta is more than three times that of the Mississippi delta. It has a mean river discharge about twice that of the Mississippi, with exceedingly high discharge taking place during the monsoon season when extreme flooding is common. Mean tidal range is large, about 4 m, and wave energy is relatively low. Sand transport is intense during the monsoon season, leading to deposition of sandy deposits similar to those in braided streams. The delta is characterized by tidal-flat environments, natural levees, and flood basins in which fine sediment is deposited from suspension. The strong tidal influence is manifested by the presence of a network of tidal sand bars and channels oriented roughly parallel to the direction of tidal current flow (Fig. 12.9). A variety of sediment types thus accumulate on the Ganges–Brahmaputra delta, including tidal-bar sands; braided, channel-fill sands; and natural levee, tidal-flat, and flood-basin muds.

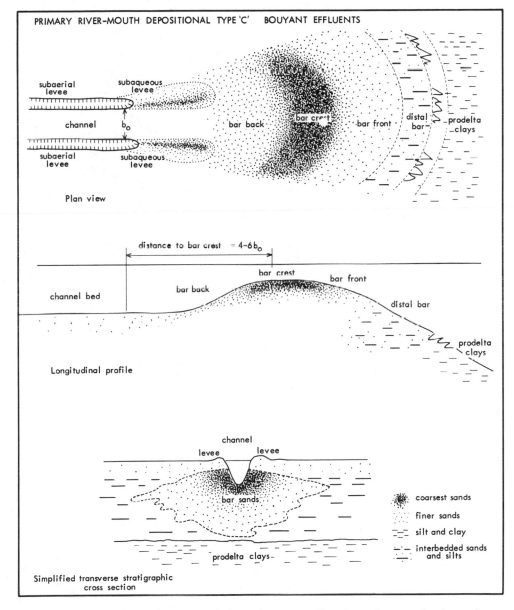

FIGURE 12.7 Depositional patterns relating to buoyant outflow from river mouths; b_0 = channel width. (From Wright, L. D., 1977, Fig. 6, p. 863.)

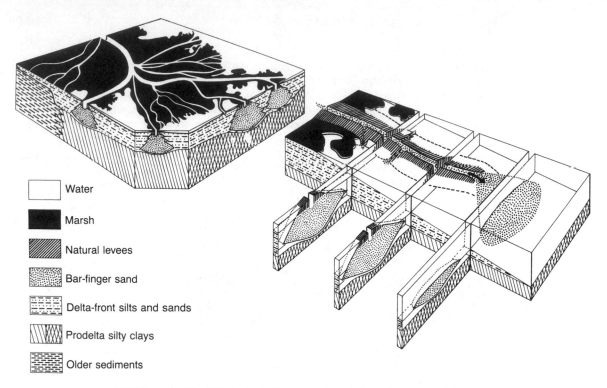

Water

Marsh

Natural levees

Bar-finger sand

Delta-front silts and sands

Prodelta silty clays

Older sediments

FIGURE 12.8 The Mississippi delta system—a fluvial-dominated delta. (From Reineck, H. E., 1970, Marine sandkörper, rezent und fossil: Geol. Rundschau, v. 60. Fig. 2, p. 305, reprinted by permission. Originally modified from H. N. Fisk, E. McFarland, C. R. Kolb, and L. J. Wilbert, 1954, Sedimentary framework of the modern Mississippi delta: Jour. Sedimentary Petrology, v. 24, Fig. 1, p. 77.)

FIGURE 12.9 The modern Ganges–Brahmaputra delta—a tide-dominated delta. (After Fisher, W. L., L. F. Brown, Jr., A. J. Scott, and J. H. McGowen, 1969, Delta systems in the exploration for oil and gas—a research colloquium: Texas Bur. Econ. Geology. Fig. 47 reprinted by permission of University of Texas, Austin.)

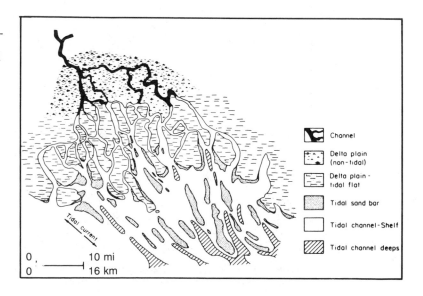

Channel

Delta plain (non-tidal)

Delta plain - tidal flat

Tidal sand bar

Tidal channel - Shelf

Tidal channel deeps

Tidal current

0   10 mi
0   16 km

404

Wave-dominated Deltas. Strong waves cause rapid diffusion and deceleration of river outflow and produce constricted or deflected river mouths. Distributary-mouth deposits are reworked by waves and redistributed along the delta front by longshore currents to form wave-built shoreline features such as beaches, barrier bars, and spits. A smooth delta front, consisting of well-developed coalescent beach ridges, may eventually be generated; delta-plain geometries may range from arcuate to cuspate (Galloway and Hobday, 1983).

The Sao Francisco delta of Brazil (Fig. 12.10) provides a good modern example of a wave-dominated delta. It is smaller in areal dimensions than the Mississippi and has lower discharge. Tidal range is about 2 m; however, wave power is reported to be about 100 times that of the Mississippi delta (Coleman, 1976). Owing to this extreme wave energy, the Sao Francisco delta is dominated by high-energy environments in which sand deposition takes place. Muds accumulate locally in marshes and floodplains, but the interdistributary bay mud deposits characteristic of the Mississippi delta are absent. Sao Francisco delta deposits are dominated by beach-ridge barrier sands that cover most of the delta surface (Fig. 12.10). A particularly unusual feature

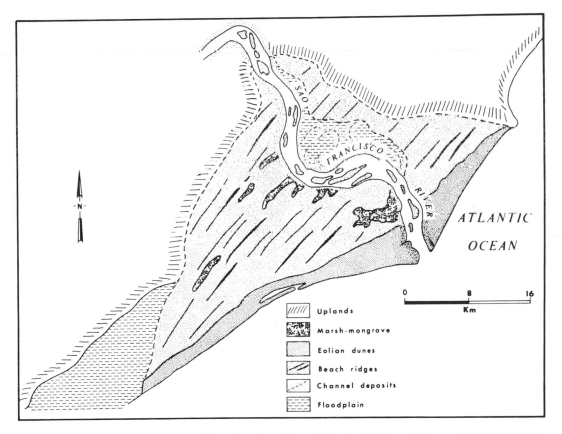

FIGURE 12.10   Map of Sao Francisco River delta, Brazil. (From Coleman, J. M., 1981, Deltas: Processes of deposition and models for exploration, 2nd ed. Fig. 4.13, p. 104, reprinted by permission of IHRDC Publications, Boston.)

of the Sao Francisco delta is the presence of a narrow belt of eolian dune sands distributed along the outer margin of the delta.

The examples discussed above illustrate some of the differences in characteristics of modern fluvial-dominated, tide-dominated, and wave-dominated deltas. The principal differences in characteristics of these types of deltas are summarized further in Table 12.2. Deltas with characteristics transitional between these "end-member" types also occur.

Constructional vs. Destructional Phases of Delta Development. During the active phases of delta outbuilding, most sedimentation processes on deltas are constructional in the sense that delta formation is dominated by sediment deposition. On the other hand, tidal currents and waves represent destructional processes to the extent that they cause erosion and redistribution of some sediment. Destructional processes become particularly important when deltas, or portions of deltas, enter an inactive phase. Channel or distributary abandonment, foundering owing to subsidence, or marine transgression may interrupt active construction of a delta. Such an interruption leads to a phase when erosion by waves and tidal currents becomes dominant as sediment influx from the river ceases. Destructional processes cause redistribution of eroded sediment, commonly producing thin, laterally persistent beds of sand or mud that contain marine faunas and are abundantly bioturbated. Such beds form important marker units for stratigraphic studies in otherwise heterogeneous deltaic sequences.

Growth of deltas thus tends to be cyclic (Coleman and Wright, 1975; Eliott, 1978a). During active, prograding phases, prodelta fine silts and clays are progressively overlain by delta-front silts and sands, distributary-mouth sands, and finally marsh, fluvial, and possibly eolian deposits as the delta builds seaward (Fig. 12.11), producing a coarsening-upward regressive sequence. Interruption of progradation by delta-lobe abandonment or marine transgression brings on a destructive phase in which erosion and redistribution of river-mouth deposits predominate. Subsequent distributary shifting or regression may bring on another phase of active progradation. A complete delta cycle may range in thickness from 50 to 150 m. Smaller scale cycles representing progradation of individual distributaries range from only about 2 m to 15 m (Miall, 1984b).

TABLE 12.2   Characteristics of fluvial-, tide-, and wave-dominated deltas.

	Fluvial-dominated	**Wave-dominated**	**Tide-dominated**
Lobe geometry	Elongate to lobate	Arcuate	Estuarine to irregular
Bulk composition	Muddy to mixed	Sandy	Muddy to sandy
Framework facies	Distributary-mouth bar and delta-front sheet sand; distributary channel-fill sand	Coastal-barrier sand; distributary-channel sand	Tidal sand ridge sand; estuarine distributary channel-fill sand
Framework orientation	Highly variable; average parallels depositional slope	Dominantly parallels depositional strike; subsidiary dip trends	Parallels depositional slope unless skewed by local basin geometry
Common channel type	Suspended load to fine mixed load	Mixed load to bedload	Variable, tidally modified geometry

Source: Galloway, W. E., and D. K. Hobday, 1983, Terrigenous clastic depositional systems. Table 5.1, p. 109, reprinted by permission of Springer-Verlag, Heidelberg.

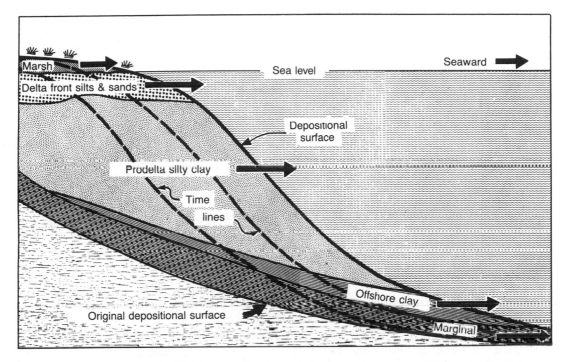

FIGURE 12.11 Highly generalized coarsening-upward sequence of facies developed by seaward progradation of a delta. Note the development of clinoform depositional surfaces of the delta front and prodelta. (From Scruton, P. C., 1960, Delta building and delta sequences, in F. P. Shepard, F. B. Phleger, and T. H. van Andel (eds.), Recent sediments of the northwest Gulf of Mexico. Fig. 9, p. 93, reprinted by permission of American Association of Petroleum Geologists, Tulsa, Okla.)

Physiographic and Sediment Characteristics of Deltaic Systems

Owing to variations in sediment input, outflow velocity, wave and current energy, and other factors, the depositional features of deltas exhibit a high degree of variability from one delta to another. Nevertheless, all deltas can be divided into subaerial and subaqueous components, each of which can be further subdivided (Fig. 12.12, Table 12.3). The subaerial component of deltas is much larger than the subaqueous component and is divided into an **upper delta plain,** which lies largely above high-tide level, and a **lower delta plain,** lying between low-tide mark and the upper limit of tidal influence. The upper delta plain is commonly the oldest part of the delta and is dominated by fluvial processes. The lower delta plain is exposed during low tide but is covered by water during high tide. Thus, it is subjected to both fluvial and marine processes. The **subaqueous delta plain** lies seaward of the lower deltaic plain below low-tide water level and is characterized by relatively open marine faunas. The uppermost part of the subaqueous delta, lying at water depths down to 10 m or so, is commonly called the **delta front.** The remaining seaward part of the subaqueous delta is called the **prodelta,** or prodelta slope.

FIGURE 12.12 Principal components of a delta system. (From Coleman, J. M., and D. B. Prior, 1982, Deltaic environment of deposition, *in* P. A. Scholle and D. Spearing (eds.), Sandstone depositional environments: Am. Assoc. Petroleum Geologists Mem. 31. Fig. 1, p. 139, reprinted by permission of AAPG, Tulsa, Okla.)

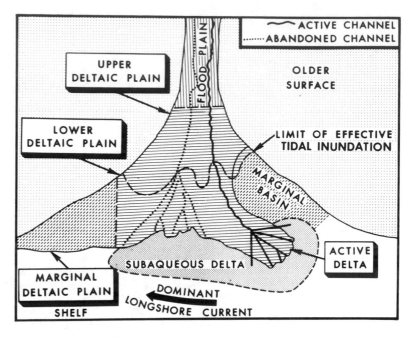

Upper Delta-Plain Sediments. The upper delta plain lies mainly above tidal influence and is little affected by marine processes. Sedimentation on the upper delta is dominated by distributary-channel migration and associated fluvial sedimentation processes such as channel and bar deposition, overbank flooding, and crevassing into lake basins. The principal depositional environments include braided channels, meandering channels, lacustrine delta fill, backswamps, and floodplain environments such as swamps, marshes, and freshwater lakes (Coleman and Prior, 1982). Therefore, upper delta-plain sediments are predominantly fluvial sands, gravels, and muds that may be closely associated with lacustrine, swamp, and marsh deposits.

Lower Delta-Plain Sediments. The lower delta plain extends from the low-tide shoreline landward to the uppermost limit of tidal influence. The width of the lower delta

TABLE 12.3 Principal categories of delta facies

Upper delta plain
Migratory channel deposits—braided-channel and meandering-channel deposits
Lacustrine delta-fill and floodplain deposits
Lower delta plain
Bay-fill deposits (interdistributary bay, crevasse splay, natural levee, marsh)
Abandoned distributary-fill deposits
Subaqueous delta plain
Distributary-mouth bar deposits (prodelta distal bar, distributary-mouth bar)
River-mouth tidal-range deposits
Subaqueous slump deposits

Source: Coleman, J. M., and D. B. Prior, 1982, p. 140, reprinted by permission of AAPG, Tulsa, Okla.

plain is thus greatest on deltas where tidal range is large. This plain includes the active distributary system of the delta, as well as abandoned distributary-fill deposits, and may be flanked by marginal-basin or bay-fill deposits. Distributary channels are numerous but environments between channels make up the largest percentage of the lower delta plain. These environments include actively migrating tidal channels, natural levees, interdistributary bays, bay fills (crevasse splays), marshes, and swamps (Coleman and Prior, 1982). The major sand bodies generated in this environment are bay-fill deposits, which may form thin sand wedges stacked one on top of the other and separated by finer grained interdistributary-bay and marsh deposits. In very arid climates, evaporites also may be deposited in some parts of the lower delta plain. Deposits of the lower delta plain also commonly include abandoned distributary deposits. These consist of locally derived sands, muds, and organic debris that gradually fill distributary channels after they have been abandoned by the main stream owing to blocking or other processes that cause channel shifting.

Subaqueous Delta Plain. The subaqueous delta plain constitutes that area of a delta that lies seaward of low tide level and actively receives fluvial sediments. It may extend outward for distances of a few kilometers to tens of kilometers and to water depths of as much as 300 m. Deposits of the subaqueous delta thus form the base over which subaerial delta deposits prograde as the delta builds seaward. The deposits typically include sands, and possibly gravels, deposited near the river mouths, forming distributary-mouth bar deposits. These deposits grade seaward to finer sands and coarse silts that settle from suspension to form the distal bar (Fig. 12.7). The uppermost part of the subaqueous delta, the delta front, may be dominated by high-energy marine processes, including waves, longshore currents, and tides in some cases. Sediment is reworked and winnowed by these processes, creating well-sorted delta-front sheet sands that are cross-bedded on a variety of scales. The finest silts and clays are transported still farther seaward and settle on the prodelta on the outermost part of the subaqueous delta. Previously deposited sediments may be reentrained, transported, and redeposited farther downslope on the subaqueous delta by gravity-driven mass-movement processes that include landslides, slumps, and mudflows. Mud diapirs are also a common feature on many subaqueous deltas (Fig. 12.13). These structures are piercement-type bodies of soft sediment that is squeezed upward and intruded into overlying sand bodies owing to sediment loading,

Recognition of Ancient Deltaic Deposits

The subenvironments of delta systems range from normal marine to totally nonmarine, and a variety of different sediment types can be deposited in these subenvironments, as the preceding discussion indicates. Deltaic sedimentary sequences thus contain no single distinctive lithofacies. Rather, they are characterized by assemblages of lithofacies, each of which can occur in other environments, such as fluvial, lacustrine, and shallow-marine environments. Recognition of ancient deltaic sediments is further complicated by the fact that we must consider three different delta models, fluvial-dominated, tide-dominated, and wave-dominated, although fluvial-dominated deltas appear to be most common. Miall (1984b) suggests that identification of ancient delta deposits is best accomplished in a series of steps, eliminating other possibilities and using distinguishing characteristics of facies types, bed geometry, and types of cyclic successions to focus gradually on the correct delta model. Some general characteristics

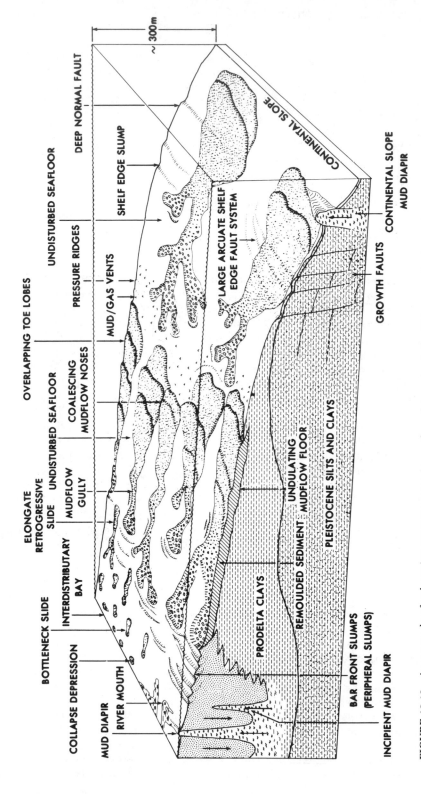

FIGURE 12.13 An example of submarine mass movements on the subaqueous delta. This example shows the major types of submarine landslides, diapirs, and contemporary faults in the Mississippi River delta. (From Coleman, J. M., and D. B. Prior, 1982, Fig. 23, p. 168, reprinted by permission of AAPG, Tulsa, Okla.)

of delta deposits that can be useful in their recognition include (1) geometry, (2) lateral facies relationships, (3) vertical sequence, and (4) sedimentary structures and fossils.

Geometry. Ideally, deltas are triangular in areal shape; however, much variation from this ideal shape can occur, particularly with tide- and wave-dominated deltas. In cross section, deltas are typically wedge- or lens-shaped bodies. Modern deltas vary in their areal dimensions from small bodies a few thousand square kilometers in size to huge deltas exceeding $125,000$ km^2 (Table 12.1).

Lateral Facies Relationships. A wedge- or lobe-shaped deposit of nonmarine to shallow-marine sand, silt, and mud that grades landward into totally nonmarine, largely fluvial sediments and basinward into finer grained, deeper water marine sediments is indicative of deltaic origin. On a smaller scale, lateral facies relationships are likely to be complex. Delta-plain deposits may range from coarse distributary-channel deposits to finer grained marsh or interdistributary-bay or lacustrine deposits. The lateral associations of delta-front sediments also can be highly variable depending upon whether deposition was dominated by fluvial, wave, or tidal processes; it is difficult to generalize about these facies relationships. Coarser, delta-front sediments may grade into prodelta silts and clays which, in turn, grade to open-shelf muds. Prodelta muds may be difficult to differentiate from open-shelf muds except perhaps by greater thickness and higher sedimentation rates. They tend also to contain more mud-turbidite units than shelf muds.

Vertical Sequence. Progradation of deltaic deposits during active delta growth produces a generally coarsening-upward sedimentary sequence. Migration of delta-front sands over prodelta silts and clays generates fairly well-defined, large-scale, coarsening-upward sequences. Progradation of subaerial delta-plain sediments over subaqueous delta-plain sediments tends to produce smaller scale, coarsening-upward units. Locally, filling of abandoned channels may even cause fining-upward sequences to develop. Major progradational coarsening-upward facies may be interrupted by thin, widespread facies generated during inactive stages of delta growth, producing cycles of progradational and channel-abandonment facies. Although the generalized coarsening-upward progradational model for deltas serves as a useful norm for comparison, students are cautioned that variations in delta behavior can produce vertical sequences that differ from this idealized model. The lithologic types, sedimentary structures, textures, and other features preserved in the progradational facies depend upon the type of delta. Examples of typical sequences of progradational facies produced in a fluvial-dominated delta—the Mississippi Delta—and a tide-dominated delta—the Ord Delta—are illustrated in Figures 12.14 and 12.15.

Sedimentary Structures and Fossils. Numerous types of sedimentary structures such as cross-bedding, ripple marks, bioturbation structures, slump structures, and mud diapirs occur in deltaic deposits. With the possible exception of mud diapirs, all of these structures are found also in many other environments; none is diagnostic of delta environments. Suites of structures may help to define a particular type of deltaic deposit, such as fluvial deposits. Paleocurrent directions in deltaic deposits can be highly variable, ranging from unidirectional patterns in fluvial-dominated portions of deltas to bidirectional patterns in tide-dominated portions. There are no specific fossil taxa that are characteristic of deltas alone. If a transition from freshwater to brackish-water

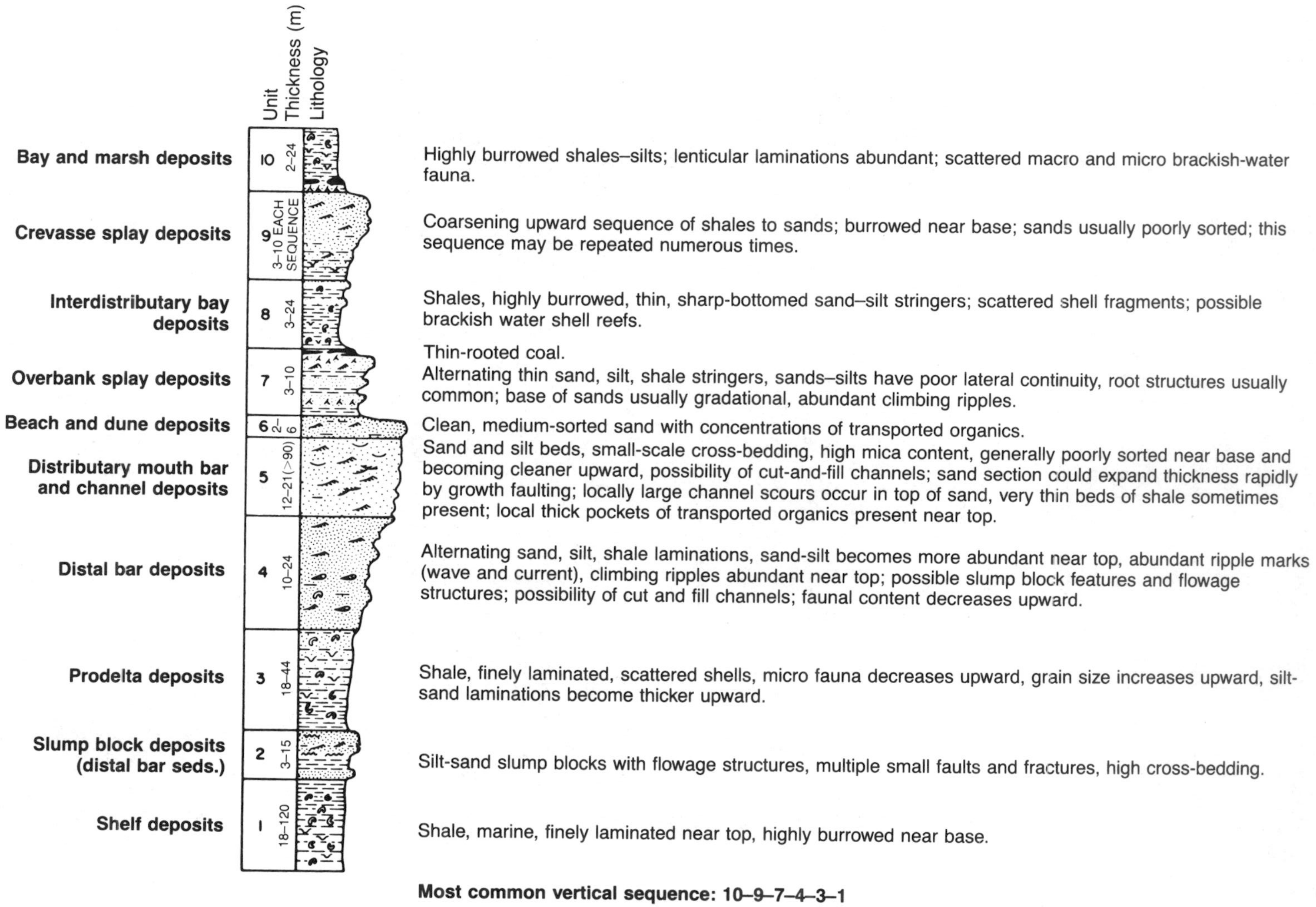

FIGURE 12.14 Idealized vertical sequence of facies in a fluvial-dominated (Mississippi) delta.

Deposit	Unit	Thickness (m)	Description
Bay and marsh deposits	10	2–24	Highly burrowed shales–silts; lenticular laminations abundant; scattered macro and micro brackish-water fauna.
Crevasse splay deposits	9	3–10 each sequence	Coarsening upward sequence of shales to sands; burrowed near base; sands usually poorly sorted; this sequence may be repeated numerous times.
Interdistributary bay deposits	8	3–24	Shales, highly burrowed, thin, sharp-bottomed sand–silt stringers; scattered shell fragments; possible brackish water shell reefs.
Overbank splay deposits	7	3–10	Thin-rooted coal. Alternating thin sand, silt, shale stringers, sands–silts have poor lateral continuity, root structures usually common; base of sands usually gradational, abundant climbing ripples.
Beach and dune deposits	6	2–6	Clean, medium-sorted sand with concentrations of transported organics.
Distributary mouth bar and channel deposits	5	12–21(<90)	Sand and silt beds, small-scale cross-bedding, high mica content, generally poorly sorted near base and becoming cleaner upward, possibility of cut-and-fill channels; sand section could expand thickness rapidly by growth faulting; locally large channel scours occur in top of sand, very thin beds of shale sometimes present; local thick pockets of transported organics present near top.
Distal bar deposits	4	10–24	Alternating sand, silt, shale laminations, sand-silt becomes more abundant near top, abundant ripple marks (wave and current), climbing ripples abundant near top; possible slump block features and flowage structures; possibility of cut and fill channels; faunal content decreases upward.
Prodelta deposits	3	18–44	Shale, finely laminated, scattered shells, micro fauna decreases upward, grain size increases upward, silt-sand laminations become thicker upward.
Slump block deposits (distal bar seds.)	2	3–15	Silt-sand slump blocks with flowage structures, multiple small faults and fractures, high cross-bedding.
Shelf deposits	1	18–120	Shale, marine, finely laminated near top, highly burrowed near base.

Most common vertical sequence: 10–9–7–4–3–1

FIGURE 12.14 Idealized vertical sequence of facies in a fluvial-dominated (Mississippi) delta. (From Coleman, J. M., 1981, Deltas: Processes of deposition and models for exploration, 2nd ed. Fig. 4.3, p. 91, reprinted by permission of IHRDC Publications, Boston.)

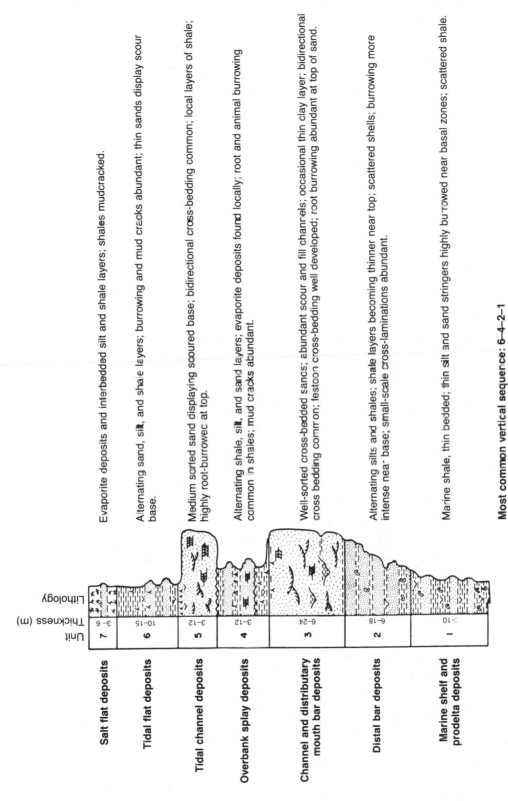

Unit	Thickness (m)	Lithology		
7	3–6		Salt flat deposits	Evaporite deposits and interbedded silt and shale layers; shales mudcracked.
6	10–15		Tidal flat deposits	Alternating sand, silt, and shale layers; burrowing and mud cracks abundant; thin sands display scour base.
5	3–12		Tidal channel deposits	Medium sorted sand displaying scoured base; bidirectional cross-bedding common; highly root-burrowed at top.
4	3–12		Overbank splay deposits	Alternating shale, silt, and sand layers; evaporite deposits found locally; root and animal burrowing common in shales; mud cracks abundant.
3	6–24		Channel and distributary mouth bar deposits	Well-sorted cross-bedded sands; abundant scour and fill channels; occasional thin clay layer; bidirectional cross bedding common; festoon cross-bedding well developed; root burrowing abundant at top of sand.
2	6–18		Distal bar deposits	Alternating silts and shales; shale layers becoming thinner near top; scattered shells; burrowing more intense near base; small-scale cross-laminations abundant.
1	<10		Marine shelf and prodelta deposits	Marine shale, thin bedded; thin silt and sand stringers highly burrowed near basal zones; scattered shale.

Most common vertical sequence: 6–4–2–1

FIGURE 12.15 Idealized vertical sequence of facies in a tide-dominated (Ord) delta. (From Coleman, J. M., 1981, Fig. 4.9, p. 97, reprinted by permission of IHRDC Publications, Boston.)

413

to saltwater fauna can be recognized in facies having the characteristics described under geometry and lateral facies environments, this is indicative of deltaic origin; however, such transitions can occur also in estuarine sediments.

Ancient Deltaic Systems

Many of the ancient delta systems that have been recognized and described appear to be fluvial-dominated deltas or deltas transitional between fluvial-dominated types and other types. Ancient examples of wave-dominated and tide-dominated deltaic systems are known also. Miall (1984b) provides a bibliography of selected examples of these various types of ancient delta systems. Deltaic sediments have been reported in stratigraphic sequences of many ages, but they are particularly common in Pennsylvanian and Tertiary rocks. Deltaic deposits have been described in several Pennsylvanian-age formations in the Appalachians and the Illinois Basin, the Atoka Formation and Bluejacket Sandstone of Oklahoma, and the Carboniferous Abottsham Formation of Great Britain. Tertiary-age deltas include the Coaledo Formation (Eocene) of Oregon and many deltas in the Gulf Coast, such as the Wilcox Group (Eocene) and Frio delta systems of Texas. Other examples of ancient delta systems include the Ordovician Queenston delta and the Devonian Catskill delta in the central and southeastern Appalachians and deltas of the Frontier Formation (Cretaceous), Wyoming. Many of these, and other, examples of ancient deltas are described in Broussard (1975), Eliott (1978a), Morgan (1970), and Shirley and Ragsdale (1966).

The Rockdale delta system of the Wilcox Group in Texas is a well-studied subsurface example of an ancient fluvial-dominated delta (Fisher and McGowen, 1967). The Rockdale system is a lobate delta characterized by interdistributary-bay muds, muddy marsh deposits containing numerous lignite coal beds, channel sands, and muddy, fine-grained delta-front sands (Fig. 12.16). Thin units of wave- and current-reworked sand, referred to by Fisher and McGowen as marine destructive units, are interbedded with the thicker delta-front muddy sands. Delta-front deposits grade laterally in a seaward direction into prodelta muds. A complete vertical sequence through the Rockdale delta system thus begins with prodelta muds at the base and grades upward through muddy, delta-front sands, with interbeds of better sorted marine destructive units, to muddy marsh deposits with lignite lenses.

Tankard and Barwis (1982) describe a wave-dominated delta system in the Devonian Bokkeveld Basin of South Africa. Figure 12.17 shows a stratigraphic section through the Bokkeveld Group in which several coarsening-upward delta cycles are evident. This figure shows also the detailed facies in an idealized cycle and the environmental interpretation of these facies. The lower 105 m of the detailed section are facies typical of river-dominated delta progradation. This unit consists of a coarsening-upward sequence of shales, siltstones, and thin sandstones interpreted as the buildup of prodelta to distal-mouth-bar deposits. Proximal, cross-bedded, distributary-mouth-bar sandstones form the top of this sequence. The uppermost part of the detailed section shows smaller cycles up to 20 m thick composed of sandstones with a few interbedded mudstones, representing the progradation of river-mouth bars and barrier and tidal sands produced by marine reworking.

These examples illustrate the characteristics of two specific ancient delta systems and are not intended to serve as general models for delta deposits. Readers should consult the references given in the preceding discussion for additional examples of the wide variety of facies and stratigraphic sequences that characterize ancient delta systems.

Mt. Pleasant Fluvial System

☐ Tributary channel facies
▤ Slightly meandering channel facies
▦ Highly meandering channel facies

Rockdale Delta System

▨ Marsh–distributary channel facies
■ Lignite
▦ Delta front facies
▧ Marine destructive units
▤ Prodelta facies

Cross section scale

160 m

12 km

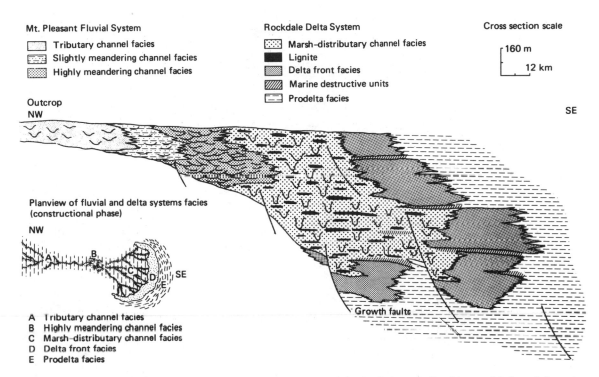

Outcrop
NW

SE

Planview of fluvial and delta systems facies
(constructional phase)

NW

SE

Growth faults

A   Tributary channel facies
B   Highly meandering channel facies
C   Marsh–distributary channel facies
D   Delta front facies
E   Prodelta facies

FIGURE 12.16 Diagrammatic cross section showing facies relationships and inferred depositional environment of the Mt. Pleasant Fluvial System and Rockdale Delta System in the Eocene Wilcox Formation of Texas. (From Fisher, W. L., and J. H. McGowen, 1967, Depositional systems in the Wilcox Group of Texas and their relationship to occurrence of oil and gas: Gulf Coast Assoc. Geol. Soc. Trans., v. 17. Fig. 3, p. 109, reprinted by permission.)

12.3   BEACH AND BARRIER-ISLAND SYSTEMS

Introduction

Mainland beaches are long, narrow accumulations of sand aligned parallel to the shoreline and attached to land. Bodies of beach sand are typically cut across here and there by headlands and sea cliffs, estuaries, river deltas, tidal inlets, bays, and lagoons. Barrier-island beaches are similar to mainland beaches but are separated from land by a shallow lagoon, estuary, or marsh. They are also commonly dissected by tidal channels or inlets. Beaches may occur within delta systems, along depositional strike from deltas, or in oceanic or even lacustrine settings that have no connection with deltas (Eliott, 1978b). They are the most dynamic of all depositional environments and are subject to both seasonal and longer range changes that keep them in a state of virtually constant flux. In contrast to deltas, which are influenced by both fluvial and marine processes, beach and barrier-island systems are generated predominantly by marine processes, aided to a minor degree by eolian sand transport.

Modern and Holocene beaches have perhaps been studied more extensively than any other depositional environment owing to their recreational use; accessibility; economic potential as a source of placer gold, platinum, and various minerals; and im-

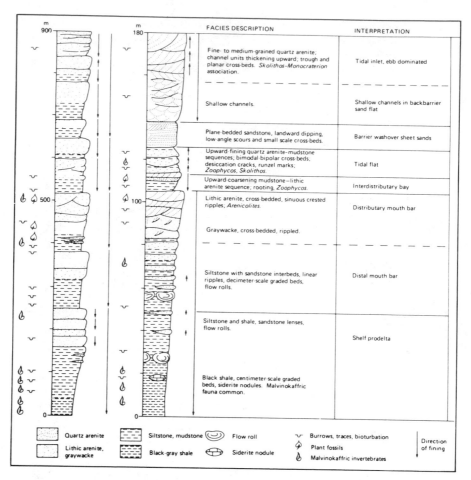

FIGURE 12.17 Stratigraphic section through the Bokkeveld Group, Cape Provence, South Africa (left), and an idealized cyclic sequence (right) showing detailed lithofacies and environmental interpretation. (From Miall, A. D., 1984, Deltas, *in* R. G. Walker (ed.), Facies models: Geoscience Canada Reprint Ser. 1. Fig. 14, p. 114, reprinted by permission of Geological Association of Canada. Modified from A. J. Tankard and J. H. Barwis, 1982, Wave-dominated deltaic sedimentation in the Devonian Bokkeveld Basin of South Africa: Jour. Sedimentary Petrology, v. 52, Fig. 12, p. 967.)

portance as an erosion buffer between the sea and the land. Much of the study of modern beaches has been carried out by coastal engineers, geographers, and geomorphologists. Geologists also have a strong scientific interest in beaches because of the insight beaches provide into ancient depositional processes and environments. Moreover, beaches are a great place to do research and relax a little on the side, which may account for some of the research appeal. Ancient beach deposits, also have been extensively studied. In addition to their significance as indicators of ancient nearshore processes and conditions, ancient beach and barrier-island sediments have considerable economic importance as reservoirs for petroleum and natural gas and as host rocks for uranium. Hundreds of research papers have been devoted to study of modern and ancient beaches, and several books have been published on this subject (e.g., Davies et al., 1971; R. A. Davis and Ethington, 1976; R. A. Davis and Fox, 1972; J. S. Fisher and Dolan, 1977; Hayes and Kana, 1976; Hails and Carr, 1975; Komar, 1976, Schwartz, 1973). Useful shorter summaries of beach and barrier-island systems are provided by R. A. Davis (1978a), Eliott (1978b), Heward (1981), McCubbin (1982), Reineck and Singh (1980), and Reinson (1984), among others. Although most beaches are composed of siliciclastic sediments, some modern beaches on carbonate shelves are made up predominantly of carbonate grains consisting of skeletal fragments, oolites, pellets, and other particles. Carbonate beach deposits are known also from the geologic record. A good discussion of carbonate beaches is given in Inden and Moore (1983).

Depositional Setting

Beach and barrier-island complexes are best developed on wave-dominated coasts where tidal range is small to moderate. Coasts are classified on the basis of tidal range into three groups: (1) **microtidal** (0–2 m tidal range), (2) **mesotidal** (2–4 m tidal range), and (3) **macrotidal** (>4 m tidal range). Hayes (1975) has shown that barrier-island and associated environments occur preferentially along microtidal coasts, where they are well developed and nearly continuous. They are less characteristic of mesotidal coasts and when present are typically short or stunted, with tidal inlets common. Barriers are generally absent on macrotidal coasts because extreme tidal range causes wave energy to be dispersed and dissipated over too great a width of shore zone to effectively form barriers.

Considerable difference of opinion exists about the origin of barrier-island complexes. As summarized by Reineck and Singh (1980), mechanisms of origin may include (1) shoal and longshore-bar aggradation, that is, upward building and eventual emergence of offshore bars; (2) spit segmentation by breaching and detachment of spits oriented parallel to the coast; (3) mainland ridge engulfment owing to submergence and drowning of shoreline-attached beaches; (4) welding or veneering of Holocene dune, beach, and foreshore sand into and over pre-Holocene topographic highs; and (5) lateral shifting of coastal sands during transgression to form the barrier islands. Submergence of mainland beaches appears to be the most favored hypothesis, but several modes of origin seem possible.

Subenvironments. The deposits of the beach and barrier-island environment can occur as (1) a **single beach** attached to the mainland; (2) a broader beach-ridge system that constitutes a **strand plain** consisting of multiple parallel beach ridges and parallel swales, but which generally lacks well-developed lagoons or marshes; or (3) a **barrier island** separated wholly or partly from the mainland by a lagoon or marsh (Fig. 12.18). A type of strand plain consisting of sandy ridges elongated along the coast and separated by coastal mudflat deposits is called a **chenier plain.** As illustrated in Figure 12.19, the barrier-island setting is not a single environment but a composite of three separate environments: (1) the sandy barrier-island chain itself (the subtidal to subaerial barrier-beach complex); (2) the enclosed lagoon, estuary, or marsh behind it (the back-barrier, subtidal–intertidal region); and (3) the channels that cut through the barrier and connect the back-barrier lagoon to the open sea (the subtidal–intertidal delta and inlet–channel complex) (Reinson, 1984). Lagoonal, estuarine, and tidal-flat environments are discussed in subsequent sections of the book and will not be considered in detail here; however, identification and interpretation of ancient barrier-island complexes require that this intimate association of facies be recognized. Barrier-island systems are not simply prograding barrier-beach complexes.

FIGURE 12.18 Morphological relationship between beaches, strand plains, and barrier islands. (From Reinson, G. E., 1984, Barrier island and associated strand-plain systems, *in* R. G. Walker (ed.), Facies models: Geoscience Canada Reprint Ser. 1. Fig. 1, p. 119, reprinted by permission of Geological Association of Canada.)

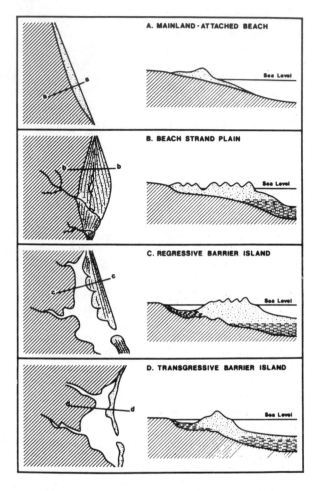

Morphology of the Beach–Nearshore Zone. The morphological features of the beach profile are similar on mainland coasts and the seaward coast of barriers. The beach is divided into the **backshore,** which extends landward from the beach berm above high-tide level and commonly includes back-beach dune deposits; the **foreshore,** which mainly encompasses the intertidal zone between low-tide and high-tide level; and the **shoreface,** also called the nearshore, that extends from about low-tide level to the transition zone between beach and shelf sediments (Fig. 12.20A). Figure 12.20A illustrates also the approximate zones of shoaling and breaking waves and the positions of the surf zone and swash zone, discussed in a succeeding section. The relationship of the beach profile to other elements of a barrier-island complex are illustrated in Figure 12.20B.

Depositional Processes on Beaches

Transport and depositional processes on beaches have been studied extensively by engineers interested in coastal processes, as well as by geologists. The published results of engineering studies tend to be expressed in mathematical terms that are not

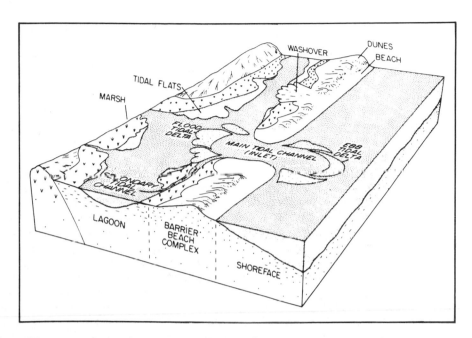

FIGURE 12.19 Subenvironments of a barrier island system. (From Reinson G. E., 1984, Fig. 3, p. 120, reprinted by permission of Geological Association of Canada.)

FIGURE 12.20 Generalized profile of (A) the beach and nearshore zone and (B) a barrier-island complex showing major environments. (A from Reinson, G. E., 1984, Fig. 5, p. 122, reprinted by permission of Geological Association of Canada. B from Richard A. Davis, Jr., Depositional systems: A genetic approach to sedimentary geology, © 1983, Fig. 12.2, p. 405. Reprinted by permission of Prentice-Hall, Englewood Cliffs, N.J.)

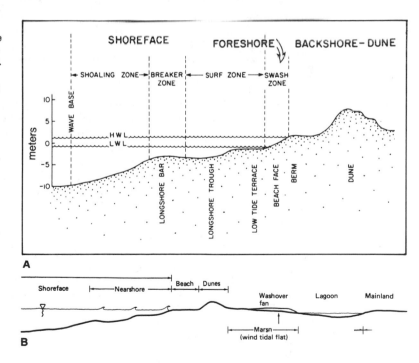

419

too well understood by the average, nonmathematically inclined geologist. Perhaps the most detailed and mathematically rigorous description of beach processes written by a geologist is that of Komar (1976). Beach processes have been summarized in less rigorous form by several other workers, including R. A. Davis (1978a), Eliott (1978b), and Reineck and Singh (1980). Only a very brief description of these processes is given here.

Wave Processes. Beaches and barrier islands are constructed primarily by wave processes. The influence of orbital wave motion in generating bedforms such as ripples and dunes is discussed in Chapter 3. As deep-water orbital waves approach shallow water where depth is about one-half the wave length, the orbital motion of the water is impeded by interaction with the bottom. Orbits become progressively more elliptical and eventually develop near the bottom a nearly horizontal to-and-fro motion that can move sediment back and forth. This to-and-fro movement is important in generating ripple bedforms as well as in producing some net sediment transport. As waves progress farther shoreward into the shallow **shoaling zone,** forward velocity of the wave slows, wave length decreases, and wave height increases. The waves eventually steepen to the point where orbital velocity exceeds wave velocity and the wave breaks, creating the **breaker zone.** Breaking waves generate turbulence that throws sediment into suspension and also brings about a transformation of wave motion to create the **surf zone.** In this zone, a high-velocity translation wave, or bore, is projected up the beach face, causing landward transport of bedload sediment and generation of a short-duration "suspension cloud" of sediment. At the shoreline, the surf zone gives way to the **swash zone,** in which a rapid, very shallow swash flow moves up the beach, followed almost immediately by a backwash flow down the beach. The backwash begins at very low velocity but accelerates quickly. The width of the surf zone is governed by the steepness of the beach face. Very steep beaches may show no surf zone at all and waves break very close to shore, whereas gently sloping beaches commonly have very wide surf zones.

Sediment transport on beaches is particularly important landward of the shoaling zone. In the high-energy breaker zone, coarse sediments move by saltation in a series of elliptical paths that move sediment parallel to the coast, while finer sediment is thrown into suspension (Fig. 12.21). So-called translation waves, which are actually currents, transport sediment through the surf and swash zone up the beach face. If waves approach the shoreline obliquely, a very common occurrence, sediment is transported alongshore in a zigzag manner owing to the fact that the upswash is directed across the beach at an angle, whereas the backswash flow is normal to the beach face. Thus, waves of normal—that is, moderate to low—energy tend to produce a net landward and alongshore transport of sediments in a largely constructive sedimentation regime in which the beach builds owing to deposition. Repeated deposition and reentrainment of sediment in the beach regime tend to winnow and remove the finest sediment, producing generally well-sorted, positively skewed deposits. During high-energy conditions created by storms, steep, long-period storm waves cause considerable erosion of the beach area and a net displacement of sediment in a seaward direction (R. A. Davis, 1978a). Great quantities of sediment are thrown into suspension during storms for transport by surf-zone currents, and sand bars on the inner beach may be planed off and displaced seaward considerable distances. Thus, it is quite common to observe marked seasonal changes on modern beaches, which often build in a landward direction during low-energy summer conditions but are eroded and reduced in size during winter storm conditions.

FIGURE 12.21 Sediment transport associated with a breaking wave. Coarser sediment moves as bedload in a series of loops at position (B). Finer sediment moves in suspension in position (A). Shoreward sediment (C) and seaward sediment (D) move toward the breaking wave as shown by the arrows. (After Ingle, J. C., 1966, The movement of beach sand: Developments in Sedimentology, v. 5. Fig. 46, p. 53, reprinted by permission of Elsevier Science Publishers, Amsterdam.)

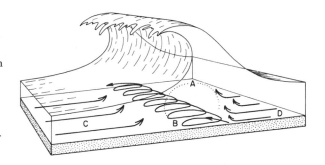

Wave-induced Currents. As breakers and winds pile water against the beach, they not only create bidirectional translation waves that move up and down the swash zone, but they also create two different types of unidirectional currents. **Longshore currents** are generated when waves that approach the shore at an angle break, and a portion of the translation wave is deflected laterally parallel to the shore. These currents move parallel to shore following longshore troughs, which are shallow troughs in the lower part of the surf zone oriented parallel to the strandline (**Fig. 12.20**). This system of parallel longshore troughs between shallow beach ridges is referred to as a ridge-and-runnel system. The velocity of longshore currents is related to wave height and the angle at which the waves approach shore. As water piles up between shallow sand bars and the shoreline with continued shoreward movement of waves, it cannot go back against incoming waves the way it came. It must find a different way to return seaward. Thus, it moves parallel to shore as a longshore current until it finds a topographic low between sand bars, where it converges and moves seaward as a narrow, near-surface current. These converging, seaward-moving currents are called **rip currents** (**Fig. 12.22**). Longshore currents play a very important role in sediment transport and deposition on beaches because they achieve velocities great enough to transport

FIGURE 12.22 Schematic representation of longshore currents and rip currents generated in the surf by obliquely approaching waves. (From McCubbin, D. G., 1982, Barrier island and strand-plain facies, in P. A. Scholle and D. Spearing (eds.), Sandstone depositional environments: Am. Assoc. Petroleum Geologists Mem. 31. Fig. 11, p. 252, reprinted by permission of AAPG, Tulsa, Okla.)

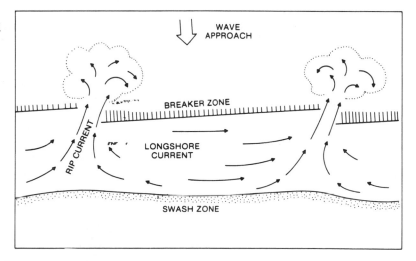

sand. Together with the processes producing swash transport, they are primary agents of alongshore sand movement. The transport paths of sand under longshore currents of different relative velocity are illustrated in Figure 12.23. Rip currents are primarily surface phenomena and thus are less important in near-bed sediment transport than are longshore currents. Nonetheless, they have been reported to entrain considerable quantities of sediment and move it through the breaker zone back out into shoal water.

Characteristics of Beach and Barrier-Island Deposits

Overall Geometry and Lithofacies. The mainland-beach and barrier-island system as a whole generates a narrow body of sediments elongated parallel to the depositional strike, or strike of the shoreline. This body of sediment is composed predominantly of sand that originates on the beach shoreface, foreshore, and backshore and is commonly tens to hundreds of meters broad, up to hundreds of kilometers long, and 10–20 m thick (Reineck and Singh, 1980). It may be interrupted in many places along its length by deltaic, estuarine, bay, and other deposits where these features cut across the beach. Where barrier islands occur, sands of the barrier beach grade landward into back-barrier sediments that may include washover sands; tidal-delta sands and muds; lagoonal silts and muds; and sandy, muddy tidal-flat and marsh deposits.

Shoreface Deposits. The shoreface environment of the beach is constantly under water and extends from mean low-tide level down to the lower limit of fair-weather wave base. Wave base is the depth below which normal waves do not react with the bottom. It is a function of wave length and wave period, and thus varies with wave conditions on the shoreface. The depth of wave base on the shoreface is commonly on the order of 10 m (Fig. 12.20A), but this depth can be lowered significantly during storms. The shoreface can be divided into lower, middle, and upper shorefaces, each of which is distinguished by characteristic facies.

  Upper shoreface deposits form within the surf zone in an environment dominated by strong bidirectional translation waves and longshore currents. Depending upon local sediment supply and energy conditions, sediment textures range from fine sand to gravel. Sedimentary structures consist predominantly of multidirectional trough cross-bed sets, but may also include low-angle, bidirectional cross-beds and subhorizontal plane beds. Bidirectional cross-beds oriented parallel to depositional strike are common also. These structures may indicate deposition under strong longshore current conditions (Reinson, 1984). Trace fossils such as *Skolithos* are common but not abundant.

  Middle shoreface deposits form under higher energy conditions owing to breaking waves and associated longshore and rip currents. This is the zone of longshore-bar development. Sediments consist of fine- to medium-grained clean sand with minor amounts of silt and shell material. Sedimentary structures can be highly complex, depending upon the presence or absence of longshore bars, and can include landward-dipping ripple cross-lamination; seaward-dipping, low-angle planar bedding; subhorizontal plane laminations; and seaward- and landward-dipping trough cross-beds. Trace fossils consisting of vertical burrows, such as those of *Skolithos* and *Ophiomorpha,* are common in this zone.

  Lower shoreface deposits form under relatively low-energy conditions and grade seaward into open-shelf deposits. They are composed predominantly of fine to very fine sand but may also contain thin intercalated layers of silt and mud. Small-scale

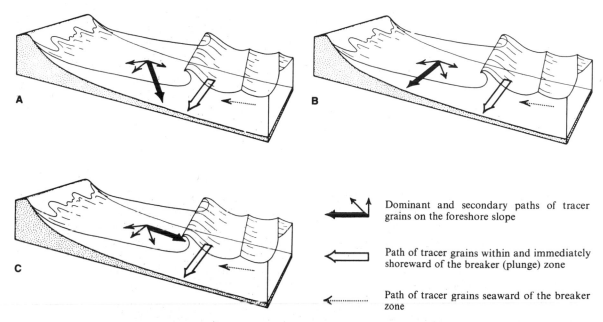

FIGURE 12.23 Transport of sand on beaches by longshore currents under different surf conditions. A. Sediment movement under surf conditions where the longshore current and the wave motion exert an equal influence. B. Sediment grain motion under conditions of a high-velocity longshore current (velocity >60 cm/sec). C. Sediment grain motion where the longshore current velocity is <30 cm/sec, and the onshore-offshore motion of waves controls the sediment grain transport. (After Ingle, J. C., 1966, The movement of beach sand: Developments in Sedimentology, v. 5. Fig. 46, p. 53, reprinted by permission of Elsevier Science Publishers, Amsterdam.)

cross-stratification formed by predominantly landward-migrating ripples occurs, but planar, nearly horizontal laminated bedding is the predominant sedimentary structure preserved. Hummocky cross-stratification may occur also in lower shoreface sands. Hummocky cross-stratification has not been observed in modern deposits, but it has now been reported in many ancient shoreface sequences. Such deposits may, in fact, constitute much of the ancient record of shoreface deposits. In some deposits, hummocky stratified beds have a basal lag deposit consisting of shells or mud clasts and may be capped by a thin layer with wave-ripple stratification, suggesting that each hummocky stratified unit was deposited by a single storm event (McCubbin, 1982). Such deposits are also reported to contain abundant plant materials, mica flakes, and other hydraulically light particles. Laminations in lower shoreface deposits tend to be obliterated by bioturbation, and suspension-feeder and deposit-feeder traces such as *Thalassinoides* may be common.

Storms may strongly modify shoreface deposits formed under normal wave-energy conditions because effective wave base can be lowered significantly by storm-generated waves, as mentioned. The effects of storm waves are particularly important on the middle and lower shoreface, causing severe erosion and redeposition of sediment. Waves scour the bottom, causing suspension of sediment, which is then redeposited farther seaward as the storm wanes (Reinson, 1984). Storm deposits tend to be thicker and more lenticular than normal shoreface deposits and, as discussed in the preceding paragraph, tend to be characterized by hummocky cross-stratification.

Foreshore and Backshore Deposits. The **foreshore,** or beach face, is the intertidal zone extending from mean low-tide level to mean high-tide level, corresponding to the zone of wave swash. Sediments of the foreshore consist predominantly of fine to medium sand, but may include scattered pebbles and pebble lenses. Sedimentary structures are mainly parallel laminations that dip gently (2–3 degrees) seaward and form during swash–backwash flow. Thin, heavy mineral laminae are commonly present, alternating with layers of quartzose sand. Thin, lenticular sets of low-angle, landward-dipping laminations, possibly formed by antidune migration during backswash, may be present. Some foreshore sands display high-angle, landward-dipping cross-beds caused by migration or foreshore ridges. The foreshore is separated from the backshore by a break in slope at the berm crest. The **backshore** is inundated only during storm conditions and is thus a zone dominated by intermittent storm-wave deposition and eolian sand transport and deposition. Faint, landward-dipping, horizontal laminations, interrupted locally by crustacean burrows, record deposition by storm waves. These beds may be overlain by small- to medium-scale eolian trough cross-bed sets, which are commonly disturbed by root growths and burrows of land-dwelling organisms.

Summary Characteristics of Beach Deposits. Beach deposits are composed predominantly of fine- to medium-grained, well-sorted sand that displays subhorizontal parallel laminations and low-angle, seaward-, landward-, and alongshore-dipping crossbeds. Bioturbation structures are common in middle and lower shoreface deposits and in sediments of the transition zone between the beach and open shelf. The characteristic features of an idealized, low-energy, prograding beach are illustrated in Figure 12.24.

Back-Barrier Deposits. Sediments are deposited in several subenvironments in the back-barrier lagoon landward of barrier beaches. **Washover deposits** occur where storm-driven waves cut through and overtop barriers, washing lobes of sandy beach sediment into the back-barrier lagoon (Fig. 12.19). Washover sediments consist predominantly of fine- to medium-grained sand that displays subhorizontal planar laminations and small- to medium-scale delta-foreset bedding. Where tidal channels cut through barriers into the inner lagoon, sediments are deposited in a number of tide-related environments, including tidal channels, tidal deltas, and tidal flats (Fig. 12.19). **Tidal-channel deposits** consist predominantly of sand, and the deposits commonly have an erosional base marked by coarse lag sands and gravels. Sedimentary structures may include bidirectional large- to small-scale planar and trough cross-beds that may display a general fining-upward textural trend. **Tidal-delta deposits** form on both the lagoonal side of the barrier (flood-tidal delta) and the seaward side of the barrier (ebb-tidal delta). They are predominantly sandy deposits up to tens of meters thick with a gross parabolic shape or geometry. They are characterized by a highly varied sequence of planar and trough cross-bed sets that may dip in either a landward or seaward direction. **Tidal-flat deposits** form along the margins of the mainland coast and the back of the barrier. They grade from fine- to medium-grained ripple-laminated sands in lower areas of the tidal flats through flaser- and lenticular-bedded fine sand and mud in middle tidal flats to layered muds in higher parts of the flats. **Lagoonal** and

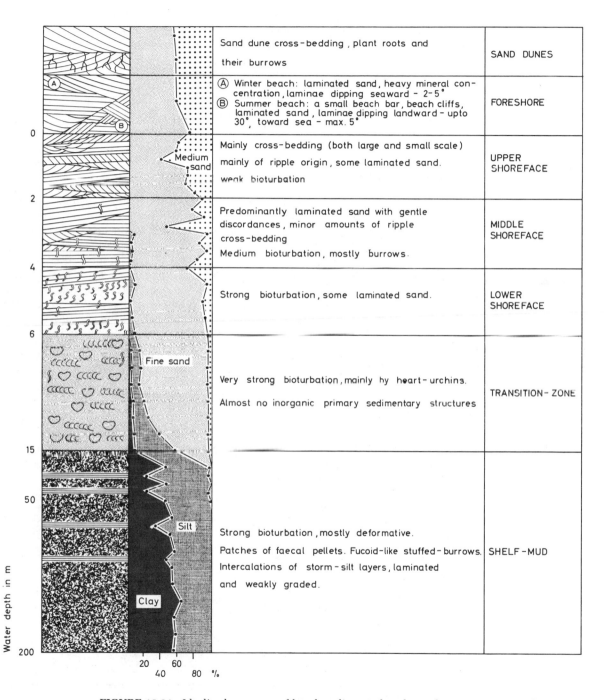

FIGURE 12.24 Idealized sequence of beach sediments found on a low-energy, prograding beach. (From Reineck, H. E., and I. B. Singh, 1980, Depositional sedimentary environments, 2nd ed. Fig. 534, p. 387, reprinted by permission of Springer-Verlag, Heidelberg.)

marsh deposits accumulate in the low-energy back-barrier lagoon and grade laterally into higher energy, sandy deposits of tidal channels, deltas, and washover lobes. They consist largely of interbedded and interfingering fine sands, silts, muds, and peat deposits that may be characterized by disseminated plant remains, brackish-water invertebrate fossils such as oysters, and horizontal to subhorizontal layering. A generalized sequence of facies deposited in a back-barrier environment is illustrated in Figure 12.25. Such a sequence might be confused with some deltaic sequences, although the absence of delta-front muddy sands and prodelta clays and mud turbidites should help to distinguish it from deltaic deposits.

Transgressive and Regressive Barrier Sequences

The vertical sequence of facies generated in beach and barrier-island systems depends upon whether changes in depositional environments with time are the result of transgression or regression. The different effects of transgression and regression are illustrated in Figure 12.26. The specific characteristics of transgressive or regressive sequences depend upon rates of sea-level change, rates of basin subsidence, and the sediment supply. Transgression, or movement of the shoreline in a landward direction, can take place as a result of rising sea level, either eustatic or relative rise, provided that the influx of terrigenous clastic sediment is not too rapid to prevent the shoreline from shifting landward. Regression, or seaward shift of the shoreline, occurs particularly as a result of falling sea level, but it can occur also during static sea level or even rising sea level if influx of clastic sediment is exceptionally great. These factors are discussed further in Chapter 14.

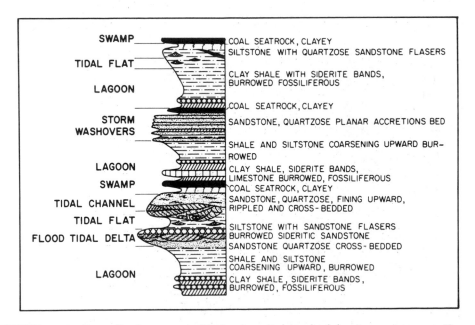

FIGURE 12.25 Generalized sequence of facies deposited in a back-barrier environment. (From Horne, J. C., J. C. Ferm, F. T. Caruccio, and B. P. Baganz, 1978, Depositional models in coal exploration and mine planning in Appalachian region: Am. Assoc. Petroleum Geologists Bull., v. 62. Fig. 4, p. 2385, reprinted by permission of AAPG, Tulsa, Okla.)

FIGURE 12.26 Effects of regression (A) and transgression (B) on barrier beach complexes. (After Galloway, W. E., and D. K. Hobday, 1983, Terrigenous clastic depositional systems. Fig. 6.10, p. 126, reprinted by permission of Springer-Verlag, Heidelberg. A originally after H. A. Bernard, C. F. Major, Jr., B. S. Parrott, and J. J. LeBlanc, Sr., 1970, Recent sediments of southeast Texas—A field guide to the Brazos alluvial and deltaic plains and the Galveston barrier island complex: Tex. Bur. Econ. Geology. B originally from J. C. Kraft and C. J. John, 1979, Lateral and vertical facies relations in a transgressive barrier: Am. Assoc. Petroleum Geologists Bull., v. 63, Fig. 12, p. 2161.)

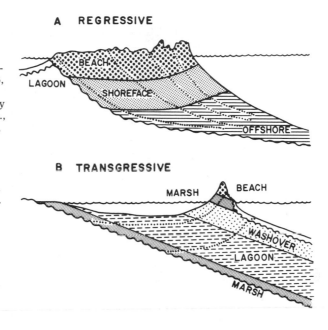

In any case, regression brings about overlap of back-barrier lagoonal and associated deposits onto barrier-beach deposits, which in turn overlap finer grained offshore deposits (Fig. 12.26A). Regressive beach and barrier complexes are thought to have relatively good preservation potential, and many ancient regressive or progradational beach and barrier complexes have been reported, particularly from Tertiary and Cretaceous sequences. Theoretically, transgressive sequences also can occur, as illustrated in Figure 12.26B, leading to overlap of back-barrier lagoonal and marsh deposits by sandy deposits of the barrier-beach complex. The preservation of beach and barrier-island deposits during transgression is, in fact, controversial. The controversy centers around disagreement about whether landward advance of the shoreline occurs owing to shoreface erosion, as might take place during slow rise of sea level, or whether it occurs by relatively sudden upward "jumps" of the shoreline during rapidly rising sea level. Beach and upper shoreface deposits would presumably be destroyed during transgression, owing to shoreface erosion, but might be preserved owing to "sudden" inundation, or drowning, during a rapid rise in sea level. Ancient beach and barrier deposits interpreted to be transgressive deposits have been reported (McCubbin, 1982), but such deposits appear to be much less common than progradational sequences.

Differences in vertical sequences of lithofacies generated by transgression and regression are illustrated in Figure 12.27. The general sequence of facies deposited on a regressive, prograding beach is shown in Figure 12.27A (see also Fig. 12.24). In general, regression produces a coarsening-upward sequence from fine-grained lower-shoreface deposits to coarser foreshore and backshore deposits. If back-barrier lagoonal and marsh deposits are preserved during regression, the upper part of this sequence will be overlain by these finer grained deposits (not shown in Fig. 12.27A). The model for a transgressive barrier-island complex (Fig. 12.27B) is more complicated than the regressive model. It is characterized by interbedded back-barrier deposits and does not display a definite fining- or coarsening-upward trend. The model illustrated in Figure

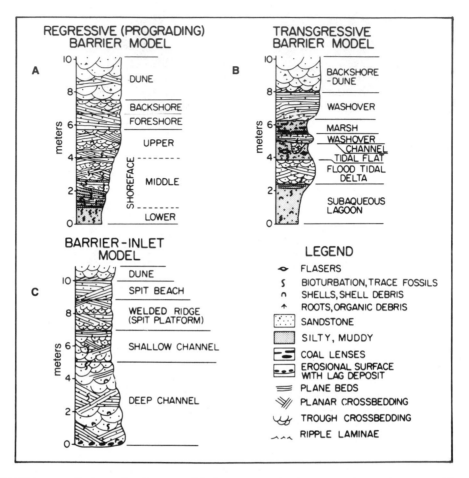

FIGURE 12.27 End-member facies models for transgressive barrier, regressive barrier, and barrier inlet stratigraphic sequences. A standard 10 m unit is shown, but thickness could range up to a few tens of meters. (From Reinson, G. E., 1984, Barrier island and associated strand-plain systems, in R. G. Walker (ed.), Facies models, 2nd ed: Geoscience Canada Reprint Ser. 1. Fig. 26, p. 133, reprinted by permission of Geological Association of Canada.)

12.27B is incomplete in the sense that it does not show the deposits of the foreshore and shoreface that should lie on top of backshore-dune deposits with continued transgression. Figure 12.27C illustrates an idealized model of the vertical sequence of facies generated in the barrier-island environment by migration of spit-beach sands over tidal-channel deposits.

Beach and Barrier-Island Deposits in the Geologic Record

Recognition of ancient beach and barrier-island complexes requires general knowledge of idealized models, such as those shown in Figure 12.27, plus an understanding of how real stratigraphic sequences can deviate from these norms. Facies developed during regression produce different vertical sequences than those formed during transgression; sequences developed on mainland beaches without barriers lack the back-barrier lagoonal and associated deposits that characterize those deposited on coasts with barrier islands; and local variations in depositional patterns within a given beach and barrier-island environment can generate different vertical patterns of facies that may include tidal-delta deposits, tidal-channel deposits, beach deposits, and back-barrier lagoonal and marsh deposits. Two examples are included here to illustrate differences in ancient regressive barrier deposits and transgressive deposits. Readers should consult the references listed in the introduction and under additional readings at the end of the chapter for further examples.

Figure 12.28 shows a regressive, or progradational, sequence in the Cretaceous Gallup Sandstone of northwestern New Mexico (McCubbin, 1982). In this sequence, sandy beach deposits overlie burrowed silty, offshore shales containing normal marine fossils. The basal unit of the sequence consists of fine- to very fine-grained sandstone with mostly subhorizontal to planar stratification and hummocky cross-stratification. Burrows are abundant. This unit is overlain above a scoured surface by fine- to medium-grained sandstones with high-angle cross-stratification in trough-shaped sets. Some thin interbeds with planar stratification are present also. Burrowing is less common in this unit than in the basal section. The uppermost part of the sequence consists of fine-grained, well-sorted sandstones characterized by nearly horizontal planar stratification and very low-angle cross-stratification; root traces are present at the top. These three divisions of the Gallup Sandstone are interpreted to represent, in ascending order, lower-shoreface, upper-shoreface, and beach-foreshore deposits. The grain-size and stratification characteristics of the Gallup Sandstone suggest that it was developed on a coast with moderate to high wave energy. Because back barrier lagoonal sediments are absent in this sequence, we can infer that it is probably a mainland beach deposit, formed on a coast that lacked barriers.

The La Ventana Sandstone in the Cliff House Formation in northwestern New Mexico has been interpreted as a transgressive barrier complex (Fig. 12.29) (McCubbin, 1982). The basal part of this unit consists of laminated dark shales with abundant plant fragments and widely spaced layers with carbonized plant roots. Coal beds up to 1.8 m thick are common, and brackish-water fossils such as oysters are present. This shale and coal unit is overlain by and interfingers with fine- to medium-grained sandstones characterized by high-angle trough cross-stratification, hummocky cross-stratification, and planar stratification, as shown in Figure 12.29. *Ophiomorpha* burrows are common. These deposits are interpreted to represent a transgressive barrier sequence consisting of basal back-barrier lagoonal sediments overlain by beach-shoreface deposits.

Siliciclastic sedimentary sequences identified as beach and barrier-island complexes are known from rocks of widely differing ages in North America. To name but a few, such sequences have been reported from several Pennsylvanian formations in the Appalachian Basin of Kentucky, Virginia, West Virginia, and Tennessee; the Lower Cretaceous Muddy Sandstone of Wyoming and Montana; the Eocene Wilcox Group of eastern Texas; and the Quaternary of California. Several ancient carbonate deposits also have been interpreted as beach complexes. The Lower Cretaceous Edwards Formation of western Texas, the Lower Cretaceous Cow Creek Formation of central Texas, the Mississippian Newman Formation in eastern Kentucky, and the Mississippian Mission Canyon Formation in the Williston Basin in the Montana area are examples of ancient stratigraphic units that contain putative carbonate beach deposits. Carbonate beach deposits are further discussed in Chapter 13.

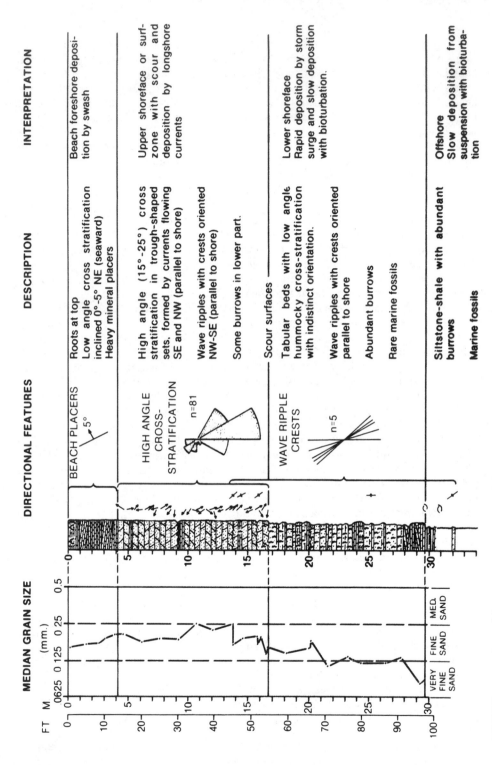

FIGURE 12.28 Vertical facies sequence in part of the Cretaceous Gallup Sandstone, northwestern New Mexico. The sequence is interpreted as a progradational beach deposit, probably formed on a nonbarred coast. (After Mc-Cubbin, D. G., 1982, Barrier island and strand-plain facies, in P. A. Scholle and D. Spearing (eds.), Sandstone depositional environments: Am. Assoc. Petroleum Geologists Mem. 31. Fig. 25, p. 260, reprinted by permission of AAPG, Tulsa, Okla.)

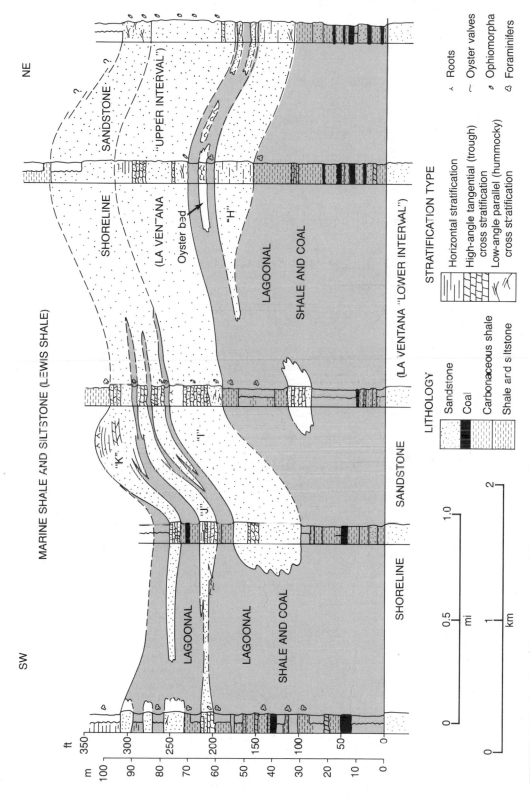

FIGURE 12.29 Stratigraphic cross section of the upper interval of the La Ventana Tongue of the Cretaceous Cliff House Formation, northwestern New Mexico. This sequence is interpreted as a transgressive barrier complex. (After McCubbin, D. G., 1982, Fig. 51, p. 273, reprinted by permission of AAPG, Tulsa, Okla.)

431

12.4   ESTUARINE AND LAGOONAL SYSTEMS

Introduction

Relatively small, semienclosed coastal embayments are loosely called coastal bays. Two broad types of coastal bays are recognized: estuaries and lagoons. **Estuaries** are generally considered to be the lower courses of rivers open to the sea; however, they have been defined somewhat differently by geologists, geographers, and chemists. In this book, I follow the usage of Fairbridge (1980, p. 7), who defines an estuary as "an inlet of the sea reaching into a river valley as far as the upper limit of tidal rise, usually divisible into three sectors: (a) a marine or lower estuary, in free connection with the open sea; (b) a middle estuary, subject to strong salt and freshwater mixing; and (c) an upper or fluvial estuary, characterized by freshwater but subject to daily tidal action." By contrast, a coastal **lagoon** is defined as a shallow stretch of seawater, such as a sound, channel, bay, or saltwater lake, near or communicating with the sea and partly or completely separated from it by a low, narrow elongate strip of land, such as a reef, barrier island, sandbank, or spit (R. L. Bates and Jackson, 1980). Lagoons commonly extend parallel to the coast, in contrast to estuaries, which are oriented approximately normal to the coast. Lagoons also differ from estuaries, which generally have a free connection with the open sea, in that they have restricted circulation with the normal marine environment. Many lagoons have no significant freshwater runoff; however, some coastal embayments that otherwise satisfy the general definition of lagoons do receive river discharge. Estuaries and lagoons may occur in close association with river deltas, barrier islands, and tidal flats.

Owing to the generally small size of coastal bays, estuarine and lagoonal sediments are volumetrically less significant in the geologic record than deltaic sediments. When present, they provide important information about shoreline conditions and environments; they may have economic significance also. Estuaries are among the most biologically productive environments known (Lauff, 1967); therefore, estuarine sediments may be important source rocks for petroleum. Furthermore, the association of well-sorted sandy facies and muddy facies in many estuaries and lagoons provides a favorable setting for stratigraphic traps for petroleum. The hydrologic characteristics and sediment-transport conditions of estuaries and lagoons are quite variable depending upon climate, tidal range, and wave energy. These characteristics, as well as other aspects of the geologic, biologic, and chemical properties of estuaries and lagoons, are described in detail in several monographs, including those of Barnes (1980), Castanares and Phleger (1969), Cronin (1975), Kjerfve (1978), Lauff (1967), Nelson (1972), Nichols and Biggs (1985), Officer (1977), Olaussen and Cato (1980), and Wiley (1976). The distinguishing characteristics of estuaries and lagoons have been summarized in shorter contributions by Biggs (1978), Boothroyd (1978), Clifton (1982), Colombo (1977), Fairbridge (1980), Hayes (1975), Nichols and Biggs (1985), and Phleger (1969).

Physiography of Estuaries and Lagoons

On the basis of physiographic characteristics of relative relief and degree of channel-mouth blocking, seven basic types of modern estuaries are recognized (Fairbridge, 1980) (Fig. 12.30). **Fjords** are high-relief estuaries with a U-shaped valley profile formed by drowning of glacially eroded valleys during Holocene sea-level rise. **Fjards,**

FIGURE 12.30 Principal types of estuaries based on physiographic characteristics. (From Fairbridge, R.W., The estuary: its definition and geodynamic cycle, *in* E. Olausson and I. Cato (eds.), Chemistry and biochemistry of estuaries, Fig. 2, p. 9, © 1980 John Wiley & Sons, Ltd. Reprinted by permission of John Wiley & Sons, Ltd., Chichester, England.)

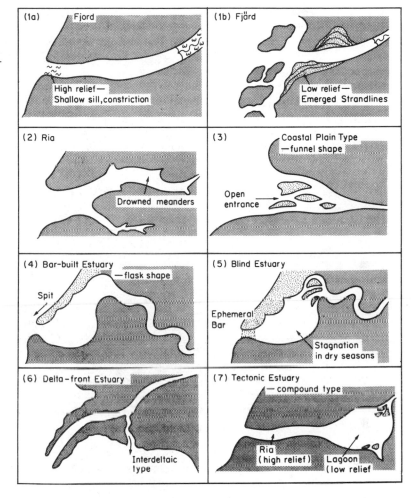

or firths, are related to fjords but have lower relief. Estuaries developed in winding valleys with moderate relief are **rias. Coastal-plain estuaries** are low-relief estuaries, funnel shaped in plan view, that are open to the sea. Low-relief estuaries that are L-shaped in plan view and have lower courses parallel to the coast are **bar-built estuaries.** Similar estuaries that are seasonally blocked by longshore drift or dune migration are called **blind estuaries.** Bar-built estuaries and blind estuaries are very similar to lagoons and might be considered lagoons by some workers. **Deltaic estuaries** occur on delta fronts as ephemeral distributaries. Flask-shaped, high-relief rias backed by a low-relief plain created by tectonic activity are called compound estuaries, or **tectonic estuaries.** Most modern lagoons are formed behind spits or offshore barriers of some type and thus are elongated bodies lying parallel to the coast and having a narrow connection to the open ocean (Fig. 12.31). Lagoons form also behind barrier reefs and atolls.

FIGURE 12.31 Lagoonal systems enclosed by barrier island chains in the United States. *a.* Long Island, New York. *b.* Pamlico and Albermarle Sounds, North Carolina. *c.* Laguna Madre and associated lagoons, Texas. *d.* A lagoon enclosed within the land: The Lagoa dos Patos between Porto Alegre and Rio Grande, Brazil. (From Barnes, R. S. K., 1980, Coastal lagoons. Fig. 1.3, p. 5, reprinted by permission of Cambridge University Press, Cambridge, England.)

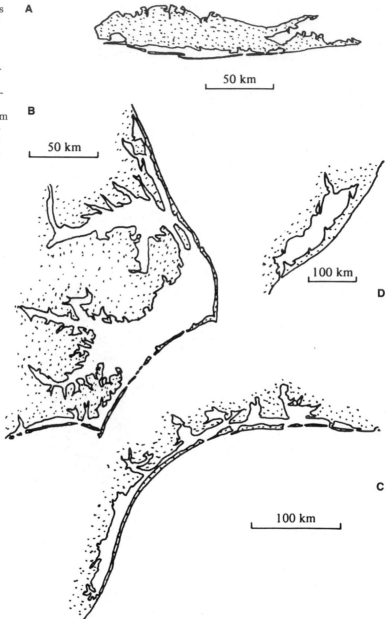

Hydrologic Characteristics

Estuaries

Water circulation patterns in estuaries are affected mainly by freshwater runoff and tidal forces. Therefore, salinity distributions and circulation patterns within estuaries are complex and differ from estuary to estuary, depending upon freshwater discharge and tidal range. In microtidal estuaries (Hayes, 1975) where tidal range is low and tidal influence minimal, high river discharge may prevent tides from entering river mouths, thereby inhibiting the development of estuarine conditions. During periods of seasonally low river discharge, or in rivers with perennial low discharge, a **stratified** estuarine circulation pattern may develop in which a wedge of saltwater extends upstream along the stream bottom. Freshwater flows outward over this wedge with little mixing of fresh- and saltwater. Inflowing flood-tidal currents are too weak to transport most sediment; thus, sediment moves mainly out of the estuary.

In mesotidal areas, where stream flow and tidal flow are about equal, a well-developed saltwater wedge may be produced in which a current of saltwater flows upstream along the bottom while a current of freshwater flows seaward over the top. Some mixing of fresh- and saltwater occurs along the freshwater–saltwater interface, producing a **partially mixed estuary,** but some vertical density stratification of the water is maintained. The velocity of incoming tidal currents during flood-tide stage may be high enough to transport sediment into the estuary. If freshwater discharge is low compared to high-tidal flow, mixing of fresh- and saltwater is much greater and a partially mixed to **well-mixed estuary** results in which the vertical gradient of salinity is much reduced.

Considerable sediment transport can take place both into and out of mesotidal estuaries. The velocities of flood- and ebb-tide currents are commonly asymmetrical, resulting in net sediment transport either into or out of the estuary. Depositional features may include flood- and ebb-tide deltas, transverse bars, and spillover lobes. Where river discharge varies seasonally, the dominant direction of sediment transport may reverse in response to discharge fluctuations. In some Oregon estuaries, for example, flood-tide velocity is dominant during low river discharge in summer and a flood-tide delta builds upsteam into the estuary. During high river discharge in winter, velocity asymmetry reverses, the flood-tide delta is eroded, and all the sand transported into the estuary during summer is transported back to the coast (Boggs and Jones, 1976).

On macrotidal coasts, tidal currents cause essentially complete mixing of salt- and freshwater, producing well-mixed estuaries. Sediment transport is bidirectional and the predominant depositional features in the main part of the estuary are linear tidal-sand ridges oriented parallel to the estuary (Fig. 12.30, 3).

Lagoons

Water-circulating patterns in lagoons are much less affected by freshwater inflow than they are in estuaries, and many lagoons receive no freshwater discharge. Also, circulation with the open ocean is restricted by the presence of some type of barrier. Consequently, the principal movement of water within lagoons occurs in the form of tidal currents that move in and out through the narrow inlets between barriers. Except within these tidal channels that extend into the lagoon, lagoons are predominantly areas of low water energy. Tidal deltas commonly develop at the ends of these tidal inlets, both within the lagoon and on the ocean sides, and sandy sediment may be deposited within the higher energy tidal channels inside the lagoon. Otherwise, sedimentation within lagoons is dominated by deposition of silt and mud, although occasional high wave activity during storms can cause washover of sediment from the barrier.

Salinity within lagoons can range from hypersaline to essentially that of freshwater, depending upon hydrologic conditions and climate. Lagoons formed in arid or semiarid coastal areas, where little freshwater influx occurs, are commonly hypersaline, with salinities well above that of normal seawater. Lagoons in more humid regions may be characterized by brackish water. Salinity within lagoons may vary in response to seasonal rainfall and evaporation rates. Also, salinity at a particular time may not be uniform throughout a lagoon. Lagoons receiving considerable freshwater inflow commonly display distinct lateral salinity zones (Fig. 12.32).

Characteristics of Estuarine and Lagoonal Sediments

Estuarine Sediments

Sediment can be brought into estuaries both by rivers and marine processes, particularly by tidal currents. The sedimentation pattern that develops in an estuary thus depends upon the relative dominance of these processes. Therefore, it is difficult to describe the characteristics of estuarine sediments in general terms. Microtidal estuaries tend to be characterized by low energy conditions. These estuaries are dominated by accumulations of mud and minor amounts of sand, contributed mainly from river sources, together with some biogenic debris such as molluscan shells, wood fragments, and organically produced mud pellets. Mesotidal and macrotidal estuaries are distinguished by higher energy conditions arising from the dominance of tidal processes, although low-energy conditions exist in portions of the estuaries. These estuaries are characterized by sand deposition in tidal channels and point bars, but muddy sediment may accumulate in associated shallow bays and tidal marshes. Fine sediment is supplied mainly by riverine processes, but much of the sand is transported into the estuary from marine sources by tidal currents. If the supply of sediment from rivers becomes dominant, estuaries change into river deltas, and a progradational deltaic sequence develops (Reineck and Singh, 1980).

   Low-energy estuaries thus tend to be characterized by deposition of muddy sediments that display mainly horizontal to subhorizontal stratification and bioturbation structures. Channels are commonly poorly developed in this environment and coarser grained channel deposits are generally absent except perhaps in the lowermost part of the estuary. The deposits of high-energy estuaries are much more complex and may include sandy sediments deposited in tidal deltas, bars, and channels, as well as muddy, silty sediment deposited on tidal flats and other lower energy areas of the estuary. In general, sediments near the estuary inlet consist predominantly of sand,

FIGURE 12.32 A hypersaline lagoon showing four salinity environments within the lagoon. Some lagoons may have only one or two of these environments. (From Barnes, R. S. K., 1980, Coastal lagoons. Fig. 3.1, p. 20, reprinted by permission of Cambridge University Press, Cambridge, England.)

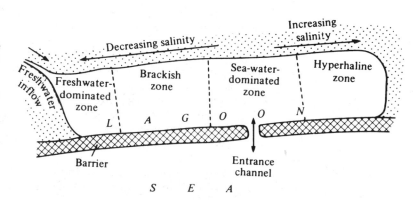

whereas those in the upper parts of the estuary are mostly mud. The sandy sediments characteristically display bedforms ranging from small-scale ripples to large dunes or sand waves. They are commonly cross-bedded, and individual cross-bed sets may dip either seaward or landward, reflecting bidirectional tidal-current transport. Cross-bedding may be destroyed locally by burrowing organisms.

During slackwater periods, fine sediment can be deposited from suspension to form thin mud layers draped over rippled sands, producing flaser bedding. Sedimentary structures in muddy sediments deposited on the floor of lower energy parts of the estuary consist mainly of nearly planar alternating laminations of silt, clay, very fine sand, and carbonaceous debris. Bioturbation by burrowing and feeding organisms may locally mix and "homogenize" these layers. Unfortunately, no trace fossils produced in estuaries are diagnostic of this environment; none can be used to distinguish between intertidal and subtidal estuarine settings (Clifton, 1982). Estuarine sediments typically contain a brackish-water fauna that may include oysters, mussels, other pelecypods, and gastropods.

As tidal channels in estuaries shift back and forth with time, vertical sequences of estuarine sediments are generated. Clifton (1982) describes sequences produced in sediments of Willapa Bay, Washington, that may be typical of many mesotidal estuaries. Migration of sandy channels produces a vertical sequence that begins at the base with a lag deposit composed of sand, shells, gravels, mud fragments, and wood fragments. This lag deposit is typically overlain by cross-bedded or ripple-bedded sand, which in turn is overlain by structureless or bioturbated intertidal-flat sand and finally supratidal mud that may show stratification or root structures (Fig. 12.33). Migration of muddy channels, which are more common in the upper reaches of estuaries, produces a vertical sequence that begins with a lag deposit overlain by laminated mud. The laminated mud in turn may either be overlain by cross-bedded mud or grade upward into broadly stratified or bioturbated intertidal-flat mud, which grades upward into supratidal muddy facies (Fig. 12.34). Channels lying in that part of the estuary between clearly sandy areas and clearly muddy areas may contain mixed sand and mud. As these channels migrate, that part of the vertical section formed by the accretionary bank consists of interbedded sand and mud, flaser bedding, some cross-bedded sand, and some laminated mud. The part of the sequence deposited on intertidal flats may be either bioturbated sand or mud, which grades upward into supratidal mud facies.

Lagoonal Sediments

The deposits of lagoons may differ from those of estuaries in several ways. First, many lagoons do not receive freshwater discharge from rivers. Thus, all the sediment in such lagoons is from marine sources. Lagoons are typically low-energy environments, although tidal currents move into lagoons through inlets between barriers, winds create some wave action along shorelines, storms provide occasional episodes of high-energy waves that wash over barriers into the lagoon, and prevailing winds may more or less continuously blow small amounts of sediment from barriers into the lagoon. Because of the dominance of low-energy conditions in lagoons, lagoonal deposits consist mainly of fine-grained sediments. Sandy sediments are confined principally to tidal deltas constructed at the mouths of the tidal inlets, some tidal channels that extend into the lagoon, washover lobes behind barriers, and some parts of the lagoonal shoreline (lagoonal beaches). Small amounts of sandy sediment blown from barriers may be scattered throughout the lagoon. Sandy sediment in tidal channels is characterized by current ripples and internal small-scale cross-bedding that may dip in either a landward or a seaward direction. Most of the lagoonal bottom is covered with silty or muddy sediments, commonly extensively bioturbated, that may contain thin intercalations of sand brought in by storms or blown in by wind. This sand is generally

FIGURE 12.33 Vertical sequence of estuarine facies produced by a migrating sandy channel. For approximate scale, channel bottom deposits can be considered 1 m thick. (From Clifton, H. E., 1982, Estuarine deposits, in P. A. Scholle and D. Spearing (eds.), Sandstone depositional environments: Am. Assoc. Petroleum Geologists Mem. 31. Fig. 19, p. 186, reprinted by permission of AAPG, Tulsa, Okla.)

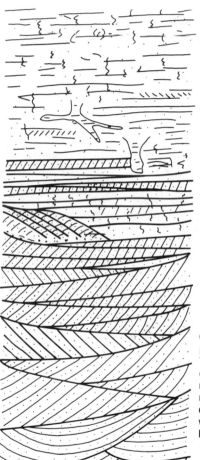

Supratidal and upper intertidal deposits. Generally interlaminated fine sand, silt, and clay often associated with rhizome structures.

Intertidal sand flat deposit. Intensely bioturbated with little physical structure preserved. May contain intertidal runoff channel deposits which consist of cross- or ripple-bedded sand.

Channel bank deposits. Composed predominantly of sand-sized material. The thickness of foreset units usually decreases upsection. Small to medium scale cross-stratification may reflect either floor or ebb currents, depending on location within the channel. Bioturbation can be intense locally.

Channel bottom deposit. Generally contains abundant shell debris often associated with gravel, sand, wood fragments and eroded mud clasts. Bioturbation is common only in the smaller, lower velocity channels.

horizontally laminated, but it may display ripple cross-laminations. The faunas that inhabit lagoons are highly variable, depending upon the salinity conditions of the lagoon, but are generally characterized by low diversity. Lagoons with normal salinity show faunas similar to those of the open ocean, whereas brackish-water faunas domi-

FIGURE 12.34 Vertical sequence of estuarine facies produced by a migrating muddy channel. For approximate scale, channel bottom deposits can be considered 1 m thick. (From Clifton, H. E. 1982, Fig. 20, p. 187, reprinted by permission of AAPG, Tulsa, Okla.)

Supratidal (meadow) deposits. Laminated silt and clay with abundant root traces.

Mud breccia formed at base of wave-cut scarps into supratidal deposits.

Salt marsh deposits characterized by rhizomes of *Triglochin* and other grasses.

Mud flat deposits. Commonly completely bioturbated, except where algal mats generate a broad alternation of silt and clay.

Intertidal creek accretionary bank deposit. Well-bedded, commonly finely laminated sand, silt, clay and carbonaceous debris. Strata generally form muddy cross-strata in units up to one meter thick. Slumped strata relatively common.

Intertidal creek channel deposit. Generally disorganized due to slumping from side banks, may contain mud clasts and concentrations of shells and wood fragments.

Accretionary bank deposit. Generally well-bedded gently inclined, alternations of silt (or fine sand) or clay. Silty layers may show intense small scale bioturbation. Deposits at similar depths tend to become progressively finer in an up-river direction.

Channel bottom deposit. Intensely bioturbated, disorganized shell lag wood fragments common. Toward river mouths, channel may contain interbedded gravel or coarse sand and mud.

nate lagoons in front of river mouths. Hypersaline lagoons commonly contain few organisms because few species are adapted to such high salinities.

In areas where the availability of siliciclastic sediment is low and climatic conditions are favorable, sedimentation in lagoons is dominated by chemical and biochemical deposition. Under very arid conditions, lagoonal sedimentation may be characterized by deposition of evaporites, which are mainly gypsum but may include some
halite, and minor dolomites. Under less hypersaline conditions, carbonate deposition
prevails, particularly in lagoons developed behind barrier reefs. Deposits in such lagoons may consist largely of carbonate muds and associated skeletal debris, although
oolites may form in more agitated parts of the lagoon. Algal mats commonly developed
in the supratidal and shallow intertidal zones may trap fine carbonate mud to form
stromatolites. Algal mats in the supratidal zone generally display mudcracks with
curled margins.

Ancient Estuarine and Lagoonal Deposits

Both estuaries and lagoons are ephemeral features. They tend to fill with sediments in
geologically short periods of time; the preservation potential of these sediments is
relatively high. Nevertheless, few estuarine deposits have been reported from the geologic record, possibly because they have not been widely recognized and distinguished
from associated fluvial, deltaic, lagoonal, or shallow-marine deposits. Recognition of
ancient estuarine deposits is difficult. They tend to have restricted faunal assemblages
that include brackish-water species, but no unique physical criterion exists for these
deposits. The most useful criteria for recognition include the fining-upward vertical
sequences described, reversing or highly variable cross-bedding in sandy deposits,
flaser bedding (also common in tidal-flat environments), and abundant silt and clayey
sediments. Clifton (1982) describes Pleistocene estuarine deposits from Willapa Bay,
Washington, analogous in many ways to the modern estuarine deposits now accumulating in the bay. Figures 12.33 and 12.34 show some of the characteristics of these
deposits and provide models for comparison with other ancient estuarine deposits.

References to ancient lagoonal deposits have been made by many workers in
discussions of barrier-island complexes. Criteria that can be used to distinguish ancient lagoonal deposits from estuarine deposits include evidence for restricted circulation, such as the presence of evaporites or anoxic facies such as black shales, lack of
strong tidal influence, slow rates of terrigenous sediment influx, low faunal diversity,
and extensive bioturbation (R. A. Davis, 1983). The St. Mary River Formation of southern Alberta, Canada, contains back-barrier deposits that illustrate some of the typical
characteristics of lagoonal sediments (Fig. 12.35). This lagoonal sequence begins with
a basal section of sandstones, siltstones, and coals that represent marsh–tidal-flat deposits (Young and Reinson, 1975). These deposits lie on the eroded surface of tidal-
inlet deposits of the underlying Blood Reserve Formation. The basal coal-bearing beds
are succeeded upward by gray shales containing oysters, which are brackish-water
faunas, and disseminated carbonaceous materials and imprints of plant remains. These
deposits are interpreted as subtidal lagoonal sediments. They are overlain by fine-
grained planar and cross-laminated sandstones, which are probably washover deposits
from the barrier beach. More subtidal shales lie above these washover sands, and the
sequence is capped by sandstones and siltstones with coal lenses of probable marsh–
tidal-flat origin. Overall, the Blood Reserve and St. Mary River formations appear to
represent a progradational barrier-island complex, with the lagoonal deposits of the
St. Mary River Formation forming the topmost part of the regressive sequence.

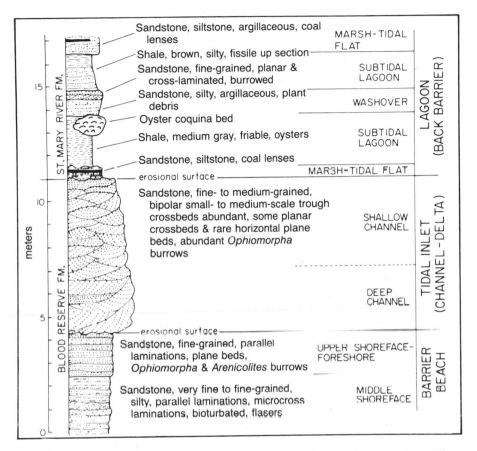

FIGURE 12.35 Composite stratigraphic section of Cretaceous formations in southern Alberta, Canada. Back-barrier lagoonal deposits of the St. Mary River Formation overlie tidal-inlet and tidal-delta deposits of the Blood Reserve Formation. (From Reinson, G. E., 1984, Barrier island and associated strand-plain systems, *in* R. G. Walker (ed.), Facies models, 2nd ed.: Geoscience Canada Reprint Ser. 1. Fig. 16, p. 128, reprinted by permission of Geological Association of Canada.)

12.5 TIDAL-FLAT SYSTEMS

Introduction

Tidal flats form primarily on mesotidal and macrotidal coasts where strong wave activity is absent. They develop either along open coasts of low relief and relatively low wave energy or behind barriers on high-energy coasts where protection is afforded from waves by barrier islands, spits, reefs, and other structures. Tidal flats thus occur within estuaries, bays, barrier-island complexes, and deltas, as well as along open coasts. I have briefly mentioned tidal-flat sediments in discussions of these other environments, but we turn now to a more detailed analysis of the depositional setting, sedimentation processes, and sediment characteristics of tidal flats.

Tidal flats are marshy and muddy to sandy areas partially uncovered by the rise and fall of tides. They constitute almost featureless plains dissected by a network of tidal channels and creeks that are largely exposed during low tide. As tide level rises, flood-tide waters move into the channels until at high tide the channels are overtopped and water spreads over and inundates the adjacent shallow flats. Ebb tide again exposes the channels and intervening flats. In temperate regions, salt marshes commonly cover the upper parts of tidal flats, and muds and silts accumulate near high-water level. At the same time, mixed mud and sand are deposited in the midtidal flat region, and sands accumulate in channels and on the lower parts of the tidal flat. In arid to semiarid regions, tidal flats may be desiccated and marked by mudcracks and the growth of gypsum and halite crystals in muds. Although there is a common tendency to regard tidal flats as primarily sites of siliciclastic deposition, carbonate sediments, and in some areas evaporites, accumulate on many modern tidal flats such as those in the Bahamas, the Persian Gulf, Florida Bay, and the western coast of Australia. Many examples of ancient carbonate tidal-flat sediments have been reported in the geologic literature.

Much of what is known about ancient tidal-flat sediments comes from research on modern tidal flats. Modern tidal flats have been studied intensively in many parts of the world since the 1950s, particularly in Germany; Holland; England; the North Sea; the Bay of Fundy, Nova Scotia; and the Gulf of California. Several important monographs on tidal sediments have now been published, including those of Ginsburg (1975), Klein (1976, 1977), Stride (1982), and Thompson (1968). Tidal environments and deposits have been summarized by numerous authors in shorter contributions, including those of Boothroyd (1978), G. Evans (1965), Frey and Basan (1978), Klein (1970), Reineck (1972; which includes an extensive bibliography of papers published prior to 1972), Weimer et al. (1982), Shinn (1983), Stride et al. (1982), and Van Straaten (1961). Oil and gas deposits have been discovered in both siliciclastic and carbonate tidal facies, and uranium has been reported in sandy tidal facies. Therefore, tidal deposits have economic significance as well as general scientific interest.

Depositional Setting

Although tidal currents may operate in the ocean to depths of 2000–2500 m, the tidal-flat environment is confined to the shallow margin of the ocean. The vertical distance between the high- and low-tide lines in modern tidal environments commonly ranges from 1 to 4 m, depending upon the locality (Reineck, 1972), although tidal ranges up to 10–15 m or more occur in some localities such as the Bay of Fundy. The total width of tidal flats may range from a few kilometers to as much as 25 km. Topographic relief within the tidal-flat environment is generally rather small, except for tidal channels, and slopes of the tidal flat are gentle, although commonly irregular.

It is convenient for the purpose of discussion to divide the tidal-flat environment into three zones. The **subtidal zone** encompasses that part of the tidal flat that normally lies below mean low-tide level. It is inundated with water most of the time and is normally subjected to the highest tidal-current velocities. Tidal influence in this part of the environment is particularly important within tidal channels, where bedload transport and deposition are predominant, although this zone is also influenced to some extent by wave processes. The **intertidal zone** lies between mean high- and low-tide levels. It is subaerially exposed either once or twice each day, depending upon local wind and tide conditions, but commonly does not support significant vegetation. Both bedload and suspension sedimentation take place in this zone. The **supratidal**

zone, or upper intertidal zone of some workers, lies above normal high-tide level but is incised by tidal channels and flooded by extreme tides. This part of the tidal flat is exposed to subaerial conditions most of the time, but may be flooded by spring tides twice each month or by storm tides at irregular intervals. Sedimentation is predominantly from suspension. On some tidal flats, the supratidal zone is a salt-marsh environment incised by tidal channels. In arid or semiarid climates the supratidal zone is commonly an environment of evaporite deposition and is often referred to as a **sabkha.** The relationship between the subtidal, intertidal, and supratidal portions of the tidal flat is illustrated schematically in Figure 12.36.

Sedimentation Processes

Physical Processes. Sedimentation on tidal flats takes place in response to both tidal processes and waves. Sedimentation in the channels of tidal flats is dominated by tidal currents, but wind-driven waves and the currents generated by these waves also play an important role in deposition on the flats between channels. Tidal currents move up the gentle slope of the tidal flat during flood tide and back down during ebb tide. The tidal velocities achieved during reversing tides are commonly asymmetrical, and the velocities of flood tides may differ significantly from those of ebb tides. Within the channels, tidal currents can reach velocities of up to 1.5 m/sec or more and velocities on the flats of 30–50 cm/sec (Reineck and Singh, 1980). These velocities are adequate to cause transport of sandy sediment and produce ripple and dune bedforms, cross-bedding, and plane bedding.

Deposition in the subtidal zone takes place mainly by lateral accretion of sandy sediment in tidal channels and point bars (Weimer et al., 1982). Tidal channels can be

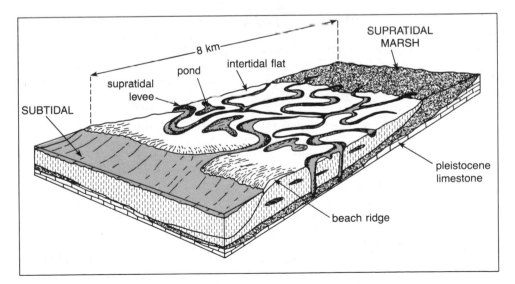

FIGURE 12.36 Schematic diagram showing relationship of subtidal, intertidal, and supratidal zones of the tidal-flat environment. (After Shinn, E. A., 1983, Tidal flat environment, *in* P. A. Scholle, D. G. Bebout, and C. H. Moore, (eds.), Carbonate depositional environments: Am. Assoc. Petroleum Geologists Mem. 33. Fig. 2, p. 172, reprinted by permission of AAPG, Tulsa, Okla.)

quite large, ranging in depth to 15 m or more. These channels migrate laterally, similarly to those of meandering river systems. Shinn (1983) compares highly channeled tidal flats to river deltas "turned wrong side out." The sea rather than the land is the sediment source, and the tidal channels with their distributaries branching in a landward direction provide the pathways for sediment delivery to the tidal flat.

The intertidal zone is a zone of mixed lateral accretion and suspension deposition. Sand or muddy sand deposition is predominant within channels, but both fine sand and mud can be deposited on the broad flats between tidal channels. Bedload transport and deposition of sand by tidal currents takes place during the higher velocity phases of a tidal cycle, during either rising or falling tide, producing ripple or dune bedforms. These bedforms may be modified by wind-generated waves and currents. During periods of lower velocity or zero-velocity flow accompanying high- or low-water slack periods, suspension deposition of clay- or silt-size sediment takes place as either discrete particles or aggregates of particles (Klein, 1977). Deposition of these muddy sediments over rippled, sandy sediments gives rise to a variety of lenticular and flaser bedding. Wave activity tends to be greatest in the lower part of the intertidal zone, leading to reworking of sediment and resuspension of fine sediment, sorting of sands, and a predominance of sandy deposits. Wave and current energy is lowest in the upper part of the intertidal zone, which is consequently dominated by deposition of muddy sediment. Periodic storms also can affect deposition in the middle and high intertidal zones. Significant quantities of sediments may be carried in during storms from farther offshore and deposited in the intertidal zone, and storm waves may erode and reentrain previously deposited sediments and modify or destroy bedforms and sedimentary structures.

The supratidal zone is only slightly affected by tidal currents and marginally affected by waves. The common presence of salt-marsh vegetation and mangrove trees in this zone further dampens wave energy. Thus, the supratidal zone is the zone of lowest energy on the tidal flat. Deposits of this zone are mainly muds, and sedimentation rates are very low. Desiccation of muds and subsequent development of mudcracks are diagnostic features of the supratidal and upper intertidal zones. Most of the mud deposited in the intertidal and supratidal zones originates in the subtidal zone, where it is stirred into suspension during storms and transported to the tidal flats by storm waves.

Chemical and Biologic Processes. The preceding discussion refers primarily to deposition on siliciclastic tidal flats. Much the same processes occur on carbonate tidal flats, with the difference that biochemical and biologic sedimentation processes are important in these settings in supplying carbonate sediment. Carbonate muds are generated mainly in the shallow subtidal zone, possibly owing to inorganic precipitation but more likely to biochemical breakdown of calcareous algae or other calcareous organisms (Chapter 4). These lime muds, together with sand-size skeletal fragments, also generated by biologic processes within the subtidal zone, are transported into the intertidal and supratidal zones by tidal currents and waves. In arid and semiarid climates, chemical precipitation of gypsum, anhydrite, and dolomite may occur in the supratidal and upper intertidal zones owing to strong evaporation.

Biogenic processes are important in other ways on both siliciclastic and carbonate tidal flats. Although tidal flats, owing to intermittent exposure, are a harsh environment for many organisms, they are inhabited by a limited number of species of organisms such as gastropods and pelecypods, crustaceans, polychaete worms, foraminifers, diatoms, and blue-green algae (cyanobacteria). Many of these organisms produce fecal pellets of mud that accumulate on the flats; they may also cause extensive bioturbation of sediment. In general, bioturbation appears to be strongest in mud flats, weaker in mixed mud and sand flats, and least in sand flats. Blue-green algae are particularly important agents in the supratidal and intertidal zones in trapping and binding fine sediment to produce stromatolitic bedding.

Characteristics of Tidal-Flat Sediments

General Morphology. The body of sediments that makes up tidal flats may range from a few kilometers to a few tens of kilometers in width. A tidal flat is elongated parallel to the coast and along open coasts may extend for tens to hundreds of kilometers, cut across here and there by major tidal channels and river estuaries. The tidal flats of bays and barrier-island systems are more restricted in size, and plan-view shape depends upon the shape of the bay shore. As mentioned, the upper surfaces of tidal flats are characterized by low relief, except for tidal channels. In cross section, tidal deposits form a gross wedge-shaped body tapering shoreward.

Lithology. Sediments on siliciclastic tidal flats are composed primarily of mud and sand. Muds predominate in the supratidal and upper intertidal zones, and the deposits of supratidal marshes are further characterized by abundant plant debris. Mixed mud and sand characterize the middle part of the intertidal zone, and sand dominates the lower intertidal zone as well as the channel and bar deposits of the shallow subtidal zone. Muds may be deposited between channels in the subtidal zone below wave base. This general distribution of siliciclastic tidal-flat facies in a modern bay tidal flat is illustrated in plan view in Figure 12.37. The proportion of muds and sand in modern tidal flats varies considerably. Some tidal flats are dominated by mud, whereas others are sand-dominated flats. The distribution of facies on carbonate tidal flats is similar.

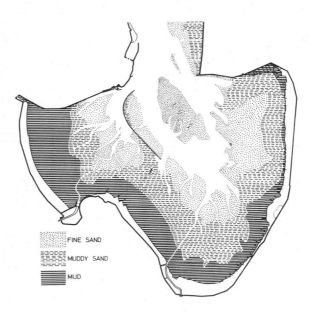

FIGURE 12.37 Distribution of sediments on the intertidal flats of Jade Bay. (From Reineck, H. E., and I. B. Singh, 1980, Depositional sedimentary environments. Fig. 591, p. 432, reprinted by permission of Springer-Verlag, Heidelberg. Originally after S. Gadow, 1970, 1. Sedimente und chemismus, *in* H. E. Reineck (ed.), Das Watt, Ablagerungs- und Lebensraum: W. Kramer, Frankfurt a. M.)

FINE SAND

MUDDY SAND

MUD

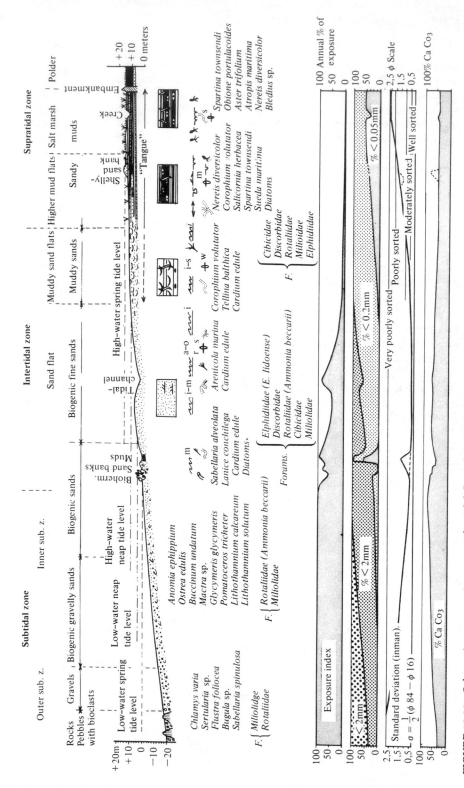

FIGURE 12.38 Schematic cross section across the tidal flats of St. Michel, France, showing sedimentation patterns, grain-size variation, faunal assemblages, and so forth. (From Larsonneur, C., 1975, Tidal deposits, Mont Saint-Michel Bay, France, in R. N. Ginsburg, (ed.), Tidal deposits. Fig. 3.2, p. 24, reprinted by permission of Springer-Verlag, Heidelberg.)

Figure 12.38 illustrates in cross section the distribution of such a facies on the tidal flat of Mont Saint-Michel Bay, France, where the clastic sediments contain a high content of biogenic carbonates (>50 percent). Figure 12.38 shows also the distribution of different species of organisms across the tidal flat, including various species of foraminifer diatoms, gastropods, and pelecypods. Carbonate tidal flats are thus dominated by muddy deposits in low-energy areas of the flat and carbonate sands composed of skeletal fragments, intraclasts, or oolites in channels and other high-energy areas. Evaporite minerals such as gypsum, anhydrite, and occasionally halite are present on all modern arid tidal flats (Shinn, 1983), commonly in association with dolomite and other carbonate minerals.

Sedimentary Structures. The predominant types of sedimentary structures in tidal-flat sediments vary in different parts of the tidal flat. Channel sands are characterized by megaripples and internal cross-bedding that may show bimodal directions of foreset dip. Reversing tides during an asymmetrical tidal cycle can cause erosion of ripple crests followed by redeposition during the next tidal cycle, producing **reactivation surfaces** (Fig. 12.39). Reactivation surfaces are most typical of tidal deposits, but they have been reported in fluvial sediments. Sandy and muddy sediments on the mixed flats are characterized by small-scale ripple cross-stratification, flaser bedding, wavy bedding, lenticular bedding, and, more rarely, finely laminated bedding. Some tidal

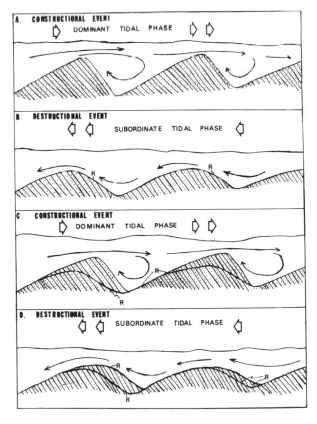

FIGURE 12.39 Reactivation surface developed owing to alternation of a dominant tidal phase (constructional event) with a subordinate phase (destructional event). A. Dunes develop by dominant-phase tidal flow and produce internal avalanche cross-stratification. B. Reversed tidal flow during subordinate phase destroys sharp asymmetry of dunes, producing a subdued, asymmetrical profile and a rounded reactivation surface (R). C. During the next constructional event, dunes build on reactivation surface (R) and develop a superimposed set of avalanche cross-stratification. D. Destruction of the dune profile occurs during reversed, subordinate tidal flow and produces a second reactivation surface (R). (From Klein, G. deV., 1970, Depositional and dispersal dynamics of intertidal sand bars: Jour. Sedimentary Petrology, v. 40. Fig. 28, p. 1118, reprinted by permission of Society of Economic Paleontologists and Mineralogists, Tulsa, Okla.)

flat sediments display **herringbone** cross-bedding in which cross-laminated sediments deposited during flood tide dip in the opposite direction to those formed almost immediately afterward during ebb tide (Fig. 12.40).

Deposits on the mud flats are characterized by thick mud layers separated by thin sand laminae, which in arid climates may be disrupted by growth of gypsum, anhydrite, or halite crystals and development of mudcracks. Anhydrite is typically characterized by chickenwire texture (Chapter 8), and distortion of gypsum and anhydrite layering is common. Muddy deposits of supratidal marshes may show some thin layering, but they are commonly bioturbated and disturbed by root growth. In addition to mudcracks, other structures indicating subaerial exposure may be present, including raindrop imprints, hail marks, and foam marks. Sedimentary structures on carbonate tidal flats are generally similar to those formed on siliciclastic tidal flats; however, megaripples, cross-stratification, and other current-generated structures tend to be less abundant, and mudcracked algal stromatolites are more characteristic of the upper intertidal zone and the supratidal zone. Bioturbation structures are prevalent in tidal-flat sediments. They are most abundant in sediments of the muddy upper intertidal and supratidal zones and least abundant in sediments of the sandy lower intertidal zone and the subtidal zone. Trace fossils produced by organic activity include not only burrows and various kinds of feeding and resting traces of organisms living in the intertidal zone, but also the tracks of birds, land animals, and insects.

Vertical Sequences. Transgression and regression cause deposits of laterally adjacent tidal-flat environments to become superimposed, generating characteristic sequences of vertical facies. Progradation produces a generalized fining-upward sequence that

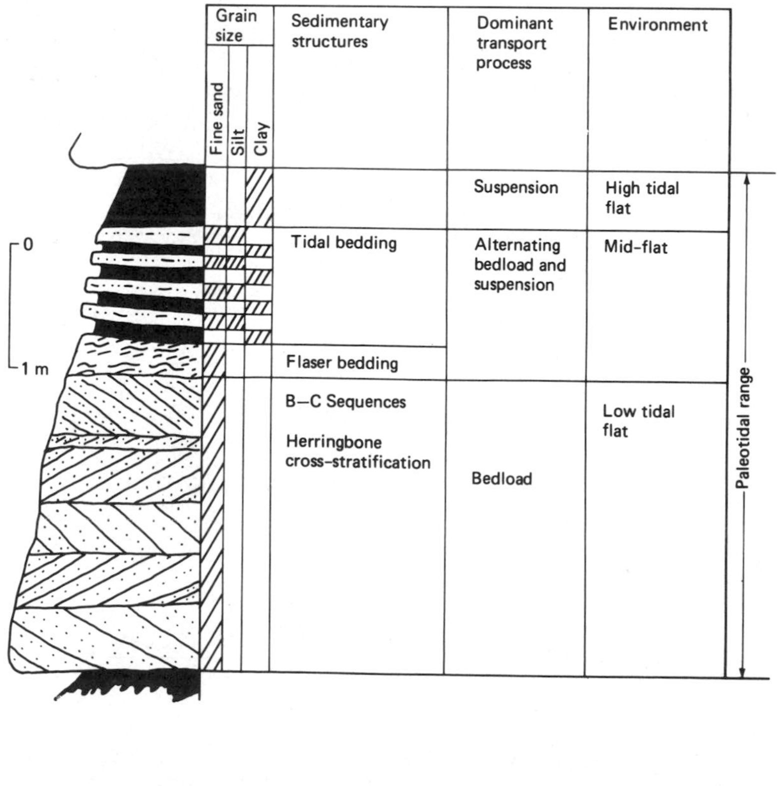

FIGURE 12.40 Progradational sequence of tidal-flat deposits. Based on Middle Member, Wood Canyon Formation (Late Precambrian), Nevada. Diagram also shows interpretation of dominant sediment-transport processes and depositional environments. (From Klein, G. deV., 1977, Clastic tidal facies. Fig. 76, p. 85, reprinted by permission of Continuing Education Publication Co., Champaign, Ill.)

begins with subtidal and lower intertidal cross-bedded sands, followed upward by mixed sand and mud deposited in the middle intertidal zone, then mud and peat from the upper intertidal and supratidal zones. A typical vertical regressive sequence developed on a siliciclastic tidal flat is illustrated schematically in Figure 12.40. This sequence may be used as a general model for progradational tidal-flat deposits. Transgression generates a coarsening-upward sequence that displays the same general facies but in reverse order. Figure 12.41 illustrates an idealized, diagrammatic model of the vertical and lateral facies relationships that might develop on a regional scale owing to transgression and regression on a coast characterized by carbonate tidal flats. On a local scale, vertical sequences may develop also as a result of lateral channel migration. Channel migration can occur during either transgression or regression and thus may generate small-scale vertical sequences that differ from the idealized sequence shown in Figure 12.41.

Ancient Tidal-Flat Sediments

Although tidal-flat deposits have several distinctive characteristics that help to differentiate them from sediments of most other environments, their overall characteristics are similar to those of estuarine deposits and it may be difficult to distinguish them from estuarine deposits in ancient sedimentary sequences. The most important criteria for recognition of ancient tidal-flat deposits are commonly regarded to include (1) bi-

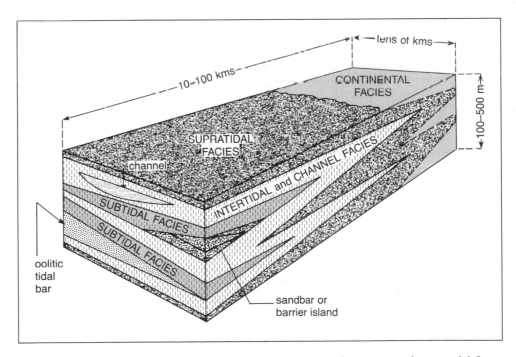

FIGURE 12.41 Idealized model of transgressive and regressive facies on a carbonate tidal flat. (After Shinn, E. A., 1983, Tidal flat environment, *in* P. A. Scholle, D. G. Bebout, and C. H. Moore (eds.), Carbonate depositional environments: Am. Assoc. Petroleum Geologists Mem. 33. Fig. 42, p. 198, reprinted by permission of AAPG, Tulsa, Okla.)

modality of current-formed cross-bedding resulting from reversing tidal currents, (2) the occurrence of facies that reflect repeated, small-scale alternations in sediment-transport conditions and the joint occurrence of large-scale (channel) and small-scale (sand- and mud-flat) structural units in superposition or juxtaposition, (3) the presence of abundant reactivation surfaces, and (4) a high frequency of erosional contacts and abrupt facies changes. Other supporting criteria include the typical vertical sequence of facies discussed above, the high degree of bioturbation of many tidal-flat sediments, the presence of mudcracked stromatolites, and other evidence of subaerial exposure, such as raindrop imprints, hail marks, and animal or bird tracks.

Tidal-flat deposits have been reported from stratigraphic units of virtually all ages from Precambrian to Holocene. Numerous examples are cited in the bibliography given by Reineck (1972) and by Klein (1977) and Weimer et al. (1982). Several examples of both ancient siliciclastic and carbonate tidal-flat deposits are discussed in the excellent compendium volume on tidal deposits edited by Ginsburg (1975). Carbonate deposits are covered in Chapter 13 and are not discussed further here.

Some of the sequences containing barrier and lagoonal deposits also include tidal deposits. For example, the upper part of the Blood Reserve Formation of Alberta, shown in Figure 12.35, contains siliciclastic tidal-channel and tidal-delta deposits composed of fine- to medium-grained, trough cross-bedded sandstones. The generalized lagoonal sequence from Carboniferous rocks of eastern Kentucky shown in Figure

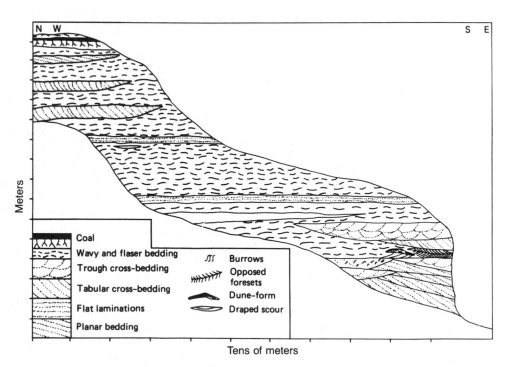

FIGURE 12.42 Tidal-flat deposits in exposure of the Jurassic Lower Coal Series of Bornholm, Denmark. (From Sellwood, B. W., 1972, Tidal-flat sedimentation in the Lower Jurassic of Bornholm, Denmark: Palaeogeography, Palaeoclimatology, Palaeoecology, v. 11. Fig. 3, p. 97, reprinted by permission of Elsevier Science Publishers, Amsterdam.)

12.25 includes tidal-flat siltstones with sandstone layers and rippled and cross-bedded tidal-channel deposits interbedded with other back-barrier deposits.

Sellwood (1972, 1975) describes an interesting sequence of tidal deposits from the Jurassic Lower Coal Series on the island of Bornholm near the mouth of the Baltic Sea. This sequence illustrates many of the general features of tidal-flat deposits (Fig. 12.42). The sequence is about 15 m thick and has a basal cross-bedded sandstone unit about 5 m thick. The basal unit consists of fine- to medium-grained sandstone that displays large-scale tabular to trough cross-bedding, herringbone cross-stratification, and planar stratification. The herringbone cross-stratification appears to be the preserved remains of megaripples with reactivation surfaces and, together with planar bedding, suggests deposition under upper-flow-regime conditions. A transition upward in the section to bedding characteristics indicative of small ripples suggests change in flow conditions to lower-flow regime. Burrows and shallow sand-filled channels are present in some of the beds. This basal unit probably reflects deposition in the tidal-dominated low-tidal-flat environment (Fig. 12.40), where energy conditions are generally high. The sequence lying above the cross-bedded unit is characterized by fine-grained sands that show wave and current ripples draped with clay. Flaser and lenticular bedding are prevalent. The characteristics of this sequence correlate well with facies of the midflat environment of Figure 12.40. The sequence is capped by coal beds up to about 30 cm thick and having root traces extending down into underlying flaser bedding. These marsh-type deposits represent the high-tidal-flat environment.

ADDITIONAL READINGS

Deltaic Systems

Broussard, M. I. (ed.), 1975, Deltas, models for exploration: Houston Geological Society, 555 p.

Coleman, J. M., 1981, Deltas, processes of deposition and models for exploration, 2nd ed.: Burgess, Minneapolis, Minn., 124 p.

Coleman, J. M., and D. B. Prior, 1980, Deltaic sand bodies: Am. Assoc. Petroleum Geologists, Education Course Notes 15, 171 p.

Morgan, J. P. (ed.), 1970, Deltaic sedimentation—modern and ancient: Soc. Econ. Paleontologists and Mineralogists, Spec. Pub. 15, 312 p.

Shirley, M. L., and J. A. Ragsdale (eds.), 1966, Deltas in their geologic framework: Houston Geological Society, 251 p.

Beach and Barrier Island Systems

Davis, R. A., Jr. (ed.), 1985, Coastal sedimentary environments, 2nd ed.: Springer-Verlag, New York, 716 p.

Davis, R. A., Jr., and R. L. Ethington (eds.), 1976, Beach and nearshore sedimentation: Soc. Econ. Paleontologists and Mineralogists, Spec. Pub. 24, 187 p.

Fisher, J. S., and R. Dolan (eds.), 1977, Beach processes and coastal hydrodynamics: Dowden, Hutchinson and Ross, Stroudsburg, Pa., 382 p.

Komar, P. D., 1976, Beach processes and sedimentation: Prentice-Hall, Englewood Cliffs, N.J., 429 p.

Leatherman, S. P. (ed.), 1979, Barrier islands from the Gulf of St. Lawrence to the Gulf of Mexico: Academic Press, New York, 325 p.

Schwartz, M. L. (ed.), 1972, Spits and bars: Benchmark Papers in Geology, v. 3, Dowden, Hutchinson and Ross, Stroudsburg, Pa., 452 p.

Schwartz, M. L. (ed.), 1973, Barrier islands: Benchmark Papers in Geology, v. 9. Dowden, Hutchinson and Ross, Stroudsburg, Pa., 451 p.

Estuarine and Lagoonal Systems

Barnes, R. S. K., 1980, Coastal lagoons: Cambridge University Press, Cambridge, 106 p.

Cronin, L. E. (ed.), 1975, Estuarine research, v. 2: Geology and Engineering: Academic Press, New York, 587 p.

Dyer, K. R. (ed.), 1979, Estuarine hydrography and sedimentation: A handbook: Cambridge University Press, Cambridge, England, 230 p.

Kennedy, V. S. (ed.), 1984, The estuary as a filter: Academic Press, Orlando, Fla., 511 p.

Ketchum, B. H., 1983, Estuaries and enclosed seas: Ecosystems of the World 26, Elsevier, Amsterdam, 500 p.

Kjerfve, B. (ed.), 1978, Estuarine transport processes: University of South Carolina Press, 331 p.

Lauff, G. H. (ed.), 1967, Estuaries: Am. Assoc. Adv. Science, Spec. Pub. 83, 757 p.

Nelson, B. W. (ed.), 1972, Environmental framework of coastal plain estuaries: Geol. Soc. America Mem. 133, 619 p.

Officer, C. B. (chm.), 1977, Estuaries, geophysics, and the environment: National Academy of Sciences, Washington, D.C., 127 p.

Olaussen, E., and I. Cato (eds.), 1980, Chemistry and biogeochemistry of estuaries: John Wiley & Sons, New York, 452 p.

Wiley, M. (ed.), 1977, Estuarine processes, v. II: Circulation, sediments, and transfer of material in the estuary: Academic Press, New York, 428 p.

Tidal-Flat Systems

Ginsburg, R. N., 1975, Tidal deposits, a casebook of recent examples and fossil counterparts: Springer–Verlag, New York, 428 p.

Klein, G. deV. (ed.), 1976, Holocene tidal sedimentation: Benchmark Papers in Geology, v. 5, Dowden, Hutchinson and Ross, Stroudsburg, Pa., 423 p.

Klein, G. deV., 1977, Clastic tidal facies: Continuing Education Publication, Champaign, Ill., 149 p.

Stride, A. H. (ed.), 1982, Offshore tidal sands: Chapman and Hall, London, 222 p.

13
The Marine Environment

13.1 INTRODUCTION

The marine environment is that part of the ocean lying seaward of the zone dominated by shoreline processes. Water depth in the marine realm ranges from a few meters to more than 10,000 m. The salinity of seawater in the open ocean averages about 35‰, although higher or lower salinities can occur locally in restricted bodies of the ocean. Marine life forms are characterized by high diversity and large populations, and most are stenohaline organisms adapted to conditions of normal salinity. The energy of the bottom water lying immediately above the ocean floor is generally low, except on the shallow continental shelf, which is affected by a variety of tidal processes and wind and storm-wave activity.

The major subdivisions of the oceanic realm are the **continental margin** and the **ocean basin.** These in turn can be further subdivided as shown in Figure 13.1. The **continental shelf** extends seaward from the shoreline at a gentle slope of about 1 degree to a point where a perceptible increase in rate of slope takes place; this is called the **shelf break.** The shelf break occurs in the modern ocean at an average distance from shore of about 75 km, although the distance ranges from a few tens of meters to more than 1000 km. Average water depth at the slope break is about 125 m. The **continental slope** descends from the shelf break to the deep seafloor with a typical slope of about 4 degrees. On passive, or divergent, continental margins, the foot of the continental slope merges with the **continental rise,** which is a gently sloping surface created by coalescing submarine fans at the base of the slope. The continental rise passes gradually into the nearly flat floor of the ocean basin—the **abyssal plain.** In some ocean basins, the abyssal plain is interrupted in the central part of the basin by a gigantic **mid-ocean ridge.** On active, or convergent, margins, the continental slope may descend into a **deep-sea trench,** and the continental rise is absent. On the basis of

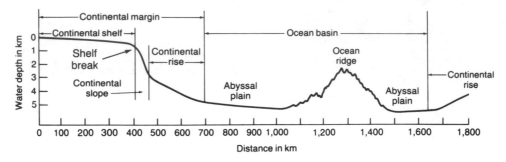

FIGURE 13.1 Diagrammatic profile of the ocean floor showing the principal features of the continental margin and ocean basin. Vertical exaggeration approximately 50×. On some active continental margins, such as the Japan Pacific margin, a deep trench lies at the foot of the continental slope and the continental rise is absent. (After D. A. Ross, Introductory oceanography, 3rd ed., © 1982, Fig. 5.17, p. 105. Reprinted by permission of Prentice-Hall, Englewood Cliffs, N.J.)

water depth, we divide the ocean into two major zones, the neritic zone and the oceanic zone. The shallow **neritic zone** extends from the shoreline to the shelf break. The **oceanic zone** extends from shelf break to shelf break and encompasses the deeper part of the ocean.

13.2   THE NERITIC ENVIRONMENT

Introduction

The neritic zone of the ocean encompasses the shallow-water areas lying shoreward of the shelf break. Although the shelf break on modern shelves occurs at an average depth of about 125 m, it may be located on some shelves at depths as shallow as 18 m or as deep as 915 m (Bouma et al., 1982). In the modern ocean, the shallow-marine environment mainly occupies the shelf area around the margin of the continents, forming what is referred to as a **pericontinental,** or marginal, sea (Heckel, 1972). At various times in the geologic past, broad, shallow **epicontinental,** or epeiric, seas occupied extensive areas within the continents (Fig 13.2), somewhat like the Hudson Bay area of the North American arctic region. The following discussion of the neritic environment is focused primarily on the pericontinental environment because we can draw on the modern continental-shelf environment as a model. Readers should keep in mind, however, that many of the shallow-marine deposits in the geologic record may have been deposited in broad epicontinental seas, for which we may have no truly representative modern analogs. We commonly assume that similar sedimentologic processes operated on continental shelves and in epicontinental seaways, but, in fact, the similarities of and differences between these two environments are not yet completely understood. Furthermore, modern continental shelves may not provide a good analog for ancient marginal seas because rapid rise of sea level following the final episode of Pleistocene glaciation has stranded coarse sediment in deeper parts of the shelves, creating conditions of sediment–water disequilibrium. Thus, sediment grain size on

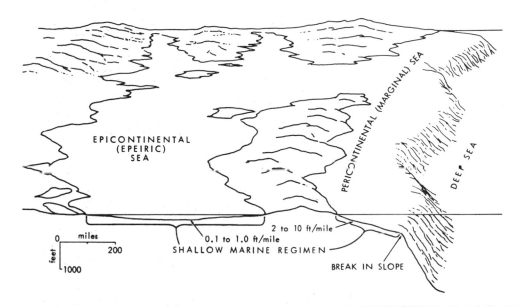

FIGURE 13.2 Schematic diagram illustrating the difference between pericontinental and epicontinental shallow-marine environments. (From Heckel, P. H., 1972, Recognition of ancient shallow marine environments, in J. K. Rigby and W. K. Hamblin (eds.), Recognition of ancient sedimentary environments: Soc. Econ. Paleontologists and Mineralogists Spec. Pub. 16. Fig. 1, p. 227, reprinted by permission of SEPM, Tulsa, Okla.)

some parts of modern shelves is not consistent with present water depth, energy conditions, and sedimentation processes on the shelves.

Both siliciclastic and carbonate sediments can accumulate in the marine-shelf environment, although most modern continental shelves are covered by siliciclastic sediments. Carbonate sediments are restricted to a few shelves in mainly tropical areas. Nonetheless, the geologic record suggests that carbonate-shelf deposition was extremely important at various times in the geologic past. Therefore, we will examine the distinguishing characteristics of both siliciclastic and carbonate shelves. To avoid possible confusion, the two systems are discussed separately, beginning with siliciclastic shelves.

Siliciclastic Shelf Systems

Physiography and Depositional Setting

The siliciclastic shelf environment is bounded by various coastal environments on the landward side and by the continental slope on the seaward side. It can be divided into the shallow **inner shelf,** which is dominated by tidal, wind, and storm-wave processes, and the deeper **outer shelf,** which may be affected by intruding major ocean currents such as the Gulf Stream (Fig. 13.3). The outer shelf may be affected also by density currents generated by temperature–salinity differences or suspended sediment differences in water bodies. The boundary between the inner and outer shelf is not well defined, and its position fluctuates with changing sea level. In fact, during greatly

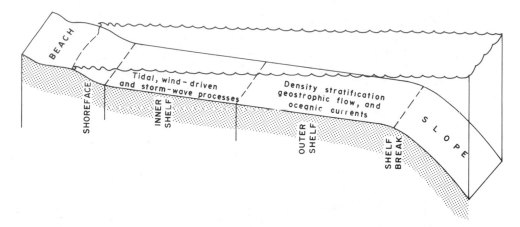

FIGURE 13.3 Subdivisions of the continental shelf. (From Galloway, W. E., and D. K. Hobday, 1983, Terrigenous clastic depositional systems. Fig. 7.1, p. 144, reprinted by permission of Springer-Verlag, Heidelberg.)

lowered sea level, the normal inner shelf is subaerially exposed and the outer shelf may be covered only by very shallow water.

The width of shelves varies according to their plate tectonic setting (Shepard, 1973). Shelves along the fore-arc region of convergent continental margins tend to be very narrow. By contrast, broad shelves and platforms occur in the back-arc basins of convergent margins, on divergent or trailing-edge continental margins, and on cratonic downwarps that open to the sea. Although shelves are fundamentally low-relief platforms, the shelf surface can vary considerably. It may be relatively smooth, or covered by a variety of small- to large-scale bedforms. It may also contain banks, islands, or shoals near its offshore edge (Bouma et al., 1982). Shepard (1977) divides the present continental shelves into five major categories based mainly on depositional setting. **Glaciated shelves** occur at high latitudes where glaciers spread from land masses onto the continental shelf. They are characterized by ice-scoured troughs and glacially deposited sediments. **Shelves with elongate sand ridges** occur particularly along the United States Atlantic coast. These subparallel sand ridges and associated bedforms are composed of Pleistocene sediment now being reworked and modified within the modern environment. **Shelves off large deltas** are low-relief platforms that build seaward as delta progradation occurs. **Shelves with coral reefs** occur in tropical waters and locally in subtropical waters, but are very limited in the modern ocean. Coral reefs are typically developed along the shelf edge or shelf break, but can occur also as patches on the inner shelf. Reef-rimmed shelves are discussed further under carbonate shelf systems. **Shelves bordered by rocky banks and islands** are characterized by rocky elevations along their outer edges. These elevations may be either islands or rock banks; within the elevated areas they may contain basins or channels partly or completely filled with sediment.

Depositional Processes

Ideas concerning transport and depositional processes on continental shelves are currently in a state of flux, and research on both modern shelf processes and ancient shallow-marine systems is very active. The current state of knowledge is summarized in an excellent paper by R. G. Walker (1984a). Walker traces the history of shallow-marine studies beginning with early ideas of a "graded shelf" (D. W. Johnson, 1919), which was believed to display progressive decrease in grain size from coarse at the shoreline to very fine at the shelf edge, in response to presumed decrease in water energy seaward. The graded-shelf concept was eventually abandoned when it became obvious with additional study that the grain-size distribution on many shelves is patchy or irregular. The concept of **relict** shelf sediments was introduced to explain such irregular sediment-distribution patterns as the presence of coarse sands and gravels in deep water. Relict sediments are deposits that are not in equilibrium with present hydrodynamic conditions. They were deposited on the shelf by fluvial or glacial processes during low stands of sea level. Relict sediments were initially believed to remain on the shelf floor without significant reworking as they were inundated by rising sea level (Shepard, 1932; Emery, 1968). It was subsequently recognized, however, that some so-called relict sediments had likely been reworked to some extent during sea-level rise. This reworking caused the sediments to be brought into partial or complete equilibrium with present shelf processes and conditions. Such reworked sediment, having partly relict and partly modern characteristics, was called **palimpsest** by Swift et al. (1971). These authors also identified and discussed four different types of shelf currents that operate on the shelf to rework and transport sediment: (1) tidal currents; (2) storm-generated currents; (3) intruding ocean currents, such as the Gulf Stream; and (4) density currents. Shallow-marine systems have subsequently been divided on the basis of dominant shelf processes into three main types: **tide-dominated shelves** (about 17 percent of the world's shelves), **storm-dominated shelves** (about 80 percent of the world's shelves), and **shelves dominated by intruding ocean currents** (about 3 percent of the world's shelves). No modern shelves are dominated by density-current processes.

Tidal Processes. Tides are generated by the gravitational attraction of the moon and sun for Earth in conjunction with the rotation of Earth. Tidal influence is manifested at any given coastal locality by daily rise and fall of the sea over an average range of 1–4 m on open coasts, but tidal range may exceed 15 m in some enclosed basins. Some localities experience **diurnal tides,** characterized by two highs and two lows each day. Others have **semidiurnal tides,** distinguished by one high and one low each day, and still others have **mixed tides.** Tidal currents on continental shelves are propagated as a large wave or tidal bulge generated in deep ocean basins (Fox, 1983). In major ocean basins, this tidal bulge rotates around a central point of no tidal movement—an **amphidromic point.** The tidal wave follows an elliptical path that is almost circular in the open ocean. In more restricted areas, the ellipse is strongly elongated, forming a narrow rectilinear pattern. The origin of tidal currents is treated in detail by Howarth (1982).

The currents generated on the shelf by tides are bidirectional, but asymmetrical with respect to velocity. Asymmetrical currents may result in sediment transport in only one direction if tidal velocity is near the threshold needed to erode and transport sediment. Tidal-current velocity decreases with water depth; thus tidal-current transport is most important in shallow water. Tidal-current velocities ranging up to about 2 m/sec have been measured in some enclosed basins such as the Bay of Fundy, Nova Scotia, but tidal velocities on most shelves are less than about 1 m/sec. On many shelves, they are so low that they are below the threshold velocities required for sediment entrainment and transport. The most outstanding example of a modern shelf dominated by strong tidal currents is the North Sea, which lies between the United

Kingdom and the coasts of Denmark and Norway. A good source of additional information on tidal currents is the volume on offshore tidal sands edited by Stride (1982); this study treats both tidal processes and tidal deposits in detail. It also provides an extensive bibliography on tidal currents and sediments.

Storm-generated Waves and Currents. The importance of storm-generated currents as agents of sediment transport and deposition on the shelf appears to have been suggested first by Hayes (1967), who described the effects of Hurricanes Carla and Cindy (1961 and 1963) on sediment erosion and transport on the Texas coast and shelf. Since that time, numerous additional papers dealing with storm processes and storm-deposited sediments have appeared in print, and this subject continues to be one of the most active and controversial areas of research dealing with continental-shelf sedimentation. R. G. Walker (1984a) reviews this problem and provides a bibliography of papers dealing with storm processes and sediments. Four types of storm-related processes may operate on shelves to transport and rework sediment: (1) waves, (2) wind-forced currents, (3) relaxation (storm-surge–ebb) currents, and (4) turbidity currents.

Waves, as previously discussed, generate a circular orbital motion of water particles; the motion is translated downward to a nearly horizontal to-and-fro motion as shoaling orbital waves feel bottom, leading to formation of oscillation ripples. Bottom-water motion generated by orbital waves can stir up bottom sediment and place it in suspension, where it can be transported by relatively weak unidirectional currents (Komar, 1976). Also, the to-and-fro motion of water along the bottom may be asymmetrical in velocity, causing some net sediment transport that tends to be shoreward in the case of fair-weather waves. Fair-weather waves can entrain sediment only to depths of about 10 m or so; however, storm waves of much longer wave length are capable of disturbing sediment on the outer shelf to depths of 200 m or more. In these deeper water areas, storm waves can stir up and rework bottom sediment, but they probably produce relatively little net transport of sand. By contrast, storm waves acting on the shoreline erode sediment from the beach and deposit it seaward on the shoreface and shelf.

Wind-forced currents are unidirectional currents generated by wind shear stress as wind blows across the water surface, gradually putting into motion deeper and deeper layers of water. If the velocity and duration of wind are great enough, water movement may extend to the seabed with enough velocity to transport sediments. Winds commonly create wind-forced currents that flow parallel to shore and therefore do not provide much offshore sediment transport; however, currents moving along the shoreline can be deflected landward owing to the Coriolis force, causing an onshore pileup of water. The Coriolis force is generated by Earth's rotation, causing moving objects to be deflected to the right in the Northern Hemisphere and to the left in the Southern Hemisphere. Piling up of water onshore creates an elevation of the water surface—a **coastal set-up**—of perhaps a meter or two, and consequently a seaward-directed pressure gradient owing to ebb of bottom water seaward (relaxation) (Fig. 13.4). As the bottom water flows back seaward, it is deflected laterally to form a **geostrophic current** that subsequently flows roughly parallel to the bathymetric contours, or isobaths; that is, roughly parallel to the shoreline. Some available bottom-current measurements from modern shelves suggest that such flows may achieve velocities of several tens of centimeters per second (R. G. Walker, 1984a), a rate adequate to transport sandy sediment and develop bedforms. Note again that such transport by geostrophic currents tends to be mostly parallel to isobaths and not directly seaward.

FIGURE 13.4 Seaward flow of water owing to coastal setup (storm surge). Bottom water flows seaward following pileup of water along the coast owing to storm waves, but is deflected to the right (Northern Hemisphere) by the Coriolis force to form a geostrophic current that flows roughly parallel to isobaths. (From Walker, R. G., 1984, Shelf and shallow marine sands, in R. G. Walker (ed.), Facies models, 2nd ed.: Geoscience Canada Reprint Ser. 1. Fig. 1, p. 142, reprinted by permission of Geological Association of Canada.)

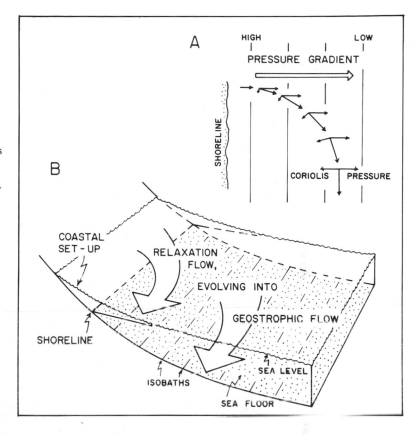

Therefore, sediment may not be transported by such currents to any great distance outward onto the shelf.

Relaxation (storm-surge–ebb) currents are presumably generated by storm surges, which are unusually high-water coastal set-ups of up to 6–7 m that form owing to strong storm winds blowing water directly onshore. As coastal water level rises, there is a return bottom flow of bottom water—a seaward-oriented relaxation flow. Relaxation flow arising from storm surges is thus somewhat similar to the seaward ebb of water from coastal set-ups caused by wind-forced currents, but it occurs on a larger scale. Hayes (1967) suggested that storm-surge currents ebbed seaward as a density current (i.e., a turbidity current), resulting in deposition of graded beds. The magnitude of some relaxation currents has been presumed to be great enough that they are capable of transporting sand considerable distances seaward below the fair-weather wave base. There is as yet no definite proof of this postulated process based on studies of modern shelves, but some geologists studying ancient continental-shelf sediments have deduced from paleocurrent patterns that sands may be transported seaward by storms and introduced into normally quiet areas of the shelf. Such transport in the geologic past has been attributed to storm surges of unusual magnitude, so-called "thousand-year events," not yet observed and documented on modern shelves (R. G. Walker, 1984a). Readers should be aware that the concept of storm-surge currents is controversial. Not all workers who are currently investigating shelf processes and shelf sediments agree that such postulated currents constitute a viable mechanism of shelf-sediment transport. The concept of storm-surge currents is presented here simply as an interesting hypothesis for shelf-sediment transport, not as a concept that is necessarily correct.

Turbidity currents of the surge type have not been observed or conclusively documented on modern shelves. Geological evidence obtained by study of ancient shelf sequences points to the possibility that turbidity currents triggered by storm activity may be agents of sediment transport and deposition on shelves. In particular, sharp-based, graded sandstone units or sharp-based hummocky cross-stratified sandstones, in association with bioturbated mudstones, have been suggested to indicate turbidity current transport. R. G. Walker (1984a) proposes that turbidity currents are generated in the nearshore zone close to wave base owing to storm-produced cyclic wave loading of rapidly deposited fine sediment. Wave loading causes liquefaction of the sediment, which then flows downslope, creating a turbidity current owing to suspension of fine sediment as a result of combined flow acceleration and expulsion of pore fluid.

Ocean Currents. Major, semipermanent ocean currents intrude onto some shelves with sufficient bottom velocity to transport sandy sediment. About three percent of modern shelves are dominated by these ocean currents, which operate most effectively on the outer shelf. Modern examples include the northwestern Gulf of Mexico, which is affected by the Gulf Stream; shelves swept by the Panama and North Equatorial currents off the northeast coast of South America; Taiwan Strait between Taiwan and mainland China, which is intruded by a branch of the Kuroshio Current flowing north from the Philippines; and the outer shelf of southern Africa, which is crossed by the southward-flowing Agulhas Current of the western Indian Ocean. These currents commonly contribute little if any new sediment to the shelf, but they are capable of transporting significant volumes of fine sediment along the shelf. Some achieve bottom velocities great enough to transport sandy sediment and create sand waves and other bedforms; for example, the Kuroshio Current (Boggs et al., 1979) and the Agulhas Current (Fleming, 1980) (Fig. 13.5).

Density Currents. Density currents are created by density differences within water masses owing to variations in temperature, salinity, or suspended sediment. Density currents are important mainly in transport of fine suspended sediment. High concentrations of suspended sediments near river mouths may create dilute density currents that move along the bottom. Conversely, plumes of warm river water carrying suspended terrigenous clay may override denser seawater and transport the clay for some distance across the shelf before mixing and flocculation cause the clays to settle. Transport of sediment within the near-bottom nepheloid layer, described in Chapter 3, is also related to water density. In arid climates, excessive nearshore evaporation may generate dense brines that flow seaward along the bottom as an underflow. Overall, density currents, other than the possible surge-type turbidity currents, are not significant agents of shelf transport, and no modern shelf is dominated by such processes.

Biologic Activities. The modern continental shelves are among the most densely populated of all depositional environments, and the geologic record suggests that ancient epeiric seas also were inhabited by large populations of organisms. The shelf floor is

FIGURE 13.5 Sediment transport by the Agulhas Current off the southeastern tip of Africa. Sand waves migrate under the influence of the Agulhas Current. The stippled pattern indicates coarse lag deposits; black streaks indicate sand ribbons. Sand-wave fields are up to 20 km long and 10 km wide, and individual sand waves are up to 17 m high. (From Fleming, B. W., 1980, Sand transport and bedforms on the continental shelf between Durban and Port Elizabeth (southeast Africa continental margin): Sed. Geology, v. 26. Fig. 15, p. 194, reprinted by permission of Elsevier Science Publishers, Amsterdam.)

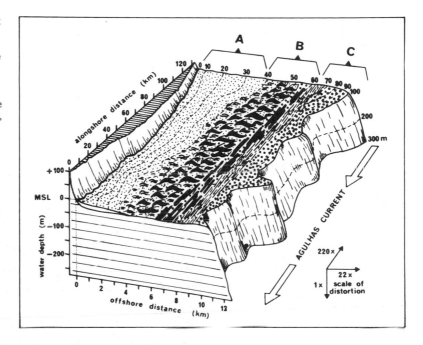

habitat for a high diversity of invertebrate organisms, including molluscs, echinoderms, corals, sponges, worms, and arthropods. Both infauna and vagrant and sessile epifauna are represented. Organisms are most abundant in lower-energy areas of the shelf, and the greatest populations of organisms occur on the inner shelf just below wave base (Howard, 1978). Organisms on siliciclastic-dominated shelves are particularly important as agents of bioturbation. Both type and abundance of bioturbation structures vary with sediment type and water depth. As discussed in Chapter 6, burrows in the nearshore high-energy zone are escape structures that tend to be predominantly vertical. The burrow style changes to oblique or horizontal feeding structures with deepening of water across the shelf. In general, muddy sediments of the shelf are more highly bioturbated than sandy sediments, and physical sedimentary structures in these sediments may be almost completely obliterated by bioturbation. By contrast, only a few species of organisms can survive in the very high-energy nearshore shelf and beach zone. Therefore, sandy sediments of the beach–shelf transition zone are dominated by physical structures such as cross-bedding rather than by bioturbation structures. Sandy layers deposited in deeper water on the shelf may be bioturbated to some degree in their upper parts. Bioturbation structures associated with various water depths on the shelf are illustrated in Figure 13.6. In addition to their importance as bioturbation agents, some organisms produce fecal pellets from muddy sediment; the pellets become hardened and coherent enough to behave as sand grains. Organisms with shells or other fossilizable hard parts also leave remains that may be preserved to become part of the sediment record.

Shelf Sediments

Source. Sediments on siliciclastic shelves of the modern ocean may include modern, relict, and palimpsest types. **Modern sediments** are deposits that are in equilibrium

FIGURE 13.6 Bioturbation structures at different water depths on
the shelf. (From Reineck, H. E.,
and I. B. Singh, 1980, Depositional
sedimentary environments. Fig.
554, p. 402, reprinted by permission of Springer-Verlag, Heidelberg. Originally after H. E. Reineck,
J. Dörjes, S. Dadow, and G. Hert-
weck, 1968, Sedimentologie,
fauenzonierung und faziesabfolge
vor der Ostküste der inneren
Deutschen Bucht: Senckenbergiana
Lethaea, v. 49, p. 261–309.)

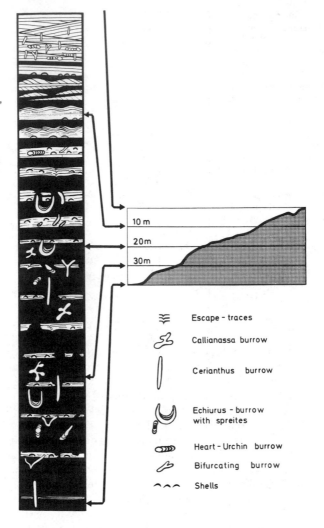

Escape - traces

Callianassa burrow

Cerianthus   burrow

Echiurus - burrow
with spreites

Heart - Urchin burrow

Bifurcating   burrow

Shells

with the hydrodynamic conditions now existing on the shelf. They were transported
onto the shelf by processes currently operating on the shelf or they were generated in
place by shelf processes. By far the greatest volume of modern sediments are **siliciclastic muds and sands,** contributed to the shelf by river outflow and shoreline erosion, together with some fine dust blown onto the shelf by winds. On shelves that
border volcanic arcs, such as the Japan continental shelf, materials contributed by
volcanism may be important components of the shelf sediment (Boggs, 1984). Under
the category of modern sediments, Emery (1968) includes also **biogenic sediments,**
composed mainly of carbonate shells and tests; **authigenic sediments,** which consist
mainly of glauconite and phosphorite; and **residual sediments,** produced by in situ
submarine weathering of bedrock.

Relict sediments are deposits that are not in equilibrium with present hydrodynamic conditions, as discussed. These sediments were derived from sources on land

and transported onto the shelves by rivers or glaciers during lowered sea level. Therefore, they may include significant quantities of gravels as well as sands. Emery (1968)
estimates that about 70 percent of the area of modern shelves is covered by relict
sediment. In making this estimate he did not take into account certain parts of the
shelves that have been extensively reworked by physical and biological processes operating today. These reworked relict sediments are the palimpsest sediments of Swift
et al. (1971).

General Characteristics and Distribution. The sediments of modern siliciclastic
shelves consist largely of muds and sands, although large patches of relict gravels
occur in many areas. The areal distribution of muds and sands varies markedly on
different shelves. On some shelves, nearshore sands grade seaward through a transition zone of mixed sand and mud to deeper-water muds, in essentially the classic
pattern of seaward-decreasing grain size visualized by D. W. Johnson (1919) in his
concept of a graded shelf. On some other shelves, coastal sands grade directly offshore
into Pleistocene relict sands or, alternatively, muds may cover the shelf floor right up
to the shoreline (Reineck and Singh, 1980). Modern shelves, with their mixture of
modern and relict sediment, may not be perfect analogs for ancient continental shelves
and epeiric seaways. We know, for example, that some types of modern shelf-sediment
bodies, such as transgressive, shoreface-detached sand ridges, have not yet been reported from the ancient geologic record (R. G. Walker, 1984a). In any case, the types
of sediments and sedimentary structures found on modern shelves are related both to
Pleistocene depositional patterns and to the dominant sedimentation processes now
operating on the shelves.

Sediments of Tide-dominated Shelves. Tide-dominated shelves are distinguished by
tidal currents with velocities ranging between 50 and 100 cm/sec. Modern examples
include the North Sea; Korea Bay of the Yellow Sea, the Gulf of Cambay, India; and
the northern Australia shelf (R. A. Davis, 1983). Tide-dominated shelves are characterized particularly by the presence of sand bodies of various types and dimensions.
Large **sand waves** a few meters in height, with wave lengths of up to 100 m or so,
typically occur in fields that may cover areas of 15,000 km^2 or more. Sand waves have
symmetrical cross-sectional shapes if produced by tidal currents with equal ebb and
flood peak speeds or asymmetrical shapes if ebb and flood velocities are unequal
(Belderson et al., 1982). **Tidal sand ridges** or **sand ribbons** up to 40 m high, 5 km
wide, 60 km long, and covering areas up to 5000 km^2 have been reported from the
North Sea shelf (Swift, 1975) (Fig. 13.7). These ridges are oriented parallel to tidal flow
and have steep sides that dip seaward. They have been referred to as "shoal retreat
massifs." They are believed to have formed initially by shoreline detachment during
Holocene transgression, but they are currently maintained by tidal flows. In addition
to sand waves and tidal sand ridges, tide-dominated shelves include **sand sheets** (and
banks) and **gravel sheets,** characterized by small-scale bedforms, and patches of bioturbated muds in areas sheltered from tidal currents and waves (Stride et al., 1982).
Many of the types of sand bodies found on modern tide-dominated shelves are illustrated in the picture atlas *Sonographs of the Sea Floor* (Belderson et al., 1972).

Sediments of Storm-dominated Shelves. Storm-dominated shelves predominate on
most of the world's coasts. They are characterized by low tidal current velocities, commonly <25 cm/sec, and fair-weather wave base is normally shallow (<~10 m). Be-

FIGURE 13.7   Sand ridges in the North Sea. Ridges are up to 35–40 m high; ridge crests are in about 10–15 m of water; steepest sides of the ridges face northeast and dip about 5 degrees. (From Swift, D. J. P., 1975, Tidal sand ridges and shoal-retreat massifs: Marine Geology, v. 18. Fig. 18, p. 129, reprinted by permission of Elsevier Science Publishers, Amsterdam. Originally after J. J. H. C. Houbolt, 1968, Recent sediments in the southern bight of the North Sea: Geol. Mijnbouw, v. 47. Fig. 9, p. 252, reprinted by permission of Royal Geological and Mining Society of the Netherlands, Martinus Nijhoff, Dordrecht.)

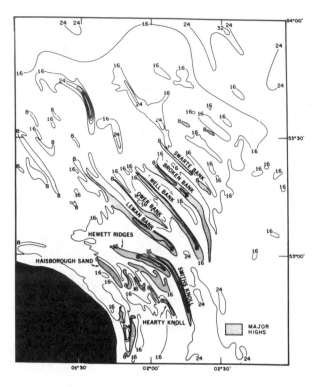

cause of these characteristics, little sediment movement occurs on these shelves except during intense storm conditions. Specific examples of storm-dominated shelves include the Atlantic shelf off the eastern coast of the United States, the Pacific shelf off Oregon and Washington, and the Bering Sea. Sedimentation patterns on storm-dominated shelves may be quite complex, depending particularly upon the extent to which the shelves are mantled by relict sediments or modern sediments. Shelves with abundant relict sediment, such as the Atlantic, are characterized particularly by sand bodies. These bodies take several different forms. **Short retreat massifs** are large sand bodies similar to those reported in the North Sea (Fig. 13.7). They formed during Holocene transgression, but are reworked and maintained by storm activity. **Linear sand ridges** up to 10 m high, 2–3 km wide, a few tens of kilometers long, and spaced 2–7 km apart, occur between or superimposed on retreat massifs. These ridges have been interpreted both as originating from relict barrier bars and as arising from modern shelf processes. **Lower-relief sand bodies,** characterized by ripple, megaripple, and sandwave bedforms, are common also. These bodies presumably form by a combination of oscillatory and unidirectional flow arising from storm activity (R. G. Walker, 1984a). Shelves with a greater component of modern versus relict sediments, such as the Pacific shelf off Oregon and Washington, are typically characterized by less relief and finer grained sediments than Atlantic-type shelves. Although sediment on these shelves may display a general trend of seaward fining, extensive "windows" occur where relict sands or gravels show through a discontinuous blanket of fine-grained modern sediment. Also, mixing of relict sands and modern muds may take place in some areas. Muds are typically thoroughly bioturbated. Summary characteristics of several storm-dominated shelves are described by H. D. Johnson (1978).

Sediments on Shelves Dominated by Intruding Ocean Currents. These shelves make up only a very small percentage of the world's shelves. Sediment on such shelves is largely relict, but is commonly reworked by intruding currents to form sand waves, sand ribbons, and coarse sand and gravel lag deposits. The best studied example of this type is the southeastern shelf of South America, which is intruded by the Agulhas Current of the Indian Ocean. Sand waves up to 17 m high, with wave lengths up to 700 m, occur in sand-wave fields up to 10 km wide and 20 km long (Fig. 13.5). Taiwan Strait between Taiwan and China is another broad shelf invaded by ocean currents that create extensive sand-wave fields.

Sedimentary Structures. Because most of the shelf is constantly covered by water, sedimentary structures in modern shelf sediments have to be studied largely by indirect methods. Small-scale bedforms can be observed and photographed by divers or remote-controlled cameras. Larger bedforms such as sand waves and sand ridges are investigated by sonar bottom profiling and side-scan sonar techniques. Small-scale internal sedimentary structures can be studied in cores of bottom sediment, and sub-bottom seismic profiling methods, described in Chapter 16, may be used to study some large-scale features such as bedding. None of these methods allow detailed examination of modern shelf structures. Therefore, relatively few data are available on the sedimentary structures of modern shelves, and geologists have been forced to turn to the geologic record to develop a fuller understanding of shelf structures. The bedding characteristics of shelf sediments, for example, are known almost exclusively from ancient shelf sediments, which typically consist of evenly bedded, laterally extensive units. On the basis of what is known from studies of modern and ancient shelf sediments, most shelf sands are characterized particularly by cross-bedding. Small-scale cross-bedding and ripple cross-lamination, produced by migration of ripples and small dunes, and large-scale cross-bedding generated by migration of sand waves and sand ridges are both common. Depending upon tidal influence, foreset dip directions of cross-lamination may be bidirectional or unidirectional. Plane beds develop also under some flow conditions on the shelf. Physical structures thus tend to dominate shelf sands, which typically display relatively few bioturbation structures. As mentioned, muddy shelf sediments tend to be highly bioturbated with few physical structures except possibly planar lamination. Some of the characteristic bioturbation structures in shelf sediments are illustrated in Figure 13.6.

Storm layers and hummocky cross-stratification are sedimentary structures that appear to be especially characteristic of shelf sediments. **Storm layers** are commonly thin layers consisting of concentrations of coarser grains interlayered or embedded in finer grained muds. The coarser material typically consists of coarse silt, fine sand, or shell fragments. The layers characteristically show vertical size grading. It is suggested that they form when storm waves resuspend surface sediment, which subsequently settles from suspension. Other origins are possible, including deposition from turbidity currents, which originated closer to shore near fair-weather wave base and then flowed outward onto the shelf, and deposition by wind-forced currents during waning stages of flow. These sandy layers are sometimes called **tempestites.** They are best developed on the inner shelf, but they have been found as far as 40 km from the coast (Reineck and Singh, 1980). Hummocky cross-stratification has apparently been identified in few, if any, modern shelf sediments, but it has been described in numerous ancient shelf sediments ranging in age from Precambrian to Pleistocene. As discussed in Chapter 6, it consists of curving, gently dipping laminations, both convex-up (hummocks) and concave-up (swales), that intersect at a low angle. It is commonly interbedded with bioturbated mudstones (Fig. 13.8). Most workers now appear to agree that hummocky cross-stratification forms owing to storm waves acting in some manner below fair-weather wave base; however, the exact mechanism of formation is still controversial.

Vertical Sequences. Several types of vertical sequences may be generated in shelf sediments depending upon whether deposition takes place during transgression or regression and upon the dominant type of shelf processes operating during deposition. It is difficult to generalize about these successions except to say that transgression tends to produce fining-upward sequences that typically begin with coarse lag deposits, and regression produces coarsening-upward sequences. Some idealized vertical shelf sequences produced under different postulated sedimentation conditions are illustrated in Figures 13.9 and 13.10. Both tide-dominated and storm-dominated transgressive sequences are shown in Figure 13.9 to illustrate differences that can develop in facies under these different shelf conditions. These sequences should be considered only as working models. Actual transgressive and regressive sequences may differ markedly in detail from these idealized profiles. The vertical and lateral facies relationships of shelf sediments are discussed in greater detail by H. D. Johnson (1978).

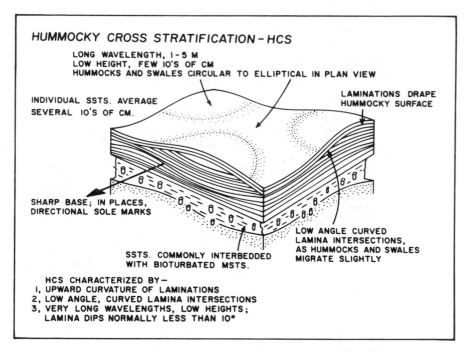

FIGURE 13.8 Schematic diagram of hummocky cross-stratification, which typically occurs interbedded with bioturbated mudstone. (From Walker, R. G., 1984, Shelf and shallow marine sands, in R. G. Walker (ed.), Facies models, 2nd ed.: Geoscience Canada Reprint Ser. 1. Fig. 11, p. 149, reprinted by permission of Geological Association of Canada. Originally after R. G. Walker, 1982, Hummocky and swaley cross stratification, in R. G. Walker (ed.), Clastic units of the Front Range between Field, B. C. and Drumheller, Alberta: Internat. Assoc. Sedimentologists, 11th Internat. Congress on Sedimentology (Hamilton, Canada), Guidebook to Excursion 21A, p. 22–30.)

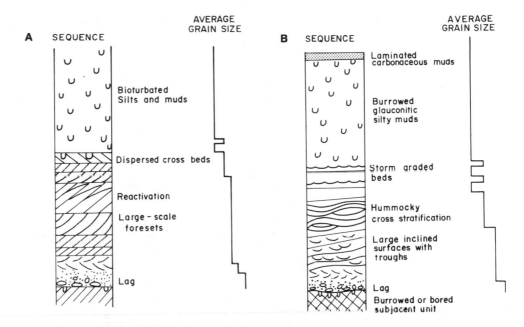

FIGURE 13.9 Schematic diagrams illustrating typical fining-upward transgressive shelf sequences. A. Tide-dominated shelf. B. Storm-dominated shelf. (After Galloway, W. E., and D. K. Hobday, 1983, Terrigenous clastic depositional systems. Figs. 7.14, 7.15, p. 159, 160, reprinted by permission of Springer-Verlag, Heidelberg.)

Ancient Siliciclastic Shelf Sediments

Although recognition of ancient shelf sediments is aided by study of modern continental shelves, modern shelves are not necessarily good analogs of ancient shelves. The prevalence of relict sediments on modern shelves may be atypical, and certain features of modern shelves, such as shoal retreat massifs, have not been recognized in the ancient rock record. Conversely, some structures believed to be diagnostic of ancient shelf sediments, notably storm-generated hummocky cross-stratification, have apparently not been recognized in modern shelf environments. Thus, no well-defined model for shelf sediments combining data from modern environments and study of ancient rocks has yet been formulated. In general, ancient shelf sediments appear to be distinguishable on the basis of the following characteristics: (1) tabular shape; (2) extensive lateral dimensions (thousands of square kilometers) and great thickness (hundreds of meters); (3) moderate compositional maturity, with quartz dominating over feldspars and rock fragments; (4) generally well developed, even, laterally extensive bedding; (5) presence of hummocky cross-stratification and/or storm layers; (6) fining-upward (retrogradational) or coarsening-upward (progradational) vertical sequences; (7) wide diversity and large numbers of normal-marine, stenohaline organisms; and (8) diagnostic associations of trace fossils. No single criterion is generally adequate to identify ancient shelf sediments, but taken together these criteria are fairly distinctive.

Ancient shelf deposits have been reported from stratigraphic units of all ages from all continents. They are probably the most extensively preserved rocks in the geologic record (R. A. Davis, 1983). Examples are so numerous that little purpose is

FIGURE 13.10 Idealized coarsening-upward regressive shelf sequence on a storm-dominated
shelf. (After Galloway, W. E., and
D. K. Hobday, 1983, Fig. 7.17, p.
162, reprinted by permission of
Springer-Verlag, Heidelberg.)

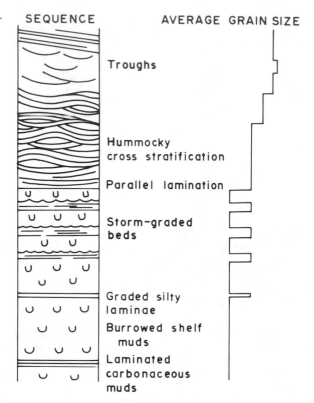

served here in trying to tabulate them. R. G. Walker (1984a) lists many of the better
known North American examples together with comments on probable depositional
setting. Space does not permit examination of ancient examples illustrating the many
different models of shelf sediments, but one example is discussed here to illustrate
deposition on a storm-dominated shelf and some of the problems involved in interpreting sediment transport and depositional mechanisms on such shelves.

The Lower Cretaceous Grayson Formation of Texas, described by Hobday and
Morton (1984), displays many of the features commonly attributed to storm deposition; however, these features can be interpreted in different ways. A 50-m section of
the Grayson Formation exposed in the Lake Texoma area of northeast Texas shows
characteristic features of the formation (Fig. 13.11). The basal part of this section consists of thick graded sandstones with well-developed hummocky cross-stratification,
separated by subordinate siltstones. The hummocky cross-stratification merges upward
into parallel lamination overlain by ripple cross-lamination. The uppermost few centimeters of the sandstones are invariably burrowed, causing ripple forms and cross-
stratification to be almost entirely obliterated in some units. The sandstones are compositionally and texturally immature and contain abundant fine plant remains. A thin
siltstone unit separates the hummocky cross-laminated basal sandstone unit from an
overlying thick, lenticular sandstone with scoured upper surface, interpreted as a
channel-mouth bar deposit. This channel sand is overlain by mudstones with plant
fragments and a sandstone unit with large wave ripples that may have formed owing

FIGURE 13.11 Vertical section through 50-m thick unit of the Lower Cretaceous Grayson Formation of Texas near Lake Texoma. This section includes many depositional characteristics attributed to storm-shelf deposition. (After Hobday, D. K., and R. A. Morton, 1984, Lower Cretaceous shelf storm deposits, northeast Texas, *in* R. W. Tillman and C. T. Siemers (eds.), Siliciclastic shelf sediments: Soc. Econ. Paleontologists and Mineralogists Spec. Pub. 34. Fig. 3, p. 208, reprinted by permission of SEPM, Tulsa, Okla.)

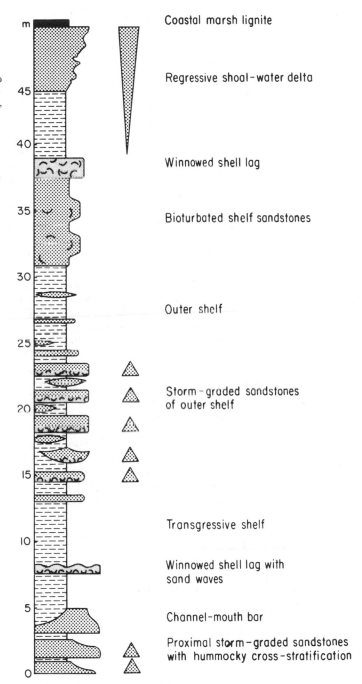

Coastal marsh lignite

Regressive shoal-water delta

Winnowed shell lag

Bioturbated shelf sandstones

Outer shelf

Storm-graded sandstones of outer shelf

Transgressive shelf

Winnowed shell lag with sand waves

Channel-mouth bar

Proximal storm-graded sandstones with hummocky cross-stratification

to storm swells. This wave-rippled layer also contains concentrations of oysters and other molluscs. The upper part of the section above the wave-rippled sand consists of siltstones and mudstones with interbeds of thinner graded sandstones and layers with shell concentrations, capped at the top by a coarsening-upward mudstone–sandstone sequence with a lignite bed above. The graded sandstones are sharp based, with sole markings and elongate fossils oriented offshore. Internal structures are mainly parallel laminations with some low-angle foresets and ripple cross-lamination. Fewer than 10 percent of the graded sandstones are intensely bioturbated at the top. The remainder contain a few horizontal burrows or no trace fossils at all. The associated mudstones intervening between successive sandstone beds contain some burrows but are not highly bioturbated. The uppermost sandstone of this sequence is highly bioturbated and contains scattered molluscs. Overlying this unit is a thick sequence of laminated mudstone, which grades upward into sandstone with clinoform foresets and thin coal partings. Root-disrupted impure lignite deposits form the top of the sequence.

Hobday and Morton (1984) interpret the hummocky cross-stratified sandstones and the wave-rippled sandstone in the base of the section as deposits formed above storm wave base at water depths less than about 30 m. They suggest that the hummocky cross-stratification was produced by interaction of unidirectional currents and storm waves. The sharp-based, graded sandstones in the overlying unit were probably deposited in quiet water, possibly below storm wave base, but the origin of graded shelf sandstones of this type has been interpreted in different ways. They resemble turbidites in many respects and have been interpreted by some authors to be shelf turbidites, deposited by turbidity currents generated in the nearshore zone owing to bottom sediment being liquefied by cyclic loading caused by storm waves (R. G. Walker, 1984a). They have been interpreted by other workers as storm-surge deposits, presumably deposited from relaxation currents generated during strong storms that move seaward as density currents. Alternatively, they might be deposited above storm wave base by settling of clouds of sediment temporarily stirred up and suspended by storm waves. Hobday and Morton believe that the characteristics of the Grayson Formation graded sandstones are best explained as the deposits of wind-forced currents. According to these authors, strong seaward-directed bottom currents related to high-energy bottom-return flow were produced by wind-forcing during the heights of storms. These currents transported the sandy sediment seaward. Upward gradation from parallel-laminated sandstones into siltstones, or from parallel lamination through cross-lamination into homogeneous mudstone, is explained by rapidly waning flow of these wind-forced currents. Such an interpretation seems to require deposition relatively close to shore, as wind-forced currents have not yet been proven to move long distances seaward. The mudstones separating the graded sandstone units appear to be suspension sediments deposited during fair-weather conditions at such a rapid rate that little bioturbation occurred. The uppermost mudstone, sandstone, lignite unit represents a change in sedimentation conditions from transgressive shelf deposition to regressive sedimentation along a prograding delta front. Thus, prodelta mudstones grade upward to shallow-water deltaic sands, capped by delta-plain marsh deposits (lignite).

Carbonate Shelf Systems

Introduction

Carbonate deposits constitute the dominant sediment cover on a few modern shelves. These shelves are located primarily at low latitudes in clear, shallow, tropical to subtropical seas (Fig. 13.12) where little terrigenous siliciclastic detritus is introduced. Carbonate sediments form also on some higher latitude, cold-water shelves, where

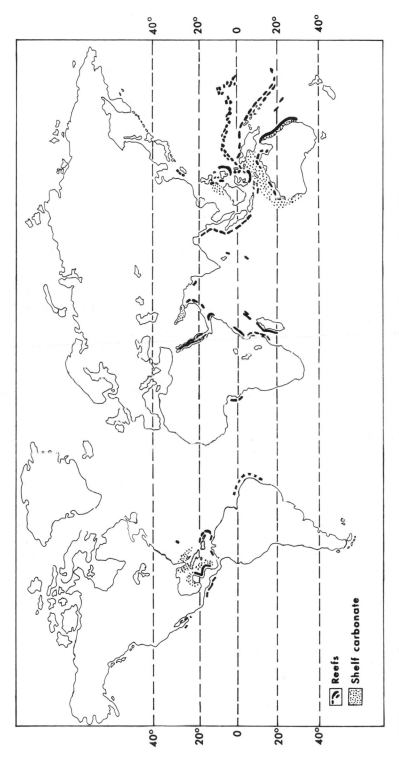

FIGURE 13.12 Distribution of shallow-marine carbonate sediments and reefs in the modern ocean. (From Wilson, J. L., 1975, Carbonate facies in geologic history. Fig. 1.1, p. 2, reprinted by permission of Springer-Verlag, Heidelberg.)

Reefs

Shelf carbonate

they consist predominantly of shell remains. Carbonate-producing environments with low terrigenous input are found today on some shelves attached to the mainland, such as Florida Bay and western Australia. They occur also in a few smaller shelf areas that surround oceanic islands where terrigenous influx is extremely low. The Bahama Platform and the narrow shelves around Pacific atolls are examples. The relatively minor importance of modern carbonate deposition is decidedly atypical of many geologic periods of the past when widespread deposition of carbonate sediments characterized sedimentation in broad epeiric seas hundreds of kilometers wide. During the middle Paleozoic, for example, carbonate deposition prevailed in shallow inland seas that spread over much of the continental interior of North America. In spite of the small areal extent of modern shelf carbonate environments, carbonate-dominated shelves nonetheless provide outstanding "laboratories" for studying the mechanisms of carbonate sedimentation. Much of what we now understand about carbonate textures and the basic processes of carbonate deposition has come from study of modern carbonate environments. On the other hand, we must turn to the ancient rock record itself for insight into the environmental conditions that typified carbonate-dominated epeiric seas.

Geologists have been greatly interested in carbonate rocks for well over a hundred years. In fact, the science of petrography was initiated by an English geologist named Henry Clifton Sorby, who began his petrographic studies about 1851 with the study of limestones. Other historically interesting early studies of carbonates include investigation of carbonate sediments in the Bahamas by Black (1933) and Cayeux's (1935) classic publication on the carbonate rocks of France. Modern study of carbonate sediments and depositional processes is generally regarded to have begun in the 1950s with the publications of Newell et al. (1951), Illing (1954), and Ginsburg (1956) dealing with modern carbonate sediments in the Bahamas and Florida Bay. Since that time, the pace of research on carbonates has accelerated at an astounding rate and dozens of books and hundreds of research papers have been devoted to both modern and ancient carbonate sediments and depositional processes. Many of these books tend to be narrowly focused on special aspects of carbonate sedimentology, but several excellent books with good general coverage of carbonate environments have been published. Probably the most outstanding of these is the symposium volume *Carbonate Depositional Environments* (Scholle, Bebout, and Moore, 1983). Other informative and useful monographs include those of Bathurst (1975), H. E. Cook et al. (1983), Friedman (1969), Frost et al. (1977), Hardie (1977), Laporte (1974), Logan et al. (1970, 1974), Milliman (1974), and J. L. Wilson (1975). Shelf carbonate depositional environments have been summarized in shorter papers by H. E. Cook (1983), Enos (1983), Hine (1983), N. P. James (1984a, b, c) Sellwood (1978), J. L. Wilson and Jordan (1983), and many others.

Depositional Setting

Carbonate sediments are deposited primarily on shallow-marine shelf platforms, including some parts of the marginal-marine environments such as beaches, lagoons, and tidal flats. The geologic record suggests that carbonate shelves can occur on the margins of cratonic blocks, in intracratonic basins, across the tops of major offshore banks, and on localized positive features (banks, for example) on wide shelves (J. L. Wilson and Jordan, 1983). Three basic types of carbonate platforms are recognized: (1) carbonate ramps, (2) rimmed carbonate shelves, and (3) isolated carbonate platforms (Hine, 1983; Read, 1982; Ginsburg and James, 1974).

FIGURE 13.13 Schematic block diagram of a homoclinal carbonate ramp. (After Read, J. F., 1982, Carbonate platforms of passive (extensional) continental margins: Tectonophysics, v. 81. Fig. 1A, p. 198, reprinted by permission of Elsevier Science Publishers, Amsterdam.)

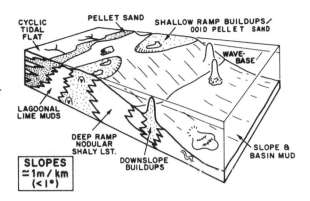

Carbonate ramps are gently sloping ($<\sim 1$ degree) platforms on which shallow-water deposits pass downslope with only a slight break in slope into deeper water facies (Fig. 13.13). The break in slope is not marked by a pronounced reef trend, but discontinuous carbonate sand shoals may be present along the shelf edge where water energy is high. Water circulation across the shelf may be adequate to allow development of a moderately high-energy beach zone along shore in addition to formation of skeletal or ooid-pellet sand shoals along the shelf edge.

Rimmed carbonate shelves are shallow platforms marked at their outer edges by a pronounced break in slope into deeper water (Fig. 13.14). They have a nearly continuous rim or barrier along the platform edge. This barrier consists of either a reef buildup or a skeletal, oolitic sand shoal, which restricts wave action and water circulation to create a low-energy "lagoon" landward of the shelf-edge barrier (Ginsburg and James, 1974). The lagoon commonly grades landward into a low-energy tidal-flat environment rather than a high-energy beach zone.

Isolated platforms (Bahama type) are shallow-water platforms tens to hundreds of kilometers wide, commonly located offshore from shallow continental shelves and surrounded by deep water that may range from several hundreds of meters to a few kilometers deep (Read, 1982; Fig. 13.15). The platforms may have gently sloping, ramplike margins, or more steeply sloping margins resembling those of rimmed shelves.

FIGURE 13.14 Schematic block diagram of an accretionary-type rimmed carbonate shelf. (After Read, J. F., 1982, Fig. 2A, p. 200, reprinted by permission of Elsevier Science Publishers, Amsterdam.)

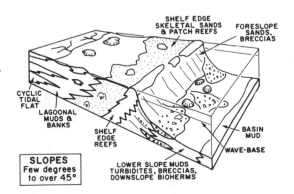

FIGURE 13.15 Schematic block diagram illustrating the principal features of an isolated carbonate platform (Bahama type). (After Read, J. F., 1982, Fig. 3, p. 204, reprinted by permission of Elsevier Science Publishers, Amsterdam.)

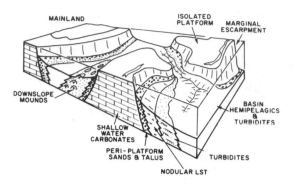

Carbonate workers subdivide the carbonate shelf environment into inner, middle, and outer shelves (Fig. 13.16). The **inner shelf** includes nearshore shallow-water to subaerial environments such as beaches, tidal flats, and possibly nearshore lagoons, which are environments not considered part of the shelf in our previous discussion of siliciclastic shelves. The **middle shelf** encompasses the shallow, subtidal zone lying between the nearshore areas and the shelf break. It is referred to as a "subtidal lagoon" by some workers, particularly in reference to rimmed carbonate shelves. The **outer shelf** is a very narrow zone that constitutes the shelf break and that encompasses reef buildups or carbonate sand shoals along the shelf edge. J. L. Wilson (1975) recognized nine "facies belts" in carbonate environments extending seaward from the supratidal zone to the slope and basin. The shelf encompasses facies belts 4 through 9 (Fig. 13.16), although the term shelf facies is most often used for sediments of the middle shelf, or so-called open shelf. Note that the above subdivision of the shelf differs from that commonly used for siliciclastic shelves (Fig. 13.3). Furthermore, the term "basin" as used by carbonate sedimentologists simply signifies a deepening of water beyond the shelf edge to below normal wave base. The term does not necessarily imply that the carbonate basin environment is analogous to the deep oceanic environment. In fact, carbonate basins generally lie well above abyssal depth, commonly less than a few hundred meters below sea level.

In contrast to most siliciclastic shelves, carbonate platforms are commonly characterized by a topographic buildup at the shelf margin of the outer shelf. This buildup may be caused by the presence of organic reefs or banks, lime-sand shoals, or small islands that create a barrier to incoming waves. Outer-shelf buildups are particularly characteristic of rimmed carbonate shelves. This outer barrier is commonly dissected by a network of tidal channels that allow high-velocity tidal currents to flow through onto the shelf. Water depth may be only a few meters over this buildup, but depth increases over the middle shelf (Fig. 13.16) to perhaps several tens of meters in some cases. The outer shelf is the highest-energy zone of the shelf. Much of the middle shelf is commonly below fair-weather wave base. Water energy is thus low over most of the middle shelf except over patch reefs, localized banks or shoals, and along the shoreline of some carbonate ramp platforms. The elevation and lateral continuity of the shelf-edge carbonate barrier control water circulation over the entire shelf. The effect of this barrier on water circulation, coupled with the width of the shelf, strongly influences the type and distribution of carbonate facies that develop on the shelf. If a well-developed barrier is present, or if the shelf is very wide, water circulation on the shelf is restricted (J. L. Wilson and Jordan, 1983). This is so because, on wide shelves, water energy is expended in friction with the bottom, leading to poor water circulation. Restricted water circulation leads to development of salinity conditions that deviate from normal (~35‰). Salinities may rise well above normal in arid or semiarid climates where evaporation rates are high or may fall below normal in areas affected by considerable freshwater runoff. Variations in salinity affect the diversity and numbers of organisms living on shelves; the organisms, in turn, owing to the extremely important role they play in carbonate sedimentation processes, strongly affect carbonate deposition. The inner shelf is especially characterized by restricted conditions.

Although carbonate environments extend from the supratidal zone to deeper basins off the shelf, the shallow basin platform that constitutes the middle, or subtidal, shelf and outer shelf is the primary site of carbonate production. N. P. James (1984a) refers to this shelf platform as the "subtidal carbonate factory" (Fig. 13.17). The sediments produced on the shelf also accumulate mainly on the shelf; however, some

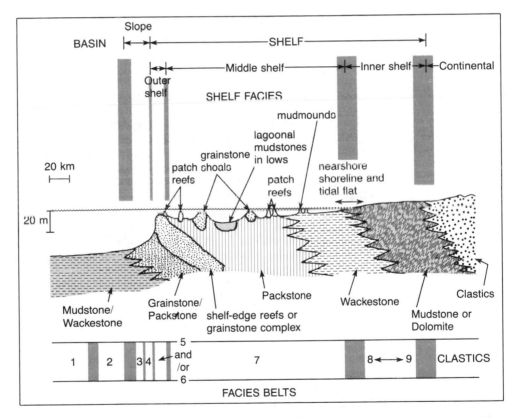

FIGURE 13.16 Schematic profile of the carbonate-shelf environment. Note subdivisions of the shelf into major subdivisions, and contrast this subdivision with that of siliciclastic shelves (Fig. 13.3). This figure shows also the approximate positions of Wilson's (1975) carbonate facies belts: 1, basin; 2, open-sea shelf; 3, deep-shelf margin; 4, foreslope; 5, organic buildups; 6, winnowed platform edge (sands); 7, open-circulation shelf; 8, restricted-circulation shelf and tidal flats; and 9, evaporite sabhkas–salinas. (After Wilson, J. L., and C. Jordan, 1983, Middle shelf environment, *in* P. A. Scholle, D. G. Bebout, and C. H. Moore (eds.), Carbonate depositional environments: Am. Assoc. Petroleum Geologists Mem. 33, Fig. 1a, p. 298, reprinted by permission of AAPG, Tulsa, Okla.)

FIGURE 13.17 The main areas of marine carbonate production. Most carbonates accumulate in water less than 30 m deep—the "subtidal carbonate factory." (From James, N. P., 1984, Introduction to carbonate facies models, *in* R. G. Walker (ed.), Facies models, 2nd ed.: Geoscience Canada Reprint Ser. 1. Fig. 2, p. 210, reprinted by permission of Geological Association of Canada.)

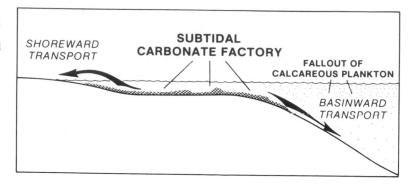

sediments are eventually transported landward onto tidal flats, beaches, and into subtidal lagoons. Others are transported seaward off the shelf onto the slope and into the deeper basin. Little carbonate sediment is generated in the deeper water basin environment off the shelf except for fallout of calcium carbonate-secreting plankton from surface waters.

Sedimentation Processes on Carbonate Shelves

Unlike the deposition of siliciclastic shelf sediments—a process controlled primarily by physical processes—carbonate deposition is controlled by a combination of physical, chemical, biological, and biochemical processes. Also in contrast to siliciclastic deposition, deposition of carbonate sediments is largely an autochthonous process in the sense that carbonate sediments are generated primarily within the same basin in which they are deposited. Siliciclastic sediments are largely allochthonous materials derived from extrabasinal sources. Carbonate muds and carbonate grains such as oolites, pellets, and fossils are formed mainly through chemical and biochemical processes within the depositional basin, aided secondarily by physical processes such as water agitation. After formation, they commonly undergo some physical transport before final deposition.

Chemical and Biochemical Processes. The principal chemical and biological/biochemical controls on carbonate deposition are discussed in Chapter 4 and need not be repeated here. Although the solubility of calcium carbonate is controlled by pH, temperature, and carbon dioxide content of seawater, simple inorganic precipitation of calcium carbonate in the modern ocean seems to be relatively unimportant. Carbonate deposition appears to be controlled primarily by organisms capable of extracting calcium carbonate from the seawater to build their shells. These organisms also contribute to the formation of carbonate sediment through their feeding and bioturbation activities, which cause breakdown of skeletal fragments and other carbonate materials and generate various kinds of trace fossils. It is worth stressing that the organisms primarily responsible for carbonate production in the modern ocean are not necessarily the same as those that were major carbonate formers in the past. Figure 13.18 shows the relative importance of some major groups of organisms as carbonate formers during Phanerozoic time. Note from this figure that the principal carbonate formers have shifted somewhat with time. For example, crinoids, byrozoans, and brachiopods were more important during the Paleozoic than the Cenozoic, whereas coccoliths, plank-

FIGURE 13.18 The approximate diversity, abundance, and relative importance of various calcareous marine organisms as sediment producers. P, Paleozoic; M, Mesozoic; C, Cenozoic. (Modified from Wilkinson, B. H., 1979, Biomineralization, paleooceanography, and evolution of calcareous marine organisms: Geology, v. 7, Fig. 1, p. 526. Published by Geological Society of America, Boulder, Co.)

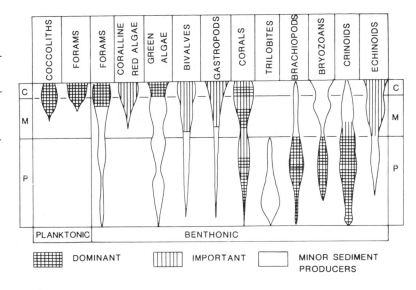

tonic foraminifers, coralline algae, and green algae were particularly important carbonate formers during the Cenozoic.

Physical Processes. Physical processes are important primarily in the reworking and transport of carbonate materials on the shelf, but they also aid in the production of carbonate sediments. Circulation of water onto the shelf brings fresh nutrients, necessary for organic growth, from deeper water. Breaking of waves against reef barriers on the outer shelf increases oxygen content in the water by interaction with the atmosphere and decreases CO_2 owing to decreased water pressure. Thus, modern reefs are best developed in wave-agitated zones, and biogenic production of carbonate sediment in general is stimulated by strong water movement. On the other hand, strong waves crashing on the reef front cause breakdown of reef rock, producing sand- and gravel-size bioclasts that subsequently undergo transport both seaward and landward from the reef.

Agitated water is important to the formation of oolites and may aid in the generation and preservation of grapestones and hardened fecal pellets by submarine accretion and cementation (J. L. Wilson, 1975). Waves and currents also winnow fine carbonate mud from coarser sediment and transport this mud off the shelf platform or into sheltered or protected areas of the shelf. Depending upon water energy, the coarser sediment itself may either remain as a winnowed lag deposit, forming sand- or gravel-covered flats, or be transported and deposited to create wave-formed bars and shoals, beaches, spits, or tidal deltas and bars. Wave- and current-transported and winnowed carbonate sand deposits are particularly common along the outer edge of the shelf platform, where water energy is highest. In resuspension and transport of sediment, storms are as important on carbonate shelves as they are on siliciclastic shelves. For example, most transport of sediment from the subtidal shelf into the intertidal (tidal-flat) environment is accomplished by storms. Absence of wave and current activity on the shelf leads to stagnant circulation, with consequent deviations from normal salinity and possibly anoxic conditions. Such restricted environments constitute unfavorable habitats for many normal marine organisms.

Characteristics of Sediments in Modern Carbonate Environments

The discussions of depositional environments presented in Chapters 11 and 12 and in preceding sections of this chapter deal mainly with deposition of siliciclastic sediments. Therefore, it may be desirable at this point to emphasize some of the differences between siliciclastic and carbonate sediments. Aside from the obvious differences in composition, carbonate sediments differ in several other ways. These differences may include their distribution with respect to climate zones, the relationship of particle size to water energy, and the genetic significance of muddy facies. These and other differences are described briefly in Table 13.1. We will now examine some of the distinguishing characteristics of sediments deposited in carbonate shelf environments. Carbonate reefs are treated as a special kind of shelf environment and are discussed separately.

TABLE 13.1   Major differences in carbonate and siliciclastic sediments

Carbonate sediments	Siliciclastic sediments
The majority of sediments occur in shallow, tropical environments.	Climate is no constraint, sediments occur worldwide and at all depths.
The majority of sediments are marine.	Sediments are both terrestrial and marine.
The grain size of sediments generally reflects the size of organism skeletons and calcified hard parts.	The grain size of sediments reflects the hydraulic energy in the environment.
The presence of lime mud often indicates the prolific growth of organisms whose calcified portions are mud-size crystallites.	The presence of mud indicates settling out from suspension.
Shallow-water lime-sand bodies result primarily from localized physicochemical or biological fixation of carbonate.	Shallow-water sand bodies result from the interaction of currents and waves.
Localized buildups of sediments without accompanying change in hydraulic regimen alter the character of surrounding sedimentary environments.	Changes in the sedimentary environments are generally brought about by widespread changes in the hydraulic regimen.
Sediments are commonly cemented on the seafloor.	Sediments remain unconsolidated in the environment of deposition and on the seafloor.
Periodic exposure of sediments during deposition results in intensive diagenesis, especially cementation and recrystallization.	Periodic exposure of sediments during deposition leaves deposits relatively unaffected.
The signature of different sedimentary facies is obliterated during low-grade metamorphism.	The signature of sedimentary facies survives low-grade metamorphism.

Source: James, N. P., 1984, Introduction to carbonate facies, in R. G. Walker (ed.), Facies models, 2nd ed.: Geoscience Canada Reprint Ser. 1. Table 1, p. 209, reprinted by permission of Geological Association of Canada.

Carbonate Shelf Environments and Sediments

This short synopsis of carbonate environments and facies cannot possibly do justice to this exceedingly interesting and important topic. It provides only an overview of the most important features of carbonate environments. Readers are referred for additional details particularly to the outstanding symposium volume edited by Scholle, Bebout, and Moore (1983); this volume contains more than a dozen beautifully illustrated individual papers, many with color plates, dealing with carbonate environments.

The plan-view setting of carbonate platforms typical of many modern rimmed carbonate shelves is illustrated in Figure 13.19. This figure graphically depicts the different subenvironments of the platform environment, each of which is characterized by specific environmental conditions and carbonate facies.

The outer shelf is commonly the highest-energy environment of the shelf. It is characterized by the development of lime-sand or gravel sheets and shoals. These deposits are lag deposits composed of skeletal fragments of shallow-water organisms such as corals, calcareous algae, bryozoans, and/or ooids and peloids. Much of the skeletal material is derived from adjacent organic reefs and is commonly abraded and rounded by wave action. Sand shoal deposits are generally well sorted and cross-bedded. Facies of ancient carbonate rocks representative of this zone are mainly bioclastic grainstones or packstones (biomicrites or biomicrudites) and oosparites.

The middle shelf (lagoon) is a zone of generally low water energy, particularly on rimmed shelves, and much of the shelf floor is below wave base. Local patch reefs, shoals, or banks extend above wave base. Water circulation ranges from open to partially restricted, depending upon the conditions of the shelf. Typically, the outer part of the middle shelf is characterized by open circulation, whereas the inner part of the shelf experiences somewhat more restricted circulation. This is a zone of high carbon-

FIGURE 13.19 Plan view of a modern carbonate platform showing the location of tidal flats both adjacent to land and in the lee of lime-sand shoals. (From James, N. P., 1984, Shallowing upward sequences in carbonate rocks, *in* R. G. Walker (ed.), Facies models, 2nd ed.: Geoscience Canada Reprint Ser. 1. Fig. 5, p. 215, reprinted by permission of Geological Association of Canada.)

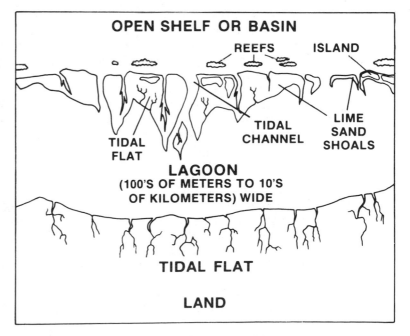

ate production: skeletal sands, lime muds, peloids, and grapestones. Organisms that populate the shelf are typically normal marine stenohaline forms, but more tolerant forms may inhabit restricted areas of the shelf. Typical shelf fauna include brachiopods, pelecypods, gastropods, crinoids, echinoids, calcareous algae, bryozoans, benthonic foraminifers, and, in older rocks, ammonites and nautiloids. Bioturbation is commonly extensive. Owing to the generally low water energy over the middle shelf, the sediments are typically poorly winnowed, with a high ratio of micrite to skeletal fragments and other grains. This condition may not be true of some carbonate ramps where higher energy conditions exist across the platform and carbonate sands may be developed over much of the shelf. Ancient facies of this zone are represented mainly by mudstones (micrites) and skeletal or pelleted wackestones (biomicrites, biopelmicrites, pelmicrites), although, depending upon specific conditions of deposition, accumulations of better winnowed grainstones or packstones also may occur. Sediments of this facies are very common in the geologic record.

The inner shelf in most carbonate environments is a low-energy, tidal-flat environment (Fig. 13.19) in which predominantly fine-grained sediments accumulate; however, on some ramp platforms a higher energy nearshore zone may be present where beaches or lime-sand shoals develop. The distinguishing physical characteristics of tidal-flat environments have already been described in Chapter 12. Carbonate sediments deposited in the subtidal and lower intertidal portions of tidal flats are similar to those formed on the inner part of the middle shelf in that they consist generally of pelleted mudstones and some skeletal wackestones that are commonly burrowed and bioturbated. Where salinities are near normal, tidal ponds, creeks, and other water bodies support a restricted but prolific population of gastropods and foraminifers (N. P. James, 1984b). These gastropods graze on blue-green algae (cyanobacteria) and keep the organisms cropped to the point that algal mats cannot flourish. In hypersaline areas where gastropods cannot survive, prolific growth of algal mats can occur.

The middle and upper intertidal zones are characterized by thin, graded, limemud layers, probably deposited by storm waves, and extensive development of algal mats (James, 1984b). Both mud layers and algal mats are disrupted by well-developed desiccation cracks. Bedding tends to be irregular, with algal mat layers alternating with graded storm layers. Trapping of fine lime mud by algal mats and repeated renewal and growth of these mats generate finely laminated stromatolites. Parts of the algal mats may rot away during burial, leaving irregular, subhorizontal elongated cavities called "laminoid fenestrae," or fenestral porosity. Mud crusts formed on intermittently exposed parts of mud flats and tidal channels may be broken off and reworked by incoming tidal currents or storm waves to form intraclasts. Carbonate sediments of the supratidal zone are similar to those of the upper intertidal zone, except that long periods of subaerial exposure cause storm-deposited layers to form lithified surface crusts several centimeters in thickness and fractured into irregular mudcracks. Sediments of the supratidal zone commonly contain some dolomite in association with calcite and aragonite. In hypersaline environments, gypsum, anhydrite, and halite may be present also. Crystallization of evaporite minerals within the sediments can cause doming up of desiccation polygons to form "teepee" structures.

In summary, tidal-flat carbonates are characterized particularly by irregularly to evenly laminated algal mats and associated fenestrae; irregularly bedded, thin, graded storm layers; and desiccation cracks. Other desiccation features, such as raindrop imprints and animal or bird tracks, may be present. Ancient carbonate facies of the inner shelf are represented by pelleted mudstones (micrites), nearly homogeneous mudstones, fenestral laminated mudstones, intraclast-rich wackestones (in channels), stromatolites, and in some deposits nodular anhydrite or gypsum. A good description of inner-shelf deposits is provided by Enos (1983). Because of their distinctive characteristics, carbonate tidal-flat deposits in the rock record are relatively easily identified.

With the exception of mud chips or intraclasts concentrated in tidal channels, coarse-grained carbonate sediments are relatively rare in the tidal-flat environment. On some carbonate ramp platforms where an outer barrier is absent, a high-energy sand-shoal beach facies may develop along the inner platform edge in lieu of the tidal-flat deposits of low-energy shelves. Carbonate grains on these beaches consist of skeletal fragments, ooids, pellets, and possibly intraclasts ripped up from layers of previously deposited, submarine cemented lime mud. Inden and Moore (1983) discuss carbonate beaches, including both barrier beaches and mainland beaches, and indicate that sediments of the lower shoreface are commonly coarse-grained, poorly sorted sands with a mud matrix and are characterized by trough cross-bedding. Sediments of the upper shoreface are much better sorted sands and gravels that display planar cross-bedding with foresets that dip mainly seaward at angles less than about 15 degrees. They point out two modern examples of mainland carbonate beaches on the Trucial coast of the Persian Gulf.

Examples of Modern Carbonate Platforms

Modern carbonate shelves of both the ramps and rimmed types have now been studied in considerable detail. The characteristics of several of these platforms are summarized by B. W. Sellwood (1978) and J. L. Wilson and Jordan (1983). Examples of open shelves or carbonate ramps include the eastern Gulf of Mexico off the Florida coast; the Yucatan Shelf, Mexico, in the southern part of the Gulf of Mexico; and the Persian Gulf. Examples of rimmed shelves include Florida Bay, the Bahama Platform (an iso-

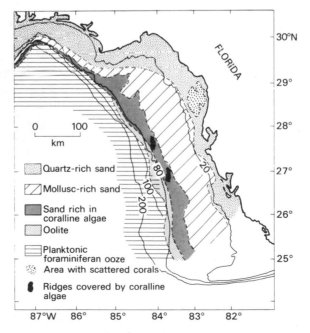

FIGURE 13.20 Example of an open-shelf or carbonate ramp—West Florida shelf in the eastern Gulf of Mexico. (From Sellwood, B. W., 1978, Shallow-water carbonate environments, in H. G. Reading (ed.), Sedimentary environments and facies. Fig. 10.17, p. 276, reprinted by permission of Elsevier Science Publishers, Amsterdam. Originally after R. N. Ginsburg and N. P. James, 1974, Holocene carbonate sediments of continental shelves, in C. A. Burk and C. L. Drake (eds.), The geology of continental margins, Fig. 6, p. 140: Springer-Verlag, New York.)

lated platform), the Belize Shelf in the western Caribbean off Guatemala, and the Great Barrier Reef area of Australia. Other important deposits of carbonate sediments in Australian waters occur along the western coast. Space does not permit discussion of these modern carbonate environments, but a sediment facies maps of three well-known carbonate shelves is included to show some of the facies-distribution patterns of these types of shelves. Figure 13.20 illustrates an open shelf or carbonate ramp (West Florida Shelf), and Figure 13.21 shows a rimmed shelf (South Florida). The best studied ex-

FIGURE 13.21   Example of a rimmed shelf, South Florida Bay. (From Sellwood, B. W., 1978, Fig. 10.21A, p. 281, reprinted by permission of Elsevier Science Publishers, Amsterdam. Originally after R. N. Ginsburg and N. P. James, 1974, Fig. 23, p. 150: Springer-Verlag, New York.)

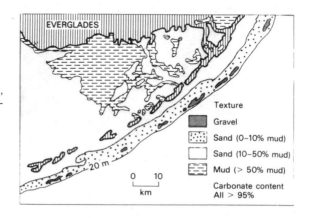

FIGURE 13.22   Sediment distribution on an isolated carbonate platform—the Great Bahama Banks. (From Sellwood, B. W., 1978, Fig. 10.21C, p. 281, reprinted by permission of Elsevier Science Publishers, Amsterdam. Originally after E. G. Purdy, 1963, Recent calcium carbonate facies of the Great Bahama Banks: Jour. Geology, v. 71, Fig. 1, p. 473.)

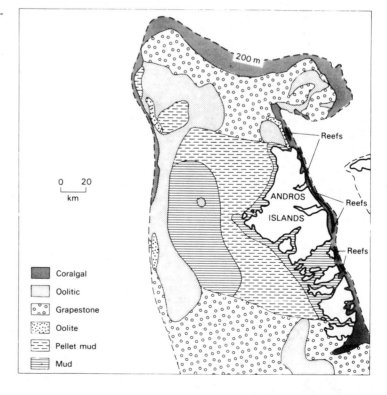

ample of a modern isolated platform is the Bahama Platform, which is also rimmed (Fig. 13.22). Note the general progression of facies on the rimmed shelves from reef buildups and shelf-edge sands on the outer shelf to carbonate muds and muddy carbonate sands on the middle and inner shelves. By contrast, most of the carbonate ramp in Figure 13.20 is covered by carbonate sand deposits mixed, on this shelf, with some terrigenous quartz sands.

Vertical Sequences in Carbonate Rocks

The rock record shows that carbonate rocks are deposited in cyclic sequences ranging from a few tens of meters to hundreds of meters thick. Many sequences begin with a high-energy carbonate sand or conglomerate unit followed upward progressively in the depositional sequence by sediments deposited in the lower energy subtidal open-marine shelf, intertidal zone, supratidal zone, and possibly nonmarine environment. Such a sequence is basically regressive; however, because rates of carbonate sedimentation commonly exceed rates of basin subsidence or sea-level rise, sediments also build upward toward sea level. Sediment is thus deposited in progressively shallower water as the sediment surface accretes toward sea level, leading N. P. James (1984b) to refer to these regressive units as "shallowing-upward sequences." Figure 13.23 shows the hypothetical sequence of vertical facies that develop on a typical low-energy rimmed shelf as tidal-flat deposits build outward and upward over low-energy, subtidal open-

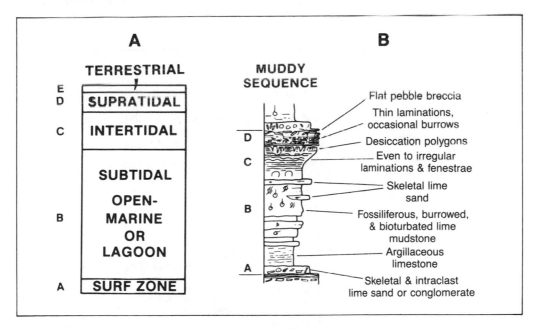

FIGURE 13.23 Hypothetical shallowing-upward sequence on a low-energy intertidal carbonate shelf. A. Subdivisions of the shallowing-upward model for carbonates. B. Sequence with a low-energy tidal-flat unit developed on a low-energy subtidal unit. (Modified from James, N. P., 1984, Shallowing upward sequences in carbonate rocks, in R. G. Walker (ed.), Facies models, 2nd ed.: Geoscience Canada Reprint Ser. 1. Figs. 2, 9, p. 214, 218, reprinted by permission of Geological Association of Canada.)

shelf deposits. Figure 13.24 illustrates the hypothetical shallowing-upward sequence that would form on a high-energy, nonrimmed, ramp-type shelf where a mainland beach occurs shoreward of the subtidal zone. These two sequences can be considered generalized models for vertical sequences formed on rimmed and nonrimmed carbonate shelves; however, considerable deviation from these norms can occur, depending upon exact depositional conditions. For example, the presence of shoaling areas within the subtidal area of rimmed shelves or extensive evaporite deposition in the supratidal zone can produce sequences that differ in detail from these models.

Repetition of large-scale shallowing-upward sequences may be the result of repeated episodes of rapid sea-level rise, flooding the carbonate platform, followed by periods of standstill during which shallowing-upward sequences develop (Wilkinson, 1982). The development of small-scale sequences is more likely tied to variations in the supply of carbonate sediment from the subtidal shelf, which is progressively reduced in area as intertidal sediments build out onto the shelf. When the area of the subtidal shelf becomes too small to supply sediments, deposition ceases until basin subsidence again creates a subtidal platform deep enough to supply sediments and begin the cycle anew (Ginsburg, 1971).

Ancient Carbonate Shelf Sequences

Carbonate shelf deposits are abundant in the geologic record in stratigraphic sequences ranging in age from middle Precambrian to Holocene. Carbonate deposition was particularly prevalent during the middle Paleozoic, late Mesozoic, and Tertiary, especially

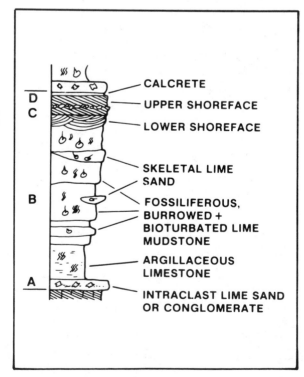

FIGURE 13.24 Hypothetical shallowing-upward carbonate-shelf sequence formed under conditions where a high-energy intertidal beach developed adjacent to a low-energy subtidal environment. Letters refer to the subdivisions of the shallowing-upward models shown in Figure 13.23. (After James, N. P., 1984, Shallowing upward sequence in carbonate rocks, in R. G. Walker (ed.), Facies models, 2nd ed.: Geoscience Canada Reprint Ser. 1. Fig. 17, p. 223, reprinted by permission of Geological Association of Canada.)

in broad epeiric seas as opposed to the small marginal-platform occurrences in the modern ocean. As indicated in the preceding discussion, shelf carbonate facies are characterized by distinctive suites of largely normal marine organisms and carbonate textures that are generally muddy, although lithofacies types range from lime mud-stones, wackestones, grainstones, and packstones to stromatolitic boundstones and patch-reef boundstones. Shelf carbonates occur in distinctive shallowing-upward sequences in which tidal-flat deposits are particularly diagnostic units. Bedding of shelf-carbonates is variable, with lens- or wedge-shaped layers common, although some shelf carbonate beds may be laterally extensive. Shelf carbonates are commonly interbedded with thin shale beds. Sedimentary structures include cross-bedding in lime-sand units, extensive bioturbation structures and burrows, and flaser and nodular bedding.

The Lower Cretaceous Edwards Formation of central Texas is a well-known carbonate sequence that illustrates many of these characteristic features of carbonate shelf deposits. This formation was deposited on an extensive carbonate platform characterized by carbonate skeletal sands and lime muds. Reefs composed of rudistid clams developed along the platform margin and some parts of the inner shelf (Fig. 13.25). Over the Llano Uplift of central Texas, the Edwards Limestone contains tidal-flat mud-stones, foraminiferal grainstones, and other skeletal carbonates. The inner shelf grades seaward over the middle shelf, where a series of high-energy grainstone bars or banks developed not far offshore (J. L. Wilson and Jordan, 1983). The banks are composed of oolitic grainstone, with skeletal wackestones to the north in front of the bank. Rudist reefs surrounded by wackestone occur farther offshore. The inner part of the middle shelf behind the banks is characterized by scattered rudist patch reefs, reef-derived skeletal sands, and muddy, pelleted skeletal packstones. These facies grade landward to carbonate sands formed in the beach environment, restricted lagoonal lime muds, and supratidal dolomites. A complete vertical sequence of the Edwards Limestone thus grades upward from oolitic, skeletal reef derived sands deposited near the shelf margin through middle-shelf pelleted, skeletal packstones to inner-shelf bioclastic beach sands, lagoonal lime muds, and supratidal dolomitic muds.

Organic Reef Buildups

Introduction

As mentioned, the outer shelf of many rimmed platforms is characterized by the presence of nearly continuous carbonate reefs that constitute an effective barrier to wave movement across the shelf. Reefs also may be developed, under some conditions, as fringing masses along the shoreline or as isolated patches within the inner shelf (Fig. 13.16). Reefs constitute a unique depositional environment that differs greatly from other parts of the shelf environments. Although they have been studied intensively for years, discussion of reefs has long been plagued by confusion over the precise meaning of the term. The nomenclature problem stems fundamentally from inability of workers to agree on whether to restrict use of the term **reef** to carbonate buildups, or bioherms, that have a rigid organic framework or core built of colonial organisms or to extend the definition to include carbonate buildups of other types that do not have a rigid-framework core. The word **bioherm** is a nongenetic term used for lenslike bodies of organic origin that are enclosed in rocks of different lithology or character. It carries no connotation of the internal structure or composition of the lens. Likewise, J. L. Wilson (1975) uses the term **carbonate buildup** for a body of locally formed, laterally

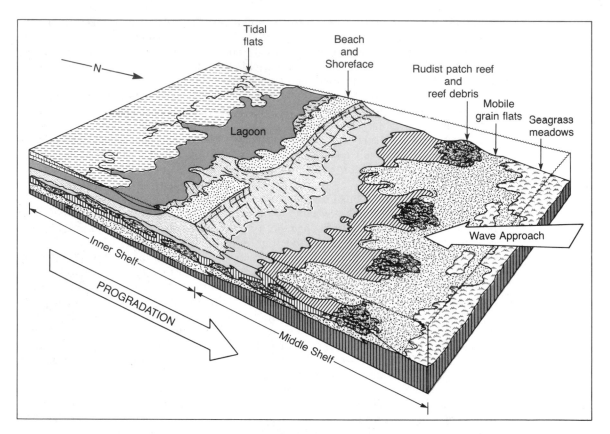

FIGURE 13.25   Depositional model of the Lower Cretaceous Edwards Limestone of Texas. The model shows progradational inner and middle shelf facies. (After Kerr, R. S., 1977, Facies, diagenesis, and porosity development in a Lower Cretaceous bank complex, Edwards Limestone, north-central Texas, *in* D. G. Bebout and R. G. Loucks (eds.), Cretaceous carbonates of Texas and Mexico, applications to subsurface exploration: Tex. Bur. Econ. Geology Rept. Inv. 89. Fig. 9, p. 223, reprinted by permission of University of Texas, Austin.)

restricted, carbonate sediment that possesses topographic relief, without regard to the internal makeup of the buildup.

Dunham (1970) attempted to solve the nomenclature dilemma by proposing two types of reefs: (1) ecologic reefs, which are rigid, wave-resistant topographic structures produced by actively building and sediment-binding organisms, and (2) stratigraphic reefs, characterized simply as thick, laterally restricted masses of pure or largely pure carbonate rock. The reef problem was reviewed again by Heckel (1974), who proposed his own definition of reef in still another effort to generate an acceptable, single meaning for the term. Longman (1981, p. 10) modified Heckel's definition slightly to arrive at the definition of a reef as "any biologically influenced buildup of carbonate sediment which affected deposition in adjacent areas (and thus differed to some degree from surrounding sediments), and stood topographically higher than surrounding sediments during deposition." He suggested (p. 10) using the term **reef complex** for "the specific type of reef having a significant rigid organic framework which generally forms in high wave energy, shallow water environments, as well as the genetically related facies associated with the framework." We have not likely heard the last word in this nomenclature wrangle, but this brief review of the nomenclature problem should at least provide readers with an understanding of the generally accepted meaning of reef, as well as some alternate usages of the term.

Modern Reefs and Reef Environments

Depositional Setting. Most modern reefs occur in shallow water. The most striking of these occurrences are the linear reefs located along platform margins, commonly called **barrier reefs.** These reefs are more or less laterally continuous, and the reef trend may extend for hundreds of kilometers—as, for example, the Great Barrier Reef of Australia, which runs for some 1900 km along the eastern shelf of Australia. In a few modern localities where shelves are very narrow, linear reefs are located hard up against the shoreline, with no intervening lagoon, and thus are called **fringing reefs.** Isolated, doughnut-shaped reefs called **atolls** occur around the tops of some Pacific seamounts. These reefs form an outer wave-resistant barrier that encloses a shallow lagoon. Small isolated reef masses commonly referred to as **patch reefs, pinnacle reefs,** or **table reefs** occur along some shelf margins or scattered on the middle shelf. In addition to these shallow-water reefs, reefs or reeflike carbonate buildups occur in deeper water. For example, organically produced mounds 100 m long and 50 m high have been reported from the Straits of Florida in water 600–700 m deep (Neuman et al., 1977). These mounds are composed of mud-cemented remains of various types of deep-water organisms such as crinoids, ahermatypic hexacorals, and sponges.

Reef Organisms. We tend to think of all reefs as coral reefs; however, many organisms in addition to corals can contribute to the formation of reefs. These organisms include coralline red algae, green algae, encrusting foraminifers, encrusting bryozoans, sponges, and molluscs (Heckel, 1974). In the geologic past, reef-building organisms also included some now-extinct groups such as the stromatoporoids, fenestellid bryozoans, and rudistid clams. Nonetheless, corals are certainly dominant constituents of modern reefs, and two types of corals are recognized. The principal corals in shallow-water reefs are hermatypic hexacorals. Hermatypic corals are unique among hexacorals in that they carry out a symbiotic relationship with a type of unicellular algae called zooxanthellae. These algae live in the tissues of the corals and aid them in secreting calcium carbonate by removing CO_2 from the tissues during photosynthesis. Because the zooxanthellae require sunlit waters, hermatypic corals are restricted to living in very shallow water. Ahermatypic hexacorals lack the symbiotic relationship and are not restricted to shallow water. They are the principal organisms today that build framework structures in deeper water carbonate buildups. The growth forms of reef-building organisms are closely related to the water energy over the reef. Organisms that live in low-energy parts of the reef tend to have delicate, branching forms or platelike forms. Those living in higher-energy zones of the reef develop hemispherical, encrusting, or tabular forms that are better able to withstand strong wave action (Fig. 13.26).

Reef Facies. Space limitations do not permit a discussion of all types of modern reefs and reef facies; therefore, we will examine the zoning and facies development of high-energy, platform-margin reefs as a general model for high-energy reef environments.

FIGURE 13.26 Growth forms of reef-building organisms and the principal types of environments in which they occur. (From James, N. P., 1983, Reef environment, *in* P. A. Scholle, D. G. Bebout, and C. H. Moore (eds.), Carbonate depositional environments: Am. Assoc. Petroleum Geologists Mem. 33. Fig. 59, p. 374, reprinted by permission of AAPG, Tulsa, Okla.)

GROWTH FORM		ENVIRONMENT	
		Wave Energy	Sedimentation
	Delicate, branching	low	high
	Thin, delicate, plate-like	low	low
	Globular, bulbous, columnar	moderate	high
	Robust, dendroid, branching	mod-high	moderate
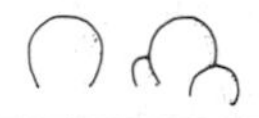	Hemispherical, domal irregular, massive	mod-high	low
	Encrusting	intense	low
	Tabular	moderate	low

Readers may wish to consult N. P. James (1983) or the symposium volume edited by Frost et al. (1977) for details of other types of modern reefs. Figure 13.27 illustrates schematically the principal facies subdivisions of a platform-margin reef. Note that the reef consists of a central core, the **reef framework,** which grades seaward into the **reef slope,** and a loose accumulation of reef debris called the **fore-reef talus.** The nearly flat uppermost, shallowest part of the reef is called the **reef flat,** which grades landward into **back-reef coral-algal sands** and **subtidal lagoonal deposits.** James (1983) subdivides reefs physiographically into **fore-reef, reef-front, reef-crest, reef-flat,** and **back-reef** zones (Fig. 13.28). Figure 13.28 shows also the types of carbonate materials formed in different zones of the reef. The words rudstone, floatstone, bafflestone, bindstone, and framestone in Figure 13.28 are terms used by Embry and Klovan (1971) as modifications of Dunham's (1962) limestone classification. Floatstone and rudstone are unbound carbonate grains, more than 10 percent of which are >2 mm in size; floatstones are mud supported; and rudstones are grain supported. Bafflestones are carbonate components bound together at the time of deposition by stalked organisms that trapped sediment by acting as baffles. Bindstones were bound during deposition by encrusting and binding organisms such as encrusting foraminifers and bryozoans, and framestones were bound by organisms such as corals, which build a rigid-framework structure.

Water energy, dominant sedimentation processes, types of organisms, percentage of framework components, and grain size and sorting of sediment vary in each zone of the reef. Table 13.2 summarizes these characteristics, following the physiographic subdivisions of reefs illustrated in Figure 13.28. Water energy is highest on the reef crest, which also contains the highest percentage of framework constituents. As water energy decreases toward both the fore reef and back reef, the percentage of framework constituents also decreases. Note that overall the framework component of reefs is commonly much smaller than the volume of nonframework constituents. Longman (1981) compares the structures of reefs to that of an apple, which has a central core, or framework, surrounded by the much larger edible fruit. The nonframework fraction of reefs consists of organisms such as echinoderms, green algae, and molluscs that do not build framework structures, together with bioclasts broken from the reef by wave activity and, in lower-energy zones of the reef, some lime mud. The fore-reef talus slope and back-reef coral-algal sand zone are made up entirely of nonframework constituents that

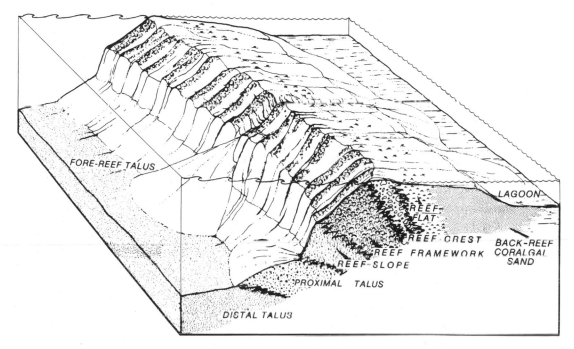

FIGURE 13.27 Idealized facies in a typical modern, mature coral reef with a well-developed reef framework. (From Longman, M. W., 1981, A process approach to recognizing facies of reef complexes, *in* D. F. Toomey (ed.), European fossil reef models: Soc. Econ. Paleontologists and Mineralogists Spec. Pub. 30, Fig. 10, p. 23, reprinted by permission of SEPM, Tulsa, Okla.)

FIGURE 13.28 Cross-section view of a hypothetical zoned marginal reef illustrating the major zones, principal types of limestones produced, and the growth forms of organisms in different parts of the reef. (From James, N. P., 1984, Reefs, *in* R. G. Walker (ed.), Facies models, 2nd ed.: Geoscience Canada Reprint Ser. 1. Fig. 9, p. 233, reprinted by permission of Geological Association of Canada.)

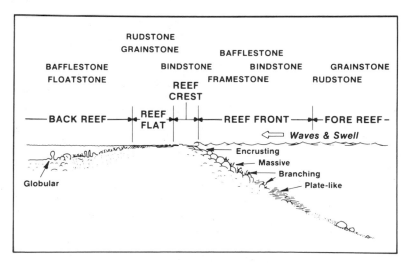

TABLE 13.2 Principal depositional processes and facies characteristics in modern reef complexes

Facies	Process of sedimentation and controls on organisms	Types of organisms likely to be preserved	Grain size	Sorting	Amount of framework (%)	Typical depth (m)	Dominant rock type
Lagoon	Low energy, much burrowing, sporadic currents and turbidity, possible terrigenous influx	Molluscs, echinoids, miliolids, forams, ostracods	Mud mixed with coarse skeletal debris	Poor	0	5–30	Wackestone
Back-reef sand	Sporadic storms and currents across reef, saltation, gravity sliding	Halimeda, miliolids, minor red algae, sparse finger corals	Coarse	Moderate to good	0	1–10	Grainstone
Reef flat	Sporadic storms, good current circulation, winnowing of mud	Finger corals, red and green algae, larger (benthic) forams, head corals	Coarse–very coarse	Moderate	0–10	1–3	Grainstone, rudstone, scattered corals
Reef crest	High wave energy, constant turbulence, good water circulation	Wave-resistant corals and algae	Very coarse	Moderate to good	0–80	0–2	Grainstone (minor bindstone)
Reef framework	Good water circulation, high wave energy—sporadic at greater depths	Abundant corals, algae, molluscs, echinoderms, forams	Framework and sand	Poor; mud in some cavities	20–80	1–30	Framestone
Reef slope	Limited light, sporadic turbulence, gravity transport of reef debris	Soft corals, flattened coral plates, sponges	Mixed	Poor	5–40	20–50	Packstone, bindstone, bafflestone
Proximal talus	Sporadic turbulence, gravity transport, little light, unstable substrate	Few living organisms	Medium to coarse	Poor to good	0	40–100	Grainstone, packstone, rudstone
Distal talus	Quiet water, no light, gravity sliding of sediments	Planktic forams	Fine	Moderate to good	0	100–200	Packstone

Source: Modified slightly from Longman, M. W., 1981. A process approach to recognizing facies of reef complexes, in D. F. Toomey (ed.), European fossil reef models: Soc. Econ. Paleontologists and Mineralogists Spec. Pub. 30. Table 3, p. 24, reprinted by permission of SEPM. Tulsa. Okla.

consist principally of reef-derived bioclasts. Relatively few organisms live in these zones.

Low-Energy Reef Facies. The facies of modern, high-energy, platform margin-type reefs thus consist fundamentally of a central framework core composed largely of corals and coralline algae; the core grades seaward through a zone of rubbly fore-reef talus to deeper water lime muds or shales and landward through back-reef coral-algal sands to finer grained lagoonal deposits. This model serves reasonably well for high-energy reefs developed in most settings; however, some reefs form under much lower energy conditions. Low-energy reefs do not develop the characteristic zoning of high-energy roofs (Fig. 13.29) and tend to be circular to elliptical in plan view. Organisms growing on such reefs are dominated by the more delicate, branching forms (Fig. 13.26). Some low-energy "reefs" do not contain the typical reef structure described but are constructed simply of carbonate sands and muds built by organisms that are very close to reef-type organisms in composition (N. P. James, 1984c). Other low-energy buildups are composed largely of nonreef-type organisms. They consist of mound-shaped piles of skeletal fragments and/or bioclastic lime muds rich in skeletal organisms, with minor amounts of organic boundstone. These structures are commonly called **reef mounds.** Heckel (1974) groups these various types of buildups under the headings of (1) loose skeletal buildups, composed largely of nonencrusting organisms and including both rooted and free-living forms; (2) abraded skeletal buildups, made up of skeletal materials that have undergone some transport, rounding, and sorting; (3) lime mud buildups, composed predominantly of lime mudstones and bioclastic wackestones; and (4) mixed buildups.

Ancient Reef Deposits

Reef Facies. Reefs as they appear in the fossil record differ in some important respects from modern reefs. First, we commonly see ancient reefs only in vertical exposures. We observe a two-dimensional limestone body composed of different components formed at different times (N. P. James, 1983). Thus, we cannot detect all of the facies

FIGURE 13.29 Schematic diagram illustrating variations in reef zonation in response to different energy conditions ranging from calm water to rough water. (From James, N. P. 1984, Reefs, in R. G. Walker (ed.), Facies models, 2nd ed.: Geoscience Canada Reprint Ser. 1. Fig. 10, p. 234, reprinted by permission of Geological Association of Canada.)

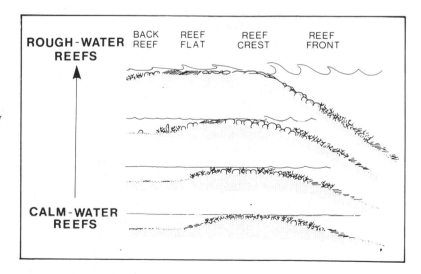

zones displayed by modern reefs. Ancient reefs can commonly be divided into only three main facies: (1) the **reef core,** consisting of the massive unbedded framework of the reef and composed of reef-building organisms cemented with a matrix of lime mud; (2) the **reef-flank facies,** consisting of bedded, commonly poorly sorted lime conglomerates (talus breccias) and/or lime sands that thin and dip away from the core; and (3) the **inter-reef facies,** composed of fine-grained subtidal lime muds or possibly siliciclastic muds. The nature of the flank and inter-reef facies exposed in outcrop obviously may differ depending upon whether the vertical exposure cuts through the reef to expose a cross section from fore reef to back reef or a longitudinal section running parallel to the reef crest. An additional factor that can further complicate recognition of ancient reef facies is diagenesis, which may cause selective dolomitization or solution that can obliterate or destroy parts of the reef complex.

The carbonate buildups in the Cretaceous deposits of northern Mexico provide a good example that illustrates the general characteristics of ancient reef complexes. The generalized biofacies and lithofacies across the Golden Lane "atoll" in Mexico are shown in Figure 13.30 (J. L. Wilson, 1975). The reef core stands tens of meters above the deeper carbonate facies and consists of rudistid clams, colonial corals, stromatoporoids, and encrusting algae. The core facies grades shoreward through reef-derived oolitic-biogenic grainstones to back-reef (inner-reef) micrites, foraminiferal grainstones, and bioturbated wackestones with a fauna indicating restricted circulation. This facies grades farther shoreward into an even more restricted facies containing evaporites. In a basinward direction, reef-core facies grade through a reef-flank facies consisting of coarse intraclastic-biogenic boulders embedded in micrite. These in turn grade into more basinward facies composed mainly of micritic limestones with a fauna of pelagic organisms.

Depending upon their age, ancient reefs may differ markedly from modern reefs in terms of the dominant reef-building organisms that make up the reef. The hermatypic corals that dominate modern coral reefs first appeared in the Mesozoic and thus are not components of older reefs. Older reefs are dominated by other kinds of frame-building organisms, such as tabulate corals, stromatoporoids, hydrozoans, sponges, encrusting bryozoans, coralline algae, and blue-green algae (cyanobacteria).

Occurrence of Ancient Reefs. "Reefs" of some type are found in carbonate rocks of all ages. Although carbonate-secreting organisms were not present during the Precambrian, carbonate buildups composed of stromatolites have been reported from rocks of that age from various localities in North America, Europe, Africa, and Australia. Phanerozoic rocks of all ages contain reefs composed of calcium carbonate-secreting organisms. Many of these reefs were built by framework-constructing or encrusting organisms. Reef development was not uniform throughout geologic time, however, and reefs are much more abundant in some parts of the rock record than in others. The middle Paleozoic, Mesozoic, and Tertiary were particularly important times for reef building. Figure 13.31 graphically depicts the distribution of reefs through time and shows also the major kinds of organisms that were responsible for reef building at different times. Specific examples of ancient reefs are too numerous to be listed and described here. Readers are referred to Heckel (1974) and N. P. James (1983) for detailed discussions of the distribution and characteristics of reefs formed during each period of geologic time. Many other examples of ancient reefs are described in detail in Frost et al. (1977), Laporte (1974), Toomey (1981), and J. L. Wilson (1975).

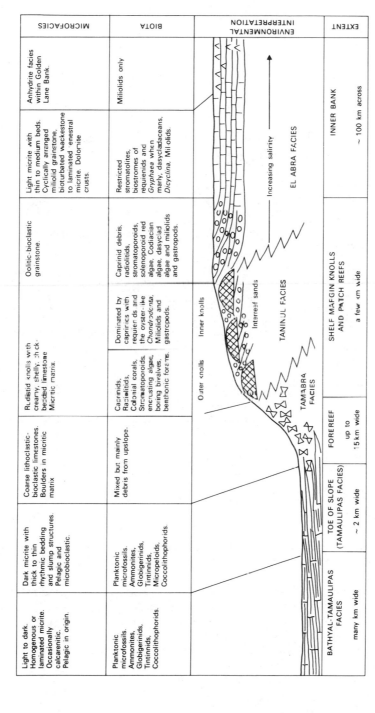

MICROFACIES	BIOTA	ENVIRONMENTAL INTERPRETATION	EXTENT
Light to dark. Homogenous or laminated micrite. Occasionally calcarenitic. Pelagic in origin.	Planktonic microfossils, Ammonites, Globigerinids, Tintinnids, Coccolithophorids.	BATHYAL-TAMAULIPAS FACIES	many km wide
Dark micrite with thick to thin rhythmic bedding and slump structures. Pelagic and microbioclastic.	Planktonic microfossils, Ammonites, Globigerinids, Tintinnids, Micropeloids, Coccolithophorids.	TOE OF SLOPE (TAMAULIPAS FACIES)	~ 2 km wide
Coarse lithoclastic-bioclastic limestones. Boulders in micritic matrix.	Mixed but mainly debris from upslope.	FOREREEF	up to 15 km wide
Rudistid knolls with creamy, shelly, chalk-bedded limestone Micritic matrix.	Caprinids, Radiolitids, Colonial corals, Stromatoporoids, encrusting algae, boring bivalves, benthonic forams. / Dominated by caprinids with requienids and the oyster-like *Chondrodonta*, Miliolids and gastropods.	TAMABRA FACIES / TANINUL FACIES (Outer knolls / Inner knolls)	SHELF MARGIN KNOLLS AND PATCH REEFS — a few km wide
Oolitic-bioclastic grainstone.	Caprinid debris, radiolitids, stromatoporoids, solenoporoid red algae, Codiacian algae, dasyciad algae and miliolids and gastropods.	(Interreef sands)	
Light micrite with thin to medium beds. Cyclically arranged miliolid grainstone, bioturbated wackestone to laminated enestral micrite. Dolomite crusts.	Restricted stromatolites, biostromes of requienids and *Gryphaea* when marly, dasycladaceans, *Dicyclina*, Miliolids.	EL ABRA FACIES — Increasing salinity	INNER BANK — ~ 100 km across
Anhydrite facies within Golden Lane Bank.	Miliolids only		

FIGURE 13.30 Generalized facies of Middle Cretaceous sediments across a large carbonate buildup in central Mexico. (From Wilson, J. L., 1975, Carbonate facies in geologic history. Fig. XI. 3, p. 323, reprinted by permission of Springer-Verlag, Heidelberg. As modified by B. W. Selwood, 1978, Shallow-water carbonate sediments, Fig. 10.42, p. 307: Elsevier, New York.)

493

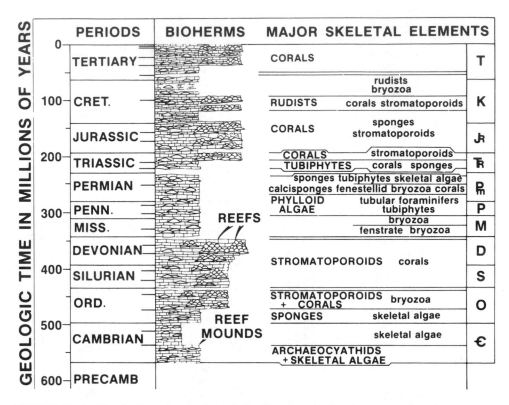

FIGURE 13.31 Distribution of reefs through the Phanerozoic showing times when there were only reef mounds, times when both reef mounds and reefs existed, and times when no reefs or bioherms were formed (gaps in the lithic column). (From James, N. P., 1983, Reef environment, *in* P. A. Scholle, D. G. Bebout, and C. H. Moore (eds.), Carbonate depositional environments: Am. Assoc. Petroleum Geologists Mem. 33. Fig. 61, p. 375, reprinted by permission of AAPG, Tulsa, Okla.)

13.3   OCEANIC SYSTEMS

Introduction

In the discussion of depositional environments to this point, I have focused on the continental, marginal-marine, and shallow-marine environments because most of the preserved sedimentary record was deposited in these environments. In terms of size of the environmental setting, however, these nonmarine to shallow-water environments actually cover a much smaller area of Earth's surface than do deep-water environments. By far the largest portion of Earth's surface lies seaward of the continental shelf in water deeper than about 200 m. Approximately 65 percent of the surface is occupied by the continental slope, continental rise, deep-sea trenches, and the deep ocean floor. Even so, most textbooks that discuss sedimentary environments typically give only slight coverage to oceanic environments. This is probably so because, as a whole, deep-water sediments are much more poorly represented in the exposed rock record than are shallow-water sediments. Deep-water deposits are less abundant than shallow-water deposits in the exposed rock record because sedimentation rates overall are slower in deeper water; thus the sediment record is thinner. Also, part of the sediment record of the deep seafloor may have been destroyed by subduction in trenches, and those deep-water sediments that have escaped subduction have required extensive faulting and uplift to bring them above sea level where they can be viewed. Deep-water sedimentary environments have not been studied as thoroughly as shallow-water environments—perhaps in part because deep-water sediments have less economic potential for petroleum—but interest in the deep ocean floor has increased considerably in recent years. Owing to the advent of seafloor spreading and global plate tectonics concepts, the deep seafloor has taken on enormous significance for geologists. Consequently, intensive research efforts have been focused on the continental margins and deep seafloor since the early 1960s. Also, the continuing need to add to our fossil fuel reserves is pushing petroleum exploration into deeper and deeper water, and the possibility of mining manganese nodules and metalliferous muds from the seafloor is increasing economic interest in the deep ocean.

Deep-sea research has been particularly stimulated by the JOIDES Deep Sea Drilling Project (discussed in Chapter 1), which began in 1968. Since that time, several hundred holes have been drilled by DSDP teams, to an average depth of 300 m (~1000 ft), throughout the ocean basins of the world. An average of about one hole for every 800,000 km^2 of the ocean floor has now been drilled (Warme et al., 1981). In addition to deep coring by DSDP, thousands of shallow piston cores have been collected from the seafloor throughout the ocean by marine geologists from major oceanographic institutions of the world. Also, hundreds of thousands of kilometers of seismic profiling lines have been run in criss-cross patterns across the ocean floor in an attempt to unravel the sub-bottom structure of the ocean. Although much of this research has been aimed at understanding the larger scale features of the ocean basins in order to illuminate the origin and evolutionary history of the ocean basins along plate tectonics lines, many data on sedimentary facies and sedimentary environments have also been collected. Much additional new information on ocean-circulation systems and sediment-transport systems has also been generated by oceanographers who study ocean-bottom currents and bottom-water masses. Thus, a significant increase in understanding of the ocean basins and the deep ocean floor has come about in the last few decades.

Detailed discussion of the deep ocean basin is outside the scope of this book. We shall concentrate our discussion here on the fundamental processes of sediment transport and deposition on continental slopes and the deep ocean floor and the principal types of facies developed in these environments. Additional details of the structure, stratigraphy, and sediment characteristics of continental margins and ocean basins are available in several symposium volumes and other monographs, including those of Bally (1981), Bouma et al. (1978), H. E. Cook et al. (1983), H. E. Cook and Enos (1977), Curray et al. (1977), Doyle and Pilkey (1979), Hay (1974), Hsü and Jenkyns (1974), Lisitzin (1972), Saxov and Nieuwenhuis (1982), Seibold and Berger (1982), Siemers et al. (1981), Stanley and Moore (1983), and Warme et al. (1981).

Depositional Setting

Continental Slope. The continental slope extends from the shelf break, which occurs at an average depth of about 125 m, to the deep seafloor (Fig. 13.32). The lower boundary is typically located at water depths ranging from 1400 to 3200 m, but locally in deep trenches it may extend to depths exceeding 10,000 m (Bouma, 1979). Continental

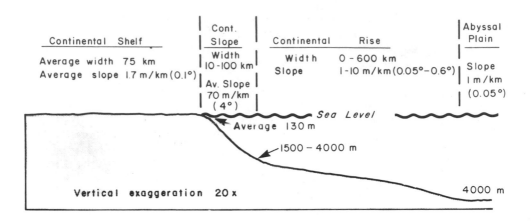

FIGURE 13.32   Principal elements of the continental margin. (After Drake, C. L., and C. A. Burk, 1974, Geological significance of continental margins, *in* C. A. Burk and C. L. Drake (eds.), The geology of continental margins. Fig. 9, p. 8, reprinted by permission of Springer-Verlag, Heidelberg. Modified by H. E. Cook, M. E. Field, and J. V. Gardner, 1982, Characteristics of sediments on modern and ancient continental slopes, *in* P. A. Scholle and D. Spearing (eds.), Sandstone depositional environments: Am. Assoc. Petroleum Geologists Mem. 31. Fig. 1, p. 329, reprinted by permission of AAPG, Tulsa, Okla.)

slopes are comparatively narrow, 10–100 km wide, and they slope seaward much more steeply than the shelf. The average inclination of modern continental slopes is about 4 degrees, although slopes may range from less than 2 degrees off major deltas to more than 45 degrees off some coral islands.

The origin and internal structure of continental slopes is not of primary interest here; however, I point out that that Emery (1977) has distinguished six types of continental slopes based on surface morphology, internal structure, and sediment cover (Fig. 13.33). Five of these types (A through E in Fig. 13.33) occur on pull-apart, or passive, margins such as the Atlantic margin. Type F occurs on seismic or converging margins. Type A consists of folded or faulted blocks or ridges of rocks covered to some degree by sediments. Type B forms by progradation over the shelf owing to delta formation or other large influx of terrigenous sediment from land. Type C slopes form on pull-apart margins owing to deposition of wedges of shallow-marine sediments over relatively low-relief prerift topography that was downwarped toward the ocean. Type D slopes are distinguished by the presence of a large carbonate reef mass developed at the shelf break position; great thickness of the reef is due to continued subsidence of the underlying crust. Type E slopes are controlled by the presence of subsurface evaporites and their resulting diapirs that form owing to salt flowage. Type F slopes are found on converging margins where subduction has caused sediment to be scraped off the descending oceanic plate and stacked against the inner trench wall to form an accretionary wedge. This type of slope descends into a deep trench.

Continental slopes may have a smooth, slightly convex surface morphology, such as that found on progradational slopes (type B), or they may be irregular on a small to

FIGURE 13.33 Types and distribution of continental slopes in the Atlantic Ocean. (From Emery, K. O., 1977, Statigraphy and structure of pull-apart margins: Am. Assoc. Petroleum Geologists Educ. Course Note Ser. 5. Fig. 5, p. B-11, reprinted by permission of AAPG, Tulsa, Okla.)

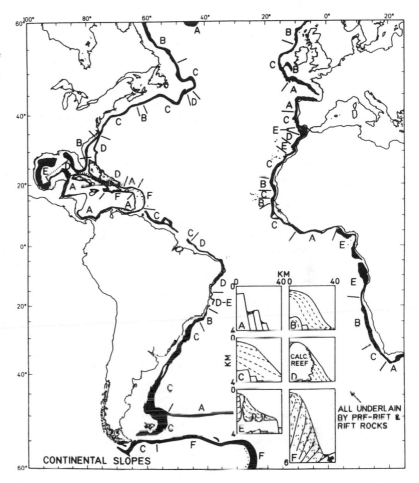

very large scale. Type F slopes tend to be particularly irregular, as illustrated by the Pacific slope off Japan. This slope, which descends to a depth of about 7000 m into the Japan Trench, is characterized by the presence of structural terraces and basins together with anticlinal welts and fault-bounded ridges arranged in an **en echelon** pattern roughly parallel to the Japan coast (Boggs, 1984). These ridges and folds form prominent structural "dams" behind which sediments are ponded. In general, structural barriers on highly irregular slopes can inhibit movement of bottom sediment across the slope and create catchment basins for sediment. Modern continental slopes are gashed to varying degrees by submarine canyons, oriented approximately normal to the shelf break, which provide accessways for turbidity currents moving across the slope. Most submarine canyons have their heads near the slope break and do not cross the shelf; however, a few major canyons on modern shelves extend onto the shelf and head up very close to shore. Some large canyons also extend seaward beyond the base of the slope to form deep-sea channels that may meander over the nearly flat ocean floor for hundreds of kilometers. The Toyama Deep Sea Channel in the Japan Sea, for example, winds its way across the sea floor from the mouth of the Toyama Trough for approximately 500 km before emptying onto the Japan Sea abyssal plain.

Continental Rise and Deep Ocean Basin. The continental rise and deep ocean basin encompass that part of the ocean lying below the base of the continental slope. Together they make up about 80 percent of the total ocean. The deeper part of the ocean seaward of the continental slope is divided into two principal physiographic components: (1) the deep **ocean floor** and (2) the **oceanic ridges.** On passive continental margins, the ocean floor can be further subdivided into the **continental rise** and the **abyssal plain** (Fig. 13.1). The continental rise is a gently sloping surface that leads gradually onto the abyssal plain and is built in part from submarine fans extending seaward from the foot of the slope. It commonly has little relief other than that due to the presence of submarine canyons or volcanic peaks (seamounts). Continental rises may not be present on convergent or seismic margins where active subduction is taking place, such as along much of the Pacific margin. On margins of this type, a long, arcuate **deep-sea trench** commonly lies at the foot of the continental slope and the rise is absent. Trenches in less active subduction zones, such as along the Oregon–Washington coast, may be filled with sediment. Abyssal plains are extensive, nearly flat areas interrupted here and there by seamounts. Some abyssal plains are cut by deep-sea channels, as mentioned. **Mid-ocean ridges** and **rises** extend across some 60,000 km of the modern ocean and overall make up about 30–35 percent of the area of the ocean. (Rises are like ridges but with more gentle slopes.) Mid-ocean ridges are particularly prominent in the Atlantic, where they rise about 2.5 km above the abyssal plains on either side. Rocks on these ridges and rises are predominantly volcanic, and the ridges are cut by numerous transverse fracture zones along which significant lateral displacements may be apparent. Ridges play a crucial role in the seafloor spreading process, but they are not particularly active areas of sedimentation. They do have a very important effect on circulation of deep bottom currents in the ocean and thus have an indirect effect on sedimentation in the deep ocean.

Transport and Depositional Processes in Deep Water

Sediment transport within the ocean basin originates primarily on the shelf (Fig. 13.34), although some sediment can be transported, or retransported, off oceanic ridges and rises, and biogenic sediments accumulate on the ocean floor owing to the rain of pelagic organisms from surface and near-surface waters. Fine sediment can move across the shelf into deeper water as freshwater surface plumes or in near-bottom nepheloid layers. Transport in the nepheloid layer appears to be particularly important. Storm waves resuspend fine bottom sediment from the shelf floor, creating a turbid near-bottom layer of suspended sediment that may reach several tens of meters in thickness, particularly on the outer shelf. This suspended sediment can then be carried across and off the shelf owing to seaward-directed wind-driven flow or the seasonal production of dense, cool, saline bottom waters that flow offshore under the influence of gravity (McGrail and Carnes, 1983).

Coarser sediment has a more difficult time crossing the shelf. Where submarine canyons cut the shelf and head up close to shore, sandy sediment may move laterally along the coast by longshore transport into the heads of these canyons (Fig. 13.34) and then down the canyons to deeper water by processes to be discussed. If submarine canyons are absent on the shelf, sandy sediments probably cannot cross the barrier created by deeper water of the outer shelf, although tidal currents, storm surges, turbidity currents, or geostrophic currents may move sands some distance seaward onto the inner shelf. Therefore, during periods of high sea level such as the present, sands tend to be trapped in the nearshore zone and prevented from moving across and off

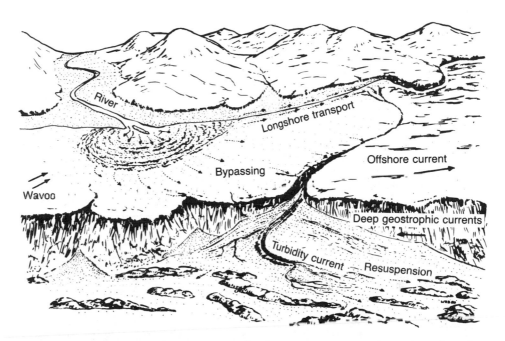

FIGURE 13.34 Transport and redistribution of sediment on a continental margin such as the west coast of North America. Coarser sediment input by rivers moves laterally along the coast by longshore transport until intercepted by a submarine canyon. Fine sediment bypasses the shelf and comes to rest on the continental slope or in deeper water. (From Seibold, E., and W. H. Berger, 1982, The sea floor. Fig. 4.1, p. 79, reprinted by permission of Springer-Verlag, Heidelberg. Redrawn from D. G. Moore, 1969, Reflecting profiling studies of the California continental borderland: Structure and Quaternary turbidite basins: Geol. Soc. America Spec. Paper 107, Fig. 5, p. B-11.)

the shelf to deeper water. By contrast, during low stands of the sea, rivers can flow across the subaerially exposed shelf and dump coarse sediment into the heads of submarine canyons that originate near the shelf break. Thus, sand transport onto the continental slope is commonly much greater during periods of low sea level.

Processes capable of transporting sediment farther into deeper water beyond the shelf edge can be grouped into: (1) suspension transport by near-surface water and by wind, (2) near-bottom nepheloid-layer transport, (3) tidal-current transport in submarine canyons, (4) sediment gravity flows and other mass transport processes, (5) transport by geostrophic contour currents, and (6) transport by floating ice. In addition, sedimentation in deep ocean basins occurs by the rain of dead pelagic organisms from near-surface waters and by airfall and submarine settling of pyroclastic particles generated by explosive volcanism within the ocean basin.

Suspension, Nepheloid, and Wind Transport. Clay and fine silt are separated from sands in the surf zone during river outflow, and the fine sediment may move outward as a freshwater plume over the denser seawater. As discussed under shelf transport, these plumes of suspended sediment can move considerable distances from shore before mixing and flocculation cause the clay particles to settle. Some plumes have been observed to extend into deeper water as much as 100 km offshore (Reineck and Singh,

1980), and small amounts of fine sediment may be carried in near-surface waters even greater distances into the basins. It is likely, however, that movement of fine sediment to great distances off the shelf takes place mainly by nepheloid transport, which involves intermittent deposition, resuspension, and seaward migration of fine sediment in near-bottom turbid layers. Fine sediment is injected into the water column owing to erosion of the seabed by bottom currents (discussed in a succeeding section) and possibly by turbidity currents. It is then transported seaward by these currents while still in suspension. Because fine sediment can remain in suspension for periods ranging up to weeks or even months, it travels much farther than coarser bottom sediment, which also may be eroded and transported by bottom currents. Nepheloid layers in the modern ocean extend seaward for hundreds of kilometers and to water depths of 6000 m or more. Winds also can transport fine suspended dust particles seaward, where they settle out over the ocean hundreds of kilometers from shore. In fact, wind transport may be the primary mechanism by which pelagic sediment is transported to the deep ocean.

Tidal Currents. Tidal currents measured in submarine canyons at depths exceeding 1000 m may be capable of transporting silt and fine sand (Shepard, 1979). Two types of currents have been detected in submarine valleys: (1) ordinary tidal currents that rarely exceed 50 cm/sec and that flow alternately up and down the valley in response to tidal reversal and (2) occasional surges of strong downcurrent flow that commonly do not exceed 100 cm/sec. Shepard (1979) interprets the surges as low-velocity turbidity currents. There are few data as yet on the quantitative importance of net downcanyon transport owing to tidal currents; however, surge currents of the magnitude reported by Shepard are certainly capable of transporting fine sediment seaward.

Turbidity Currents and Other Mass-Transport Processes. Catastrophic, or surge-type, high-velocity turbidity currents generated on the shelf or upper slope are probably the single most important mechanism for transporting sands and gravels to deeper water through submarine channels. On passive margins and in back-arc basins, the deposits of these flows spread out from the mouths of the canyons onto the deep seafloor to form deep-sea fans (Fig. 13.34) and contribute in part to building of the continental rise. In the fore-arc region of seismic margins, submarine canyons discharge turbidites into fore-arc basins on the slope or into deep-sea trenches, where they spread out along the canyon axis (Boggs, 1984). Down-canyon grain flow of beach sands, swept into the heads of nearshore submarine canyons during storms, and submarine debris flows may be locally important. In addition to these processes, other mass transport processes such as creep, gliding (sliding), and slumping appear to be responsible for large-scale en masse sediment transport on oversteepened continental slopes and ridge slopes. Some of these slump masses can be of enormous size. Blocks up to 300 m thick and 100 km long have been reported. The importance of these mass-transport processes as agents of slope modification and basin filling is recognized in a recent volume entitled *Marine Slides and other Mass Movements*, edited by Saxov and Nieuwenhuis (1982).

Contour Currents. Density differences in surface ocean water owing to temperature or salinity variations create vertical circulation of water masses in the ocean. This is commonly referred to as **thermohaline circulation.** Circulation is initiated primarily at high latitudes as cold surface waters sink toward the bottom, forming deep water masses that flow along the ocean floor as bottom currents. The path of these bottom currents is influenced by the position of oceanic ridges and rises and other topographic features such as narrow passages through fracture zones. Owing to density stratification of ocean water, bottom currents tend to flow parallel to depth contours, or isobaths, and thus are often called **contour currents.** The movement of these currents is affected also by the Coriolis force, which likewise tends to deflect them into paths parallel to depth contours; thus, they are sometimes also called geostrophic contour currents. Because contour currents are best developed in areas of steep topography where the bottom topography extends through the greatest thickness of stratified water column (Kennett, 1982), they are particularly important on the continental slope and rise. Photographs of the deep seafloor have revealed the presence of current ripples in some areas and suspended sediment clouds with lineations in others, both of which suggest that some contour currents can achieve velocities great enough to transport sediment on or near the seafloor. Contour currents are believed to have had a particularly important role in shaping and modifying continental rises (Fig. 13.35).

Floating Ice. During glacial episodes of the Pleistocene when sea level was low and many land areas were covered by ice, rafting of sediment of all sizes into deeper water by icebergs was a particularly important transport process. Ice transport is still going on today on a more limited scale at high latitudes in the Arctic and Antarctic regions. Melting of the floating ice dumps sediments of mixed sizes, commonly referred to as glacial-marine sediment, onto the shelf and deep ocean floor. The overall quantitative significance of iceberg transport through geologic time has probably not been significant, but locally and at certain times it was quite important.

Pelagic Rain. Calcareous and siliceous planktonic organisms settle through the ocean water column to the seafloor upon death, a process called **pelagic rain.** The geographic

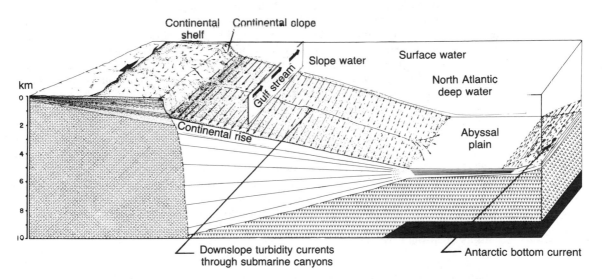

FIGURE 13.35 Schematic diagram illustrating the shaping of the continental rise of eastern North America by contour currents. (After Heezen, B. C., C. D. Hollister, and W. F. Ruddiman, 1966, Shaping the continental rise by deep geostrophic contour currents: Science, v. 152. Fig. 4, p. 507, reprinted by permission of American Association for the Advancement of Science, Washington, D.C.)

distribution of these organisms in surface waters is affected by nutrients and prevailing ocean currents. After the dead organisms settle onto the ocean floor, they may be re-transported by turbidity currents or contour currents. These pelagic organisms form extensive biogenic deposits, or **oozes,** in some areas of the modern ocean floor. They are also important contributors to ancient deep-ocean sediments, particularly in Jurassic and younger rocks.

Explosive Volcanism. Volcanism within the marine basin may contribute important quantities of sediment to both the shelf and deeper water, particularly near volcanic arcs. Volcanic ash, lapilli, and bombs can be ejected both subaerially and subaqueous-ly. Coarse material ejected subaerially tends to be deposited by airfall close to the eruption column on all sides of the vent. If strong prevailing winds are blowing during eruption, fine ash will be carried downwind for considerable distances before settling. Pyroclastic particles ejected beneath the sea as well as airfall particles can be dispersed still more widely within the ocean basin by the various transport processes discussed. Pumice may even be dispersed to some extent by floating on the ocean surface.

Principal Kinds of Deep-Sea Sediments

Shepard (1973) recognizes four types of deep-sea sediments based on origin: (1) lithogenous (terrigenous or volcanic particles), (2) biogenous (skeletal materials), (3) hydrogenous (inorganic precipitates such as carbonates, phosphorites, manganese nodules, zeolites, and pyrite), and (4) cosmogenous (micrometeorite spherules). Of these types, cosmogenous particles are minor constituents of ocean sediments, and the volume of true hydrogenous sediments is very small compared to those of lithogenous and biogenous sediments. Several attempts have been made to classify deep-sea sediments more fully under such largely genetic headings as pelagic, hemipelagic, glacial-marine, turbidites (e.g., Shepard, 1973; Berger, 1974). A descriptive classification for fine-grained deep-sea sediments has been proposed by Dean et al. (1985), who use a three-component classification system in which calcareous-biogenic, siliceous-biogenic, and nonbiogenic components form the three end members. None of these classifications is totally satisfactory as a general classification for all deep-sea sediment. Berger's (1974) classification (Table 13.3) and Dean et al.'s classification are useful for the fine-grained sediments that dominate the deep seafloor; however, these classifications cannot be applied to a variety of other sediments such as turbidite sands and gravels, glacial-marine sediments, slide and slump deposits, and grain-flow deposits. Regardless of the details of the classification schemes used, deep-sea sediments are divided by most workers into two rather broad groups: **pelagic** and **terrigenous.** Unfortunately, this grouping does not adequately recognize an important category of fine- to very coarse-grained, shelf-derived carbonate sediments, partially but not wholly biogenic, that accumulate on the continental slope and basin floor owing to turbidity current transport, debris flows, and slumping and sliding. I refer to these deposits simply as **allochthonous deep-sea carbonates.**

Pelagic Sediments. The term "pelagic sediment" has been defined in different ways, but is generally taken to mean fine-grained sediment deposited far from land influence by slow settling of particles suspended in the water column. Pelagic sediments may be composed predominantly of clay-size particles of terrigenous or volcanogenic ori-

TABLE 13.3 Classification of deep-sea sediments

I. (Eu-) pelagic deposits (oozes and clays)
  <25% of fraction >5 μm of terrigenic, volcanogenic, or neritic origin
  Median grain size <5 μm (except authigenic minerals and pelagic fossils):
  A. Pelagic clays. $CaCO_3$ and siliceous fossils <30%
    1. $CaCO_3$: 1–10%; (slightly) calcareous clay
    2. $CaCO_3$: 10–30%; very calcareous (or marl) clay
    3. Siliceous fossils: 1–10%; (slightly) siliceous clay
    4. Siliceous fossils: 10–30%; very siliceous clay
  B. Oozes. $CaCO_3$ or siliceous fossils >30%
    1. $CaCO_3$: >30%; <⅔ $CaCO_3$: marl ooze; >⅔ $CaCO_3$: chalk ooze
    2. $CaCO_3$: <30%; >30% siliceous fossils: diatom or radiolarian ooze
II. Hemipelagic deposits (muds)
  >25% of fraction >5 μm is of terrigenic, volcanogenic, or neritic origin
  Median grain size >5 μm (except authigenic minerals and pelagic fossils):
  A. Calcareous muds. $CaCO_3$ > 30%
    1. <⅔ $CaCO_3$: marl mud; >⅔ $CaCO_3$: chalk mud
    2. Skeletal $CaCO_3$: >30%: foram; nanno; coquina
  B. Terrigenous mud. $CaCO_3$ <30%. Quartz, feldspar, mica dominant
    Prefixes: quartzose, arkosic, micaceous
  C. Volcanogenic muds. $CaCO_3$ <30%. Ash, palagonite, and so on dominant
III. Pelagic or hemipelagic deposits
    1. Dolomite–sapropelite cycles
    2. Black (carbonaceous) clay and mud: sapropelites
    3. Silicified claystones and mudstones: chert
    4. Limestone

Source: Berger, W. H., 1974, Deep-sea sedimentation, in C. A. Burk and C. L. Drake (eds.), The geology of continental margins. Table 1, p. 214, reprinted by permission of Springer-Verlag, Heidelberg.)

gin, or they may contain significant amounts of fine-size planktonic biogenic remains (Table 13.3) **Pelagic clays** are siliciclastic muds, commonly red to red brown owing to oxidation by oxygen-bearing deep waters in areas of very slow sedimentation. These clays cover vast areas of the deeper parts of the ocean below about 4500 m. Pelagic sediments that contain significant quantities of biogenic remains are called **oozes.** Most workers have defined oozes as fine-grained, deep-sea sediments containing more than 30 percent biogenic components; however, Dean et al. (1985) restrict the name ooze to sediments containing more than 50 percent biogenic remains. Oozes composed predominantly of $CaCO_3$ tests are **calcareous oozes;** those composed mainly of siliceous tests are **siliceous oozes.** Calcareous oozes are dominated by the tests of foramininers and nannofossils such as coccoliths, but may include somewhat larger fossils such as petropods, which are planktonic molluscs. Calcareous oozes are widespread in the modern deep ocean at depths shallower than about 4500–5000 m, the calcium carbonate compensation depth, particularly in the Atlantic Ocean (Fig. 13.36). In deeper ocean basins such as the Pacific, such oozes may occur on the shallow tops of ridges and rises. Lithified equivalents of oozes are called chalks. Excellent discussions of pelagic carbonate environments and sediments are given by Scholle, Arthur, and Ekdale (1983) and Jenkyns (1978). Siliceous oozes are particularly abundant in the modern ocean at high latitudes in a belt about 200 km wide stretching across the ocean at water depths greater than about 4500 m (Fig. 13.36). They occur also in some equatorial regions of upwelling where nutrients are abundant and productivity of siliceous organisms is high. They are composed primarily of the remains of diatoms and radiolarians but may include other siliceous organisms such as silicoflagellates and sponge spicules. Diatom oozes occur mainly in high-latitude areas and along some continental margins, whereas radiolarian oozes are more characteristic of equatorial areas. Planktonic sediments that settle onto steep slopes on seamounts, ridges, or other structures may be retransported to adjacent basins by turbidity currents or slumping and sliding.

Terrigenous Sediments. A wide variety of deep-sea siliciclastic sediments are grouped as terrigenous sediments, including turbidites and other sediment gravity-flow deposits, contourites, glacial-marine sediments, hemipelagic muds, volcanogenic sediments (although locally derived volcaniclastic sediments are not strictly terrigenous materials), and slumps and slides.

Turbidites. The general characteristics of turbidites have been described elsewhere in this book. Turbidites may occur in the lower reaches of submarine canyons and farther

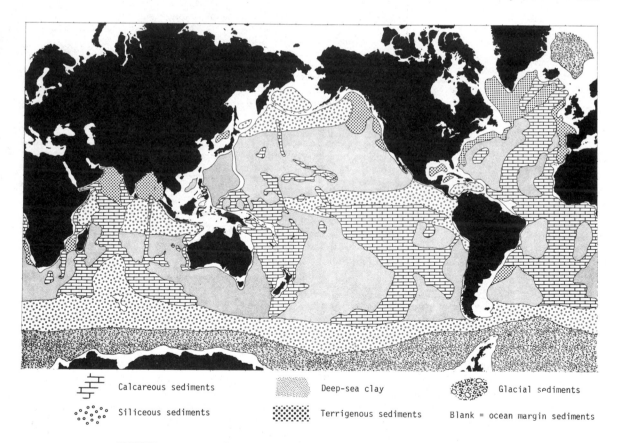

FIGURE 13.36 Distribution and dominant types of deep-sea sediments in the modern ocean. (From Davies, T. A., and D. S. Gorsline, 1976, Oceanic sediments and sedimentary processes, *in* J. P. Riley and R. Chester (eds.), Chemical oceanography, v. 5, 2nd ed., Fig. 24.7, p. 26, reprinted by permission of Academic Press, Orlando, Fla.)

seaward in deep-sea channels, but most are deposited in broad, cone-shaped fans. These **turbidite fans** spread outward on the seafloor from the mouths of canyons. Where submarine canyons are closely spaced along the slope, the fans at the base of the slope may coalesce to build a broad, gently sloping rise. On active margins where a trench is present, turbidites commonly occur on the trench floor throughout the length of the trench owing to deposition from turbidity currents flowing longitudinally through the trench. Most turbidites are composed of sands, silty sands, or gravelly sands interbedded with pelagic clays. They commonly display normal size grading and may or may not consist of complete Bouma sequences (Fig. 13.37). Many turbidites lack either the basal or upper part of the Bouma sequence, or both. Sole markings, including flute casts, groove casts, and load casts, are common on the bases of many turbidite sequences. Mud turbidites occur on many parts of the modern ocean floor. These turbidites are composed of normally graded silt and clay that may be either laminated or massive and that commonly lack extensive bioturbation. Turbidites are widely distributed in the modern ocean on passive margins and in both the back-arc and fore-arc regions of seismic margins.

Because the main depositional environments of turbidites are submarine fans, and most modern turbidites occur in such fans, geologists have displayed considerable interest in developing a general model for submarine-fan sedimentation. Several fan models have been published, but perhaps the most widely used is that of Normark (1978), shown in Figure 13.38. His model is based on the characteristics of several

FIGURE 13.37 Graded volcani-clastic-carbonate turbidite from a DSDP core of Eocene sediment in the Daito Basin, Philippine Sea. Large-scale divisions are in centimeters. (From Scholle, P. A., M. A. Arthur, and A. A. Ekdale, 1983, Pelagic environment, in P. A. Scholle, D. G. Bebout, and C. H. Moore (eds.), Carbonate depositional environments: Am. Assoc. Petroleum Geologists Mem. 33. Fig. 54, p. 658, reprinted by permission of AAPG, Tulsa, Okla.)

modern fans. A somewhat similar model by R. G. Walker (1978) also shows characteristic facies in different parts of the fan (Fig. 13.39). A salient feature of these models is subdivision of the fan into three major elements: the upper fan, midfan and lower fan. The **upper fan,** or inner fan, is characterized by channel-fill turbidite conglomerates that may show normal or inverse grading, debris-flow deposits, and slumps. Turbidity current flow on the upper fan is commonly through a single, deep, leveed channel. The levee deposits themselves consist of much finer grained, thin-bedded turbidites. The **midfan** is built up of overlapping lobes of turbidite sediments, referred to as suprafan lobes. The different lobes, which are fed by shallower, branching chan-

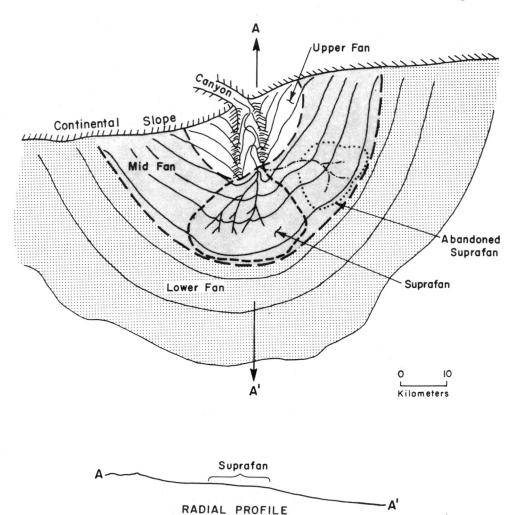

FIGURE 13.38 Schematic representation of submarine fan model proposed by Normark. This model emphasizes active and abandoned depositional lobes called suprafans. (From Normark, R. W., 1978, Fan valleys, channels, and depositional lobes on modern submarine fans: Characters for recognition of sandy turbidite environments: Am. Assoc. Petroleum Geologists Bull., v. 62. Fig. 1, p. 914, reprinted by permission of AAPG, Tulsa, Okla.)

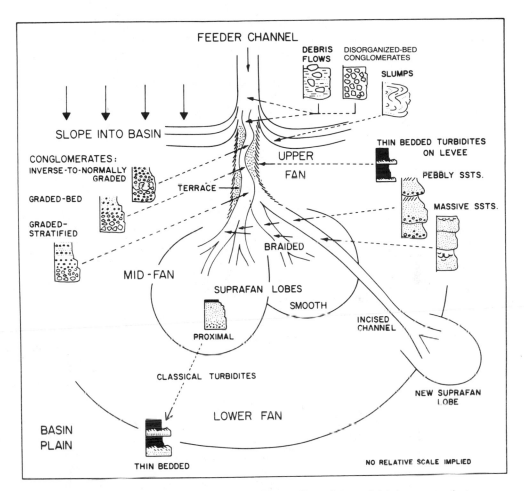

FIGURE 13.39 Submarine fan model proposed by Walker. This model incorporates features such as terraces, inner-fan meandering channel, and levees which, although common, may not occur on all fans. Facies defined in ancient rocks are shown in their inferred positions on the fan. Note also an incised channel, indicating a phase of downcutting, fan extension, and new lobe development. (From Walker, R. G., 1978, Deep-water sandstone facies and ancient submarine fans: Models for exploration for stratigraphic traps: Am. Assoc. Petroleum Geologists Bull., v. 62. Fig. 13, p. 946, reprinted by permission of AAPG, Tulsa, Okla.)

nels, apparently form owing to lateral channel switching as the upfan channel aggrades and fills up, causing a levee break. Midfan turbidites are commonly thick-bedded graded sandstones that may have pebbly bases. The **lower fan,** or outer fan, is the most distal part of the fan and forms outward from the midfan. It consists of finer grained, thin-bedded turbidites. The outer fan is characterized by a topographically smooth surface that grades into a basin plain. Various modern fans may differ in detail from these models, but the models nevertheless constitute a useful norm to which ancient submarine fan deposits can be compared.

Because most turbidites are deposited in fans, progradation of fans as they build outward from canyon mouths produces a vertical sequence of turbidite facies. R. G. Walker (1984c) suggests that fan deposits can display both thickening-upward sequences, accompanied in some cases by a coarsening-upward trend, and fining-upward sequences. Coarsening-upward sequences result from progradation of fan lobes seaward. Thinning-upward sequences are interpreted to represent gradual channel filling and abandonment. A generalized hypothetical sequence of turbidite deposits developed during overall fan progradation is illustrated in Figure 13.40.

FIGURE 13.40 Generalized hypothetical turbidite sequence on a prograding submarine fan. CT, classical turbidites; MS, massive sandstones; PS, pebbly sandstones; CGL, conglomerate; DF, debris flows; SL, slump. (From Walker, R. G., 1984, Turbidites and associated coarse clastic deposits, *in* R. G. Walker (ed.), Facies models, 2nd ed.: Geoscience Canada Reprint Ser. 1. Fig. 20, p. 182, reprinted by permission of Geological Association of Canada.)

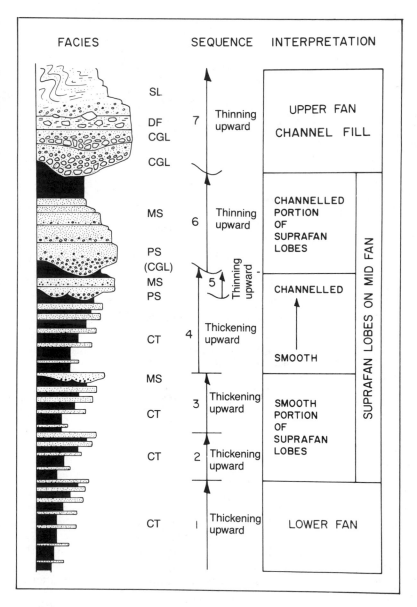

Contourites. Contourites are sediments believed to be the deposits of contour currents. They occur principally on continental rises. Coring of continental rises has shown that turbidites make up only part of continental-rise sediments. Rise sediments also include numerous thin layers of clean quartz silt, which is commonly cross-bedded, well sorted, and poorly graded, and which may include occasional thin, heavy-mineral layers (Reineck and Singh, 1980). These thin, quartz-rich layers are considered to be contourites, which differ from turbidites in several important respects. Among other things, they tend to be better sorted, thinner bedded, less regularly graded, better laminated, and more compositionally mature than turbidites (Table 13.4). In the modern ocean, contourites appear to be particularly well developed on the continental rise off eastern North America (Fig. 13.41).

Glacial-marine Sediments. Sediments ice-rafted to deep water are typically poorly sorted gravelly sands or gravelly muds that show crude to well-developed stratification. The coarse fraction may include angular, faceted, and striated pebbles. Significant areas of the modern ocean floor at high latitudes are covered by these glacial-marine sediments, particularly the subpolar North Atlantic, circum-Antarctic, Arctic Ocean, North Pacific, and Norwegian Sea (Ruddiman, 1977) (Fig. 13.36). These deposits are discussed in further detail under glacial systems in Chapter 11.

Hemipelagic Muds. Hemipelagic muds are fine-grained sediments that contain more than 25 percent terrigenous, volcanogenic, and/or neritic sediments (Table 13.3). These muds range in color from gray to green, and more rarely reddish brown. Textures range from clay to silty, sandy mud. They are commonly composed of fine terrigenous quartz, feldspar, micas, and clay minerals and/or volcanogenic sediments such as ash, fine pumice, and palagonite. Volcanic ash or glass may be intermixed with other hemipelagic sediments or concentrated into distinct **tephra** layers that range in thickness from less than 1 cm to more than 25 cm (Boggs, 1984). Granules and pebbles of pumice may occur as isolated fragments, pockets, or distinct layers associated with hemipelagic muds. Hemipelagic muds may contain the remains of siliceous organisms, particularly diatoms, and calcareous organisms such as foraminifers and nannofossils, as well as fine lime muds swept off carbonate platforms into deeper water. They are poorly laminated to massive, and generally are moderately to highly bioturbated. Hemipelagic muds are deposited closer to shore than pelagic muds. They are widely distributed on continental slopes of volcanic arcs, such as those of the Western Pacific. They also occur in back-arc basins, on inner trench walls, and on the tops of some rises.

Slump and Slide Deposits. These deposits consist of previously sedimented pelagic or terrigenous deposits that have been emplaced downslope owing to mass-movement processes. During the transport process, consistency of the slump masses is disturbed, resulting in faulted, contorted and chaotic bedding and internal structure. Studies of the ocean floor with side-scan sonar and bottom and sub-bottom acoustical profiling show that slump and slide deposits are particularly common on continental slopes with high rates of deposition, such as off the Mississippi and Rhone deltas, and on slopes with glacial-marine deposits. Nardin, Edwards, and Gorsline (1979) studied mass-flow deposits in the Santa Cruz Basin, California borderland, by use of seismic reflection profiling and identified four types of gravity deposits based on acoustical signatures on reflection records. The first type is recognized as slides, in which failure

TABLE 13.4 Important characteristics of turbidite and contourite sand or silt

		Turbidite	Contourite	Conclusions
Size sorting		Moderate to poorly sorted (std. dev. <1.50φ, Folk, 1974)	Well to very well sorted (std. dev. <0.75 φ, Folk, 1974)	Contourite is better sorted.
Bed thickness		Usually 10–100 cm	Usually <5 cm	Contourite has thinner bedding.
Primary sedimentary structure	Grading	Normal grading ubiquitous, bottom contacts sharp, upper contacts poorly defined	Normal and reverse grading, bottom and top contacts sharp	Contourite tends to be less regularly graded and has sharp upper contacts.
	Cross-laminations	Common, accentuated by concentrations of lutite	Common, accentuated by concentrations of heavy minerals	Contourite contrasts sharply with turbidite in that heavy mineral placers are in the form of small-scale stratification.
	Horizontal laminations	Common in upper portion only, accentuated by concentrations of lutite	Common throughout, accentuated by concentrations of heavy minerals or foraminifer shells	
Massive bedding		Common, particularly in lower portion	Absent	Contourite is ubiquitously laminated.
Grain fabric		Little or no preferred grain orientation in massive graded portions	Preferred grain orientation parallel to the bedding plane ubiquitous throughout bed	Contourite has better grain orientation.
Principal constituents of sand and silt beds	Matrix (<2 μ)	10–20%	0–5%	Contourite has less matrix.
	Microfossils	Common and well preserved, sorted by size throughout bed	Rare and usually worn or broken, often size sorted in placers	Contourite shows more evidence of reworking
	Plant and skeletal remains	Common and well preserved, sorted by size throughout bed	Rare and usually worn or broken	Contourite shows more evidence of reworking
Classification (Williams, Turner, Gilbert, 1982)		Lithic wacke	Lithic wacke, feldspathic wacke, quartz arenite	Contourite is more "mature."

Source: Modified from Bouma, A. H., and C. D. Hollister, 1973, Deep ocean basin sedimentation, in G. V. Middleton and A. H. Bouma (eds.), Turbidites and deep water sedimentation: Soc. Econ. Paleontologists and Mineralogists Pacific Section Short Course. Table 1, p. 95, reprinted by permission of SEPM, Tulsa, Okla.

FIGURE 13.41 Core of contourite sediment from the North American Atlantic continental rise off New England in 4746 m water. Note intercalations of hemipelagic claystone and coarser, cross-laminated contourite siltstone. Core is about 13 cm long. (From Heezen, B. C., and C. D. Hollister, 1971, The face of the deep. Fig. 9.66, p. 398, reprinted by permission of Oxford University Press, New York. Photograph courtesy of C. D. Hollister.)

took place elastically and only minor internal deformation of strata occurred. The other three types of failed masses were emplaced plastically and are characterized by various degrees of internal deformation. Much more study is needed to better understand the characteristics and volumetric importance of slump and slide deposits; however, emplacement of sediment along the modern lower continental slopes by mass-transport processes appears to be the dominant sediment transport process on some slopes.

Allochthonous Deep-Sea Carbonates. Although we tend to think of carbonate sediments as strictly shallow-water deposits, deep-water carbonates have been identified in several areas of the modern ocean, such as the slope and adjacent basin floor around

the Bahama Platform. They have been reported also from many Phanerozoic-age stratigraphic sequences. As shown in Figure 13.17, carbonate sediments are generated primarily on the shelf. No source of carbonate sediments exists within deep water except
that provided by the rain of calcareous pelagic organisms. Therefore, with the exception of calcareous oozes, carbonate sediments in deep water are derived from the shelf
by transport processes that include storm waves, turbidity currents, debris and grain
flows, slumping, sliding, and rockfalls. Very commonly, carbonate sediment deposited
on the slope and basin by these processes consists of bioclastic debris and limestone
blocks derived from the talus slopes off reef fronts. Also, sediments may be transported
downslope from carbonate-sand shoals or lime-mud deposits on the platform margin.
Excellent summaries of carbonate slope sedimentation and deposits are given by Mc-
Ilreath and James (1984), Enos and Moore (1983), and H. E. Cook and Mullins (1983).

McIlreath and James (1984) divide autochthonous slope carbonates into five
principal groups. **Hemipelagic sediments,** sometimes called "periplatform ooze," settle
from the water column more or less constantly. They are augmented by episodic contributions of fine material swept off the shelf during storms or by warm, sediment-rich
waters that float off the shelf over cooler basinal waters during tidal exchange. **Periplatform talus** consists of a debris apron of limestone blocks, skeletons of reef organisms, and lime sands and muds that accumulate directly seaward of reefs or platform-
edge lime shoals. They accumulate mainly by rockfall and grain flow from shallow
water. **Carbonate breccias** and **conglomerates** are coarse breccias derived from either
shallow water or the slope, or both. They consist of petromict carbonate clasts of various sizes, some of enormous dimensions, that commonly have a matrix of lime mud,
lime sand, or argillaceous lime mud. The internal fabric of these deposits may range
from chaotic to imbricated, horizontal, vertical, or even-graded. These deposits are
transported on the slope by submarine debris flows or other mass-transport mechanisms. **Graded calcarenites** are the carbonate equivalents of siliciclastic turbidites.
They commonly have sharp, planar bases, but sole markings are rare. Complete Bouma
sequences may occur, but more commonly only the A division (and in some deposits
the B and C divisions) is present. Graded calcarenites are deposited both on the slope
and on the deeper seafloor at the base of the slope. These carbonate turbidites are
derived mainly from loose carbonate sand and gravel accumulations near the platform
margin, but may be derived also from finer grained sediment lower on the slope. **Nongraded calcarenites** are massive to cross-bedded and ripple-marked lime-sand deposits
that generally have sharp bases and lenticular to irregular geometry. They contain various amounts of lime mud and may range from clean grainstones to fine- to coarse-
grained wackestones. They are possibly formed by some type of grain-flow mechanism.
Alternatively, they may be previously deposited sediments that were reworked and
winnowed by bottom contour currents.

The distribution of carbonate facies that develop on carbonate slopes and adjacent basins depends upon the character of the shelf-margin sediments that serve as a
source for slope carbonates, the nature of the platform edge, and the relief between the
platform and basin. Margins characterized by a gentle transition from platform to slope
are called "depositional margins." Figure 13.42 schematically depicts a depositional
margin capped by lime-sand shoals and illustrates the distribution of facies on the
slope and the hypothetical sequence of deposits formed in the adjacent basin. By contrast, Figure 13.43 illustrates the facies and deposition sequence developed on a "by-
pass margin" with a very abrupt transition from platform to slope owing to the presence of a cliff-fronted reef. Allochthonous deep-water carbonate facies thus range from
hemipelagic lime muds to relatively clean lime sands to graded turbidites to chaotic

FIGURE 13.42 Depositional margin characterized by shallow-water lime-sand shoals. Schematic model shows the hypothetical sequence of deposits within the adjacent basin. (From McIlreath, I. A., and N. P. James, 1984, Carbonate slopes, *in* R. G. Walker (ed.), Facies models, 2nd ed.: Geoscience Canada Reprint Ser. 1. Fig. 13, p. 252, reprinted by permission of Geological Association of Canada.)

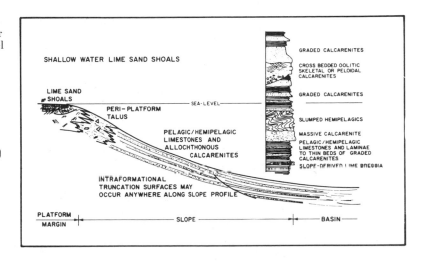

petromict conglomerates and breccias. The gradation of upper-slope petromict conglomerates and hemipelagic muds seaward into carbonate turbidites and eventually pelagic muds is diagnostic.

Ancient Deep-Sea Sediments

As mentioned, deep-sea sedimentary deposits are not as abundant in the rock record as shallow-water sediments because the potential for preservation and uplift of these sediments above sea level is much lower. Nonetheless, they are known from stratigraphic units of all ages. Typically, deep-sea sedimentary rocks consist of pelagic and hemipelagic shales, turbidite sandstones and conglomerates, bedded cherts (recrystallized siliceous oozes), chalks and marls (lithified, clayey pelagic oozes), and petromict

FIGURE 13.43 By-pass carbonate margin characterized by shallow-water reefs. Schematic model shows a hypothetical sequence of deposits formed in the adjacent basin. (From McIlreath, I. A., and N. P. James, 1984, Fig. 14, p. 253, reprinted by permission of Geological Association of Canada.)

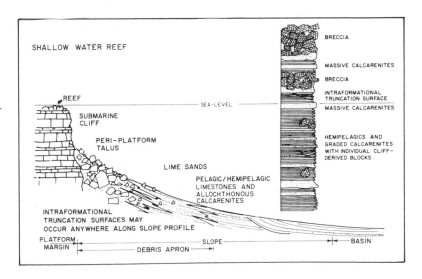

limestone breccias (slope deposits) and carbonate turbidites. Except for turbidites and carbonate breccias, which may be very coarse deposits, these deep-sea deposits are distinguished in general by their fine grain size. Other than turbidites, most deep-sea deposits do not show well-developed vertical facies sequences. Physical sedimentary structures consist predominantly of thin, horizontal laminations, although rippled bedding and graded bedding are common in turbidites, and cross-lamination occurs in some contourites. The bedding of many deep-sea deposits is well developed, even, and laterally persistent. Deep-water turbidites commonly display repetitive, well-bedded sequences of thin, graded units that are referred to as **rhythmites.** Such sequences are also called **flysch** facies, although the meaning of this term is somewhat controversial. Colors of deep-water sediments are typically dark gray to black. Red pelagic shales are much rarer. Muds may be well bioturbated or essentially nonbioturbated, and they are commonly characterized by distinctive deep-water trace-fossil associations. Fine-grained deep-water sediments are characterized also by the presence of much greater concentrations of planktonic organisms than occur in shallow-water sediments. These organisms include diatoms, radiolarians, foraminifers, coccoliths and, in older rocks, graptolites. Deep-water sedimentary rocks occur in extensive tabular- or blanket-shaped deposits, and may be associated with ocean crustal rocks such as submarine basalts and ophiolite assemblages consisting of serpentized peridotite, dunite, gabbros, sheeted dikes, and pillow lavas.

The Cretaceous Rosario Group in the San Diego area, California, provides a good example of an ancient basin-plain and submarine-fan sequence that illustrates some of the characteristics of deep-water siliciclastic deposits. This sequence has been described in detail by Nilsen and Abbott (1981). The Rosario Group in the San Diego area consists of the Point Loma Formation and the overlying Cabrillo Formation (Fig. 13.44). The basal Point Loma Formation is composed of a thick, massive mudstone unit with thin interbeds of graded siltstone and very fine-grained sandstone. The mudstones are thoroughly bioturbated, rich in carbonaceous materials, and are inferred to be probable slope and basin-plain deposits. Overlying the mudstones is a distinctive unit composed of interbedded turbidite sandstones and mudstones, forming a series of thickening- and coarsening-upward sequences. The thicker and coarser upper sandstone beds are locally scoured and channeled, whereas the thinner and finer lower beds are planar and more highly bioturbated. Abundant carbonaceous matter, mica, and megafossil fragments are present in this facies. Nilsen and Abbott interpret this unit as the deposits of outer-fan sandstone lobes. The uppermost part of the Point Loma Formation consists predominantly of well-defined, repetitive, thinning- and fining-upward sandstone sequences with some mudstone interbeds. The thicker lower sandstones are locally pebbly and may contain dish structures, parallel laminae, wood fragments, and mudstone rip-up clasts. They range from nongraded to normally graded to reversely graded. The thinner upper sandstone beds contain abundant parallel laminae, convolute laminae, ripple markings, flute and groove casts, mica flakes, carbonaceous matter, and bioturbation structures. These deposits are thought to be largely channel-fill sandstones deposited in a midfan environment. The basal part of the overlying Cabrillo Formation consists of thick beds of conglomerate with some interbedded sandstone. These coarse-grained sediments are interpreted as inner-fan channel deposits that prograded over the midfan sandstone channel deposits of the Point Loma Formation. The inner-fan conglomerates are overlain by a retrogradational sequence of midfan channel sandstones. The inferred paleogeographic setting in which the Rosario Group was deposited is shown in Figure 13.45. Lateral progradation and retrogradation of the turbidite fan generated the vertical sequence of facies illustrated in Figure 13.44.

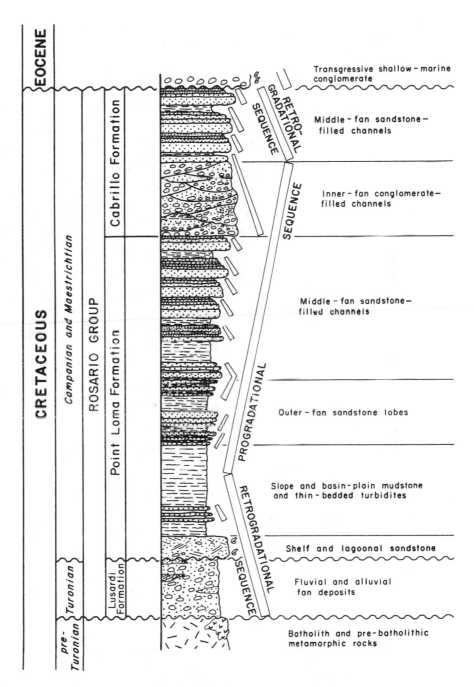

FIGURE 13.44 Composite columnal section of the Late Cretaceous strata in the San Diego area, California, showing progradational and retrogradational cycles of submarine fan deposition. (From Nilsen, T. H., and P. A. Abbott, 1981, Paleogeography and sedimentology of Upper Cretaceous turbidites, San Diego, California: Am. Assoc. Petroleum Geologists Bull., v. 65. Fig. 19, p. 1280, reprinted by permission of AAPG, Tulsa, Okla.)

FIGURE 13.45 Paleogeographic map of the San Diego area, California, in Late Cretaceous time, showing the inferred depositional setting of the submarine fan and associated deposits shown in Figure 13.44. (From Nilsen, T. H., and P. A. Abbott, 1981, Fig. 20, p. 1281, reprinted by permission of AAPG, Tulsa, Okla.)

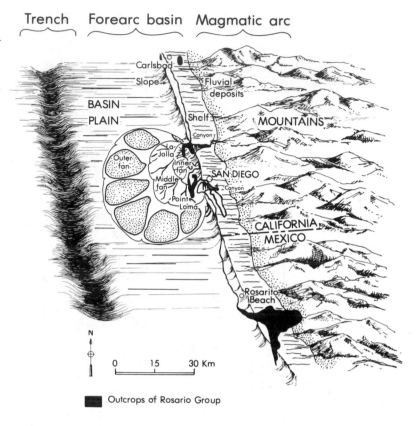

Jurassic carbonates of the central High Atlas Mountains, Morocco, described by I. Evans and Kendall (1977), illustrate a somewhat similar basin-plain submarine-fan setting in which carbonate rather than siliciclastic deposition took place. The Jurassic carbonates of this area consist of micrites and carbonate turbidites, and both a basin sequence and a slope sequence are recognized. The inferred paleogeographic setting of the basin is shown in Figure 13.46. The oldest exposed deposits are laminated micrites, which record an early supratidal to subtidal environment prior to the onset of deep-water conditions. The laminated micrites are overlain by massively bedded micrites, 10–20 m thick, containing sponges, brachiopods, and calcareous algae. These massively bedded bodies are referred to by Evans and Kendall as lithoherms and were presumably formed by trapping of carbonate mud by sponges and algae in water a few hundred meters deep. The massive micrites pass upward into a rhythmic alternation of micrites and marls that contain coccoliths and ammonites but lack the skeletal remains of benthonic organisms. The micrites also contain a low-diversity suite of trace fossils. The micrites are interpreted as pelagic deposits and the marls as the distal suspension loads of turbidites, both deposited in a basin environment. These deep-water deposits are overlain by a progressively shallowing-upward sequence of micrites and marls that are capped by coral reefs, recording the filling of the trough. A sequence of carbonate turbidite deposits, inferred to be essentially coeval with the basin facies described, is exposed on the southern slope of the High Atlas Mountains. The sequence is nearly 1 km thick and is distinguished by two different facies. One facies is

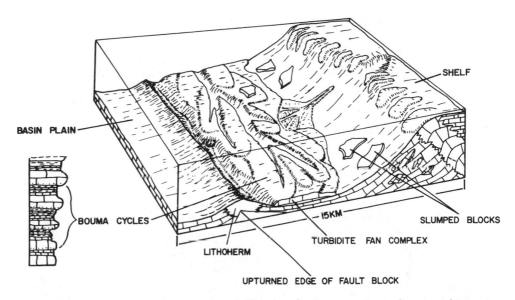

BASIN PLAIN

BOUMA CYCLES

LITHOHERM

UPTURNED EDGE OF FAULT BLOCK

TURBIDITE FAN COMPLEX

SLUMPED BLOCKS

SHELF

15KM

FIGURE 13.46   Schematic representation of the inferred paleoenvironmental setting of Jurassic carbonate submarine-fan deposits of the central High Atlas Mountains of Morocco. (From Evans, I., and C. G. St. C. Kendall, 1977, An interpretation of the depositional setting of some deep-water Jurassic carbonates of the central High Atlas Mountains, Morocco, in H. E. Cook and P. Enos (eds.), Deep-water carbonate environments: Soc. Econ. Paleontologists and Mineralogists Spec. Pub. 25. Fig. 12, p. 257, reprinted by permission of SEPM, Tulsa, Okla.)

composed of irregularly bedded black micrites interbedded with normally graded to nongraded beds, 2–3 m thick, containing sand- to pebble-size carbonate grains that include intraclasts, ooids, algally coated grains, and bioclastic debris in a micrite matrix. These deposits are channeled in places and display few Bouma cycles. They are interpreted as midfan channel-fill deposits and levee and overbank deposits. The second facies consists of carbonate turbidites containing mud clasts, pisolites, and fragments of corals, bryozoans, and brachiopods and show abundant Bouma cycles and deep-water trace fossils. This facies is interpreted to represent outer-fan deposition by turbidity currents that may have flowed parallel to the shelf margin (Fig. 13.46).

Numerous other ancient deep-water siliciclastic and carbonate deposits have been described in the literature. Readers are referred to H. E. Cook and Enos (1977), H. E. Cook and Mullins (1983), Jenkyns (1978), Rupke (1978), Scholle, Arthur, and Ekdale (1983), and Siemers et al. (1981) for discussion of some of these deep-water sequences.

ADDITIONAL READINGS

Siliciclastic Shelf Systems

Burk, C. A., and C. L. Drake (eds.), 1974, The geology of continental margins: Springer–Verlag, New York, 1009 p.

Greenwood, B., and R. A. Davis, Jr. (eds.), 1984, Hydrodynamics and sedimentation in wave-dominated coastal environments: Elsevier, New York, 473 p.

Hails, J., and A. Carr (eds.), 1975, Nearshore sediment dynamics and sedimentation: John Wiley & Sons, New York, 316 p.

Moslow, T. F., 1985, Depositional models of shelf and shoreline sandstones: Am. Assoc. Petroleum Geologists Education Course Notes Ser. 27, 102 p.

Nittrouer, C. A. (ed.), 1981, Sedimentary dynamics of continental shelves: Elsevier, Amsterdam, 449 p.

Scholle, P. A., and D. Spearing (eds.), 1982, Sandstone depositional environments: Am. Assoc. Petroleum Geologists Mem. 31, 410 p.

Stanley, D. J. (ed.), 1969, New concepts of continental margin sedimentation: Am. Geol. Inst. Short Course Notes: Washington, D.C., 400 p.

Stride, A. H. (ed.), 1982, Offshore tidal sands: Processes and deposits: Chapman and Hall, London, 222 p.

Swift, D. J. P., D. B. Duane, and O. H. Pilkey (eds.), 1972, Shelf sediment transport: Process and pattern: Dowden, Hutchinson and Ross, Stroudsburg, Pa., 656 p.

Tillman, R. W., and C. T. Siemers (eds.), 1984, Siliciclastic shelf sediments: Soc. Econ. Paleontologists and Mineralogists Spec. Pub. 34, Tulsa, Okla., 268 p.

Van Straaten, L. M. J. U. (ed.), 1964, Deltaic and shallow marine deposits: Developments in sedimentology, v. 1. Elsevier, New York, 464 p.

Carbonate Shelf and Reef Systems

Bathurst, R. G. C., 1975, Carbonate sediments and their diagenesis: Elsevier, Amsterdam, 658 p.

Bhattacharyya, A., and G. M. Friedman (eds.), 1983, Modern carbonate environments: Benchmark Papers in Geology, v. 74. Hutchinson and Ross, Stroudsburg, Pa., 376 p.

Cook, H. E., A. C. Hine, and H. T. Mullins (eds.), 1983, Platform margin and deep water carbonates: Soc. Econ. Paleontologists and Mineralogists Short Course Lecture Notes No. 12.

Friedman, G. M. (ed.), 1969, Depositional environments in carbonate rocks: Soc. Econ. Paleontologists and Mineralogists Spec. Pub. 14, 209 p.

Frost, S. H., M. P. Weiss, and J. B. Saunders (eds.), 1977, Reefs and related carbonates—Ecology and sedimentology: Am. Assoc. Petroleum Geologists, Studies in Geology No. 4, 421 p.

Laporte, L. F., (ed.), 1974, Reefs in time and spaces: Soc. Econ. Paleontologists and Mineralogists Spec. Pub. 18, 256 p.

Scholle, P. A., D. G. Bebout, and C. H. Moore, 1983, Carbonate depositional environments: Am. Assoc. Petroleum Geologists Mem. 33, 708 p.

Toomey, D. F. (ed.), 1981, European fossil reef models: Soc. Econ. Paleontologists and Mineralogists Spec. Pub. 30, 546 p.

Wilson, J. L., 1975, Carbonate facies in geologic history: Springer-Verlag, Berlin, 471 p.

Continental Slope and Deep-Sea Systems

Cook, H. E., and P. Enos (eds.), 1977, Deep-water carbonate environments: Soc. Econ. Paleontologists and Mineralogists Spec. Pub. 25, 336 p.

Cook, H. E., M. E. Field, and J. V. Gardner, 1982, Characteristics of sediments on modern and ancient continental slopes, in P. A. Scholle and D. Spearing (eds.), Sandstone depositional environments: Am. Assoc. Petroleum Geologists Mem. 31, p. 329–364.

Dott, R. H., and R. H. Shaver (eds.), 1974, Modern and ancient geosynclinal sedimentation: Soc. Economic Paleontologists and Mineralogists Spec. Pub. 19, 380 p.

Doyle, L. J., and D. H. Pilkey (eds.), 1979, Geology of continental slopes: Soc. Econ. Paleontologists and Mineralogists Spec. Pub. 27, 374 p.

Hay, W. W. (ed.), 1974, Studies in paleooceanography: Soc. Econ. Paleontologists and Mineralogists Spec. Pub. 20, 218 p.

Heezen, B. C., and Hollister, C. D., 1971, The face of the deep: Oxford University Press, New York, 659 p.

Hsü, K. J., and H. C. Jenkyns (eds.), 1974, Pelagic sedimentation on land and under the sea: Internat. Assoc. Sedimentologists Spec. Pub. 1, Blackwell, London, 447 p.

Lisitzin, A. P., 1072, Sedimentation in the world ocean: Soc. Econ. Paleontologists and Mineralogists Spec. Pub. 17, 218 p.

Middleton, G. V., and Bouma, A. H. (eds.), 1973, Turbidites and deep water sedimentation: Soc. Econ. Paleontologists and Mineralogists, Pacific Section Short Course, Anaheim, Calif., 157 p.

Nelson, C. H., and T. H. Nilsen, 1984, Modern and ancient deep-sea fan sediments: Soc. Econ. Paleontologists and Mineralogists Short Course Notes No. 14, 404 p.

Riley, J. P., and R. L. Chester (eds.), 1976, Chemical oceanography, v. 5, 2nd ed.: Academic Press, New York, 401 p.

Saxov, S., and J. K. Nieuwenhuis (eds.), 1982, Marine slides and other mass movements: Plenum Press, New York, 353 p.

Seibold, E., and W. H. Berger, 1982, The sea floor: Springer-Verlag, Berlin, 288 p.

Stanley, D. J., and G. T. Moore (eds.), 1983, The shelfbreak: Critical interface on continental margins: Soc. Econ. Paleontologists and Mineralogists Spec. Pub. 33, 467 p.

Stow, D. A. V., and D. J. W. Piper (eds.), 1984, Fine-grained sediments: Deep-water processes and facies: Geol. Soc. Spec. Pub. 15, Blackwell, Oxford, 659 p.

Warme, J. E., R. G. Douglas, and E. L. Winterer, 1981, the Deep Sea Drilling Project: A decade of progress: Soc. Econ. Paleontologists and Mineralogists Spec. Pub. 32, 564 p.

PART SEVEN
Principles of Stratigraphy

The emphasis in the preceding part of this book has been on sedimentary processes, the environments in which these processes take place, and the properties of sedimentary rocks generated in these environments. In this final part of the book, we focus on a different aspect of sedimentary rocks. Our concern here is not so much sedimentary processes and detailed rock properties, but rather the larger scale vertical and lateral relationships between units of sedimentary rock that are defined on the basis of lithologic or physical properties, paleontological characteristics, geophysical properties, age relationships, and geographic position and distribution. It is the study of these characteristics of layered rocks that we consider to encompass the discipline of stratigraphy. Understanding the principles and terminology of stratigraphy is essential to geologic study of sedimentary rocks because stratigraphy provides the framework within which systematic sedimentologic studies can be carried out. It allows the geologist to bring together the details of sediment composition, texture, structure, and other features into an environmental and temporal synthesis from which we can interpret the broader aspects of Earth history.

Prior to the 1960s, the discipline of stratigraphy was concerned particularly with stratigraphic nomenclature, the more classical concepts of lithostratigraphic, biostratigraphic, and chronostratigraphic successions in given areas, and correlation of these successions between areas. **Lithostratigraphy** deals with the lithology of physical properties of strata and their organization into units based on lithologic character. **Biostratigraphy** is the study of rock units based on the fossils they contain. **Chronostratigraphy** deals with the ages of strata and their time relations. This classical approach to stratigraphy is particularly exemplified in Weller's (1960) textbook, *Stratigraphic Principles and Practice*. These established principles are still the backbone of stratigraphy; however, today's students must go beyond these basic principles. They must also have a thorough understanding of depositional systems and be able to apply stratigraphic and sedimentological principles to interpretation of strata within the context of global plate tectonics. This means, among other things, becoming familiar with new branches of stratigraphy that have developed since the early 1960s as new concepts and methods of studying sedimentary rocks and other rocks by remote sensing techniques have unfolded. Two new offshoots of stratigraphy that have made particularly important contributions to our understanding of the physical stratigraphic relationships, ages, and environmental significance of subsurface strata and oceanic sediments are **magnetic stratigraphy,** which deals with stratigraphic relationships based on the magnetic properties of sedimentary rocks and layered volcanic rocks, and **seismic stratigraphy,** which is the study of stratigraphic and depositional facies as interpreted from seismic data.

14
Lithostratigraphy

14.1 INTRODUCTION

Lithostratigraphy deals with the study and organization of strata on the basis of their lithologic characteristics. The term **lithology** is used by geologists in two different but related ways. Strictly speaking, it refers to study and description of the physical character of rocks, particularly in hand specimens and outcrops. It is used also as a term that refers to these physical characteristics. Rock type, color, mineral composition, and grain size are all lithologic characteristics. For example, we may refer to the lithology of a particular stratigraphic unit as sandstone, shale, limestone, and so forth. Thus, lithostratigraphic units are rock units defined or delineated on the basis of their physical properties, and lithostratigraphy deals with the study of the stratigraphic relationships among strata that can be identified on the basis of lithology.

In this chapter, we begin study of stratigraphic principles by briefly discussing the nature of lithostratigraphic units, followed by an explanation of the various types of contacts that separate these units. We then explore the extremely important concepts of sedimentary facies and depositional sequences. The essentials of stratigraphic nomenclature and classification as they apply to lithostratigraphic units are discussed next, including examination of the new North American Code of Stratigraphic Nomenclature. Finally, correlation of lithostratigraphic units is explained and the various methods of correlation described.

14.2 TYPES OF LITHOSTRATIGRAPHIC UNITS

Lithostratigraphic units are bodies of sedimentary, extrusive igneous, metasedimentary, or metavolcanic rock distinguished on the basis of lithologic characteristics. A

523

lithostratigraphic unit generally conforms to the law of superposition, which states that in any sequence of strata not subsequently disturbed or overturned since deposition younger rocks lie above older rocks. Lithostratigraphic units are commonly stratified and tabular in form. They are recognized and defined on the basis of observable rock characteristics. Boundaries between different units may be placed at clearly identifiable or distinguishable contacts or be drawn arbitrarily within a zone of gradation. Definition of lithostratigraphic units is based on a stratotype, or type section, consisting of readily accessible rocks, where possible, in natural outcrops, excavations, mines, or boreholes. Lithostratigraphic units are defined strictly on the basis of lithic criteria as determined by descriptions of actual rock materials. They carry no connotation of age. They cannot be defined on the basis of paleontologic criteria, and they are independent of time concepts. They may be established in subsurface sections as well as rock units exposed at the surface, but they must be established on the basis of lithic characteristics and not on geophysical properties or other criteria. Geophysical criteria, described in Chapter 16, may be used to aid in fixing boundaries of subsurface lithostratigraphic units, but the units cannot be defined exclusively on the basis of remotely sensed physical properties.

Wheeler and Mallory (1956) introduced the term **lithosome** to refer to masses of rock of essentially uniform character and having intertonguing relationships with adjacent masses of different lithology. Thus, we speak of shale lithosomes, limestone lithosomes, sand–shale lithosomes, and so forth. Krumbein and Sloss (1963) explain the meaning of lithosomes by asking readers to imagine the body of rock that would emerge if it were possible to preserve a single rock type, such as sandstone, and dissolve away all other rock types. The resulting sandstone body would thus appear as a roughly tabular mass with intricately shaped boundaries. These irregular boundaries would represent surfaces of contact with erosion surfaces and with other rock masses of differing constitution above, below, and to the sides. Lithosomes have no specified size limits and may range in gross shape from thin sheetlike or blanketlike units to thick prisms, or narrow, elongated shoestrings.

Of course, stratigraphic units of a single lithology rarely exist as isolated bodies. They are commonly in contact with other rock bodies of different lithology. An important part of lithostratigraphy is identifying and understanding the nature of contacts between vertically superposed or laterally adjacent bodies. Another important aspect is the identification of single lithosomes, groups of lithosomes, or subdivisions of lithosomes that are so distinctive that they form lithostratigraphic units that can be distinguished from other units that may lie above, below, or adjacent. The fundamental lithostratigraphic unit of this type is the formation. A **formation** is a lithologically distinctive stratigraphic unit that is large enough in scale to be mappable at the surface or traceable in the subsurface. It may encompass a single lithosome, or part of an intertonguing lithosome, and thus consist of a single lithology. Alternatively, a formation can be composed of two or more lithosomes, and thus include rocks of different lithology. Some formations may be divided into smaller distinctive units called **beds,** which are the smallest formal lithostratigraphic units. Formations having some kind of stratigraphic unity can be combined to form **groups,** and groups can be combined to form **supergroups.** All formal lithostratigraphic units are given names, which are derived from some geographic feature in the area where they are studied.

Subdivision of thick units of strata into smaller lithostratigraphic units such as formations is essential for tracing and correlation of strata both in outcrop and in the subsurface. We will come back to discussion of these formal stratigraphic units near the end of this chapter and include discussion of some of the problems involved in naming stratigraphic units. First, however, we need to examine the nature of contacts between stratigraphic units and the lateral and vertical facies relationships that characterize strata.

14.3   STRATIGRAPHIC CONTACTS

Different lithologic units are separated from each other by **contacts,** which are plane or irregular surfaces between different types of rocks. Vertically superposed strata are said to be either conformable or unconformable depending upon continuity of deposition. Conformable strata are characterized by unbroken depositional sequences, generally deposited in parallel order, in which layers are formed one above the other by more or less uninterrupted deposition under the same general conditions. The surface that separates conformable strata is a **conformity;** that is, a surface that separates younger strata from older rocks, but along which there is no physical evidence of nondeposition. A conformable contact indicates that no significant break, or hiatus, in deposition has occurred. A **hiatus** is defined as the total interval of geologic time represented by missing strata at a specific position along a stratigraphic surface.

Unconformable strata are strata, in a vertical sequence, that do not succeed underlying rocks in immediate order of age or do not fit together with them as part of a continuous whole. The contacts between such strata are called unconformities. An **unconformity** is a surface of erosion or nondeposition, separating younger strata from older rocks, that represents a significant hiatus. Unconformities indicate a lack of continuity in deposition and correspond to periods of nondeposition, weathering, or erosion, either subaerial or subaqueous, prior to deposition of younger beds. Unconformities thus represent a substantial break in the geologic record that may correspond to periods of erosion or nondeposition lasting millions or even hundreds of millions of years.

Contacts are present also between laterally adjacent lithosomes. These contacts are formed between rock units of equivalent age that developed different lithologies owing to different conditions in the depositional environment. I exclude from discussion here contacts between laterally adjacent bodies that arise from postdepositional faulting. Contacts between laterally adjacent bodies may be gradational, where one rock type grades gradually into another, or they may be intertonguing. This latter type of contact is explained more fully in a succeeding section.

Contacts between Conformable Strata

Contacts between conformable strata may be either abrupt or gradational. **Abrupt contacts** occur as a result of sudden, distinct changes in lithology. Most abrupt contacts coincide with primary depositional bedding planes that formed as a result of changes in local depositional conditions, as discussed in Chapter 6. In general, bedding planes represent minor interruptions in depositional conditions. Such minor depositional breaks, involving only short hiatuses in sedimentation with little or no erosion before deposition is resumed are called **diastems.** Abrupt contacts may be caused also by postdepositional chemical alteration of beds, producing changes in color owing to oxidation or reduction of iron-bearing minerals, changes in grain size owing to recrystallization or dolomitization, or changes in resistance to weathering owing to cementation by silica or carbonate minerals.

Conformable contacts are said to be **gradational** if the change from one lithology to another is gradual, reflecting gradual change in depositional conditions with time. Gradational contacts may be of either the progressive gradual type or the intercalated type. **Progressive gradual contacts** occur where one lithology grades into another by progressive, more or less uniform changes in grain size, mineral composition, or other characteristic. Examples include sandstone units that become progressively finer grained upward until they change to mudstones (Fig. 14.1A) or quartz-rich sandstones that become progressively enriched upward in lithic fragments until they change to lithic arenites. **Intercalated contacts** are gradational contacts that occur owing to an increasing number of interbeds of another lithology that appear upward in the section (Fig. 14.1B).

Contacts between Unconformable Strata

As mentioned, the surfaces that separate unconformable strata are called unconformities. Four types of unconformities—(1) angular unconformity, (2) disconformity, (3) paraconformity, and (4) nonconformity—are recognized on the basis of the presence or absence of an angular relationship between the unconformable strata, the presence or absence of a marked erosional surface separating these strata, and the nature of the rocks underlying the surface of unconformity. The first three types of unconformities occur between bodies of sedimentary rock. The last type occurs between sedimentary rock and metamorphic or igneous rock.

Angular Unconformity. An angular unconformity is a type of unconformity in which younger sediments rest upon the eroded surface of tilted or folded older rocks; that is, the older rocks dip at a different, commonly steeper, angle than do the younger rocks (Fig. 14.2A). Angular unconformities confined to limited geographic areas are called local unconformities. Those that extend for tens or even hundreds of kilometers are regional unconformities. Some angular unconformities are clearly visible in a single outcrop (Fig. 14.3). By contrast, regional unconformities between stratigraphic units of

FIGURE 14.1 Types of gradational vertical contacts. A. Progressive gradual. B. Intercalated.

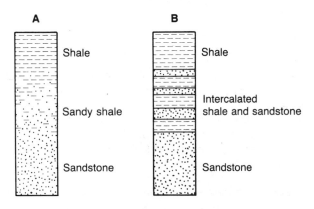

FIGURE 14.2 Four basic types of unconformities. A. Angular unconformity. B. Disconformity. C. Paraconformity. D. Nonconformity. (After Dunbar, C. O., and J. Rodgers, Principles of stratigraphy. © 1957 by John Wiley & Sons, Inc. Fig. 57, p. 117, reprinted by permission of John Wiley & Sons, New York.)

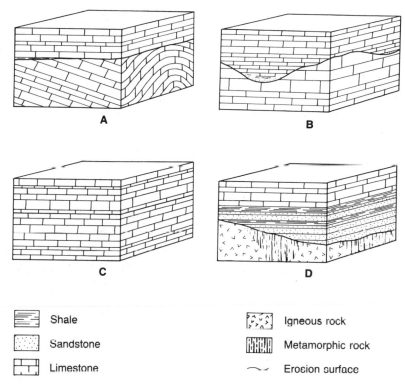

	Shale		Igneous rock
	Sandstone		Metamorphic rock
	Limestone		Erosion surface

very low dip may not be apparent in a single outcrop and may require detailed mapping over a large area before they can be identified.

Disconformity. An unconformity in which the bedding planes above and below the unconformable surface are essentially parallel and the contact between younger and older beds is marked by a visible, irregular, or uneven erosional surface is a disconformity (Fig. 14.2B). Disconformities are most easily recognized by this erosional surface, which may have relief ranging up to tens of meters. Disconformity surfaces, as well as angular unconformity surfaces, may be marked also by "fossil" soil zones or may include lag-gravel deposits lying immediately above the unconformable surface and containing pebbles of the same lithology as the underlying unit. Disconformities are presumed to form as a result of a significant period of erosion during which older rocks remained essentially horizontal during nearly vertical uplift and subsequent downwarping.

Paraconformity. A paraconformity is an obscure unconformity in which the beds above and below the unconformity contact are parallel and in which no erosional surface or other physical evidence of unconformity is discernable. The contact may even appear to be a simple bedding plane (Fig. 14.2C). Paraconformities are not easily recognized and must be identified on the basis of missing strata as determined from paleontologic evidence such as absence of faunal zones or abrupt faunal changes.

FIGURE 14.3 Angular unconformity between steeply dipping, well-bedded marine sandstones and shales of the Cretaceous Hudspeth Formation and nearly horizontal beds of volcaniclastic sediments in the overlying Clarno Formation (Eocene). The line of boulders marks the position of the unconformity. West of Mitchell, north-central Oregon.

Nonconformity. An unconformity developed between sedimentary rock and older igneous or massive metamorphic rock that has been exposed to erosion prior to being covered by sediments is a **nonconformity** (Fig. 14.2D).

The presence of unconformities has considerable significance in sedimentological studies. Many stratigraphic sequences are bounded by unconformities, indicating that these sequences are incomplete records of past sedimentation. Unconformities not only show that some part of the stratigraphic record is missing, they also indicate that an important geologic event took place during the time period represented by the unconformity—an episode of uplift and erosion or, less likely, an extended period of nondeposition.

Contacts between Laterally Adjacent Lithosomes

In the preceding discussion we examined the kinds of stratigraphic contacts or boundaries that separate sedimentary units into distinct vertical lithologic sequences. Stratigraphic units also have finite lateral boundaries. They do not extend indefinitely laterally but must eventually terminate, either abruptly as a result of erosion or more gradually by change to a different lithology. Lateral changes may be accompanied by progressive thinning of units to extinction—**pinchouts** (Fig. 14.4A); lateral splitting of a lithologic unit into many thin units that pinch out independently—**intertonguing** (Fig. 14.4B); or **progressive lateral gradation,** similar to progressive vertical gradation (Fig. 14.4C).

FIGURE 14.4 Lateral relationship of sedimentary units: A. Pinchout. B. Intertonguing. C. Lateral gradation.

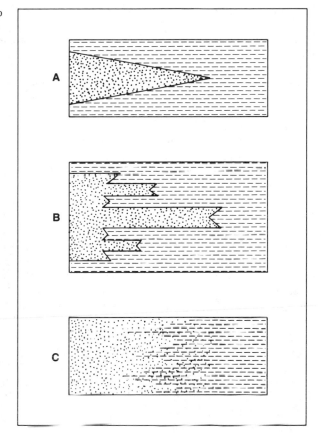

14.4 VERTICAL SUCCESSIONS OF STRATA

Conformities and unconformities divide sedimentary rocks into vertical successions of beds, each characterized by a particular lithologic aspect. Different types of beds can succeed each other vertically in a great variety of ways, and distinctions can be drawn between rock units characterized by (1) lithologic uniformity, (2) lithologic heterogeneity, and (3) cyclic successions (Weller, 1960). Rock units that have complete lithologic uniformity are rare, although many beds may display a high degree of uniformity in color, grain size, composition, or resistance to weathering. Beds that are most likely to be uniform are fine-grained sediments deposited slowly under essentially uniform conditions in deeper water or coarser sediments that have been deposited rapidly by some type of mass sediment transport mechanism such as grain flow. Heterogeneous bodies of sedimentary strata are characterized by internal variations or irregularities in properties. Heterogeneous units may include strata such as extremely poorly sorted tillites or debris-flow deposits, as well as thick units broken internally by thinner beds characterized by differences in grain size or bedding features.

Many stratigraphic successions display repetitions of strata that reflect a sequence of related depositional processes and conditions that occurred in the same order. Such repetitious events are called **cyclic sedimentation,** or **rhythmic sedimentation.** Cyclic sedimentation leads to the formation of vertical sequences of sedimentary strata that display repetitive orderly arrangement of different kinds of sediments. The term cyclic sediment has been used for a wide variety of repetitious strata, including such small-scale features as presumed annually deposited varves in glacial lakes, as well as large-scale sediment cycles, or cyclothems, caused by long-period, cyclic migration of depositional environments. Other common examples of cyclic deposits include rhythmically bedded turbidites, laminated evaporite deposits, limestone–shale rhythmic sequences, and black shale deposits. Cyclic sequences occur on all continents in essentially every stratigraphic system. They are produced by processes that range in geographic scope and duration from very local, short-term events such as seasonal climatic changes that generate varves to global changes in sea level that may involve entire geologic periods. Many problems related to the origin of cyclic deposits are still unresolved. Cyclic sedimentation is discussed in greater detail by Merriam (1964), Duff et al. (1967), Schwarzacher (1975), Vail et al. (1977b), and Einsele and Seilacher (1982).

14.5   LATERAL SUCCESSIONS OF STRATA

In the preceding chapters dealing with depositional environments, many examples are pointed out of sediments of one type grading laterally into sediments of a different type deposited in laterally contiguous parts of a given depositional setting. For example, sandy sediments of the beach shoreface may grade seaward to muddy sediments of the shallow inner shelf; delta-front sands and silts commonly grade seaward to prodelta muds; and shelf-edge skeletal or oolitic carbonate sands grade toward the open shelf to pelleted carbonate muds. I have already referred to such laterally equivalent bodies of sediment with distinctive characteristics as facies. Thus a deposit may be characterized by shale facies, sandstone facies, limestone facies, and so forth. The concept of facies is so important in stratigraphy that a more detailed explanation of the meaning and significance of facies is necessary at this point.

Sedimentary Facies

The term **facies** was introduced into the geological literature by Nicolas Steno in 1669 (Teichert, 1958); however, modern scientific use of the term is credited to the Swiss geologist Amanz Gressly, who used the term in 1838 in his description of Upper Jurassic strata in the region of Solothurn in the Jura Mountains to describe marked changes in lithology and paleontology of these strata. Krumbein and Sloss (1963) maintain that Gressly intended to confine usage of the term to lateral changes within a stratigraphic unit such as those illustrated in Figure 14.5. Other workers have interpreted Gressly's usage to include vertical changes in the character of rock units as well (Teichert, 1958). Subsequently, the term has been used with numerous meanings, many of which bear little resemblance to Gressly's original meaning. These various meanings have been summarized and discussed by Moore (1949), Teichert (1958), Weller (1958), Markevich (1960), and others. The extended meanings of facies have included referring to all strata of a particular type as a certain facies, such as referring

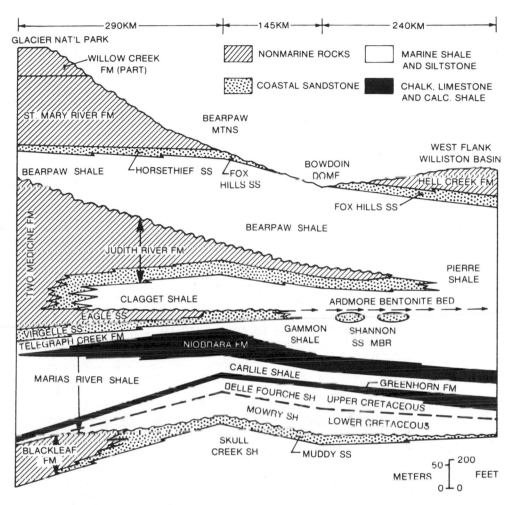

FIGURE 14.5 Facies relationships in Upper Cretaceous strata of the Rocky Mountains in Montana. (From Swift, D. J. P., and D. D. Rice, 1984, Sand bodies on muddy shelves: A model for sedimentation in the western interior Cretaceous seaway, North America, *in* R. W. Tillman and C. T. Siemers (eds.), Siliciclastic shelf sediments: Soc. Econ. Paleontologists and Mineralogists Spec. Pub. 34. Fig. 2, p. 45, reprinted by permission of SEPM, Tulsa, Okla.)

to all redbeds as the "redbed facies," and even such nonstratigraphical usage as "metamorphic facies," "igneous facies," and "tectonic facies." Because of these rather loose and inconsistent usages of the term, the real meaning of facies has become considerably clouded.

A commonly accepted definition of facies in the United States is that of Moore (1949, p. 32), who described facies as "any areally restricted part of a designated stratigraphic unit which exhibits characters significantly different from those of other parts of the unit." Facies comprise "one or any two or more different sorts of deposits which are partly or wholly equivalent in age and which occur side by side or in somewhat close neighborhood." According to Moore's definition, facies are restricted in areal extent but the same facies could be found at different levels within the same stratigraphic unit. A different usage of facies that is closer to that of Gressly usage—and to that of many European geologists—is to consider facies simply as stratigraphic units distinguished by lithological, structural, and organic aspects detectable in the field. The areal distribution of facies thus designated may not be well known (Blatt et al., 1980), in contrast to the restricted areal distribution required by Moore's definition. Regardless of the exact definition followed in defining facies, it is now common practice to designate facies identified on the basis of lithologic characteristics as **lithofacies** and facies distinguished by paleontologic characteristics (fossil content) without regard to lithologic character as **biofacies.**

An important objective of facies studies is to ultimately make environmental interpretations from the facies. Thus some geologists designate facies on the basis of assumed depositional environment and speak of "continental facies," "fluvial facies," "delta facies," and so forth. Such generic usage involves subjective judgments that may not always be justified. It is better to make the usage of facies purely descriptive and objective and then make subjective interpretations of environment on the basis of these descriptive facies (Hallam, 1981).

14.6 COMBINED VERTICAL AND LATERAL STRATIGRAPHIC RELATIONSHIPS

Walther's Law

It is implicit in the concept of facies that different facies represent different depositional environments. As laterally contiguous environments in a given region shift with time in response to shifting shorelines or other geologic conditions, facies boundaries also shift so that eventually the deposits of one environment may lie above those of another environment. This deceptively simple idea embodies one of the single most important concepts in stratigraphy—the concept that a direct environmental relationship exists between lateral facies and vertically stacked or superimposed successions of strata. This concept was first formally stated by Johannes Walther in 1894 and is now called the **law of the correlation (or succession) of facies,** or simply **Walther's Law.** This law has often been misstated as "the same facies sequences are seen laterally as vertically." The correct statement of the law as translated by Middleton (1973, p. 979) is

> The various deposits of the same facies-area and similarly the sum of the rocks of different facies-areas are formed beside each other in space, though in a cross-section we see them lying on top of each other . . . it is a basic statement of far-reaching significance that only those facies and facies-areas can be superimposed primarily which can be observed beside each other at the present time (Walther, 1884).

Walther's Law is thus interpreted to mean that facies that occur in **conformable** vertical successions of strata also occurred in laterally adjacent environments. Middleton (1973) is careful to point out that the law does not state that vertical successions always reproduce the horizontal sequence of environments, but only that those facies can be superimposed that can now be seen developing side by side. For example, the beach and barrier-island environmental setting discussed in Chapter 12 may include several laterally adjacent environments such as beach, back-barrier lagoon, marsh, tidal flat, tidal channel, and tidal delta. Depending upon the manner in which these lateral environments shift with time, the vertical sequence produced by deposition in a particular barrier-island setting might consist only of beach sands overlain by lagoonal muds and capped by marsh peats. The entire lateral sequence of deposits formed in the contiguous environments is not preserved, but those deposits that are preserved in the vertical sequence originally occurred side by side.

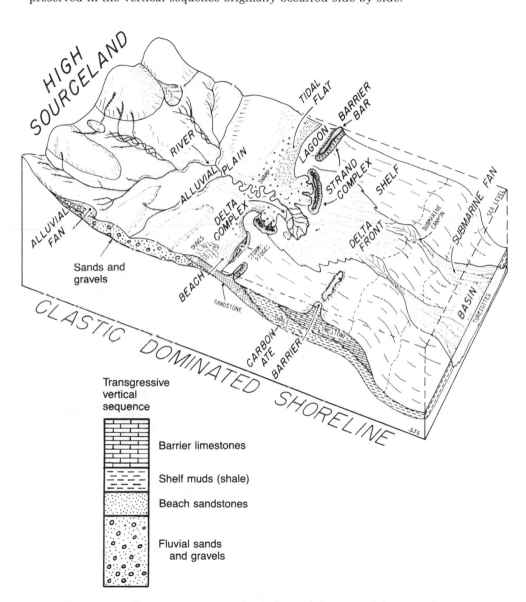

FIGURE 14.6 Lateral environments on a clastic-dominated coast and the vertical sequences of facies that would likely develop owing to transgression. (Modified from Fichter, L. S., and D. J. Poché, 1979, Ancient environments and the interpretation of geologic history. Fig. following p. 168, reprinted by permission of Burgess Publishing Co., Minneapolis, Minn.)

Transgressions and Regressions

The principles embodied in Walther's Law are illustrated in Figure 14.6. The top of the block diagram in this figure shows the lateral sequence of environments typical of a clastic-dominated shelf undergoing transgression. As discussed in Chapter 12, transgression refers to movement of a shoreline in a landward direction, also called retrogradation. The left side of the block in Figure 14.6 shows how sediments from different laterally adjacent environments have become superimposed vertically as the shoreline advanced landward, resulting in sediments from more distant offshore environments being deposited progressively on top of sediments previously laid down nearer to shore. Transgression of shorelines thus produces vertical sequences of sedimentary units in which deeper water, fine-grained sediments are superimposed on coarser grained nearshore sediments, creating fining-upward successions of strata. Transgressions occur during a relative rise in sea level when the influx of terrigenous sediments from land sources is low enough (Fig. 14.7A) to allow deeper water marine sediments to encroach landward over nearshore deposits, a process called coastal en-

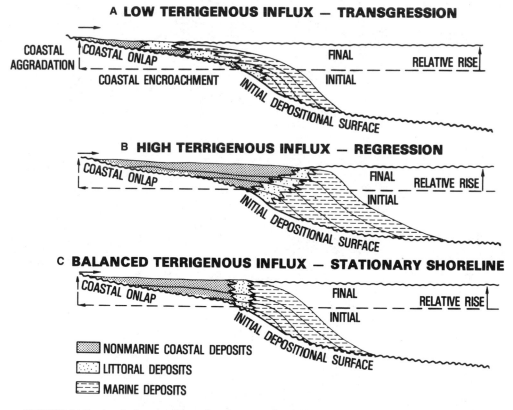

FIGURE 14.7   A relative rise in sea level can produce (A) transgression, (B) regression, or (C) a stationary shoreline, depending upon the rate of terrigenous influx. (After Vail, P. R., R. M. Mitchum, Jr., and S. Thompson, III, 1977, Seismic stratigraphy and global changes of sea level. Part 3: Relative changes of sea level from coastal onlap, *in* C. E. Payton (ed.), Seismic stratigraphy—Applications to hydrocarbon exploration: Am. Assoc. Petroleum Geologists Mem. 26. Fig. 3, p. 66, reprinted by permission of AAPG, Tulsa, Okla.)

croachment. It is important to stress here that transgression will not occur during rising sea level if the influx of terrigenous sediments is so high that outbuilding of the shoreline takes place.

Under some conditions, shorelines and environments may shift in a seaward direction. Seaward movement of a shoreline is called **regression,** or **progradation.** Regression leads also to vertical superposition of contiguous lateral facies, but in this case coarse-grained nearshore sediments become progressively stacked on top of finer grained, deeper water sediments, leading to coarsening-upward successions. Regression may occur during a relative rise in sea level or during static sea level if the influx of terrigenous clastics is high (Fig. 14.7B), or it may occur during a relative fall in sea level (Fig. 14.8). Transgression followed by regression tends to produce a **wedge** of sediments in which deeper water sediments are deposited on top of shallower water sediments in the basal part of the wedge, and shallower water sediments are deposited on top of deeper water sediments in the top part of the wedge (Fig. 14.9). Note the marked coastal overlap illustrated in Figure 14.9. The initial depositional surface at the base of a transgressive sequence is commonly an unconformity. The bounding surface at the top of the regressive sequence can also be an unconformity if regression occurs as a result of a relative fall in sea level accompanied by erosion.

Transgressions and regressions thus lead to deposition of vertically stacked successions of fining-upward and coarsening-upward deposits. I point out, however, that under at least two types of depositional conditions transgression or regression may not occur for a geologically significant period of time. If during a relative rise in sea level the shoreline is stationary for a long period of time as a result of a balance in terrigenous influx, so that neither sediment outbuilding (progradation) nor coastal encroachment (retrogradation) occurs, lateral facies do not become vertically superimposed (Fig. 14.7C). During a relative standstill of the sea, when relative sea level is neither rising nor falling and terrigenous influx is sufficiently high, a type of deposition called **coastal toplap** (Vail et al., 1977a) occurs. Coastal sediments cannot build above the effective wave base and aggrade because of the standstill of sea level, so onlap cannot be produced. Instead, each unit of strata laps out in a landward direction, but the successive terminations lie progressively seaward (Fig. 14.10).

Effects of Climate on Coastal Sedimentation Patterns

It should be clear from the preceding discussion that both the rate of influx of terrigenous clastic sediments and change in relative sea level exert controls on sedimentation patterns in coastal areas and the continental shelf. In turn, terrigenous influx is itself influenced by tectonism and climatic conditions. Tectonism produces changes in elevation of sediment source areas and thus affects rates of erosion, which generally increase with increase in land elevation. Also, source areas at higher elevations tend to

FIGURE 14.8 Rapid fall in relative sea level indicated by downward shift in coastal onlap. (After Vail, P. R., R. M. Mitchum, Jr., and S. Thompson, III, 1977, Fig. 8, p. 72, reprinted by permission of AAPG, Tulsa, Okla.)

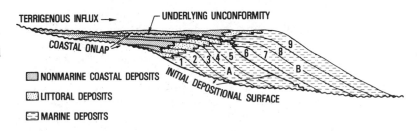

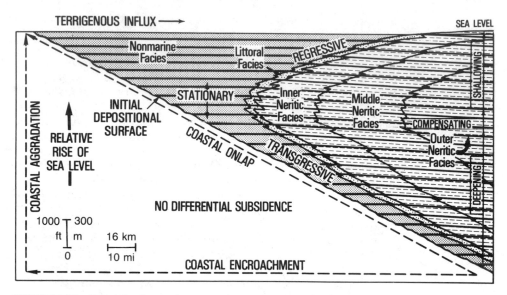

FIGURE 14.9   Coastal onlap owing to marine transgression and regression. During relative rise in sea level, littoral facies may be transgressive, stationary, or regressive. Neritic (shallow shelf) facies may be deepening, shallowing, or compensating (maintaining a given depth). (From Vail, P. R., R. M. Mitchum, Jr., and S. Thompson, III, 1977, Fig. 4, p. 67, reprinted by permission of AAPG, Tulsa, Okla.)

shed coarser sediment than those at lower elevations. Climate regulates sediment influx by controlling rates of weathering and erosion, sediment-transport conditions, and sediment mechanisms. For example, significantly greater terrigenous influx will occur during long periods of rainy climatic conditions, when erosion rates are accelerated and stream transport is increased, than during prolonged dry periods. On a shorter time scale, more sediment, and coarser sediment, may be eroded and transported during a single unusually large, high-velocity flood that occurs only once every hundred years than during all the smaller floods that may have occurred during the preceding hundred years. Thus, the rates of sediment influx and the grain sizes of sediments delivered to coastal areas from continents has varied throughout geologic time in response to these variables of tectonism and climate.

Effects of Sea-Level Changes on Sedimentation Patterns

Changes in sea level also control sedimentation patterns in coastal areas. Two kinds of sea-level changes are possible. **Relative** sea-level changes occur simply as a result of tectonic uplift or downwarping of land masses. Such tectonic changes affect sea level locally but may have little or no effect on worldwide sea levels. **Eustatic** sea-level changes are worldwide changes that affect sea level on all continents essentially simultaneously. Eustatic sea-level changes have been attributed to a variety of causes. The best known cause is tied to continental glaciation. Sea level drops during glacial stages, when seawater is locked up on land as ice, and rises during interglacial stages as continental ice sheets melt. Other possible causes of eustatic change include in-

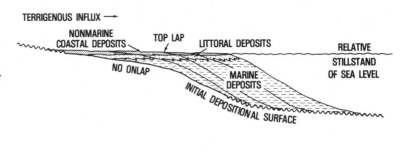

FIGURE 14.10 Coastal toplap. Coastal toplap indicates relative standstill of sea level. During a relative standstill of sea level, no relative rise of base level occurs; therefore, nonmarine coastal and/or littoral deposits cannot aggrade and no onlap is produced. Instead, sediment bypassing takes place, producing toplap. (From Vail, P. R., R. M. Mitchum, Jr., and S. Thompson III, 1977, Fig. 6, p. 70, reprinted by permission of AAPG, Tulsa, Okla.)

crease in volume of ocean water, owing to generation of juvenile water (water derived directly from magmas) at mid-ocean ridge axes and in island arcs, and variations in sediment influx. Sediment infill of an ocean basin, for example, would cause sea level to rise. Changes in the volume of the mid-ocean ridge system may be another cause. It has been suggested that changes in volume of mid-ocean ridges occur as a result of changes in rates of seafloor spreading. The idea is that an increase in rate of seafloor spreading causes an increase in volume of mid-ocean ridges and consequent rise in sea level, and a decrease in spreading rate generates a decrease in ridge volume and a corresponding fall in worldwide sea level. Pitman (1978) suggests, for example, that change in the rate of seafloor spreading from 2 cm/yr to 6 cm/yr in the modern ocean could produce a rise in sea level of more than 100 m over a period of 70 million years (Fig. 14.11). Correspondingly, a decrease in spreading rate back to 2 cm/yr for the next 70 million years would cause sea level to drop by more than 100 m. Pitman further proposes that transgressive or regressive events may not be simply indicated by rise and fall of sea level but by changes in the rate of sea-level rise or fall. Thus, a decrease in rate of sea-level rise or an increase in rate of sea-level fall may result in regression, whereas an increase in rate of sea-level rise or a decrease in rate of sea-level fall may produce transgression (Fig. 14.11).

Application of the Facies Concept

To return to Walther's Law of the correlation of facies, this law not only offers a rational explanation for vertical sequences of facies, but also has practical application in the study of ancient sedimentary facies and environments. For example, it is possible on the basis of Walther's Law to study the vertical sequence of beds in an outcrop section or well boring located along the edge of a basin and to predict from this sequence what the lateral succession of facies will be as a particular stratigraphic unit is followed into the basin. Such predictions are of particular importance to petroleum geologists, who must try to determine possible facies changes in petroleum reservoir beds that extend from outcrop areas or drilled regions into prospective undrilled regions.

Utilization of Walther's Law can also aid greatly in understanding depositional environments of strata in a vertical succession. Visher (1965) points out that only a limited number of depositional sequences or associations have been found in the study of modern and ancient sediments. For example, the environments recognized on an inner-marine shelf undergoing regression include tidal flat, lagoon–bay, littoral

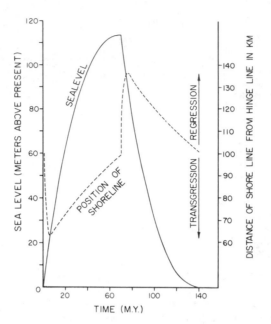

FIGURE 14.11 Change in sea level (solid line) owing to change in spreading rate of a ridge 10,000 km long that has been spreading at a rate of 2 cm/yr for a long time. At 0 m.y., the spreading rate changes to 6 cm/yr and stays at this rate until 10 m.y., when the rate is reduced to 2 cm/yr. It stays at this lower rate until 140 m.y. The dashed line shows the position of the shoreline as a function of rate of sea-level change. (From Pitman, W.C., III, 1978, Relation between eustacy and stratigraphic sequences of passive margins: Geol. Soc. America Bull., v. 89, Fig. 3, p. 1392.)

(beach), wave zone, shoreface, and the neritic zone below wave base. Sediments deposited in each of these environments have distinctive properties of grain size, sorting, sedimentary structures, and geometry (Fig. 14.12) that characterize that particular environment. Knowing that this lateral succession of environments and facies exists in a marine environment undergoing regression, a geologist studying ancient regressive marine sediments can interpret the depositional environment of various parts of the vertical succession with greater confidence and can also recognize the absence of facies that may have been removed from a particular environment by erosion. The regressive marine facies model discussed above is only one of many possible examples of the interrelationship of lateral facies and vertical successions. Sedimentary processes operating in other depositional settings—fluvial, deltaic, and barrier island, for example—also produce distinctive vertical successions of strata that are caused by shifts in laterally contiguous depositional environments, as discussed in Chapters 11–13.

14.7 DEPOSITIONAL SEQUENCES

Definition. The term "sequence" is commonly used informally by geologists, and has been so used in this book, to refer to any grouping or succession of strata. Sequence is used also in a more restricted and formal sense to identify distinctive stratigraphic units that are commonly bounded by unconformities. Sloss (1963) considered sequences to be major rock-stratigraphic units of interregional scope, separated and delimited by interregional unconformities. On the North American craton, he recognized and named six major sequences, each separated by demonstrable regional unconformities that can be traced from the Cordilleran region of western North America to the Appalachian Basin in the east. Each succession or sequence represents a major cycle

REGRESSIVE MARINE MODEL					
	Grain size	Sorting	Lithology	Sedimentary structures	Geometry
Tidal flat	Fine-medium ○—○	Poor-fair	Silt-clay Sand	Laminated, ripple cross-beds scour & fill, mudcracks raindrop-scuffed ripples	
Lagoon-bay	Fine ○	Poor	Silt-clay (sand?)	Bored & churned plant remains	
Dune	Fine-medium ○—○	Very good	Sand	Festoon & planar cross-bedding	
Littoral	Coarse ◯			Swash & rill marks parallel to wavy bedding	
Wave zone	◯			Parallel bedding ripples	
Shoreface	○			Graded bedding current structures thin bedded	
Below wave zone	Very fine ○	Poor	Clay-silt	Bored & churned laminated (?)	

FIGURE 14.12 General characteristics of sediments deposited in a coastal marine environment during regression. (After Visher, G. S., 1965, Use of vertical profiles in environmental reconstruction: Am. Assoc. Petroleum Geologists Bull., v. 49. Fig. 2, p. 45, reprinted by permission of AAPG, Tulsa, Okla.)

of transgression and regression, that is, advance and retreat of shorelines. Recognition of the sequences is based on physical relationships among rock units, although Sloss indicates that the sequences also have time-stratigraphic significance.

The sequence concept has subsequently been extended and redefined by Mitchum, Vail, and Thompson (1977). These authors define (p. 53) a **depositional sequence** as "a stratigraphic unit composed of a relatively conformable succession of genetically related strata and bounded at its top and base by unconformities or their correlative conformities." Sequences as thus defined differ from Sloss's sequences in that they may be much smaller rock units. Also, because they are bounded by interregional unconformities and their equivalent conformities, they may be traceable over major areas of ocean basins as well as continents. Distinct, related groups of depositional sequences superposed one on another are designated by Mitchum and his coworkers as **supersequences.** These supersequences are of the same general order of magnitude as Sloss's original sequences. The basic concept of depositional sequences is illustrated in Figure 14.13. Because sequences are defined on the basis of physical relationships of the strata—that is, bounded at the top and base by unconformities or their

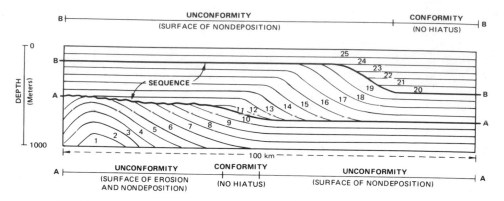

FIGURE 14.13 Illustration of the concept of depositional sequences. A depositional sequence is composed of relatively conformable genetically related strata bounded at its base (A) and top (B) by unconformities that pass laterally to correlative conformities. Individual units of strata 1 through 25 are traced by following stratification surfaces; they are assumed to be conformable where successive strata are present. Where units of strata are missing, hiatuses are present. (From Mitchum, R. M., Jr., P. R. Vail, and S. Thompson, III, 1977, Seismic stratigraphy and global changes of sea level. Part 2: The depositional sequence as a basic unit for stratigraphic analysis, in C. E. Payton (ed.), Seismic stratigraphy—Applications to hydrocarbon exploration: Am. Assoc. Petroleum Geologists Mem. 26. Fig. 1, p. 54, reprinted by permission of AAPG, Tulsa, Okla.)

correlative conformities—they are not primarily dependent for recognition upon determination of rock types, fossils, or depositional processes. The criteria for recognizing sequence boundaries is explained below. Depositional sequences generally range in thickness from tens to hundreds of meters.

Time Significance. Mitchum, Vail, and Thompson (1977) recognize that depositional sequences have time-stratigraphic significance in the sense that all strata within a sequence were deposited during a given broad interval of time, although the age range of individual strata within the sequence may differ from place to place. Surfaces separating depositional sequences may be either unconformities or, in the case of correlative conformities, stratal surfaces, or bedding planes (Fig. 14.13). The hiatus represented by unconformities may range from millions to hundreds of millions of years. On the other hand, the physical surfaces that separate groups of strata or individual beds and laminae within sequences were produced during a relatively short period of time and are essentially synchronous.

Internal Relationships. The strata that make up a depositional sequence may be either **concordant,** that is, essentially parallel to the sequence boundary, or **discordant,** lacking parallelism with respect to the sequence boundaries. Concordant relations can occur at either the upper or lower boundary of a sequence and may be expressed as parallelism to an initially horizontal, inclined, or uneven surface (Fig. 14.14). Discordance is the most important physical criterion used in determining sequence boundaries. Two main types of discordance are recognized, depending upon the manner in which the strata terminate against the sequence boundaries. **Truncation** is the lateral termination of strata owing to being cut off from their original depositional limits by erosion. Truncation occurs at the upper boundary of a sequence and may be of either

UPPER BOUNDARY

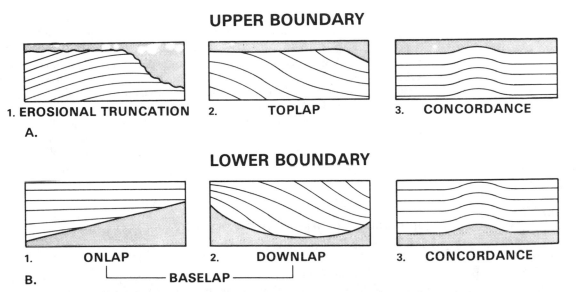

1. **EROSIONAL TRUNCATION**   2.   **TOPLAP**   3.   **CONCORDANCE**

A.

LOWER BOUNDARY

1.   **ONLAP**   2.   **DOWNLAP**   3.   **CONCORDANCE**

B.   └────── **BASELAP** ──────┘

FIGURE 14.14   Relations of strata to the (A) upper boundary and (B) lower boundary of a depositional sequence. A1. Erosional truncation: strata terminate against the upper boundary mainly owing to erosion. A2. Toplap: initially inclined strata terminate against an upper boundary created mainly as a result of nondeposition (e.g., foreset strata terminating against an overlying horizontal surface where no erosion or deposition took place). A3. Top concordance: strata at the top of a sequence do not terminate against an upper boundary. B1. Onlap: strata terminate updip against an inclined surface. B2. Downlap: initially inclined strata terminate downdip progressively against initially horizontal or inclined surface. B3. Base-concordance: strata at base of a sequence do not terminate against lower boundary. (From Mitchum, R. M., Jr., P. R. Vail, and S. Thompson, III, 1977, Fig. 2, p. 58, reprinted by permission of AAPG, Tulsa, Okla.)

local or regional extent. **Lapout** is the lateral termination of strata against a boundary at their original depositional limit. Lapout relationships are further divided into two types, baselap and toplap, depending upon the specific nature of the discordant relationship with the upper or lower sequence boundary.

Baselap occurs at the lower boundary of a depositional sequence and may itself be of two types. **Onlap** is baselap in which an initially horizontal or inclined stratum terminates against a surface of greater inclination. **Downlap** is baselap in which an initially inclined stratum terminates downdip against an initially horizontal or inclined surface. Onlap and downlap indicate nondepositional hiatuses (Fig. 14.14) and not erosional breaks in deposition.

Toplap is lapout at the upper boundary of a depositional sequence; for example, the lateral termination updip of the foreset beds of a deltaic complex. Toplap is also evidence of a nondepositional hiatus. Mitchum, Vail, and Thompson (1977) suggest that toplap results from a depositional base level, such as sea level, being too low to permit the strata to extend farther updip, thus allowing sedimentary bypassing and possibly minor erosion to occur above base level while prograding strata are deposited below base level.

Toplap and baselap relations of the strata in a depositional sequence should not be confused with the foreset bedding of cross-laminated units, which form parts of beds rather than of sequences. Figure 14.15 diagrammatically illustrates in a regional setting the relationships to sequence boundaries of strata in depositional sequences.

Identification of Depositional Sequences. Depositional sequences can be identified in both outcrop sections and subsurface sections. Subsurface identification depends largely upon the use of instrumental well logs (such as electric logs that measure resistivity of rock units) and seismic data (Chapter 16) to locate and trace unconformities, truncations, and lapout relationships. Sequence boundaries are identified on the basis of physical stratigraphic relationships, but where fossil information is available, as from well-core data, the geologic ages of the sequences are also determined. An example of sequences defined in a subsurface section by correlation of strata using electric log data is given in Figure 14.16.

14.8   NOMENCLATURE AND CLASSIFICATION OF LITHOSTRATIGRAPHIC UNITS

The general, informal term lithosome was introduced in Section 14.2 for bodies of sedimentary strata of distinctive lithology. To bring order to strata and to understand to the fullest extent the geologic history recorded in these strata, it is necessary also to have a formal system for defining, classifying, and naming geologic units. Such a stratigraphic procedure promotes systematic study of the physical properties and sequence relationships of sedimentary strata and is an essential requirement for interpretation of depositional environments and other aspects of Earth history. The need for systematic organization of strata was recognized as early as the latter half of the eighteenth century by European scientists such as Johann Gottlob Lehman, Giovanni Arduino, and Georg Christian Füchsel, who made early attempts to organize strata on the basis of relative age (Krumbein and Sloss, 1963). The gradual evolution of these efforts to organize and classify strata continued through the eighteenth and nineteenth centuries and eventually culminated in formulation of the internationally used Geologic Time Scale and the Geologic, or Stratigraphic, Column. This evolution is one of the more fascinating chapters in the history of stratigraphic study. Succinct summaries of these early efforts at stratigraphic classification are given by Weller (1960), Krumbein and Sloss (1963), and Dunbar and Rodgers (1967).

FIGURE 14.15 Terminology for relations that define unconformable boundaries of a depositional sequence. (After Mitchum, R. M., Jr., P. R. Vail, and J. B. Sangree, 1977, Seismic stratigraphy and global changes of sea level. Part 6: Stratigraphic interpretation of seismic reflection patterns in depositional sequences, in C. E. Payton (ed.), Seismic stratigraphy—Applications to hydrocarbon exploration: Am. Assoc. Petroleum Geologists Mem. 26. Fig. 1, p. 118, reprinted by permission of AAPG, Tulsa, Okla.)

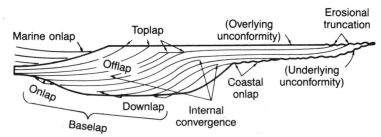

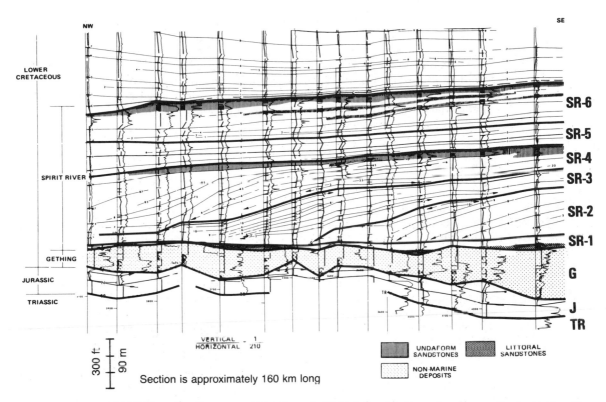

FIGURE 14.16 Sequences (SR-1 through SR-6) defined in the subsurface on the basis of well-log marker correlation. (From Mitchum, R. M., Jr., P. R. Vail, and S. Thompson, III, 1977, Fig. 9, p. 60, reprinted by permission of AAPG, Tulsa, Okla.)

Development of the Stratigraphic Code

Local study of rock strata requires subdivision of the stratigraphic column into smaller units that are systematically arranged on the basis of inherent properties and attributes. The purpose of stratigraphic classification is thus to promote understanding of the geometry and sequences of rock bodies. To insure uniform usage of stratigraphic nomenclature and classification, attempts have been under way for several decades to adopt a code of stratigraphic nomenclature that formulates views on stratigraphic principles and practices designed to promote standardized classification and formal nomenclature of rock materials. In the United States, such codes have been drafted by the Committee on Stratigraphic Nomenclature, 1933, and its successors, the American Commission on Stratigraphic Nomenclature, 1961, and the North American Commission on Stratigraphic Nomenclature, 1983. The Code of Stratigraphic Nomenclature published by the American Commission on Stratigraphic Nomenclature in 1961, and revised slightly in 1970, standardized terminology and practices used in stratigraphy in the United States at that time and has been widely accepted by North American geologists. New concepts and techniques, particularly the concept of global plate tectonics, have developed in the past few decades. These developments have revolutionized the earth sciences and necessitated revision of the 1961 Code. An International

Stratigraphic Guide published by the International Subcommission on Stratigraphic Classification in 1976 (Hedberg, 1976) provided a comprehensive treatment of stratigraphic classification, terminology, and procedures from an international point of view. In order to incorporate new concepts and techniques and to recognize the contribution of international stratigraphic organizations, the North American Commission on Stratigraphic Nomenclature published a new North American Stratigraphic Code in May 1983 (North American Commission on Stratigraphic Nomenclature, 1983). For the convenience of readers, this code is reproduced in full in the Appendix.

Major Types of Stratigraphic Units

The 1983 North American Code of Stratigraphic Nomenclature is a new code, not a revision of the 1961 Code. Some categories of the stratigraphic units included in the 1961 Code have disappeared and others are new. In general, the new code has been prepared to be as consistent as possible with the International Stratigraphic Guide, but several differences exist. Stratigraphic units are divided in the 1983 Code into two broad categories—a material category based on content or physical limits of strata and one based on geologic age (Table 14.1). "Content of strata" refers to physical, or lithic, characteristics of rocks—composition, texture, fabric, structure, and color—as well as chemical composition and biologic content or properties. Subcategories of stratigraphic units defined on the basis of material content or physical limits include: (1) **lithostratigraphic units** that conform to the law of superposition and are distinguished on the basis of lithic characteristics and lithostratigraphic position; (2) **lithodemic units** consisting of predominantly intrusive, highly metamorphosed, or intensely deformed rock that generally does not conform to the law of superposition; (3) **magnetopolarity units,** which are bodies of rock identified by remanent magnetic polarity; (4) **biostratigraphic units** defined and characterized by their fossil content; (5) **pedostratigraphic units,** consisting of one or more pedologic (soil) horizons developed in one or more lithic units now buried by a formally defined lithostratigraphic or allostratigraphic unit or units; and (6) **allostratigraphic units,** which are mappable strati-

TABLE 14.1 Categories of stratigraphic units defined by the 1983 North American Stratigraphic Code

Material categories based on content or physical limits
Lithostratigraphic units
Lithodemic units
Magnetopolarity units
Biostratigraphic units
Pedostratigraphic units
Allostratigraphic units
Categories expressing or related to geologic age
Material categories to define temporal spans
Chronostratigraphic units
Polarity-chronostratigraphic units
Temporal (non-material) categories
Geochronologic units
Polarity-chronologic units
Diachronic units
Geochronometric units

Source: After North American Commission on Stratigraphic Nomenclature, 1983, North American Stratigraphic Code: Am. Assoc. Petroleum Geologists Bull., v. 67. Table 1, p. 848, reprinted by permission of AAPG, Tulsa, Okla.

form bodies defined and identified on the basis of bounding discontinuities. In the International Stratigraphic Guide, stratigraphic units based on magnetic reversals and seismic velocity are included under the general heading of **geophysical units.** The 1983 North American Code of Stratigraphic Nomenclature does not formally recognize units based on seismic velocity.

Categories of stratigraphic units that express or are related to geologic age are of two kinds. **Materials categories used to define temporal spans** refers to categories of stratigraphic units that serve as standards for recognizing and isolating materials of a particular age. Two types of units are included in this category. **Chronostratigraphic units** are bodies of rock established to serve as the material reference for all rocks formed during the same span of time. **Polarity-chronostratigraphic units** are bodies of rock that contain a primary magnetopolarity record (Chapter 15) imposed when the rock was deposited or crystallized. **Temporal (non-material) categories** are not material units but conceptual units; they are divisions of time. Prior to publication of the 1983 Code, the only standard of reference for temporal units was the geochronologic unit. **Geochronologic units** are divisions of time distinguished on the basis of the rock record as expressed by chronostratigraphic units. Three new temporal units have been added by the 1983 Code. **Polarity-chronologic units** are divisions of geologic time distinguished on the basis of the record of magnetopolarity as embodied in polarity-chronostratigraphic units. **Diachronic units** comprise the unequal spans of time represented by one or more specific diachronous rock bodies, which are bodies with one or two bounding surfaces that are not time synchronous and thus "transgress" time. **Geochronometric (chronometric) units** are isochronous units (units having equal time duration) that are direct divisions of geologic time expressed in years. Geochronometric units have no material referents.

Categories and ranks of all stratigraphic units defined in the 1983 Code are shown in Table 14.2. Procedures and requirements for defining formal stratigraphic units are set forth in detail in the Code (Appendix). These procedures include requirements for picking a name, designating a stratotype or type section, describing the units, specifying the boundaries between units, and publication of appropriate descriptions of the units in a recognized scientific medium. Our immediate concern here is with subdivision and nomenclature of lithostratigraphic units. Other types of stratigraphic units are described in appropriate parts of the text.

Formal Lithostratigraphic Units

The concept of formations and other formal lithostratigraphic units is introduced in Section 14.2. In terms of size, the hierarchy of lithostratigraphic units in descending order is supergroup, group, formation, member, and bed (Table 14.2). Although a formation is not the largest lithostratigraphic unit, it is nonetheless the fundamental unit of lithostratigraphic classification. All other lithostratigraphic units are defined either as assemblages or subdivisions of formations. A **formation** is defined in Article 24 of the Stratigraphic Code (Appendix) as a body of rock, identified by lithic characteristics and stratigraphic position, that is prevailingly but not necessarily tabular and is mappable at Earth's surface and traceable in the subsurface. Formations may be defined on a single lithic type, repetitions of two or more lithic types, or extreme lithic heterogeneity, where such heterogeneity constitutes a form of unity when comparison is made with adjacent units. For example, a formation might be composed entirely of shale, entirely of sandstone, or of an intimate mixture of sandstone and shale beds that is distinctive because of the mixed lithology. Boundaries of formations, as with all lith-

TABLE 14.2 Categories and ranks of stratigraphic units defined in the 1983 North American Stratigraphic Code

A. Material Units

LITHOSTRATIGRAPHIC	LITHODEMIC	MAGNETOPOLARITY	BIOSTRATIGRAPHIC	PEDOSTRATIGRAPHIC	ALLOSTRATIGRAPHIC
Supergroup	Supersuite (Complex)				
Group	Suite (Complex)	Polarity Superzone			Allogroup
Formation	*Lithodeme*	*Polarity zone*	*Biozone* (Interval, Assemblage or Abundance)	*Geosol*	*Alloformation*
Member (or Lens, or Tongue)		Polarity Subzone	Subbiozone		Allomember
Bed(s) or Flow(s)					

B. Temporal and Related Chronostratigraphic Units

CHRONO-STRATIGRAPHIC	GEOCHRONOLOGIC GEOCHRONOMETRIC	POLARITY CHRONO-STRATIGRAPHIC	POLARITY CHRONOLOGIC	DIACHRONIC
Eonothem	Eon	Polarity Superchronozone	Polarity Superchron	
Erathem (Supersystem)	Era (Superperiod)			
System (Subsystem)	*Period* (Subperiod)	*Polarity Chronozone*	*Polarity Chron*	*Episode*
Series	Epoch			Phase (Diachron)
Stage (Substage)	Age (Subage)	Polarity Subchronozone	Polarity Subchron	Span (Diachron)
Chronozone	Chron			Cline

***Fundamental units are italicized.**

Source: North American Commission on Stratigraphic Nomenclature, 1983, Table 2, p. 852, reprinted by permission of AAPG, Tulsa, Okla.

ostratigraphic units, are placed at the position of lithic change. Boundaries between different formations may, therefore, occur both vertically and laterally. That is, a formation may be located above or below another formation or be positioned laterally adjacent to another formation where lateral facies changes occur. Illustrations of different types of formation boundaries are given in Figure 2, Appendix. A formation must be of sufficient areal extent and thickness to be mappable at the scale of mapping commonly used in the region where it occurs.

A **member** is the formal lithostratigraphic unit next in rank below a formation and is always a part of some formation. Formations may be divided completely into members, have only certain parts designated as members, or contain no members. A geographically restricted member that terminates on all sides within a formation may be called a **lens,** or lentil, and a wedge-shaped member that extends beyond the main boundary of a formation or that wedges or pinches out within another formation may be called a **tongue.**

Beds are subdivisions of members. A bed is the smallest formal lithostratigraphic unit of sedimentary rock. A **flow** is the smallest formal lithostratigraphic unit of volcanic flow rock.

A **group** is of higher rank than a formation and consists of assemblages of formations, but groups need not be composed entirely of named formations. A **super-**

group is a formal assemblage of related or superposed groups or of groups and formations.

Formal lithostratigraphic units are assigned names that consist of a geographic name combined with the appropriate rank (formation, member, and so forth) or an appropriate lithic term, such as limestone, or both. Formation names consist of a geographic name followed by either the word formation or a lithic designation. For example, a particular formation might be called the Otter Point Formation (geographic name only) or the Eureka Quartzite (geographic name plus lithic designation). The names of members include a geographic name and the word member, or the name may have an intervening lithic designation such as Eau Claire Sandstone Member. A group name combines a geographic name with the word "group," as in Arbuckle Group. The first letters of all words used in formal names of lithostratigraphic units are capitalized.

Informal names may be used for lithostratigraphic units when there is insufficient need, insufficient information, or an inappropriate basis to justify designation as a formal unit (Hedberg, 1976). Informal names may be applied to such units as oil sands, coal beds, mineralized zones, quarry beds, and key or marker beds. Informal names are not capitalized. Examples of informally designated names are "shaly zone," "coal-bearing zone," "pebbly beds," and "siliceous-shale member." The depositional sequences described above are also informal stratigraphic units. As originally defined by Sloss (1963), sequences are considered to be "rock-stratigraphic units of higher rank than group, megagroup, or supergroup . . . "; however, as redefined by Mitchum, Vail, and Thompson (1977), depositional sequences are an order of magnitude smaller than Sloss's sequences. Thus, a depositional sequence might include a single formation, groups of formations, or, on a smaller scale, subdivisions of formations. The International Stratigraphic Guide (Hedberg, 1976) proposes the term **synthem** for unconformity-bounded units, and the North American Stratigraphic Code of 1983 introduces the name **allostratigraphic unit** for mappable stratiform bodies of sedimentary rock defined on the basis of bounding, laterally traceable discontinuities. Neither of these formal names appears to be completely applicable to depositional sequences, which can be recognized where the boundaries are conformities as well as unconformities.

14.9 CORRELATION OF LITHOSTRATIGRAPHIC UNITS

Introduction

In the simplest sense, stratigraphic correlation is the demonstration of equivalency of stratigraphic units. Correlation is a fundamental part of stratigraphy, and much of the effort by stratigraphers in creating formal stratigraphic units has been aimed at finding practical and reliable methods of correlating these units from one area to another. Without correlation, treatment of stratigraphy on anything but a purely local level would be impossible.

The concept of correlation goes back to the very roots of stratigraphy. The fundamental principles of correlation have been presented in numerous early textbooks on geology and stratigraphy; especially interesting reviews of these general principles are given in Dunbar and Rodgers (1957), Weller (1960), and Krumbein and Sloss (1963). The continued strong interest in correlation is demonstrated by more recent publication of several books and numerous articles dealing with correlation, particularly statistical methods of correlation; for example, Gill and Merriam (1979), Harbaugh and Merriam (1968), Schwarzacher (1975), Mann (1981), Merriam (1981), and Cubitt and Reyment (1982).

The fundamental concepts of stratigraphic correlation were already firmly established by the 1950s and 1960s. These basic principles are still important today; however, the emergence of new concepts and more advanced analytical tools has changed our perception of correlation to some degree, as well as adding new methods for correlation. The development of the field of magnetostratigraphy since the late 1950s, for example, has provided an extremely important new tool for global time-stratigraphic correlation based on magnetic polarity events. Also, rapid advances in computer technology and availability and the application of computer-assisted statistical methods to stratigraphic problems have added a new quantitative dimension to the field of stratigraphic correlation. I will attempt in this chapter to bring out some of these new developments, along with discussion of the more "classical" concepts of stratigraphic correlation.

Definition of Correlation

In spite of the fact that the concept of correlation goes back to the early history of stratigraphy, disagreement has persisted over the exact meaning of the term. Historically, two points of view have prevailed. One view rigidly restricts the meaning of correlation to demonstration of time equivalency, that is, to demonstration that two bodies of rock were deposited during the same period of time (Dunbar and Rodgers, 1957; Rodgers, 1959). From this point of view, establishing the equivalence of two lithostratigraphic units on the basis of lithologic similarity does not constitute correlation. A broader interpretation of correlation allows that equivalency may be expressed in lithologic, paleontologic, or chronologic terms (Krumbein and Sloss, 1963). In other words, two bodies of rock can be correlated as belonging to the same lithostratigraphic or biostratigraphic unit even though these units may be of different ages. It is clear from a pragmatic point of view that most geologists today accept the broader view of correlation. Petroleum geologists, for example, routinely correlate subsurface formations on the basis of lithology of the formations, the specific "signatures" recorded within the formations by instrumental well logs, or the reflection characteristics on seismic records. The 1983 North American Stratigraphic Code (Appendix) recognizes three principal kinds of correlation: (1) **lithocorrelation,** which links units of similar lithology and stratigraphic position, (2) **biocorrelation,** which expresses similarity of fossil content and biostratigraphic position, and (3) **chronocorrelation,** which expresses correspondence in age and in chronostratigraphic position.

Even though our concern in this chapter is correlation based on lithology, it is important to clarify the relationship between chronocorrelation and lithocorrelation. Chronocorrelation can be established by any method that allows matching of strata by age equivalence. Correlation of units defined by lithology may also yield chronostratigraphic correlation on a local scale, but when traced regionally many lithostratigraphic units transgress time boundaries. As mentioned, stratigraphic units deposited during major transgressions and regressions are notably time-transgressive. Perhaps the most famous North American example of a time-transgressive formation is the Cambrian Tapeats Sandstone in the Grand Canyon region. This sandstone is apparently all Early Cambrian in age at the west end of the canyon and all Middle Cambrian in age at the eastern end (Fig. 14.17). Thus, the Tapeats Sandstone, which can be traced continuously through the canyon, correlates from one end of the canyon to the other as a

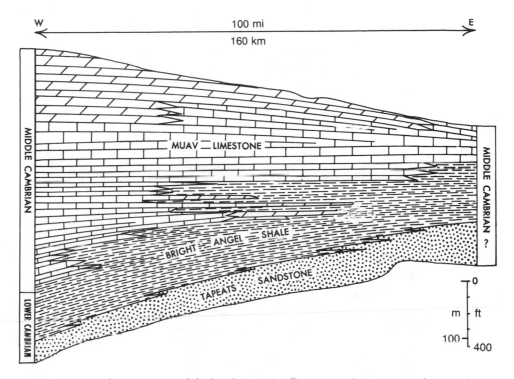

FIGURE 14.17 Changes in age of the basal Cambrian Tapeats Sandstone across the Grand Canyon region. (From Clark, T. H., and C. W. Stern, 1968, Geological evolution of North America, 2nd ed. Fig. 7.10, p. 138, reprinted by permission of John Wiley & Sons, Inc. Originally from E. D. McKee, 1954, Cambrian history of the Grand Canyon region. Part 1. Stratigraphy and ecology of the Grand Canyon Cambrian: Carnegie Inst. Washington Pub. 563, Washington, D.C.)

lithostratigraphic unit but not as a chronostratigraphic unit. The important point stressed here is that the boundaries defined by criteria used to establish time correlation of stratigraphic units need not be the same as those defined by criteria used to establish lithologic correlation. Because of this fact, different methods of correlation may yield different results when applied to the same stratigraphic sequence.

Another point that requires some clarification is the difference between matching of stratigraphic units and correlation of these units. Matching has been defined simply as correspondence of serial data without regard to stratigraphic units (Schwarzacher, 1975; B. R. Shaw, 1982). For example, two rock units identified in stratigraphic sections at different localities as having essentially identical lithology—say two black shales—can be matched on the basis of lithology; however, these units may have neither time equivalence nor lithostratigraphic equivalence. For example, physical tracing of the units between the localities may show that one unit lies stratigraphically above the other. Matching by lithologic characteristics in this particular case does not constitute demonstration of equivalence. B. R. Shaw (1982) states that the process of correlation is the demonstration of geometric relationships between rocks, fossils, or sequences of geologic data for interpretation and inclusion in facies models, paleontologic reconstructions, or structural models. The object of correlation is the establishment of equivalency of stratigraphic units between geographically separated parts of a geologic unit. Implicit in this definition is the concept that *correlation is made between stratigraphic units,* that is, lithostratigraphic units, biostratigraphic units, or chronostratigraphic units. The difference between correlation and matching is illustrated in Figure 14.18. Figure 14.18A shows two stratigraphic sections that appear to be perfectly matched. The actual correlation is shown in Figure 14.1B. The tie lines in Figure 14.1A do not constitute correlation because they do not encompass equivalent stratigraphic units.

Correlation can be regarded as either direct (formal) or indirect (informal) (B. R. Shaw, 1982). **Direct correlation** is correlation that can be established physically and unequivocally. Physical tracing of continuous stratigraphic units is the only unequivocal method of showing correspondence of a unit in one locality to that in another. **Indirect correlation** can be established by numerous methods such as visual comparison of instrumental well logs, polarity reversal records, or fossil assemblages; however, such comparisons have different degrees of reliability and can never be totally unequivocal. If indirect correlation is made on the basis of a single unique set of physical or biological attributes which are deemed to be both necessary and sufficient to establish equivalence, the correlation is termed **monothetic.** If demonstration of equivalence is determined statistically on the basis of the greatest number of shared characteristics, when no single characteristic is essential or adequate for correlations, it is called **polythetic.** Polythetic correlation commonly requires systematic measures involving statistical applications rather than simple visual comparisons. The differences between matching, formal correlation, and indirect correlations are illustrated in Figure 14.19.

Lithocorrelation

We turn now to the methods used for correlating strata on the basis of lithology. Methods of biocorrelation and chronocorrelation are discussed in appropriate sections of Chapters 15–18.

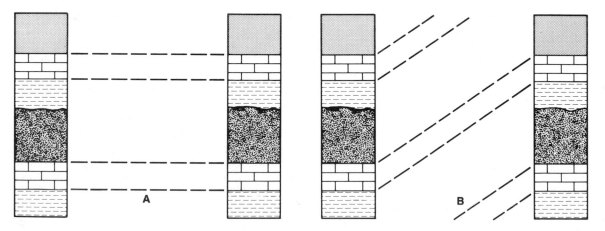

FIGURE 14.18 Illustration of the difference between matching and correlation. A. Matching of similar-appearing strata. B. Actual correlation. (Modified from Shaw, A. B., 1964, Time in stratigraphy. Fig. 30.1, p. 214, © 1964. Reprinted by permission of McGraw-Hill Book Company, New York.)

FIGURE 14.19 The relationship of formal (direct) correlation, indirect correlation, and matching. (From Shaw, B. R., The correlation of geologic sequences, *in* J. M. Cubitt and R. A. Reyment (eds.), Quantitative stratigraphic correlation. Fig. 2, p. 11, © 1982 by John Wiley & Sons, Ltd. Reprinted by permission of John Wiley & Sons, Ltd., Chichester, England.)

Correlation	Formal	Physical tracing of stratigraphic units		
	Indirect	Arbitrary	Systematic	
		Visual comparisons	Monothetic	Polythetic
			Numeric equivalence	Statistical equivalence
Matching		Comparisons of nonstratigraphic units		

Continuous Lateral Tracing of Lithostratigraphic Units

Direct, continuous tracing of a lithostratigraphic unit from one locality to another is the only correlation method that can establish the equivalence of such a unit without doubt. This correlation method can be applied only where strata are continuously or nearly continuously exposed. If outcrops of strata are interrupted by large areas of soil or vegetation cover, or if continuity of the strata has been broken by erosion, as across a large valley, or by faulting, physical tracing of beds may not be possible. Under these circumstances, other (indirect) correlation techniques must be used. The most straightforward way of tracing lithostratigraphic units laterally is by walking out the beds. A geologist who traces a stratigraphic unit continuously from one locality to another by walking along the top of a particular bed can be quite confident that correlation has been established. Thus, the application of field boots and a bit of physical effort yields the satisfaction of achieving a virtually unequivocal correlation. Another useful, but somewhat more equivocal, method of tracing stratigraphic units laterally is to follow the beds on aerial photographs. In areas where surface exposures are abundant and visibility is little hampered by soil or vegetation cover, lateral tracing of thick, distinctive stratigraphic units on aerial photographs can be done rapidly and effectively. This method is limited to tracing of distinctive beds that are thick enough to show up on photographs of a suitable scale (Fig. 14.20).

Although physical tracing of beds is the only unequivocal method of correlation, it is not without limitations. The most serious of these is the fact that in most areas where geologists do mapping, beds cannot be traced continuously for more than a very short distance before encountering covered areas, structural complications (faults), or erosional terminations. In fact, it is often impossible to trace a given stratigraphic unit more than a few hundred meters before the unit is lost for one of these reasons. An additional problem may arise if the beds being traced pinch out or merge with others laterally, a very common occurrence in nonmarine strata. In such a case, tracing of an individual bed or bedding plane will be impossible. Therefore, in practice, geologists commonly trace a gross lithostratigraphic unit (e.g., a member or a formation) consisting of beds of like character, rather than trying to trace individual beds.

Lithologic Similarity and Stratigraphic Position

Lithologic Similarity. As pointed out, direct lateral tracing of stratigraphic units cannot be successfully accomplished in a great many areas owing to discontinuous out-

FIGURE 14.20 Evenly bedded, laterally extensive Permo-Pennsylvanian formations exposed along the Colorado River at Dead Horse Point, near Moab, Utah.

crops. Geologists working on these areas must depend for correlation of lithostratigraphic units upon methods that involve matching strata from one area to another on the basis of lithologic similarity and stratigraphic position. Because matching of strata does not necessarily indicate correlation, correlation by lithologic similarity has varying degrees of reliability. The success of such correlation depends upon the distinctiveness of the lithologic attributes used for correlation, the nature of the stratigraphic sequence, and the presence or absence of lithologic changes from one area to another. Facies changes that take place in lithostratigraphic units between two areas under study obviously complicate the problem of lithologic correlation.

Lithologic similarity can be established on the basis of a variety of rock properties. These include gross lithology (sandstone, shale, or limestone, for example), color, heavy mineral assemblages or other distinctive mineral assemblages, primary sedimentary structures such as bedding and cross-lamination, and even thickness and weathering characteristics. The greater the number of properties that can be used to establish a match between strata the stronger the likelihood of a reliable match. A single property such as color or thickness may change laterally within a given stratigraphic unit, but a suite of distinctive lithologic properties is less likely to change. I caution again that matching of strata on the basis of lithology is not a guarantee that correlation has been established. Strata with very similar lithologic characteristics can form in similar depositional environments widely separated in time or space. It may be quite possible, for example, to obtain an excellent lithologic match between a clean, well-sorted, cross-bedded, eolian sandstone unit of Triassic age and a virtually lithologically identical sandstone of Jurassic age, yet these sandstones do not correlate either as lithostratigraphic or chronostratigraphic units. Correlation on the basis of lithologic identity is particularly difficult between cyclic sequences, such as the Pennsylvanian cyclothems of the United States midcontinent region. Very similar-appearing sequences of units can be repeated over and over in the stratigraphic section owing to the fact that similar environmental conditions can reappear in a region time after time during repeated transgressive-regressive cycles of deposition.

The most reliable lithologic correlations are made when it is possible to match not just one or two distinctive beds or rock types but a sequence of several distinctive units. For example, the Triassic and Jurassic formations of the Colorado Plateau in the western United States consist of a highly distinctive sequence of largely nonmarine red to green siltstone and mudstone units (the Moenkopi, Chinle, Kayenta, Summerville, Morrison formations) interstratified with red to white, cross-bedded eolian(?) sandstones (the Wingate, Navajo, Entrada formation). This sequence of formations is so distinctive that it can be recognized and correlated lithologically with considerable confidence over wide areas on the Colorado Plateau (Fig. 14.21).

FIGURE 14.21 Correlation of strata between two localities on the Colorado Plateau based on similar lithology of distinctive stratigraphic units. (From Mintz, I., W., 1981, Historical geology, The science of a dynamic Earth, 3d ed. Fig. 10.1, p. 241, reprinted by permission of Charles E. Merrill Publishing Co., Columbus OH.)

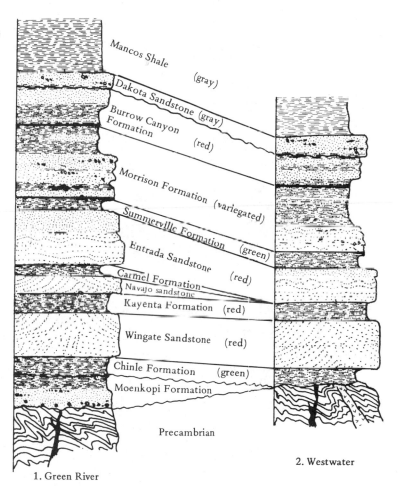

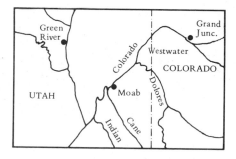

Stratigraphic Position in a Sequence. The preceding illustration points out the importance of position in a stratigraphic sequence when correlating units by lithologic identity. Several of the Colorado Plateau formations are lithologically similar, but because they occur in a sequence of strata distinctive enough to be correlated from one area to another, individual formations can be correlated also by their positions in this sequence. Another way in which position in a stratigraphic sequence is important has to do with establishing correlation of strata by relation to some highly distinctive and easily correlated unit or units. Such distinctive beds serve as control units for correlation of other strata above and below. For example, a thin, ash-fall unit or bentonite bed may be present and easily recognized throughout a particular region. If it is the only such bed in the stratigraphic sequence in the region, and thus cannot be confused with any other bed, it can serve as a **key bed,** or **marker bed,** to which other strata are related. Strata immediately above or below this control unit can be correlated with a reasonable degree of confidence with strata that are in a similar stratigraphic position with respect to the control unit in other areas. If two or more marker beds are present in a sequence, this gives even greater reliability to correlation of units that lie between the marker beds. Obviously, correlation becomes more equivocal with increasing stratigraphic separation above or below the control units.

Correlation by Instrumental Well Logs

To generations of petroleum geologists weaned on use of well logs for subsurface correlations and other applications in the petroleum industry, a discussion of instrumental well logs may seem like bringing coals to Newcastle. Readers who lack this firsthand experience with logs may find a brief explanation of well logs useful. Well logs are simply analog traces or curves sketched on paper charts. These traces record variations in such rock properties as electrical resistivity, transmissibility of sound waves, or adsorption and emission of nuclear radiation in the rocks surrounding a borehole. These variations are a reflection of changes in gross lithology, mineralogy, fluid content, porosity, and other characteristics of the subsurface formations. Thus, correlation by use of well logs is not based totally on lithology. Nonetheless, most of the rock properties measured by well logs are closely related to lithology.

Very briefly, well logs are obtained by the following procedure. After an exploratory well is drilled by a petroleum company, the well is logged before being completed as an oil or gas producer or abandoned as a dry hole. The logging procedure begins with lowering an instrument called a **sonde** to the bottom of the well bore. The sonde may be designed to measure the electrical resistivity of a rock unit, natural or induced gamma radiation emitted by the unit, velocity of sound waves passing through the rock, or other rock properties. As the sonde is slowly withdrawn from the borehole through a succession of stratigraphic units, it continuously measures the particular property of the rock that it is designed to analyze and electrically transmits this information to a chart recorder located in a logging truck at the surface.

One common type of well log is the **electric log,** which records resistivity of rock units as the sonde passes up the borehole in contact with the wall of the hole. Resistivity is affected by the lithology of the rock units and the amount and nature of pore fluids in the rock. For example, a marine shale that has its pore spaces filled with saline formation water will have a much lower electrical resistivity (higher conductivity) than a porous sandstone or limestone filled with oil or gas. With experience in a given geological province, petroleum geologists can recognize the particular signatures represented by the analog traces on the log and can relate these signatures to particular types of lithostratigraphic units or to a specific formation. Lithology cannot be read directly from such logs, but the characteristics of the log traces are a reflection of lithology (and fluid content). Other types of logging sondes may measure the natural gamma radiation in rock units, yielding **gamma ray logs,** or the velocity with which a sound signal passes through rock units. Logs that measure sound velocity are called **sonic logs.** In addition to their usefulness in correlation, sonic logs can be used to determine the porosity of subsurface formations, owing to the fact that sound waves are slowed in their passage through rocks by the presence of fluid-filled pores.

Other types of well logs are in general use, but all have the common characteristics that they consist of electrically produced signatures, or traces, that represent some particular property of subsurface lithostratigraphic units that is related in some way to lithology, fluid content, bed thickness, or other property. Most logs actually consist of two different types of traces arranged on either side of a central column that represents the well bore. This central column is calibrated in feet or meters to show depth below the surface. Figure 14.22 illustrates a section of a well log showing a sonic curve opposite a gamma ray curve. The curve shapes generated by a particular lithostratigraphic unit are not unique, but a trained, experienced well-log analyst can learn to recognize the signature of a particular formation or sequence of formations and can match up the signatures in logs from one area to those from nearby wells.

Characteristically, the well-log curves of adjacent wells are very similar, but the degree of similarity decreases in more distant wells. By working with a series of closely spaced wells, however, a geologist can carry a correlation across an entire sedimentary basin, even when pinchouts or facies changes occur. In fact, one of the reasons petroleum geologists find correlation of well logs so useful in petroleum exploration is that correlation permits recognition of pinchouts and facies changes that may be potential traps for oil and gas. Figure 14.23 is an example of correlation by electric logs across a portion of a basin. Geologists often add lithologic information obtained from drill cores or cuttings to the well logs; however, I stress again that correlation by well logs is not necessarily correlation based entirely on lithologic identity because the shapes of the curves can represent a variety of rock properties, including porosity and fluid content. Correlation by well logs is actually based more on the position of each unit in a succession of units represented on the logs rather than on the character of any individual unit reflected in the curves. Correlation by well logs is thus the approximate subsurface equivalent of correlation of surface sections by position in the sequence.

Correlation by instrumental well logs can be a laborious process where large numbers of logs are involved and is subject also to considerable subjectivity owing to the similarity of the log curves or traces in different parts of a logged stratigraphic section. Differences between stratigraphic units may be manifested only by very subtle differences in the analog traces and can be difficult to discern visually. The availability of computers and sophisticated statistical techniques now makes it possible to apply automated approaches to stratigraphic correlation of well logs, removing some of the subjectivity in correlation. These approaches involve digitizing the analog traces, or use of digital tapes, to segment the logs for use in computational systems. These systems then provide a statistical match for correlation purposes. Details of automated well-log correlation are given by B. R. Shaw and Cubitt (1978) and Griffiths (1982).

FIGURE 14.22 A section of a sonic–gamma ray log. The gamma ray curve is shown on the left. The sonic log (interval transit time log) is on the right. Depths (in feet) below the surface are shown in the central column. Four distinct correlatable units are indicated.

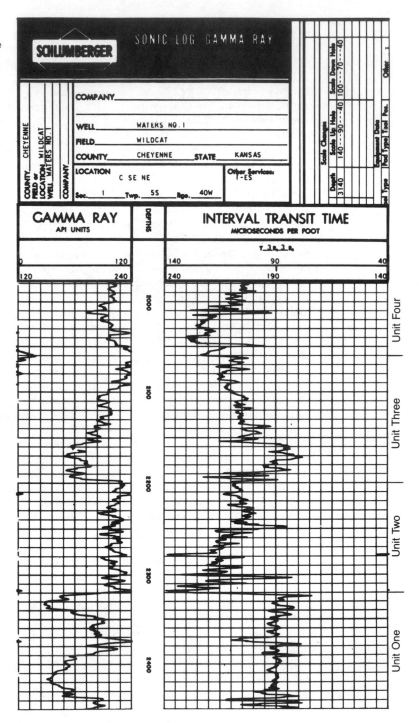

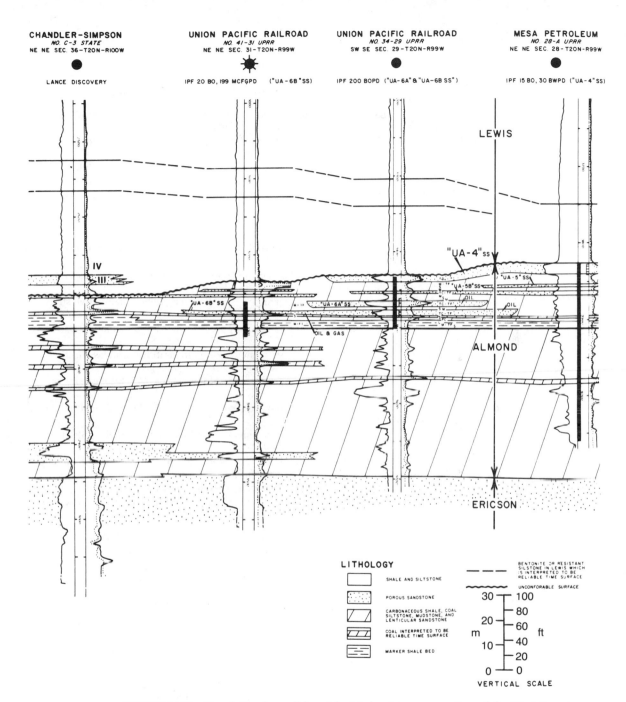

FIGURE 14.23 Correlation by well logs (electric logs) across a portion of the West Desert Springs oil field in Wyoming. Lithologic information has been added to the well logs, but correlation is based on the "signatures" of the well-log curves. (After Weimer, R. J., J. D. Howard, and D. R. Lindsay, 1982, Tidal flats and associated tidal channels, *in* P. A. Scholle and D. Spearing (eds.), Sandstone depositional environments: Am. Assoc. Petroleum Geologists Mem. 31. Fig. 44, pp. 224, 225, reprinted by permission of AAPG, Tulsa, Okla.)

ADDITIONAL READINGS

Childs, O. E., 1983, Correlation of stratigraphic units of North America, COSUNA, 1977–83: Am. Assoc. Petroleum Geologists and U.S. Geol. Survey, 49 p.

Conklin, B. M., and J. E. Conklin, 1984, Stratigraphy: Foundations and concepts: Benchmark Papers in Geology, v. 82, Van Nostrand Reinhold, New York, 365 p.

Cubitt, J. M., and R. A. Reyment (eds.), 1982, Quantitative stratigraphic correlation: John Wiley & Sons, New York, 301 p.

Donovan, D. T., 1966, Stratigraphy—An introduction to principles: John Wiley & Sons, New York, 199 p.

Dunbar, C. O., and J. Rodgers, 1957, Principles of stratigraphy: John Wiley & Sons, New York, 356 p.

Einsele, G., and A. Seilacher, 1982, Cyclic and event stratification: Springer-Verlag, Berlin, 536 p.

Gill, D., and D. F. Merriam (eds.), 1979, Geomathematical and petrophysical studies in sedimentology: Pergamon, New York, 267 p.

Gradstein, F. M., F. P. Atterberg, J. C. Brower, and W. J. Swarzacher, 1985, Quantitative stratigraphy: D. Reidel, Dordrecht, Holland.

Hallam, A., 1981, Facies interpretation and the stratigraphic record: W. H. Freeman, San Francisco, 287 p.

Hedberg, H. D., 1976, International stratigraphic guide: A guide to stratification classification, terminology, and procedure: John Wiley & Sons, New York, 200 p.

Krumbein, W. C., and L. L. Sloss, 1963, Stratigraphy and sedimentation: W. H. Freeman, San Francisco, 660 p.

Merriam, D. F. (ed.), 1981, Computer applications in the earth sciences—An update of the 70s: Plenum, New York, 385 p.

North American Commission on Stratigraphic Nomenclature, 1983, North American Stratigraphic Code: Am. Assoc. Petroleum Geologists Bull., v. 67, p. 841–875.

Schwarzacher, W. J., 1975, Sedimentation models and quantitative stratigraphy: Developments in Sedimentology 19, Elsevier, Amsterdam, 382 p.

Visher, G. S., 1984, Exploration stratigraphy: Pennwell, Tulsa, Okla., 334 p.

Weller, J. M., 1960, Stratigraphic principles and practice: Harper and Brothers, New York, 725 p.

15
Magnetostratigraphy

15.1 INTRODUCTION

Magnetic stratigraphy, or **magnetostratigraphy,** is a new branch of stratigraphy developed largely since about the middle 1960s. The principles of magnetic stratigraphy were initially applied to study of volcanic rocks and sediments younger than about 5 million years. Magnetic stratigraphic techniques have now been extended to much older rocks, and a magnetic polarity time scale has been developed for rocks as old as the Jurassic. Magnetic stratigraphy came about through the discovery that magnetic iron oxide minerals in igneous and sedimentary rocks can preserve the orientation or field direction of the Earth's magnetic field at the time the rocks were formed. During the cooling of molten rock, iron oxide minerals become magnetized in alignment with the Earth's magnetic field as they cool through a critical temperature of about 500–600°C—the **Curie point.** As they approach this temperature, the influence of the magnetic field begins to exert itself and small atomic groups within the minerals begin to line up parallel to one another and to the direction of the magnetic lines of force around Earth. With further cooling these atoms become locked into this orientation and each mineral in essence becomes a small magnet, with polarity parallel to the Earth's magnetic field. During deposition of sediments, small magnetic minerals are able to rotate in the loose, unconsolidated sediment on the depositional surface and thus align themselves mechanically with the Earth's magnetic field. This preferred orientation of magnetic minerals in igneous and sedimentary rocks imparts bulk magnetic properties to the rocks. These properties are retained for geologically long periods of time unless the rocks are again heated above the Curie point. Therefore, this residual magnetism is called **remanent magnetism.** Because sediment grains can be disturbed by bioturbating organisms or by physical and chemical processes during burial and diagenesis, the magnetization of sedimentary rocks is less stable, as well as being weaker, than that of volcanic lavas. The study of remanent magnetism in rocks

559

of various ages to determine the intensity and direction of Earth's magnetic field in the geologic past is called **paleomagnetism.**

Remanent magnetism is measured by instruments called magnetometers. Early magnetometers were capable of making paleomagnetic measurements only in igneous rocks and highly magnetized iron-bearing red sediments. Modern superconducting magnetometers with a sensitivity at least 100 times that of previous types can measure the magnetism in much more weakly magnetic sediments, including carbonates. Remanent magnetism is complex and can include secondary magnetism caused by prolonged effects of the Earth's present magnetic field or by chemical changes owing to alteration of one magnetic mineral to another. Demagnetization techniques are available for destroying this secondary magnetic effect in the laboratory so that primary magnetization can be measured. It is this primary magnetic component, recording the Earth's geomagnetic field at the time volcanic or sedimentary rocks formed, that is of interest in stratigraphic studies.

The significance of primary remanent magnetism for stratigraphic studies stems from the fact that the Earth's magnetic field has not remained constant throughout geologic history but has experienced frequent reversals. The geomagnetic field is generated by the motion of highly conducting nickel–iron fluids in the outer part of the Earth's core; this motion is controlled by thermal convection and by the Coriolis force generated by the Earth's rotation. Studies of the remanent magnetism in igneous and sedimentary rocks show that the dipole (main) component of the Earth's magnetic field has reversed its polarity at irregular intervals since Precambrian time, apparently owing to instabilities in the outer core. When the Earth's magnetic field has the present orientation, it is said to be normal. When this orientation changes 180 degrees, it is referred to as reversed. Figure 15.1 illustrates diagrammatically the magnetic lines of force around Earth during normal and reversed polarity epochs and shows what the orientation of a compass needle would be at points in the Northern and Southern hemispheres at such times. Reversals of the Earth's magnetic field are recorded in sediments and igneous rocks by patterns of normal and reversed remanent magnetism. The direction of magnetization of a rock is defined by its **north-seeking magnetization.** If the north-seeking magnetization points toward the Earth's present magnetic north pole, the rock is said to have **normal magnetization.** If the north-seeking magnetization points toward the present-day south magnetic pole, the rock has **reversed magnetization,** or reversed polarity. Thus, sedimentary and igneous rocks that display bulk magnetic properties of the same magnetic polarity as the present magnetic field of Earth have **normal polarity,** whereas those that have the opposite magnetic orientation have **reverse polarity.**

These geomagnetic reversals are contemporaneous worldwide phenomena. Thus, they provide unique stratigraphic markers in igneous and sedimentary rocks. The process of reversal is thought to take place over a period of 1000–2000 years, during which a decrease in intensity of the magnetic field by 60–80 percent occurs over a period of about 10,000 years preceding reversal, followed by a buildup of intensity for the next 10,000 years (Cox, 1969). The last unquestioned reversal of the magnetic field took place approximately 700,000 years ago, although a reversal may have occurred about 20,000 years ago. Intervals of reversed or normal polarity lasting 100,000 years or more are called "epochs" and those having a duration of 10,000–100,000 years are called "events." Magnetic stratigraphy in pre-Pleistocene rocks is based on these changes in polarity recorded in sediments or volcanic rocks. Recognizable patterns of alternating polarity stratigraphic units can be used for chronological and correlation purposes.

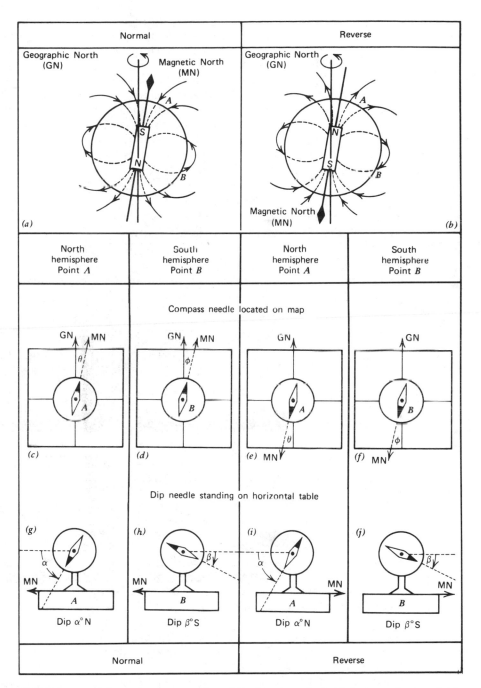

FIGURE 15.1 Schematic representation of Earth's magnetic field during episodes of (a) normal and (b) reversed polarity. The diagram also shows behavior during these episodes of a horizontal compass needle (c–f) and a vertical compass needle (g–j) at two different points on Earth. (From Wyllie, P. J., The way the Earth works, Fig. 9.1, p. 120, © 1976, John Wiley & Sons, Inc. Reprinted by permission of John Wiley & Sons, Inc., New York.)

The principle of developing a polarity time scale is illustrated in Figure 15.2. This figure shows three lava flows. The oldest erupted about 1.9 million years ago when Earth's magnetic field was normal. Thus, when it cooled, it acquired normal magnetic polarity (Fig. 15.2A). The second erupted about 1.5 million years ago during an episode of reversed magnetic polarity (Fig. 15.2B), and the youngest erupted 0.5 million years ago after the magnetic field had reversed back to normal polarity (Fig. 15.2C). Although the lavas have subsequently undergone weathering and erosion, each has retained its original magnetic polarity (Fig. 15.2D). By measuring the remanent magnetization in each lava flow and then determining its age by radiometric dating methods, a polarity time scale for these lavas can be constructed and is shown at the left of the figure.

15.2 DEVELOPMENT OF THE MAGNETIC POLARITY TIME SCALE

The concept of remanent magnetism is well known to students of geology today; however, only a few studies of rock magnetism had been made prior to the 1960s. The

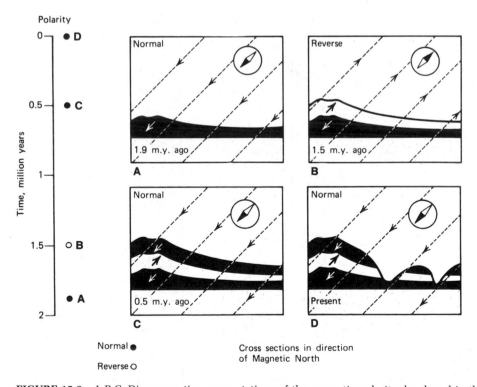

FIGURE 15.2 A,B,C. Diagrammatic representations of the magnetic polarity developed in three lava flows that erupted during the last 2 million years. Each of these lavas became magnetized at the time of eruption, with the direction of magnetization oriented parallel to the magnetic lines of force (dashed arrows) around Earth. D. Present time, showing retention of magnetic polarity. By measuring the radiometric ages of the lavas, a polarity time scale (left) can be constructed. (After Wyllie, P. J., The way the Earth works, Fig. 9.2, p. 122, © 1976, John Wiley & Sons, Inc. Reprinted by permission of John Wiley & Sons, Inc., New York.)

basic principles of magnetostratigraphy were developed in the early and middle 1960s in the remarkably short time of about five years by two groups of scientists working independently and competitively—one group in northern California and one in Australia. The initial development of a magnetic polarity sequence by these groups of scientists is summarized by Cox (1973a), Glen (1982), McDougall (1977), and Watkins (1972).

The use of magnetic polarity reversals as a stratigraphic tool is based on identification of characteristic patterns of reversals. Each reversal is known as a **magnetic**

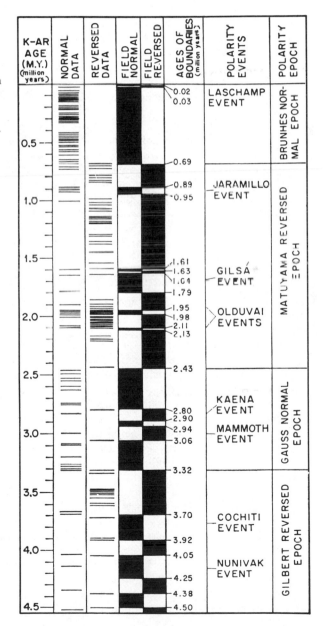

FIGURE 15.3 The geomagnetic time scale for the last 4.5 million years. Each short horizontal line represents the magnetic polarity and potassium–argon date of one volcanic cooling unit. The duration of events is based in part on paleomagnetic data from deep-sea sediments. (From Cox, A., 1969, Geomagnetic reversals: Science, v. 163. Fig. 4, p. 240, reprinted by permission of American Association for the Advancement of Science, Washington, D.C.)

anomaly. Magnetic anomalies are significant deviations from Earth's magnetic background on either a local or regional scale. If the absolute age of each anomaly can be established by independent means, a quantitative time scale for reversals can be set up. The first such polarity scales were achieved by measuring the ages and magnetic polarities of young volcanic rocks on land using potassium–argon techniques to estimate the ages of the rocks (Cox, 1969). These polarity scales were developed only for the last 4.5 million years (Fig. 15.3) because extension of the time scale was limited by lack of resolution of the potassium–argon dating method. For ages greater than about 5 m.y., the typical value of ±2% for the precision of a K–Ar date is equivalent to ±0.1 m.y., which is longer than the duration of many of the shorter polarity intervals (Channell, 1982). The polarity time scale was subsequently extended to about 7 million years by use of stratigraphically related Icelandic lavas (McDougall et al., 1977). Note from Figure 15.3 that the polarity time scale is subdivided into polarity "epochs," each named for a distinguished scientist who contributed to development of the field of geomagnetism, and shorter "events," named for localities where definitive study of the paleomagnetic characteristics of specific groups of rocks has been carried out.

In addition to study of the polarity of volcanic rocks on land, a second very important source of information about magnetic reversal sequences is provided by the linear anomaly patterns discovered in volcanic rocks of the ocean floor over mid-ocean ridges and first reported by Vine and Matthews (1963). These linear "stripes" of normal- and reversed-polarity magnetic rocks (Fig. 15.4) are roughly parallel to ridge crests and are typically 5–50 km wide and hundreds of kilometers long. They were produced by reversals in the Earth's magnetic field as successive flows of lava erupted along ridge crests and cooled below the Curie point. Previously magnetized volcanic rock was pushed or pulled aside from the ridges as new volcanic rock formed and became magnetized. Vine and Matthews discovered that the linear magnetic anomaly patterns on the ocean floor correlate with normal and reversed polarity intervals in the geomagnetic scale established on land, allowing age of the anomalies to be estimated. The fact that the magnetic anomalies are roughly symmetrical about spreading ridges was a critically important piece of evidence for developing the concept of seafloor spreading.

The discovery of these linear magnetic anomalies on the seafloor provided the necessary tool for extending and calibrating the magnetic polarity time scale developed on land. Geophysicists assign numbers to particularly characteristic magnetic anomalies, beginning with number 1 at ridge axes, as illustrated in Figure 15.5. These ocean-floor magnetic anomalies do not in themselves determine an independent reversal time scale because the ages of oceanic volcanic rocks are not usually known (Cox, 1969); however, once the anomalies are calibrated against known points on the radiometrically dated polarity scale, they provide a nearly continuous record of magnetic polarity intervals. The seafloor magnetic anomaly record is particularly valuable because it is continuous, unlike the on-land record, and may include polarity events that exist within gaps in the on-land radiometric data. The principal problem with the oceanic record is that it is very difficult to date directly. Paleontologic ages on the oldest sediments overlying anomalies are available where basement has been reached by DSDP drill holes, but large uncertainties are often associated with these age determinations.

A polarity time scale for Mesozoic and Cenozoic oceanic events has been constructed by extrapolating ages on the basis of rates of seafloor spreading. Heirtzler et al. (1968) assumed that magnetic anomaly profiles above the ocean ridges and basins

FIGURE 15.4 Linear magnetic anomaly patterns in the northeastern Pacific. Positive anomalies (normal polarity) are black. (After Mason, R. G., and A. D. Raff, 1961, Magnetic survey of the west coast of North America, 32°N. latitude to 42°N. latitude: Geol. Soc. America Bull., v. 72, Fig. 1, p. 1260; and A. D. Raff and R. G. Mason, 1961, Magnetic survey of the west coast of North America, 40°N. latitude to 52°N. latitude: Geol. Soc. America Bull., v. 72, Fig. 1, p. 1268; as modified by K. C. Conde, 1982, Plate tectonics and crustal evolution, 2nd ed., Fig. 4.22, p. 65, reprinted by permission of Pergamon Press, New York.)

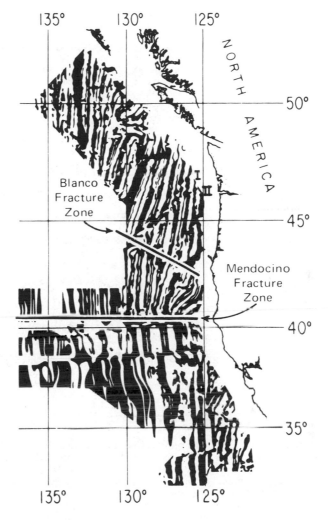

were manifestations of earlier reversals in the polarity of the Earth's magnetic field. By further assuming a constant rate (1.9 cm/yr) of seafloor spreading since Late Cretaceous time in the South Atlantic, ages were assigned to the magnetic reversal time scale by extrapolation from a date of 3.35 m.y. for the older reversal boundary correlated to the Gilbert–Gauss magnetic polarity epoch (Figs. 15.3, 15.5). The assumption of nearly constant spreading rate has proven to be surprisingly good, although subsequent work has shown that the calculated spreading rate may be about 7 percent too low (LaBrecque et al., 1977).

A third method of obtaining information on magnetic polarity is based on study of the record of polarity events in land sections of sedimentary rocks and in oceanic cores. Bulk remanent magnetism is produced in sediments by mechanical alignment of magnetic iron oxide minerals during slow settling through water or shortly after deposition while the sediment is still highly water saturated. Study of polarity reversals in sediments has been hampered by several factors, including gaps in the strati-

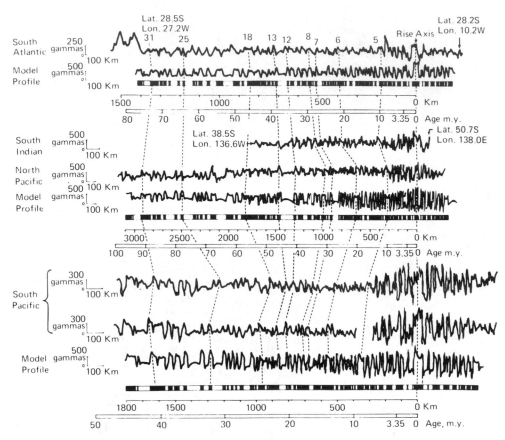

FIGURE 15.5 Magnetic profiles from the Atlantic, Indian, and Pacific ocean basins. Model profiles are also given from the South Atlantic, South Pacific, and North Pacific based on the normal (black) and reversely (white) magnetized bands beneath each model profile. Suggested correlations of anomalies are shown by dashed lines. Each time scale was constructed assuming an age of 3.35 million years for the end of the Gauss epoch. (From Heirtzler, J. R., G. O. Dickson, E. M. Herron, W. C. Pitman, III, and X. Le Pichon, Marine magnetic anomalies, geomagnetic field reversals, and motions of the ocean floor and continents: Jour. Geophysical Research, v. 73. Fig. 1, p. 2120, © 1968, American Geophysical Union, Washington, D.C.)

graphic record; variable rates of deposition; and chemical alteration (authigenesis) of magnetic iron oxide minerals, causing secondary magnetism. Also, many sediments or sedimentary rocks are weakly magnetized; early magnetometers were not sensitive enough to measure their magnetic polarity. The development of modern superconducting magnetometers and improvements in laboratory techniques for removal of unstable components of magnetic overprinting, called secondary magnetism, have alleviated some of these problems and allowed extension of paleomagnetic studies to many types of sedimentary rocks.

The major advantage of paleomagnetic studies of sediments and sedimentary rocks is that even though gaps exist in the sedimentary record, sedimentary sequences are stratigraphically far more continuous than most volcanic sequences. Furthermore, ages of magnetic anomalies can be estimated on the basis of associated fossils. Conventional piston-coring techniques in the oceans have provided paleontologic control on reversal sequences as far back as the early Miocene (McElhinney, 1978). Cores recovered during DSDP drilling have also provided some useful ages, although their usefulness is limited by incomplete recovery and physical distortion of cores and by disturbance of magnetism caused by drilling. Although, owing to gaps in the stratigraphic record, many land-based sections of sedimentary rock are inappropriate for detailed magnetic polarity studies, some land sections are now providing good paleontologic calibration of the polarity time scale for the early Tertiary and Mesozoic. An essentially complete section of Middle Cretaceous to Paleocene calcareous, pelagic sediments exposed at Gubbio in the Umbrian Apennines of Italy has yielded well-defined reversal stratigraphy that can be tied to detailed foraminiferal biostratigraphy (Arthur and Fischer, 1977; Lowrie and Alvarez, 1977; Alvarez et al., 1977).

Land section magnetostratigraphy can be correlated with reversal stratigraphy derived from oceanic anomalies, thereby making possible paleontological dating of the oceanic anomalies. This procedure allows the establishment of dated calibration points in the oceanic geomagnetic reversal time scale. Interpolation of ages of anomalies between these points is made by extrapolation, assuming constant rates of seafloor spreading. As many as eleven calibration points have now been established for the Late Cretaceous and Cenozoic (Lowrie and Alvarez, 1981); these data have allowed extension and dating of the magnetic polarity time scale beyond 100 million years (Fig. 15.6). Note from this figure the extended period of normal polarity that occurred during the Cretaceous. Such periods are commonly referred to as "quiet zones" in contrast to "disturbed intervals," which are characterized by frequent reversals. Extension of the oceanic polarity time scale beyond the Late Cretaceous is still somewhat tenuous, although Larson and Hilde (1975) have developed a time scale of magnetic reversals for the Early Cretaceous and Jurassic based on northwest Pacific Ocean reversal patterns that were dated by paleontologic ages in DSDP drill holes that reached volcanic basement. This scale is incorporated into Figure 15.6.

A detailed magnetic polarity time scale for rocks older than the Jurassic has not yet been established because the oceanic geomagnetic time scale cannot be extrapolated beyond about 200 m.y., the age of the oldest oceanic crust; however, magnetic reversals are known to occur in land sections in rocks at least as old as 1.5 billion years (Conde, 1982). Study of reversals in older rocks shows that the percentage of normal and reversed magnetization for any increment of geologic time has varied with time. In general, the Cretaceous, Jurassic, and late Ordovician–Silurian are characterized by normal polarities and the early and late Paleozoic by reversed polarities. The distribution of magnetic polarities during the Phanerozoic is illustrated in Figure 15.7, which shows overlapping 50-m.y. averages of polarity ratios from land-based paleomagnetic data. A summary of the overall pattern of Phanerozoic magnetic anomalies has emerged from paleomagnetic studies on land and from oceanic magnetic anomalies and is given in Figure 15.8. This figure shows the polarity bias depicted in Figure 15.7, polarity zones for the Mesozoic and Cenozoic, and epoch events for the last 5 million years of geologic time.

15.3   NOMENCLATURE AND CLASSIFICATION OF MAGNETOSTRATIGRAPHIC UNITS

Magnetostratigraphy, as we see from the preceding discussion, is the study of remanent magnetism in sedimentary and volcanic rocks. It is that element of stratigraphy that deals with the magnetic characteristics of rocks. In a broad sense, magnetostratig-

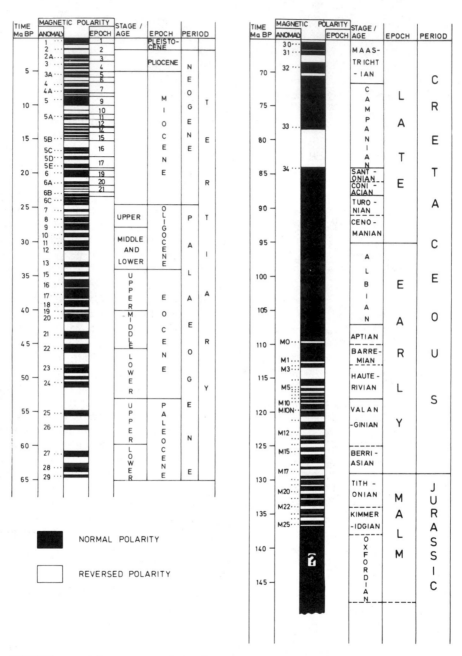

FIGURE 15.6   Polarity time scale for the Mesozoic–Cenozoic. The magnetic polarity sequence is mainly from oceanic magnetic anomaly data, with correlation to stage boundaries mainly from paleontologically dated land sections. (From Channell, J. E. T., Palaeomagnetic stratigraphy as a correlation technique, in G. S. Odin (ed.), Numerical dating in stratigraphy, Fig. 9, p. 97, © 1982, John Wiley & Sons, Ltd. Reprinted by permission of John Wiley & Sons, Ltd., Chichester, England.)

FIGURE 15.7 Overlapping 50-million year averages of polarity ratios from land-based paleomagnetic data, through the Phanerozoic. Values with normal polarity bias lie above the horizontal line. Those with reverse polarity bias lie below. (From Irving, E., and G. Pullaiah, 1976, Reversals in the geomagnetic field, magnetostratigraphy and relative magnitude of paleosecular variations in the Phanerozoic: Earth Science Reviews, v. 12. Fig. 13, p. 55, reprinted by permission of Elsevier Science Publishers, Amsterdam.)

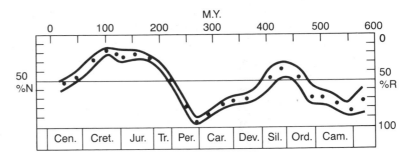

raphy encompasses all aspects of the study of remanent magnetism. Therefore, it includes study of magnetic susceptibility (the ratio of induced magnetism to the strength of the field causing magnetism), dipole-field position (variations in Earth's magnetic field intensity and pole positions, from which is derived polar wandering), nondipole components (secular or short-term variations on the order of 5,000–10,000 years in the direction and intensity of Earth's magnetic field), and magnetic inclination (the angle at which magnetic field lines dip), as well as study of magnetic field reversals. Field reversals are of special interest, however, because reversals in magnetic polarity have proven to be the most useful property for systematically subdividing strata on the basis of their remanent-magnetic properties.

Although units of rock having uniform magnetic properties, such as a uniform direction of magnetic polarity, are not necessarily coincident with lithostratigraphic units, remanent magnetism is a physical attribute of rocks and may be used to characterize a body of rock and separate it stratigraphically from other bodies with different magnetic properties. Early workers in the field of magnetic stratigraphy set up a magnetic polarity scale based on informal "epochs" lasting more than about 100,000 years and "events" lasting between 10,000 and 100,000 years. Serious efforts have been under way since the early 1970s to develop a formal system of classification and nomenclature for stratigraphic bodies defined on the basis of magnetic properties. These efforts have been spearheaded by the International Subcommission on Stratigraphic Classification (ISSC) of the International Union of Geological Sciences (IUGS) and the American Commission on Stratigraphic Nomenclature, which is now the North American Commission on Stratigraphic Nomenclature. The work of these two organizations has culminated in separate publications, each of which sets forth proposed terminology and procedures for establishing stratigraphic units based on magnetic polarity.

The IUGS International Subcommission on Stratigraphic Nomenclature and IUGS/IAGA Subcommission on a Magnetic Polarity Time Scale (1979) published a supplementary chapter, "Magnetostratigraphy Polarity Units," to the ISSC International Stratigraphic Guide. This supplement presents the recommendations of the International Subcommission on Stratigraphic Classification for naming and defining formal stratigraphic units based on magnetic properties. Rock units characterized by any type of magnetic property (magnetic susceptibility, magnetic field intensity, direction of natural remanent magnetism) are classified under the general heading of **magneto-**

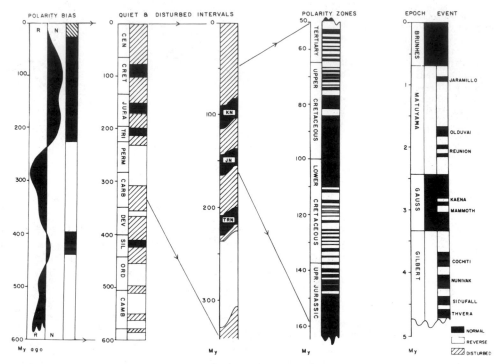

FIGURE 15.8   Patterns of geomagnetic fields through the Phanerozoic. Left to right: Polarity bias, quiet and disturbed intervals, Cretaceous and Late Jurassic magnetic anomaly sequence, polarity time scale for the last 5 million years. (From McElhinny, M. W., 1978, Magnetic polarity time scale: Prospects and possibilities in magnetostratigraphy, *in* G. V. Cohee, M. F. Glaessner, and H. D. Hedberg (eds.), The geologic time scale: Am. Assoc. Petroleum Geologists Studies in Geology 6. Fig. 6, p. 63, reprinted by permission of AAPG, Tulsa, Okla.)

stratigraphic units, or magnetozones, which are "bodies of rock strata unified by similar magnetic characteristics which allow them to be differentiated from adjacent strata" (p. 579). Rock units defined specifically on the basis of their magnetic polarity are called **magnetostratigraphic polarity units,** which are defined as "bodies of rock in original sequences, unified by their magnetic polarity which allows them to be differentiated from adjacent strata" (p. 579).

  Because magnetostratigraphic polarity units are of primary interest in stratigraphy, these units are formally subdivided into polarity superzones, polarity zones, and polarity subzones (Table 15.1). The **polarity zone** is the fundamental polarity unit. A polarity zone may consist of strata with a single direction of polarization throughout, be composed of an intricate alternation of normal and reversed units, or be predominantly either normal or reversed but with minor subdivisions of the opposite polarity. A **polarity superzone** consists of two or more polarity zones, and a **polarity subzone** is a subdivision of a polarity zone. The principal polarity zone names now in use are well-established names like Brunhes, Matuyama, Gauss, and Gilbert, which are used for the last 5 million years of Earth's history. Historically, these units have been called epochs (Fig. 15.3); however, it is now recommended that these "epochs" be called the Brunhes, Matuyama, Gauss, and Gilbert polarity zones. Similarly, the so-called "events" such as the Jaramillo, Gilsá, and Olduvai should now be referred to as the Jaramillo, Gilsá, and Olduvai polarity subzones.

TABLE 15.1 Nomenclature of magnetostratigraphic units

Magnetostratigraphic units	Description	Example	
		New name	Old name
Polarity superzone	Magnetostratigraphic unit composed of two or more polarity zones		
Polarity zone	Magnetostratigraphic unit distinguished by a single direction of magnetic polarization, or by a distinctive alternation of normal and reversed polarities	Brunhes Normal Polarity Zone	Brunhes Normal Epoch
Polarity subzone	Subdivision of a polarity zone	Jaramillo Polarity Subzone Gilsá Polarity Subzone Olduvai Polarity Subzone	Jaramillo Event Gilsá Event Olduvai Event

In the 1983 North American Stratigraphic Code, the North American Commission on Stratigraphic Nomenclature follows approximately the same scheme of nomenclature for remanent-magnetic stratigraphic units as that proposed by IUGS, and defines magnetostratigraphic units and magnetopolarity units in roughly the same way (Appendix). With regard to magnetopolarity units, the 1983 Code further states that the upper and lower limits of a magnetostratigraphic unit are defined by boundaries marking a change in polarity and that such boundaries may represent either a depositional discontinuity **(polarity-reversal horizon)** or a magnetic field transition **(polarity transition-zone).** According to the Code, a polarity-reversal horizon is either a single, clearly definable surface or a thin body of strata constituting a transitional interval across which a change in magnetic polarity is recorded. Polarity-reversal horizons describe transitional intervals of 1 m or less. If the change in polarity takes place over a stratigraphic interval greater than 1 m, the term "polarity transition-zone" should be used. Polarity-reversal horizons and polarity transition-zones provide the boundaries for polarity zones, although they may also be contained within a polarity zone, where they mark an internal change subsidiary in rank to those at its boundaries.

The North American Commission on Stratigraphic Nomenclature also follows the IUGS in considering the polarity zone as the fundamental unit of magnetostratigraphic classification and defines a polarity zone as "a unit of rock characterized by the polarity of its magnetic signature" (Appendix, Article 46). Polarity zones may be grouped into polarity superzones or subdivided into polarity subzones.

Names, numbers, and letters are currently in use for designating magnestostratigraphic units. Both the IUGS and the North American Commission on Stratigraphic Nomenclature recommend that magnetostratigraphic units be given formally designated names, consisting preferably of a geographic name followed by the term polarity zone, polarity subzone, or polarity superzone, as appropriate. The name may be modified by including the words normal, reversed, or mixed, as in Deer Park Reversed Polarity Zone.

The IUGS guide and the North American Commission on Stratigraphic Nomenclature both recommend that a stratotype, or type section, be established for magnetostratigraphic units and that boundaries be defined in terms of recognizable lithostratigraphic or biostratigraphic units, in much the same manner as stratotypes are established for lithostratigraphic units. A special problem arises with selecting stratotypes for magnetopolarity units, however, because the best sequential record of reversals of the Earth's magnetic field available for the past 150 m.y. is that preserved in the linear seafloor-spreading anomalies described in Section 15.2. These anomalies have been dated by extrapolation and interpolation from radiometric and paleontologic evidence in on-land sections. Because they are deduced from remotely obtained shipboard magnetometer records rather than by magnetic polarity determinations on outcropping or cored volcanic or sedimentary rocks, it is not possible to designate any satisfactory type intervals or type boundaries for them, as required of conventional stratotypes. Instead, the standard reference for these ocean-floor magnetic anomalies is magnetometer profiles such as those shown in Figure 15.5.

15.4 APPLICATIONS OF MAGNETOSTRATIGRAPHY

Correlation

The primary application of magnetostratigraphy lies in its use as a tool for global correlation of marine strata. Magnetostratigraphic correlation is particularly important where paleontologic or lithologic correlation is difficult. It has special significance for international correlation because geomagnetic reversals are contemporaneous, synchronous worldwide phenomena. They have worldwide scope owing to the fact that reversals of the Earth's magnetic field affect the magnetic field everywhere on Earth at the same time. Because the polarity time scale can be calibrated radiometrically or paleontologically, polarity events thus provide a precise tool for chronostratigraphic correlation. The first significant application of magnetostratigraphic techniques to correlation and age determinations of rocks was correlation of linear ocean-floor magnetic anomalies to on-land sections of volcanic strata whose ages had been determined by radiometric methods. These correlation techniques were subsequently extended to cores of oceanic sediments.

Until very recently, correlation of sediment cores by use of magnetic polarity events had its greatest application in the study of marine sediments younger than about 6–7 million years. Correlation was previously restricted to very young rocks because the magnetic time scale had not been developed beyond about 7 m.y., and because most gravity and piston cores of ocean-floor sediment did not penetrate deep enough to sample older sediments. As mentioned, the magnetic time scale has subsequently been extended to about 170 m.y. Furthermore, deeper coring by use of hydraulic piston cores now makes it possible to obtain undisturbed cores of sediments as old as about Middle Miocene. Longer cores obtained during the Deep Sea Drilling Project by rotary coring methods have included rocks as old as the Jurassic; however, DSDP cores are commonly too badly disturbed to provide reliable paleomagnetic data. Because paleomagnetic methods have now been extended to correlation of on-land sections as old as Middle Cretaceous, this development opens up the possibility of even more extensive future use of paleomagnetic methods for correlating on-land stratigraphic sections. Magnetostratigraphy thus becomes a potentially important tool for international correlation of older, on-land strata as the magnetic polarity time scale is extended farther back into geologic time.

Figure 15.9 provides an easily visualized example of paleomagnetic correlation in cores of young oceanic sediment. Beginning with the Brunhes Normal Epoch at the

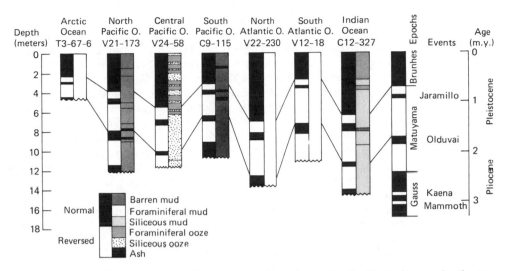

FIGURE 15.9 Paleomagnetic correlation of cores from the Arctic, Pacific, Indian, and Atlantic oceans. Cores have different lithologies and fossil assemblages. (From Opdyke, N., Paleomagnetism of deep-sea cores: Reviews Geophysics and Space Physics, v. 10. Fig. 20, p. 244, © 1972, American Geophysical Union, Washington, D.C.)

top of the cores, the correlation can be carried downward on the basis of the patterns of reversed and normal events. With longer cores and older sediment, correlation becomes more difficult because the magnetostratigraphic record consists of many sets of reversals (Figs. 15.3 and 15.6) that may look very much alike. Correlation of these reversal patterns may require independent radiometric or paleontologic age evidence to first establish stratigraphic position. Paleomagnetic events are particularly useful for correlating long distances across biogeographic boundaries where correlation by fossils, even planktonic fossils, may be difficult owing to the fact that different biogeographic provinces are marked by different fossil assemblages. Figure 15.9 illustrates that paleomagnetic correlation can be carried across the Arctic, Pacific, Indian, and Atlantic ocean basins, each of which is characterized by sediments composed of different lithologies with different fossil assemblages.

Geochronology

Although magnetostratigraphic sequence in itself does not normally provide unequivocal ages for geologic events preserved in strata, correlation of magnetic polarity zones or anomalies from areas where the ages of magnetic events have been established by radiometric methods or paleontologic data to areas where the ages of the strata are unknown provides a means of estimating the ages of events in the new areas. Magnetostratigraphic chronometry can also provide absolute ages for sediments that have been zoned by fossils and whose ages have been estimated from fossil data. Magnetostratigraphy is thus a useful, if somewhat limited, chronological tool. At the present time, the chronological framework is most accurate for the last 5–7 million years, but establishment of correlation points in older sequences of strata whose ages can be determined has extended the usefulness of geomagnetic chronometry to rocks as old as 100 million years or more.

The paleomagnetism of sediment cores can be used also as a tool for estimating the ages of volcanic eruptions that took place either on land or in the ocean. This is done by determining the ages of ash fallout from these eruptions where such fallout is preserved as ash beds in oceanic sediments. An example of the use of paleomagnetic stratigraphy for determining the ages of ash layers, a technique called **tephrochronology,** is shown in Figure 15.10. This figure demonstrates that ages of ash layers in piston cores in the Antarctic Ocean can be estimated on the basis of their paleomagnetic characteristics. A related application is in determination of rates of sedimentation for deep-sea sediments. Paleomagnetic correlation of deep-sea cores with rocks on land whose ages have been determined radiometrically allows absolute ages to be assigned to the boundaries between different magnetic events in the cores. The thickness of sediments between horizons within the cores whose ages are thus determined can then be used to calculate the sedimentation rate. For example, if we assume that 1,000 cm (10 m) of sediment were deposited in a given area of the ocean during the time represented in a core by the Matuyama Reversed Polarity Zone extending from 2,430,000 to 700,000 years B.P., a time interval of 1,730,000 years, the sedimentation rate for this area of the ocean can be calculated as

$$1,000 \text{ cm}/1,730,000 \text{ yrs} = 0.58 \text{ cm}/1,000 \text{ yrs, or } 5.8 \text{ m/m.y.}$$

Paleoclimatology

Ages of sediment cores determined by paleomagnetic methods have also been used to study paleoclimate oscillations during the Quaternary and late Pliocene. For example, the development of the paleomagnetic stratigraphy of deep-sea sediments provided a means of estimating the ages of ice-rafted debris in piston cores. It also furnished a method of studying the history of deposition of siliceous oozes as a reflection of increased biologic productivity owing to increased oceanic upwelling and resulting increase in nutrients during cooler periods. Quantitative determinations of variations in microfossil assemblages, particularly planktonic foraminiferal assemblages, have also been studied in relation to climatic cycles. Some microfossil species are much more abundant during cooling trends, whereas others are more abundant during warming periods. Use of paleomagnetic methods to estimate the ages of these climate-related biologic oscillations and variations in distribution of ice-rafted material in the oceans has added significantly to our knowledge of climatic fluctuations on land. It has also radically changed our ideas about the number of climatic cycles that occurred during the Quaternary. We now know, for example, that many more cycles of cooling and warming took place than the four major glacial advances and retreats postulated from land-based studies.

Definition of Stratigraphic Boundaries

It has been suggested that magnetostratigraphy can be used to help resolve the definition of boundaries between major systems of rock. For example, studies of the Precambrian–Cambrian boundary in some areas suggests that a quiet reversed zone is present just below the possible boundary, followed by a disturbed zone of mixed polarity (McElhinney, 1978). The top of this quiet zone may thus prove to be a definitive boundary for the Precambrian and Cambrian that can be used on a worldwide basis. A related idea is that major faunal changes, which have been used by geologists to define the boundaries of major stratigraphic subdivisions, may be linked to paleomag-

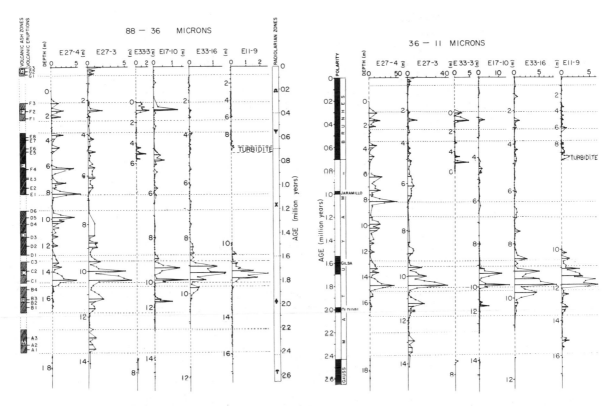

FIGURE 15.10 Use of magnetostratigraphy in tephrochronology of a suite of piston cores from the Antarctic Ocean. Periods of maxima on volcanic-glass accumulation rates are labelled A to G, with members 1 to 8 indicating possible single eruptions or closely spaced series of eruptions. Black = normal polarity; clear = reversed polarity. (From Huang, T. C., N. D. Watkins, and D. M. Shaw, 1975, Atmospherically transported volcanic glass in deep-sea sediments: Volcanism in subantarctic latitudes of the South Pacific during Pliocene and Pleistocene time: Geol. Soc. America Bull., v. 86, Fig. 3, p. 1307.)

netic reversals. Dramatic extinctions of organisms during Phanerozoic time occurred at the end of the Cambrian, Ordovician, Devonian, Permian, Triassic, and Cretaceous periods; less dramatic extinctions have occurred at other times. It has been suggested (Uffen, 1963) that significant evolutionary changes, involving extinctions or new appearances of species, should occur at every polarity reversal. Uffen postulatates that, during a reversal, the magnetic field intensity is much reduced for a relatively short time, and as a result cosmic radiation increases at the Earth's surface, leading to high mutation rates and rapid evolution. Alternatively, it has been suggested by other investigators that magnetic reversals may in some way cause changes in worldwide climates that might in turn produce rapid faunal changes.

Several detailed studies of the possible relationship of faunal extinctions and magnetic reversals have now been carried out. Some, but not all, of these studies have indeed demonstrated that extinctions have occurred within time intervals marked also by reversal of the Earth's magnetic field. Compelling evidence has not as yet been advanced to prove that paleoclimatic changes of sufficient magnitude at the time of reversal could have caused these extinctions or that the changes are due to excessive radiation at the time of polarity change. In fact, several later workers have now suggested that the radiation increase at the Earth's surface at the time of magnetic reversals is negligible and that an increase in mutation rate is unlikely to increase evolutionary change in the absence of other environmental pressures. Furthermore, the extinctions that have occurred within time intervals of magnetic reversals have not affected entire taxonomic groups, such as phyla, as might be expected. Instead, only certain orders of organisms within a given phyla were affected, suggesting that random extinctions or other factors are responsible for the observed faunal changes.

Polar Wandering and Suspect Terranes

Finally, in passing, I mention that one of the earliest uses of paleomagnetism was to test the idea of polar wandering. Rocks that became magnetized when they formed preserve a "fossil" record of the direction of the Earth's magnetic field at the time and place of their origin. Thus, when we measure this direction in an ancient rock, it shows the attitude or direction in the rock of the magnetic lines of force at the time the rock was originally magnetized. In very young rocks this direction is very nearly the same as that of the present lines of force about Earth. In older rocks, the direction may show considerable discrepancy with respect to Earth's present magnetic field. This discrepancy suggests that either the magnetic poles of Earth have shifted with time or the rocks themselves have shifted with respect to the poles. Magnetized rocks record not only the polarity of the magnetic field—that is, normal or reversed—at the time they were formed, but also the direction of magnetization. This direction points toward the ancient magnetic pole. To illustrate, let's assume we find an outcrop of a magnetized rock such as a lava flow and select a piece of this rock to be collected and analyzed for remanent magnetism. First we use a compass to determine the direction of the present geographic pole and mark this direction on the specimen as a line with an arrow pointing toward the pole, say the north pole. We also mark a horizontal line on the rock, a line we will assume is parallel to the bedding surface of the flow. The specimen is then broken off the outcrop with a hammer and taken to the laboratory, where the direction of the weak magnetization is determined with a sensitive magnetometer. A piece of the specimen is used to determine the radiometric age of the rock. The specimen is placed on a horizontal table with its original horizontal plane parallel to the table. To insure that the specimen is in the same relative geographic position that it occupied before it was collected, the north-pole direction marked on the specimen is now lined up, by use of a compass, with the present geographic north-pole direction. The measured direction of remanent magnetization in the rock now points toward the ancient pole position, and the angle between this direction and the direction of the present geographic pole is the declination between the present pole and the ancient pole.

If we plot on a map the location where the rock specimen was collected and then from this point draw a line that has the same orientation as the measured direction of remanent magnetization in the rock, this line will indicate the direction to the ancient pole. To actually locate the position of the ancient pole on this line, we have to know how far away it was from where the rock was formed. We determine this distance by measuring the paleomagnetic inclination in the rock. This inclination is the vertical angle between a line representing the direction of remanent magnetization in the rock and a horizontal plane. If a compass needle is hinged on a horizontal pin rather than a vertical pin, the needle can swing freely up and down in a vertical plane.

At the equator, the needle will assume a horizontal position. As we move away from the equator to the north or south, the needle will dip from the horizontal at a progressively steeper angle until at the pole it will point straight down, 90 degrees from the horizontal. This vertical angle of dip is the inclination just mentioned, and the latitude of any position on Earth can be determined if we know the magnetic inclination. By measuring the remanent magnetic inclination in an ancient rock, we can thus determine the paleolatitude at which it formed. The direction in which the inclination dips tells us if the pole position is in the Northern or Southern hemisphere, and the amount of dip tells the paleolatitude. If we assume that a given paleolatitude was roughly the same distance from the pole as the present latitude, the paleomagnetic inclination thus tells how far away the pole was at that time.

Now, assume that we determine ancient pole positions from many different rocks of different ages on a particular continent, plot these positions on a map, and connect the points with a line. This line shows the apparent wandering path of the pole through geologic time. Such apparent polar-wandering curves for North America

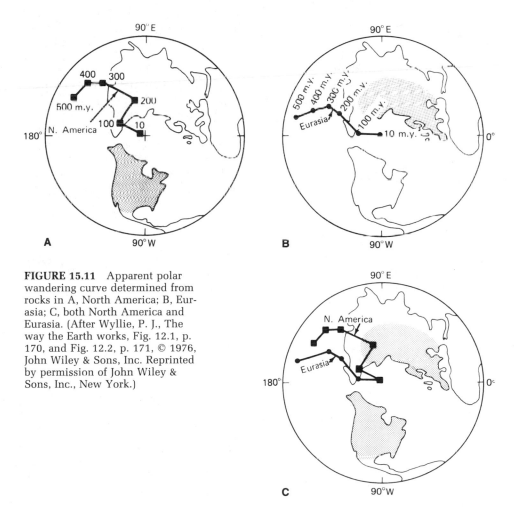

FIGURE 15.11 Apparent polar wandering curve determined from rocks in A, North America; B, Eurasia; C, both North America and Eurasia. (After Wyllie, P. J., The way the Earth works, Fig. 12.1, p. 170, and Fig. 12.2, p. 171, © 1976, John Wiley & Sons, Inc. Reprinted by permission of John Wiley & Sons, Inc., New York.)

and Eurasia are shown in Figure 15.11A,B. If the apparent polar-wandering curves for both Eurasia and North America are plotted on a single map (Fig. 15.11C), it becomes immediately apparent that the two curves do not coincide. Because it seems unlikely that more than one pole was wandering around out there in the past, some enterprising researchers decided to see what would happen to the ancient pole positions if the present continents were shoved back together and reassembled. When they did this and replotted the pole positions, they found that most of the differences in the pole positions charted from different continents disappeared, suggesting that migration of the continents and not the poles accounts for the observed differences. It is now believed that the axis of Earth's rotation has remained fixed in space through time and that the magnetic pole has not shifted far from the pole of rotation. The apparent wandering of the poles indicated by paleomagnetic data is in reality due to the movements of continents with respect to the geographic pole of rotation. To an observer in space who viewed Earth over a period of hundreds of millions of years, the geographic pole would thus appear fixed and the continents would slip about over the face of Earth with respect to the pole.

The magnetic inclination of ancient magnetized rocks is now being used extensively as a tool to examine these presumed movements of continental masses and smaller blocks. By measuring the remanent magnetic inclination and declination in ancient rocks, it is possible to reconstruct the original geographic position of these rocks at the time they formed. These studies have shown not only that major continents have shifted in their positions with time, but that many smaller blocks of rock have also moved from their original locations. That is, these blocks are now located in different latitudes than those in which they formed. Quite commonly the blocks have different lithologies and structural attitudes than those of adjoining areas. These small, exotic blocks are often called **suspect terranes,** referring to the probability that they are not now in the geographic position in which they were originally formed. There is growing evidence that large portions of many continental margins are made up of a collage of these suspect terranes, assembled by seafloor spreading and subduction processes over long periods of time from different parts of Earth.

Other Applications

Other applications of paleomagnetism include dating of archaeological materials, tracing the source of provenance of these materials, and study of magnetic fabrics in sedimentary, igneous, and metamorphic rocks. These and other applications are discussed by Tarling (1983).

ADDITIONAL READINGS

Conde, K. C., 1982, Plate tectonics and crustal evolution, 2nd ed.: Pergamon, New York, 310 p.

Cox, A. (editor), 1973, Plate tectonics and geomagnetic reversals: W. H. Freeman, San Francisco, 702 p.

Glen, W., 1982, The road to Jaramillo: Critical years of the revolution in the earth sciences: Stanford University Press, 459 p.

IUGS International Subcommission on Stratigraphic Classification and IUGS/IAGA Subcommission on a Magnetic Polarity Time Scale, 1979, Magnetostratigraphic polarity units—A supplementary chapter of the ISSC International Stratigraphic Guide: Geology, v. 7, p. 578–583.

Kennett, J. P. (ed.), 1980, Magnetic stratigraphy of sediments: Benchmark Papers in Geology 54, Dowden, Hutchinson and Ross, Stroudsburg, Pa., 438 p.

Tarling, D. H., 1983, Paleomagnetism: Principles and applications in geology, geophysics, and archaeology: Chapman and Hall, London, 379 p.

Wyllie, P. J., 1976, The way the Earth works: John Wiley & Sons, New York, 296 p.

16

Seismic Stratigraphy

16.1 INTRODUCTION

Seismology is the study of earthquakes and the structure of Earth on the basis of characteristics of seismic waves. Although the broad subject of seismology lies outside the scope of this book, some aspects of seismology have a very important application to stratigraphy. The emphasis of this chapter is focused on what is commonly referred to as exploration seismology and more specifically on application of the techniques of exploration seismology to stratigraphic study. Exploration seismology deals with the use of artificially generated seismic waves to obtain information about the geologic structure, stratigraphic characteristics, and distributions of rock types. The techniques of exploration seismology were developed initially to locate structural traps for petroleum deposits, and they are still used extensively for that purpose. Seismic methods are now being used also for many other scientific purposes to increase knowledge of Earth's structure and stratigraphy.

In this chapter, we examine the application of seismic methods to stratigraphic problems. Seismic methods can be used to delineate rock bodies with distinctive geometries and internal structures, which can be thought of as constituting "seismic facies." Study of the lateral and vertical variations in seismic facies patterns is used also as a basis for interpreting lithology, depositional environments, and geologic history of subsurface stratigraphic units. **Seismic stratigraphy** is thus the study of seismic data for the purpose of extracting stratigraphic information. Like magnetostratigraphy, seismic stratigraphy is a relatively new science, born in the early 1960s. Owing to its wide applicability to subsurface study both on land and at sea, where other types of stratigraphic data are few, it has already achieved an important position alongside the more traditional branches of stratigraphy.

16.2 EARLY DEVELOPMENT OF SEISMIC METHODS

Much of the theory of elasticity and propagation of seismic waves through rock materials was developed in the early part of the nineteenth century. An English seismologist, Robert Mallet, was the first scientist to measure the velocity of seismic waves in subsurface materials. He initiated experimental seismology in 1848 by measuring the speed of seismic waves through comparatively near-surface materials. He used black powder as an energy source to create a disturbance in the rocks and the surface of a bowl of mercury as the detector for the arriving seismic waves. The possibility of using seismic techniques to define the characteristics of subsurface rocks was apparently first put forward by a scientist named Milne in 1898. Two other interesting applications of the principles of seismology were experimented with in the early part of the twentieth century. These were a method for detecting icebergs—after the sinking of the Titanic by an iceberg in 1912—and the use of mechanical seismographs to detect the position of large enemy guns during World War I.

Refraction Seismic Methods

Practical use of seismology in detection of rock structures began immediately after the end of World War I both in the United States and in Europe, especially Germany and England. The first applications were in petroleum exploration in Germany and the Gulf Coast region of the United States, particularly in exploration for petroleum traps associated with salt domes. These early exploration efforts used the **refraction** seismic method for determining the structure of subsurface formations. This method is based on the principle that artificially generated seismic waves are refracted or bent at discontinuity surfaces as they travel downward below the surface. When waves generated from a point source meet a subsurface discontinuity, such as an unconformity, some of the waves will strike the discontinuity at a critical angle to the wave paths. Thus, they are refracted, or bent, into the underlying layer, in a manner somewhat analogous to the way light rays are bent as they pass from air to water. The velocity of seismic waves tends to increase with depth in the subsurface; therefore, the refracted waves will travel along the top of the layer underlying the discontinuity surface at a greater velocity than that at which they passed downward through the overlying layer.

The rock above the discontinuity interface is subjected to oscillating stresses from the faster moving waves below, which, as the waves travel, generate new disturbances along the boundary. These disturbances spread out from points all along the boundary and move back toward the surface along ray paths that form the same critical angle with the discontinuity as that formed by the downward-moving waves. If a seismic signal detector is located on the surface close to the sound source, the first seismic waves to reach the detector will be those that travel horizontally through the rocks near the surface. If the detector is located farther from the source, the first waves to reach the detector will be those that were refracted at the discontinuity and then traveled along the discontinuity at a faster velocity before being refracted back to the surface. In other words, the longer ray path of these refracted waves is more than made up for by the faster velocity with which they travel through the rocks underlying the discontinuity. Therefore, in refraction surveys the detectors are placed several times farther from the shot point, or energy source, than the vertical distance to the underlying formations being investigated. The time required for seismic waves to travel from the surface to a subsurface formation, advance along the discontinuity surfaces, and be refracted back to the surface where the detectors are located can be used to compute depths to the formations. Such calculations require that the velocities with which the sound waves pass through the rock overlying the discontinuity and along the discontinuity be known. By shooting a pattern or spread of refraction profiles in a given region and computing depths to subsurface formations at many locations, it is possible to detect the presence of subsurface structures such as anticlines and salt domes.

Reflection Seismic Method

Although several shallow petroleum deposits in salt domes were discovered during the early years of seismic exploration by the refraction method, this method did not work well for deeper structures because of the excessive distances required between shot points and detectors. Therefore, it was soon largely supplanted in petroleum exploration by the **reflection** seismic method. In the reflection method, waves created by an explosion are reflected back to the surface directly from subsurface rock interfaces without being refracted and traveling laterally along discontinuity surfaces. Because detectors can be located at relatively short distances from the shot points, reflection seismic techniques can be used for delineating very deep structures. Of course, refracted waves also are generated by the explosions used in reflection shooting but, owing to the closer location of the detectors to the source, are not picked up. After the reflection method was introduced about 1930, it quickly became the primary tool in the petroleum industry for locating buried anticlines and other structural traps. The principles of reflection seismology are explained in greater detail in Section 16.3.

Marine Seismic Surveys and Subsequent Developments

Early seismic surveys were carried out mainly on land, although some operations in very shallow water began in the late 1920s and 1930s. Extensive marine operations did not begin until about the middle 1940s. In these initial marine operations, explosive charges were detonated at the water surface to provide the energy source, and detectors were dragged along the bottom on a cable. Early marine operations were severely hampered by problems of accurately locating the shot point and detector positions and had to be carried out within sight of land so that locations could be determined by land-based surveying methods. The development about 1949–50 of radio-navigation methods and the floating streamer cable that allowed detectors to be towed in a floating cable behind the ship came about as a result of advancements during World War II. These technical advances made possible rapid progress in marine seismic surveying methods in the years following.

Subsequent progress in seismic surveying techniques have included improvements in energy sources, including sources that do not require explosives, development of new detection equipment and procedures, and advances in treatment and analysis of seismic data. In particular, computer analysis of seismic data has allowed a quantum leap forward in filtering and enhancing seismic signals and in methods of displaying and interpreting seismic data. Additional details of the early history of seismic exploration are given by Sheriff and Geldart (1982).

Seismic Stratigraphy

The early development of seismic methods was aimed primarily at detecting and delineating subsurface structures. Petroleum geologists were particularly interested in identifying anticlines, faults, and salt domes that were potential petroleum traps. It was not until about the early part of the 1960s that geologists began to take a strati-graphic-facies approach to seismic exploration when they began using reflection seis-mic profiles to interpret lithofacies and, subsequently, depositional systems. Applica-tions of geophysical principles and techniques to stratigraphic study thus initiated seismic stratigraphy as an offshoot of exploration seismology. Most of the pioneering work that went into the development of seismic stratigraphy was done by geologists of major petroleum companies, and much of the work was proprietary and confiden-tial. Because many of these proprietary data have not yet been released, and because seismic stratigraphy is a relatively new application of geophysical principles, rela-tively little has been published in the open literature about seismic stratigraphy. The most authoritative publications dealing with the principles and applications of seismic stratigraphy are the two memoir volumes *Seismic Stratigraphy—Applications to Hy-drocarbon Exploration*, edited by C. E. Payton (1977), and *Seismic Stratigraphy II—An Integrated Approach*, edited by O. R. Berg and D. G. Wolverton (1985). The first volume contains more than two dozen articles, written by a small group of experts in the field, that deal with the fundamentals of seismic stratigraphy. The second volume is somewhat more technical and has a greater emphasis on application of seismic stra-tigraphy principles to specific geologic problems. Much of the information about seis-mic stratigraphy that follows is drawn from this these memoirs and from the publica-tions of Brown and Fisher (1980) and Sheriff (1980). Other informative publications on seismic stratigraphy include those of Anstey (1982) and T. L. Davis (1984).

16.3 PRINCIPLES OF REFLECTION SEISMIC METHODS

Readers who are not familiar with seismic methods may find useful a short discussion of the principles of the reflection seismic method before we proceed to a discussion of the applications of reflection seismology to stratigraphic analysis. Only a very brief summary of basic principles is given here. Additional details of the physical principles upon which reflection methods are based are given by Anstey (1982), Dobrin (1976), and Sheriff and Geldart (1982).

The reflection seismic method for delineating the structure of subsurface rock units is based on the principle that elastic or seismic waves travel at known velocities through rock materials. These velocities vary with the type of rock, but where the subsurface lithology is known relatively well from drill-hole information, it is possible to make accurate calculations of the time required for a seismic signal to travel from the surface to a given depth and then be reflected back to the surface. The reflection technique involves first generating elastic waves at the surface at a point source, orig-inally called a shot point because explosives were first used to create the seismic waves. Seismic detectors, called geophones, are laid out in arrays extending outward from the point source. Seismic waves reflected back from subsurface discontinuities are picked up by these detectors and fed electronically to a recording device. The principal discontinuities that reflect seismic waves are bedding planes and uncon-formities. By multiplying one-half of the travel time elapsed from initiation of the elastic waves at the point source to their arrival at the detector by the velocity of travel,

FIGURE 16.1 (opposite) Diagram illustrating the equipment and procedures used in seismic exploration in the 1940s. (From Nettleton, L. L., Geophysical prospecting for oil. Fig. 155, fol-lowing p. 332, © 1940. Reprinted by permission of McGraw-Hill Book Company, New York.)

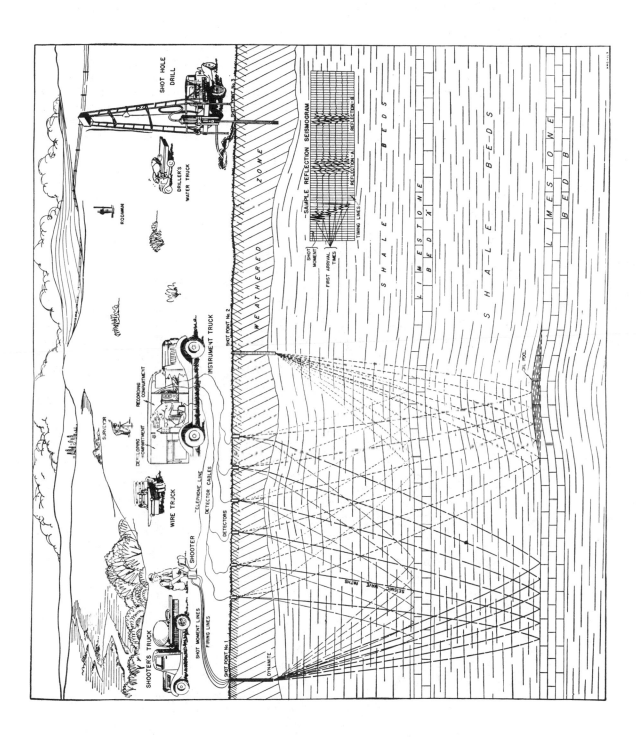

depths to the discontinuities can be accurately calculated. This procedure thus allows the subsurface position of the discontinuities to be determined. The data obtained in this manner can then be displayed as seismic sections or profiles that depict the structure of the major rock units as they appear in cross section. Alternatively, the data may be used to prepare structure contour maps on the tops of particular reflecting horizons.

In practice, the technique for on-land reflection seismic profiling involves (1) selecting locations for shot points and emplacing the energy source, either explosive or other device; (2) laying out and burying the geophones in a predetermined array and connecting the geophones to the recording equipment by long cables; (3) triggering the energy source; (4) recording on magnetic tape the seismic signals picked up by the detectors; and (5) computer processing of the tapes and preparation of visual analog displays of the seismograms. In the initial stages of seismic exploration, seismic waves were generated by explosives placed in shallow shot holes drilled through the near-surface weathered zone to bedrock. Nonexplosive sources located on the surface are now also in common use. These nonexplosive energy sources include vibratory devices that produce continuous vibrations at the surface (Vibroseis) or devices that drop heavy weights onto a metal plate placed on the ground surface (Thumper or Weight Dropper).

The general principles of on-land reflection seismic "shooting" are illustrated in the interesting old diagram (Fig. 16.1) from Nettleton's 1940 *Geophysical Prospecting for Oil.* The equipment and techniques for surveying locations, shooting, recording, and processing seismic data have changed and significantly improved since the 1940s; however, the basic principles illustrated in this figure still apply. As seismic waves pass downward and outward from the point energy source through the subsurface formations, they are reflected from successively deeper formations back to the surface, where they are picked up by the electronic detectors. The signal from the detectors is then amplified, filtered to remove excess "noise," and fed to a recording truck to be recorded on magnetic tape. The data recorded on the magnetic tape must then be presented in visual form for monitoring and interpretation. Prior to the use of magnetic tape for recording, the visual seismic records, or seismograms, were mechanically produced wiggly-trace records such as that shown on the right in Figure 16.1. Photographic or dry-paper recording methods are now used for visual display, and several modes of displaying the amplitude of arriving seismic waves against arrival time are in use. A common type of display, called a **variable-density mode,** is generated by a technique by which light intensity is varied to display differences in wave amplitude (height) by producing alternating light and dark areas on film or paper, thereby accentuating the amplitude of waves from a particular reflecting surface (Fig. 16.2).

Marine seismic operations employ the same principles as those used on land, but differ in the speed at which they take place and the specific details of the shooting and detection processes. Sound sources and detectors are towed behind the survey ship, which can operate at a speed of 6 knots or more on a continuous 24 hr/day basis. In the early years of marine operations, a half-pound block of TNT was tossed over the ship's side every 3 minutes to provide a continuous seismic record. This method was potentially dangerous and also very damaging to fish and other ocean life. It has now been largely replaced by techniques that use acoustic sources such as airguns, which produce sound energy by releasing highly compressed air. The detectors, called hydrophones, are contained in long streamer cables, or eels, that may be up to several kilometers in length. The position of the survey ships at sea can be accurately and continuously fixed by precise satellite navigation methods. The general principles of marine seismic profiling are illustrated in Figure 16.3.

16.4 APPLICATION OF REFLECTION SEISMIC METHODS TO STRATIGRAPHIC ANALYSIS

Introduction

The science of seismic stratigraphy was developed largely by petroleum companies out of pragmatic necessity to locate petroleum deposits in deep, unexplored basins

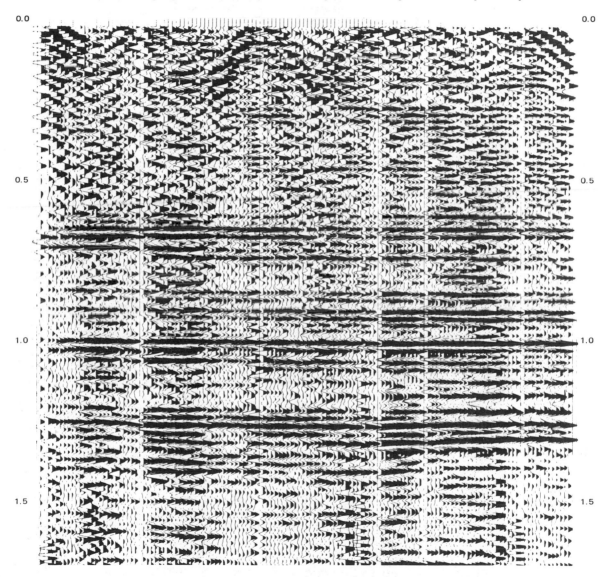

FIGURE 16.2 Example of a seismic record displayed using the variable density method of printing. (From Neidell, N. S., 1979, Stratigraphic modeling and interpretation: Geophysical principles and techniques: Am. Assoc. Petroleum Geologists Education Short Course Notes Ser. 13. Fig., p. 49, reprinted by permission of AAPG, Tulsa, Okla.)

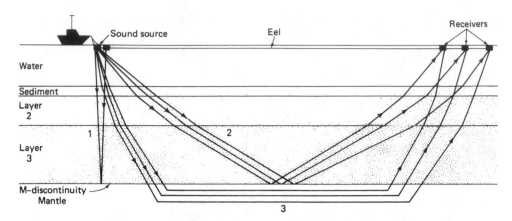

FIGURE 16.3 Diagram illustrating the principle of marine seismic surveying: 1, vertical inci-
dence reflector, 2, wide-angle reflector, and 3, refracted waves. (From Kennett, J., Marine geology, © 1982, Fig. 2.14, p. 40. Reprinted by permission of Prentice-Hall, Englewood Cliffs, N.J.)

both on land and offshore. Geologists have not yet discovered a successful geochemical method for directly detecting the presence of oil or gas in deep subsurface formations, although some progress has been made in direct detection of hydrocarbon deposits by seismic methods. Therefore, the successful search for petroleum still requires
that explorationists locate and drill petroleum traps such as anticlines and salt domes.
Because most shallow petroleum traps were located and tested long ago during the
earlier phases of petroleum exploration, petroleum companies have been forced to
extend exploration efforts to deeper formations and to frontier basins, which are undrilled or sparsely drilled basins, on land and offshore. Inasmuch as successful oil
finding depends upon a knowledge of stratigraphic relationships as well as structural
anomalies, and because poorly explored basins lack sufficient well control for stratigraphic analysis, new techniques had to be developed to allow stratigraphic information to be extracted from seismic data. Thus, seismic stratigraphy was born in the
1960s as a tool that made possible the integration of stratigraphic concepts with geophysical data—that is, a geologic approach to stratigraphic interpretation of seismic
data.

Seismic reflections are generated by physical surfaces in subsurface rocks. In the
conventional structural application of seismic data, seismic reflections are used to
identify and map the structural attitudes of subsurface sedimentary layers. By contrast,
seismic stratigraphy uses seismic reflection correlation patterns to identify depositional sequences, to predict the lithology of seismic facies by interpreting depositional
processes and environmental settings, and to analyze relative changes in sea level as
recorded in the stratigraphic records of coastal regions. Seismic stratigraphy thus
makes possible many types of stratigraphic interpretations such as geologic time correlations, definition of genetic depositional units, and thickness and depositional environment of genetic units (Vail and Mitchum, 1977).

Parameters Used in Seismic
Stratigraphic Interpretation

To accomplish the objective of interpreting stratigraphy and depositional facies from
seismic data, characteristic features of seismic reflection records (seismograms) must

be identified and related to the geologic factors responsible for the reflections. An understanding of the factors that generate seismic reflections is therefore critical to the entire concept of seismic stratigraphy. Fundamentally, primary seismic reflections occur in response to the presence of significant density–velocity changes at either unconformity or bedding surfaces. Reflections are generated at unconformities because unconformities separate rocks having different structural attitudes or physical properties, particularly different lithologies. The density–velocity contrast along unconformities may be further enhanced if the rocks below the unconformity have been altered by weathering. Reflections are generated at bedding surfaces because, owing to lithologic or textural differences, a velocity–density contrast exists between some sedimentary beds; however, not every bedding surface will generate a seismic reflection. Also, a given reflection event identified on a seismic record may not necessarily be caused by reflection from a single surface, but may represent the sum or average of reflections from several bedding surfaces, particularly if beds are thin.

The seismic records produced as a result of primary reflections from unconformities or bedding surfaces have distinctive characteristics that can be related to depositional features such as lithology, bed thickness, bed spacing, and continuity. The principal parameters that are useful in seismic stratigraphy for interpreting geologic information are reflection configuration, continuity, amplitude, and frequency; interval velocity; and external form and association of seismic facies units (Table 16.1). These parameters are explained in the following paragraphs.

Reflection Configuration

Reflection configuration refers to the gross stratification patterns identified on seismic records. Four basic types of configurations are recognized (Mitchum, Vail, and Sangree, 1977). **Parallel patterns,** including subparallel and wavy patterns (Fig. 16.4A,B),

TABLE 16.1 Seismic reflection parameters commonly used in seismic stratigraphy, and the geologic significance of these parameters

Seismic facies parameters	Geologic interpretation
Reflection configuration	Bedding patterns Depositional processes Erosion and paleotopography Fluid contacts
Reflection continuity	Bedding continuity Depositional processes
Reflection amplitude	Velocity-density contrast Bed spacing Fluid content
Reflection frequency	Bed thickness Fluid content
Interval velocity	Estimation of lithology Estimation of porosity Fluid content
External form and areal association of seismic facies units	Gross depositional environment Sediment source Geologic setting

Source: Mitchum, R. M., Jr., P. R. Vail, and J. B. Sangree, 1977, Seismic stratigraphy and global changes of sea level, Part 6: Stratigraphic interpretation of seismic reflection patterns in depositional sequences, *in* C.E. Payton (ed.), Seismic stratigraphy—Applications to hydrocarbon exploration: Am. Assoc. Petroleum Geologists Mem. 26. Table 2, p. 122, reprinted by permission of AAPG, Tulsa, Okla.

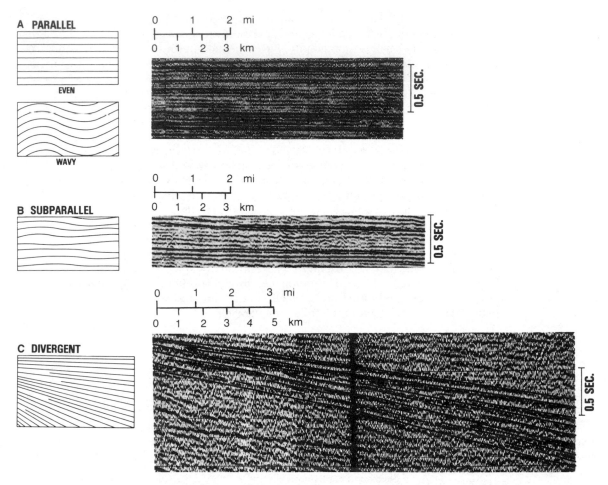

FIGURE 16.4 Principal types of seismic reflection configurations. A. Parallel (even or wavy). B. Subparallel. C. Divergent. (After Mitchum, R. M., Jr., P. R. Vail, and J. B. Sangree, 1977, Figs. 4 & 5, p. 123, 124, reprinted by permission of AAPG, Tulsa, Okla.)

are generated by strata that were probably deposited at uniform rates on a uniformly subsiding shelf or in a stable basin setting. **Divergent configurations** are characterized by a wedge-shaped unit in which lateral thickening of the entire unit is caused by thickening of individual reflection subunits within the main unit (Fig. 16.4C). Divergent configurations are interpreted to signify lateral variations in rates of deposition or progressive tilting of the sedimentary surface during deposition. **Prograding reflection configurations** are reflection patterns generated by strata that were deposited by lateral

FIGURE 16.5 (opposite) Examples of reflection patterns interpreted as prograding clinoforms. A. Sigmoid. B. Mostly tangential. C. Mostly parallel oblique. D. Complex sigmoid-oblique. E. Shingled. F. Hummocky clinoforms. (After Mitchum, R. M., Jr., P. R. Vail, and J. B. Sangree, 1977, Figs. 6–8, p. 125, 126, 127, reprinted by permission of AAPG, Tulsa, Okla.)

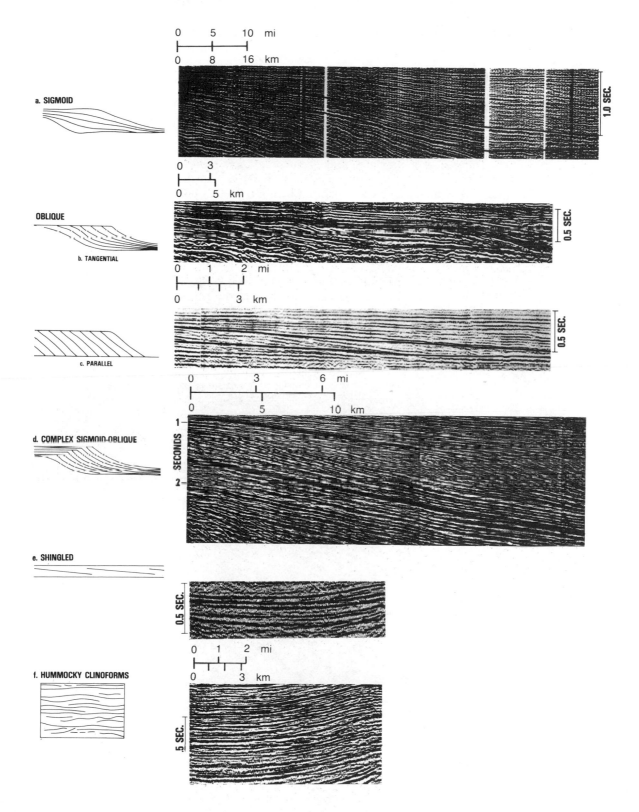

a. SIGMOID

OBLIQUE

b. TANGENTIAL

c. PARALLEL

d. COMPLEX SIGMOID-OBLIQUE

e. SHINGLED

f. HUMMOCKY CLINOFORMS

591

outbuilding or progradation to form gently sloping depositional surfaces called clinoforms (explained in the next paragraph). As represented on seismic records, prograding reflection configurations may have a variety of patterns, including **sigmoid** (superposed S-shaped reflectors), **oblique,** or **hummocky** (Fig. 16.5). These stratal
configurations are all caused in some way by outbuilding of strata—commonly from
shallow water into deeper water, as along the front of a delta—or by infilling of channels. Differences in configurations of the clinoforms represent variations in sediment
supply or rates of basin subsidence or changes in sea level, water energy of the depositional environment, or water depth. **Chaotic reflection patterns** (Fig. 16.6) are interpreted to represent a disordered arrangement of reflection surfaces owing to penecontemporaneous soft-sediment deformation, or possibly to deposition of strata in a
variable, high-energy environment. Some chaotic reflections may be related to overpressured or geopressured zones in deep formations. Reflection-free areas on seismic
records may represent homogeneous, nonstratified units, such as igneous masses or
thick salt deposits, or highly contorted or very steeply dipping strata.

The terms undaform, clinoform, and fondoform were introduced by Rich (1951)
to describe depositional environments in relation to wave base (Fig. 16.7). The **unda-**

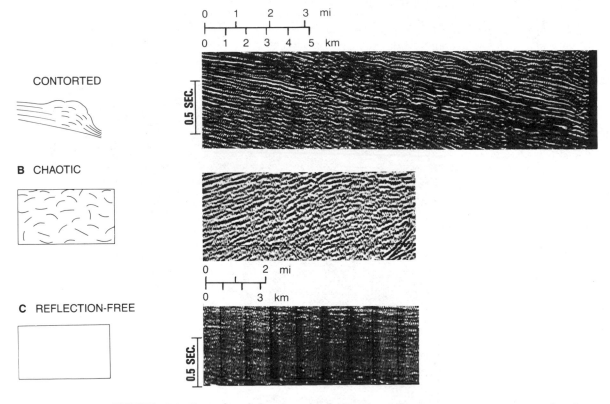

FIGURE 16.6 Examples of chaotic and reflection-free reflection patterns. A. Interpreted as the
reflection from contorted stratal surfaces that are still recognizable after penecontemporaneous
(soft sediment) deformation. B. Reflections are so chaotic no stratal pattern can be reliably interpreted. C. Largely reflection-free, where no or very few reflections occur in seismically homogenous strata. (After Mitchum, R. M., Jr., P. R. Vail, and J. B. Sangree, 1977, Figs. 9, 10, p. 128,
129, reprinted by permission of AAPG, Tulsa, Okla.)

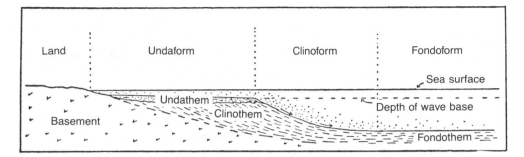

FIGURE 16.7 Sketch illustrating the meaning of the terms undaform, clinoform, and fondo-form as used by Rich. (After Rich, J. L., 1951, Three critical environments of deposition and criteria for recognition of rocks deposited in each of them: Geol. Soc. America Bull., v. 62, Fig. 1, p. 3.)

form is the more or less flat topographic surface that exists above wave base in an aqueous environment where bottom sediments are moved or stirred by waves and currents, particularly during storms. The **clinoform** is the sloping surface extending from wave base down to the generally flat floor, called the **fondoform,** of the water body. The rock units formed in each of these environmental settings are called unda-them, clinothem, and fondothem.

Reflection Continuity

Reflection continuity depends upon the continuity of the density–velocity contrast along bedding surfaces or unconformities. It is closely associated with continuity of strata, and it provides information about depositional process and environment. Continuous reflectors (Fig. 16.8) characteristically indicate widespread uniformly stratified

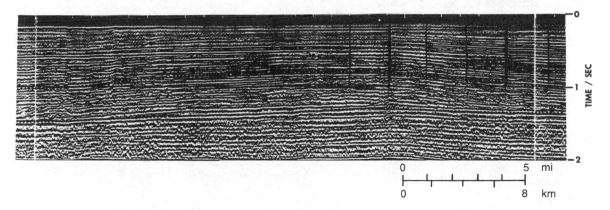

FIGURE 16.8 Illustration of broad, low-relief seismic mound facies. This example is from off-shore Africa. It shows mounded external form with reflections concordant at the top and down-lapping in opposite directions at the base. (From Sangree, J. B., and J. M. Widmier, 1977, Seismic stratigraphy and global changes in sea level, Part 9: Seismic interpretation of clastic depositional facies, *in* C. E. Payton (ed.), Seismic stratigraphy—Applications to hydrocarbon exploration: Am. Assoc. Petroleum Geologists Mem. 26. Fig. 4, p. 172, reprinted by permission of AAPG, Tulsa, Okla.)

deposits, although this is not necessarily true of all widespread reflectors. In some depositional sequences, such reflectors may, for reasons that are still unknown, reflect isochronous horizons that cut across some stratal surfaces. In contrast to continuous reflectors, reflection patterns showing reflection terminations (Fig. 16.9) indicate stratigraphic relationships such as onlap, downlap, and toplap that occur in coastal regions in response to transgressions and regressions.

Reflection Amplitude

Reflection amplitude has to do with seismic wave height and is a function of the energy of seismic waves. On a seismic record, amplitude is measured as the distance from the midposition of a wave to the extreme position. Amplitude is thus equal to one-half the height of the wave above the adjacent trough (Fig. 16.10). The amplitude of reflected seismic waves is controlled principally by the velocity and density contrast along individual reflecting surfaces and increases with increasing velocity–density contrast. It is affected also by the spacing between reflecting surfaces. Where bed spacing is optimum, lower energy responses are phased together constructively (constructive interference) to intensify or amplify the reflected energy and thus increase amplitude. For example, if bed thickness is less than the wave length of the seismic wave—for example, one-fourth of a wave length—the reflections from the top and base of the bed can be phased together to give exceptionally large amplitudes. When a bed is very thick, the reflections from the top and base of the bed are completely separate

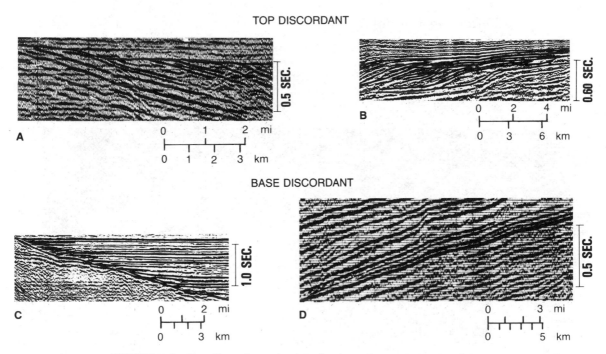

FIGURE 16.9 Top-discordant seismic reflection patterns: A. erosional truncation. B. toplap. Base-discordant seismic reflection patterns: C. onlap. D. downlap. (After Mitchum, R. M., Jr., P. R. Vail, and J. B. Sangree, 1977, Figs. 2 & 3, p. 119, 120, reprinted by permission of AAPG, Tulsa, Okla.)

FIGURE 16.10 Amplitude of seismic waves. (After Neidell, N. S.,
1979, Stratigraphic modeling and
interpretation: Geophysical principles and techniques: Am. Assoc.
Petroleum Geologists Education
Short Course Note Ser. 13. Fig., p.
31, reprinted by permission of
AAPG, Tulsa, Okla.)

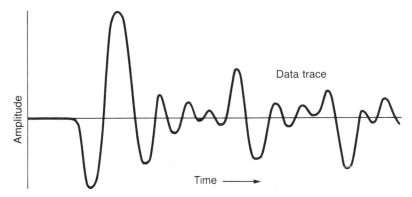

(Sheriff, 1980). Figure 16.11 shows the effect of bed thickness on amplitude in the case
of a sandstone with a gradational base.

The amplitude of reflected seismic waves can be affected also by fluid content
of sedimentary beds or by accumulations of gas in the beds. The presence of hydrocarbons in beds can produce a marked increase in amplitude of waves, showing up on
seismic records as so-called "bright spots." Bright-spot analysis was introduced in the
petroleum industry in the early 1970s and is now used as a method for direct detection
of hydrocarbon deposits.

Reflection Frequency

Reflection frequency refers to the number of vibrations or oscillations of seismic waves
per second. It is numerically equal to wave velocity divided by wave length. The frequency of a seismic wave is commonly expressed in hertz (Hz) or kilohertz (KHz). A
hertz is a unit of frequency equal to one cycle per second; a kilohertz is 1000 hertz.
The frequency of seismic waves affects both the depth of penetration of the waves into

FIGURE 16.11 Seismic response
for a sand with a gradational base.
The 30-foot thickness is about ⅛
wave length. (After Neidell, N. S.,
and E. Poggiagliolmi, 1977, Stratigraphic modeling and interpretations—Geophysical principles and
techniques, in C. E. Payton (ed.),
Seismic stratigraphy—Applications
to hydrocarbon exploration: Am.
Assoc. Petroleum Geologists Mem.
26. Fig. 27, p. 413, reprinted by
permission of AAPG, Tulsa, Okla.)

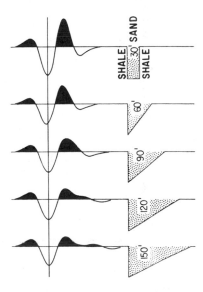

the subsurface and the resolution of the seismic records; that is, the sharpness with which details of the seismograms can be distinguished. Lower frequencies give greater depth of penetration but less resolving power.

The frequency of seismic waves is induced by the particular energy sound source used to create the waves. As the waves pass downward through subsurface formations and are reflected back to the surface, the initial induced frequency is modified by bed thickness, which controls the spacing of reflectors. Thus, differences between the initial induced frequency of seismic waves and the final frequencies of the reflected waves provide information about bed thickness. Frequency is affected also by lateral changes in fluid content of beds—the presence of hydrocarbon accumulations, for example—and by lateral thickness changes in beds.

Interval Velocity

Interval velocity refers to the average velocity of seismic waves between reflectors. Seismic wave velocity is affected by several factors, especially porosity, density, external pressure, and pore (fluid) pressure. Porosity has a particularly significant effect on velocity, which increases as porosity decreases. Thus, because porosity commonly decreases with depth, velocity increases with depth. Velocity increases also with density of the rocks and with increasing overburden pressure. It decreases with increasing interstitial fluid pressure; the presence of gas at low saturations in the pore spaces of the rocks also causes a decrease in velocity.

Seismic velocity is of particular interest because of the possibility that different rock types, which are characterized by different densities, porosities, pore fluid pressures, and other characteristics, can be differentiated on the basis of seismic velocity. Figure 16.12 shows, for example, that the velocity of seismic waves is lower in terrig-

FIGURE 16.12 Characteristic velocity–depth relations for terrigenous clastic sedimentary rocks, carbonate rocks, and salt. Younger rocks tend to have lower velocities than older rocks because they generally have higher porosities, are less cemented, and have undergone less deformation. (From Sheriff, R. E., 1976, Inferring stratigraphy from seismic data: Am. Assoc. Petroleum Geologists Bull., v. 60. Fig. 6, p. 533, reprinted by permission of AAPG, Tulsa, Okla.)

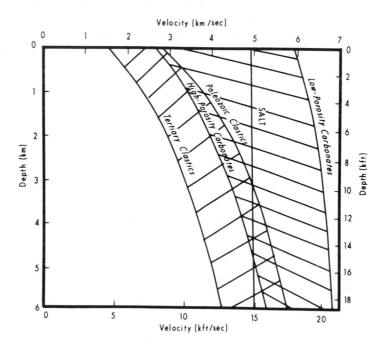

enous siliciclastic rocks such as sandstones than in carbonate rocks and salt. Thus, where velocity contrasts in different rocks are large, velocity can be used as an indicator of gross lithology. Unfortunately, there is appreciable overlap in the velocities of seismic waves in various rock types, owing mainly to variations in porosity, so that velocity alone is not sufficient to unequivocally distinguish rock types. For example, the velocity of seismic waves in a low-porosity sandstone can be approximately the same as in a high-porosity carbonate rock. Therefore, in order to predict lithology, velocity information must be combined with other types of seismic data to allow interpretation of depositional processes and environmental settings. Nonetheless, velocity is one of the most critical factors in seismic data processing and interpretation.

External Form

The external form or geometry of seismic facies units and their areal associations can be used to identify "seismic facies," which may be interpreted in terms of depositional environments of the lithologic analogs of these seismic facies. Seismic facies analysis also provides information on sediment source and geologic setting, including major facies changes. It is an extremely important aspect of seismic stratigraphy and is discussed in greater detail in the following paragraphs.

Procedures in Seismic Stratigraphic Analysis

The significance and appeal of the seismic stratigraphic approach to study of subsurface sedimentary rocks lies in the fact that it permits geologists and geophysicists to interpret stratigraphic relationships and depositional processes as well as to use seismic data for conventional structural mapping. Interpretation is a subjective process, but when seismic stratigraphic analysis is pursued in a logical manner and interpretation is based upon analogy with established stratigraphic and depositional models that have been generated by other types of studies, it becomes an extremely valuable tool. Seismic stratigraphy can thus provide insight into such stratigraphic and depositional factors as lithofacies and facies changes, relief and topography of unconformities, paleobathymetry (depth relationships and topography of ancient oceans), geologic time correlations, depositional history, and subsidence and tilting history (burial history). The procedures for interpreting stratigraphy from seismic data involve three principal stages: (1) seismic sequence analysis, (2) seismic facies analysis, and (3) sea level analysis (Mitchum and Vail, 1977).

Seismic Sequence Analysis

Seismic sequence analysis involves identification of major reflection "packages" that can be delineated by recognizing surfaces of discontinuities. The concept of depositional sequence was introduced in Chapter 14, where sequences are defined as stratigraphic units separated by unconformities or their correlative conformities (Mitchum, Vail, and Thompson, 1977). The mapping of unconformities is thus the key to seismic sequence analysis. Unconformities are generally good reflectors and also commonly separate rock units having different dips, at least on a regional scale. Unconformities may thus be recognized by interpreting systematic patterns of reflection terminations along the unconformity surface. These patterns are categorized (Chapter 14) as onlap, downlap, toplap, and truncation.

Seismic resolution is generally not adequate to delineate minor sedimentary sequences, that is, $<\sim$30–50 m (T. L. Davis, 1984); however, major depositional units or systems such as progradational delta-slope systems, carbonate shelf-margin systems, or marine offlap–onlap systems can be identified. Once basinwide correlation of depositional sequences has been made, these sequences provide a first-order stratigraphic framework within which more detailed seismic facies studies can be carried out. Figure 16.13, a seismic section from the Beaufort Sea (north of Alaska), provides an example of how sequences can be delineated by picking unconformities that separate different depositional units.

The procedure for carrying out seismic sequence analysis thus involves (1) picking unconformities in a given area by recognizing reflection terminations along their surfaces; (2) extending or extrapolating these boundaries over the complete section, including areas where the reflectors are conformable, to define the sequences completely; (3) repeating the process of delineating sequence boundaries on seismic records from other parts of a basin or region and correlating the sequences throughout the seismic grid to produce a three-dimensional framework of successive stratified seismic sequences separated by unconformities or correlative conformities; and (4) mapping sequence units on the basis of thickness, geometry, orientation, or other feature to see how each sequence relates to neighboring sequences. Readers desiring more details on seismic sequence mapping and its application to geologic problems may wish to consult Hubbard et al. (1985a,b).

Seismic Facies Analysis

Seismic facies analysis takes the interpretation process one step beyond seismic sequence analysis by examining within sequences smaller reflection units that may be the seismic response to lithofacies. In seismic facies analysis, seismic parameters such as reflection configuration, continuity, amplitude, and interval velocity are used to identify seismic facies units. A **seismic facies unit** is a mappable, areally definable, three-dimensional unit composed of seismic reflections whose characteristics, or elements—reflection configuration, amplitude, continuity, frequency, and interval velocity—differ from those of adjacent units. This unit is considered to represent or express the gross lithologic aspect and stratification characteristics of the depositional unit that generates the reflections.

Procedures for Interpreting Seismic Facies. The objective of seismic facies analysis is regional interpretation of lithology, depositional environments, and geologic history. Mitchum and Vail (1977) describe several distinct steps in the interpretation process.

The first step is recognizing and delineating seismic facies units within each sequence on all the seismic sections in the region being mapped. Each seismic facies unit is distinguished from adjacent units on the basis of (1) the dominant type of reflection configuration (parallel, divergent, sigmoid, or oblique); (2) reflection continuity; (3) bounding relationships, that is, relation of reflectors to upper and lower sequence boundaries and types of reflection terminations and lateral changes such as onlap, downlap, toplap, truncation, and concordance; (4) reflection amplitude and frequency; (5) interval velocity; and (6) external geometry of the seismic facies units. All of these factors except interval velocity and external geometry can be estimated visually from two-dimensional seismic profiles. Estimation of interval velocity requires special computer processing techniques and is done by a trained geophysical processor. Determination of external geometry must be done by mapping. The three-dimensional geometry of several types of seismic facies is shown in Figure 16.14.

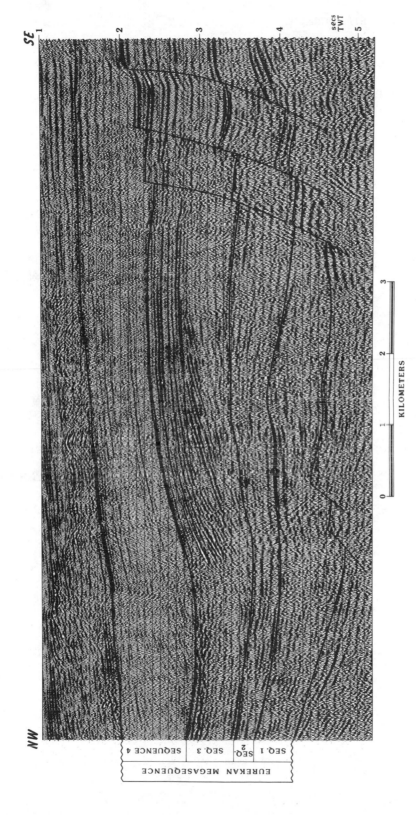

FIGURE 16.13 Depositional sequences as defined from seismic records. In this example from the Beaufort Sea, seismic sequence boundaries are shown by solid black lines. The three heavy, sloping lines in the right-hand part of the seismic record indicate large faults. The vertical scale on this record is given in seismic wave travel time rather than depth. (From Hubbard, R. J., J. Pape, and D. G. Roberts, 1985, Depositional sequence mapping as a technique to establish tectonic and stratigraphic framework and evaluate hydrocarbon potential on a passive continental margin, in O. R. Berg and R. G. Wolverton (eds.), Seismic stratigraphy II—An integrated approach: Am. Assoc. Petroleum Geologists Mem. 39, p. 84, reprinted by permission of AAPG, Tulsa, Okla.)

After distribution (geometry) and thickness of the reflection packages have been mapped, the next step in seismic facies analysis is to combine this information with any other distinctive seismic information, such as interval velocity, and any nonseismic data, such as well and outcrop data, that are available.

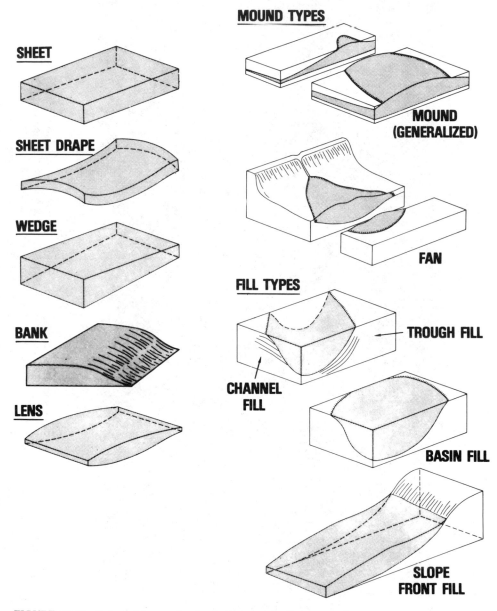

FIGURE 16.14 External forms of seismic facies units. (From Mitchum, R. M., Jr., P. R. Vail, and J. B. Sangree, 1977, Fig. 12, p. 131, reprinted by permission of AAPG, Tulsa, Okla.)

The final step is interpretation of the seismic facies in terms of depositional setting, that is, major environments, paleobathymetry and related factors, and gross lithology.

Brown and Fisher (1980) suggest that stratigraphic interpretation of seismic facies is a process of elimination. Because of obvious inconsistencies with available data or because of personal knowledge of the basin or region under study, the interpreter may be able to immediately eliminate certain lithofacies or depositional environments. Additional analysis involving study of relationships to other units, reflection characteristics, or other feature commonly allows further reduction of the remaining options until only one or two depositional or lithofacies models fit the available data. Lateral facies equivalents must be given special attention in the interpretation process and the interpreter must be experienced and have a good knowledge of depositional process and systems—lithofacies composition, geometry, and spatial relationships. Even so, it may not always be possible to arrive at a final, unique interpretation; the interpreter may have to settle for the best conclusion that can be made consistent with available data.

Principal Types of Seismic Facies. Seismic facies units are recognized on the basis of reflection configuration (Figs. 16.4, 16.5, 16.6), bounding relationships (Fig. 14.14), and external geometry (Fig. 16.14). Brown and Fisher (1980) use these parameters to group seismic facies units into four basic types: (1) parallel/divergent, (2) progradational, (3) mounded/draped, and (4) onlap/fills. Each of these facies types is described below in greater detail.

Parallel to subparallel and **divergent configurations** (Fig. 16.4) are the most common types of reflections within basins and are also the most difficult types to interpret because they do not represent unique depositional environments. Depositional settings that generate parallel/divergent facies include shelf or platform, delta platform, alluvial plain/distal fan delta, and basinal plain. Although all of these environmental settings generate similar appearing seismic reflectors, they are characterized by different lithofacies that range from shallow neritic limestones, shales, and sandstones on the shelf to hemipelagic clays and siltstones in basinal environments (Table 16.2). Lithofacies interpretation of parallel/divergent seismic facies cannot be done from reflection configuration alone and requires integration of all available data on seismic properties, including lateral changes and bounding relationships, reflection amplitude, continuity, and frequency.

Seismic facies characterized by **progradational configurations** display reflections in dip sections (sections oriented normal to strike) that are inclined with respect to underlying and overlying reflectors (Fig. 16.5). The reflection configurations that dominate in these facies include sigmoid, oblique, shingled, and complex composites of these configurations. Progradational seismic facies are particularly characteristic of slopes associated with a prograding shelf platform or in prodelta/slope settings (Table 16.3). Mitchum, Vail, and Sangree (1977) suggest that sigmoid configurations are generally associated with progradation of shelf or platform systems and that oblique progradations are associated with delta systems. Shingled configurations are most common in seismic facies interpreted as depositional units prograding into shallow water.

Mounded/draped configurations are characterized by geometric form ranging from lenses to mounds and by sheet to blanket geometry exhibiting drape over underlying surfaces (Table 16.4, Fig. 16.15). Reflection configurations may be mounded (wavy), chaotic, reflection-free, or, in the case of drape structures, parallel. Brown and

TABLE 16.2 Summary of seismic facies characterized by parallel and divergent reflection configurations

Properties of seismic facies	Depositional environments/settings			
	Shelf/platform	Delta platform: *delta front/ delta plain*	Alluvial plain/ distal-fan delta	Basinal plain
Reflection configuration	Parallel/slightly divergent; highly divergent near rare growth faults	Parallel/slightly divergent on shelf; highly divergent near growth faults in deep-water deltas	Parallel, generally grades basinward into delta plain or into shelf/platform facies	Parallel/slightly divergent; may grade laterally into divergent fills or mounds
Lithofacies and composition	Alternating neritic limestone and shale; rare sandstone; undaform deposits	Shallow marine delta-front sandstone/shale grading upward into subaerial delta-plain shale, coal, sandstone channels; prodelta facies excluded except where toplap is absent; undaform deposits	Meanderbelt and channel-fill sandstone and floodbasin mudstone; marine reworked fan delta sandstones/ profan shale; undaform deposits	Alternating hemipelagic clays and siltstone; calcareous and terrigenous composition; fondoform deposits
Geometry and structure	Sheetlike to wedge shaped or tabular; very stable setting; uniform subsidence	Sheetlike to wedge shaped or tabular on shelf, prismatic to lenticular basinward of subjacent shelf edge with growth faults and roll-over anticlines; relatively stable, uniform subsidence on shelf; rapid subsidence and faulting in deep-water delta	Sheetlike to wedge shaped (individually elongate ribbons or lobes), commonly tilted and eroded	Sheetlike to wedge-shaped; may be slightly wavy or draped over subjacent mounds; generally stable to uniform subsidence; may grade laterally into active structural areas
Lateral relationships	May grade landward into coastal facies and basinward into shelf-margin carbonate facies; local carbonate mounds	May grade landward into alluvial systems and basinward into prodelta/slope clinoforms (on shelf) or growth-faulted prodelta/ slope facies (deep-water setting)	Grade landward into reflection-free, high-sandstone facies; alluvial facies grade basinward into upper delta plain; fandelta facies grade basinward into shelf/platform or into slope clinoforms	Commonly grades shelfward into mounded turbidites or slope clinoforms; may grade laterally into deep-water mounds or fills
Nature of upper/ lower boundaries	Concordant, coastal onlap and/ or baselap over upper surface; upper surface may be eroded by submarine canyons; basal surface concordant, low-angle baselap or (rare) toplapped by subjacent clinoforms	Normally concordant at top but may be rarely onlapped or baselapped; upper surface may be eroded by submarine canyons; basal surface generally toplapped by prodelta/slope clinoforms (on shelf); rarely concordant with prodelta on shelf but common in deep-water, roll-over anticlines	Upper surface may be onlapped by coastal facies; top may be angular unconformity; base is generally concordant; fan deltas rarely overlie clinoforms (toplap)	Generally concordant at top and base; may onlap eroded slope clinoforms or eroded mounds; upper surface rarely eroded

TABLE 16.2 *continued*

Amplitude	High	High in delta front and coal/lignite or marine transgressive facies within delta plain; low/moderate in most delta plain and in prodelta where in continuity with delta front	Variable—low–high	Low to moderate
Continuity	High	High in delta-front, coal/lignite and marine transgressive facies; low/moderate in remainder of delta plain and prodelta where in lateral continuity with delta front	Discontinous; continuity decreases landward	High
Frequency (cycle breadth)	Broad or moderate; little variability	Variable; broader in delta front; coal/lignite and marine transgressive facies moderate; narrower in other delta plain and prodelta where in continuity with delta front	Variable; generally narrower cycles than shelf/platform	Generally narrower than shelf/platform; commonly very uniform breadth throughout

Source: Brown, L. F., Jr., and W. L. Fisher, 1980, Seismic stratigraphic interpretation and petroleum exploration: Geophysical principles and techniques: Am. Assoc. Petroleum Geologists Education Course Note Ser. 16. Table 1, p. 42, reprinted by permission of AAPG, Tulsa, Okla.

TABLE 16.3 Summary of seismic facies characterized by progradational reflection configurations.

Properties of seismic facies	Depositional environments/settings	
	Slope: *associated with prograding shelf/platform*	Prodelta/slope: *associated with prograding-shelf delta or shelf-margin delta; or slope: associated with prograding neritic shelf supplied periodically by shelf delta/fan delta*
Reflection configuration	Sigmoid clinoforms Progradational in dip profile; parallel to disrupted and mounded in strike profile	Oblique clinoforms Progradational in dip profile; hummocky, progradational to mounded in strike profile; mounds more common in deep-water slope than in prodelta/slope on shelf
Lithofacies and composition	Hemipelagic slope facies in upper/mid-clinoform; submarine fans common in lower clinoform; generally calcareous clay, silt, and some sand (base of clinoform); clinoform deposited in deep water beyond shelf edge	*On shelf:* prodelta (upper) and shallow slope facies (mid-clinoform and lower clinoform); deposited on submerged shelf; composition generally terrigenous clay, silt and sand; sand concentrated in submarine fans at base of clinoform *Beyond shelf edge:* (1) prodelta and deep-water slope associated with shelf-margin delta; may be growth-faulted; clay, silt and sand (in basal submarine fans); and (2) deep-water slope associated with prograding neritic shelf supplied periodically by shelf deltas/fan deltas; clay, silt, and sand (in basal submarine fans)

TABLE 16.3 *continued*

Geometry and structure	Lens-shaped slope system; poorly defined individual submarine fans and point sources; strike profile may intersect facies to define parallel to slightly mounded configurations; rarely affected by growth faults; represents low rate of sedimentation under relatively uniform sea-level rise and/or subsidence rate	Complex fan geometry with apices at shelf-edge point sources; each submarine fan resembles a bisected cone; total slope system lens- to wedge-shaped; strike profiles intersect fans or cones to display complex mounds; seismic facies deposited rapidly relative to subsidence and/or sea-level rise; highly unstable slopes associated with deep-water deltas (growth faults, roll-over anticlines)
Lateral relationships	Grades updip through shelf/platform edge facies into parallel/divergent shelf/platform (undaform) reflections; may grade downdip into basinal plain (fondoform) or mound/drape seismic facies; grades along strike to similar facies; may change landward to oblique facies	Terminates updip against base of delta platform or shelf/platform (undaform) facies and may grade downdip into basinal plain (fondoform), or mound/drape facies; may change basinward into sigmoid facies; grade along strike into mounded facies and locally submarine canyon-fill facies
Nature of upper/lower boundaries	Generally concordant at top and downlap (baselap) terminations at base; upper surface of outer or distal sigmoids may be eroded by submarine erosion and submarine canyons; eroded surface commonly onlapped by continental rise facies	Toplap termination at top and downlap (baselap) termination at base; may contain local or minor submarine erosion/onlap sequences; outer or distal oblique clinoforms commonly eroded by submarine erosion and submarine canyon cutting; eroded surface generally onlapped by continental rise facies
Amplitude	Moderate to high; uniform	Moderate to high in upper clinoform; moderate to low in lower clinoform; highly variable
Continuity	Generally continuous	Generally continuous in upper clinoform; discontinuous in mid-clinoform and lower clinoform; may exhibit better continuity near base
Frequency (cycle breadth)	Broadest in mid-clinoform where beds thickest; uniform along strike	Broadest at top, generally decreasing downdip as beds thin; variable along strike

Source: Brown, L. F., Jr., and W. L. Fisher, 1980, Table 2, p. 49, reprinted by permission of AAPG, Tulsa, Okla.

TABLE 16.4 Summary of seismic facies characterized by mounded and draped reflection configurations

Properties of seismic facies	Depositional environments/settings		
	Reefs and banks: *shelf/platform margin, back-shelf patch reefs and pinnacle/barrier reefs*	Submarine canyon and lower slope: *proximal turbidites, slumped clastics*	Hemipelagic clastics: *proximal basin and lower slope*
Reflection configuration	Mounded, chaotic, or reflector-free; pull-up or pull-down common	Mounded; complex and variable	Parallel; mirrors underlying surface

TABLE 16.4 *continued*

Lithofacies and composition	Shallow-water carbonate biogenic buildups; may or may not exhibit reef-forming framework	Sand and shale submarine fans; complex gravity-failure fans or mounds; turbidity flow; other grain flows, submarine landslides/ debris flows; clinoform/fondoform deposits	Terrigenous and calcareous clays (commonly alternating); pelagic oozes; deposition from suspension plumes and nepheloid clouds; fondoform deposits
Geometry and structure	Elongate lens-shaped (shelf/platform edge and barrier reefs); elongate to subcircular lens-shaped (patch and pinnacle reefs/banks); form on stable structural elements	Irregular fan-shaped to mounded geometry; common but not restricted to unstable basins	Sheet to blanket geometry exhibiting drape over underlying surface; common in deep, subsiding basins
Lateral relationships	Shelf/platform edge facies grade updip into parallel/divergent shelf/platform facies; grade downdip into talus and sigmoid clinoform facies; patch reef/bank facies grade updip and downdip into parallel/divergent shelf/platform facies; pinnacle and barrier facies grade downdip into talus clinoforms and to basinal plain (fondoform) facies	May grade shelfward into progradational clinoforms (normally oblique) or canyon onlap fill, or may pinch out against eroded slope; may grade basinward and laterally into basinal plain (fondoform); onlap fills or drapes	Commonly grades laterally or basinward into basinal plain (fondoform) facies; may grade shelfward into submarine canyon onlap fill; may onlap eroded slope
Nature of upper and lower boundaries	Upper surface concordant or may be onlapped by flank reflections; basal surface concordant, base-lapping, or may overlie clinoform toplap; pull-up or pull-down of basal surface common	Upper surface commonly erosional and onlapped, base-lapped, or concordant (with drape); basal surface irregularly baselapping; may appear concordant (low resolution), or may onlap (mounded onlap fill)	Upper surface commonly concordant, but may be onlapped or baselapped; basal surface generally concordant but may onlap eroded mound or slope
Amplitude	High along boundaries; may be moderate to low internally; commonly reflector-free	Variable; generally low; some higher internal amplitudes may be thin hemipelagic drapes	Low to moderate; some high amplitude reflections (well defined on high-frequency, shallow data)
Continuity	High along boundaries; internally discontinuous to reflector-free	Discontinuous to chaotic	High
Frequency (cycle breadth)	Broad; cycle may diverge into massively bedded buildup	Highly variable; commonly narrow	Narrow, uniform

Source: Brown, L. F., Jr., and W. L. Fisher, 1980, Table 3, p. 64, reprinted by permission of AAPG, Tulsa, Okla.)

Fisher (1980) suggest that mounded/draped configurations occur in response to facies deposited either in carbonate shelf/platform environments or in deep-water clastic slope/basin environments. Clastic mounds that occur in shelves are normally thick delta systems that show up as a mounded configuration in seismic profiles oriented parallel to the strike of the beds. Clastic mounds in deeper water form by sediment slumping, flow of dense turbidity currents, or transport by other gravity-flow mechanisms. They are particularly common as parts of submarine fan systems at the distal ends of submarine canyons. Some mounded clastic deposits in deep water have been attributed to transport and deposition by contour currents (Fig. 16.15).

Carbonate mounds are formed by biogenic buildup of skeletal material and deposition of associated carbonate deposits. They occur as reefs or banks that are commonly located along the shelf edge, but they may be located also on the inner shelf or along the shelf slope. Carbonate buildups large enough to be resolved on seismic records include elongate shelf or platform-margin deposits, local patch and barrier reefs,

FIGURE 16.15 Schematic diagrams illustrating mounded seismic facies units. (From Mitchum, R. M., Jr., P. R. Vail, and J. B. Sangree, 1977, Fig. 13, p. 132, reprinted by permission of AAPG, Tulsa, Okla.)

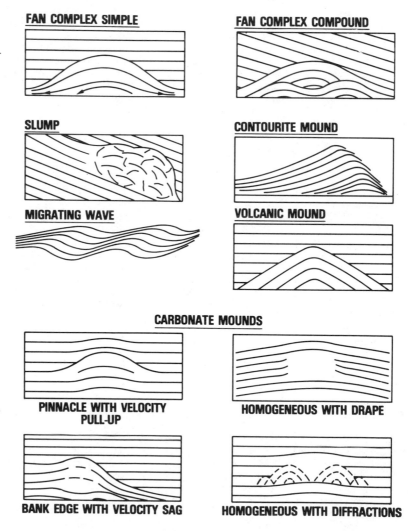

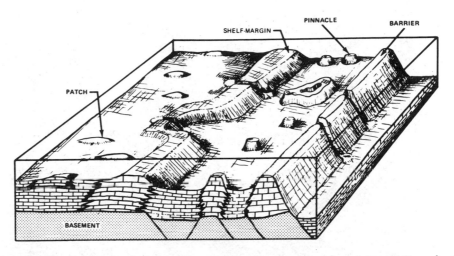

FIGURE 16.16 Types of carbonate buildups that can be recognized from interpretation of seismic data. (From Bubb J. N., and W. G. Hatelid, 1977, Seismic stratigraphy and global changes of sea level, Part 10: Seismic recognition of carbonate buildups, *in* C. E. Payton (ed.), Seismic stratigraphy—Applications to hydrocarbon exploration: Am. Assoc. Petroleum Geologists Mem. 26. Fig. 1, p. 186, reprinted by permission of AAPG, Tulsa, Okla.)

and elongate barrier reefs (Fig. 16.16). Carbonate mounds can be distinguished from clastic mounds on seismic sections by boundary outlines, drape of strata over carbonate buildups, and reflection configurations that are commonly chaotic or reflection-free (Bubb and Hatelid, 1977).

Draped configurations in deep water result from long periods of hemipelagic sedimentation that produce a sediment blanket that mirrors the bathymetric surface. Draped sediments are commonly composed of terrigenous clay, but may also include calcareous or siliceous pelagic oozes.

Onlap/fill configurations are discussed by Mitchum, Vail, and Sangree, 1977. They interpret fill reflection patterns as strata filling negative features in the underlying strata. The underlying strata may show either erosional truncation or concordance along the basal surfaces of the fill units. The external shapes of some fill units (channel fill, trough fill, basin fill, slope-front fill) are shown in Figure 16.14. Internal reflection configurations of fill units range from chaotic fill to mounded onlap fill and onlap fill (Fig. 16.17). Seismic fill units occur in environments such as erosional channels, canyons, and structural troughs or in depressions associated with fans or slumps. Figure 16.18 shows an example of sediment fill in major troughs or canyons, with the fill overlapping adjacent ridges.

FIGURE 16.17 Schematic diagrams illustrating some fill seismic facies units. (From Mitchum, R. M., Jr., P. R. Vail, and J. B. Sangree, 1977, Fig. 15, p. 133, reprinted by permission of AAPG, Tulsa, Okla.)

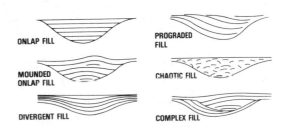

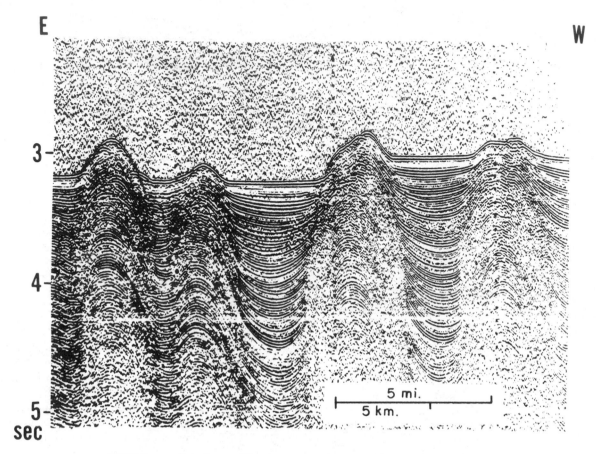

FIGURE 16.18 Sediment fill in major troughs as shown by seismic records. Sediments fill troughs and onlap ridges. This example shows folded Miocene sediments of the Mexican Ridge province with Pliocene–Pleistocene sediments filling the troughs. (From Stuart, C. J., and C. A. Caughey, 1977, Seismic facies and sedimentology of terrigenous Pleistocene deposits in northwest and central Gulf of Mexico, *in* C. E. Payton (ed.), Seismic stratigraphy—Applications to hydrocarbon exploration: Am. Assoc. Petroleum Geologists Mem. 26. Fig. 10, p. 260, reprinted by permission of AAPG, Tulsa, Okla.)

Although Vail et al. (1977a) do not designate coastal onlap reflectors as seismic facies, Brown and Fisher (1980) use coastal onlap (Fig. 14.7) to define proximal delta, fan delta, and shallow/platform seismic facies. Onlap is also recognized on continental rises along passive continental margins lacking an arc–trench system. Seismic facies characterized by onlap and fill reflections in coastal, continental rise, submarine canyon, and deep-water basins are summarized in Table 16.5.

Sea-Level Analysis

In addition to seismic sequence analysis and seismic facies analysis, the principles of seismic stratigraphy have been applied also to study and interpretation of ancient sea

TABLE 16.5 Summary of seismic facies characterized by onlap and fill reflection configurations

Properties of seismic facies	Depositional environments/settings			
	Coastal (paralic) onlap facies	Continental rise: *slope-front fill and onlap clastics*	Submarine canyon-fill deposits	Other deep-water fill deposits: *mounded, chaotic, structurally active basins*
Reflection configuration	Parallel; coastal onlap	Parallel/divergent; platform or shelfward onlap	Parallel/divergent; landward and lateral onlap	Parallel/divergent; chaotic, mounded onlap
Lithofacies and composition	Delta/alluvial plain and medial-fan delta sands and shales; supratidal clastic/carbonate facies; rarely beach/shoreface clastic facies	Sand and shale deposited in submarine fans by turbidity flows; hemipelagic terrigenous/calcareous clays; distal pelagic oozes	Sand and shale deposited by turbidity flow in submarine fans near base; hemipelagic and neritic shale/calcareous clays in middle and upper sequence, respectively; locally may contain coarse proximal turbidites	Sand and shale deposited by turbidity flow in submarine fans; hemipelagic terrigenous/calcareous clays; pelagic oozes; locally proximal turbidites
Geometry and structure	Sheetlike or tabular; uniform subsidence during deposition; periodic tilting and erosion; deposited near basinal hingeline during subsidence and/or sea-level rise	Wedge-shaped lens; may be fan shaped or lobate in plan view; slow subsidence	Elongate; lens-shaped in transverse section; may bifurcate updip; pinches out updip; slow subsidence	Variable lens shaped; commonly irregular; reflects bathymetric configuration of structural depression; slow to rapid subsidence
Lateral relationships	Pinches out landward; grades basinward into lower delta plain, distal fan-delta, or shelf/platform facies; may grade laterally into marine embayment facies	Pinches out updip; grades basinward into basinal plain or hemipelagic drape facies; continuous laterally for tens of kilometers	Pinches out updip and laterally; grades downdip into continental rise, mounded turbidites, or large submarine fans	Pinches out in every direction
Nature of upper/lower boundaries	Upper surface commonly tilted, eroded, and onlapped by similar deposits; base of facies onlaps unconformity, commonly angular	Upper surface commonly baselapped by prograding clinoforms; basal surface onlaps updip against eroded slope (and commonly outer shelf); may show baselap basinward against mounds or bathymetric highs	Upper surface may be concordant with overlying shelf or platform reflections or commonly baselapped by prograding prodelta and slope facies; basal surface onlaps updip and laterally; baselap onto basin floor rarely observed	Upper surface may be concordant with hemipelagic drape or baselapped by prograding clinoforms; basal surface onlaps in all directions

TABLE 16.5 *continued*

Amplitude	Variable; locally high but normally low to moderate	Variable; hemipelagic facies moderate to high; clastics low to moderate	Variable; generally low to moderate	Variable; generally low to moderate
Continuity	Low in clastics; higher in carbonate facies; decreases landward	Moderate to high; continuous reflections in response to hemipelagic facies	Variable; generally low to moderate	Variable; poor in chaotic or mounded fill; high in low-density turbidites and hemipelagics
Frequency (cycle breadth)	Variable; generally moderate to narrow	Narrow; uniform	Variable but generally narrow	Variable; commonly narrow; may increase breadth toward axis of fill

Source: Brown, L. F., Jr., and W. L. Fisher, 1980, Table 4, p. 80, reprinted by permission of AAPG, Tulsa, Okla.

levels and changes in sea level throughout geologic time. Studies of sea-level changes have special relevance with respect to analysis of cyclic sequences in the stratigraphic record. Sea-level changes through time have been studied particularly intensively by P. R. Vail and his associates at the Exxon Research Laboratory in Houston (Vail et al., 1977a, b). These authors used seismic data to integrate occurrences of coastal onlap, marine (deep-water) onlap, baselap, and toplap into a model involving asymmetric cycle oscillations of relative sea level.

Vail and his group constructed relative sea-level charts or curves by plotting the relative magnitude of a sea-level rise from measurements of coastal aggradation as determined from seismic sections and by determining the relative amount of a sea-level fall on the basis of differences in level between a unit that is displaced a significant distance seaward and the top of the next older unit. The specific procedures used in constructing these sea-level curves are outlined by Vail et al. (1977a) and are illustrated in Figure 16.19.

The first step in constructing the curve is analysis of coastal and marine sequences such as those shown as units A through E of Figure 16.19A. Sequence boundaries, areal distributions, and the presence or absence of coastal onlap and toplap are determined by tracing reflections on the seismic sections. Available age control from well data is used to establish the geologic-time range of each sequence. An environmental analysis is made from seismic and other available data to distinguish coastal facies from marine facies.

The second step is to construct a chronostratigraphic (time-stratigraphic) chart of the sequences. Both stratal surfaces and unconformities give time-stratigraphic information. Because they are depositional surfaces, stratal surfaces are chronostratigraphic reflectors. In addition, because seismic reflectors are isochronous they can cross lithologic boundaries. That is, the seismic reflections from a given surface may extend laterally through a variety of lithofacies. Seismic reflectors may be traced continuously, for example, through a shelf system, over the shelf edge, and downward through an equivalent slope system. Unconformities are not isochronous surfaces; however, all strata below an unconformity are older than all strata above it. Therefore, strata between unconformities constitute time-stratigraphic units. After determining

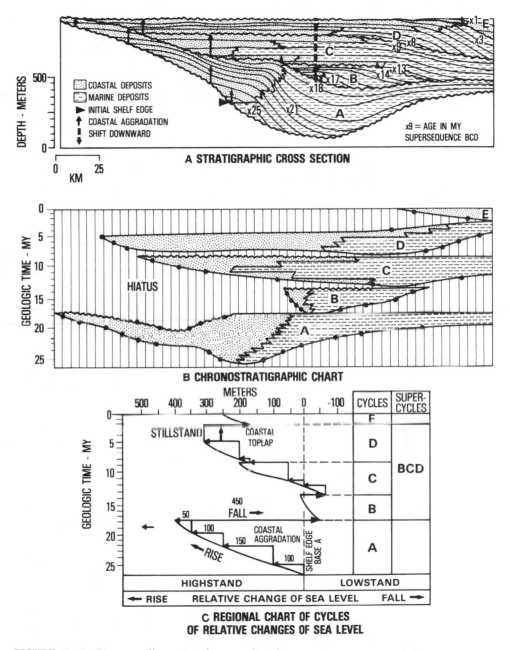

A STRATIGRAPHIC CROSS SECTION

B CHRONOSTRATIGRAPHIC CHART

C REGIONAL CHART OF CYCLES
OF RELATIVE CHANGES OF SEA LEVEL

FIGURE 16.19 Diagrams illustrating the procedure for constructing a regional chart of cycles of relative changes of sea level. (From Vail, P. R., R. M. Mitchum, Jr., and S. Thompson, III, 1977, Seismic stratigraphy and global changes of sea level, Part 3: Relative changes of sea level from coastal onlap, in C. E. Payton (ed.), Seismic stratigraphy—Applications to hydrocarbon exploration: Am. Assoc. Petroleum Geologists Mem. 26. Fig. 13, p. 78, reprinted by permission of AAPG, Tulsa, Okla.)

the ages of depositional sequences (such as those shown on the stratigraphic cross section in Fig. 16.19A) from well control or other information, the stratigraphic information is plotted against geologic time to construct a chronostratigraphic correlation chart (Fig. 16.19B).

The final step in the procedure for constructing a curve of relative sea-level change is to identify cycles of relative rise and fall of sea level in each seismic sequence, measure the magnitude of the rises and falls, and plot them and the sea-level standstills against geologic time, as shown in Figure 16.19C. Vail et al. (1977a) suggest that coastal aggradation is the best measure of a relative rise in sea level. A relative standstill is indicated by coastal toplap (Fig. 14.10). The plots of rise, fall, and standstill of sea level are repeated for each sequence (cycles A through E of Fig. 16.19C) to complete the relative sea-level chart.

The time interval occupied by a relative rise and fall of sea level, as illustrated in Figure 16.19C, constitutes a **cycle** of relative sea-level change in a region. Correlative regional cycles can be composited and used to construct charts of global cycles of relative change in sea level, but the cycles of sea-level change determined from seismic records are subject to different interpretations. Vail et al. (1977b) visualize a cycle as consisting of a gradual relative rise in sea level, requiring perhaps a few million years (Fig. 16.20); a period of standstill; and a relatively rapid fall in sea level, in perhaps a few thousand years. They consider these cycles to be worldwide and apparently to be controlled by absolute or eustatic sea-level changes. The Vail et al. (1977b) model for global sea-level change appears to imply that eustatic rates of change are greater than rates of basin subsidence.

Brown and Fisher (1980) challenge this basic premise as well as the concept that an abrupt seaward shift in onlap patterns is evidence of rapid sea-level fall. In contrast to Vail and his associates, Brown and Fisher visualize periods of sea-level rise interrupted by occasional standstills of sea level, but only rarely a sea-level fall. In their alternative model for sea-level change, gradual eustatic change is superimposed on basin subsidence and tilting. Gradually rising sea level, and/or basin subsidence, causes slope (marine) onlap, followed by coastal onlap and extensive marine transgression. Diminishing sea-level rise, or basin subsidence, and adequate sediment supply allow initiation of progradational coastal systems over the submerged shelf. During sea-level standstills, or minimum basin subsidence, extensive progradation of coastal systems and/or the shelf edge occurs, along with slope (marine) offlap. The principal difference between the alternative concept of Brown and Fisher and that of Vail and his coworkers thus lies in the implications of marine onlap.

Brown and Fisher are not the only workers to challenge the concept of Vail and his associates that coastal onlap and offlap reflect worldwide episodes of eustatic sea-level rise and fall. The controversy regarding sea-level changes is summarized by Kerr (1984). In their 1977 papers, Vail and his group assumed on the basis of seismic profiles that rising sea level laid down sediment layers higher and higher on the continental margin. When the sea withdrew, it laid down a series of retreating sediment layers.

FIGURE 16.20 (opposite)   Global cycles of relative change in sea level during Jurassic to Tertiary time according to Vail and his associates. Cretaceous cycles (hachured area) not released for publication at the time this diagram was published. (From Vail, P. R., R. M. Mitchum, Jr., and S. Thompson, III, 1977, Seismic statrigraphy and global changes of sea level, Part 4: Global cycles of relative changes of sea level, in C. E. Payton (ed.), Seismic stratigraphy—Applications to hydrocarbon exploration: Am. Assoc. Petroleum Geologists Mem. 26. Fig. 2, p. 85, reprinted by permission of AAPG, Tulsa, Okla.)

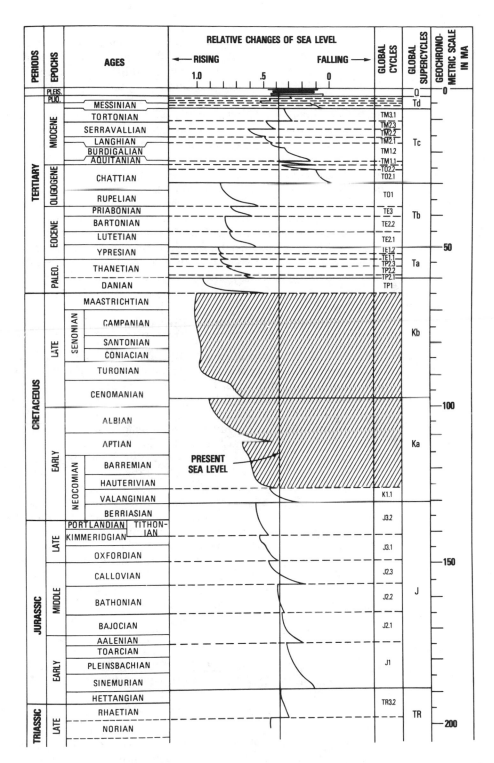

Exposure of the shelf then caused the development of unconformities. Based on this interpretation of the seismic profiles, the sea appeared to rise gradually and then drop almost instantaneously, giving the sea-level curves the saw-toothed character shown in Figure 16.20. Other workers in addition to Brown and Fisher have pointed out that there were other things moving around in addition to sea level. For example, if during sea-level fall the margin of a continent subsided more rapidly than falling sea level, the seismic data would suggest that sea level was still rising, whereas it was actually falling. Furthermore, variations in the rate of influx of terrigenous sediment can cause shorelines to shift independently of sea level. Vail and his group apparently now agree that factors other than sea-level change can affect the seismic facies patterns observed in seismic records (Vail and Todd, 1981) and that the interplay of sea level, subsidence, and sediment supply determines the configuration of sediment layers. Nonetheless, they apparently see no alternative to rapid, global sea-level changes to explain the apparent global synchroneity of major unconformities indicated by the seismic records. Hallam (1984) also reviews in detail the problem of pre-Quaternary sea-level changes. He presents a eustatic sea-level curve for Phanerozoic time (Fig. 16.21A) that is considerably smoother than Vail et al.'s. 1977 curve (Fig. 16.21B), suggesting a slower rate of sea-level drop than that shown in the Vail curve.

16.5   CORRELATION BY SEISMIC EVENTS

As discussed in Section 16.2, seismic methods are predicated on the fact that elastic waves transmitted downward from a point source at the surface are reflected back to the surface from discontinuities. These discontinuities are either bedding planes or unconformities. Mitchum, Vail, and Sangree (1977) emphasize the fact that both bedding planes and unconformities have chronostratigraphic significance. The physical surfaces that separate individual beds, laminae, or groups of strata are essentially synchronous surfaces in contrast to the boundaries of major lithostratigraphic units. The latter boundaries may or may not transgress time boundaries. Although the hiatus represented by an unconformity may not be of the same duration everywhere, unconformities nonetheless have time-stratigraphic significance because the rocks that lie above an unconformity are everywhere younger than those that lie below.

Because elastic waves are reflected from bedding planes and unconformities rather than from the boundaries between lithostratigraphic units (formation and other boundaries), seismic reflection patterns can be used for large-scale time-stratigraphic correlation throughout a region even where lithologic units are markedly diachronous. I pointed out, for example, that seismic reflectors may be traced continuously in some cases through a shelf system, over the shelf edge, and downward through an equivalent slope system. Figure 16.22 provides an example of time-stratigraphic correlations of regional scope based on lateral tracing of seismic reflection horizons.

16.6   FORMAL NOMENCLATURE AND CLASSIFICATION OF SEISMIC STRATIGRAPHIC UNITS

The terminology of seismic stratigraphic units used in this chapter was developed principally by Vail and associates at the Exxon Research Laboratory in Houston. At this time, no formal system of nomenclature equivalent to that established for litho-

FIGURE 16.21 Eustatic sea-level
curves for Phanerozoic time. A.
Hallam, 1984, B. Vail et al., 1977b.
(From Hallam, A., Pre-Quaternary
sea-level changes: Ann. Rev. Earth
and Planetary Sciences, v. 12. Fig.
5, p. 220, reprinted by permission.)

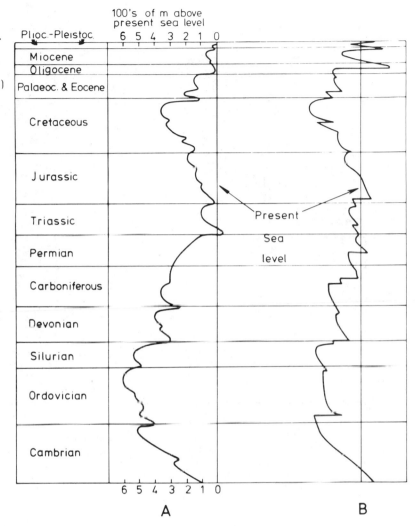

stratigraphic and magnetostratigraphic units has been proposed for seismic stratigraphic units by either the International Subcommission on Stratigraphic Nomenclature or the North American Commission on Stratigraphic Nomenclature. The 1976
International Stratigraphic Guide does suggest that seismic units can be treated as (informal) zones.

The degree to which seismic reflections can be segregated into identifiable sequences and facies depends upon the resolving power of the seismic technique and
the equipment in use. Packages of seismic reflections may represent stratigraphic intervals encompassing lithostratigraphic units as large as groups or supergroups or as
small as very thick beds. Seismic sequences, for example, probably represent packages
of lithofacies containing several formations or groups of formations that together may
make up the rocks of an entire geologic system, such as the Jurassic. Seismic facies
units, on the other hand, may include formations or lithostratigraphic units of lesser

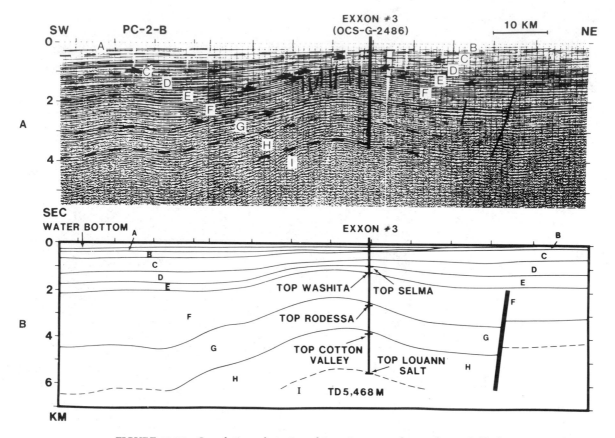

FIGURE 16.22   Correlation of stratigraphic units across the northeast Gulf of Mexico on the basis of seismic reflections. A. Original seismic section in which the vertical scale is calibrated in seconds. B. Time has been converted to depth and the geologist's interpretation of the correlation of units A–I is shown more clearly. (From Addy, S. K., and R. T., Buffler, 1984, Seismic stratigraphy of shelf and slope, northwestern Gulf of Mexico: Am. Assoc. Petroleum Geologists Bull., v. 68. Fig. 2, p. 1786, reprinted by permission of AAPG, Tulsa, Okla.)

rank, such as members. I emphasize the point that there is no direct equivalent relationship between seismic sequences and facies and lithostratigraphic units such as groups, formations, and members. This lack of relationship is especially true because seismic units are time-stratigraphic units that may transgress the boundaries of lithologic units.

ADDITIONAL READINGS

Anstey, N. A., 1982, Simple seismics: International Human Resources Development Corporation, Boston, 168 p.

Berg, O. R., and D. G. Wolverton (eds.), 1985, Seismic stratigraphy II: An integrated approach: Am. Assoc. Petroleum Geologists Mem. 39, 276 p.

Brown, L. F., Jr., and W. L. Fisher, 1980, Seismic stratigraphic interpretation and petroleum exploration: Geophysical principles and techniques: Am. Assoc. Petroleum Geologists Education Course Note Ser. 16, 56 p.

Neidell, N. S., 1979, Stratigraphic modeling and interpretation: Geophysical principles and techniques: Am. Assoc. Petroleum Geologists, Education Course Note Ser. 13, 141 p.

Payton, C. E. (ed.), 1977, Seismic stratigraphy—Applications to hydrocarbon exploration: Am. Assoc. Petroleum Geologists Mem. 26, 516 p.

Sheriff, R. E., 1980, Seismic stratigraphy: International Resources Development Corporation, Boston, 227 p.

Sheriff, R. E., and L. P. Geldart, 1982, Exploration seismology: History, theory, and data acquisition: Cambridge University Press, Cambridge, England, 253 p.

17
Biostratigraphy

17.1 INTRODUCTION

The preceding three chapters focus on stratigraphic relations in sedimentary sequences that exist owing to the physical characteristic of sedimentary rocks, either lithology or physical properties that can be remotely sensed by magnetic or seismic instrumentation. We turn in the present chapter to examination of the exceedingly important role that fossil organisms play in stratigraphy. First, fossils provide an additional and highly useful method for subdividing sedimentary rocks into identifiable stratigraphic units called biostratigraphic units. In addition, they make possible the ordering and relative-age dating of strata and their correlation on both a continental and a global scale. The characterization and correlation of rock units on the basis of their fossil content is called **biostratigraphy.** Stratigraphy based on the paleontologic characteristics of sedimentary rocks is referred to also as **stratigraphic paleontology,** the study of fossils and their distributions in various geologic formations.

Separation of rock units on the basis of fossil content may or may not yield stratigraphic units whose boundaries coincide with the boundaries of lithic stratigraphic units. In fact, lithostratigraphic units such as formations commonly can be subdivided by distinctive fossil assemblages into several smaller biostratigraphic units. Indeed, one of the primary objectives of biostratigraphy is to make possible differentiation of strata into small-scale subunits or zones that can be dated and correlated over wide geographic areas, allowing interpretation of Earth history within a precise framework of geologic time. On the other hand, it is quite possible for biologically defined stratigraphic units to span the boundaries of formally defined lithostratigraphic units. Some biostratigraphic units may thus include parts of two members or formations, for example, or even encompass two or more entire members or formations (Fig. 17.1).

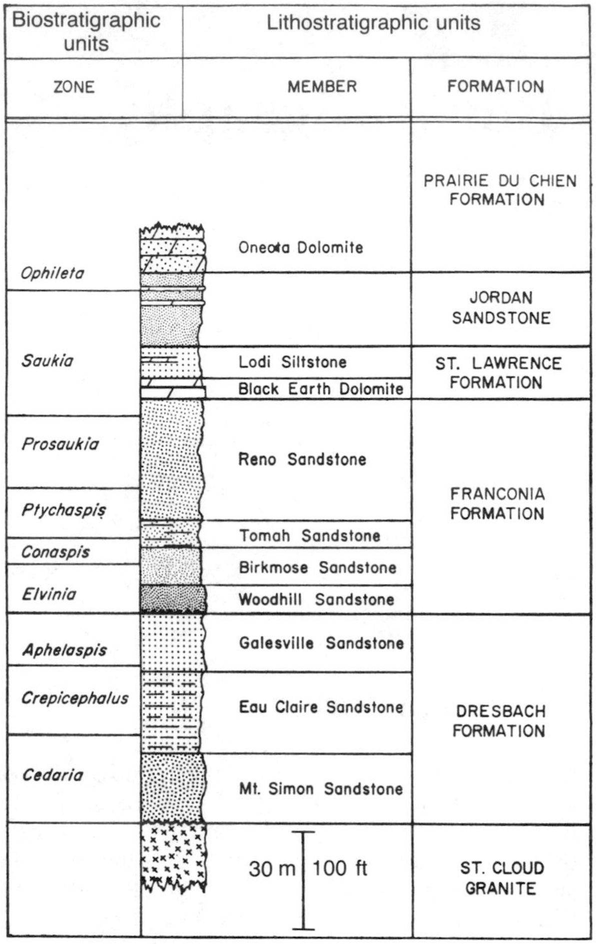

FIGURE 17.1 Relationship of biostratigraphic units to lithostratigraphic units. (Modified from Berg, R. R., C. A. Nelson, and W. C. Bell, 1956, Upper Cambrian rocks in southeastern Minnesota, *in* Lower Paleozoic of the upper Mississippi Valley: Geol. Soc. America Guidebook Ser., Minneapolis meeting, Field Trip 2.)

Biostratigraphic units	Lithostratigraphic units		
ZONE		MEMBER	FORMATION
			PRAIRIE DU CHIEN FORMATION
Ophileta		Oneota Dolomite	
			JORDAN SANDSTONE
Saukia		Lodi Siltstone	ST. LAWRENCE FORMATION
		Black Earth Dolomite	
Prosaukia		Reno Sandstone	FRANCONIA FORMATION
Ptychaspis			
		Tomah Sandstone	
Conaspis		Birkmose Sandstone	
Elvinia		Woodhill Sandstone	
Aphelaspis		Galesville Sandstone	
Crepicephalus		Eau Claire Sandstone	DRESBACH FORMATION
Cedaria		Mt. Simon Sandstone	
		30 m 100 ft	ST. CLOUD GRANITE

The concept of biostratigraphy is based on the observation that organisms have undergone successive changes throughout geologic time. Thus, any unit of strata can be dated and characterized by its fossil content. That is, any stratigraphic unit can be differentiated on the basis of its contained fossils from stratigraphically younger and older units. Biostratigraphy is obviously closely allied to paleontology, and a skilled biostratigrapher must also be a well-trained paleontologist. In fact, the study of biostratigraphy is for specialists who have intimate knowledge of large groups of organisms and their temporal and spatial distribution. Because stratigraphic paleontology is such a complex field, comprehensive treatment of this subject is beyond the scope of this book. The aim of this chapter is to introduce some very basic concepts and principles of biostratigraphy. Readers who wish more in-depth treatment of biostratigraphy should consult standard reference works on paleontology as well as more specialized, biostratigraphically oriented monographs such as those of Berry (1968), Cubitt and Reyment (1982), Dodd and Stanton (1981), Gray and Boucot (1979), Hallam (1973b), Kauffman and Hazel (1977), Middlemass et al. (1971), A. B. Shaw (1964), and S. M. Stanley (1979).

We begin discussion of biostratigraphy by examining the concept that fossils constitute a valid basis for stratigraphic subdivision. As part of this examination, the origin and development of methods for biostratigraphic zonation are traced, and the stratigraphic procedures currently in use for classifying, naming, and describing biostratigraphic units are discussed. Organic evolution and the distribution of organisms in both time and space is explored next. We conclude the chapter with a discussion of the extemely important role that biostratigraphy plays in correlation of stratigraphic units. Biochronology, the use of fossils for calibrating the geologic time scale, is discussed in Chapter 18.

17.2 FOSSILS AS A BASIS FOR STRATIGRAPHIC SUBDIVISION

Principle of Faunal Succession

An English surveyor and civil engineer named William Smith, who worked in England and Wales in the late eighteenth century, is credited with discovering the fundamental principle of biostratigraphy. Previous workers had recognized that fossils are the remains of once-living organisms, and some workers had even suggested the possibility that certain species of marine shelled organisms had become extinct. Smith was evidently the first to utilize fossils as a practical tool for characterizing, subdividing, and correlating strata from one area to another. In his work as a surveyor and canal builder, he had discovered by about 1796 that the strata in and around Bath in Somerset and for some distance outward were always found in the same order of superposition—the order in which rocks are placed above one another. Furthermore, he noted that each layer in the stratigraphic succession was characterized by the same distinctive fossil assemblage wherever it was found throughout the region. Soon, Smith was able to assign any fossil-bearing rock to its proper superpositional interval by comparing its fossils with others whose stratigraphic position he knew from previous study. He thus discovered that strata occur in a definite and determinable order. On the basis of Smith's discovery, we now know that rocks formed during any particular interval of geologic time can be recognized and distinguished by their fossil content from rocks formed during other time intervals. This concept has consequently become known as the **principle of faunal succession,** or the **law of faunal succession.** Even without assigning names to fossils, Smith was successful in using them to establish a stratigraphic succession and to subdivide the rocks into mappable units by a combination of lithologic characteristics and fossil assemblages.

It is important to stress that Smith did not subdivide rock successions on the basis of fossils alone. His strata were first delineated and named according to their lithology. Then, their characteristic fossils were collected and studied. The use of fossils alone to subdivide thick, essentially lithologically homogenous formations did not come about for another 15 years. The French scientist George Cuvier, a contemporary of Smith's, recognized the desirability of using fossils to subdivide rocks but did not attempt this process himself. Subdivision of rock successions on the basis of fossils was first carried out on sediments of Tertiary age in the early 1830s. Deshayes in France in 1830, Bronn in Germany in 1831, and Lyell in England in 1833 all proposed subdivisions of Tertiary strata based on fossils (Hancock, 1977). Lyell's subdivisions are historically noteworthy. He split the Tertiary strata into four units on the basis of proportions of living to extinct species in the rocks (Table 17.1). Thus, we see here for

TABLE 17.1 Lyell's subdivisions of the Tertiary

Name of subdivision	Extant species in the rocks (%)
Pliocene (more recent)	
Newer Pliocene	90
Older Pliocene	33–50
Miocene (less recent)	18
Eocene (dawn of Recent)	3.5

apparently the first time the use of fossils as an essential part of the definition of units of geologic time and the possibility of biostratigraphy freed from lithologic control.

Concept of Stage

Although Smith's principle of faunal succession was to be the cornerstone for all subsequent biostratigraphy, his own work on biostratigraphic successions led to only vaguely defined time units (Berry, 1968). Lyell's subdivisions were similarly vague and, in any case, were confined to the Tertiary. A closer look at fossil successions was needed to refine their use in dating and correlation. This important step came with introduction of the concept of stage, credited to the French paleontologist Alcide d'Orbigny. About 1842, d'Orbigny came up with the idea of erecting major subdivisions of strata, each systematically following the other and each bearing a unique assemblage of fossils. Like Smith, d'Orbigny recognized that similarity of fossil assemblages was the key to correlating rock units, but he went a step further to propose that strata characterized by distinctive and unique fossil assemblages might include many formations (lithostratigraphic units) in one place or only a single formation or part of a formation in another place. He defined as **stages** groups of strata containing the same major fossil assemblages. He named these stages after geographic localities with particularly good sections of rock containing the characteristic fossils on which the stages are based. Using the stage concept, he was able to divide the rocks of the Jurassic system into ten stages and the Cretaceous rocks into seven stages, each characterized strictly by its fossil fauna.

The boundaries of d'Orbigny's stages were defined at intervals marked by the last appearance, or disappearance, of distinctive assemblages of life forms and their replacement in the rock record by other assemblages. He conceived these stages as having worldwide extent and to be the result of repeated catastrophic destruction of life on Earth followed by new creations. His ideas on catastrophic destruction and special new creation, like those of Cuvier who preceded him, failed to gain lasting acceptance among geologists, and subsequent study has shown that his stages and their characteristic faunas are local rather than worldwide. Nonetheless, d'Orbigny's concept of stages as major bodies of strata characterized by large assemblages of fossils unique to that part of the total stratigraphic column was a significant and lasting contribution to the growing discipline of biostratigraphy. Somewhat different interpretations of the meaning of stage have been used by subsequent workers, but d'Orbigny's basic concept is still valid.

Concept of Zone

The stage concept of d'Orbigny permitted subdivision of strata into major successions based on fossils. What it did not provide was a method by which fossiliferous strata could be divided into small-magnitude, clearly delimited units. Friedrich Quenstedt in Germany was particularly critical of d'Orbigny's stages because, according to him, d'Orbigny's method "centered around the acceptance, as the diagnostic faunal aggregate, of species of many strata in many localities, lumped together without enough regard for their precise stratigraphic ranges" (Berry, 1968, p. 125). Quenstedt maintained that only by extremely detailed study of strata on essentially a centimeter-by-centimeter basis could full understanding of the succession of faunas be developed. Quenstedt's own work did not bring this notion of detailed biostratigraphic subdivision into full fruition. It remained for his student, Albert Oppel, to expand, synthesize, and meld Quenstedt's ideas into the concept of the **zone.**

Oppel introduced the concept of zone in 1856 and thereby altered for all time the practice of biostratigraphy. Working with Jurassic rocks in various parts of Germany, he conceived the idea of small-scale units defined by the **stratigraphic ranges** of fossil species irrespective of lithology of the fossil-bearing beds. Oppel noted that the vertical ranges of some species were very short, that is, the species existed for only a very short time geologically. Others were quite long, but most were of some intermediate length (Berry, 1968). Oppel noted also that the assemblages of fossils that characterized the strata were made up of **overlapping** ranges of fossils. He defined his zones by exploring the vertical range of each separate species. Each zone was characterized by the joint occurrence of species not found together above or below this zone. Thus, the range of some species began at the base of a zone (the first appearance of a species), others ended at the top of a zone (the last appearance of a species), whereas still others ranged throughout the zone or even extended beyond it. Using species ranges, Oppel discovered that he could delineate the boundaries between small-scale rock units and distinguish a succession of unique fossil assemblages. Each of these assemblages was bounded at its base by the appearance of distinctive new species and at its top—that is, the base of the succeeding section—by the appearance of other new species. It is, however, the overlapping stratigraphic ranges of the species that make up the fossil assemblage that typifies a zone (Fig. 17.2). Because a zone represents the time between the appearance of species chosen as the base of the zone and the appearance of other species chosen as the base of the next succeeding zone, recognition of zones thus permits delineation of clear-cut, small-scale **time units.** Each of Oppel's zones was named after a particular fossil species, called an **index fossil,** or **index species,** which is but one fossil species in the assemblage of species that characterize the zone.

The concept of zone thus allowed subdivision of stages into two or more smaller, distinctive biostratigraphic units that could be recognized and correlated over long distances. Oppel was able, for example, to subdivide the Jurassic rocks of western Europe into 33 zones. It should be noted that Oppel did not start with d'Orbigny's stages and subdivide them into zones. Instead, he delineated zones on the basis of fossil ranges and then combined the zones into stages, all of which did not necessarily fit into one of d'Orbigny's stages as then defined (Hancock, 1977).

Zones were slow to be adopted into stratigraphic practice, especially in the United States, but with minor modifications of Oppel's method they have now become the common denominator of biostratigraphic study. Zones have been extended to all parts of the fossil record, not just the Jurassic, and to all areas of the world. It is now recognized, however, that there are definite geographic limits beyond which most zones cannot be traced. The area within which a zone can be recognized is a **biogeographic province.** Because zones constitute the basic unit of biostratigraphic classifi-

FIGURE 17.2 Diagrammatic illustration of an Oppel zone defined by the overlapping ranges of two or more taxa.

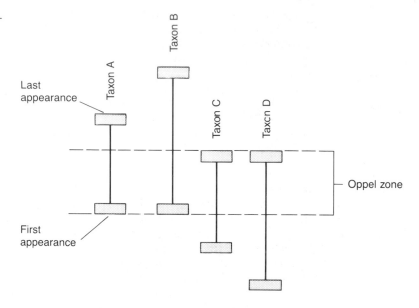

cation, considerable work has gone into efforts to standardize their usage. The current usage of zone and the different kinds of zones now recognized are described in the following section.

17.3 BIOSTRATIGRAPHIC UNITS

As the preceding discussion makes clear, a biostratigraphic unit is a body of rock strata characterized by its fossil content that distinguishes and differentiates it from adjacent strata. Furthermore, the zone, or **biozone,** is the fundamental biostratigraphic unit. Zones do not have any prescribed thickness or geographic extent. They may range in thickness from thin beds to units thousands of meters thick and in geographic extent from local units to those with nearly worldwide distribution. Recent attempts to standardize nomenclature and usage of zones have been made by the International Subcommission on Stratigraphic Classification (Hedberg, 1976) in the International Stratigraphic Guide and by the North American Commission on Stratigraphic Nomenclature (1983) in the North American Stratigraphic Code. The usage in this book follows that recommended in the North American Stratigraphic Code, but reference is made to usage in the International guide where appropriate.

Principal Categories of Zones

The North American Stratigraphic Code (Articles 49–52, Appendix) subdivides biostratigraphic units into three principal kinds of biozones: interval zones, assemblage zones, and abundance zones. Each of these zones is distinguished by different criteria.

An **interval zone,** or subzone, is the body of strata between two specific, documented lowest and/or highest occurrences of single taxa. Taxa is the plural of taxon, which is a term used to signify a taxonomic group or entity, for example, a species. The boundaries of interval zones are defined by lowest and/or highest occurrences of single taxa, and three basic types of interval zones are recognized:

1. *The interval between the lowest and highest occurrences of a single taxon* (Fig. 4A, Appendix) constitutes the simplest type of interval zone in that it involves a single species or other taxon. This type of interval zone is called a **taxon range zone** in the International Stratigraphic Guide. It is the body of strata representing the total range of occurrence of specimens of a particular taxon, as opposed to the local range.

2. *The interval between the documented lowest occurrence of one taxon and the documented highest occurrence of another taxon* (Fig. 4B, Appendix) defines a type of interval zone that is called a concurrent range zone in some cases. For example, the International Stratigraphic Guide uses the term **concurrent range zone** for such an interval zone when the occurrence results in stratigraphic overlap of the taxa (Fig. 4B1, Appendix). When such occurrences do not result in stratigraphic overlap (Fig. 4B2, Appendix), the interval zone may be called a **partial range zone.** An example of the use of concurrent range zones for subdividing the Cretaceous (Maastrichtian) chalks of northwestern Europe is shown in Figure 17.3.

3. *The interval between documented successive lowest occurrences or successive highest occurrences of two taxa* (Fig. 4C, Appendix) forms the final type of interval zone. When the interval is between successive documented lowest occurrences within an evolutionary lineage (Fig. 4C1, Appendix), it is the **lineage zone** of the International Stratigraphic Guide. When the interval is between successive lowest occurrences of unrelated taxa or between successive highest occurrences of either related or unrelated taxa (Fig. 4C2, Appendix), it corresponds to the **interval zone** of the International Stratigraphic Guide. Note that the interval-zone, defined in the International Stratigraphic Guide as an interval between two distinctive biostratigraphic horizons, is thus much more restrictive than the general definition of an interval zone given by the North American Stratigraphic Code.

An **assemblage zone** is defined by the North American Stratigraphic Code as a biozone characterized by the association of three or more taxa. An assemblage zone may consist of a geographically or stratigraphically restricted assemblage or may incorporate two or more contemporaneous assemblages with shared characterizing taxa (Fig. 5C, Appendix), in which case it may be referred to as a **composite assemblage zone.** The Code suggests that two concepts can be used to define assemblage zones. If the zone is characterized by taxa without regard to their range limits (Fig. 5A, Appendix), it is called an assemblage zone. If, on the other hand, it is characterized by more than two taxa and is further characterized by having boundaries based on two or more documented first and/or last occurrences of the included characterizing taxa (Fig. 5B, Appendix), it is called an **Oppel zone.** The inclusion of the Oppel zone as a subcategory of an assemblage zone differs from usage in the International Stratigraphic Guide. In the Guide, an Oppel zone is defined as a zone characterized by an association or aggregation of selected taxa of restricted and largely concurrent range, chosen as indicative of approximate contemporaneity. It is considered in the Guide to be a subcategory of a concurrent-range-zone. These differences may cause readers some confusion, but that's the way it is.

An **abundance zone** is a biozone characterized by quantitatively distinctive maxima of related abundance of one or more taxa (Article 52, Appendix). The abundance zone is equivalent to the **acme-zone** of the International Stratigraphic Guide. The

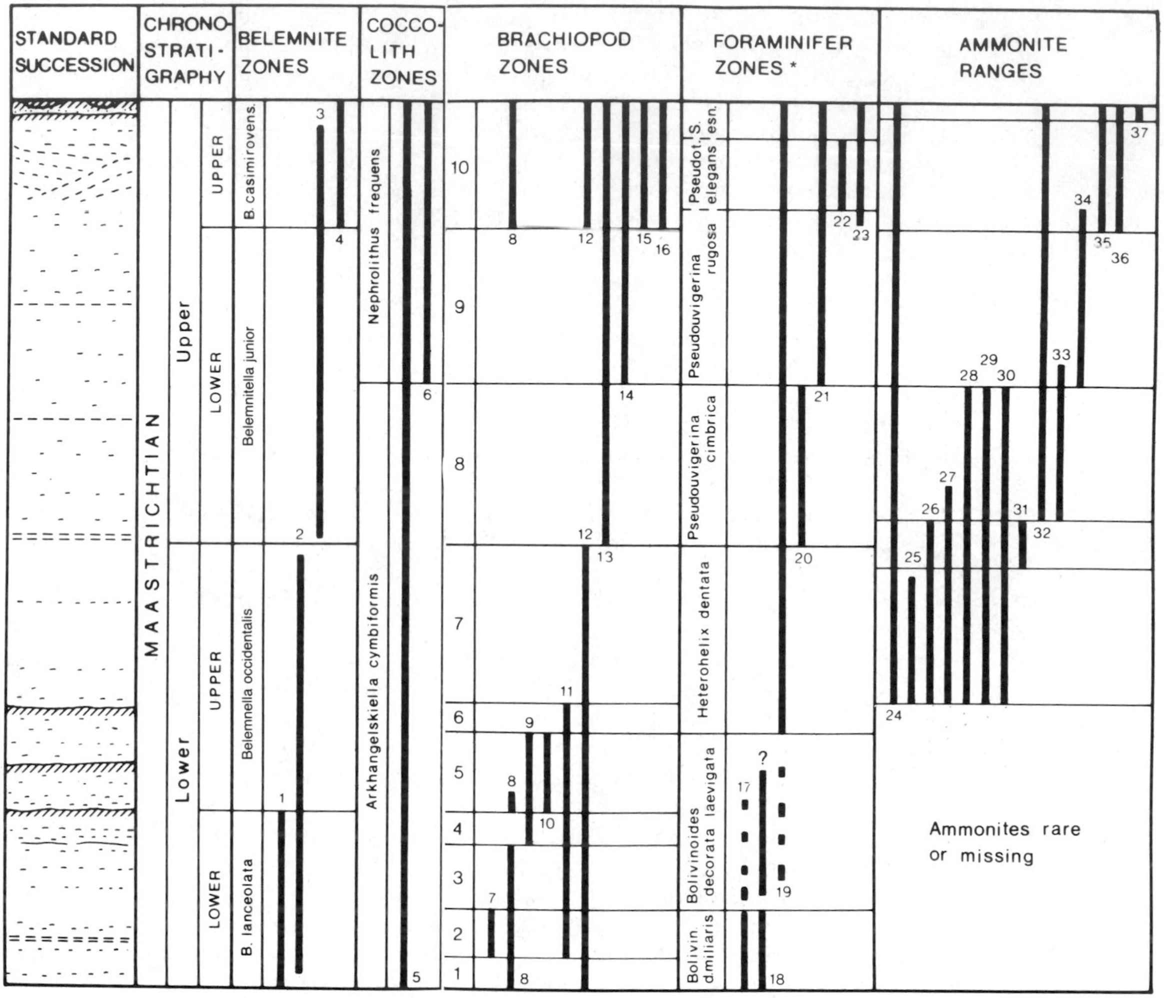

*S.esn = Stensioeina esnehensis; Pseudot. = Pseudotextularia

FIGURE 17.3 Subdivision of the Cretaceous (Maastrichtian) Chalk in northwest Europe using concurrent range zones (biozones). Numbers besides range bars refer to individual species. (From Surlyk, F., and T. Birkelund, 1977, An integrated stratigraphical study of fossil assemblages from the Maastrichtian white chalk of northwestern Europe, *in* E. G. Kauffman and J. E. Hazel (eds.), Concepts and methods of biostratigraphy. Fig. 13, p. 280. Copyright 1977, 1983 by Van Nostrand Reinhold. All rights reserved.)

Guide suggests that an acme-zone represents the maximum development, which commonly means maximum abundance or frequency of occurrence, of some species, genus, or other taxon, but not its total range. Figure 17.4 shows Cretaceous strata in northwest Germany subdivided on the basis of abundance zones of foraminiferal species.

FIGURE 17.4 Abundance zones illustrated by the vertical range and frequency of the foraminifer *Gaudryina dividens* and four species of *Spiroplectinata* from the Cretaceous of northwest Germany. (From Thenius, E., 1973, Fossils and the life of the past. Fig. 54, p. 89, reprinted by permission of Springer-Verlag, Heidelberg. After B. Grabert, 1959, Senckenburg naturf. Gesell. Abh., 496, Frankfurt.)

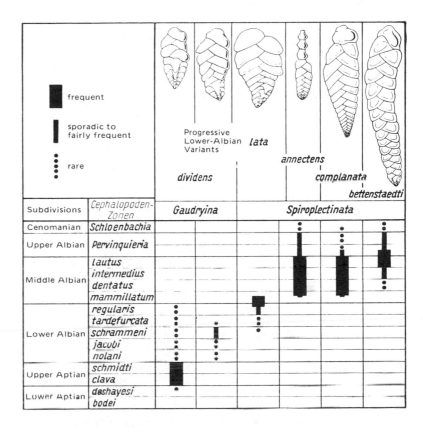

Rank of Biostratigraphic Units

The zone, or biozone, is the fundamental unit of biostratigraphic classification. Other biostratigraphic units are formed by either grouping or subdividing zones. The International Stratigraphic Guide suggests that some kinds of biozones, such as assemblage zones and Oppel zones, may be subdivided into subzones and/or grouped into superzones. The North American Stratigraphic Code provides that a biozone may be completely or partly divided into formally designated subbiozones (subzones) if such divisions serve a useful purpose.

Naming Biostratigraphic Units

The formal procedure for establishing and naming biostratigraphic units is outlined in the North American Stratigraphic Code (Appendix). Each biozone is given a unique name, which is compound and which designates the kind of biozone; for example, the *Exus albus* assemblage zone or the *Rotalipora cushmani* taxon range zone. The name may be based on one or two characteristic and common taxa that are restricted to the biozone or have their total stratigraphic overlap within the biozone. These names most commonly are those of genera or subgenera, binomial designations of species, or trinomial designations of subspecies.

17.4   THE BASIS FOR BIOSTRATIGRAPHIC ZONATION: CHANGES IN ORGANISMS THROUGH TIME

Evolution

The practicality of biostratigraphy as a tool for characterizing and correlating strata had been clearly established by Smith, d'Orbigny, and Oppel by the middle of the nineteenth century. None of these workers fully understood why fossil assemblages changed from one stratigraphic layer to another, although d'Orbigny apparently gave considerable thought to the origin of his stage boundaries. Charles Darwin provided the answer to this puzzle a few years after Oppel conceived the zone concept. In his monumental work on the origin of species, published in 1859, Darwin demonstrated the existence of organic evolution and thereby greatly changed subsequent geological and philosophical thought, although his ideas were by no means accepted by all of his contemporaries.

Darwin was not the first to conceive the general idea of evolution, but previous workers had marshalled little supporting evidence for their ideas. By contrast, Darwin drew together data on the fossil records of extinct organisms, the results of selective breeding of domestic animals, observations on ecological adaptations and variations among living organisms, and details on comparative anatomy. He then wove these data into a powerful argument for organic evolution. Darwin pointed out that all organisms have high reproductive rates, yet populations of these organisms remain essentially constant over the long run. He explained this observation by suggesting that not all organisms of the same kind (species) are equally well equipped to survive, and therefore many individuals die before reproducing. Each individual of a species differs from other individuals as a result of variations that arise within an organism entirely by chance. Some of these chance variations may be an advantage to the organism in coping with its environment in its struggle for existence. Others may be a disadvantage. Successful variations help organisms survive and extend their environment and range. Unsuccessful variations result in extinction.

Darwin termed this process of weeding out the unfit and survival of the fittest **natural selection.** Furthermore, he proposed that these favorable variations are inheritable and can be transmitted from one generation to the next. Darwin's fundamental contribution to understanding evolution was thus recognition that natural selection was the process by which new species arise. New species appear because the composition of populations changes with time owing to the fact that those individuals that undergo favorable adaptations will stand a better chance of surviving and reproducing. He did not understand how variations arose or how these traits were passed on from one generation of organisms to the next. The concept of spontaneous changes in genes that we now call **mutation** was not known at the time Darwin published *The Origin of Species.*

Taxonomic Classification and the Importance of Species

Organisms can be classified in a variety of ways, including habitat (planktonic, nektonic, benthonic) and environmental distribution (littoral, neritic, bathyal, and so forth); however, **taxonomic classification** based on genetic relationships is most pertinent to recognition of evolution and biostratigraphic zonation. The basic system of taxonomic classification now in use was introduced in 1735 by the Swedish naturalist

TABLE 17.2 Taxonomic systems for classifying organisms

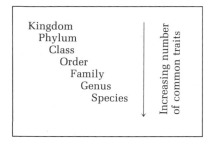

Kingdom
Phylum
Class
Order
Family
Genus
Species

Increasing number of common traits

Linnaeus, who grouped organisms into a hierarchy of different categories based on the number of distinctive characteristics shared in common. Organisms in the lowest, or least inclusive, category have the greatest number of common characteristics; those in the next highest category have fewer common characteristics; and so on until the highest, or most inclusive, category is reached. In the last category, organisms share only a very few common characteristics, or traits. Linnaeus's system of classification, as modified by some later additions, is illustrated in Table 17.2.

The Linnaean system of taxonomic classification brought to light the fact that degrees of similarities among organisms differ at different levels of classification. Differences among groups of organisms are greatest at the kingdom level and least at the species level. Species have thus become the fundamental entity of biostratigraphy. Biologists define species as *a breeding community that preserves its genetic identity by its ability to exchange genes with other breeding communities.* In other words, all members of a given species have the ability to interbreed, but they do not normally breed with members of a different species.

Changes in Species through Time

The importance of species in biostratigraphic study lies in the fact that species do not remain immutable for all times. If environmental conditions remained absolutely constant through time, perhaps species would never change. The fact is, environments do change, and as they change species also change, although environments do not directly cause species to change. Both gene mutation, or gene pool combinations, and shifting environmental conditions are essential to the evolution of species. Most species are well adjusted to their normal environments, but if an appropriate variation appears in a species just at the time when it is becoming inadaptive to a changing environment, the force of natural selection may preserve this novel variant (A. B. Shaw, 1964). Thus, species have evolved through time as a result of natural selection of those random, chance mutations that brought the species into better adjustment with changing environmental conditions.

All indications from the geologic record suggest that species variations are one-directional and nonreversible. Once a species has become extinct, it does not reappear in the fossil record. As members of a new species increase in numbers, they may eventually become abundant and widespread enough to show up in the geologic record as the **first appearance** of the species. When the species is no longer able to adjust to shifting environmental conditions, its members decrease in number and eventually disappear—the extinction, or **last appearance,** of the species. Some species exist for only a fraction of a geologic period. Others may persist for longer periods of time. Organisms that were abundant, geographically widespread, and had relatively short

FIGURE 17.5 The most important macrofossil groups of marine invertebrate organisms for biostratigraphic zonation. The white columns show the time span of distribution, the black columns the time span in which the organisms are important as index fossils. (From Thenius, E., 1973, Fossils and the life of the past, Fig. 50, p. 79, reprinted by permission of Springer-Verlag, Heidelberg.)

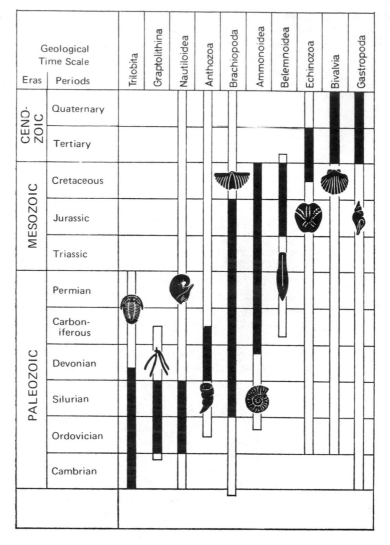

ranges have the greatest time-stratigraphic utility, that is, the greatest usefulness for biostratigraphic study (Fig. 17.5).

Rates of Evolution. Rates of evolution among different groups of organisms have varied through time. Some species seem to have changed dramatically in a short period of time, whereas others evolved very little over long stretches of time. There is currently considerable controversy among paleontologists concerning the "tempo" of change in organic evolution. Two principal points of view prevail. One view states that evolution proceeds mainly as a gradual change by slow, steady transformation of well-established lineages—**phyletic evolution,** or **gradualism.** The second view holds that most species arise very rapidly from small populations of organisms that have become isolated from the parental range, and then subsequently change very little after their successful origin. This latter view represents evolution by speciation or branching of lineages, the so-called **punctuated equilibria** model of Eldredge and Gould (1972). Differences in these two postulated modes of evolution are illustrated diagrammatically in Figure 17.6.

The gradualist concept has been the traditional view of species evolution, and many prominent workers still hold this view. Perusal of recent paleontologic literature suggests, however, that an increasing number of scientists now consider species evolution to be mainly punctuational. In the punctuational model, **speciation,** or branching of species, is viewed as a very rapid process, requiring only tens of thousands of years or possibly as little as a few hundred years (S. M. Stanley, 1979) after a population becomes reproductively isolated from the parent population. Although the duration of species from first appearance to extinction may be measured in millions of years (Table 17.3), species are believed by the punctuationalists to change morphologically very little and only very slowly after initial speciation. This concept is stated very succinctly by Eldredge and Gould (1977, p. 40), who emphasize the importance of speciation (splitting) and claim that "most morphological differences between two species appear in conjunction with the speciation process itself, whereas most of a species' history involves little further change, at least of a progressive nature." A third style of evolution, called **reticulate speciation,** was suggested by Sylvester-Bradley (1977). This evolutionary style combines on a small scale the mechanisms of phyletic evolution and punctuated equilibrium. It was proposed to account for the behavior of some modern species that appear to have remained stable for millions of years before experiencing rapid speciation following the Pleistocene ice age. Reticulated speciation has not yet been demonstrated in fossil groups.

A related and also controversial question about evolution concerns the factors that control the rates of evolution and extinction. The principal question here is this. Is it chiefly environmental factors external to organisms that control organic change, or does change arise from some independent and internal dynamic control, some key biologic innovation, within organisms themselves (T. J. M. Schopf, 1977)? Most workers seem to agree that environment is the crucial factor controlling diversity of organisms, but the exact way in which environment regulates evolution is less clear, although competition appears to be the key element of the model.

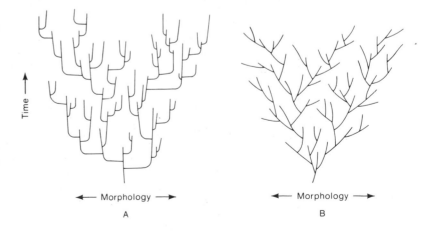

FIGURE 17.6 Diagrammatic sketch of hypothetical phylogenies (lines of direct descent in a group of organisms) representing the punctuational model (A) and the gradualistic model (B). Note that some phyletic evolution is indicated in model A and that some speciation events in model B display accelerated evolution. (From Macroevolution, patterns and process, by S. M. Stanley. W. H. Freeman and Company. Fig. 2.4, p. 17, © 1979.)

Time

← Morphology →

A

← Morphology →

B

TABLE 17.3 Mean duration of species and genera based on empirical data

| Group | Duration in millions of years | |
	Species	Genera
Brachiopods		Shortest lived not <10 Mode 16–20
Bryozoans	10–20	
Bivalves	~7	
Ammonites	Mean ~0.5–1.0 Range 0.2–25	
Graptolites	Mean 1.9 Range <1–8	Mean ~5
Mammals	~1	~2 (for rodents)
Mammals	~1	

Source: Schopf, T. J. M., 1977, Patterns of evolution: A summary and discussion, *in* A. Hallam (ed.), Patterns of evolution, as illustrated by the fossil record. Table IX, p. 559, reprinted by permission of Elsevier Science Publishers, Amsterdam.

Boucot (1975) suggests that worldwide population size is inversely related to rates of evolution and that population size forms the first-order control related to rates of evolution. Thus, there is a general increase in rate of evolution related to smaller population size. Boucot recognizes that there may be many second-order controls that affect population size and thus rates of evolution. These lower-order controls on marine invertebrates may include (1) size and number of biogenographic entities; (2) size and number of organic communities; (3) number, size, and scatter of reef environments; (4) strength or weakness of climatic gradients acting as one of the major barriers involved in presence or absence of biogeographic units; (5) areas of the continental platform covered by shallow seas during intervals of transgression and regression; and (6) food supply.

Rates of evolution have been tied also to physical size of organisms. Hallam (1975, p. 496) suggests that "organisms that increase in size more rapidly should become extinct more quickly, since extreme specialization would occur earlier and hence render them more vulnerable to extinction, whilst given food resources are likely to become scarcer, the number of individuals will be smaller, and hence larger species will once again be more liable to extinction." Kauffman (1977, p. 120) maintains that the major controls on rates of evolution are externally manifest and include "population size, degree of isolation, rate of isolation, diversity of niches, variations in selective pressures, size and mobility of organisms and ecological controls on their distribution, trophic relationships, and breadth of ecological tolerances."

Deterministic vs. Probabilistic Evolution. An interesting side issue to the problem of evolutionary controls relates to the question of whether or not such evolutionary events as adaptive radiation and periods of mass extinction are deterministic or probabilistic. That is to say, are evolutionary events explainable only in terms of causal factors, or are there statistical laws or generalizations that can explain these events on the basis of random variations or processes? Probabilistic evolutionary models are called **stochastic models.** Van Valen (1973, p. 1) asserted, for example, that "all groups for which data exist go extinct at a rate that is constant for a given group." Such statements should not be taken to mean that extinctions occur without cause. Extinction of a species may be the result of any number of specific causes—predation, aging, starvation, or genetic deformity, for example—and it may therefore be invalid to attribute the death of individuals to chance. If frequency of death is considered at population levels, however, it may be mathematically valid to describe the frequency as being governed by random stochastic processes. Therefore, the observed frequency of death can be used to compute the probability of extinction if the population size is known (Raup, 1977). Thus, in the stochastic approach, the pattern of evolution as a whole is perceived to be a random process, although individual fluctuations in this pattern can be explained by cause and effect.

Raup (1977, p. 76) states that "Stochastic models serve to separate those features of the evolutionary record which are amenable to deterministic explanations from those where the search for a specific cause is not warranted," but he points out also that, owing particularly to the questionable validity of some of the underlying assumptions used in the models, there are many pitfalls in application of stochastic models to evolutionary studies. Furthermore, not all geologists and paleontologists agree with the stochastic approach to evolution. Some take the view, for example, that the apparent correlation of episodes of massive extinction with key environmental events argues that these extinction waves are deterministic (Valentine, 1977a). S. M. Stanley (1979, p. 279) also favors deterministic models, stating that "rapid radiation and decline of large clades (clusters of lineages) normally relates to nonrandom causal factors: adaptive innovations and the agents of species selection."

Mass Extinctions. The geologic record shows that many groups of organisms became extinct or suffered dramatic reductions in numbers and diversity at particular times in the geologic record. Examples of such times are the Late Cambrian, Late Ordovician, Late Devonian, Late Permian, Late Triassic, and Late Cretaceous (Table 17.4). As Table 17.4 indicates, extinctions affected both terrestrial and marine forms. Many of these mass extinction episodes are so dramatic that it is indeed difficult to accept the stochastic model as an explanation for the demise of these groups of organisms, and many scientists feel compelled to seek specific causal factors to explain these extinction waves. The extinction of the dinosaurs at the end of the Cretaceous and extinction of the fusulinid foraminifers at the end of the Permian are but two examples of abrupt disappearance of major groups of organisms. These dramatic extinctions have taxed the imagination of paleontologists to provide an acceptable causal explanation. The Late Permian extinction phase has received particular attention because of the number of major groups affected and the sharpness of the change with which these groups disappeared from the geologic record between the Late Permian and the Triassic.

Hallam (1981) mentions several factors that may have been responsible for episodes of mass extinction (Table 17.5). He suggests that the eustatic sea-level control hypothesis provides the most promising way of accounting for alternating extinction and radiation episodes for shallow-marine invertebrates. On the other hand, he points out that it does not fully account for the mass extinctions of large land vertebrates or the mass extinction of calcareous phytoplankton and zooplankton species such as occurred at the end of the Cretaceous. Planktonic groups, in particular, should not have been seriously affected by transgressions and regressions. In any event, worldwide extinctions of major groups of organisms, while very interesting, play only a limited role in biostratigraphy. Change in local environmental conditions is probably a more significant cause of extinction of individual species, which form the most important basis for biostratigraphy.

TABLE 17.4 Major extinctions of organisms during the Phanerozoic

Extinction episode	Major animal groups strongly affected	Percentage of families extinct
Late Cretaceous	Ammonites* Belemnites Rudistid bivalves* Corals Echinoids Bryozoans Sponges Planktonic foraminifers Dinosaurs* Marine reptiles*	26
Late Triassic	Ammonites Brachiopods Conodonts* Reptiles Fish	35
Late Permian	Ammonites Rugose corals* Trilobites* Blastoids* Inadunate, flexibiliate, and camerate crinoids* Productid brachiopods* Fusulinid foraminifers* Bryozoans Reptiles	50
Late Devonian	Corals Stromatoporoids Trilobites Ammonoids Bryozoans Brachiopods Fish	30
Late Ordovician	Trilobites Brachiopods Crinoids Echinoids	24
Late Cambrian	Trilobites Sponges Gastropods	52

*Last appearance of group.

Source: Facies interpretation and the stratigraphic record, by A. Hallam. W. H. Freeman and Company. Table 10.1, p. 216, 217, © 1981.

17.5 DISTRIBUTION OF ORGANISMS IN SPACE: PALEOBIOGEOGRAPHY

When d'Orbigny introduced the concept of stage, he believed that the fossil assemblages upon which his stages were based had worldwide distribution. We now know that the fossil species and assemblages that characterize biostratigraphic units are not necessarily found everywhere that rocks of the appropriate ages occur. Few species are distributed throughout the entire world. Most, in fact, are restricted in their geographic range, although some fossil groups ranged widely throughout whole ecological

TABLE 17.5 Causes of mass extinction of organisms

Cause	Description
Geomagnetic reversal events	Weakening of Earth's magnetic field during magnetic reversals has been postulated to allow excessive levels of cosmic radiation to reach Earth, causing extinction of some groups of organisms. Conclusive evidence to support this hypothesis is still lacking.
Climate and atmospheric changes	Marked changes in climate have been invoked to explain extinction of such groups as dinosaurs and certain foraminifers. Climate changes may be the result of mountain building, shifting of pole positions, or other factors. One current theory attributes climate changes to the impact of a large meteorite or other cosmic body that could throw huge clouds of dust into the atmosphere, blocking the sun's rays and significantly altering global climates.
Salinity changes in the ocean	Decrease in salinity of the oceans from normal values (~35‰) to perhaps less than 30‰ owing to withdrawal of large quantities of salt from the oceans to form massive evaporite deposits has been postulated for certain geologic periods such as the Late Permian and Cretaceous.
Haug effect	The extinction of dinosaurs at the end of the Cretaceous has been attributed to exceptionally low topographic diversity, with consequent reduction of highlands and ensuing loss of habitat variety, owing to widespread transgression
Anoxic events	Episodes of widespread reducing (anoxic) conditions in the ocean, indicated by the presence of extensive black shale deposits in the Devonian, Jurassic, etc., is suggested to lead to greatly reduced habitable areas for neritic (shallow-shelf) organisms.
Sea-level changes	Major episodes of regression lead to shrinkage of areas of epicontinental sea habitats and thus to intense competition for food and space, ultimately causing extinction of less adaptive groups. Episodes of extinction are followed by emergence of new species as a result of adaptive radiation of the survivors during expansion of habitats accompanying the succeeding transgression.

Source: Based on *Facies interpretation and the stratigraphic record*, by A. Hallam. W. H. Freeman and Company. © 1981.

realms at times in the geologic past (Valentine, 1977a). The region within which a particular group or groups of plants or animals is distributed is called a **biogeographic province.** Biogeographic provinces are separated by physical or climatic barriers. Land areas are barriers to marine organisms; open marine water is a barrier to land animals and plants; deep water is a barrier to shallow-water, shelf-dwelling organisms; cold water is a barrier to warm-water organisms; freshwater is a barrier to organisms adapted to saline marine conditions; and so forth. A particular type of barrier may be impenetrable by one species of organism but not by another. For example, benthonic organisms that do not have a long-lived juvenile, planktonic larval stage find deep water a barrier to dispersal. By contrast, planktonic organisms that live in near-surface waters in the ocean are distributed widely throughout the oceans in both shallow and deep water.

Dispersal of Organisms

Valentine (1977b) emphasizes the point that species are the fundamental biologic units in nature. They are the basic units that undergo evolution; the species niche is the basic functional unit in ecological interactions; and species are the fundamental units of biogeography, biostratigraphic zonation, and correlation. Because of their central importance in biostratigraphy, it is essential that we understand the factors that control the dispersal and distribution of species. Obviously, different factors affect the dispersal of land organisms and plants than control the dispersal of marine organisms. Also, the distribution of invertebrate marine organisms is controlled by different factors than those that control the distribution of vertebrate marine groups. Because of the overriding importance of marine invertebrates in biostratigraphic studies, we shall confine our discussion of dispersal here to invertebrate organisms in the marine realm (Fig. 17.7).

Marine invertebrate organisms can be divided into three basic types based upon habitat: plankton, nekton, and benthos (Table 17.6). **Plankton** are mainly microscopic-size organisms that live suspended at shallow depths within the water column and have very weak or limited ability to direct their own movements. They are distributed more or less passively by currents and wave action and may be dispersed widely into all types of open-ocean environments. Because they reflect the habitat of the pelagic realm and not the bottom environment into which they fall upon death, their presence in ancient sedimentary rocks is of limited value in environmental interpretation. A

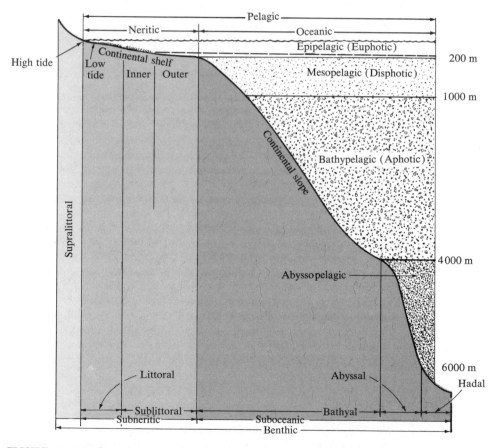

FIGURE 17.7 Bathymetric subdivisions of the marine realm. (After Thurman, H. V., 1985, Introductory oceanography, 4th ed. Fig. 13–10, p. 327, reprinted by permission of Merrill Publishing Co., Columbus, Ohio.)

TABLE 17.6 Classification of organisms by habitat or life style

Classification	Description	Example
Planktonic	Organisms that live suspended in the upper water column, and which have only a very weak or limited ability to direct their own movements	
Phytoplankton	Have the ability to carry on photosynthesis; primary food producers, or autotrophs	Diatoms, dinoflagellates, coccolithophoridae
Zooplankton	Do not carry on photosynthesis and thus cannot produce their own food (heterotrophs); feed on phytoplankton	Foraminifers, radiolarians, graptolites
Meroplankton	Spend only their juvenile stage as plankton; later become free-swimming or bottom-dwelling organisms	Larva of most benthonic organisms such as molluscs
Pseudoplankton	Organisms distributed by waves and currents as a result of attachment to floating seaweed, driftwood, etc.	Mussels, barnacles, etc.
Benthonic	Bottom-dwelling organisms that live either on or below the ocean floor	
Sessile benthos	Benthos that attach themselves to the substrate (epifauna)	Crinoids, oysters, brachiopods
Vagrant benthos	Benthos that either creep or swim over the bottom (epifauna) or burrow into the bottom (infauna)	Starfish, echinoids, crabs Clams, worms
Nektonic	Organisms able to swim freely and thus move about largely independently of waves and currents	Mobile cephalopods, fish, sharks

few plankton such as graptolites do have some value as indicators of bottom environments. Graptolites were too fragile to survive in high-energy shallow-water environments and are thus preserved mainly in the facies of quiet-water environments. Thus, they constitute "facies fossils." Planktonic organisms are exceptionally useful fossils for biostratigraphic zonation and correlation because of their widespread distribution.

Nekton include all animals that are able to swim freely. Modern nekton are distributed in the ocean at depths ranging from the surface to thousands of meters and encompass many advanced groups of animals such as fish, whales, and mammals. Nekton are less abundant in the fossil record than planktonic and benthonic organisms and thus overall appear to have somewhat less value in biostratigraphic studies. Nektonic fossils include fish remains in some deep sea clays, belemnites and other mobile cephalopods, and probably conodonts. Conodonts are an interesting type of fossil, with considerable biostratigraphic significance, that occur in rocks ranging in age from Cambrian to Triassic. They are tiny, toothlike phosphate fossils whose origin remained an enigma until 1982. A complete specimen of the conodont-bearing animal has been found in Lower Carboniferous rocks of the Edinburgh district, Scotland (Briggs et al., 1983). The specimen is an elongate, soft-bodied animal 40 mm long and 1.8 mm wide. The conodont apparatus occurs in the head or anterior region of the body and may have served as teeth or possibly some type of internal support.

Benthos are bottom-dwelling organisms that live either on or below the ocean floor. Benthonic organisms with preservable hard parts are particularly important for environmental interpretation because their remains are commonly preserved in the same environment in which they lived. Because most benthos live in shallow water and have limited ability to move long distances along the bottom, they tend to be more provincial, and of somewhat less biostratigraphic significance, than plankton. Nonetheless, benthos can be dispersed outside their local environment because many benthonic species have a planktonic, juvenile larval stage during which they can be dispersed by currents. Some workers originally questioned the importance of larval transport as a mechanism for dispersal of shallow-water benthos over long distances, such as across ocean basins. It was initially believed that the duration of the larval stage was so short that the larvae would change to the adult phase while the organisms were still over deep water, causing them to perish when they settled to the bottom. More recent work (Scheltema, 1977) has demonstrated, however, that there are many shoal-water or continental-shelf benthonic invertebrates whose larvae have a pelagic stage lasting from six months to over one year. Scheltema refers to such larvae as **teleplanic,** or "far wandering." Such long-lived larval species could cross the modern Atlantic Ocean, for example, by way of the main surface currents. Many workers interpret this ability to become dispersed by pelagic mechanisms as being favorable for interpopulation migration and, therefore, gene flow. Valentine (1977b, p. 145) summarizes the importance of these long-lived larval forms as follows:

> Species with the more long-lived and hardy pelagic larva have the greater chance to be widely dispersed after reproduction. . . .Therefore, species with such attributes would commonly be able to colonize habitats that lie at some distance from their parental ranges, and would usually be able to maintain gene flow to such outlying populations. Species with shorter planktonic development periods, smaller broods, or more restricted larval requirements would tend to colonize only localities that are fairly close to their parental regions. If a population became established at any considerable distance from others, gene exchange might be sporadic or lacking altogether, leading to divergence between the colonists and the parental population, and a reduction in their usefulness in correlation.
>
> For a species in a given locality, then, a geographic range exists for which colonization is essentially obligatory, as the region lies within the normal migratory range of the population; thus, by some standard time, occupation is virtually assured. This can be called the **local range.**

Barriers to Dispersal

Dodd and Stanton (1981) point out that each species has a potential geographic range that is determined by its habitat requirements. Few species actually occur throughout their potential range. Their distribution is restricted owing either to the presence of barriers of some type that prevent their expansion into all areas of suitable habitat or because the species may not have had time to spread to all suitable areas, especially if barriers are present. At any given time, there are many regions in the world that could be colonized by species if they could reach them in appropriate numbers, but they are barred from reaching them by intervening inhospitable areas. Many species eventually find ways to broach narrow barriers and perhaps in time even to cross wider barriers. Once barriers are crossed, or barriers disappear, the migrant species may find itself in competition for environmental niches with similar species or similarly adapted species in the new province. In the face of this competition, either the indigenous species or the migrant species may become extinct. Alternatively, the less well-adapted species could evolve and become adapted to a different environmental niche. Once a barrier is surmounted, the colonizers typically fill out their new local range, expanding until the new location is circumscribed by other barriers. The intruding species may subsequently broach still other barriers, hopping from one habitable region to another across barriers of varying difficulty of penetration, episodically expanding their total range (Valentine, 1977b).

The broaching of barriers thus leads to expansion of the total range of a species, although in some cases it may lead to extinction of the species in the new region or to its evolution to a more adaptable species. On the other hand, if the opposite situation prevails and a barrier "suddenly" appears and divides a once continuous area of suitable habitat, the result is the segregation of the species into different populations separated by the barrier. The separated populations would gradually evolve into different species, each with a more restricted geographic range than the parent species (Dodd and Stanton, 1981).

Numerous ecological factors can act as barriers to dispersal of organisms. Valentine (1977b) suggests that the factors of primary importance in range restriction can be grouped under two major categories: habitat failure—as when shelf habitats give way to deep-sea conditions or to land—and temperature. Hallam (1981) places the major controls on faunal provinciality as climate and plate movements. Dodd and Stanton (1981) indicate that the principal factors controlling geographic distribution of species are depth–elevation—that is, water depth and land elevation—and temperatures. These possible controls on provinciality are discussed in further detail in the following paragraphs.

Temperature. Temperature is clearly a major barrier to migration of species and it commonly affects larvae more than adult organisms. Because the distribution of worldwide temperatures is latitudinally controlled, temperature barriers are most important latitudinally, although seasonal and even diurnal temperature changes are also important. The boundaries of all modern biotic provinces are in part temperature controlled, and ancient biotic provinces were undoubtedly similarly controlled. Warm-water taxa are restricted primarily to the equatorial zone of the ocean because no other large parts of the ocean, either at the surface or at depth, are warm enough to sustain these tropical species. Cold-water taxa, on the other hand, can extend their range closer to the equatorial region by migrating down the bathymetric gradient into deeper and colder water, the phenomenon of **submergence,** if they are capable of adapting to greater depths. Also, if a polar species can manage to find a way of breaking through the temperature barrier and crossing the equatorial region, it can find suitable cold-water habitats at or near the surface in the higher latitudes of the other hemisphere.

Some species of organisms are adapted to a wide range of temperatures and are said to be **eurythermal.** Such temperature-tolerant species may thus be distributed through a much wider range of temperature zones than the less tolerant **stenothermal** species. Nonetheless, even eurythermal species are sensitive to temperature variations and do not occur throughout all temperature zones. It must be recognized also that marine temperature zones have changed throughout geologic time as world climatic zones have shifted in response to plate movements and episodes of glaciation. A given geographic region of the world may thus record a succession or colder or warmer water faunas through time in response to these shifting climatic conditions.

Geographic Barriers. The terms "habitat failure," "plate movements," and "depth-elevation" are all different ways of expressing the concept of geographic barriers. These geographic barriers arise out of the distribution pattern of land masses and oceans and variations in water depths of the oceans. All organisms have limited water depths at which they can survive. Thus, water that is either too deep or too shallow can constitute a barrier to a particular species of organism. Land masses constitute barriers to the dispersal of marine organisms, and the open ocean is a barrier to migration of land animals and plants from one continent to another. The most important factors influencing geographic barriers appear to be changes in sea level and changes in the nature and geographic distribution of land masses and the ocean floor brought about by plate movements (discussed in a following section).

Sea-Level Changes. Causes of major cycles of sea-level change are discussed in Chapter 14. Changes in sea level cause significant interruptions in biogeographic provinces because of changes in water depths on the continental shelves. During a major drop in sea level, water is withdrawn from the continental shelves, exposing much of the inner shelf. The habitable area of shallow water is greatly reduced, leading to crowding and increased competition among shallow water species that cannot move seaward into deeper water, and probable extinction of less adaptable groups. During major rises in sea level, water depths on the outer continental shelf are increased but the total area of shallow water along continental margins is also vastly increased owing to spread of the seas over the edges of the continents. The available environmental niches for shallow-water organisms are correspondingly increased, resulting in less competition among species for available space and food. These conditions lead to expansion of the local ranges of species as they move into favorable habitats, and also probably to rapid emergence of new species (speciation) as a result of adaptive radiation of groups that survived the preceding episode of lowered sea level.

Hallam (1981) states that an analysis of genera across the world indicates a clear inverse relationship between endemism and the area of continents covered by sea. **Endemism** is the tendency of species or other taxa to have a very limited geographic range, as contrasted with **pandemism,** which is the tendency of species to have worldwide distribution. He suggests that at times of low sea level and restriction of seas, faunal migration between continental shelf areas would be rendered more difficult, resulting in less gene flow. Thus, there would be more local speciation among the dispersible organisms that occupied shallower water habitats.

Plate Movements. Tectonism is the major factor controlling the distribution of land masses and ocean basins. Major changes in the environmental framework of the marine realm occur as the geographic positions, configurations, and sizes of continents and ocean basins are changed by global plate tectonic processes. Plate movements can greatly affect topographic barriers by producing changes in oceanic widths and depths. As previously discussed, changes in rates of seafloor spreading may have a major effect on sea level. Plate movements can also alter latitudinal temperature gradients by shifting the geographic position of continents and even affect the distribution patterns of major ocean currents. The creation or destruction of migration barriers may thus be tied closely to plate tectonics events.

Provinciality of species is greatest during times when plate motion has produced a maximum number of separate continents, such as at the present time (Dodd and Stanton, 1981). Barriers are fewer and provinciality is less when plate motions have welded continents together, as at the end of the Paleozoic. Several generalities between plate tectonics and biogeography can be stated (Fig. 17.8 and Table 17.7): (1) When spreading ridges—the Mid-Atlantic Ridge, for example—lie parallel to continents, they produce deep and ever-widening ocean basins, and thus are barriers to shallow-marine species and terrestrial organisms. (2) Transform faults—the San Andreas fault, California, for example—that run parallel to continental margins are also usually associated with deep-water barriers. (3) Subduction zones—the Peru–Chile Trench, for example—parallel to and dipping toward the continent form deep-water barriers. (4) Subduction zones dipping away from continents—the subduction zone running from

FIGURE 17.8 Schematic illustration of the relationships between
crustal plates and continents as
they affect the distribution of organisms. Table 17.7 shows the biogeographical implications for the
continental shelf for each case.
(After Valentine, J. W., 1971, Plate
tectonics and shallow marine diversity and endemism, an actualistic model: Systematic Zoology, v.
20. Fig. 4, p. 261, reprinted by permission of Society of Systematic
Zoology, Norman, Okla.)

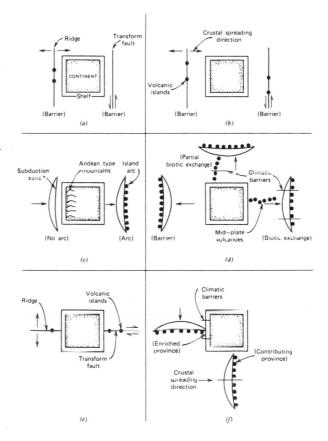

Burma to New Hebrides, for example—may have island arcs that aid in breaking down
barriers. (5) Mid-plate volcanoes—the Hawaiian Islands, for example—help break
down deep-water barriers. (6) Subduction zones, spreading ridges, and associated island arcs at high angles to continents—the Aleutian Islands, for example—may provide migration pathways breaking down barriers (Valentine, 1971).

Hallam (1973b) proposes the distinction of three patterns of changing faunal distributions with time that are related to plate movements. **Convergence** refers to the
phenomenon whereby the degree of resemblance of faunas in different regions increases from an earlier to a later time. **Divergence** refers to the reverse process, that is,
decrease in degree of resemblance of faunas in different regions from an earlier to a
later time. Table 17.8 lists several examples of this phenomenon, including examples
of both nonmarine and marine invertebrates. Note from this table that plate tectonics
movements that cause closure of ocean basins have tended to produce convergence of
shallow-water benthonic organisms. Distinctly different faunal provinces on either
side of a wide ocean are gradually brought together by closure of the intervening ocean
between contiguous continents, eventually producing a single faunal province. Opening of oceans has commonly produced divergence. Hallam also identifies a third biogeographic pattern, which he calls complementarity. **Complementarity** is the distri-

TABLE 17.7 The relationship of crustal plates to continental margins (Figure 17.8) and the effect of this relationship on biogeography

Geometry of relation	Character of margin	Distance	Biogeographic implications for continental shelf	Fig 17.8
Parallel	Ridge or transform	Near	Barrier but with depauperate provincial outliers on isolated islands	a
		Far	Barrier	b
	Subduction zone	Near	Barrier if truly marginal with no island arc; source of rich biota and dispersal route if arc present	c
		Far	No effect unless intervening region bridged by midplate volcanoes, then source of rich biota and dispersal route if no climatic barriers intervene	d
High angle	Ridge or transform	Near	Little effect, with depauperate provincial outliers on isolated islands	e
		Far	Not a case	
	Subduction zone	Near	N–S shelf, E–W arc: arc system a source of rich biota for local province. E–W shelf, N–S arc: proximal province of arc system a source of rich biota for entire shelf	f
		Far	Not a case	

Source: After Valentine, J. W., 1971, Table 1, p. 262, reprinted by permission of Society of Systematic Zoology, Norman, Okla.

butional change of contiguous marine and terrestrial organisms that occurs when one group exhibits convergence and the other divergence. Creation of a land connection between two previously isolated areas of continents, for example, allows convergence of the terrestrial faunas to take place, while at the same time the land connection creates a barrier to marine organisms, causing divergence as a result of genetic isolation.

Other Barriers. Other less important barriers than temperature and geography may also help to define the boundaries of biogeographic provinces. Salinity differences constitute an important boundary between freshwater and marine provinces; however, salinity is a relatively unimportant barrier within the marine realm itself. Marked salinity increases can occur in some small, restricted arms of the ocean where evaporation rates are high. Conversely, lower than normal salinities may ensue in some coastal areas where freshwater runoff is high. These salinity variations can control local communities of organisms but not the distribution of organisms on a provincial level. In the open ocean, salinity tends to be highest in the equatorial region, where evaporation rates are at a maximum, and lowest in the higher latitudes, where some dilution occurs as a result of freshwater runoff from the continents. Even so, the salinity in these regions varies only a few parts per thousand from the average ocean salinity (35‰), a variation not adequate to seriously affect the dispersal of organisms in the open ocean.

TABLE 17.8 Correlation of plate tectonic events and changes in faunal distribution patterns

Plate tectonic event	Convergence	Divergence
Closure of Proto-Atlantic (Ordovician, Silurian)	Trilobites, graptolites, corals, brachiopods, conodonts, anaspids, and thelodonts of the two continents flanking the Proto-Atlantic	
Closure of Urals Seaway	Post-Permian continental vertebrates of Eurasia	
Opening of Atlantic (Cretaceous, Tertiary)		Cretaceous bivalves and benthic foraminifers of Caribbean and Mediterranean; Upper Cretaceous ammonites of USA and W. Europe–N. Africa; post-Lower Eocene mammals of North America and Europe; Tertiary mammals of Africa and South America
Opening of Indian Ocean (Cretaceous)		Bivalves of East African and Indian shelves
Closure of Tethys (late Cretaceous) (mid-Tertiary)	?Ammonites of Eurasia and Africa–Arabia; mammals of Eurasia and Africa	Molluscs, foraminifers, etc., of Indian Ocean and Mediterranean–Atlantic

Source: *Facies interpretation and the stratigraphic record*, by A. Hallam. W. H. Freeman and Company. Table 10.2, p. 237, © 1981.

Currents aid in the dispersal of planktonic species and the larvae of benthonic species, but they help also, in some parts of the ocean, to maintain the temperature gradients that create barriers to dispersal. Thus, currents may act as either a barrier or an aid to dispersal. The long-term pattern of currents is itself affected by plate movements, as discussed.

17.6 COMBINED EFFECTS OF THE DISTRIBUTION OF ORGANISMS IN TIME AND SPACE

Eicher (1976) points out that both environmental and temporal records are important for interpretation of geologic history, where temporal relates to time as distinguished from space. If organisms throughout geologic time had been spread over the world and not confined to specific biogeographic provinces and environments, worldwide correlation of strata on the basis of fossils would be greatly facilitated, assuming that evolutionary changes have been simultaneous and worldwide. Under these conditions, however, fossils would provide little or no help in working out ancient depositional environments because more or less the same organisms would have lived in all environments. Conversely, if organisms were distributed in biogeographic provinces as they are today but organic evolution never occurred, we would be able to interpret local ancient environments with great confidence because ancient sedimentary rocks would contain the same species as modern environments. By the same token, these species would be of no value in correlation and the unraveling of local chronologies because the same species would have existed throughout geologic time.

The real fossil record reflects the fact that both segregation into biogeographic provinces and organic evolution took place. Owing to organic evolution, we are able to correlate strata of a given age from one area to another and to work out the relative chronology of strata in a given area. Because many organisms were confined to biogeographic provinces in the past, however, we cannot always correlate time-equivalent strata from different environments because the organisms that existed in different biogeographic provinces during the same period of time were different. Thus, correlation between biogeographic provinces is difficult, and it is commonly not possible to make worldwide correlations. On the other hand, because different groups of organisms were confined to different provinces and different environments, the provinciality of ancient organisms provides an invaluable tool for interpreting ancient sedimentary environments.

On the debit side again, the provinciality of organisms creates special problems from the standpoint of determining the total vertical stratigraphic range of a species. A species may exist in one province for long periods of time before broaching a particular barrier and spreading into a nearby province. After migration into the new province, the species may die out in the old province while continuing to thrive for some time in the new region. Therefore, the *local* vertical range of a species in a given province, sometimes called the **teil zone,** may be much shorter than the *total* range of the species. Paleontologists must be extremely careful about recognizing this possibility when using fossils for time correlations. This problem is illustrated in Figure 17.9, which shows some of the major factors that can affect the range of a species. This diagram demonstrates that the range of a species is affected both by evolutionary changes and by the presence of barriers that can regulate the times of migration into and first appearance in nearby provinces.

17.7   BIOCORRELATION

Biostratigraphic units are observable, objective stratigraphic units identified on the basis of their fossil content. As such, they can be traced and matched from one locality

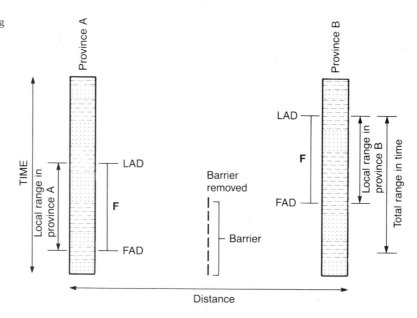

FIGURE 17.9   Diagram illustrating the difference in local range and total range of a hypothetical species (F). Species F first appears in Province A and is restricted to Province A by a barrier. Later removal of the barrier allows migration to Province B, where the species persists for a time after it has died out in Province A. FAD = first appearance datum; LAD = last appearance datum.

to another just as lithostratigraphic units are traced. Biostratigraphic units may or may not have time significance. For example, assemblage zones and abundance zones may cross time lines when traced laterally. On the other hand, interval zones, particularly those defined by first appearances of taxa, yield correlation lines that generally coincide with time lines. Biostratigraphic units may be correlated, irrespective of their time significance, by use of much the same principles employed in correlation of lithostratigraphic units—matching by identity and position in the stratigraphic sequence, for example. We will first examine correlation by assemblage zones and abundance zones, which can be correlated as biostratigraphic units even though they may not have time-stratigraphic significance. We will then discuss biocorrelation methods based on interval zones and other zones that yield time-stratigraphic correlations.

Correlation by Assemblage Zones

Assemblage zones, as opposed to Oppel zones, are based on distinctive groupings of three or more taxa without regard to their range limits (Fig. 5A, Appendix). They are defined by different successions of faunas or floras and they succeed each other in a stratigraphic section without gaps or overlaps. Assemblage zones have particular significance as an indicator of environment, which may vary greatly regionally. Therefore, they tend to be of greatest value in local correlations. Nonetheless, some assemblage zones based on marine planktonic assemblages may be used for correlation over much wider areas. The principle of correlation by assemblage zones is illustrated graphically in the very simple example shown in Figure 17.10.

A. B. Shaw (1964) points out that the boundary between assemblage zones is inherently fuzzy because above and below the limits of this zone will be transition zones in which part of the characteristic fossil assemblage will be missing because it

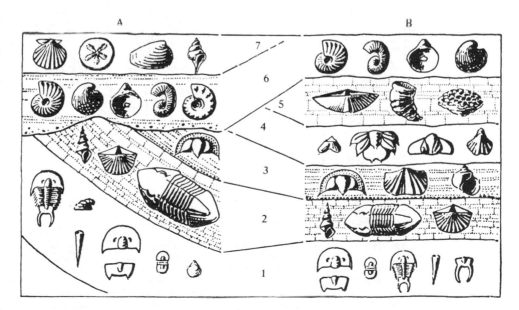

FIGURE 17.10 Generalized diagram illustrating the principle of correlation by fossil assemblages. (From Moore, R. C., C. G. Lalicker, and A. G. Fischer, Invertebrate fossils, © 1952. Fig. 1.3, p. 8, reprinted by permission of McGraw-Hill Book Company, New York.)

has not yet appeared or has already vanished. Therefore, there are practical limits to the accuracy that can be achieved by assemblage zone correlations. Part of the problem in correlation by assemblage zones stems from the fact that the number of fossil taxa that a biostratigrapher must work with is so large that it is difficult to visually assimilate the data and draw meaningful zone boundaries (Fig. 17.11). To overcome this problem, earlier workers tended to reduce the number of taxa whose distributions would be studied, or they tried to composite the samples. A more recent solution to this problem is to apply the techniques of multivariate statistical analysis to recognition and delineation of assemblage zones. These techniques provide a rational statistical basis for delineating zones based on large numbers of taxa without taking the decision making out of the hands of the biostratigrapher. Details of these multivariate techniques are given in Hazel (1977) and Brower (1981).

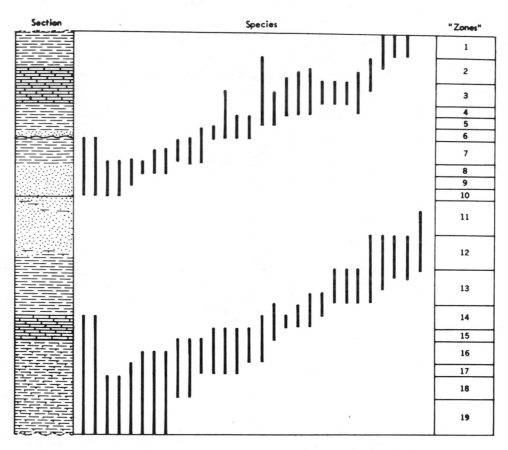

FIGURE 17.11 Hypothetical stratigraphic section illustrating the large numbers of fossil taxa that may be involved in correlation by assemblage zones. Vertical black lines represent the composite ranges of the species found at various local sections. The column at the right shows one interpretation that could be drawn from these fossil data. (From Hazel, J. E., 1977, Use of certain multivariate and other techniques in assemblage zonal biostratigraphy: Examples utilizing Cambrian, Cretaceous, and Tertiary benthic invertebrates, *in* G. G. Kauffman and J. E. Hazel (eds.), Concepts and methods of biostratigraphy. Fig. 1, p. 189. Copyright 1977, 1983 by Van Nostrand Reinhold. All rights reserved.)

Correlation by Abundance Zones

As mentioned, abundance zones, or acme zones, are defined by the quantitatively distinctive maxima of relative abundance of one or more species, genus, or other taxon rather than by the range of the taxon. They represent a time or times when a particular taxon was at the peak, or acme, of its development with respect to numbers of individuals. Some biostratigraphers previously used abundance zones for time-stratigraphic correlation under the assumption that there is a time in the history of every taxon when it reaches its maximum abundance and that this abundance peak occurs everywhere at the same time. The current prevailing opinion among biostratigraphers is that abundance zones are unreliable and unsatisfactory for time-stratigraphic correlation. This opinion is based on the apparent fact that not all species achieve a maximum abundance, or if they do that this peak is not necessarily recorded by layers of abundant specimens. Furthermore, peak abundances that are recorded in the stratigraphic record may be related to favorable ecological conditions that can occur at different times in different areas and that may persist in one area much longer than in another. Maximum abundance may thus represent local, sporadically favorable environments, sudden unfavorable environments that caused mass mortality, or mechanical concentrations of the shells of organisms after death. Some of the problems of correlating by abundance zones are illustrated in Figure 17.12. In short, abundance zones may be used for biostratigraphic correlation but they do not provide a reliable means of time-stratigraphic correlation. Although they are sometimes used locally for correlation within provinces, biostratigraphers usually prefer correlations based on assemblage zones or interval zones.

Chronocorrelation by Fossils

Chronostratigraphic correlation is the matching up of stratigraphic units based on time equivalence. Establishing the time equivalence of strata is the backbone of global stratigraphy and is considered by most stratigraphers to be the most important type of correlation. Methods for establishing time-stratigraphic correlation fall into two broad general categories: biological and physical/chemical. As mentioned, time-stratigraphic

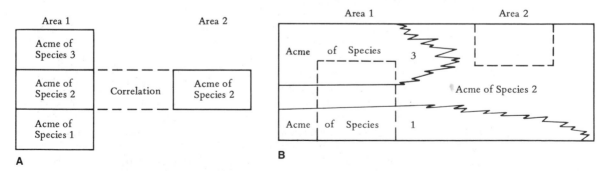

FIGURE 17.12 Problems that can develop in correlation by interval (acme) zones when correlation is incorrectly assumed to be time correlation. A. Inferred time correlation. B. Strata preserved in Area 2 are actually equivalent in time to the youngest strata in Area 1 (containing the acme of Species 3) rather than to the middle strata (containing the acme of Species 2). (From Mintz, L. W., 1981, Historical geology, the science of a dynamic Earth, 3d ed. Fig. 10.17, p. 253, reprinted by permission of Merrill Publishing Co., Columbus, Ohio.)

correlation by biological methods is based mainly on use of concurrent range zones and other interval zones. Biological correlation methods also include statistical treatment of range–zone data and correlation by biogeographical acme zones, which are biological events related to climate fluctuations. A variety of physical and chemical methods are available for chronostratigraphic correlation, and are discussed in the following chapter. Logically, this discussion of chronostratigraphic correlation by fossils also belongs in the next chapter; however, I am including it here to keep all material relating to correlation by fossils in a single unit.

Correlation by Biologic Interval Zones

Interval zones are biozones that constitute the strata that fall between the highest and/ or lowest occurrence of taxa. Several kinds of interval zones are recognized, including those formed by overlapping ranges of taxa. Figure 4 of the Appendix illustrates several ways that the first and last appearances of taxa may be used to define interval zones: (1) the interval between the first and last appearance of a single taxon (taxon range zone), (2) the interval between the first appearances or last appearances of two different taxa, (3) the interval between the first appearance of one taxon and the last appearance of another, and (4) intervals defined by overlapping range zones (concurrent range zones). These different interval zones have varying degrees of usefulness in time-stratigraphic correlation, as described in the following paragraphs.

Taxon Range Zones. Taxon range zones may be very useful for time correlation if the taxa upon which they are based have very short stratigraphic ranges. They are of little value if the taxa range through an entire geologic period or several periods. Correlation by taxon range zone is often referred to as correlation by **index fossils.** Index fossils are considered to be those taxa that have very short stratigraphic ranges, were geographically widespread, were abundant enough to show up in the stratigraphic record, and are easily identifiable. Unfortunately, the term index fossil has been used in other ways and can have other connotations. Therefore, it is less confusing when speaking of correlation based on the entire range of a taxon to refer to it simply as correlation by taxon range zone. Correlation by taxon range zone is illustrated diagrammatically in Figure 17.13.

Other Interval Zones. When individual taxon range zones are very long and correlation by taxon range zone is thus not suitable, much finer scale correlation is possible by use of other types of interval zones. Interval zones defined by the first appearance of two taxa, for example, are particularly useful in time-stratigraphic correlation because they are based on evolutionary changes, along phyletic lineages, that tend to occur very rapidly. Thus, the interval between the first documented appearance of two taxa may represent a very short span of time, and the age of the strata in this interval may be nearly synchronous throughout their extent. Interval zones defined on the last appearances of taxa are commonly considered to have less time significance than those based on first appearances because extinctions of taxa do not occur with the same suddenness that new species appear through phyletic evolution.

Figure 17.13 illustrates some of the various methods that can be used for correlating between two sections on the basis of interval zones. Note from this illustration that interval zones can be identified that represent much shorter spans of time than those represented by the range zones of most individual taxa. To avoid confusion in this illustration, I have shown only a few of the possible biostratigraphic correlation lines that could have been drawn between the two stratigraphic sections.

Graphic Method for Correlating by Taxon Range Zone. Although interval zones can be used to define units of strata deposited during relatively short periods of time, they do not necessarily yield precise time-stratigraphic correlations. Organisms may migrate laterally and appear in other areas at somewhat later times than their true first appearance (Fig. 17.9), or they may migrate out of a local area before their final extinction elsewhere. These variables of behavior make the boundaries between interval zones inherently fuzzy. The exact boundary between biozones can never be known because such boundaries are determined empirically. Additional collecting in a new area always holds the possibility of extending the known range of previously defined species or taxa, because they may have appeared earlier or persisted longer in the new area than in originally defined areas. One way to minimize the problem of fuzzy zonal boundaries is to treat range data statistically, utilizing the first and last appearances of all the species present in a stratigraphic section rather than the ranges of just one or two species. A graphical method for establishing time equivalence of strata in two stratigraphic sections by plotting first and last appearances of *all* the species in one section against the first and last appearances of the same species in another section

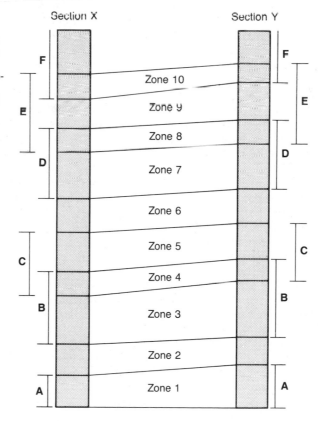

FIGURE 17.13 Correlation between two hypothetical sections based on interval zones. Note that several types of interval zones are used here for correlation. For example, Zone 1 is defined by the total vertical range of Species A, Zone 2 is an interval zone defined by the last appearance of Species A and the first appearance of Species B, Zone 4 is formed by the overlapping ranges of Species B and C, and so forth.

was first proposed by A. B. Shaw in 1964 and is now widely used by stratigraphers for detailed time-stratigraphic correlation between stratigraphic sections.

Shaw's method, as further elaborated by F. X. Miller (1977), involves first selecting a single stratigraphic section as a reference section to which other sections can be compared and correlated. This reference section should be the thickest section available, should be free of faulting or other structural complications, and should have a large and varied fossil content. The reference section is measured and sampled as completely as possible and the first and last appearances of all species are documented in terms of their positions in the stratigraphic section above an arbitrarily chosen reference point; that is, number of meters above the base of the measured section. The species ranges recorded by the first and last appearances in this local reference section may not be the true (total) ranges for all of the species; however, this fact does not preclude using them to help establish correlation, as we shall see. A second stratigraphic section is then chosen to be compared with the reference section, and the first and last appearances of the same species, and any other species, are determined in this section.

From two such stratigraphic sections (Fig. 17.14), a graph is constructed in which distance above the base of the reference section, say section A, is indicated on the horizontal axis and distance above the base of the second measured section, section B, is plotted on the vertical axis (Fig. 17.15). The first and last appearances of each species in the reference section can then be plotted against the first and last appearances of the same species in the second measured section. In Figure 17.14, for example, Species 1 first appears in reference Section A at 93 m above the base of the section and in measured Section B at about 47 m above the base. A single point can be plotted on the graph to represent these values. Similarly, additional points are plotted to represent the first and last appearances of all the species in the two sections. This procedure yields a series of points that tend to cluster around a straight line (Fig. 17.15). This line can be drawn visually to yield a "best-fit" line or can be drawn by use of statistical regression methods. The x and y coordinates of any point on this line provide a precise time-stratigraphic correlation between the two sections. In Figure 17.15, for example, the bed at 60 m in Section A correlates with the bed at 30 m in Section B and the bed at 100 m in Section A correlates with the bed at about 49 m in Section B.

First and last appearances of species represented by points that plot well off the best-fit line in Figure 17.15 indicate species that appear or disappear in Section A at distinctly different times than in Section B. Such species are either environmentally controlled (facies dependent), or their migration between Section A and B was impeded by biogeographic barriers, causing them to appear in the two sections at different times.

Shaw's graphic correlation method can take advantage of physical events such as ashfalls, or stable isotopic events that have time-stratigraphic significance, to verify the position of the best-fit line. For example, ashfalls occur over wide geographic areas almost instantaneously. Their presence in two stratigraphic sections constitutes a precise time marker that provides a very reliable point for the best-fit line and should fall exactly on this line.

In addition to its usefulness in correlating between two stratigraphic sections, Shaw's method also provides a powerful tool for evaluating differences in rates of sedimentation between two sections or the presence of a hiatus in a section. The slope of the best-fit line indicates the relative rates of sedimentation between the areas. If an

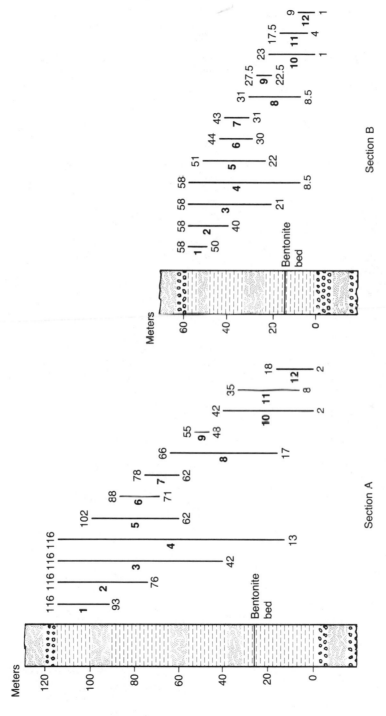

FIGURE 17.14 Two stratigraphic sections with ranges of fossil species (Species 1 through 12) graphed in meters above the base of the section. Sections A and B contain identical fossils with identical time spans; however, Section B represents only half the rate of sediment accumulation. Use of these fossil ranges in A. B. Shaw's (1964) graphic correlation method is illustrated in Figure 17.15. (After Eicher, D. L., Geologic time, 2nd ed., © 1976, Fig. 5.8, p. 112. Reprinted by permission of Prentice-Hall, Englewood Cliffs, N.J.)

FIGURE 17.15 Illustration of A.
B. Shaw's (1964) graphic correlation method using the data shown
in Figure 17.14. (After Eicher, D.
L., Geologic time, 2nd ed., © 1976,
Fig. 5.9, p. 113. Reprinted by permission of Prentice-Hall, Englewood Cliffs, N.J.)

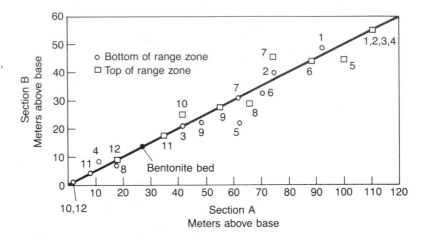

abrupt change occurs in this slope (Fig. 17.16), this change suggests a sudden relative
increase or decrease in sedimentation rates in the sections. The change in slope at
about 75 m in Section B of Figure 17.16, for example, indicates an increase in the rate
of sedimentation in Section A compared to that in Section B. The presence of a hiatus
in deposition in one section shows up as a horizontal line segment in the best-fit curve
(Fig. 17.17).

The graphic correlation method can be used not only for correlating between any
two local sections, but also can be expanded by correlating one section after another
to compile what A. B. Shaw (1964) refers to as a **composite standard reference section**
that can be used for regional or even worldwide time-stratigraphic correlations. In any
reference section chosen for study, the ranges of some fossil species will be at their
total stratigraphic maximum. Others fossils will have incomplete ranges owing to the

FIGURE 17.16. Increase in rate of
sedimentation in Section A compared to that in Section B shown
by a "dogleg" in the correlation
line determined by A. B. Shaw's
(1964) graphic correlation method.
(After Eicher, D. L., Geologic time,
2nd ed., © 1976, Fig. 5.10, p. 113.
Reprinted by permission of Prentice-Hall, Englewood Cliffs, N.J.)

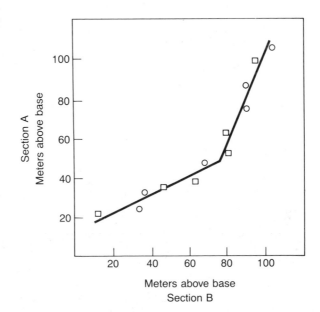

FIGURE 17.17 A hiatus in deposition in Section A shows up as a horizontal line in the graphic correlation plot, offsetting the correlation line. (After Eicher, D. L., Geologic time, 2nd ed., © 1976, Fig. 5.11, p. 114. Reprinted by permission of Prentice-Hall, Englewood Cliffs, N.J.)

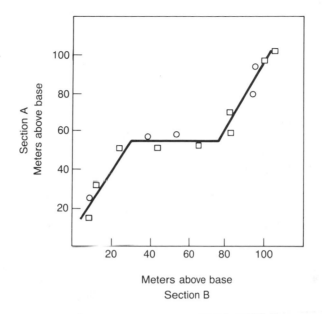

environmental or biogeographical factors described or to accidents of preservation. The purpose of creating a composite standard reference section is to establish the **total composite standard** range of each species or taxon by compounding information from other correlatable sections into the composite standard reference section. The tops and bases of the stratigraphic ranges of each taxon are adjusted in the standard composite reference section by correlating with other sections until a point is reached where the tops have been adjusted upward as high as they occur in any correlatable section and the bases downward as low as they occur. These adjusted bases and tops thus represent, as nearly as it is possible to determine, the times of speciation (evolutionary first appearance) and global extinction (last appearance anywhere on Earth) of the taxa. Once the total composite standard range of each taxon has been determined and the composite standard reference section established, it is possible to make time-stratigraphic correlations on a global scale.

Correlation by Biogeographical Acme Zones

Under the heading of biocorrelation, I discussed correlation by fossil abundance (acme) zones and pointed out that acme zones are unreliable for time-stratigraphic correlation because they are affected by environmental conditions and other factors that can cause them to be diachronous. A different approach to the use of acme zones yields correlations that have time-stratigraphic significance; this approach is correlation based on the maximum abundance of a taxon that results from geographical shifts of an environmentally sensitive fossil assemblage (Haq and Worsley, 1982). Owing to latitudinally related temperature differences in the ocean, some species or other taxa are restricted to biogeographic provinces that are defined by latitude. Thus, low-latitude taxa are ecologically excluded from high latitudes and vice versa; however, changes in climate can allow shifts of these taxa into a different biogeographic province. During major glacial stages, for example, high-latitude taxa can expand into lower latitudes, and during warming trends between major glacial stages low-latitude taxa can expand into higher latitudes. From a geochronological point of view, the spreading out (prochoresis) of certain planktonic species in response to major climatic fluctuations is essentially isochronous.

Climate-related shifts in planktonic taxa at specific times thus provide biogeographical acme events that can be correlated from one area to another. In each core or outcrop section studied, climatic curves are constructed on the basis of percentages of warm-climate to cool-climate taxa or on relative abundance of a particular taxon. These curves can then be used to identify episodes of warming and cooling that can be correlated from one section to another. Figure 17.18, constructed from this type of information, illustrates how climatically controlled latitudinal shifts in calcareous nannoplankton assemblages in the North Atlantic during Miocene time can be used for time-stratigraphic correlation in DSDP cores.

A related approach is time-stratigraphic correlation based on the coiling ratios of planktonic foraminifers, as described by Eicher (1976). The multichambered shells of some foraminifers are known to coil in one direction when the species lives in areas

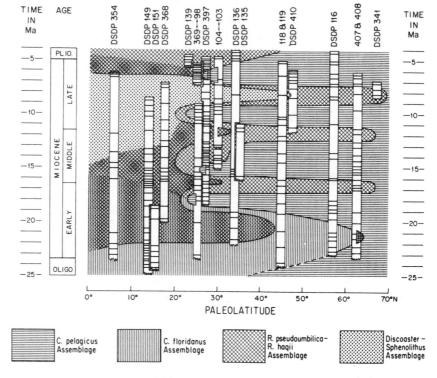

FIGURE 17.18 Use of biogeographical acme zones as a means of time correlation. Cycles of latitudinal shifts of calcareous nannoplankton assemblages in the North Atlantic Ocean during the Miocene are interpreted in response to major fluctuations in climate. The major shifts of relatively warmer, midlatitude assemblages into higher latitudes can be used for the refinement of the biochronological scale in the higher latitudes from which marker, low-latitude taxa are normally excluded. (From Haq, B. V., and T. R. Worsley, Biochronology–Biologic events in time resolution, their potential and limitations, *in* G. S. Odin (ed.), Numerical dating in stratigraphy, © 1982 by John Wiley & Sons, Ltd. Fig. 4, p. 27, reprinted by permission of John Wiley & Sons, Ltd.)

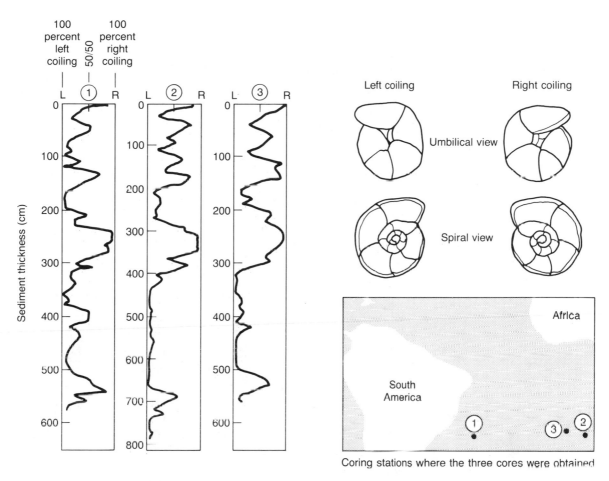

FIGURE 17.19 Biogeographical acme zone correlation based on coiling ratios of foraminifers. Correlation is based on coiling ratios of *Globorotalia truncatulinoides* in three South Atlantic Ocean cores. The depositional time represented by the cores is about 1.5 million years. (After Eicher, D. L., Geologic time, 2nd ed., © 1976, Fig. 5.12, p. 115. Reprinted by permission of Prentice-Hall, Englewood Cliffs, N.J. Original data from D. B. Ericson, and G. Woolin, 1968, Pleistocene climates and chronology in deep-sea sediments: Science, v. 162, p. 1227–1234.)

of warm water and in the opposite direction when it lives in areas of cold water. The foraminifer *Globorotalia truncatulinoides,* for example, has predominantly right-handed coils in warm water and left-handed coils in cold water. Figure 17.19 shows that during times of glacial cooling of the ocean in the Pleistocene, predominantly right-coiled populations of *Globorotalia truncatulinoides* were replaced in middle and low latitudes by predominantly left-coiled populations. These changes in coiling ratios of foraminiferal species provide a means of correlating short-term fluctuations of climatic change in the Pleistocene that are essentially synchronous throughout at least a part of an ocean basin.

The major drawback to correlation methods based on biologic response to climate fluctuations is that their use is restricted mainly to correlating sediments deposited during the Quaternary and Late Tertiary, when several episodes of cooling and warming in the world ocean took place. Nonetheless, they provide a useful supplement to correlation methods based on oxygen isotopes (Chapter 18), which also involve climate fluctuations in the Late Tertiary and Quaternary.

ADDITIONAL READINGS

Berry, W. B. N., 1968, Growth of a prehistoric time scale: W. H. Freeman, San Francisco, 158 p.

Boucot, A. J., and R. S. Carney, 1981, Principles of marine benthic paleoecology: Academic Press, New York.

Dodd, J. R., and R. J. Stanton, Jr., 1981, Paleoecology, concepts and applications: John Wiley & Sons, New York, 559 p.

Eicher, D. L., 1968, Geologic time, 2nd ed.: Prentice-Hall, Englewood Cliffs, N.J., 152 p.

Hallam, A. (ed.), 1977, Patterns of evolution as illustrated by the fossil record: Elsevier, New York, 591 p.

Hedberg, H. D. (ed.), 1976, International Stratigraphic Guide: John Wiley & Sons, New York, 200 p.

Hughes, N. F. (ed.), 1973, Organisms and continents through time: The Paleontological Association, London, 334 p.

Kauffman, E. G., and J. E. Hazel (eds.), 1977, Concepts and methods of biostratigraphy: Dowden, Hutchinson and Ross, Stroudsburg, Pa., 658 p.

Middlemass, F. A., P. F. Rawson, and G. Newell (eds.), 1971, Faunal provinces in space and time: Seel House Press, Liverpool, 236 p.

Nitecki, M. H. (ed.), 1984, Extinctions: University of Chicago Press, Chicago, 354 p.

Shaw, A. B., 1964, Time in stratigraphy: McGraw-Hill, New York, 365 p.

Stanley, S. M., 1979, Macroevolution, pattern and process: W. H. Freeman, San Francisco, 332 p.

18
Chronostratigraphy and Geologic Time

18.1 INTRODUCTION

The stratigraphic units described in the preceding chapters are rock units distinguished by lithology, magnetic characteristics, seismic reflections characteristics, or fossil content. As such, they are observable or measurable material reference units that depict the descriptive stratigraphic features of a region. Definition of these units allows the vertical and lateral relationships between rock units to be recognized and provides a means of correlating the units from one area to another. As Krumbein and Sloss (1963) point out, however, descriptive stratigraphic units do not lend themselves to interpretation of the local stratigraphic column in terms of Earth history. To do this requires that stratigraphic units be related to geologic time and that this relationship be expressed through some acceptable system of classification and terminology.

Geologists relate stratigraphic units to time through the use of **geologic time units.** In this chapter, we examine the concept of geologic time units and explore the relationship of time units to other types of stratigraphic units. We will see also how geologic time units are used to create the Geologic Time Scale and discuss methods of calibrating the time scale.

18.2 GEOLOGIC TIME UNITS

Geologic time units are conceptual units rather than material units, or actual rock units, although most geologic time units are based on material units. In fact, we recognize two distinct types of formal stratigraphic units that can be distinguished by geologic age: (1) units called **stratotypes,** based on actual rock sections and (2) units independent of material referents (Fig. 18.1). Ideally, the reference rock bodies for

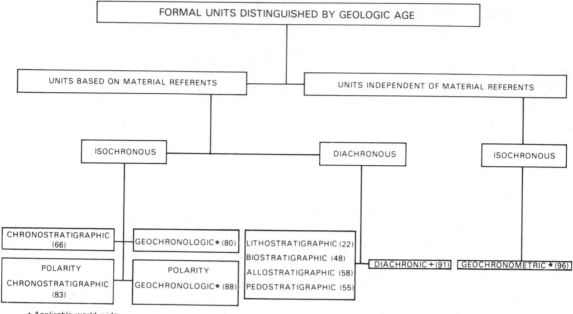

★ Applicable world-wide.
+Applicable only where material referents are present.
( )Number of article in which defined.

FIGURE 18.1   Major types of geologic time units and their relation to the kinds of rock-unit referents on which most are based. (From North American Commission on Stratigraphic Nomenclature, 1983, North American Stratigraphic Code: Am. Assoc. Petroleum Geologists Bull., v. 67. Fig. 1, p. 850, reprinted by permission of AAPG, Tulsa, Okla.)

geologic time units are **isochronous units.** That is, they are rock units formed during the same span of time and everywhere bounded by synchronous surfaces, which are surfaces on which every point has the same age. Although all geologic time units have traditionally been considered isochronous units, the 1983 North American Stratigraphic Code (Appendix) now formally recognizes certain diachronous rock bodies as referents for geologic time units. A diachronous geologic time unit is a unit that comprises the unequal spans of time represented by one or more specific diachronous rock bodies, that is, rock units that vary in age in different areas and thus cut across time planes. The wedge of sediments that typically forms during a transgressive-regressive depositional cycle (Fig. 14.9) is an example of a diachronous rock body.

18.3   ISOCHRONOUS TIME UNITS

The International Stratigraphic Guide (Hedberg, 1976) recognizes two fundamental types of isochronous geologic time units: chronostratigraphic units and geochronologic units. **Chronostratigraphic units** are tangible bodies of rock that are selected by geologists to serve as reference sections, or material referents, for all rocks formed during the same interval of time. **Geochronologic units,** by contrast, are divisions of time distinguished on the basis of the rock record as expressed by chronostratigraphic units. They are not in themselves stratigraphic units. Readers who find the distinction between these two types of units somewhat confusing may benefit from the following illustration. Chronostratigraphic units have been likened to the sand that flows through an hourglass during a certain period of time. By contrast, corresponding geochronologic units can be compared to the interval of time during which the sand flows (Hedberg, 1976). The duration of the flow measures a certain interval of time, such as an hour, but the sand itself cannot be said to be an hour.

Traditional internationally accepted chronostratigraphic units were previously based primarily on the time spans of lithostratigraphic or biostratigraphic units. The 1983 North American Stratigraphic Code now also formally recognizes polarity chronostratigraphic units, which are geologic time units based on the remanent magnetic fields in rocks.

Chronostratigraphic Units

As suggested, a chronostratigraphic unit is an isochronous body of rock that serves as the material reference for all rocks formed during the same span of time (Article 66, Appendix). Its boundaries are defined in a designated stratotype, or type section, on the basis of paleontologic or physical properties of the rock. A chronostratigraphic unit is always based upon some material referent unit or stratotype, that is, on an actual section of rock. It may be based on the time span of a biostratigraphic unit, a lithic unit, a magnetopolarity unit, or any other feature of the rock record that has a time range. The boundaries of chronostratigraphic units should be clearly defined in a designated stratotype on the basis of one or more of these types of units. In principle, chronostratigraphic units have worldwide extent and can be recognized throughout the world. In practice, worldwide use of chronostratigraphic units depends upon the extent to which the time-diagnostic features that characterize the units can be recognized everywhere.

Chronostratigraphic units are subdivided into several different ranks of smaller units. The hierarchy of these ranks is illustrated in Table 18.1, which shows also the geochronologic unit that corresponds to each chronostratigraphic rank. A short description of each chronostratigraphic rank is given below. These descriptions are drawn primarily from the 1983 North American Stratigraphic Code (Appendix).

Eonothem. An eonothem is the highest ranking chronostratigraphic unit. Two eonothems have commonly been recognized: the **Phanerozoic Eonothem,** which encompas-

TABLE 18.1 Conventional hierarchy of chronostratigraphic and geochronologic terms commonly used by stratigraphers

Chronostratigraphic hierarchy	Geochronologic hierarchy
Eonothem	Eon
Erathem	Era
System	Period
Series	Epoch
Stage	Age
Chronozone	Chron

Source: Hedberg, H. D. (ed.), International Stratigraphic Guide, Table 2, p. 68, © 1976, John Wiley & Sons, Inc. Reprinted by permission of John Wiley & Sons, Inc., New York.

ses the Paleozoic, Mesozoic, and Cenozoic erathems, and the **Precambrian Eonothem**, encompassing all older rocks. Recently, the IUGS Precambrian Subcommission, which was set up to study the problems of subdividing the Precambrian, has recommended that the Precambrian be divided into two eonothems: the **Proterozoic** and the **Archean** (Harrison and Peterman, 1980). Thus, current practice defines three eonothems.

Erathem. Erathems are subdivisions of an eonothem and commonly consist of several adjacent systems. The names for the Phanerozoic erathems were initially chosen to reflect major changes in the development of life on Earth. Thus Paleozoic refers to "old life," Mesozoic to "intermediate life," and Cenozoic to "recent life." The names and ranks of Precambrian erathems have not yet been standardized and agreed upon worldwide; different names are currently in use in different parts of the world.

System. System is next in rank below erathem and is the primary chronostratigraphic unit of worldwide major rank. That is, all of the internationally accepted systems have a time span sufficiently great that they serve as worldwide chronostratigraphic reference units. They are the fundamental reference units for the geologic time scale; as such, the names of the systems are familiar to all geologists. Thus, a stratigrapher who says that he or she is studying rocks of the Jurassic System or the Permian System will be readily understood by other geologists anywhere in the world. Systems are divided completely into units of the next lower rank (series). Under some circumstances, systems may be subdivided into **subsystems** or grouped into **supersystems.**

Series. Series is next in rank below system and is always a subdivision of a system. Most systems are divided into three series; however, the number of series in a system may range from two to six. Some series have formal names that are different from the system name. Most series take their names from the system by simply adding the appropriate adjective "Lower," "Middle," or "Upper" to the system name; for example, Lower Jurassic Series, Middle Jurassic Series, Upper Jurassic Series. Subdivision of systems into series is still a controversial issue with stratigraphers, and the numbers and names of the series in some systems have not yet been agreed upon internationally. Series are particularly useful for chronostratigraphic correlation within provinces; however, most series can be recognized worldwide and some series originally defined in Europe have now been adopted for dividing systems on other continents.

Stage. Stage is a stratigraphic unit of smaller scope and rank than series and represents a relatively short interval of geologic time. Because of its smaller scope and rank, it is particularly well suited to the practical needs and purposes of intraregional and intracontinental classification and correlation. Nonetheless, stages have the potential for worldwide recognition; they are the smallest units in the hierarchy of chronostratigraphic units that have this potential. Many European stages have now been recognized in North America and other parts of the world. If desirable, stages may be subdivided into **substages.**

Chronozone. A chronozone is the smallest of the formal chronostratigraphic units. The International Stratigraphic Guide (Hedberg, 1976) suggests that a chronozone is the lowest ranking division in the hierarchy of chronostratigraphic terms; however, the 1983 North American Stratigraphic Code (Appendix) regards chronozone as a non-hierarchical, but commonly small, formal chronostratigraphic unit. Its boundaries may be independent of those of ranked stratigraphic units. It may be based on a biostratigraphic unit, a lithostratigraphic unit, or a magnetopolarity unit.

Geochronologic Units

Geochronologic units are divisions of time traditionally distinguished on the basis of the rock record as expressed by chronostratigraphic units (Article 80, Appendix). A geochronologic unit is not an actual unit of rock, but it corresponds to the interval of time during which an established chronostratigraphic unit was deposited or formed. Thus, the beginning of a geochronologic unit corresponds to the time of deposition of the bottom of the chronostratigraphic unit upon which it is based, and the ending of a geochronologic unit corresponds to the time of deposition of the top of the referent unit.

Table 18.1 shows the hierarchical ranking of geochronologic units and the relationship between the ranks of geochronologic units and chronostratigraphic units. An eon of time is thus represented by the rocks that constitute an eonothem, era by an erathem, period by a system, epoch by a series, age by a stage, and chron by a chronozone. Names for periods and lower rank geochronologic units are identical with their corresponding chronostratigraphic units. Thus, we refer to the Jurassic system rocks that were deposited during the Jurassic Period or the Lower Jurassic series rocks that were deposited during the Early Jurassic Epoch of time. Note that the adjectives Early, Middle, and Late, rather than Lower, Middle, and Upper, must be used when referring to epochs because epochs are units of time, not units of rock. Most names for the eons and eras are the same as the names of the corresponding eonothems and erathems.

Polarity-chronostratigraphic and Polarity-geochronologic Units

Polarity-chronostratigraphic units are recognized as formal chronostratigraphic units for the first time in the 1983 North American Stratigraphic Code. A polarity-chronostratigraphic unit is defined by the Code (Article 83, Appendix) as a body of rock that exhibits the primary magnetic-polarity record imposed when the rock was deposited, or crystallized, during a specific interval of geologic time. The fundamental unit of polarity-chronostratigraphic classification is the **polarity-chronozone,** which consists of rocks of a specified primary polarity. Polarity-chronostratigraphic units, like all chronostratigraphic units, depend for definition upon actual sections of rock, or measurements on individual rock units. They are thus based on material units, specifically the polarity zones of magnetopolarity classification discussed in Chapter 15.

Polarity-geochronologic units are the chronologic units corresponding to polarity-chronostratigraphic units. They are divisions of geologic time distinguished on the basis of the record of magnetopolarity as embodied in polarity-chronostratigraphic units (Article 88, Appendix). The **polarity chron** is the fundamental unit of geologic time that corresponds to the time span of a polarity chronozone (Table 18.2). The name for a polarity chron is the same as that for the corresponding chronozone except that the term chron is substituted for chronozone.

Geochronometric Units

All of the geologic time units described in the preceding discussion are related to actual sections of rock. They are either actual units of rock or they are based upon

TABLE 18.2   Nomenclature of polarity-chronostratigraphic and polarity-geochronologic units

Polarity-chronostratigraphic units	Polarity-geochronologic units	Corresponding magnetopolarity units
Polarity superchronozone	Polarity superchron	Polarity superzone
Polarity chronozone	Polarity chron	Polarity zone
Polarity subchronozone	Polarity subchron	Polarity subzone

such units. Geochronologic units, for example, are abstractions in the sense that they are not actual material units, but are based on the time span of material units (chrono-stratigraphic units). Geologists also recognize a type of geologic time unit that is independent of a material reference section. Such an independently derived time unit is called a **geochronometric unit.** Geochronometric units are thus direct divisions of geologic time. They are not based on the time span of designated chronostratigraphic stratotypes. They are simply time divisions of an appropriate magnitude or scale. Their boundaries are arbitrarily chosen ages expressed in some convenient multiple of years. Geologic ages are commonly expressed in millions of years (m.y. or Ma), but they may also be expressed in thousands of years (Ka) or billions of years (Ga).

Geochronometric units have been used particularly in attempts to develop a time scale for Precambrian rocks (Fig. 18.2). These rocks have not yet proven generally susceptible to analysis and subdivision by superposition or by application of other lithologic or biologic principles that are commonly used in subdividing the Phanero-zoic rocks. Subdivision has proven extremely difficult owing to the nature of Precambrian igneous, metamorphic, and orogenic events; the complex rock assemblages that constitute the Precambrian record in most areas; and the generally unfossiliferous character of Precambrian rocks. The 1983 North American Stratigraphic Code suggests that geochronologic rank terms (eon, era, period, epoch, age, and chron) may be used for geochronometric terms when these geochronologic units are formalized. Thus, the Archean Eon (oldest Precambrian time) and the Proterozoic Eon (younger Precambrian time), which have been tentatively formalized for the North American Precambrian, can be considered geochronometric units, with the boundary between them chosen arbitrarily at 2.5 billion years.

The geologic time table, which we will discuss in Section 18.5, is calibrated in terms of absolute ages to the extent presently possible. In a sense, this calibration constitutes a geochronometric scale (Harland, 1978); however, I stress again that geochronometric units are not defined by chronostratigraphic units, although they may have corresponding chronostratigraphic units (eonothem, erathem, system, series, stage, and chronozone).

18.4   DIACHRONOUS TIME UNITS

The 1983 North American Stratigraphic Code introduces the term diachronic unit (Article 91, Appendix) for a stratigraphic unit that comprises the unequal spans of time represented by a specific stratigraphic unit, or assemblage of units, such as a lithostratigraphic or biostratigraphic unit. The stated purpose of introducing diachronic units as formal geologic time units is to provide

(1) a means of comparing the spans of time represented by stratigraphic units with diachronous boundaries at different localities, (2) a basis for broadly establishing in time the beginning and ending of deposition of diachronous stratigraphic units at different sites, (3) a basis for inferring the rate of change in areal extent of depositional processes, (4) a means of determining and comparing the rates and durations of deposition at different localities, and (5) a means of comparing temporal and spatial relations of diachronous stratigraphic units.

One or both boundaries of a diachronic unit are demonstrably time-transgressive (diachronous); thus, the duration and age of a diachronic unit differ from place to place. This difference is illustrated in Figure 18.3, which depicts also the relationship of diachronic units to chronostratigraphic and geochronologic units. All diachronic units are based on a diachronous reference unit of some type. These reference units

FIGURE 18.2 Proposed chronometric time scale for the Precambrian of the United States and Mexico. (After Harrison, J. E., and Z. E. Peterman, 1980, North American Commission on Stratigraphic Nomenclature Note 52—A preliminary proposal for a chronometric time scale for the Precambrian of the United States and Mexico: Geol. Soc. America Bull., v. 91, Fig. 1, p. 378. Subsequent revisions, J. E. Harrison, personal communication, 1985.)

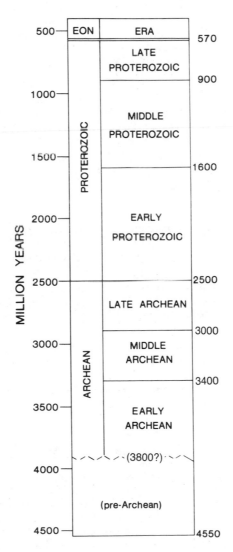

FIGURE 18.3 Comparison of geochronologic units, chronostratigraphic units, and diachronic units. (From North American Stratigraphic Commission, 1983, Fig. 10, p. 871, reprinted by permission of AAPG, Tulsa, Okla.)

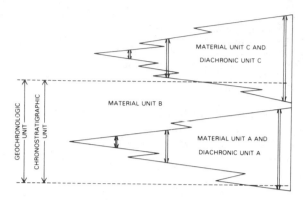

may be a lithostratigraphic unit; a biostratigraphic unit; an allostratigraphic unit, which is a mappable stratiform body of sedimentary rock that is defined and identified on the basis of its bounding discontinuities; or a pedostratigraphic, or soil-stratigraphic, unit. The fundamental diachronic unit is the **diachron.** A diachron is a nonhierarchical unit, but if a hierarchy of diachronic units is needed, the Code recommends the terms episode, phase, span, and cline, in order of decreasing rank.

18.5 THE GEOLOGIC TIME SCALE

Purpose and Scope

Classifying rocks on the basis of time involves systematic organization of strata into named units, each corresponding to specific intervals of geologic time. These units provide a basis for time correlation and a reference system for recording and systematizing specific events in the geologic history of Earth. Thus, the ultimate aim of creating a standardized geologic time scale is to establish a hierarchy of chronostratigraphic units of international scope that can serve as a standard reference to which the ages of rocks everywhere in the world can be related. Establishing the relative ordering of events in Earth's history is the main contribution that geology makes to our understanding of time.

Harland (1978) suggests that a standard geologic time scale should (1) express any age in any place, (2) express broad and general ages as well as detailed and particular ages, (3) be understandable, clear, and unambiguous, (4) be independent of opinion and therefore have some objective reference that is accessible, (5) be stable, that is, not subject to frequent change, and (6) be agreed to and used internationally in all languages.

Development of the Geologic Time Scale

Chronostratigraphic Scale. Geologists have been trying for almost 200 years to develop a systematic scheme for time-stratigraphic classification of rock units. This slow process has evolved through two fundamental stages of development: (1) determining time-stratigraphic relationships from local stratigraphic sections by applying the principle of superposition, supplemented by fossil control and, more recently, radiometric ages and (2) using these local stratigraphic sections as a basis for establishing a composite international chronostratigraphic scale, which serves as the material reference for constructing a standardized international geologic time scale.

An international chronostratigraphic scale ideally consists of a systematically arranged hierarchy of named and defined chronostratigraphic units that encompass the entire stratigraphic sequence or geologic column without gaps or overlaps. This scale serves as a standard framework for expressing the ages of all rock strata and their position with respect to Earth history. The procedure of naming units of geologic time to represent spans of time during which well-described units of rock were deposited or formed has long been standard practice in the study of Phanerozoic rocks. It has not yet been possible to extend this practice to classification of Precambrian rocks owing to the generally unfossiliferous character and structural complexity of these rocks.

Because the fundamental tool stratigraphers use for putting events in the Earth's history in chronologic order is the principle of positional relationships, or relative ages of rocks, the geologic time scale is thus based on actual sections of rocks. These chronostratigraphic sections are the material referents for all rocks formed during the same span of time. Ideally, the material reference for the geologic time scale should be a single stratigraphic section containing a complete sequence of strata that represent all geologic time, with no gaps in the stratigraphic record. Unfortunately, there is nowhere on Earth a continuous record of strata representing all intervals of geologic time. The currently used geologic systems that constitute the reference sections for the geologic time scale were not proposed by any one geologist or group of geologists in any one locality, nor were they developed as a result of any specific organized approach. They developed gradually over a long period of time, without a specific plan, through the efforts of many different geologists working independently in various parts of the world, particularly Europe.

The first attempts at placing rocks into some type of systematic sequence began as early as about 1750. The historical development of the geologic time scale, which eventually culminated in the establishment and naming of the geologic systems shown in Table 18.3, has been traced by numerous workers (for example, Dunbar and Rodgers, 1957; Krumbein and Sloss, 1963; Berry, 1968; and Eicher, 1976). The geologic time table evolved slowly over nearly two centuries, during which time many different schemes for subdividing the rock record were proposed. Many of these schemes were ultimately rejected; however, by the early to middle part of the twentieth century the system names shown in Table 18.3 had been adopted more or less internationally. As this table suggests, the criteria used by various workers to define the systems were not the same in all cases. Some systems were established on the basis of distinctive lithologies and the presence of major bounding discontinuities. Other system boundaries were defined on the basis of fossils. All of the system boundaries have subsequently been defined by distinctive fossils, in some cases in localities other than the original type locality, making it possible to recognize the geologic systems worldwide.

The geologic systems listed in Table 18.3 form a fundamental part of the conventional hierarchy of chronostratigraphic units, a hierarchy that constitutes the basis for the geologic time scale. These systems are grouped into erathems (Paleozoic, Mesozoic, Cenozoic), and each system, in turn, can be subdivided into smaller units (series, stage). Although the systems are accepted by the international geologic community as the basic reference sections for the geologic time scale, considerable controversy still exists regarding the exact placement of system boundaries and the subdivision of some systems. For example, some systems have been subdivided into units considered by one group of geologists to be series and by another group to be stages. Therefore, the

TABLE 18.3   The internationally accepted geologic systems and their type localities

System name	Type locality	Named or proposed by	Date proposed	Remarks
Cambrian	Western Wales	Adam Sedgwick	1835	Defined mainly on lithology
Ordovician	Western Wales	Charles Lapworth	1879	Set up as an intermediate unit between the Cambrian and Silurian to resolve boundary dispute; boundary defined by fossils
Silurian	Western Wales	Roderick I. Murchison	1835	Defined by lithology and fossils
Devonian	Devonshire, southern England	Roderick I. Murchison and Adam Sedgwick	1840	Boundaries based mainly on fossils
Carboniferous	Central England	William Conybeare and William Phillips	1822	Named for lithologically distinctive coal-bearing strata, but recognizable by distinctive fossils
Mississippian	Mississippi Valley, U.S.A.	Alexander Winchell	1870	The Mississippian and Pennsylvanian are subdivisions of the Carboniferous; not used outside the United States
Pennsylvanian	Pennsylvania, U.S.A.	Henry S. Williams	1891	
Permian	Province of Perm, Russia	Roderick I. Murchison	1841	Identified by distinctive fossils
Triassic	Southern Germany	Fredrick von Alberti	1843	Defined lithologically on the basis of a distinctive threefold division of strata; also defined by fossils
Jurassic	Jura Mountains, northern Switzerland	Alexander von Humboldt	1795	Defined originally on the basis of lithology
Cretaceous	Paris Basin	Omalius d'Halloy	1882	Defined initially on the basis of strata composed of distinctive chalk beds
Tertiary	Italy	Giovanni Arduino	1760	Originally defined by lithology; redefined with type section in France on the basis of distinctive fossils
Quaternary	France	Jules Desnoyers	1829	Defined by lithology, including some unconsolidated sediment

chronostratigraphic units that form the material referents for the geologic time scale are still in a state of flux, and revision of the boundaries of these units continues to the present time.

Table 18.4 shows the complete hierarchy of chronostratigraphic units in general use throughout most of the world. This chronostratigraphic and geochronometric scale

TABLE 18.4 (p. 667–669)   Nomenclature of chronostratigraphic units generally used throughout the world. Some North American stage names are also shown, together with a chronometric scale.

Source: Salvador, A., 1985, Chronostratigraphic and geochronometric scales in COSUNA stratigraphic nomenclature charts of the United States: Am. Assoc. Petroleum Geologists Bull., v. 69. Figs. 1–3, p. 182–184, reprinted by permission of AAPG, Tulsa, Okla.

GLOBAL CHRONOSTRATIGRAPHIC UNITS					NORTH AMERICAN CHRONOSTRATIGRAPHIC UNITS	NUMERICAL TIME SCALE (Ma)	
ERATHEM	SYSTEMS		SERIES / STAGES		SERIES / STAGES		
C E N O Z O I C	QUATERNARY		HOLOCENE / PLEISTOCENE		NORTH AMERICAN PLEISTOCENE GLACIAL STAGES ONLY WHEN APPLICABLE AND NECESSARY	0.01 / 1.7 to 2.8	
	T E R T I A R Y	**N E O G E N E**	PLIOCENE	PIACENZIAN		4.6 / 5.3	5
				ZANCLEAN		6.7	
			MIOCENE — UPPER	MESSINIAN			
				TORTONIAN		10.8	10
			MIDDLE	SERRAVALLIAN		15.4	15
				LANGHIAN		17	
			LOWER	BURDIGALIAN			20
				AQUITANIAN		23	
						25	25
		P A L E O G E N E	OLIGOCENE — UPPER	CHATTIAN	PACIFIC AREA STAGES	30	
						33	
			LOWER	RUPELIAN	OR		35
					MAMMALIAN STAGES ONLY	38	
			EOCENE — UPPER	PRIABONIAN	WHEN APPLICABLE	41	40
				BARTONIAN	AND NECESSARY		45
			MIDDLE	LUTETIAN		45	
			LOWER	YPRESIAN		50	50
						55	55
			PALEOCENE — UPPER	THANETIAN		62	60
			LOWER	DANIAN			65
						67	

ERATHEM	SYSTEMS	GLOBAL CHRONOSTRATIGRAPHIC UNITS SERIES / STAGES		NORTH AMERICAN CHRONOSTRATIGRAPHIC UNITS SERIES / STAGES	NUMERICAL TIME SCALE (Ma)	
M E S O Z O I C	CRETACEOUS	UPPER	MAASTRICHTIAN	SAME AS GLOBAL	67	65
			CAMPANIAN		72	70
			SANTONIAN		80	80
			CONIACIAN		85	
			TURONIAN		90	90
			CENOMANIAN		92	
		LOWER	ALBIAN		100	100
			APTIAN		108	110
			BARREMIAN		115	120
			HAUTERIVIAN		125	
			VALANGINIAN		130	130
			BERRIASIAN		135	
	JURASSIC	UPPER	TITHONIAN	SAME AS GLOBAL	140	140
			KIMMERIDGIAN		145	150
			OXFORDIAN		155	
		MIDDLE	CALLOVIAN		160	160
			BATHONIAN		165	170
			BAJOCIAN		170	
			AALENIAN		175	
		LOWER	TOARCIAN		180	180
			PLIENSBACHIAN		185	
			SINEMURIAN		190	190
			HETTANGIAN		195	
	TRIASSIC	UPPER	RHAETIAN	SAME AS GLOBAL	200	200
			NORIAN		215	210
			CARNIAN		220	220
		MIDDLE	LADINIAN		230	230
			ANISIAN		240	240
		LOWER	SCYTHIAN		245	
					250	250

GLOBAL CHRONOSTRATIGRAPHIC UNITS				NORTH AMERICAN CHRONOSTRATIGRAPHIC UNITS		NUMERICAL TIME SCALE (Ma)	
ERATHEM	SYSTEMS	SERIES / STAGES		SERIES / STAGES			
PALEOZOIC	PERMIAN	UPPER	TATARIAN	OCHOAN		250 / 255	250
			KAZANIAN	GUADALUPIAN			260
			KUNGURIAN			270 / 275	270
		LOWER	ARTINSKIAN	LEONARDIAN			280
			SAKMARIAN	WOLFCAMPIAN		285	
			ASSELIAN			290	290
	CARBONIFEROUS	UPPER	STEPHANIAN / GZHELIAN	PENNSYLVANIAN SUB-SYSTEM	VIRGILIAN		300
			STEPHANIAN / KASIMOVIAN		MISSOURIAN		
					DESMOINESIAN	310	310
		MIDDLE	WESTPHALIAN / MOSCOVIAN		ATOKAN	315	
			"NAMURIAN" / BASHKIRIAN		MORROWAN		320
			SERPUKHOVIAN			330	330
		LOWER		MISSISSIPPIAN SUB-SYSTEM	CHESTERIAN	340	340
			VISEAN		MERAMECIAN		350
					OSAGEAN	355	
			TOURNAISIAN		KINDERHOOKIAN	360	
						365	
	DEVONIAN	UPPER	FAMENNIAN	CHAUTAUQUAN	CONEWANGOAN		370
					CASSADAGAN		
			FRASNIAN	SENECAN	CHEMUNGIAN	380	380
		MIDDLE	GIVETIAN	ERIAN	FINGERLAKESIAN	385	
			EIFELIAN				390
			EMSIAN		ESOPUSIAN	395	
		LOWER	SIEGENIAN	ULSTERIAN	DEERPARKIAN	400	400
			GEDINNIAN		HELDERBERGIAN	405	
	SILURIAN	UPPER	PRIDOLIAN	CAYUGAN			410
			LUDLOVIAN	NIAGARAN	LOCKPORTIAN	415	
		LOWER	WENLOCKIAN		CLIFTONIAN / CLINTONIAN		420
			LLANDOVERIAN	ALEXANDRIAN		420 / 425	
	ORDOVICIAN	UPPER	ASHGILLIAN	CINCINNATIAN	RICHMONDIAN		430
					MAYSVILLIAN		440
			CARADOCIAN		EDENIAN		450
						455	
				SHERMANIAN / KIRKFIELDIAN / ROCKLANDIAN	BLACKRIVERIAN	460	460
		MIDDLE	LLANDEILIAN		CHAZYAN		470
			LLANVIRNIAN	CHAMPLAINIAN		475	
					WHITEROCKIAN		480
			ARENIGIAN			485	
		LOWER	TREMADOCIAN	CANADIAN		490	490
						500	500
	CAMBRIAN	UPPER		TREMPEALEAUAN			
				FRANCONIAN			510
				DRESBACHIAN		515	
							520
		MIDDLE					530
							540
						540	
		LOWER					550
							560
						570	570

was compiled by Salvador (1985) as part of the COSUNA (Correlation of Stratigraphic Units of North America) project. Some of the provincial stage names commonly used in North America are shown also; however, there now appears to be a general movement among North American stratigraphers to abandon these provincial stage names and adopt the European stages as standards for North America. Stratigraphers in Europe and many other parts of the world have for many years subdivided the Tertiary into two subsystems, the **Paleogene** and the **Neogene,** with the top of the Oligocene Series as the dividing boundary between the two. Geologists in North America have now also adopted this practice. They have likewise adopted the European usage of **Carboniferous** as a system name, with subdivision into the **Mississippian** and **Pennsylvanian** subsystems. Readers are reminded that other versions of the chronostratigraphic scale (for example, Harland et al., 1982) may differ somewhat from this scale, particularly in naming of series and stages, as well as in ages of series and stage boundaries. The geologic community has not yet achieved the ideal of a truly international chronostratigraphic scale that is accepted and used by all geologists worldwide.

Geochronologic Scale. Table 18.4 is a chronostratigraphic scale with units and boundaries based on physical divisions of the rock record, but it is not in itself a time scale. To function as a geologic time scale for expressing the age of a rock unit or a geologic event, the chronostratigraphic scale must be converted to a geochronologic scale consisting of units that represent intervals of time rather than bodies of rock that formed during a specified time interval. The geologic time scale is derived from the chronostratigraphic scale by substituting for chronostratigraphic units the corresponding geochronologic units. Thus, the geologic time scale is expressed in eras, periods, epochs, and ages rather than erathems, systems, series, and stages. The subdivision boundaries of the geologic time scale are calibrated in absolute ages; however, the geologic time scale differs from a true chronometric scale. As explained above, a chronometric scale is based purely on time, without regard to the rock record. By contrast, the subdivisions of the Phanerozoic time scale are of unequal length, owing to the fact that they are based on chronostratigraphic units that were deposited during unequal intervals of time.

The geologic time scale has been in existence for several decades, and during that time it has continued to evolve, with refinements being made particularly in subdivision of epochs and ages and absolute-age calibration of the boundaries between periods, epochs, and ages. Figure 18.4 shows a recent version of the geologic time scale published in North America by the Geological Society of America as part of the efforts involved in publication of the 27 synthesis volumes of the *Geology of North America* for the Decade of North American Geology (DNAG) (Palmer, 1983). This DNAG time scale is subdivided into ages based on the European stages, with boundaries between ages calibrated in absolute time. Absolute ages are given in millions of years (Ma) before the present, where the present refers to 1950. Methods for absolute-age calibration of the geologic time scale are discussed below. Note that the magnetic polarity scale for the most recent approximately 170 million years is also included in the time scale. Note also the use of a chronometric scale for the Precambrian, with the dividing boundary between the Archean and the Proterozoic set arbitrarily at 2500 million years, as discussed.

The traditional procedure followed in establishing the geologic time scale has been first to define and describe the chronostratigraphic reference units (system, series,

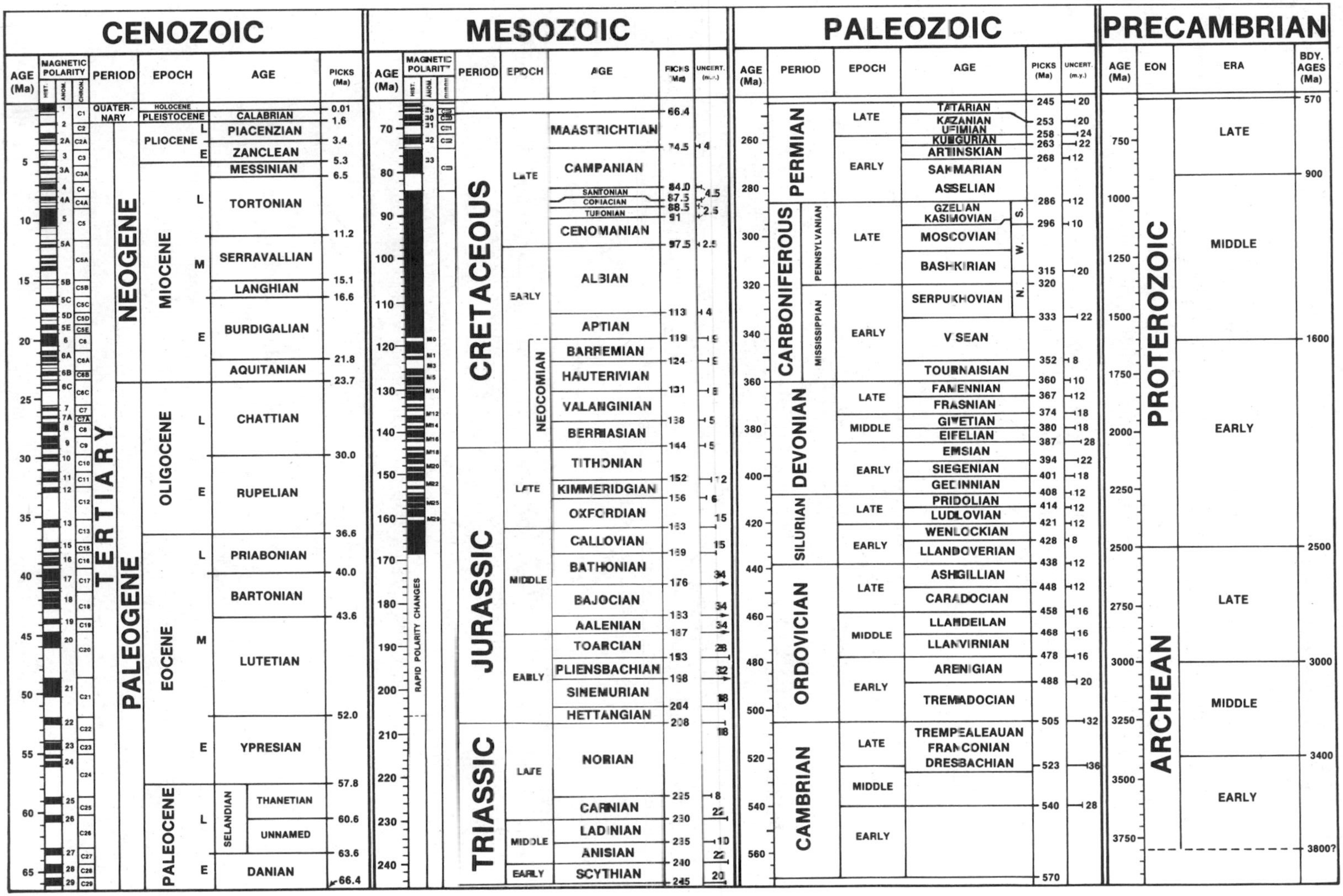

FIGURE 18.4 Geologic time table calibrated in absolute ages. The magnetic polarity scale for the last 170 million years is also shown. (From Palmer, A. R. (comp.), 1983, The Decade of North American Geology 1983 geologic time scale: Geology, v. 11, p. 504.)

stage) and then derive the corresponding geochronologic unit (period, epoch, age) from these already established rock units. In other words, geologic time intervals are conceived as being the time equivalents of rock referent units already defined. Harland (1978) and Harland et al. (1982) suggest that this procedure be reversed. They recommend that the geochronologic units be established first by selecting well-defined reference points in stratigraphic sections. Each point represents a specific time of deposition. Pairs of such points thus define the intervening time span. The period, for example, is first defined in this way by picking upper and lower reference points that can be calibrated by absolute age. The system in turn becomes the rock formed during that particular defined period. This proposed procedure has not yet been widely accepted; however, stratigraphers and other geologists throughout the world are actively working together to achieve the ultimate goal of an international geologic time scale. Names that have been suggested for this ideal geologic time scale include **Standard Stratigraphic(al) Scale, Global Chronostratigraphic (Geochronologic) Scale,** and **Geochronostratic Scale.**

18.6   CALIBRATING THE GEOLOGIC TIME SCALE

As mentioned, the geologic time scale has evolved slowly over a long period of time. To develop the scale to its present level of usefulness for fixing the position in time of a particular rock unit or geologic event, two types of information had to be available to stratigraphers: (1) some method of arranging rocks in an orderly sequence on the basis of their relative positions in time, or relative ages, and (2) a method of determining the ages of the boundaries between rock units on the basis of their absolute positions in time with respect to some fixed time horizon; for example, the present.

Placing strata in stratigraphic order in terms of their relative ages has been the guiding principle used by stratigraphers in constructing the geologic time scale. Relative ordering was determined by applying the principle of superposition, aided by use of fossils. As stated previously, the principle of superposition means simply that in a normal sequence of strata that have not been tectonically overturned since deposition, the youngest strata are on top and the age of the strata increases with depth. Application of this simple but highly important principle was of key importance in building the early chronostratigraphic scale and is still useful for determining the relative ages of strata in any local stratigraphic sequence.

Owing to the irreversible nature of evolution, fossil organisms succeed each other in the stratigraphic record in an orderly fashion—the principle of faunal succession. This organic continuum makes possible the determination of relative ages of strata because each fossiliferous unit of rock in a stratigraphic succession contains distinctive fossils that distinguish it from younger strata above and older strata below. These distinctive fossil assemblages also allow correlation of strata from one area to another, as discussed in Chapter 17, making possible determination of relative ages of strata outside the boundaries of original areas of study. Most of the divisions in the current global chronostratigraphic scale are based on fossils, and early efforts to create an international chronostratigraphic scale before methods of absolute-age determinations were developed would have been impossible without the use of fossils.

Fortunately, methods are now available not only for determining the relative ages of strata but also for fixing within reasonable limits of uncertainty the absolute ages of some strata. Development of these methods of absolute-age estimation have made it possible to place approximate absolute ages on boundaries of the chronostratigraphic scale initially established by relative-age determination methods. Absolute-age data can be used also for determining ages of poorly fossiliferous Precambrian rocks that cannot be placed in stratigraphic order by relative-age determination methods. The principal method for determining the absolute ages of rocks is based on isotopic decay of radioactive elements in minerals. Other methods of determining the absolute passage of geologic time include counting lake-sediment varves, which are presumed to represent annual sediment accumulations; growth increments in the shells of some invertebrate organisms; and growth rings in trees. These alternative methods are useful only for marking the passage of short periods of time in local areas and are not of importance in calibrating the geologic time scale.

Thus, the major tools for finding ages of sediments are relative-age determinations by use of fossils—biochronology—and absolute-age estimates based on isotopic decay—radiochronology. These tools may be used both for calibrating the chronostratigraphic scale directly and for calibrating the sequence of reversals of the Earth's magnetic field; this sequence constitutes the magnetostratigraphic scale discussed in Chapter 15. We shall now discuss each of these dating methods, beginning with biochronology.

Dating by Fossils: Biochronology

Biochronology is the organization of geologic time according to the irreversible process of evolution in the organic continuum (Berggren and Van Couvering, 1978). Useful fossil horizons are more widespread and abundant in Phanerozoic rocks than are horizons whose ages can be estimated by radiochronology, and biologic events can commonly be correlated in time more precisely than can radiometric data in all but very young rocks. Because of these factors, fossils have conventionally provided the most readily available tool for dating and long-distance correlations of Phanerozoic rocks. It is necessary, however, to make a clear distinction between biochronology and biostratigraphy. Biostratigraphy aims simply at recognizing the distinctive fossils that characterize a known stratigraphic level in a sedimentary section without regard to the inherent time significance of the fossils. For example, William Smith was able to use fossils very effectively for identifying and correlating strata even though he had little or no idea of the time relationships or time significance of the fossils. Biochronology, on the other hand, is concerned with the recognition of fossils as having an age that falls at a known point in the span of evolutionary time, as measured by fossils of a reference biostratigraphic section. Therefore, by establishing identifiable horizons in reference sections based on fossils, biochronology provides a tool both for international correlation and worldwide age determination.

The aim of biochronology is to make possible correlation and dating of the geologic record beyond the limits of local stratigraphic sections. To do this most effectively, stratigraphers use features or events in the paleontologic record that are widespread and easily identifiable and that occurred during short periods of geologic time. These events are considered to be biochronologic **datum events** because they mark a particular short period of time in the geologic past. The datum events most commonly used are the immigrations (first appearances) and extinctions (last appearances) of a fossil species or taxon. The first appearance of a species as a result of immigration from another area commonly occurs very rapidly after its initial appearance owing to evolution from its ancestral morphotype. Appearance is so rapid, in fact, that geologically speaking we consider speciation and immigration as essentially synchronous events. Extinction of a taxon also may occur very rapidly, although commonly not as rapidly as speciation.

Stratigraphers speak of the first and last appearances of a taxon as the **first appearance datum** (FAD) and the **last appearance datum** (LAD). These FADs and LADs are not totally synchronous owing to the fact that even though immigrations and extinctions can take place quite rapidly, they are not actually instantaneous events. Some planktonic species can spread worldwide in 100–1000 years (Fig. 18.5); however, bioturbation of sediment after deposition can mix fossils through a zone several centimeters thick, and accidents in preservation as well as bias in collection and analytical methods can combine to create uncertainties in the age of the FADs and LADs that can amount to thousands of years. Nevertheless, the duration of the FADs of planktonic species may be as little as 10,000 years; that is, the ages of the first appearance datum of a species will not vary by more than 10,000 years in different parts of the world (Berggren and Van Couvering, 1978). The error caused by an age discrepancy of this magnitude becomes insignificant when applied to estimation of the ages of rocks that are millions to hundreds of millions of years old. Thus, the FADs and LADs of many fossil species can be considered essentially synchronous for the utilitarian purposes of biochronology.

FADs and LADs are the most easily utilized and communicated types of fossil information upon which to base biochronology, and they can be used over great distances within the range of the defining taxa. Therefore, they have come to dominate global biochronological subdivision. The procedure for establishing the biochronology of any fossil group on the basis of FADs and LADs involves the following steps (Haq and Worsley, 1982): (1) Identify and locate in local biostratigraphic units the FADs and LADs of distinctive fossil taxa that have wide geographic distribution. (2) If possible, assign ages to these events by direct or indirect calibration through radiochronology or magnetostratigraphy. If ages can be assigned to any two events, the sedimentation rates for strata between these events can be calculated by dividing the age difference between the two by the thickness of sediment separating them. The sedimentation rates can then be used to calculate the approximate age of each event enclosed within the dated sequence. (3) If radiometric or magnetostratigraphic calibration of FADs and LADs in the local section cannot be accomplished, then the ages of the datum levels must be found in a different way. Under these conditions, ages of the FADs and LADs are estimated on the basis of their stratigraphic position with respect to calibrated datum levels of other fossil groups that occur in the sedimentary sequence and whose ages have been found by studying one or more sequences elsewhere. An example of biochronologic calibration is illustrated in Figure 18.6, which shows the use of calcareous nannoplankton to establish a biochronology for the Pleistocene by direct correlation with magnetostratigraphic units.

Absolute Ages: Radiochronology

General Principles

Fossils provide an exceptionally useful tool for determining relative ages of rocks and geologic events; however, prior to the beginning of the twentieth century scientists had no accurate method for determining the absolute passage of geologic time. Many early attempts were made to estimate the age of Earth, but with highly inconsistent results. For example, in 1897 Lord Kelvin, an outstanding physicist of the nineteenth century, used an assumed rate of cooling of a presumed molten Earth as a basis for estimating the Earth's age. He used this assumed rate of cooling, together with estimates of the time during which the sun would remain hot without a major internal energy source, to calculate an age for Earth of 20–40 million years. Two years later, a geologist named John Joly, on the basis of present salinity of ocean water and an as-

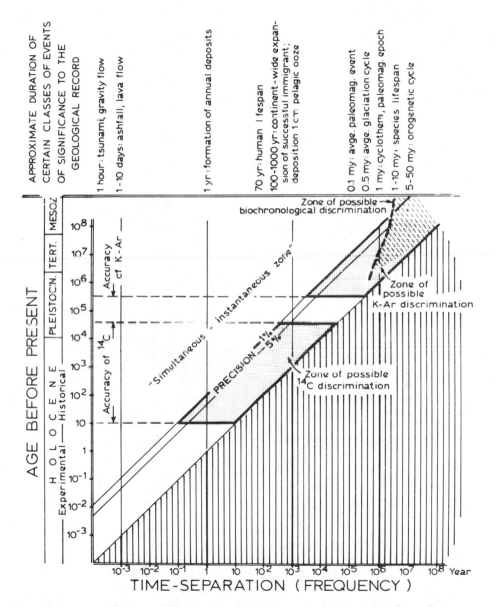

FIGURE 18.5 Resolving power of geochronologic systems in the Cenozoic. The evolutionary (biochronologic) precision is based on the lifespan of species. This system is capable of discrimination with almost undiminished precision between events of about 1 m.y. frequency to the limit of Phanerozoic time. (From Berggren, W. A., and J. A. Van Couvering, 1978, Biochronology, in G. V. Cohee, M. F. Glaessner, and H. D. Hedberg (eds.), The geologic time scale: Am. Assoc. Petroleum Geologists Studies in Geology 6. Fig. 1, p. 43, reprinted by permission of AAPG, Tulsa, Okla.)

FIGURE 18.6 An example of biochronological dating by use of nannofossil datum events correlated with magnetic polarity events. (After Gartner, S., 1977, Calcareous nannofossil biostratigraphy and revised zonation of the Pleistocene: Marine Micropaleontology, v. 2. Fig. 5, p. 12, reprinted by permission of Elsevier Science Publishers, Amsterdam.)

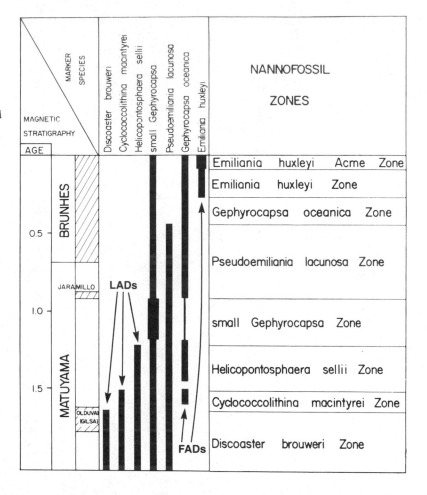

sumed annual rate of delivery of salts to the ocean by rivers, calculated the age of the oceans to be 90 million years. Between about 1860 and 1910, several geologists calculated the age of Earth by dividing the estimated thickness of the total stratigraphic record by assumed rates of sedimentation. These calculations yielded ages for Earth ranging from 3 million years to 1,584 million years, although most calculations gave ages of less than 100 million years.

We now know that the results of these early calculations are several orders of magnitude too small and that the methods used to obtain them were based on faulty assumptions and incorrect measurements. The breakthrough in developing an accurate method for determining absolute ages of rocks began about 1896 with the discovery by the French physicist Henri Becquerel that uranium has the ability to spontaneously emit rays that cause fogging of photographic plates in total darkness—a property he called **radioactivity.** This discovery led B. B. Boltwood, an American chemist at Yale University, to suggest in 1905 that radioactive breakdown of uranium leads ultimately to the production of lead. From chemical analyses of uranium minerals from many parts of the world, Boltwood was able to show by about 1907 that uranium minerals from older rocks contain more radiogenically produced lead than do uranium minerals from younger rocks. Using rough estimates of the decay rate of uranium to lead, he obtained ages based on uranium/lead ratios that ranged from 340 million years for Carboniferous rocks to 1640 million years for Precambrian rocks.

Significant improvements have been made since Boltwood's time in analytical techniques used for measuring radiometric ages, and the accuracy and precision of the ages have improved correspondingly; however, the basic principles of estimating absolute ages remain the same. That is, the age of a radiogenic mineral is usually calculated from the measured ratio of parent radionuclide to daughter product in the mineral, using the known decay rate of the parent material. The decay rate is measured in the laboratory by special counters and is commonly expressed as the half-life of the radioactive isotope (that is, the time required for one-half of the parent material to decay to daughter product). The most useful radionuclides for estimating absolute ages and the minerals, rocks, and organic materials most suitable for age determination are shown in Table 18.5. Details of radiochronologic methods and discussions of errors and uncertainties in radiometric age determinations are given in several published volumes (e.g., Mahaney, 1984; Odin, 1982b; Berger and Suess, 1979; Eicher, 1976; York and Farquhar, 1972; Hamilton and Farquhar, 1968; Faul, 1966) and will not be repeated here. We shall confine our discussion to the application of radiochronologic methods to calibration of the geologic time scale.

TABLE 18.5 Principal methods of radiometric age determination

Parent nuclide	Daughter nuclide	Half-life (years)	Approximate useful dating range (years B.P.)	Materials commonly dated
Carbon-14	*Nitrogen-14	5,730	**<~40,000	Wood, peat, CaCO$_3$ shells, charcoal
Protactinium-231 (daughter nuclide of uranium-235)	*Actinium-227	33,000	<150,000	Deep-sea sediment, aragonite corals
Thorium-230 (daughter nuclide of uranium-238/234)	*Radium-226	77,000	<250,000	Deep-sea sediment, aragonite corals
Uranium-238	Lead-206	4,510 million	>5 million	Monazite, zircon, uraninite, pitchblende
Uranium-238	Spontaneous fission tracks	—	**<~65 million	Volcanic glass, zircon, apatite
Uranium-235	Lead-207	713 million	>60 million	Monazite, zircon, uraninite, pitchblende
Potassium-40	Argon-40	1,300 million	>~100,000	Muscovite, biotite, hornblende, glauconite, sanidine, whole volcanic rock
Rubidium-87	Strontium-87	47,000 million	>5 million	Muscovite, biotite, lepidolite, microcline, glauconite, whole metamorphic rock

*Not used in calculating radiometric ages
**Can be used for dating older rocks under favorable circumstances

Radiometric Methods for Calibrating the Geologic Time Scale

Although radiochronologic methods can be applied to a variety of rock materials and organic substances (Table 18.5), they have limited application to the direct estimation of ages of sedimentary rocks. Most of the potentially usable minerals in sedimentary rocks are terrigenous minerals that when analyzed yield the age of the parent source rock, not the time of deposition of the sedimentary rock. Therefore, much of the geologic time scale has been calibrated by indirect methods of estimating ages of sedimentary rocks on the basis of their relationships to igneous or metamorphic rocks whose ages can be estimated by radiochronology. The types of rocks that are most useful for isotopic calibration of the geologic time scale are described in Table 18.6. We will now examine in greater detail the methods used to find ages of the sedimentary rocks of the international chronostratigraphic scale.

Finding Ages of Sedimentary Rocks by Analyzing Interbedded "Contemporaneous" Volcanic Rocks. Lava flows and pyroclastic deposits such as ashfalls can be incorporated very quickly into an accumulating sedimentary sequence without significantly interrupting the sedimentation process. Volcanic materials may be erupted onto "soft" unconsolidated sediment and then buried during subsequent, continued sedimentation, leading to a sequence of interbedded sedimentary rocks and volcanic rocks that are essentially contemporaneous in age. Thus, estimates of ages of such associated volcanic rocks also established the ages of contemporaneous sedimentary rocks.

Ages of whole volcanic rock can be estimated relatively easily by the potassium–argon method, and ages of minerals in these rocks can be determined by either potassium–argon or rubidium–strontium methods. Potassium–argon methods can be used to study rocks ranging in age from about 50,000 years to the age of Earth and rubidium–strontium methods are useful for studying rocks older than about 5 million years

TABLE 18.6   Categories of rocks most useful for geochronologic calibration of the geologic time scale

Type of rock	Stratigraphic relationship	Reliability of age data
Volcanic rock (lava flows and ashfalls)	Interbedded with "contemporaneous" sedimentary rocks	Give actual ages of sedimentary rocks in close stratigraphic proximity above and below volcanic layers
Plutonic igneous rocks	Intrude (cut across) sedimentary rocks	Give minimum ages for the rocks they intrude
	Lie unconformably beneath sedimentary rocks	Give maximum ages for overlying sedimentary rocks
Metamorphosed sedimentary rocks	Constitute the rocks whose ages are being determined	Give minimum ages for metamorphosed sedimentary rocks
	Lie unconformably beneath non-metamorphosed sedimentary rocks	Give maximum ages for the overlying nonmetamorphosed sedimentary rocks
Sedimentary rocks containing contemporary organic remains (fossils, wood)		Give actual ages of sedimentary rocks
Sedimentary rocks containing authigenic minerals such as glauconite		Give minimum ages for sedimentary rocks

(Table 18.5). Volcanic rocks that occur in association with nearly contemporaneous sedimentary rocks whose ages can also be determined by fossils provide extremely useful reference points for calibration. In fact, establishing the absolute ages of fossiliferous sedimentary rocks by association with contemporaneous volcanic flows whose ages can be radiometrically estimated has probably been the single most important method of calibrating the geologic time scale.

For this method to work, the contemporaneity of the interbedded volcanic and sedimentary rocks must first be established. If a pyroclastic flow such as an ashfall or a lava flow erupts over an older, exposed sedimentary rock surface where erosion is taking place or sedimentation is inactive, the flow is not contemporaneous with the underlying sedimentary rock. The age calculated for such a flow indicates only that the rock below the flow is older and the rock above younger than the flow. A geologist can establish contemporaneity by determining if fossils in sedimentary layers above and below the flow belong to the same biostratigraphic zone or by looking, along the basal contact of the flow unit, for physical evidence that may show that the underlying sediment was still soft at the time of the volcanic eruption; for example, ashfall material mixed by bioturbation into underlying sediment, mixing of soft sediment into the base of a submarine lava flow, or other such relationships.

Bracketed Ages from Associated Igneous or Metamorphic Rocks. The radiometric ages of igneous rocks that are not contemporaneous with associated sedimentary rocks can be used to estimate the ages of associated sedimentary rocks if two or more igneous bodies bracket the sedimentary unit. In this case, the age of the sedimentary unit can be established only as lying between the ages of the bracketing igneous bodies. The sedimentary unit will be older than an igneous body that intrudes it, but younger than an igneous body upon which it rests unconformably (Fig. 18.7A). For example, a sedimentary sequence deposited on the eroded, weathered surface of a granite batholith may subsequently be intruded by a dike or sill. The sedimentary unit is obviously younger than the batholith but older than the dike or sill. Unfortunately, there is no way to determine how much younger or older unless other evidence is available. Because erosional and depositional processes are relatively slow, the time represented by a bracketed age may be so long as to be of relatively little use in calibrating the geologic time scale. Only a few points on the time scale have been calibrated by this method.

Metamorphic minerals that develop in sedimentary rocks owing to regional or contact metamorphism can be studied also to provide a method of bracketing the ages of sedimentary rocks (Fig. 18.7B). The radiometric age of metamorphic minerals gives a minimum age for the metamorphosed sediment; that is, the metamorphosed sedimentary rocks are older than the time of metamorphism. If a sequence of metamorphic rocks is overlain unconformably by nonmetamorphosed sedimentary rocks, the nonmetamorphosed rocks are obviously younger than the age of the metamorphism.

Direct Radiochronology of Sedimentary Rocks

The calibration methods discussed above allow estimation of ages of sedimentary rocks only through their association in some manner with igneous or metamorphic rocks whose ages can be determined by radiometric methods. Clearly, the uncertainties involved in finding ages of sedimentary rocks by these indirect methods could be avoided if these ages could be estimated directly. As mentioned, terrigenous minerals

FIGURE 18.7 Determining the ages of sedimentary rocks indirectly by (A) bracketing between two igneous bodies, and (B) bracketing between regionally metamorphosed sedimentary rocks and an intrusive igneous body.

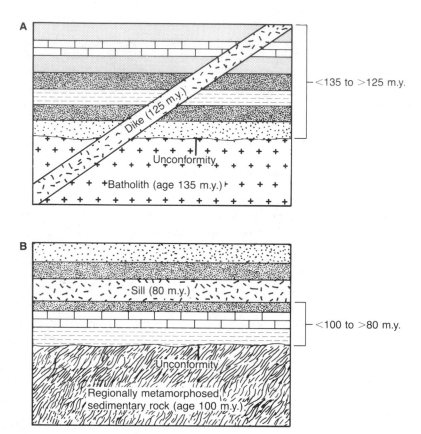

in sedimentary rocks are not useful for radiochronology because they yield ages for the parent rocks, not the time of deposition of the sediment. The only materials in sedimentary rocks that can be used for direct radiochronology are organic remains (wood, calcium carbonate fossils, and other such remains) that were deposited with the sediment and authigenic minerals that formed while the sediment was still on the seafloor or shortly after burial. The principal methods that have been used for direct radiochronology of sedimentary rocks are (1) the carbon-14 technique for organic materials, (2) potassium–argon and rubidium–strontium techniques for glauconites, (3) the thorium-230 technique for ocean floor sediments, and (4) the thorium-230/protactinium-231 technique for fossils and sediment. A short discussion of the advantages and disadvantages of each of these methods follows.

Carbon-14 Method. The carbon-14 method can be applied to the radiochronology of materials such as wood, peat, charcoal, bone, leaves, and the $CaCO_3$ shells of marine organisms. The method has been used extensively for estimating ages of archeological materials, but it has had limited application in geology owing to the very short useful age range of the method. Carbon-14 decays rapidly, with a half-life of only 5730 years. Consequently, the carbon-14 method commonly can be used only for materials less than about 40,000 years old—older materials contain too little carbon-14 to be determined by standard analytical methods. Special techniques that make use of mass spectrometers that allow analysis of smaller amounts of carbon-14, or special proportional counters with high counting efficiencies, make it possible to extend the usable ages to as much as 60,000–70,000 years (Stuiver et al., 1979). These special methods are expensive and have not been widely used in the past. Also, they are exceptionally subject to systematic error because of contamination of samples with young carbon.

The carbon-14 method has been used successfully for such applications as estimating ages of very young sediment in cores of deep-sea sediment and unraveling recent glacial history by analysis of wood in glacial deposits. Its extremely short range renders the method of little value in calibrating the geologic time scale except for very recent Quaternary events. For example, the Holocene–Pleistocene boundary has been set at about 10,000 years B.P. (before present) by use of carbon-14 dates on calcium carbonate microfossils in cores of deep-sea sediment.

Potassium-40/Argon-40 and Rubidium-87/Strontium-87 Radiochronology of Glauconites. The term **glauconite** is used loosely for a group of green clay minerals, all of which are complex potassium–aluminum–iron silicates that commonly occur in sediments as small rounded grains or pellets. The term **glaucony** is used as a "facies" name for these green pellets, especially by European geologists. The origin of glauconite is still not thoroughly understood; however, it appears to form authigenically on the seafloor by alteration of substrate materials such as skeletal debris, the fecal pellets or coprolites of organisms, and various types of mineral grains, particularly micas. The glauconization process requires exchange with seawater; therefore, authigenic growth of glauconite grains must take place within the top few centimeters of muddy sediment or the top few meters of coarse sandy sediment in order for such exchange to occur (Odin and Dodson, 1982).

Radioactive ^{40}K is incorporated into the glauconite grains as they evolve by alteration processes on the seafloor. When the glauconite grains are fully formed, they theoretically become closed systems with respect to gain or loss of potassium or argon; that is, no additional radioactive potassium is taken into the grains and the ^{40}Ar that forms by gradual decay of potassium remains trapped within the glauconite grains. Measurement of the $^{40}K/^{40}Ar$ ratio in the glauconite grains thus allows the age of the grains to be estimated. The half-life of potassium-40 is 1300 million years; therefore, it is theoretically possible to apply the K–Ar method to radiochronology of rocks ranging in age from 50,000 years to the age of Earth. As we shall see, however, uncertainties in the radiochronology of glauconites reduce the usable range of K–Ar ages for sedimentary rocks.

The wide distribution of glauconite in sediments of all ages and their undoubted early authigenic origin give them significant potential for use in estimating ages of sedimentary rocks. They have been studied more intensively for direct-age determinations of sedimentary rocks than has any other sedimentary mineral; however, owing to concern that heating of glauconites during burial may cause loss of argon, considerable difference of opinion has existed regarding the reliability of K–Ar glauconite ages. Comparison of glauconite ages with ages obtained by other radiometric methods has led several workers to suggest that, owing to argon loss, glauconite ages are commonly 10–20 percent too young. On the other hand, calculated glauconite ages may be too old in some cases owing to the presence of inherited radiogenic argon that was already in sediment at the time the glauconite grains formed. More recent work has shown that some of the uncertainties in estimating ages of glauconites can be removed by using only glauconite grains that contain more than about 7 percent K_2O; such grains appear to be less likely to contain significant amounts of inherited argon. Nonetheless, uncertainties in glauconite ages may range from thousands of years to almost a million years (Fig. 18.8). These uncertainties are less important with older rocks because plus or minus 1 million years is a less serious uncertainty when estimating the ages of rocks 100 million years old than when estimating ages of rocks 1 million years old. Therefore, the use of glauconite ages for calibrating the geologic time scale commonly is not recommended for rocks younger than about the Oligocene.

Obviously, the formation of glauconite grains and their closure to loss of argon do not occur simultaneously with deposition of the enclosing sediment. Glauconites, therefore, must yield a slightly younger age than the sediment in which they occur, even if uncertainties about inherited or lost argon are not a problem. Odin and Dodson (1982) suggest that the time required for glauconites to evolve and become closed systems may range up to 25,000 years or more. This means that in relation to biostratigraphic zonation, the glauconite K–Ar ages are closer to those of fossils in the horizon immediately above the glauconites than to the fossils deposited with the glauconites.

The ages of glauconites can be estimated also by the rubidium–strontium method (Table 18.5). Radioactive rubidium (^{87}Rb) is incorporated into glauconites as they form, along with potassium-40. Rubidium-87 decays to strontium-87 (^{87}Sr), with a half-life of 47,000 million years. This long half-life limits the use of the rubidium–strontium method to radiochronology of rocks older than about 5 million years. Like K–Ar ages for glauconites, the ages determined by the Rb–Sr method are commonly believed to be 10–20 percent too young and are thus minimum ages for the glauconites. Also, there appear to be somewhat greater analytical problems and uncertainties associated with the radiochronology of glauconites by the Rb–Sr method, and it has not been as widely used as the K–Ar method. Details of the Rb–Sr method as applied to the radiochronology of sedimentary rocks are given by Clauer (1982).

Estimating Ages of Sedimentary Rocks By Use of Other Authigenic Minerals. In addition to glauconites, several other authigenic minerals have been used in direct radiochronology of sedimentary rocks by the K–Ar and Rb–Sr methods. These minerals include other types of clay minerals such as illite, montmorillonite, and chlorite; zeolites; carbonate minerals; and siliceous minerals such as chert and opal. Owing to uncertainties about their origin—that is, authigenic or detrital—and time of closure to seawater interactions, none of the clay minerals except glauconite has so far proven to yield reliable ages. Zeolites, carbonate minerals, and siliceous minerals have been used for direct radiochronology of sedimentary rocks with some success, but the overall usefulness and reliability of methods based on these minerals have not yet been adequately investigated.

Thorium-230 and Thorium-230/Protactinium-231 Methods for Estimating Ages of Recent Sediments. Uranium-238 decays through several intermediate daughter products, including uranium-234, to thorium-230. Uranium-238 is fairly soluble in seawater and is present there in detectable amounts. By contrast, the thorium-230 daughter product precipitates quickly from seawater by adsorption onto sediment or inclusion in certain authigenic minerals and becomes incorporated into accumulating sediment on the seafloor. Thorium-230 is an unstable isotope and itself decays with a half-life of 75,000 years to still another unstable daughter product, radium-226. Owing to this fairly rapid decay of ^{230}Th, cores of sediment taken from the ocean floor exhibit a measurable decrease in ^{230}Th content with increasing depth in the cores. Assuming that sedimen-

FIGURE 18.8 Stratigraphic and genetic uncertainties related to the use of major types of chronometers for dating sedimentary rocks. The best chronometers are those that plot nearest to the x–y intercept in the figure. (From Odin, G. S., Introduction: Uncertainties in evaluating the numerical time scale, *in* G. S. Odin (ed.), Numerical dating in stratigraphy, Fig. 3, p. 10, © 1982, John Wiley & Sons, Ltd. Reprinted by permission of John Wiley & Sons, Ltd., Chichester, England.)

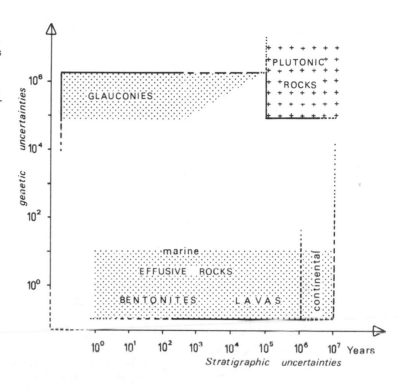

tation rates and the rates of precipitation of ^{230}Th have remained fairly constant through time, the concentration of ^{230}Th should decrease exponentially with depth. The ages of the sediments at various depths in a core can be calculated by comparing the amount of remaining ^{230}Th at any depth to the amount in the top layer of the core (surface sediment). This method can be applied to the dating of sediments younger than about 250,000 years, making the method useful for bridging the gap between maximum carbon-14 ages and minimum potassium–argon ages.

Protactinium-231 is the unstable daughter product of uranium-235 and itself decays with a half-life of about 34,000 years to actinium-227. Protactinium-231, like thorium-230, precipitates quickly from seawater and becomes incorporated into sediment along with thorium-230. Because protactinium-231 decays about twice as rapidly as thorium-230, the ^{231}Pa/^{230}Th ratio in sediments changes with time. Thus, in a sediment core, this ratio is largest in the surface layer of the core and decreases progressively with depth in the core. The age of the sediment at any depth in the core is determined by comparing the ^{231}Pa/^{230}Th ratio at that depth to the ratio in the surface sediment. The reliability of the ages determined by this method rests on the assumption that protactinium-231 and thorium-230 are produced everywhere in the ocean at a constant rate and that the starting ratio of these two isotopes in surface sediment is constant throughout the ocean.

An alternative method for calculating ages of sediment by use of protactinium-231 and thorium-230 involves measuring the ratio of these daughter products to their parent isotopes in the skeletons of marine invertebrates such as corals. Dissolved uranium-238 and uranium-235 in seawater are incorporated into corals as they grow, whereas seawater contains no appreciable protactinium-231 and thorium-231, owing to the rapid precipitation of these daughter products. Therefore, any protactinium-231 or thorium-230 present in corals results from decay of the parent uranium isotopes within the corals. The ratio of parent isotope to daughter product decreases systematically with time, providing a method for dating the corals. These ratios approach an equilibrium value with increasing passage of time, owing to the fact that the daughter products themselves continue to decay. Thorium-230 reaches a steady state after about 250,000 years and protactinium-231 after about 150,000 years. Thus, these methods can be used only for radiochronology of rocks younger than these ages. Owing to the fact that corals and other skeletal materials tend to recrystallize with burial and diagenesis, the ^{231}Pa–^{230}Th method has severe limitations. Recrystallization may open the initially closed system and allow escape of the daughter isotopes or of parent isotopes. Therefore, this method cannot be applied to estimating ages of skeletal materials that have undergone recrystallization.

Summary

Radiochronology of sedimentary rocks whose relative positions in the stratigraphic column are already established can be accomplished by several methods. The choice of method depends upon the age of the rocks and the types of materials present in the rocks. In general, calibration of the time scale by estimating ages of volcanic rocks associated with essentially contemporaneous sedimentary rocks that can be easily correlated by marine fossils is the most useful and reliable approach. Radiochronology of sedimentary glauconites or bracketing the ages of sedimentary rocks from associated plutonic intrusive rocks may also yield usable ages—the only ages available in some cases. Therefore, different methods may have to be applied to estimating ages of rocks in each geologic system. Details of the methods used for estimating ages of boundaries between and within the different systems are given in Odin (1982a), Cohee et al. (1978), and Harland et al. (1982).

Table 18.5 shows the calibration of the Decade of North American Geology 1983 Geologic Time Scale based on absolute ages obtained from a number of different sources (Palmer, 1983). Readers should be aware, however, that other recently calibrated geologic time scales have slightly different values for some of these boundaries (for example, Odin, Curry, Gale, and Kennedy, 1982; and Harland et al., 1982), indicating differences in opinion about the ages of the boundaries. Calibration of the geologic time table has changed steadily over the years as radiochronologic methods have improved, decay constants have been revised, and more absolute ages have become available. Although the ages now used to calibrate the major boundaries of the geologic time scale are unlikely to undergo major revision in the future, it is safe to assume that refinements in these ages will continue for some time.

18.7   CHRONOCORRELATION

Chronostratigraphic units are extremely important in stratigraphy because they form the basis for provincial to global correlation of strata on the basis of age equivalence. We have already established that chronostratigraphic correlation is correlation that expresses correspondence in age and chronostratigraphic position of stratigraphic units. To many geologists, correlation on the basis of age equivalence is by far the most important type of correlation and, in fact, is the only type of correlation possible on a global basis. Methods of establishing the age equivalence of strata by magneto-stratigraphic, seismic, and biologic techniques have already been discussed, and readers should refer back to appropriate chapters of the text for details of these methods. Several other methods of time-stratigraphic correlation are in common use, including correlation by short-term depositional events, correlation based on transgressive-regressive events, correlation by stable isotope events, and correlation by absolute ages. These methods are discussed in the following paragraphs.

Correlation by Short-term Depositional Events

In the discussion of seismic stratigraphy, I pointed out that seismic methods allow chronostratigraphic correlation of subsurface units by tracing of reflections generated from individual bedding surfaces or groups of closely spaced bedding surfaces. In outcrop sections, it is generally impossible to trace most individual beds physically for more than a few tens of hundreds of meters because they die out, merge with other beds, or become covered between outcrops by soil and vegetation. In some cases, however, key beds or marker beds may be present in outcrop or subsurface sections that can be traced by lithologic identity for long distances. These marker beds are useful for time-stratigraphic correlation, as well as for lithostratigraphic correlation, if they were deposited as a result of a geologic events that took place essentially "instantaneously." Correlation on the basis of such short-term geologic event markers has been called **event correlation.**

The most striking short-term depositional event is ashfall from volcanic eruptions. Beds formed from ashfalls are called ash layers, tephra layers, bentonite beds (if the ash alters to bentonite clays), or tuff layers. The ashfall from a single eruption may produce ash layers several centimeters in thickness that can cover thousands to hundreds of thousands of square kilometers. For example, ash from the eruption of Mt. Mazama in southeastern Oregon about 6500 to 7000 years ago—an eruption that subsequently led to the formation of the Crater Lake caldera—was carried northeastward by winds and deposited as far away as Saskatchewan and Manitoba, Canada. Ash from the May 1980 eruption of Mt. St. Helens also spread over thousands of square kilometers east and north of Mt. St. Helens in Washington and Idaho. Other historic examples of widespread ashfalls include the 1932 eruption of Quizapú in Chile—an eruption that distributed volcanic ash eastward for 1500 km across South America and into the Atlantic Ocean—and the eruption of Perbuatan Volcano at Krakatoa Island, Indonesia, in 1883—an eruption that spread volcanic dust around the world.

Tephra layers make extremely useful reference points in stratigraphic sections. They provide a means for reliable time-stratigraphic correlation if they are of sufficient lateral and vertical extent and if they can be identified as the product of a particular volcanic eruption. Identification of individual ash layers or bentonite beds can often be made on the basis of petrographic characteristics—types of mineral grains, rock fragments, glass shards, or other components—or trace-element composition. Ages of these layers may be determined also by radiometric methods, allowing the layers to be identified and correlated by contemporaneous age. Tephra layers are particularly useful in correlating across marine basins, but attempts have been made by some workers to correlate ash layers in marine basins with well-dated lava flows or ash layers on land, thereby extending marine correlations onto land.

Turbidity currents constitute another type of "instantaneous" geologic event that can produce thin, widespread deposits. Turbidites may have chronostratigraphic significance if a particular turbidite bed or sequence can be differentiated from other turbidite units and traced laterally. Unfortunately, turbidites commonly consist of rhythmic or cyclic sequences of units that have very similar appearance and are very difficult to differentiate. Thus, in practice, the usefulness of turbidites in time-stratigraphic correlation is rather limited. Figure 18.9 shows an example of a distinctive tuff bed or tephra layer that provides a chronostratigraphic marker horizon that can be

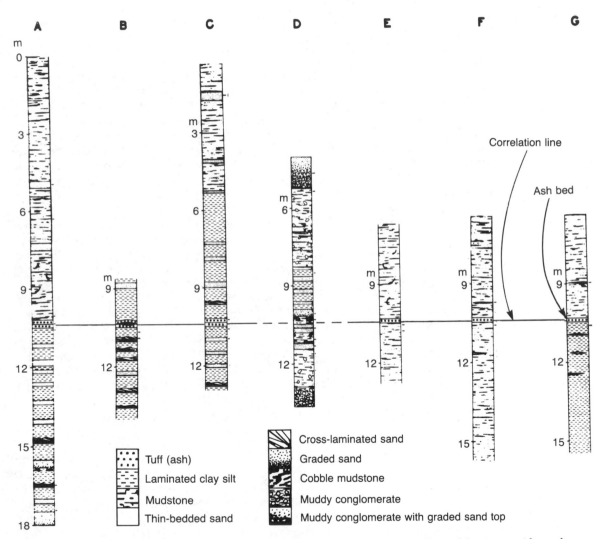

FIGURE 18.9 The Late Pliocene Bailey Ash in the Ventura Basin, California, provides a chronostratigraphic marker bed that can be correlated through outcrop sections (A through G) across the basin, allowing mudstone and sandstone beds above the marker unit to be correlated in relation to the marker. The ash bed is missing in Section D. (After Hsü, K. J., K. Kelts, and J. W. Valentine, 1980, Resedimented facies in Ventura Basin, California, and model of longitudinal transport of turbidity currents: Am. Assoc. Petroleum Geologists Bull., v. 64. Fig. 3, p. 1038, 1039, reprinted by permission of AAPG, Tulsa, Okla.)

recognized and correlated in several wells that penetrated deep-water conglomerates, sandstones, and mudstones in the Ventura Basin, California. Turbidite units and other deep-water facies can be correlated by their positions in the sequence with respect to the tuff horizon. Without this widespread marker bed, the turbidite units could not otherwise be correlated.

Other types of "catastrophic" short-term geologic events include dust storms that produce fine-grained loess deposits on land or silt–sand layers in marine basins. Storms at sea can stir up and transport sediment on the continent shelf to produce thin "storm layers" of sand or silt, as discussed. Even the formation of microtektite or micrometeorite horizons in the deep sea has been suggested to have time-stratigraphic significance (Glass and Zwart, 1979).

Slower, noncatastrophic depositional conditions also may generate thin, distinctive, widespread stratigraphic marker beds under some depositional conditions. Deposition of these beds does not necessarily take place "instantaneously." Nevertheless, they can be used for time-stratigraphic correlation if they formed as a result of deposition that took place over a large part of a basin during a relatively short period of time under essentially uniform depositional conditions. For example, a thin, widespread limestone bed within a predominantly shale or silt sequence implies deposition of the limestone under conditions that existed essentially simultaneously throughout a geologic province. Such a thin limestone bed within a sequence of nonmarine clastic units may represent a brief incursion of marine conditions into a nonmarine environment or temporary ponding of freshwater to form a large, shallow lake. Thin limestone units in a thick sequence of marine clastic deposits may indicate shelf-carbonate deposition during brief periods when clastic detritus was temporarily trapped in estuaries or deltaic environments and thus was prevented from escaping onto the shelf. By contrast, thin interbeds of sand, clay, or silt in a thick carbonate or evaporite sequence may represent temporary incursions of clastic detritus into a carbonate or evaporite basin. Such incursions may be due to a sudden increase in the supply of detritus as a result of tectonic events, periodic flooding on land, or deposition by windstorms or turbidity currents. Widespread, thin, continuous evaporite beds may also have time-stratigraphic significance because they appear to represent nearly simultaneous deposition throughout a large evaporite basin.

Correlation Based on Transgressive-Regressive Events

A different approach to event correlation is represented by local correlation based on position within a transgressive-regressive sequence or cycle (Ager, 1981). According to Ager, event correlation in this case is based on the correlation of corresponding peaks of symmetric sedimentary cycles that are presumed to be synchronous. The events represented in this type of correlation are the result of transgressions and regressions that may represent either worldwide, simultaneous eustatic changes in sea level or more local changes owing to uplift, subsidence, or fluctuation in sediment supply.

The principle of correlation based on transgressive-regressive events is illustrated in Figure 18.10. The deposits formed during any transgressive-regressive cycle contain one particular time plane that represents the time of maximum inundation of the sea—that is, the time at which water depth was greatest at any particular locality. Rocks lying stratigraphically below this time plane were deposited during transgression and those above during regression. This time plane can be identified by use of fossil data to determine depth zonation and maximum water depth of various localities, as illustrated in Figure 18.10. The position of the time plane can be established

FIGURE 18.10 Time correlation by position in a transgressive-regressive cycle. The line connecting points of deepest water conditions is a time line. (After Israelski, M. C., 1949, Oscillation chart: Am. Assoc. Petroleum Geologists Bull., v. 33. Fig. 3, p. 98, reprinted by permission of AAPG, Tulsa, Okla.)

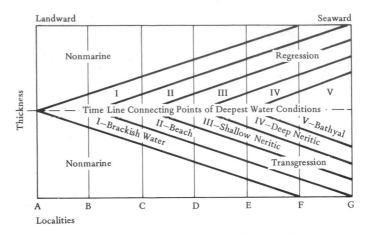

also from lithologic evidence by determining in the vertical stratigraphic section at each locality the position within the section where the rocks are symmetrically distributed with respect to the most basinward facies present. A surface connecting the most basinward rocks in each of the vertical sections defines the approximate position of the time plane and thus the time-stratigraphic correlation between the sections. Figure 18.11 further illustrates the method. Note from this illustration how time-equivalent points on the cycle are related, resulting in a correlation in which glauconitic clays at the east end of the sequence are equated to laminated beds at the west end. Correlation is expressed, as Ager (1981) puts it, in terms of degrees of "marineness."

Correlation by Stable Isotope Events

Variations in the relative abundance of certain stable, nonradioactive isotopes in marine sediments and fossils can be used as a tool for chronostratigraphic correlations of marine sediments. Geochemical evidence shows that the isotopic composition of oxygen, carbon, and sulfur in the ocean has undergone large fluctuations, or "excursions," in the geologic past—fluctuations that have been recorded in marine sediments. Because the mixing time in the oceans is about 1000 years or less, marine isotopic excursions are considered to be essentially isochronous throughout the world. Variations in isotopic compositions of sediments or fossils allow geochemists to construct **isotopic composition curves** that can be used as stratigraphic markers for correlation purposes. To be useful for correlation, fluctuations in isotopic composition must be recognizable on a global scale and must be of sufficiently short duration to show up as a shift on isotopic composition curves. Also, stratigraphers must be able to fix the relative stratigraphic positions of these fluctuations in relation to biostratigraphic, paleomagnetic, or radiometric scales. Of the various potentially useful isotopes, oxygen isotopes seem most nearly to meet these requirements and have proven to be particularly useful for chronostratigraphic correlation of Quaternary and Late Tertiary sediments. Carbon and sulfur isotopes are useful for correlating rocks of certain ages.

Oxygen Isotopes. The natural isotopes of oxygen are shown in Table 18.7. Most of the oxygen in the oceans occurs as oxygen-16. Oxygen-18 is much rarer, but is present in measurable amounts. The ratio of $^{18}O/^{16}O$ in the ocean at any given time in the past

FIGURE 18.11 Transgressive-regressive cyclic sedimentation and event correlation in the Eocene of the Isle of Wight in southern England. (From Ager, D. V., The nature of the stratigraphical record, 2nd ed., Fig. 7.2, p. 70, © 1981. Reprinted by permission of Macmillan, London and Basingstoke.)

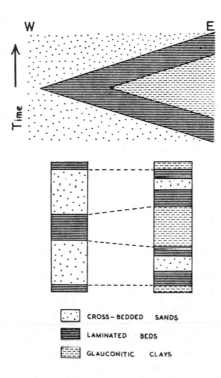

has been built into contemporaneous marine carbonate minerals and the calcium carbonate shells of marine organisms as a permanent record of the isotopic composition of the ocean at those times. Fluctuations in oxygen isotope ratios in the ocean with time thus show up as fluctuations in the isotopic ratios of these marine carbonates and fossils. Classification of deep-sea sediments on the basis of oxygen isotope ratios in the shells of calcareous marine organisms, particularly foraminifers, has given rise to

TABLE 18.7 Abundances and natural isotope ranges of carbon, oxygen, and sulfur stable isotopes

Element	Atomic number	Isotope	Geochemical abundances (ppm)	Relative isotopic abundances (%)	Range of natural variations (‰)
C	6	12	230	98.99	
		13*		1.108	90
		14		10^{-12}	
O	8	16	470,000	99.759	
		17		0.037	
		18*		0.203	110
S	16	32	470	95.0	
		33		0.76	
		34*		4.22	140
		36		0.014	

Source: Odin, C. S., M. Renard, and C. V. Grazzini, Geochemical events as a means of correlation, in G. S. Odin (ed.), Numerical dating in stratigraphy, Table 2, p. 55, © 1982, John Wiley & Sons, Ltd. Reprinted by permission of John Wiley & Sons, Ltd., Chichester, England.

a new stratigraphy for Quaternary sediments. This stratigraphic method is commonly referred to as **oxygen isotope stratigraphy.** It was first used by Emiliani (1955), who studied the isotopic composition of foraminifers in deep-sea cores and used oxygen isotope ratios to subdivide the core sediments. Oxygen isotope stratigraphy has now developed into a major new tool for correlating Quaternary marine sequences, as explained below.

The $^{18}O/^{16}O$ ratio in biogenic marine carbonates reflects both the temperature and the $^{18}O/^{16}O$ ratio of the water in which these carbonates formed. The relationships of ocean paleotemperature (T) to oxygen isotopic composition has been shown by Shackleton (1967) to be

$$T(°C) = 16.9 - 4.38 (\delta_c - \delta_w) + 0.10 (\delta_c - \delta_w)^2$$

where δ_c = equilibrium oxygen isotopic composition of calcite and δ_w = oxygen isotopic composition of the water from which the calcite was precipitated. The δ_c and δ_w notations do not refer to the actual oxygen isotopic abundances in calcite and water but to the per mil (parts per thousand) deviation of the $^{18}O/^{16}O$ ratio in calcite and water from that of an arbitrary standard. The commonly used standard for oxygen isotopes is the University of Chicago PDB standard. The PDB refers to a particular fossil belemnite from the Pee Dee Formation of South Carolina. The per mil deviation, referred to as $\delta^{18}O$, is expressed by the relationship

$$\delta^{18}O = \frac{[(^{18}O/^{16}O) \text{ sample} - (^{18}O/^{16}O) \text{ standard}]}{(^{18}O/^{16}O) \text{ standard}} \times 1000$$

Oxygen isotope stratigraphy is based on the fact that $\delta^{18}O$ values in biogenic marine carbonates reflect both the temperature and the isotopic composition of the water from which the calcite precipitates. These factors are both, in turn, functions of the climate. When water evaporates at the surface of the ocean, the lighter ^{16}O isotopes are preferentially removed in the water vapor, leaving the heavier ^{18}O in the ocean. This isotopic fractionation process thus causes water vapor to be depleted in ^{18}O with respect to the seawater from which it evaporates. When water vapor condenses to form rain or snow, the water containing heavy oxygen will tend to precipitate first, leaving the remaining vapor depleted in ^{18}O compared to the initial vapor. Thus, the precipitation that falls near the coast and runs back quickly to the ocean will contain heavier oxygen than that which falls in the interior of continents or in polar regions, where it returns more slowly to the ocean. There is a correlation also between air temperature and the $^{18}O/^{16}O$ ratio of precipitates—the colder the air the lighter the rain or snow (Odin, Renard, and Grazzini, 1982). For example, the overall average oxygen isotope composition of seawater is −0.28‰ (per mil); however, the precipitation that falls in the crests of the Greenland ice sheet is about −35‰ and in relatively inaccessible parts of the Antarctic ice sheet is as negative as −58‰.

The ^{18}O-depleted moisture that falls in polar regions is locked up as ice on land and is thus prevented from quickly returning to the ocean. Owing to this retention of light-oxygen water in the ice caps, the ocean becomes progressively enriched in ^{18}O as ^{18}O-depleted ice caps build up during a glacial stage. Marine carbonates that precipitate in the ocean during a glacial stage, particularly biogenic carbonates such as foraminifers, will be enriched in ^{18}O relative to those that precipitate during times when the climate is warmer and ice caps are absent, or are much smaller, on land. Changes in the $\delta^{18}O$ content of biogenic marine calcite thus reflect changes in the volumes of ice on land.

Decrease in temperature of the seawater in which biogenic calcite precipitates also causes an increase in the $\delta^{18}O$ values that are built into the calcite. Thus, during glacial periods, both decrease in temperature of ocean water and changes in isotopic composition of ocean water owing to buildup of ice caps on the continents combine to cause an increase in the $\delta^{18}O$ content of biogenic calcites. Conversely, melting of polar ice caps, with consequent return of light-oxygen water to the oceans, and increase in ocean temperature will be reflected in a decrease in $\delta^{18}O$ values in marine biogenic carbonates.

These oxygen isotope fluctuations between maximum and minimum $\delta^{18}O$ values in response to waxing and waning of glaciers during the Quaternary form the basis for correlating cores of deep-sea sediments. Figure 18.12 illustrates how oscillations in maximum and minimum $\delta^{18}O$ values in biogenic carbonates in two deep-sea cores from the Pacific can be used to make detailed correlations between the cores. Note the large numbers of glacial-interglacial fluctuations indicated by these isotopic data. The numerous advances and retreats of glaciers during the Pleistocene indicated by these data are in sharp contrast to the commonly accepted number of four major glacial stages interpreted from on-land evidence.

Carbon Isotopes. Carbon-12 and carbon-13 are the nonradioactive isotopes of carbon. Carbon-12 is much more abundant than carbon-13 and makes up most of the carbon in seawater (Table 18.7). The isotopic ratio of $^{13}C/^{12}C$ can be expressed in terms of per mil deviation ($\delta^{13}C$) from the PDB belemnite standard, just as oxygen isotope ratios are expressed. The $\delta^{13}C$ values in marine carbonates reflect the $^{13}C/^{12}C$ ratio of CO_2 dis-

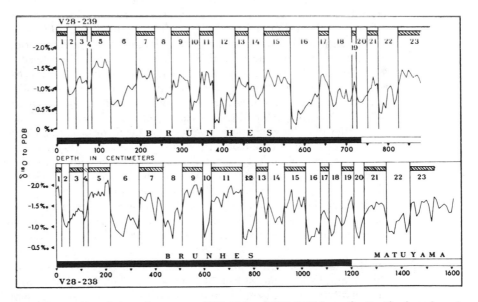

FIGURE 18.12   Correlation of two cores (V28-239 and V28-238) from the Pacific; based on the oxygen isotope signatures in the cores. The paleomagnetic record is also shown. (From Shackleton, N. J., and N. D. Opdyke, 1976, Isotope and paleomagnetic stratigraphy of Pacific core V-28-239, late Pliocene to latest Pleistocene, in R. M. Cline and J. D. Hays (eds.), Investigation of late Quaternary paleooceanography and paleoclimatology: Geol. Soc. America Mem. 145, Fig. 2, p. 453.)

solved in deep ocean water; the ratio is, in turn, a reflection of the source of carbon in the CO_2. Carbon dioxide dissolves in the ocean by interchange with the atmosphere and is generated also by decay of organic matter that originates both in the ocean and on land. Carbon dioxide derived from decaying organic matter is sharply depleted in ^{13}C compared to that derived from the atmosphere. Thus, water runoff from the continents brings organic-rich waters with low $^{13}C/^{12}C$ ratios into the ocean, significantly lowering the $\delta^{13}C$ content of surface ocean waters near the continents.

Another factor that influences the $\delta^{13}C$ content of ocean water, and thus the $\delta^{13}C$ content in the shells of marine organisms that live in these waters, is the residence time of deep-water masses in the ocean. Carbon-13 is depleted in deep-water masses that have long residence times near the ocean bottom, owing to oxidation of low $\delta^{13}C$ marine organic matter that sinks from the surface. Oxidation of this low $\delta^{13}C$ organic matter leads to production of low $\delta^{13}C$ dissolved bicarbonate (HCO_3^-), which is then used by organisms to build shells. Respiration by bottom-dwelling organisms also apparently causes a decrease in $\delta^{13}C$ of deep-bottom waters (Kennett, 1982).

Because the $\delta^{13}C$ in the calcareous shells of marine organisms is a function of the $\delta^{13}C$ content of the waters in which they live, changes in the $\delta^{13}C$ content of fossil marine organisms indicate changes in ocean water masses. Abrupt decreases in the $\delta^{13}C$ in fossil marine calcareous organisms may reflect changes in deep-ocean paleocirculation and upwelling patterns that caused low $\delta^{13}C$ deep waters to spread upward and outward into other parts of the ocean. Or such decrease may reflect changes in surface circulation patterns that brought low $\delta^{13}C$ surface ocean waters from continental margins into deeper basins. Significant increases in the total biomass produced on the continents during any particular geologic time interval could cause an increase in runoff of low $\delta^{13}C$ to the oceans, an increase that may be reflected also as an episode of low $\delta^{13}C$ surface water. Increased rates of erosion of organic-rich sediments, such as dark shales and limestones, on land could produce much the same effect as increasing runoff of low $\delta^{13}C$ to the ocean. These abrupt changes in circulation patterns or organic carbon runoff from the continents may have affected the area of a single ocean basin, such as the Pacific, or in some cases the entire ocean. On the other hand, increased rates of sediment burial in the ocean may have the opposite effect of removing fine organic matter containing low $\delta^{13}C$ from interaction with seawater. This would have the effect of increasing $\delta^{13}C$ in ocean water.

Because these changes in $\delta^{13}C$ content of the ocean are reflected in the $\delta^{13}C$ of marine calcareous organisms, they can be used for correlation. For example, a distinct shift (decrease) in $\delta^{13}C$ has been detected in marine sediments of Miocene age at about 5.9–6.2 Ma; the shift appears to be of global extent (Fig. 18.13). Other events have been recorded in the Middle Miocene, near the Eocene–Oligocene boundary, at the Paleocene–Eocene boundary, and the Cretaceous–Tertiary boundary (Odin, Renard, and Grazzini, 1982). These carbon isotopic excursions are essentially synchronous events that can be correlated over wide areas of the ocean in DSDP cores.

Sulfur Isotopes. Sulfur has four stable isotopes (Table 18.7); sulfur-32 is the most abundant, followed by sulfur-34. The ratio of $^{34}S/^{32}S$ is used in most stratigraphic studies involving sulfur isotopes and is expressed in terms of $\delta^{34}S$, which is per mil deviation of the $^{34}S/^{32}S$ ratio relative to a meteorite standard—troilite from the Canyon Diablo meteorite. Figure 18.14 shows the $\delta^{34}S$ values in various materials relative to the Canyon Diablo meteorite.

The major means of sulfur isotope fractionation in the oceans is by bacterial reduction of sulfate (SO_4^{2-}) in seawater to sulfides (H_2S, HS^-, HSO_4^-). Bacterial re-

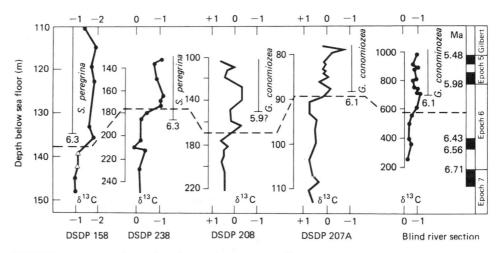

FIGURE 18.13 Correlation based on carbon isotope (δ^{13}C) oscillations in a suite of late Miocene marine sections plotted against depth. The dashed line represents an inferred isochron (based on carbon isotopic shift in each record) that is paleomagnetically dated in the Blind River section (at right) as 6.1 m.y. (From Kennett, J., Marine geology, © 1982, Fig. 3.13, p. 86. Reprinted by permission of Prentice-Hall, Englewood Cliffs, N.J.)

duction of dissolved seawater sulfate at the sediment–seawater interface causes isotopic fractionation of the sulfate, resulting in enrichment of the remaining seawater sulfate in δ^{34}S by about $+20‰$ and depletion in the reduced sulfide by about $-9‰$ (T. J. M. Schopf, 1980). Precipitation of evaporites from dissolved marine sulfates introduces an additional fractionation ($\sim +1.65‰$), causing the δ^{34}S of evaporites to be higher than that of dissolved marine sulfates (Thode and Monster, 1965). Other minor factors that can influence the δ^{34}S content of seawater include oxidation of bacterial H_2S, producing sulfates depleted in δ^{34}S relative to original sulfates and local emanations of sulfate or sulfide through volcanic activity.

Marine sulfates in the present ocean have a mean δ^{34}S of about $+21‰$; however, the δ^{34}S of ancient marine evaporites ranges from about $+10$ to $+30$ (Fig. 18.14). Study of ancient evaporite deposits has shown that the sulfur isotope ratios in the surface waters of the world ocean have undergone major changes, or excursions, at various times, as recorded in the δ^{34}S of ancient marine evaporites (Fig. 18.15). These major excursions, called "catastrophic chemical events" by Holser (1977), are characterized by sharp rises in δ^{34}S in the surface waters of the world ocean and by greater "overshoots" locally. Three major events of sharply increased δ^{34}S have been recognized and named for the evaporite formations in which the δ^{34}S increase is most strongly manifested: the **Yudomski event** in very late Precambrian time (about 635 Ma), the **Souris event** in the Late Devonian (approximately 370 Ma), and the **Rot event** in Early to Middle Triassic (approximately 240 Ma). In addition to these excursions of increased δ^{34}S, reverse events occurred also during the Late Permian and Late Paleogene (Fig. 18.15). Many of these sulfur isotope excursions appear to have affected the ocean worldwide.

Holser (1977) suggests that chemical events characterized by sharply increased δ^{34}S are caused by catastrophic mixing of deep δ^{34}S-rich brines with surface waters. Brines generated by evaporite deposition are stored in deep basins. Underneath the brines, bacterial reduction of sulfates to form pyrite builds up a store of brine heavy

FIGURE 18.14 Compilation of $\delta^{34}S$ values in various materials relative to the Canyon Diablo meteorite. (From Degens, E. T., Geochemistry of sediments, © 1965, Fig. 36, p. 163. Reprinted by permission of Prentice-Hall, Englewood Cliffs, N.J.)

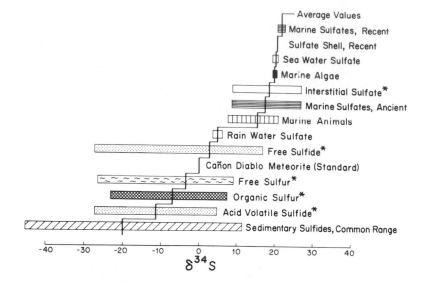

in $\delta^{34}S$ sulfate. Catastrophic mixing of these $\delta^{34}S$-rich brines with surface waters, owing to destruction of the storage basin by tectonism, causes a sharp rise in the $\delta^{34}S$ of surface ocean waters and consequently in the evaporite deposits formed from these surface waters. Gradual decrease in the $\delta^{34}S$ of surface ocean waters with time after a catastrophic event is attributed to on-land erosion of predominantly sulfide materials into the ocean in an amount that exceeds evaporite deposition (Holser, 1977). The cause of sharp reverse $\delta^{34}S$ events, such as the Late Paleogene event shown in Figure 18.15, is not fully understood.

In any case, these sulfur isotope excursions constitute a sulfur isotope age curve because each catastrophic chemical event occurred within a very short interval of geologic time. Each major event thus represents a synchronous stratigraphic marker that can be correlated in evaporite deposits from one area to another. Some of the events can be correlated on a global basis. Thus, they provide an important method for international chronostratigraphic correlation of evaporite deposits, which commonly cannot be correlated by other means because they do not contain fossils or other datable materials.

The field of stable isotope geochemistry is very complex and many questions remain with regard to the observed isotope excursions in the geologic record. A discussion of these problems and the details of correlation by stable isotopes is outside the scope of this book. Additional information on this subject is available in numerous publications such as those of Arthur et al. (1983), Claypool et al. (1980), Holland (1984), Holland and Trendall (1984), Holser (1984), Holser et al. (1986), and Veizer et al. (1980).

Correlation by Absolute Ages

As discussed, radiometric dating of igneous and sedimentary rocks provides a basis for detailed calibration of the geologic time scale. In spite of the usefulness of radiochronology for that purpose, absolute- (isotopic-) age data are still not widely used for chronostratigraphic correlation of sedimentary rocks. We might logically assume that

FIGURE 18.15 Sulfur isotopic ratios in sulfates collected from Phanerozoic formations. Dotted figures are for European samples, half-black figures are for North American samples, open figures are for samples from other parts of the world. (From Odin, G. S., M. Renard, and C. V. Grazzini, Geochemical events as a means of correlation, *in* G. S. Odin (ed.), Numerical dating in stratigraphy. Fig. 11, p. 59, © 1982, John Wiley & Sons, Ltd. Reprinted by permission of John Wiley & Sons, Ltd., Chichester, England.)

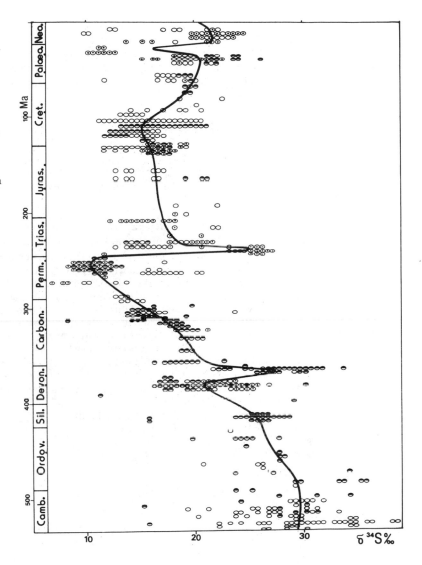

absolute-age data should constitute the primary basis for chronostratigraphic correlation; however, many sedimentary rocks do not contain datable materials. Furthermore, the precision of radiometric ages obtained from sedimentary rocks that can be dated is often not good enough for correlation purposes.

Carbon-14 and uranium-238 (^{230}Th/^{231}Pa) disequilibrium ages can be used for correlating very young rocks; however, geologists have been less successful in correlating older sedimentary rocks by absolute age owing to uncertainties in the direct-dating methods used for older rocks. Figure 18.5 shows, for example, that uncertainties in glauconite ages can range from thousands of years to as much as a million years. An error of a million years is statistically small when dating rocks hundreds of millions of years old, but it is nonetheless a large error for correlation purposes.

On the other hand, radiometric ages provide the only available tool for correlating most Precambrian rocks. Lithologic and structural complexity and lack of fossils in these rocks preclude their correlation by most other methods. Absolute-age data are a valuable asset also in stratigraphic studies of volcanic rocks and have proven particularly useful in correlating Cenozoic volcanic sequences.

ADDITIONAL READINGS

Arthur, M. A., T. F. Anderson, I. R. Kaplan, J. Veizer, and L. S. Land, 1983, Stable isotopes in sedimentary geology: Soc. Econ. Paleontologists and Mineralogists Short Course No. 10.

Bandy, O. L. (ed.), 1970, Radiometric dating and paleontologic zonation: Geol. Soc. America Spec. Paper 124, 247 p.

Cohee, G. V., M. F. Glaessner, and H. D. Hedberg, 1978, Contributions to the Geologic Time Scale: Am. Assoc. Petroleum Geologists, Tulsa, Okla., 388 p.

Eicher, D. L., 1976, Geologic time, 2nd ed.: Prentice-Hall, Englewood Cliffs, N. J., 152 p.

Faure, G., 1977, Principles of isotope geology: John Wiley & Sons, New York, 464 p.

Harland, W. B., A. V. Cox, P. G. Llewellyn, C. A. G. Pickton, A. G. Smith, and R. Walters, 1982, A geologic time scale: Cambridge University Press, Cambridge, England, 131 p.

Hedberg, H. D. (ed.), 1976, International stratigraphic guide: A guide to stratigraphic classification, terminology, and procedure: International Subcommission on Stratigraphic Classification of IUGS Commission on Stratigraphy: John Wiley & Sons, New York, 200 p.

Mahaney, W. C. (ed.), 1984, Quarternary dating methods: Developments in paleontology and stratigraphy, 7. Elsevier, Amsterdam, 431 p.

North American Stratigraphic Commission on Stratigraphic Nomenclature, 1983, North American Stratigraphic Code: Am. Assoc. Petroleum Geologists Bull., v. 67, p. 841–875.

Odin, G. S. (ed.), 1982, Numerical dating in stratigraphy; Parts I and II: John Wiley & Sons, New York, 1040 p.

York, D., and R. M. Farquhar, 1972, The Earth's age and geochronology: Pergamon, New York, 178 p.

References

Adams, A. E., W. S. Mackenzie, and G. Guilford, 1984, Atlas of sedimentary rocks under the microscope: John Wiley & Sons, New York, 104 p.

Ager, D. V., 1981, The nature of the stratigraphical record, 2nd ed.: John Wiley & Sons, New York, 122 p.

Ahlbrandt, T. S., and S. G. Fryberger, 1981, Sedimentary features and significance of interdune deposits, in F. G. Ethridge and R. O. Flores (eds.), Recent and ancient nonmarine depositional environments: Models for exploration: Soc. Econ. Paleontologists and Mineralogists Spec. Pub. 31, p. 293–314.

———1982, Introduction to eolian deposits, in P. A. Scholle and D. Spearing (eds.), Sandstone depositional environments: Am. Assoc. Petroleum Geologists Mem. 31, p. 11–47.

Aitken, J. D., 1967, Classification and environmental significance of cryptalgal limestones and dolomites, with illustrations from the Cambrian and Ordovician of southwestern Alberta: Jour. Sed. Petrology, v. 37, p. 1163–1179.

Allen, D. R., and G. V. Chilingarian, 1975, Mechanics of sand compaction, in G. V. Chilingarian and K. H. Wolf (eds.), Compaction of coarse-grained sediments, I: Developments in Sedimentology 18A, Elsevier, Amsterdam, p. 43–77.

Allen, J. R. L., 1963, The classification of cross-stratified units with notes on their origin: Sedimentology, v. 2, p. 93—114.

———1964, Studies in fluviatile sedimentation: Six cyclothems from the Lower Old Red Sandstone, Anglo-Welsh Basin: Sedimentology, v. 3, p. 163–198.

———1965a, A review of the origin and characteristics of recent alluvial sediments: Sedimentology, v. 5, p. 89–191.

———1965b, The sedimentation and paleogeography of the Old Red Sandstone of Anglesey, North Wales: Yorkshire Geol. Soc. Proc., v. 35, p. 139–185.

———1968, Current ripples: their relation to patterns of water and sediment motion: North Holland Publishing, Amsterdam, 433 p.

———1970a, Physical processes of sedimentation: George Allen & Unwin, London, 248 p.

———1970b, Studies in fluviatile sedimentation: A comparison of fining-upward cyclothems, with special reference to coarse-member composition and interpretation: Jour. Sed. Petrology, v. 40, p. 298–323.

———1982, Sedimentary structures: their character and physical basis, v. 1–2: Elsevier, Amsterdam, 664 p.

————1984, Laminations developed from upper-stage plane beds: A model based on the larger coherent structures of the turbulent boundary layer: Sed. Geology, v. 39, p. 227–242.

————1985, Loose-boundary hydraulics and fluid mechanics: Selected advances since 1961, in P. J. Brenchley and B. P. J. Williams (eds.), Sedimentology: Recent developments and Applied Aspects: Geol. Soc., Blackwell, Oxford, p. 7–30.

Aller, R. C., 1982, The effects of macrobenthos on chemical properties of marine sediments and overlying water, in P. L. McCall and M. J. S. Tevesz (eds.), Animal-sediment relations: The biogenic alteration of sediments: Plenum, New York, p. 53–102.

Alvarez, W., M. A. Arthur, A. G. Fischer, W. Lowrey, G. Napoleone, I. Premoli Silva, and M. W. Roggenthen, 1977, Upper Cretaceous-Paleocene magnetic stratigraphy at Gubbio, Italy: V. Type section for the Late Cretaceous-Paleocene geomagnetic reversal time scale: Geol. Soc. America Bull., v. 88, p. 383–389.

Anderson, R. Y., and D. W. Kirkland, 1966, Intrabasin varve correlation: Geol. Soc. America Bull., v. 77, p. 241–256.

Anderton, R., 1985, Clastic facies models and facies analysis, in P. J. Brenchley and B. P. J. Williams (eds.), Sedimentology, recent developments and applied aspects: Geol. Soc., Blackwell, Oxford, p. 31–48.

Anstey, N. A., 1982, Simple seismics: International Human Resources Development Corporation, Boston, 168 p.

Arthur, M. A., T. F. Anderson, I. R. Kaplan, J. Veizer, and L. S. Land, 1983, Stable isotopes in geology: Soc. Econ. Paleontologists and Mineralogists Short Course 10.

Arthur, M. A., and A. G. Fischer, 1977, Upper Cretaceous-Paleocene magnetic stratigraphy at Gubbio, Italy: I. Lithostratigraphy and sedimentology: Geol. Soc. America Bull., v. 88, p. 367–371.

Ashley, G. M., N. D. Smith, and J. D. Shaw, 1985, Glacial sedimentary environments: Soc. Econ. Paleontologists and Mineralogists Short Course Notes No. 16, 246 p.

Aston, S. R. (ed.), 1983, Silicon geochemistry and biochemistry: Academic Press, London, 248 p.

Bagnold, R. A., 1954, The physics of blown sand and desert dunes: Methuen, London, 265 p.

————1956, The flow of cohesionless grains in fluids: Royal Soc. London, Philos. Trans., Ser. A., v. 249, p. 235–297.

————1962, Auto-suspension of transported sediment: Turbidity currents: Royal Soc. London Proc. (A), v. 265, p. 315–319.

Baker, P. A., and M. Kastner, 1981, Constraints on the formation of sedimentary dolomite: Science, v. 213, p. 214–216.

Baker, P. A., M. Kastner, J. D. Byerlee, and D. A. Lockner, 1980, Pressure solution and hydrothermal recrystallization of carbonate sediments—An experimental study: Marine Geology, v. 38, p. 185–203.

Bally, A. W. (comp.), 1981, Geology of passive continental margins: Am. Assoc. Petroleum Geologists Short Course Notes No. 19.

Bandy, O. L. (ed.), 1970, Radiometric dating and paleontologic zonation: Geol. Soc. America Spec. Paper 124, 247 p.

Barnes, R. S. K., 1980, Coastal lagoons: Cambridge University Press, Cambridge, England, 106 p.

Barrett, P. J., 1980, The shape of rock particles, a critical review: Sedimentology, v. 27, p. 291–303.

Basan, P. B. (ed.), 1978, Trace fossil concepts: Soc. Econ. Paleontologists and Mineralogists Short Course No. 5, 181 p.

Basu, A., S. W. Young, L. J. Suttner, W. C. James, and G. H. Mack, 1975, Reevaluation of the use of undulatory extinction and polycrystallinity in detrital quartz for provenance interpretation: Jour. Sed. Petrology, v. 45, p. 873–882.

Bates, C. C., 1953, Rational theory of delta formation: Am. Assoc. Petroleum Geologists Bull., v. 37, p. 2119–2161.

Bates, R. L., and J. A. Jackson (comps.), 1980, Glossary of geology, 2nd ed.: American Geological Institute, Falls Church, Va., 749 p.

Bathurst, R. G. C., 1975, Carbonate sediments and their diagenesis, 2nd ed.: Developments in Sedimentology 12, Elsevier, Amsterdam, 658 p.

Baturin, G. N., 1982, Phosphorites on the sea floor: Origin, composition, and distribution: Developments in Sedimentology 33. Elsevier, Amsterdam, 343 p.

Belderson, R. H., M. A. Johnson, and N. H. Kenyon, 1982, Bedforms, in A. H. Stride (ed.), Offshore tidal sands: Chapman and Hall, London, p. 27–57.

Belderson, R. H., N. H. Kenyon, A. H. Stride, and A. R. Stubbs, 1972, Sonographs of the sea floor: Elsevier, Amsterdam, 185 p.

Bentor, Y. K., 1980a, Phosphorites—The unsolved problem, in Y. K. Bentor (ed.), Marine phosphorites—Geochemistry, occurrence, genesis: Soc. Econ. Paleontologists and Mineralogists Spec. Pub. 29, p. 3–18.

Bentor, Y. K. (ed.), 1980b, Marine phosphorites—Geochemistry, occurrence, genesis: Soc. Econ. Paleontologists and Mineralogists Spec. Pub. 29, 249 p.

Berg, O. R., and D. B. Wolverton (eds.), 1985, Seismic stratigraphy II, an integrated approach: Am. Assoc. Petroleum Geologists Mem. 39, 276 p.

Berger, R., and H. E. Suess (eds.), 1979, Radiocarbon dating: University of California Press, Berkeley, 787 p.

Berger, W. H., 1974, Deep-sea sedimentation, in C. A. Burk and C. L. Drake (eds.), The geology of continental margins: Springer-Verlag, New York, p. 213–241.

Berggren, W. A., and J. A. Van Couvering, 1978, Biochronology, in G. V. Cohee, M. F. Claoooncr, and II. D. Hedberg (eds.), Contributions to the geologic time scale: Am. Assoc. Petroleum Geologists, Tulsa, Okla., p. 39–55.

Berner, R. A., 1971, Principles of chemical sedimentology: McGraw-Hill, New York, 240 p.

———1975, The role of magnesium in crystal growth of aragonite from sea water: Geochim. et Cosmochim. Acta, v. 39, p. 489–505.

———1980, Early diagenesis: A theoretical approach: Princeton University Press, Princeton, N. J., 224 p.

Berner, R. A., T. Baldwin, and G. R. Holdren, Jr., 1979, Authigenic iron sulfides as paleosalinity indicators: Jour. Sed. Petrology, v. 49, p. 1345–1350.

Berner, R. A., J. T. Westrich, R. Graber, J. Smith, and C. S. Martens, 1978, Inhibition of aragonite precipitation from supersaturated seawater: A laboratory and field study: Am. Jour. Sci., v. 278, p. 816–837.

Berry, W. B. N., 1968, Growth of a prehistoric time scale: W. H Freeman, San Francisco, 158 p.

Bhattacharyya, A., and G. M. Friedman (eds.), 1983, Modern carbonate environments: Benchmark Papers in Geology, v. 74, Hutchinson and Ross, Stroudsburg, Pa., 376 p.

Bien, G. S., D. E. Contois, and W. H. Thomas, 1959, The removal of silica from fresh water entering the sea, in H. A. Ireland (ed.), Silica in sediments: Soc. Econ Paleontologists and Mineralogists Spec. Pub. 7, p. 20–35.

Bigarella, J. J., 1972, Eolian environments: Their characteristics, recognition and importance, in J. K. Rigby and W. K. Hamblin (eds.), Recognition of ancient sedimentary environments: Soc. Econ. Paleontologists and Mineralogists Spec. Pub. 16, p. 12–62.

Biggs, R. B., 1978, Coastal bays, in R. A. Davis, Jr., (ed.), Coastal sedimentary environments: Springer-Verlag, New York, p. 69–99.

Birkland, P. W., 1974, Pedology, weathering, and geomorphological research: Oxford University Press, New York, 285 p.

Black, M., 1933, The precipitation of calcium carbonate on the Great Bahama Bank: Geol. Mag., v. 70, p. 455–466.

Blatt, H., 1967, Original characteristics of quartz grains: Jour. Sed. Petrology, v. 37, p. 401–424.

———1970, Determination of mean sediment thickness in the crust: A sedimentological model: Geol. Soc. America Bull., v. 81, p. 255–262.

———1979, Diagenetic processes in sandstones, in P. A. Scholle and P. R. Schluger (eds.), Aspects of diagenesis: Soc. Econ. Paleontologists and Mineralogists Spec. Pub. 26, p. 141–157.

———1982, Sedimentary petrology: W. H. Freeman, San Francisco, 564 p.

Blatt, H., G. V. Middleton, and R. Murray, 1980, Original of sedimentary rocks, 2nd ed.: Prentice-Hall, Englewood Cliffs, N. J., 782 p.

Blatt, H., and M. W. Totten, 1981, Detrital quartz as an indicator of distance from shore in marine mudrocks: Jour. Sed. Petrology, v. 51, p. 1259–1266.

Boggs, S., Jr., 1967a, Measurement of roundness and sphericity parameters using an electronic particle size analyzer: Jour. Sed. Petrology, v. 37, p. 908–913.

———1967b, A numerical method for sandstone classification: Jour. Sed. Petrology, v. 37, p. 548–555.

———1968, Experimental study of rock fragments: Jour. Sed. Petrology, v. 38, p. 1326–1339.

———1969, Relationship of size and composition in pebble counts: Jour. Sed. Petrology, v. 39, p. 1243–1247.

———1975, Seabed resources of the Taiwan continental shelf: Acta Oceanographica Taiwanica, v. 5, p. 1–18.

————1984, Quarternary sedimentation in the Japan arc-trench system: Geol. Soc. America Bull., v. 95, p. 669–685.

Boggs, S., Jr., and C. A. Jones, 1976, Seasonal reversal of flood-tide dominant sediment transport in a small Oregon estuary: Geol. Soc. America Bull., v. 87, p. 419–426.

Boggs, S., Jr., W. C. Wang, and F. S. Lewis, 1979, Sediment properties and water characteristics of the Taiwan shelf and slope: Acta Oceanographica Taiwanica, v. 10, p. 10–49.

Bohn, H. L., B. L. McNeal, and G. A. O'Connor, 1979, Soil chemistry: John Wiley & Sons, New York, 329 p.

Bohor, B. F., and H. J. Gluskoter, 1973, Boron in illites as an indicator of paleosalinity of Illinois coals: Jour. Sed. Petrology, v. 43, p. 945–956.

Boothroyd, J. C., 1978, Mesotidal inlets and estuaries, in R. A. Davis, Jr. (ed.), Coastal sedimentary environments: Springer-Verlag, New York, p. 287–360.

————1985, Tidal inlets and tidal deltas, in R. A. Davis, Jr. (ed.), Coastal sedimentary environments, 2nd ed.: Springer-Verlag, New York, p. 445–532.

Borchert, H., and R. O. Muir, 1964, Salt deposits: The origin, metamorphism, and deformation of evaporites: Van Nostrand, London, 338 p.

Bostock, N. H., 1979, Microscopic measurement of level of catagenesis of solid organic matter in sedimentary rocks to aid exploration for petroleum and to determine former burial temperatures—A review, in P. A. Scholle and P. R. Schluger (eds.), Aspects of diagenesis, Soc. Econ. Paleontologists and Mineralogists Spec. Pub. 26, p. 17–44.

Boucot, A. J., 1975, Evolution and extinction rate controls: Developments in Paleontology and Stratigraphy, I, Elsevier, New York, 427 p.

Boucot, A. J., and R. S. Carney, 1981, Principles of marine benthic paleoecology: Academic Press, New York, 463 p.

Bouma, A. H., 1969, Methods for the study of sedimentary structures: John Wiley & Sons, New York, 457 p.

————1979, Continental slopes, in L. J. Doyle and O. H. Pilkey (eds.), Geology of continental slopes: Soc. Econ. Paleontologists and Mineralogists Spec. Pub. 27, p. 1–15.

Bouma, A. H., H. L. Berryhill, R. L. Brenner, and H. J. Knebel, 1982, Continental shelf and epicontinental seaways, in P. A. Scholle and D. Spearing (eds.), Sandstone depositional environments: Am. Assoc. Petroleum Geologists Mem. 31, p. 281–327.

Bouma, A. H., G. T. Moore, and J. M. Coleman (eds.), 1978, Framework, facies, and oil-trapping characteristics of the upper continental margin: Am. Assoc. Petroleum Geologists Studies in Geology 7, 326 p.

Braitsch, O., 1971, Salt deposits: Their origin and composition: Springer-Verlag, Berlin, 279 p.

Brenchley, P. J., and B. P. J. Williams (eds.), 1985, Sedimentology: Recent developments and applied aspects: Geol. Soc. Spec. Pub. 12, Blackwell, Oxford, 320 p.

Bricker, O. P., 1971, Carbonate cements: Johns Hopkins University Press, Baltimore, Md., 376 p.

Briggs, D. E. G., E. N. K. Clarkson, and R. J. Aldridge, 1983, The conodont animal: Lethaea, v. 16, p. 1–14.

Brookfield, M. E., 1984, Eolian sands, in R. G. Walker (ed.), Facies models, 2nd ed.: Geoscience Canada Reprint Ser. 1, p. 91–104.

Brookfield, M. E., and T. S. Ahlbrandt (eds.), 1983, Eolian sediments and processes: Elsevier, Amsterdam, 660 p.

Broussard, M. L. (ed.), 1975, Deltas: Models for exploration: Houston Geological Society, 555 p.

Brower, J. C. 1981, Quantitative biostratigraphy, 1830–1980, in D. F. Merriam (ed.), Computer applications in the earth sciences: An update of the 70s: Plenum, New York, p. 63–103.

Brown, L. F., Jr., and W. L. Fisher, 1980, Seismic stratigraphic interpretation and petroleum exploration: Geophysical principles and techniques: Am. Assoc. Petroleum Geologists Continuing Education Course Note Ser. 16, 56 p.

Bruce, C. H., 1984, Smectite dehydration—Its relation to structural development and hydrocarbon accumulation in northern Gulf of Mexico Basin: Am. Assoc. Petroleum Geologists Bull., v. 68, p. 673–683.

Bubb, J. N., and W. G. Hatelid, 1977, Recognition of carbonate buildups, in C. E. Payton (ed.), Seismic stratigraphy—Application to hydrocarbon exploration: Am. Assoc. Petroleum Geologists Mem. 26, p. 185–204.

Bull, W. B., 1972, Recognition of alluvial-fan deposits in the stratigraphic record, in J. K. Rigby and W. K. Hamblin (eds.), Recognition of ancient sedimentary environments: Soc. Econ. Paleontologists and Mineralogists Spec. Pub. 16, p. 63–83.

Burk, C. A., and C. L. Drake (eds.), 1974, The geology of continental margins: Springer-Verlag, New York, 1009 p.

Burst, J. F., 1965, Subaqueously formed shrinkage cracks in clay: Jour. Sed. Petrology, v. 35, p. 348–353.

———1976, Argillaceous sediment dewatering: Ann. Rev. Earth and Planetary Sci., v. 4, p. 293–318.

Bushman, J. R., 1983, Twelve fallacies of uniformitarianism: Comment: Geology, v. 11, p. 312–313.

Calvert, S. E., 1974, Deposition and diagenesis of silica in marine sediments, in K. J. Hsü and H. C. Jenkyns (eds.), Pelagic sediments: On land and under the sea: Internat. Assoc. Sedimentologists Spec. Pub. 1, p. 273–300.

———1983, Sedimentary geochemistry of silicon, in R. R. Aston (ed.), Silicon geochemistry and biogeochemistry: Academic Press, London, P. 143–186.

Campbell, C. V., 1967, Lamina, laminaset, bed and bedset: Sedimentology, v. 8, p. 7–26.

Cant, D. J., 1982, Fluvial facies models and their application, in P. A. Scholle and D. Spearing (eds.), Sandstone depositional environments: Am. Assoc. Petroleum Geologists Mem. 31, p. 115–138.

Cant, D. J., and Walker, R. G., 1976, Development of a braided fluvial model for the Devonian Battery Point Sandstone, Quebec: Canadian Jour. Earth Sci., v. 13, p. 102–119.

Carlson, W. D., 1983, The polymorphs of $CaCO_3$ and the aragonite-calcite transformation, in R. J. Reeder (ed.), Carbonates: Mineralogy and chemistry: Reviews in Mineralogy, v. 11, Mineralog. Soc. America, p. 191–225.

Carr, T. R., 1982, Log-linear models, Markov chains and cyclic sedimentation: Jour. Sed. Petrology, v. 52, p. 905–912.

Carroll, D., 1970, Rock weathering: Plenum, New York, 203 p.

Carver, R. E. (ed.), 1971, Procedures in sedimentary petrology: John Wiley & Sons, New York, 653 p.

Castanares, A. A., and F. B. Phleger (eds.), 1969, Coastal lagoons—A symposium: Universidad Nacional Autonoma de Mexico/UNESCO, Mexico City, 686 p.

Cayeux, L., 1931, Introduction á l'étude pétrographique des roches sédimentaires: Imp. Nat., Paris, 524 p.

———1935, Les roches sédimentares de Frances: Roches carbonatées: Masson et Cie, Paris, 447 p.

Chambre Syndical de la Recherche et de la Production due Pétrole et du Gaz Naturel (eds.), 1981, Evaporite deposits: Illustration and interpretation of some environmental sequences: Editions Technip, Paris, and Gulf Publishing, Houston, 266 p.

Channell, J. E. T., 1982, Palaeomagnetic stratigraphy as a correlation technique, in G. S. Odin (ed.), Numerical dating in stratigraphy: John Wiley & Sons, New York, p. 81–106.

Childs, O. E., 1983, Correlation of stratigraphic units of North America, COSUNA, 1977–83: Am. Assoc. Petroleum Geologists and U.S. Geol. Survey, 49 p.

Chilingarian, G. V., H. J. Bissell, and R. W. Fairbridge (eds.), 1967, Carbonate rocks. Part A: Origin, occurrence, and classification, 471 p.: Part B: Physical and chemical aspects, 413 p.: Elsevier, New York.

Chilingarian G. V., and K. H. Wolf (eds.), 1976a, Compaction of coarse-grained sediments, I: Elsevier, New York, 552 p.

———1976b, Compaction of coarse-grained sediments, II: Elsevier, New York, 808 p.

Chilingarian, G. V., and T. F. Yen, 1978, Bitumens, asphalts and tar sands: Elsevier, 331 p.

Chowns, T. M., and J. E. Elkins, 1974, The origin of quartz geodes and cauliflower cherts through the silicification of anhydrite nodules: Jour. Sed. Petrology, v. 44, p. 885–903.

Clarke, F. W., 1924, The data of geochemistry, 5th ed.: U.S. Geol. Survey Bull. 770, 841 p.

Clauer, N., 1982, The rubidium-strontium method applied to sediments: Certitudes and uncertainties, in G. S. Odin (ed.), Numerical dating in stratigraphy: John Wiley & Sons, New York, p. 245–276.

Claypool, G. E., W. T. Holser, I. R. Kaplan, H. Sakai, and I. Zak, 1980, The age curves of sulfur and oxygen isotopes in marine sulfate and their mutual interpretation: Chem. Geology, v. 28, p. 199–260.

Clayton, R. N., and E. T. Degens, 1959, Use of carbon isotope analyses of carbonates in differentiating freshwater and marine sediments: Am. Assoc. Petroleum Geologists Bull., v. 43, p. 890–897.

Clifton, H. E., 1969, Beach laminations: Nature and origin: Marine Geology, v. 7, p. 553–559.

———1976, Wave-formed sedimentary structures: A conceptual model, in R. A. Davis and R. L. Ethington (eds.), Beach and nearshore sedimentation: Soc. Econ. Paleontologists and Mineralogists Spec. Pub. 24, p. 126–148.

————1982, Estuarine deposits, *in* P. A. Scholle and D. Spearing (eds.), Sandstone depositional environments: Am. Assoc. Petroleum Geologists Mem. 31, p. 179–189.

Clifton, H. E., and J. R. Dingler, 1984, Wave-formed structures and paleoenvironmental reconstruction: Marine Geology, v. 60, p. 165–198.

Cloud, P. E., 1973, Paleoecological significance of banded iron formations: Econ. Geology, v. 68, p. 1135–1143.

Cohee, G. V., M. F. Glaessner, and H. D. Hedberg, 1978, Contributions to the geologic time scale: Am. Assoc. Petroleum Geologists, Tulsa, Okla., 388 p.

Cole, G. A., 1975, Textbook of limnology: C. V. Mosby, St. Louis, Mo., 283 p.

Coleman, J. M., 1976, Deltas: Processes of deposition and models for exploration: Continuing Education Publication Co., Champaign, Ill., 102 p.

————1981, Deltas: Processes of deposition and models for exploration, 2nd ed.: Burgess, 124 p.

Coleman, J. M., and D. B. Prior, 1980, Deltaic sand bodies: Am. Assoc. Petroleum Geologists Education Short Course Notes 15, 171 p.

————1982, Deltaic environments of deposition, *in* P. A. Scholle and D. Spearing (eds.), Sandstone depositional environments: Am. Assoc. Petroleum Geologists Mem. 31, p. 139–178.

————1983, Deltaic influences on shelfedge instability processes, *in* D. J. Stanley and G. T. Moore (eds.), The shelfbreak: Critical interface on continental margins: Soc. Econ. Paleontologists and Mineralogists Spec. Pub. 33, p. 121–127.

Coleman, J. M., and L. D. Wright, 1975, Modern river deltas: Variability of process and sand bodies, *in* M. L. Broussard (ed.), Deltas: Models for exploration: Houston Geological Society, p. 99–149.

Collins, A. G., 1975, Geochemistry of oilfield waters: Developments in petroleum science 1, Elsevier, Amsterdam, 496 p.

————1980, Oilfield brines, *in* G. D. Hobson (ed.), Developments in petroleum geology 2: Applied Science Publishers, London, p. 139–187.

Collinson, J. D., 1978a, Alluvial sediments, *in* H. G. Reading (ed.), Sedimentary environments and facies: Elsevier, New York, p. 15–60.

————1978b, Deserts, *in* H. G. Reading (ed.), Sedimentary environments and facies: Elsevier, New York, p. 80–96.

————1978c, Lakes, *in* H. G. Reading (ed.), Sedimentary environments and facies: Elsevier, New York, p. 61–79.

Collinson, J. D., and J. Lewin (eds.), 1983, Modern and ancient fluvial systems: Internat. Assoc. Sedimentologists Spec. Pub. 6, Blackwell, Oxford, 575 p.

Collinson, J. D., and D. B. Thompson, 1982, Sedimentary structures: George Allen & Unwin, London, 194 p.

Colombo, G., 1977, Lagoons, *in* R. S. K. Barnes (ed.), The coastline: John Wiley & Sons, New York, p. 63–81.

Conde, K. C., 1982, Plate tectonics and crustal evolution, 2nd ed.: Pergamon, New York, 310 p.

Conklin, B. A., and J. E. Conklin (eds.), 1984, Stratigraphy: Foundations and concepts: Benchmark Papers in Geology, v. 82, Van Nostrand Reinhold, New York, 365 p.

Conybeare, C. E. B., and K. A. W. Crook, 1968, Manual of sedimentary structures: Department of National Development, Bureau of Mineral Resources, Geology and Geophysics Bull. 102, 327 p.

Cook, H. E., 1983, Introductory perspectives, basic carbonate principles, and stratigraphic and depositional models, *in* H. E. Cook, A. C. Hine, and H. T. Mullins (eds.), Platform margin and deep water carbonates: Soc. Econ. Paleontologists and Mineralogists Short Course Notes No. 12, p. 1-1–1-89.

Cook, H. E., and P. Enos (eds.), 1977, Deep-water carbonate environments: Soc. Econ. Paleontologists and Mineralogists Spec. Pub. 25, 336 p.

Cook, H. E., M. E. Field, and J. V. Gardner, 1982, Characteristics of sediments on modern and ancient continental slopes, *in* P. A. Scholle and D. Spearing (eds.), Sandstone depositional environments: Am. Assoc. Petroleum Geologists Mem. 31, p. 329–364.

Cook, H. E., A. C. Hine, and H. T. Mullins (eds.), 1983, Platform margin and deep water carbonates: Soc. Econ. Paleontologists and Mineralogists Short Course Notes No. 12.

Cook, H. E., and H. T. Mullins, 1983, Basin margin, *in* P. A. Scholle, D. G. Bebout, and C. H. Moore (eds.), Carbonate depositional environments: Am. Assoc. Petroleum Geologists Mem. 33, p. 539–618.

Cook, P. J., 1976, Sedimentary phosphate deposits, *in* K. H. Wolf (ed.), Handbook of strata-bound and stratiform ore deposits, v. 7: Elsevier, New York, p. 505–536.

Coombs, D. S., 1971, Present status of the zeolite facies: Advances in Chemistry Ser. 101 (Molecular sieve zeolites, Am. Chem. Soc.-I), p. 317–327.

Corliss, J. B., J. Dymond, L. I. Gordon, J. M. Edmond, R. P. von Herzen, R. D. Ballard, K. Green, D. Williams, A. Bainbridge, K. Crane, and T. H. van Andel, 1979, Submarine thermal springs on the Galápagos Rift: Science, v. 203, p. 1073–1083.

Couch, E. L., 1971, Calculation of paleosalinities from boron and clay mineral data: Am. Assoc. Petroleum Geologists Bull., v. 55, p. 1829–1837.

Cox, A., 1973a, Plate tectonics and geomagnetic reversals: Introduction and reading list, in A. Cox (ed.), Plate tectonics and geomagnetic reversals: W. H. Freeman, San Francisco, p. 138–153.

Cox, A. (ed.), 1973b, Plate tectonics and geomagnetic reversals: W. H. Freeman, San Francisco, 702 p.

Cox, A., 1969, Geomagnetic reversals: Science, v. 163, p. 237–245.

Crelling, J. C., and R. R. Dutcher, 1980, Principles and applications of coal petrology: Soc. Econ. Paleontologists and Mineralogists Short Course Notes No. 8, 127 p.

Cressman, E. R., 1962, Nondetrital siliceous sediments: U.S. Geol. Surv. Prof. Paper 440-T, 22 p.

Crimes, T. P., 1975, The stratigraphical significance of trace fossils, in R. W. Frey (ed.), The study of trace fossils: Springer-Verlag, New York, p. 109–130.

Crimes, T. P., and J. C. Harper (eds.), 1970, Trace fossils: Seel House Press, Liverpool, 547 p.

———1977, Trace fossils 2: Seel House Press, Liverpool, 351 p.

Cronin, L. E. (ed.), 1975, Estuarine research, v. 2: Geology and engineering: Academic Press, New York, 587 p.

Crosby, E. J., 1972, Classification of sedimentary environments, in J. K. Rigby and W. K. Hamblin (eds.), Recognition of ancient sedimentary environments: Soc. Econ. Paleontologists and Mineralogists Spec. Pub. 16, p. 4–11.

Cubitt, J. M., and R. A. Reyment (eds.), 1982, Quantitative stratigraphic correlation: John Wiley & Sons, New York, 301 p.

Curran, H. A. (ed.), 1985, Biogenic structures: Their use in interpreting depositional environments: Soc. Econ. Paleontologists and Mineralogists Spec. Pub. 35, 347 p.

Curray, J. R., W. R. Dickinson, W. G. Dow, K. O. Emery, D. R. Seeley, P. R. Vail, and H. Yarbough, 1977, Geology of continental margins: Am. Assoc. Petroleum Geologists Short Course Notes No. 5, variously paginated.

Curray, R. R., 1966, Observations of alpine mudflows in the Tenmile Range, central Colorado: Geol. Soc. America Bull., v. 77, p. 771–776.

Dapples, E. C., 1979, Diagenesis of sandstones, in G. Larsen and G. V. Chilingar (eds.), Diagenesis in sediments and sedimentary rocks: Developments in Sedimentology 25A, Elsevier, Amsterdam, p. 31–141.

Davies, D. K., and F. G. Ethridge, 1975, Sandstone composition and depositional environments: Am. Assoc. Petroleum Geologists Bull., v. 59, p. 239–264.

Davies, D. K., F. G. Ethridge, and R. R. Berg, 1971, Recognition of barrier environments: Am. Assoc. Petroleum Geologists Bull., v. 55, p. 550–565.

Davies, T. A., and D. S. Gorsline, 1976, Oceanic sediments and sedimentary processes, in J. P. Riley and R. Chester (eds.), Chemical oceanography, v. 5, 2nd ed.: Academic Press, New York, p. 1–80.

Davis, R. A., Jr., 1968, Algal stromatolites composed of quartz sandstone: Jour. Sed. Petrology, v. 38, p. 953–955.

———1978, Beach and nearshore zone, in R. A. Davis, Jr. (ed.), Coastal sedimentary environments: Springer-Verlag, New York, p. 237–285.

———1983, Depositional systems—A genetic approach to sedimentary geology: Prentice-Hall, Englewood Cliffs, N.J., 669 p.

Davis, R. A., Jr. (ed.), 1985, Coastal sedimentary environments, 2nd ed.: Springer-Verlag, New York, 716 p.

Davis, R. A., Jr., and R. L. Ethington (eds.), 1976, Beach and nearshore sedimentation: Soc. Econ. Paleontologists and Mineralogists Spec. Pub. 24, 187 p.

Davis, R. A., Jr., and W. T. Fox, 1972, Four-dimensional model for beach and inner nearshore sedimentation: Jour. Geology, v. 80, p. 484–493.

Davis, T. L., 1984, Seismic-stratigraphic facies model, in R. G. Walker (ed.), Facies models, 2nd ed.: Geoscience Canada Reprint Ser. 1, p. 311–317.

Dean, W. E., G. R. Davies, and R. Y. Anderson, 1975, Sedimentological significance of nodular and laminated anhydrite: Geology, v. 3, p. 367–372.

Dean, W. E., M. Leinen, and D. A. V. Stow, 1985, Classification of deep-sea, fine-grained sediments: Jour. Sed. Petrology, v. 55, p. 250–256.

Dean, W. E., and B. C. Schreiber, 1978, Marine evaporites: Soc. Econ. Paleontologists and Mineralogists Short Course Notes No. 4, 193 p.

Degens, E. T., 1965, Geochemistry of sediments: Prentice-Hall, Englewood Cliffs, N.J., 342 p.

Degens, E. T., and S. Epstein, 1964, Oxygen and carbon isotope ratios in coexisting calcites and dolomites from recent and ancient sediments: Geochim. et Cosmochim. Acta, v. 28, p. 23–44.

Dickey, P. A., 1969, Increasing concentration of subsurface brines with depth: Chem. Geology, v. 4, p. 361–370.

Dickinson, W. R., 1982, Composition of sandstones in cirum-pacific subduction complexes and fore-arc basins: Am. Assoc. Petroleum Geologists Bull., v. 66, p. 121–137.

Dickinson, W. R., L. S. Beard, G. R. Brakenridge, J. L. Erjavec, R. C. Ferguson, K. F. Inman, R. A. Knepp, F. A. Lindberg, and P. T. Ryberg, 1983, Provenance of North American Phanerozoic sandstones in relation to tectonic setting: Geol. Soc. America Bull., v. 94, p. 222–235.

Dickinson, W. R., and C. A. Suczek, 1979, Plate tectonics and sandstone composition: Am. Assoc. Petroleum Geologists Bull., v. 63, p. 2164–2182.

Didyk, B. M., B. T. R. Simoreit, S. C. Brassel, and G. Eglinton, 1978, Organic geochemical indicators of paleoenvironmental conditions of sedimentation: Nature, v. 272, p. 216–222.

Diessel, C. F. K., and R. Offler, 1975, Change in physical properties of coalified and graphitized phytoclasts with grade of metamorphism: Neues Jahrg. Mineralogie Monatsh., Jahrg. 1975, p. 11–26.

Dill, D. F., 1966, Sand flows and sand falls, in R. W. Fairbridge (ed.), Encyclopedia of oceanography: Reinhold, New York, p. 763–765.

Dimroth, E., 1976, Aspects of the sedimentary petrology of cherty iron-formation, in K. H. Wolf (ed.), Handbook of strata-bound and stratiform ore deposits, v. 7: Elsevier, New York, p. 203–254.

———1979, Models of physical sedimentation of iron formations, in R. G. Walker (ed.), Facies models: Geoscience Canada Reprint Ser. 1, p. 159–174.

Dimroth, E., and M. M. Kimberley, 1976, Precambrian atmospheric oxygen: Evidence in the sedimentary distribution of carbon, sulfur, uranium and iron: Canadian Jour. Earth Sci., v. 13, p. 1161–1185.

Dobkins, J. E., and R. L. Folk, 1970, Shape development on Tahati-Nui: Jour. Sed. Petrology, v. 40, p. 1167–1203.

Dobrin, M. B., 1976, Introduction to geophysical prospecting: McGraw-Hill, New York, 630 p.

Dodd, J. R., and R. J. Stanton, 1975, Paleosalinities within a Pliocene bay, Kettleman Hills, California: A study of the resolving power of isotope and faunal techniques: Geol. Soc. America Bull., v. 86, p. 51–64.

———1981, Paleoecology, concepts and applications: John Wiley & Sons, New York, 559 p.

Donovan, D. T., 1966, Stratigraphy—An introduction to principles: John Wiley & Sons, New York, 199 p.

Dott, R. H., Jr., and J. Bourgeois, 1982, Hummocky stratification: Significance of its variable bedding sequences: Geol. Soc. America Bull., v. 93, p. 663–680.

Dott, R. H., and R. H. Shaver (eds.), 1974, Modern and ancient geosynclinal sedimentation: Soc. Econ. Paleontologists and Mineralogists Spec. Pub. 19, 380 p.

Dowdeswell, J. A., 1982, Scanning electron micrographs of quartz sand grains from cold environments examined using Fourier shape analysis: Jour. Sed. Petrology, v. 52, p. 1315–1326.

Doyle, L. J., W. J. Cleary, and D. U. Pilkey, 1968, Mica: Its use in determining shelf-depositional regime: Marine Geology, v. 6, p., 381–389.

Doyle, L. J., and O. H. Pilkey (eds.), 1979, Geology of continental slopes: Soc. Econ. Paleontologists and Mineralogists Spec. Pub. 27, 374 p.

Drever, J. I., 1971, Magnesium iron replacements in clay minerals in anoxic marine sediments: Science, v. 172, p. 1334–1336.

———1974, Geochemical model for the origin of Precambrian banded iron formations: Geol. Soc. America Bull., v. 85, p. 1099–1106.

Drever, J. I. (ed.), 1985, The chemistry of weathering: D. Reidel, Hingham, Mass., 336 p.

Duff, P. McL. D., A. Hallam, and E. K. Walton, 1967, Cyclic sedimentation: Developments in Sedimentology 10, Elsevier, Amsterdam, 280 p.

Duke, W. L., 1985, Hummocky cross-stratification, tropical hurricanes and intense winter storms: Sedimentology, v. 32, p. 167–194.

Dunbar, C. O., and J. Rodgers, 1957, Principles of stratigraphy: John Wiley & Sons, New York, 356 p.

Duncan, D. C., 1976, Geologic setting of oil shale deposits and world prospects, in T. F. Yen and G. V. Chilingarian (eds.), Oil shale: Elsevier, New York, p. 13–26.

Dunham, R. J., 1962, Classification of carbonate rocks according to depositional textures, in W. E. Ham (ed.), Classification of carbonate rocks. Am. Assoc. Petroleum Geologists Mem. 1, p. 108–121.

————1970, Stratigraphic reef versus ecologic reefs: Am. Assoc. Petroleum Geologists Bull., v. 54, p. 1931–1932.

Dunoyer De Segonzac, G., 1970, The transformation of clay minerals during diagenesis and low-grade metamorphism: A review: Sedimentology, v. 15, p. 281–346.

Dyer, K. R., 1973, Estuaries: A physical introduction: John Wiley & Sons, New York, 140 p.

Dyer, K. R. (ed.), 1979, Estuarine hydrography and sedimentation: A handbook: Cambridge University Press, Cambridge, England, 230 p.

Dzulynski, S., and E. K. Walton, 1965, Sedimentary features of flysch and greywackes: Developments in Sedimentology 7, Elsevier, Amsterdam, 274 p.

Easterbrook, D. J., 1982, Characteristic features of glacial sediments, in P. A. Scholle and D. Spearing (eds.), Sandstone depositional environments: Am. Assoc. Petroleum Geologists Mem. 31, p. 1–10.

Eckel, E. C., 1904, On the chemical composition of American shales and roofing slates: Jour. Geology, v. 12, p. 25–29.

Edmond, J. M., 1980, Ridge crest hot springs: The story so far: EOS, v. 61, p. 129–131.

Edwards, M. B., 1978, Glacial environments, in H. G. Reading (ed.), Sedimentary environments and facies: Elsevier, New York, p. 416–438.

Ehrlich, R., and B. Weinberg, 1970, An exact method for characterization of grain shape: Jour. Sed. Petrology, v. 40, p. 205–212.

Eicher, D. L., 1976, Geologic time, 2nd ed.: Prentice-Hall, Englewood, Cliffs, N.J., 152 p.

Eichler, J., 1976, Origin of the Precambrian banded iron-formations, in K. H. Wolf (ed.), Handbook of strata-bound and stratiform ore deposits, v. 7: Elsevier, New York, p. 157–201.

Einsele, G., and A. Seilacher, 1982, Cyclic and event stratification: Springer-Verlag, Berlin, 536 p.

Ekdale, A. A., R. G. Bromley, and S. G. Pemberton, 1984, Ichnology, trace fossils in sedimentology and stratigraphy: Soc. Econ. Paleontologists and Mineralogists Short Course No. 15, 317 p.

Eldredge, N., and S. J. Gould, 1972, Punctuated equilibria: An alternative to phyletic gradualism, in T. J. M. Schopf (ed.), Models in paleobiology: Freeman, Cooper, San Francisco, p. 82–115.

————1977, Evolutionary models and biostratigraphic strategies, in E. G. Kauffman and J. E. Hazel (eds.), Concepts and methods of biostratigraphy: Dowden, Hutchinson and Ross, Stroudsburg, Pa., p. 25–40.

Eliott, T., 1978a, Deltas, in H. G. Reading (ed.), Sedimentary environments and facies: Blackwell, Oxford, p. 97–142.

————1978b, Clastic shorelines, in H. G. Reading (ed.), Sedimentary environments and facies: Elsevier, New York, p. 143–177.

Embry, A. F., and J. E. Klovan, 1971, A late Devonian reef tract on the northeastern Banks Island, N.W.T.: Canadian Petroleum Geology Bull., v. 19, p. 730–781.

Emery, K. O., 1968, Relict sediments on continental shelves of the world: Am. Assoc. Petroleum Geologists Bull., v. 52, p. 445–464.

————1977, Structure and stratigraphy of divergent continental margins: Am. Assoc. Petroleum Geologists Continuing Education Course Note Ser. 5, Geology of continental margins, p. B1–B20.

Emiliani, C., 1955, Pleistocene temperatures: Jour. Geology, v. 63, p. 538–578.

Enos, P., 1983, Shelf environment, in P. A. Scholle, D. G. Bebout, and C. H. Moore (eds.), Carbonate depositional environments: Am. Assoc. Petroleum Geologists Mem. 33, p. 267–296.

Enos, P., and C. H. Moore, 1983, Fore-reef slope, in P. A. Scholle, D. G. Bebout, and C. H. Moore (eds.), Carbonate depositional environments: Am. Assoc. Petroleum Geologists Mem. 33, p. 507–538.

Epstein, A. G., J. B. Epstein, and L. D. Harris, 1977, Conodont color alteration—An index to organic metamorphism: U. S. Geol. Survey Prof. Paper 995, 27 p.

Ernst, W., 1970, Geochemical facies analysis: Elsevier, Amsterdam, 152 p.

Eslinger, E. V., S. M. Savin, and H-W. Yeh, 1979, Oxygen isotope geothermometry of diagenetically altered shales, in P. A. Scholle and P. R. Schluger (eds.), Aspects of diagenesis: Soc. Econ. Paleontologists and Mineralogists Spec. Pub. 26, p. 113–124.

Ethridge, F. G., and R. M. Flores (eds.), 1981, Recent and ancient nonmarine depositional environments: Models for exploration: Soc. Econ. Paleontologists and Mineralogists Spec. Pub. 31, 349 p.

Eugster, H. P., and L. A. Hardie, 1978, Saline lakes, in A. Lerman (ed.), Lakes: Chemistry, geology, physics: Springer-Verlag, New York, p. 237–294.

Evans, G., 1965, Intertidal flat sediments and their environments of deposition in the Wash: Geol. Soc. London Quart. Jour., v. 121, p. 209–241.

Evans, I., and C. G. St. C. Kendall, 1977, An interpretation of the depositional setting of some deep-water Jurassic carbonates of the central High Atlas Mountains, Morocco, in H. E. Cook and P. Enos (eds.), Deep-water carbonate environments: Soc. Econ. Paleontologists and Mineralogists Spec. Pub. 25, p. 249–261.

Evenson, E. B., Ch. Schlüchter, and J. Rabassa, 1983, Till and related deposits: A. A. Balkema, 454 p.

Ewers, W. E., 1983, Chemical factors in the deposition and diagenesis of banded iron-formation, in A. F. Trendall and R. C. Morris (eds.), Developments in Precambrian Geology 6: Elsevier, Amsterdam, p. 491–512.

Ewing, M., and E. M. Thorndike, 1965, Suspended matter in deep-ocean water: Science, v. 147, p. 1291–1294.

Eyles, N. (ed.), 1983, Glacial geology: An introduction for engineers and earth scientists: Pergamon, Oxford, 409 p.

Eyles, N., and A. D. Miall, 1984, Glacial facies, in R. G. Walker (ed.), Facies models, 2nd ed.: Geoscience Canada Reprint Ser. 1, p. 15–38.

Fairbridge, R. W., 1980, The estuary: Its definition and geodynamic cycle, in E. Olausson and I. Cato (eds.), Chemistry and biochemistry of estuaries: John Wiley & Sons, New York, p. 1–35.

Faul, H., 1966, Ages of rocks, planets, and stars: McGraw-Hill, New York, 109 p.

Faure, G., 1977, Principles of isotope geology: John Wiley & Sons, New York, 464 p.

———1982, The marine-strontium geochronometer, in G. S. Odin (ed.), Numerical dating in stratigraphy, Part I: John Wiley & Sons, p. 73–80.

Fisher, J. S., and R. Dolan (eds.), 1977, Beach processes and coastal hydrodynamics: Dowden, Hutchinson and Ross, Stroudsburg, Pa., 382 p.

Fisher, W. L., and J. H. McGowen, 1967, Depositional systems in the Wilcox Group of Texas and their relationship to occurrences of oil and gas: Gulf Coast Assoc. Geol. Soc. Trans., v. 17, p. 105–125.

Fleming, B. W., 1980, Sand transport and bedform patterns on the continental shelf between Durban and Port Elizabeth (southeast Africa continental margin): Sed. Geology, v. 26, p. 179–205.

Flint, R. F., 1971, Glacial and Quaternary geology: John Wiley & Sons, New York, 892 p.

Folk, R. L., 1955, Student operator error in determination of roundness, sphericity, and grain size: Jour. Sed. Petrology, v. 25, p. 297–301.

———1959, Practical petrographic classification of limestones: Am. Assoc. Petroleum Geologists Bull., v. 43, p. 1–38.

———1962, Spectral subdivision of limestone types, in W. E. Ham (ed.), Classification of carbonate rocks: Am. Assoc. Petroleum Geologists Mem. 1, p. 62–84.

———1965, Some aspects of recrystallization in ancient limestones, in L. C. Pray and R. C. Murray (eds.), Dolomitization and limestone diagenesis: Soc. Econ. Paleontologists and Mineralogists Spec. Pub. 13, p. 14–48.

———1974, Petrology of sedimentary rocks: Hemphill, Austin, Tex. 182 p.

Folk, R.L., P.B. Andrews, and D.W. Lewis, 1970, Detrital sedimentary rock classification and nomenclature for use in New Zealand: New Zealand Jour. Geol. and Geophysics, v. 13, p. 937–968.

Folk, R. L., and L. S. Land, 1975, Mg/Ca ratio and salinity: Two controls over crystallization of dolomite: Am. Assoc. Petroleum Geologists Bull., v. 59, p. 60–68.

Fouch, T. D., and W. E. Dean, 1982, Lacustrine and associated clastic depositional environments, in P. A. Scholle and D. Spearing (eds.), Sandstone depositional environments: Am. Assoc. Petroleum Geologists Mem. 31, p. 87–114.

Fox, W. T., 1983, At the sea's edge, Ch. 4, Tides, p. 93–124: Prentice Hall, Englewood Cliffs, N.J., 317 p.

Frey, R. W. (ed.), 1975, The study of trace fossils: Springer-Verlag, New York, 562 p.

Frey, R. W., 1978, Behavioral and ecological implications of trace fossils, in P. B. Basan (ed.), Trace fossil concepts: Soc. Econ. Paleontologists and Mineralogists Short Course No. 5, p. 43–66.

Frey, R. W., and P. B. Basan, 1978, Coastal salt marshes, in R. A. Davis, Jr. (ed.), Coastal sedimentary environments: Springer-Verlag, New York, p. 101–170.

Frey, R. W., and S. G. Pemberton, 1984, Trace fossil facies models, in R. G. Walker (ed.), Facies models: Geoscience Canada Reprint Ser. 1, p. 189–207.

Frey, R. W., S. G. Pemberton, and J. A. Fagerstrom, 1984, Morphological, ethological and environmental significance of the ichnogenera Scoyenia and Ancorichnus: Jour. Paleontology, v. 58, p. 511–528.

Frey, R. W., and A. Seilacher, 1980, Uniformity in marine invertebrate ichnology: Lethaea, v. 13, p. 183–207.

Friedman, G. M., 1961, Distinction between dune, beach, and river sands from their textural characteristics: Jour. Sed. Petrology, v. 31, p. 514–529.

———1967, Dynamic processes and statistical parameters compared for size frequency distribution of beach and river sands: Jour. Sed. Petrology, v. 37, p. 327–354.

Friedman, G. M. (ed.), 1969, Depositional environments in carbonate rocks: Soc. Econ. Paleontologists and Mineralogists Spec. Pub. 14, 209 p.

Friedman, G. M., 1979, Address of the retiring president of the International Association of Sedimentologists: Differences in size distributions of populations of particles among sands of various origins: Sedimentology, v. 26, p. 3–32.

Friedman, G. M., and J. E. Sanders, 1978, Principles of sedimentology: John Wiley & Sons, New York, 792 p.

Frost, S. H., M. P. Weiss, and J. B. Saunders (eds.), 1977, Reefs and related carbonates—Ecology and sedimentology: Am. Assoc. Petroleum Geologists Studies in Geology 4, 421 p.

Fryberger, S. G., and G. Dean, 1979, Dune forms and wind regime, in E. D. McKee (ed.), 1979, A study of global sand seas: U.S. Geol. Survey Prof. Paper 1052, p. 137–169.

Füchtbauer, H., 1974, Zur diagenese fluviatiler sandsteine: Geol. Rundschau, v. 63, p. 904–925.

Gains, A. M., 1980, Dolomitization kinetics: Recent experimental studies, in D. H. Zenger, J. B. Dunham, and R. L. Ethington (eds.), Concepts and models of dolomitization: Soc. Econ. Paleontologists and Mineralogists Spec. Pub. 28, p. 81–86.

Galehouse, J. S., 1971, Sedimentation analysis, in R. E. Carver (ed.), Procedures in sedimentary petrology: John Wiley & Sons, New York, p. 69–94.

Gall, J. C., 1983, Ancient sedimentary environments and the habitats of living organisms: Springer-Verlag, Berlin, 219 p.

Galloway, W. E., 1975, Process framework for describing the morphologic and stratigraphic evolution of deltaic depositional systems, in M. L. Broussard (ed.), Deltas: Models for exploration: Houston Geological Society, p. 87–98.

Galloway, W. E., and D. K. Hobday, 1983, Terrigenous clastic depositional systems: Springer-Verlag, New York, 423 p.

Garde, R. J., and K. G. Ranga Raju, 1978, Mechanics of sediment transport and alluvial stream problems: Halsted, New York, 483 p.

Garrels, R. M., and C. L. Christ, 1965, Solutions, minerals, and equilibrium: Harper & Row, New York, 450 p.

Garrels, R. M., and F. T. McKenzie, 1971, Evolution of sedimentary rocks: W. W. Norton, New York, 397 p.

Garrison, R. E., 1974, Radiolarian cherts, pelagic limestone, and igneous rocks in eugeosynclinal settings, in K. J. Hsü and H. C. Jenkyns (eds.), Pelagic sediments on land and under the sea: Internat. Assoc. Sedimentologists Spec. Pub. 1, p. 367–399.

Garrison, R. E., R. B. Douglas, K. E. Pisciotto, C. M. Isaacs, and J. C. Ingle (eds.), 1981, The Monterey Formation and related siliceous rocks of California: Soc. Econ. Paleontologists and Mineralogists, Pacific Section, Los Angeles, Calif., 327 p.

Ghent, E. D., 1979, Problems in zeolite facies geothermometry, geobarometry and fluid compositions, in P. S. Scholle and P. R. Schluger (eds.), Aspects of diagenesis, Soc. Econ. Paleontologists and Mineralogists Spec. Pub. 26, p. 81–87.

Gill, D., and D. F. Merriam (eds.), 1979, Geomathematical and petrophysical studies in sedimentology: Pergamon, New York, 267 p.

Ginsburg, R. N., 1956, Environmental relationships of grain size and constituent particles in some south Florida carbonate environments: Am. Assoc. Petroleum Geologists Bull., v. 40, p. 2384–2427.

——1971, Landward movement of carbonate mud: New model for regressive cycles in carbonates (abs.): Am. Assoc. Petroleum Geologists Bull., v. 55, p. 340.

Ginsburg, R. N. (ed.), 1975, Tidal deposits, a casebook of Recent examples and fossil counterparts: Springer-Verlag, New York, 428 p.

Ginsburg, R. N., and N. P. James, 1974, Holocene carbonate sediments of continental shelves, in C. A. Burk and C. L. Drake (eds.), The geology of continental margins: Springer-Verlag, New York, p. 137–155.

Glaister, R. P., and H. W. Nelson, 1974, Grain-size distributions, an aid to facies identifications: Canadian Petroleum Geology Bull., v. 22, p. 203–240.

Glass, B. P., and M. J. Zwart, 1979, North American microtektites in Deep Sea Drilling Project cores from the Caribbean Sea and Gulf of Mexico: Geol. Soc. America Bull., v. 90, p. 595–602.

Glen, W., 1982, The road to Jaramillo: Critical years of the revolution in the earth sciences: Stanford University Press, Stanford, Calif., 459 p.

Glennie, K. W., 1970, Desert sedimentary environments: Developments in Sedimentology 14: Elsevier, Amsterdam, 222 p.

——1972, Permian Rotliegendes of northwest Europe interpreted in light of modern desert sedimentation studies: Am. Assoc. Petroleum Geologists Bull., v. 56, p. 1048–1071.

Goldthwait, R. P. (ed.), 1971, Till: A symposium: Ohio State University Press, Columbus, Ohio, 402 p.

——1975, Glacial deposits: Dowden, Hutchinson and Ross, Stroudsburg, Pa., 464 p.

Gould, H. R., 1972, Environmental indicators—A key to the stratigraphic record, in J. K. Rigby and W. K. Hamblin (eds.), Recognition of ancient sedimentary environments: Soc. Econ. Paleontologists and Mineralogists Spec. Pub. 16, p. 1–3.

Grabau, A. W., 1913, Principles of stratigraphy: A. G. Seiler, 1185 p.

Gradstein, F. M., F. P. Atterberg, J. C. Brower, and W. S. Schwarzacher, 1985, Quantitative stratigraphy: D. Reidel, Dordrecht, Holland, 598 p.

Gray, J., and A. J. Boucot (eds.), 1979, Historical biogeography, plate tectonics, and the changing environment: Oregon State University Press, Corvallis, Ore., 500 p.

Greenwood, B., and R. A. Davis, Jr. (eds.), 1984, Hydrodynamics and sedimentation in wave-dominated coastal environments: Elsevier, New York, 473 p.

Griffith, J. C. 1967, Scientific methods in analysis of sediments: McGraw-Hill, New York, 508 p.

Griffiths, C. M., 1982, A proposed geologically significant segmentation and reassignment algorithm for petrophysical borehole logs, in J. M. Cubitt and R. A. Reyment (eds.), Quantitative stratigraphic correlation: John Wiley & Sons, New York, p. 287–298.

Griggs, D. T., 1936, The factor of fatigue in rock exfoliation: Jour. Geology, v. 44, p. 781–796.

Gross, G. A., 1980, A classification of iron formations based on depositional environments: Canadian Mineralogist, v. 18, p. 215–222.

Gulbrandsen, R. A., and C. E. Roberson, 1973, Inorganic phosphorites in seawater: Environmental phosphorus handbook: John Wiley & Sons, Ch. 5, p. 117–140.

Guy, H. P., D. B. Simons, and E. V. Richardson, 1966, Summary of alluvial channel data from flume experiments, 1956–1961: U.S. Geol. Survey Prof. Paper 462–I, 96 p.

Hails, J., and A. Carr (eds.), 1975, Nearshore sediment dynamics and sedimentation: John Wiley & Sons, New York, 361 p.

Hails, J. R., 1976, Placer deposits, in K. H. Wolf (ed.), Handbook of strata-bound and stratiform ore deposits, v. 3: Elsevier, New York, p. 213–244.

Hákanson, L., and M. Jansson, 1983, Lake sedimentation: Springer-Verlag, Berlin, 320 p.

Hallam, A., 1973a, A revolution in the earth sciences: From continental drift to plate tectonics: Oxford University Press, London, 127 p.

Hallam, A. (ed.), 1973b, Atlas of paleobiogeography: Elsevier, Amsterdam, 531 p.

Hallam, A., 1973c, Distributional patterns in contemporary terrestrial and marine animals, in N.

F. Hughes (ed.), Organisms and continents through time: The Paleontological Association, London, p. 93–105.

——1975, Evolutionary size increase and longevity in Jurassic bivalves and ammonites: Nature, v. 258, p. 493–496.

Hallam, A. (ed.), 1977, Patterns of evolution as illustrated by the fossil record: Elsevier, New York, 591 p.

Hallam, A., 1981, Facies interpretation and the stratigraphic record: W. H. Freeman, San Francisco, 291 p.

——1984, Pre-Quaternary sea-level changes: Ann. Rev. Earth and Planetary Sci., v. 12, p. 205–243.

Ham, W. E. (ed.), 1962, Classification of carbonate rocks: Am. Assoc. Petroleum Geologists Mem. 1, 279 p.

Ham, W. E., and L. C. Pray, 1962, Modern concepts and classifications of carbonate rocks, in W. E. Ham (ed.), Classification of carbonate rocks: Am. Assoc. Petroleum Geologists Mem. 1, p. 2–19.

Hamblin, W. K., 1965, Internal structures of "homogeneous" sandstones: Kansas Geol. Survey Bull. 175, pt. 1, p. 1–37.

Hamilton, E. I., and R. M. Farquhar (eds.), 1968, Radiometric dating for geologists: John Wiley & Sons, New York, 506 p.

Hancock, J.M., 1977, The historic development of biostratigraphic correlation, in E. G. Kauffman and J. E. Hazel (eds.), Concepts and methods of biostratigraphy: Dowden, Hutchinson and Ross, Stroudsburg, Pa., p. 3–22.

Haney, W. D., and L. I. Briggs, 1964, Cyclicity of textures in evaporite rocks of the Lucas Formation, in D. F. Merriam (ed.), Symposium on cyclic sedimentation: Kansas Geol. Survey, p. 191–197.

Hanor, J. S., 1979, The sedimentary genesis of hydrothermal fluids, in H. L. Barnes (ed.), Geochemistry of hydrothermal ore deposits: John Wiley & Sons, New York, p. 137–172.

Hanshaw, B. B., W. Back, and R. G. Deike, 1971, A geochemical hypothesis for dolomitization by ground water: Econ. Geology, v. 66, p. 710–724.

Häntzschel, W., 1975, Trace fossils and problematica, 2nd ed.: Treatise on invertebrate paleontology, pt. W, Misc., Suppl. 1, C. Teichert (ed.): Geol. Soc. America and Univ. Kansas, Boulder, Colo., and Lawrence, Kan., v. XXi +, 269 p.

Haq, B. U., and T. R. Worsley, 1982, Biochronology—Biological events in time resolution, their potential and limitations, in G. S. Odin (ed.), Numerical dating in stratigraphy, pt I: John Wiley & Sons, New York, p. 19–36.

Harbaugh, J. W., and D. F. Merriam, 1968, Computer applications in stratigraphic analysis: John Wiley & Sons, New York, 282 p.

Harder, H., 1970, Boron content of sediments as a tool in facies analysis: Sed. Geology, v. 4, p. 153–175.

Hardie, L. A. (ed.), 1977, Sedimentation on the modern carbonate tidal flats of northwest Andros Island, Bahamas: Johns Hopkins University Press, Baltimore, Md., 202 p.

Harland, W. B., 1978, Geochronologic scales, in G. V. Cohee, M. F. Glaessner, and H. D. Hedberg (eds.), Contributions to the geologic time scale: Am. Assoc. Petroleum Geologists, Tulsa, Okla., p. 9–32.

Harland, W. B., A. V. Cox, P. G. Llewellyn, C. A. G. Pickton, A. G. Smith, and R. Walters, 1982, A geologic time scale: Cambridge University Press, Cambridge, England, 131 p.

Harms, J. C., and R. K. Fahnestock, 1965, Stratification, bed forms and flow phenomena (with examples from the Rio Grande), in G. V. Middleton (ed.), Primary sedimentary structures and their hydrodynamic interpretation: Soc. Econ. Paleontologists and Mineralogists Spec. Pub. 12, p. 84–155.

Harms, J. C., J. B. Southard, D. R. Spearing, and R. G. Walker, 1975, Depositional environments as interpreted from primary sedimentary structures and stratification sequences: Soc. Econ. Paleontologists and Mineralogists Short Course No. 2, 161 p.

Harms, J. C., J. B. Southard, and R. G. Walker, 1982, Structures and sequences in clastic rocks: Soc. Econ. Paleontologists and Mineralogists Lecture Notes for Short Course No. 9.

Harper, C. W., Jr., 1984, Improved methods of facies sequence analysis, in R. G. Walker (ed.), Facies models, 2nd ed.: Geoscience Canada Reprint Ser. 1, p. 11–13.

Harris, A. G., 1979, Conodont color alteration, an organo-mineral metamorphic index, and its application to Appalachian Basin geology, in P. A. Scholle and P. A. Schulger (eds.), Aspects of diagenesis: Soc. Econ. Paleontologists and Mineralogists Spec. Pub. 26, p. 3–16.

Harrison, J. E., and Z. E. Peterman, 1980, North American Commission on Stratigraphic Nomen-clature Note 52–A preliminary proposal for a chronometric time scale for the Precambrian of the United States and Mexico: Geol. Soc. America Bull., v. 91, p. 377–380.

Harrison, W. E., 1979, Levels of graphitization of kerogen as a potentially useful method of as-sessing paleotemperatures, in P. A. Scholle and P. R. Schluger (eds.), Aspects of diagenesis: Soc. Econ. Paleontologists and Mineralogists Spec. Pub. 26, p. 45–53.

Haworth, E. Y., and J. W. G. Lund (eds.), 1984, Lake sediments and environmental history: Uni-versity of Minnesota Press, Minneapolis, 411 p.

Hay, W. W. (ed.), 1974, Studies in paleooceanography: Soc. Econ. Paleontologists and Mineralo-gists Spec. Pub. 20, 218 p.

Hayes, M. O., 1967, Hurricanes as geological agents: Case studies of Hurricanes Carla, 1961, and Cindy, 1963: Texas Bur. Econ. Geology Rep. Inv. 61, 54 p.

——1975, Morphology of sand accumulations in estuaries, in L. E. Cronin (ed.), Estuarine re-search, v. 2: Geology and engineering: Academic Press, New York, p. 3–22.

Hayes, M. O., and T. W. Kana (eds.), 1976, Terrigenous clastic depositional environments: Some modern examples. Univ. South Carolina Tech. Rept. 11-CRD, 315 p.

Hazel, J. E., 1977, Use of certain multivariate and other techniques in assemblage zonal biostra-tigraphy: Examples utilizing Cambrian, Cretaceous, and Tertiary benthic invertebrates, in E. G. Kauffman and J. E. Hazel (eds.), Concepts and methods of biostratigraphy: Dowden, Hutchinson and Ross, Stroudsburg, Pa., p. 187–212.

Heath, G. R., 1974, Dissolved silica and deep-sea sediments, in W. W. Hay (ed.), Studies in paleo-oceanography: Soc. Econ. Paleontologists and Mineralogists Spec. Pub. 20, p. 77–94.

Heckel, P. H., 1972, Recognition of ancient shallow marine environments, in J. K. Rigby and W. K. Hamblin (eds.), Recognition of ancient sedimentary environments: Soc. Econ. Paleon-tologists and Mineralogists Spec. Pub. 16, p. 226–286.

——1974, Carbonate buildups in the geologic record: A review, in L. F. Laporte (ed.), Reefs in time and space: Soc. Econ. Paleontologists and Mineralogists Spec. Pub. 18, p. 90–154.

Hedberg, H. D. (ed.), 1976, International Stratigraphic Guide: A guide to stratigraphical classifi-cation, terminology, and procedure: International Subcommission on Stratigraphic Classi-fication of IUGS Commission on Stratigraphy: John Wiley & Sons, New York, 200 p.

Heezen, B. C., and C. D. Hollister, 1971, The face of the deep: Oxford University Press, New York, 659 p.

Heirtzler, J. R., G. O. Dickson, E. M. Herron, W. C. Pitman, III, and X. Le Pichon, 1968, Marine magnetic anomalies, geomagnetic field reversals, and motions of the ocean floor and con-tinents: Jour. Geophys. Research, v. 73, p. 2119–2136.

Heward, A. P., 1981, A review of wave-dominated clastic shoreline deposits: Earth Sci. Rev., v. 17, p. 223–276.

Hine, A. C., 1983, Modern shallow water carbonate platform margins, in H. E. Cook, A. C. Hine, and H. T. Mullins (eds.), Platform margin and deep water carbonates: Soc. Econ. Paleon-tologists and Mineralogists Short Course Notes No. 12, p. 3-1–3-100.

Hiscott, R. N., and G. V. Middleton, 1980, Fabric of coarse deep-water sandstones, Tourelle For-mation, Quebec, Canada: Jour. Sed. Petrology, v. 50, p. 703–722.

Hobday, D. K., and R. A. Morton, 1984, Lower Cretaceous shelf storm deposits, northeast Texas, in R. W. Tillman and C. T. Siemers (eds.), Siliciclastic shelf sediments: Soc. Econ. Paleon-tologists and Mineralogists Spec. Pub. 34, p. 205–213.

Hoffman, J., and J. Hower, 1979, Clay mineral assemblages as low grade metamorphic geother-mometers: Applications to the thrust faulted disturbed belt of Montana, U.S.A., in P. A. Scholle and P. R. Schluger (eds.), Aspects of diagenesis: Soc. Econ. Paleontologists and Mineralogists Spec. Pub. 26, p. 55–79.

Holland, H. D., 1984, The chemical evolution of the atmosphere and oceans: Princeton University Press, Princeton, N. J., 582 p.

Holland, H. D., and F. F. Trendall (eds.), 1984, Patterns of change in earth evolution: Springer-Verlag, Berlin, 432 p.

Holser, W. T., 1977, Catastrophic chemical events in the history of the ocean: Nature, v. 267, p. 403–408.

——1984, Gradual and abrupt shifts in ocean chemistry during Phanerozoic time, in H. D. Holland and A. F. Trendall (eds.), Patterns of change in Earth evolution: Springer-Verlag, Berlin, p. 123–143.

Holser, W. T., M. Mordeckai, and D. L. Clark, 1986, Carbon-isotope stratigraphic correlations in the Late Permian: Am. Jour. Sci. (in press).

Howard, J. D., 1978, Sedimentology and trace fossils, in P. B. Basan (ed.), Trace fossil concepts: Soc. Econ. Paleontologists and Mineralogists Short Course No. 5, p. 13–47.

Howarth, M. J., 1982, Tidal currents of the continental shelf, in A. H. Stride (ed.), Offshore tidal sands: Processes and deposits: Chapman and Hall, London, p. 10–26.

Hsü, K. J., and H. C. Jenkyns (eds.), 1974, Pelagic sediments on land and under the sea: Internat. Assoc. Sedimentologists Spec. Pub. 1, Blackwell, London, 447 p.

Hubbard, R. J., J. Pape, and D. G. Roberts, 1985a, Depositional sequence mapping as a technique to establish tectonic and stratigraphic framework and evaluate hydrocarbon potential on a passive continental margin, in O. R. Berg and D. G. Wolverton (eds.), Seismic stratigraphy II: An integrated approach: Am. Assoc. Petroleum Geologists Mem. 39, p. 79–91.

——1985b, Depositional sequence mapping to illustrate the evolution of a passive continental margin, in O. R. Berg and D. G. Wolverton (eds.), Seismic stratigraphy II: An integrated approach: Am. Assoc. Petroleum Geologists Mem. 39, p. 93–115.

Hughes, N. F. (ed.), 1973, Organisms and continents through time: The Palaeontological Association, London, 334 p.

Hunt, J. M., 1979, Petroleum, geochemistry and geology: W. H. Freeman, San Francisco, 617 p.

Hunter, R. E., 1977, Basic types of stratification in small eolian dunes: Sedimentology, v. 24, p. 361–387.

Hutton, J., 1888, Theory of the Earth, or an investigation of the laws observable in the composition, dissolution and restoration of land upon the globe: Royal Soc. Edinburgh Trans. v. 1, p. 109–304.

Iijima, A., J. R. Hein, and R. Siever (eds.), 1983, Siliceous deposits in the Pacific region: Elsevier, Amsterdam, 472 p.

Iijima, A., H. Inagaki, and Y. Kakuwa, 1979, Nature and origin of the Paleogene cherts in the Setogawa Terrain, Shizuoka, Central Japan: Jour. Fac. Sci., Univ. Tokyo, v. 20, p. 1–30.

Iller, R. K., 1979, Chemistry of silica: Wiley-Interscience, New York, 866 p.

Illing, L. V. (1954), Bahamian calcareous sands: Am. Assoc. Petroleum Geologists Bull., v. 38, p. 1–95.

Inden, R. F., and C. H. Moore, 1983, Beach, in P. A. Scholle, D. G. Bebout, and C. H. Moore (eds.), Carbonate depositional environments: Am. Assoc. Petroleum Geologists Mem. 33, p. 211–266.

Inman, D. L., and Nordstrom, C. E., 1971, On the tectonic and morphologic classification of coasts: Jour. Geology, v. 79, p. 1–21.

International Committee for Coal Petrology, 1963, International handbook of coal petrology, 2nd ed.: Centre National de la Recherche Scientifique, Paris. Supplements published in 1971, 1975.

Ireland, H. A., 1959, Silica in sediments: Soc. Econ. Paleontologists and Mineralogists Spec. Pub. 7, 185 p.

IUGS International Subcommission on Stratigraphic Classification and IUGS/IAGA Subcommission on a Magnetic Polarity Time Scale, 1979, Magnetostratigraphic polarity units—A supplementary chapter of the ISSC International Stratigraphic Guide: Geology, v. 7, p. 578–583.

James, H. L., 1966, Chemistry of the iron-rich sedimentary rocks: Data of geochemistry, 6th ed.: U.S. Geol. Survey Prof. Paper 440-W, 61 p.

James, H. L., and P. K. Sims (eds.), 1973, Precambrian iron formations of the world: Econ. Geology, v. 68, p. 913–1179.

James, N. P., 1983, Reef environment, in P. A. Scholle, D. G. Bebout, and C. H. Moore (eds.), Carbonate depositional environments: Am. Assoc. Petroleum Geologists Mem. 33, p. 345–440.

——1984a, Introduction to carbonate facies models, in R. G. Walker (ed.), Facies models: Geoscience Canada Reprint Ser. 1, p. 209–212.

——1984b, Shallowing-upward sequences in carbonates, in R. G. Walker (ed.), Facies models: Geoscience Canada Reprint Ser. 1, p. 213–228.

——1984c, Reefs, in R. G. Walker (ed.), Facies models: Geoscience Canada Reprint Ser. 1, p. 229–244.

Jenkyns, H. C., 1978, Pelagic environments, in H. G. Reading (ed.), Sedimentary environments and facies: Elsevier, New York, p. 314–371.

Johansson, E. E., 1976, Structural studies of frictional sediments: Geograf. Annaler, v. 58A, p. 200–300.

Johnson, D. W., 1919, Shore processes and shoreline development: John Wiley & Sons, New York, 584 p.

Johnson, H. D., 1978, Shallow siliciclastic seas, in H. G. Reading (ed.), Sedimentary environments and facies: Elsevier, New York, p. 207–258.

Jones, B. F., and C. J. Bowser, 1978, The mineralogy and related chemistry of lake sediments, in A. Lerman (ed.), Lakes: Chemistry, geology, physics: Springer-Verlag, New York, p. 179–235.

Jopling, A. V., and B. C. McDonald, 1975, Glaciofluvial and glaciolacustrine sedimentation: Soc. Econ. Paleontologists and Mineralogists Spec. Pub. 23, 320 p.

Jopling, A. V., and R. G. Walker, 1968, Morphology and origin of ripple-drift cross-lamination with examples from the Pleistocene of Massachusetts: Jour. Sed. Petrology, v. 38, p. 971–984.

Jordan, C. F., G. E. Freyer, and E. H. Hemmen, 1971, Size analysis of silt and clay by hydrophotometer: Jour. Sed. Petrology, v. 41, p. 489–496.

Journal Geological Society (London), 1980, v. 136, pt. 6 (an issue devoted to phosphatic and glauconitic sediments), p. 657–805.

Kastner, M., and J. M. Gieskes, 1983, Opal-A to opal-CT transformation: A kinetic study, in A. Iijima, J. R. Hein, and R. Siever (eds.), Siliceous deposits in the Pacific region: Developments in Sedimentology 36, Elsevier, Amsterdam, p. 211–228.

Kauffman, E. G., 1977, Evolutionary rates and biostratigraphy, in E. G. Kauffman and J. E. Hazel (eds.), Concepts and methods of biostratigraphy: Dowden, Hutchinson and Ross, Stroudsburg, Pa., p. 109–142.

Kauffman, E. G., and J. E. Hazel (eds.), 1977, Concepts and methods of biostratigraphy: Dowden, Hutchinson and Ross, Stroudsburg, Pa., 658 p.

Keith, M. L., G. M. Anderson and R. Eichler, 1964, Carbon and oxygen isotopic composition of mollusk shells from marine and fresh-water environments: Geochim. et Cosmochim. Acta, v. 28, p. 1757–1786.

Keith, M. L., and J. N. Weber, 1964, Carbon and oxygen isotope composition of selected limestones and fossils: Geochim. et Cosmochim. Acta, v. 28, p. 1787–1816.

Kelts, K., and K. J. Hsü, 1978, Calcium carbonate sedimentation in freshwater lakes and the formation of non-glacial varves in Lake Zurich, in A. Lerman (ed.), Lakes: Chemistry, geology, physics: Springer-Verlag, New York, p. 295–324.

Keller, W. D., 1955, The principles of chemical weathering: Lucas Brothers, Columbia, Mo., 88 p.

Kendall, A. C., 1979, Subaqueous evaporites, in R. G. Walker (ed.), Facies models: Geoscience Canada Reprint Ser. 1, p. 159–174.

Kennedy, V. S., 1984, The estuary as a filter: Academic Press, Orlando, Fla., 511 p.

Kennett, J. P., 1982, Marine geology: Prentice-Hall, Englewood Cliffs, N. J., 812 p.

Kennett, J. P. (ed.), 1980, Magnetic stratigraphy of sediments: Benchmark Papers in Geology 54, Dowden, Hutchinson and Ross, Stroudsburg, Pa., 438 p.

Kerr, A., B. J. Smith, W. B. Whalley, and J. P. McGreevy, 1984, Rock temperatures from southeast Morocco and their significance for experimental rock-weathering studies: Geology, v. 12, p. 306–309.

Kerr, R. A., 1984, Vail's sea-level curves aren't going away: Science, v. 226, p. 677–678.

Kersey, D. G., and K. J. Hsü, 1976, Energy relations and density current flows: An experimental investigation: Sedimentology, v. 23, p. 761–790.

Ketchum, B. H. (ed.), 1983, Estuaries and enclosed seas: Ecosystems of the World 26: Elsevier, Amsterdam, 500 p.

Kirkland, D. W., and R. Evans (eds.), 1973, Marine evaporites: Origin, diagenesis and geochemistry: Dowden, Hutchinson, and Ross, Stroudsburg, Pa., 426 p.

Kjerfve, B. (ed.), 1978, Estuarine transport processes: Univ. South Carolina Press, 331 p.

Klein, G. deV., 1963, Analysis and review of sandstone classifications in the North American geological literature, 1940–1960: Geol. Soc. America Bull. v. 74, p. 555–576.

———1965, Diverse origins of graded bedding: Geol. Soc. America Spec. Paper 82, 109 p.

———1970, Depositional and dispersal dynamics of intertidal sand bars: Jour. Sed. Petrology, v. 40, p. 1095–1127.

———1977, Clastic tidal facies: Continuing Education Publication Co., Champaign, Ill., 149 p.

———1985, Intertidal flats and intertidal sand bodies, in R. A. Davis, Jr. (ed.), Coastal sedimentary environments, 2nd ed.: Springer-Verlag, New York, p. 187–224.

Klein, G. deV. (ed.), 1976, Holocene tidal sedimentation: Benchmark Papers in Geology 5: Dowden, Hutchinson and Ross, Stroudsburg, Pa., 423 p.

Knauth, L. P., 1979, A model for the origin of chert in limestone: Geology, v. 7, p. 274–277.

Kolodny, Y., 1980, The origin of phosphorite deposits in the light of occurrences of Recent seafloor phosphorites, in Y. K. Bentor (ed.), Marine phosphorites: Soc. Econ. Paleontologists and Mineralogists Spec. Pub. 29, p. 249.

———1981, Phosphorites, in C. Emiliani (ed.), The ocean lithosphere: The sea, v. 7: John Wiley & Sons, New York, p. 981–1023.

Kolodny, Y., and I. R. Kaplan, 1970, Uranium isotopes in sea floor phosphorites: Geochim. et Cosmochim. Acta, v. 34, p. 3–24.

Komar, P. D., 1976, Beach processes and sedimentation: Prentice-Hall, Englewood Cliffs, N.J., 429 p.

Komar, P. D., and M. C. Miller, 1975, Sediment threshold under oscillatory waves: 14th Conference on Coastal Engineering Proc., p. 756–775.

Koster, E. H., and R. J. Steel (eds.), 1984, Sedimentology of gravels and conglomerates: Canadian Soc. Petroleum Geologists Mem. 10, 441 p.

Krauskopf, K. B., 1959, The geochemistry of silica in sedimentary environments, in H. A. Ireland (ed.), Silica in sediments: Soc. Econ. Paleontologists and Mineralogists Spec. Pub. 7, p. 4–19.

———1979, Introduction to geochemistry, 2nd ed.: McGraw-Hill, New York, 617 p.

Krinsley, D., 1962, Applications of electron microscopy to geology: New York Acad. Sci. Trans., v. 25, p. 3–22.

Krinsley, D., and J. Doornkamp, 1973, Atlas of quartz sand surface textures: Cambridge Press, Cambridge, England, 91 p.

Krumbein, W. C., 1934, Size frequency distribution of sediments: Jour. Sed. Petrology, v. 4, p. 65–77.

———1941, Measurement and geological significance of shape and roundness of sedimentary particles: Journ. Sed. Petrology, v. 11, p. 64–72.

Krumbein, W. C., and F. A. Graybill, 1965, An introduction to statistical models in geology: McGraw-Hill, New York, 475 p.

Krumbein, W. C., and F. J. Pettijohn, 1938, Manual of sedimentary petrography: Appleton-Century Crofts, New York, 549 p.

Krumbein, W. C., and L. L. Sloss, 1963, Stratigraphy and sedimentation: W. H. Freeman, San Francisco, 660 p.

Kuenen, Ph. H., 1958, Experiments in geology: Geol. Society Glasgow Trans., v. 23, p. 1–28.

———1959, Experimental abrasion, part 3: Fluviatile action on sand: Am. Jour. Sci., v. 257, p. 172–190.

———1960, Experimental abrasion, part 4: Eolian action: Jour. Geology, v. 68, p. 427–449.

———1964, Experimental abrasion, part 6: Surf action: Sedimentology, v. 3, p. 29–43.

La Brecque, J. L., D. V. Kent, and S. C. Cande, 1977, Revised magnetic polarity time scale for late Cretaceous and Cenozoic time: Geology, v. 5, p. 330–335.

Landis, C. A., 1971, Graphitization of dispersed carbonaceous material in metamorphic rocks: Contr. Mineralogy and Petrology, v. 30, p. 34–45.

Laporte, L. F., 1968, Ancient environments: Prentice-Hall, Englewood Cliffs, N. J., 116 p.

Laporte, L. F. (ed.), 1974, Reefs in time and space: Soc. Econ. Paleontologists and Mineralogists Spec. Pub. 18, 256 p.

Larsen, G., and G. V. Chilingar (eds.), 1967, Diagenesis in sediments: Developments in Sedimentology 8, Elsevier, Amsterdam, 551 p.

———1979, Diagenesis in sediments and sedimentary rocks: Elsevier North Holland, New York, 579 p.

———1983, Diagenesis in sediments and sedimentary rocks, 2: Elsevier, New York, 572 p.

Larsen, R. L., and T. W. C. Hilde, 1975, A revised time scale of magnetic reversals for the Early Cretaceous and Late Jurassic: Jour. Geophys. Research, v. 80, p. 2586–2594.

Lauff, G. H. (ed.), 1967, Estuaries: Am. Assoc. Adv. Sci. Spec. Pub. 83, 757 p.

Leatherman, S. P. (ed.), 1979, Barrier Islands from the Gulf of St. Lawrence to the Gulf of Mexico: Academic Press, New York, 325 p.

LeBlanc, R. J., 1975, Significant studies of modern and ancient deltaic sediments, in M. L. Broussard (ed.), Deltas: Models for exploration: Houston Geological Society, p. 13–85.

Leeder, M. R., 1982, Sedimentology: Process and product: George Allen & Unwin, London, 344 p.

Legget, R. F. (ed.), 1976, Glacial till: An inter-disciplinary study: Royal Soc. Canada Spec. Pub. 12, 412 p.

Leopold, L. B., and M. G. Wolman, 1957, River channel patterns, braided, meandering, and straight: U.S. Geol. Survey Prof. Paper 282-B, p. 39–85.

Lepp, H. (ed.), 1975, Geochemistry of iron: Benchmark Papers in Geology, v. 18, Dowden, Hutchinson, and Ross, Stroudsburg, Pa., 464 p.

Lepp, H., and S. S. Goldich, 1964, Origin of Precambrian iron formation: Econ. Geology, v. 58, p. 1025–1061.

Lerman, A. (ed.), 1978, Lakes: Chemistry, geology, physics: Springer-Verlag, New York, 363 p.

Lewan, M. D., 1978, Laboratory classification at very fine-grained sedimentary rocks: Geology, v. 6, p. 745–748.

Lewis, D. W., 1984, Practical sedimentology: Hutchinson and Ross, Stroudsburg, Pa., 229 p.

Lippman, F., 1973, Sedimentary carbonate minerals: Springer-Verlag, New York, 228 p.

Lisitzin, A. P., 1972, Sedimentation in the world ocean: Soc. Econ. Paleontologists and Mineralogists Spec. Pub. 17, 218 p.

Logan, B. W., G. F. Davies, J. F. Read, and D. Cebulski, 1970, Carbonate sedimentation and environments, Shark Bay, Western Australia: Am. Assoc. Petroleum Geologists Mem. 13, 223 p.

Logan, B. W., J. F. Read, G. M. Hagen, P. Hoffman, R. G. Brown, P. J. Woods, and C. D. Gebelien, 1974, Evolution and diagenesis of Quaternary carbonate sequences, Shark Bay, Western Australia: Am. Assoc. Petroleum Geologists Mem. 22, 358 p.

Logan, B. W., R. Rezak, and R. N. Ginsburg, 1964, Classification and environmental significance of algal stromatolites: Jour. Geology, v. 72, p. 68–83.

Longman, M. W., 1981, A process approach to recognizing facies of reef complexes, in D. F. Toomey (ed.), European fossil reef models: Soc. Econ. Paleontologists and Mineralogists Spec. Pub. 30, p. 9–40.

———1982, Carbonate diagenesis as a control on stratigraphic traps: Am. Assoc. Petroleum Geologists Education Course Notes Ser. 21, 159 p.

Loughnan, F. C., 1969, Chemical weathering of silicate minerals: Elsevier, New York, 154 p.

Lowe, D. R., 1976, Subaqueous liquefied and fluidized sediment flows and their deposits: Sedimentology, v. 23, p. 285–308.

———1982, Sedimentary gravity flows: II. Depositional models with special reference to the deposits of high-density turbidity currents: Jour. Sed. Petrology, v. 52, p. 279–297.

Lowe, D. R., and R. D. LoPiccolo, 1974, The characteristics and origin of dish and pillar structures: Jour. Sed. Petrology, v. 44, p. 484–501.

Lowrie, W., and W. Alvarez, 1977, Upper Cretaceous-Paleocene magnetic stratigraphy at Gubbio, Italy. III. Upper Cretaceous magnetic stratigraphy: Geol. Soc. America Bull. v. 88, p. 374–377.

———1981, One hundred million years of geomagnetic polarity history: Geology, v. 9, p. 392–397.

Lowrie, W., W. Alvarez, G. Napeleone, K. Perch-Neilsen, I. P. Silva, and M. Toumarkine, 1982, Paleogene magnetic stratigraphy in Umbrian pelagic carbonate rocks: The Contessa sections, Gubbio: Geol. Soc. America Bull. v. 93, p. 414–432.

Lundegard, P. D., and N. D. Samuels, 1980, Field classification of fine-grained rocks: Jour. Sed. Petrology, v. 50, p. 781–786.

Lyell, Sir Charles, 1830–33, Principles of geology, v. 1: John Murray, London, 511 p.

MacQueen, R. W., 1983, Carbonate sedimentology for the mineral explorationist: Geol. Assoc. Canada, Cordilleran Section, Short Course 1.

Mahaney, W. C. (ed.), 1984, Quaternary dating methods: Elsevier, New York, 431 p.

Maiklem, W. R., D. G. Bebolt, and R. P. Glaister, 1969, Classification of anhydrite—A practical approach: Canadian Petroleum Geology Bull., v. 17, p. 194–233.

Mann, C. J., 1981, Stratigraphic analysis: Decades of revolution (1970–1979) and refinements (1980–1989); in D. F. Merriam (ed.), Computer applications in earth sciences—An update of the 70s: Plenum, New York, p. 211–242.

Markevich, V. P., 1960, The concept of facies: Internat. Geol. Rev., v. 2, p. 376–379, 498–507, 582–604.

Martens, C. S., and R. C. Harris, 1970, Inhibition of apatite precipitation in marine environment by magnesium ions: Geochim. et Cosmochim. Acta, v. 34, p. 621–625.

Matter, A., and M. E. Tucker (eds.), 1978, Modern and ancient lake sediments: Internat. Assoc. Sedimentologists Spec. Pub. 2, 290 p.

McBride, E. F., 1963, A classification of common sandstones: Jour. Sed. Petrology, v. 33, p. 664–669.

McBride, E. F. (ed.), 1979, Silica in sediments: Nodular and bedded chert: Soc. Econ. Paleontologists and Mineralogists Reprint Ser. 8, 184 p.

McBride, E. F., R. G. Shepard, and R. A. Crawley, 1975, Origin of parallel, near-horizontal laminae by migration of bed forms in a small flume: Jour. Sed. Petrology, v. 45, p. 132–139.

McCall, P. L., and M. J. S. Tevesz, 1982, Animal-sediment relations. The biogenic alteration of sediments: Plenum, New York, 336 p.

McCubbin, D. J., 1982, Barrier-island and strand-plain facies, in P. A. Scholle and D. Spearing (eds.), Sandstone depositional environments: Am. Assoc. Petroleum Geologists Mem. 31, p. 247–279.

McDonald, D. A., and R. C. Surdam, 1984, Clastic diagenesis: Am. Assoc. Petroleum Geologists Mem. 37, 434 p.

McDonald, K. C., K. B. F. N. Spiess, and R. D. Ballard, 1980, Hydrothermal heat flux of the "black smoker" vents on the East Pacific Rise: Earth and Planetary Sci. Letters, v. 48, p. 1–7.

McDougall, I., 1977, The present status of the geomagnetic polarity time scale, in M. W. McElhinney (ed.), The Earth: Its origin, structure, and evolution (a volume in honor of J. C. Jaeger and A. L. Hales): Academic Press, New York, p. 543–566.

McDougall, I., K. Saemundsson, H. Johannesson, N. D. Watkins, and L. Kristjansson, 1977, Extension of the geomagnetic polarity time scale to 6.5 m.y.: K-Ar dating, geological and paleomagnetic study of a 3500-m lava succession in western Iceland: Geol. Soc. America Bull., v. 88, p. 1–15.

McElhinney, M. W., 1978, The magnetic polarity time scale: Prospects and possibilities in magnetostratigraphy, in G. V. Cohee, M. F. Glaessner, and H. D. Hedberg (eds.), Contributions to the geologic time scale: Am. Assoc. Petroleum Geologists Studies in Geology 6, p. 57–66.

McGowen, J. H., and C. G. Groat, 1971, Van Horn Sandstone, West Texas: An alluvial fan model for mineral exploration: Texas Bur. Econ. Geology Rept. Inv. 72, Austin, Tex., 57 p.

McGrail, D. W., and M. Carnes, 1983, Shelfedge dynamics and the nepheloid layer in the northwest Gulf of Mexico; in D. J. Stanley and G. T. Moore (eds.), The shelfbreak: Critical interface on continental margins: Soc. Econ. Paleontologists and Mineralogists Spec. Pub. 33, p. 251–264.

McIlreath, I. A., and N. P. James, 1984, Carbonate slopes, in R. G. Walker (ed.), Facies models, 2nd ed.: Geoscience Canada Reprint Ser. 1, p. 245–257.

Mckee, E. D., 1965, Experiments on ripple lamination, in G. V. Middleton (ed.), Primary sedimentary structures and their hydrodynamic interpretation. Soc. Econ. Paleontologists and Mineralogists Spec. Pub. 12, p. 66–83.

McKee, E. D. (ed.), 1979a, A study of global sand seas: U. S. Geol. Survey Prof. Paper 1052, 429 p.

McKee, E. D., 1979b, Sedimentary structures in dunes, in E. D. McKee (ed.), A study of global sand seas: U.S. Geol. Survey Prof. Paper 1052, p. 83–134.

———1979c, Introduction to a study of global sand seas, in E. D. McKee (ed.), A study of global sand seas: U.S. Geol. Survey Prof. Paper 1052, p. 1–19.

McKee, E. D., J. R. Douglass, and S. Rittenhouse, 1971, Deformation of lee-side laminae in eolian dunes: Geol. Soc. America Bull., v. 82, p. 359–378.

McKee, E. D., and G. W. Weir, 1953, Terminology for stratification and cross-stratification in sedimentary rocks: Geol. Soc. America Bull., v. 64, p. 381–390.

McKelvey, V. E., J. S. Williams, R. P. Sheldon, E. R. Cressman, and T. M. Channey, 1959, The Phosphoria, Park City and Shedhorn formations in the Western Phosphate Field: U.S. Geol. Survey Prof. Paper 313-A.

Melnik, Y. P., 1982, Precambrian banded iron-formations: Developments in Precambrian Geology 5. Elsevier, Amsterdam, 310 p.

Merino, E., 1975a, Diagenesis in Tertiary sandstones from Kettleman North Dome, California—I. Diagenetic mineralogy: Jour. Sed. Petrology, v. 45, p. 320–336.

———1975b, Diagenesis in Tertiary sandstones from Kettleman North Dome, California—II. Interstitial solutions: Distribution of aqueous species at 100°C and chemical relation to the diagenetic mineralogy: Geochim. et Cosmochim. Acta, v. 39, p. 1629–1645.

Mero, J. L., 1965, The mineral resources of the sea: Elsevier, New York, 312 p.

Merriam, D. F., 1964, Symposium on cyclic sedimentation: Geol. Survey Kansas Bull. 169, v. 1 and 2, 636 p.

Merriam, D. F. (ed.), 1981, Computer applications in the earth sciences—An update of the 70s: Plenum, New York, 385 p.

Meybeck, M., 1981, River transport of organic carbon to the ocean, *in* Carbon dioxide effects research and assessment program: Flux of organic carbon by rivers to the ocean: Committee on Flux of Organic Carbon to the Ocean, G. E. Likens (chm.), Div. Biol. Sci., Natl. Research Council, U.S. Dept. Energy, Off. Energy Research, Washington, D.C., Reference: CONF-80009140 UC, 219 p.

Miall, A. D., 1973, Markov chain analysis applied to an ancient alluvial plain succession: Sedimentology, v. 20, p. 347–365.

———1977, A review of the braided-river depositional environment: Earth Science Rev., v. 13, p. 1–62.

Miall, A. D. (ed.), 1978, Fluvial sedimentology: Canadian Soc. Petroleum Geologists Mem. 5, 589 p.

Miall, A. D., 1982, Analysis of fluvial depositional systems: Am. Assoc. Petroleum Geologists Education Course Note Ser. 20, 75 p.

———1984a, Principles of sedimentary basin analysis: Springer-Verlag, New York, 490 p.

———1984b, Deltas, *in* R. G. Walker (ed.), Facies models: Geoscience Canada Reprint Ser. 1, 2nd ed., p. 105–118.

Middlemass, F. A., P. F. Rawson, and G. Newall (eds.), 1971, Faunal provinces in space and time: Geology Jour. Spec. Issue 4, Seel House Press, Liverpool, 236 p.

Middleton, G. V., 1973, Johannes Walther's Law of the Correlation of Facies: Geol. Soc. America Bull., v. 84, p. 979–988.

———1976, Hydraulic interpretation of sand size distributions: Jour. Geology, v. 84, p. 405–426.

———1978, Facies, *in* R. W. Fairbridge and J. Bourgeois (eds.), Encyclopedia of sedimentology: Dowden, Hutchinson, and Ross, Stroudsburg, Pa., p. 323–325.

Middleton, G. V. (ed.), 1965, Primary sedimentary structures and their hydrodynamic interpretation: Soc. Econ. Paleontologists and Mineralogists Spec. Pub. 12, 265 p.

Middleton, G. V., and A. H. Bouma (eds.), 1973, Turbidites and deep water sedimentation: Soc. Econ. Paleontologists and Mineralogists, Pacific Section, Short Course, Anaheim, Calif., 157 p.

Middleton, G. V., and M. A. Hampton, 1976, Subaqueous sediment transport and deposition by sediment gravity flows, *in* D. J. Stanley and D. J. P. Swift (eds.), Marine sediment transport and environmental management: John Wiley & Sons, New York, p. 197–218.

Middleton, G. V., and J. B. Southard, 1978, Mechanics of sediment movement: Soc. Econ. Paleontologists and Mineralogists Short Course Notes No. 3, variously paginated.

———1984, Mechanics of sediment movement, 2nd ed.: Eastern Section, Soc. Econ. Paleontologists and Mineralogists Short Course 3, 401 p.

Miller, F. X., 1977, The graphic correlation method in biostratigraphy, *in* E. G. Kauffman and J. E. Hazel (eds.), Concepts and methods in biostratigraphy: Dowden, Hutchinson and Ross, Stroudsburg, Pa., p. 165–186.

Miller, M. F., A. A. Ekdale, and M. D. Picard (eds.), 1984, Trace fossils and paleoenvironments: Marine carbonate, marginal marine terrigenous and continental terrigenous settings: Jour. Paleontology, v. 58, p. 283–597.

Milliman, J. D., 1974, Marine carbonates: Springer-Verlag, New York, 375 p.

Milner, H. B., 1962, Sedimentary petrography, v. 2: Principles and applications: MacMillan, New York, 715 p.

Mitchum, R. M., Jr., and P. R. Vail, 1977, Seismic stratigraphic interpretation procedures, *in* C. E. Payton (ed.), Seismic stratigraphy—application to hydrocarbon exploration: Am. Assoc. Petroleum Geologists Mem. 26, p. 135–143.

Mitchum, R. M., Jr., P. R. Vail, and J. B. Sangree, 1977, Stratigraphic interpretation of seismic reflection patterns in depositional sequences, *in* C. E. Payton (ed.), Seismic stratigraphy—Application to hydrocarbon exploration: Am. Assoc. Petroleum Geologists Mem. 26, p. 117–133.

Mitchum, R. M., Jr., P. R. Vail, and S. Thompson, III, 1977, The depositional sequence as a basic unit for stratigraphic analysis, *in* C. E. Payton (ed.), Seismic stratigraphy—Applications to hydrocarbon exploration: Am. Assoc. Petroleum Geologists Mem. 26, p. 53–62.

Molnia, B. F., 1983, Glacial-marine sedimentation: Plenum, New York, 844 p.

Moore, R. C., 1949, Meaning of facies: Geol. Soc. America Mem. 39, p. 1–34.

Morey, G. W., R. O. Fournier, and J. J. Rowe, 1962, The solubility of quartz in water in the temperature interval from 25°C to 300°C: Geochim. et Cosmochim. Acta, v. 26, p. 1029–1043.

———1964, The solubility of amorphous silica at 25°C: Jour. Geophys. Research, v. 69, p. 1995–2002.

Morgan, J. P. (ed.), 1970, Deltaic sedimentation—Modern and ancient: Soc. Econ. Paleontologists and Mineralogists Spec. Pub. 15, 312 p.

Moslow, T. F., 1985, Depositional models of shelf and shoreline sandstones: Am. Assoc. Petroleum Geologists Education Course Notes Ser. 17, 102 p.

Mount, J., 1985, Mixed siliciclastic and carbonate sediments: A proposed first-order textural and compositional classification: Sedimentology, v. 32, p. 435–442.

Mucci, A., and J. W. Morse, 1983, The incorporation of Mg^{2+} and Sr^{2+} into calcite overgrowths: Influence of growth rates and solution composition: Geochim. et Cosmochim. Acta, v. 47, p. 217–233.

Nanz, R. H., 1953, Chemical composition of Precambrian slates with notes on the geochemical evolution of lutites: Jour. Geology, v. 61, p. 51–64.

Nardin, T. R., B. D. Edwards, and D. S. Gorsline, 1979a, Santa Cruz Basin, California borderland: Dominance of slope processes in basin sedimentation, in L. J. Doyle and O. H. Pilkey (eds.), Geology of continental slopes: Soc. Econ. Paleontologists and Mineralogists Spec. Pub. 27, p. 209–221.

Nardin, T. R., F. J. Hein, D. S. Gorsline, and B. D. Edwards, 1979b, A review of mass movement processes, sediment and acoustic characteristics, and contrasts in slope and base-of-slope systems versus canyon-fan-basin floor system, in L. J. Doyle and O. R. Pilkey (eds.), Geology of continental slopes: Soc. Econ. Paleontologists and Mineralogists Spec. Pub. 27, p. 61–73.

Neidell, N. S., 1979, Stratigraphic modeling and interpretation: Geophysical principles and techniques: Am. Assoc. Petroleum Geologists Education Course Notes 13, 141 p.

Nelson, B. W. (ed.), 1972, Environmental framework of coastal plain estuaries: Geol. Soc. America Mem. 133, 619 p.

Nelson, C. H., and T. H. Nilsen, 1984, Modern and ancient deep-sea fan sedimentation: Soc. Econ. Paleontologists and Mineralogists Short Course 14, 404 p.

Neumann, A. C., J. W. Kofoed, and G. H. Keller, 1977, Lithoherms in the Straits of Florida: Geology, v. 5, p. 4–11.

Neumann, A. C., and J. S. Land, 1975, Lime mud deposition and calcareous algae in the Bight of Abaco, Bahamas: A budget: Jour. Sed. Petrology, v. 45, p. 763–786.

Newell, N. D., J. K. Rigby, A. J. Whitman, and J. S. Bradley, 1951, Shoal-water geology and environments, eastern Andros Island, Bahamas: Am. Mus. Nat. History Bull., v. 97, p. 1–29.

Newton, R. S., 1968, Internal structure of wave-formed ripple marks in the nearshore zone: Sedimentology, v. 11, p. 275–292.

Nichols, M. M., and R. B. Biggs, 1985, Estuaries, in R. A. Davis, Jr. (ed.), Coastal sedimentary environments, 2nd ed.: Springer-Verlag, New York, p. 77–186.

Nilsen, T. H., 1980, Modern and ancient submarine fans: Discussion of papers by R. G Walker and W. R. Normark: Am. Assoc. Petroleum Geologists Bull., v. 64, p. 1094–1112.

———1982, Alluvial fan deposits, in P. A. Scholle and D. Spearing (eds.), Sandstone depositional environments: Am. Assoc. Petroleum Geologists Mem. 31, p. 49–86.

Nilsen, T. H. (ed.), 1985, Modern and ancient alluvial fan deposits: Van Nostrand Reinhold, New York, 372 p.

Nilsen, T. H., and P. L. Abbott, 1981, Paleogeography and sedimentology of Upper Cretaceous turbidites, San Diego, California: Am. Assoc. Petroleum Geologists Bull., v. 65, p. 1256–1284.

Nisbet, E. G., and I. Price, 1974, Siliceous turbidites: Bedded cherts as redeposited, ocean ridge-derived sediments, in K. J. Hsü and H. C. Jenkyns (eds.), Pelagic sediments: On land and under the sea: Internat. Assoc. Sedimentologists Spec. Pub. 1, p. 351–366.

Nitecki, M. H. (ed.), 1984, Extinctions: University of Chicago Press, 354 p.

Nittrouer, C. A., 1981, Sedimentary dynamics of continental shelves: Elsevier, Amsterdam, 449 p.

Normark, W. R., 1978, Fan valleys, channels, and depositional lobes on modern submarine fans: Characters for recognition of sandy turbidite environments: Am. Assoc. Petroleum Geologists Bull., v. 62, p 912–931.

North American Commission on Stratigraphic Nomenclature, 1983, North American Stratigraphic Code: Am. Assoc. Petroleum Geologists Bull., v. 67, p. 841–875.

Nriagu, J. O., and P. B. Moore (eds.), 1984, Phosphate minerals: Springer-Verlag, New York, 434 p.

Odin, G. S. (ed.), 1982a, Numerical dating in stratigraphy: John Wiley & Sons, New York. Pt. 1, p. 1–630, pt. II, p. 631–1040.

Odin, G. S., 1982b, Introduction: Uncertainties in evaluating the numerical time scale, in G. S. Odin (ed.), Numerical dating in stratigraphy: John Wiley & Sons, New York, p. 3–16.

———1982c, Zero isotopic ages of glauconies, in G. S. Odin (ed.), Numerical dating in stratigraphy: John Wiley & Sons, New York, p. 277–305.

Odin, G. S., D. Curry, N. H. Gale, and W. J. Kennedy, 1982, The Phanerozoic time scale in 1981; in G. S. Odin (ed.), Numerical dating in stratigraphy: John Wiley & Sons, New York, p. 957–960.

Odin, G. S., and M. H. Dodson, 1982, Zero isotopic ages of glauconies, in G. S. Odin (ed.), Numerical dating in stratigraphy: John Wiley & Sons, New York, p. 277–306.

Odin, G. S., M. Renard, and C. V. Grazzini, 1982, Geochemical events as a means of correlation, in G. S. Odin (ed.), Numerical dating in stratigraphy: John Wiley & Sons, New York, p. 37–72.

Officer, C. B. (chm.), 1977, Estuaries, geophysics, and the environment: Nat. Acad. Sci., Washington, D.C., 127 p.

Okada, H., 1971, Classification of sandstone: Analysis and proposal: Jour. Geology, v. 79, p. 509–525.

Olausson, E., and I. Cato (eds.), 1980, Chemistry and biogeochemistry of estuaries: John Wiley & Sons, New York, 452 p.

Ollier, C., 1969, Weathering: American Elsevier, New York, 304 p.

Otto, G. H., 1938, The sedimentation unit and its use in field sampling: Jour. Geology, v. 46, p. 569–582.

Palmer, A. R. (comp.), 1983, The Decade of North American Geology 1983 Geologic Time Scale: Geology, v. 11, p. 503–504.

Pantin, H. M., 1979, Interaction between velocity and effective density in turbidity flow: Phase plane analysis with criteria for autosuspension: Marine Geology, v. 31, p 59–99.

Park, W. C., and E. H. Schot, 1968, Stylolitization in carbonate rocks, in C. Müller and G. M. Friedman (eds.), Recent developments in carbonate sedimentology in central Europe: Springer-Verlag, New York, p. 66–74.

Parkash, B., and G. V. Middleton, 1970, Downcurrent textural changes in Ordovician turbidite graywackes: Sedimentology, v. 14, p. 259–293.

Parker, A., and B. W. Sellwood (eds.), Sediment diagenesis: D. Reidel, Boston, 427 p.

Parker, G., 1982, Conditions for the ignition of catastrophically erosive turbidity currents: Marine Geology, v. 46, p. 307–327.

Passega, R., 1957, Texture as a characteristic of clastic deposition: Am. Assoc. Petroleum Geologists Bull., v. 41, p. 1952–1984.

———1964, Grain size representation by CM patterns as a geological tool: Jour. Sed. Petrology, v. 34, p. 830–847.

———1977, Significance of CM diagrams of sediments deposited by suspensions: Sedimentology, v. 24, p. 723–733.

Payton, C. E. (ed.), 1977, Seismic stratigraphy—Applications to hydrocarbon exploration: Am. Assoc. Petroleum Geologists Mem. 26, 516 p.

Peterson, M. N., and C. C. von der Borch, 1965, Chert: Modern inorganic deposition in a carbonate-precipitating locality: Science, v. 149, p. 1501–1503.

Petrakis, L., and D. W. Grandy, 1980, Coal analysis, characterization and petrography: Jour. Chem. Education, v. 57, p. 689–694.

Pettijohn, F. J., 1941, Persistence of minerals and geologic age: Jour. Geology, v. 49, p. 610–625.

———1975, Sedimentary rocks, 3rd ed.: Harper & Row, New York, 628 p.

Pettijohn, F. J., and P. E. Potter, 1964, Atlas and glossary of primary sedimentary structures: Springer-Verlag, New York, 370 p.

Pettijohn, F. J., P. E. Potter, and R. Siever, 1973, Sand and sandstone: Springer-Verlag, New York, 618 p.

Phleger, F. B., 1969, Some general characteristics of coastal lagoons, in A. A. Castanares and F. B. Phleger (eds.), Coastal lagoons—A symposium: Universidad Nacional Autonoma de Mexico/UNESCO, Mexico City, p. 5–26.

Picard, M. D., 1971, Classification of fine-grained sedimentary rocks: Jour. Sed. Petrology, v. 41, p. 179–195.

Picard, M. D., and L. R. High, Jr., 1972, Criteria for recognizing lacustrine rocks, in J. K. Rigby and W. K. Hamblin (eds.), Recognition of ancient sedimentary environments: Soc. Econ. Paleontologists and Mineralogists Spec. Pub. 16, p. 108–145.

———1973, Sedimentary structures of ephemeral streams: Elsevier, New York, 223 p.

———1981, Physical stratigraphy of ancient lacustrine deposits, in F. G. Ethridge and R. M. Flores (eds.), Recent and nonmarine depositional environments: Models for exploration: Soc. Econ. Paleontologists and Mineralogists Spec. Pub. 31, p. 233–259.

Pitman, W. C., III, 1978, Relationship between eustacy and stratigraphic sequences of passive margins: Geol. Soc. America Bull., v. 89, p. 1389–1403.

Plummer, P. S., and V. A. Gostin, 1981, Shrinkage cracks: Desiccation or synaeresis? Jour. Sed. Petrology, v. 51, p. 1147–1156.

Porrenga, D. H., 1967, Glauconite and chamosite as depth indicators in the marine environment, in A. Hallam (ed.), Depth indicators in marine sedimentary rocks: Marine Geology, Spec. Issue 5, no. 5/6, p. 495–502.

Potter, P. E., 1962, Late Mississippian sandstones of Illinois Basin: Illinois Geol. Survey Circ. 340, 36 p.

———1967, Sand bodies and sedimentary environments. A review: Am. Assoc. Petroleum Geologists Bull., v. 51, p. 337–365.

Potter, P. E., J. B. Maynard, and W. A. Pryor, 1980, Sedimentology of shale: Springer-Verlag, New York, 306 p.

Potter, P. E., and F. J. Pettijohn, 1977, Paleocurrents and basin analysis, 2nd ed.: Springer-Verlag, New York, 460 p.

Potter, P. E., N. F. Shimp, and J. Witters, 1963, Trace elements in marine and fresh-water argillaceous sediments: Geochim. et Cosmochim. Acta, v. 27, p. 669–694.

Powers, D. W., and R. G. Easterling, 1982, Improved methodology for using embedded Markov chains to describe cyclic sediments: Jour. Sed. Petrology, v. 56, p. 913–923.

Powers, M. C., 1953, A new roundness scale for sedimentary particles: Jour. Sed. Petrology, v. 23, p. 117–119.

Pray, L. C., and R. C. Murray (eds.), 1965, Dolomitization and limestone diagenesis: A symposium: Soc. Econ. Paleontologists and Mineralogists, Tulsa, Okla., 180 p.

Prospero, J. M., 1981, Eolian transport to the world ocean, in C. Emiliani (ed.), The oceanic lithosphere: The sea, v. 7, John Wiley & Sons, New York, p. 801–874.

Rachocki, A., 1981, Alluvial fans: John Wiley & Sons, New York, 161 p.

Rahmani, R. A., and R. M. Flores (eds.), 1985, Sedimentology of coal and coal bearing sequences: Internat. Assoc. Sedimentologists Spec. Pub. 7, Blackwell, Oxford, 412 p.

Raup, D. M., 1977, Stochastic models in evolutionary palaeontology, in A. Hallam (ed.), Patterns of evolution as illustrated by the fossil record: Elsevier, New York, p. 59–78.

Rautman, C. A., and R. H. Dott, Jr., 1977, Dish structures formed by fluid escape in Jurassic shallow marine sandstones: Jour. Sed. Petrology, v. 47, p. 101–106.

Read, J. F., 1982, Carbonate platforms of passive (extensional) continental margins: Types, character and evolution: Tectonophysics, v. 81, p. 195–212.

Reading, H. G., 1978a, Facies, in H. G. Reading (ed.), Sedimentary environments and facies: Elsevier, New York, p. 4–14.

Reading, H. G. (ed.), 1978b, Sedimentary environments and facies: Elsevier, New York, 557 p.

Reddy, M. M., and K. K. Wang, 1980, Crystallization of calcium carbonate in the presence of metal ions. I. Inhibition by magnesium ions at pH 8.8 and 25°C: Jour. Crystal Growth, v. 50, p. 470–480.

Reed, W. R., R. LeFever, and G. J. Moir, 1975, Depositional environment interpretation from settling-velocity (psi) distributions: Geol. Soc. America Bull., v. 86, p. 1321–1328.

Reeder, R. J. (ed.), 1983, Carbonates: Mineralogy and chemistry: Rev. in Mineralogy, v. 11, 394 p.

Reeves, C. C., Jr., 1968, Introduction to paleolimnology: Elsevier, New York, 228 p.

Reinson, G. E., 1984, Barrier-island and associated strand-plain systems, in R. G. Walker (ed.), Facies models: Geoscience Canada Reprint Ser. 1, 2nd ed., p. 119–140.

Reineck, H. E., 1972, Tidal flats, in J. K. Rigby and W. K. Hamblin (eds.), Recognition of ancient depositional environments: Soc. Econ. Paleontologists and Mineralogists Spec. Pub. 16, p. 146–159.

Reineck, H. E., and I. B. Singh, 1980, Depositional sedimentary environments, 2nd ed.: Springer-Verlag, Berlin, 549 p.

Rich, J. L., 1951, Three critical environments of deposition and criteria for recognition of rocks deposited in each of them: Geol. Soc. America Bull., v. 62, p. 1–20.

Riecke, H. H., III, and G. V. Chilingarian (eds.), 1974, Compaction of argillaceous sediments: Elsevier, New York, 424 p.

Rigby, J. K., and W. K. Hamblin (eds.), 1972, Recognition of ancient sedimentary environments: Soc. Econ. Paleontologists and Mineralogists Spec. Pub. 16, 340 p.

Riley, J. P., and R. L. Chester (eds.), 1976, Chem. Oceanography, v. 5, 2nd ed.: Academic Press, New York, 401 p.

Roberts, H. H., and C. H. Moore, Jr., 1971, Recently cemented aggregates (grapestones), Grand Cayman Island, BWI, in O. P. Bricker (ed.), Carbonate cements: Johns Hopkins University Press, Baltimore, Md., p. 88–90.

Roedder, E., 1976, Fluid-inclusion evidence on the genesis of ores in sedimentary and volcanic rocks, in K. H. Wolf (ed.), Handbook of strata-bound and stratiform ore deposits: Elsevier, Amsterdam, v. 4, no. 2, p. 67–110.

————1979, Fluid inclusion evidence on the environments of sedimentary diagenesis, a review, in P. A. Scholle and P. R. Schluger (eds.), Aspects of diagenesis: Soc. Econ. Paleontologists and Mineralogists Spec. Pub. 26, p. 89–107.

Rodgers, J., 1959, The meaning of correlation: Am. Jour. Sci., v. 257, p. 684–691.

Rona, P. A., and R. P. Lowell (eds.), 1980, Seafloor spreading centers: Hydrothermal systems: Benchmark Papers in Geology, v. 56, Dowden, Hutchinson and Ross, Stroudsburg, Pa., 424 p.

Ronov, A. B., V. E. Khain, A. N. Balukhovsky, and K. B. Seslavinsky, 1980, Quantitative analysis of Phanerozoic sedimentation: Sed. Geology, v. 25, p. 311–325.

Ross, C. A., and J. R. P. Ross (eds.), 1984, Geology of coal: Benchmark Papers in Geology 77: Hutchinson and Ross, Stroudsburg, Pa., 349 p.

Rothe, P., J. Hoefs, and V. Sonne, 1974, The isotopic composition of Tertiary carbonates from the Mainz Basin: An example of isotopic fractionation in "closed basins": Sedimentology, v. 21, p. 373–395.

Rouse, H., and J. W. Howe, 1953, Basic mechanics of fluids: John Wiley & Sons, New York, 245 p.

Ruddiman, W. F. (1977), Late Quaternary deposition of ice-rafted sand in the subpolar North Atlantic (lat. 40° to 65° N): Geol. Soc. America Bull., v. 88, p. 1813–1827.

Rupke, N. A., 1978, Deep clastic seas, in H. G. Reading (ed.), Sedimentary environments and facies: Elsevier, New York, p. 372–415.

Russell, R. D., and R. E. Taylor, 1937, Roundness and shape of Mississippi River sands: Jour. Geology, v. 45, p. 225–267.

Sagoe, K-M. O., and G. S. Visher, 1977, Population breaks in grain-size distributions of sand—A theoretical model: Jour. Sedimentary Petrology, v. 47, p. 285–310.

Salvador, A., 1985, Chronostratigraphic and geochronometric scales in COSUNA Stratigraphic Correlation Charts of the United States: Am. Assoc. Petroleum Geologists Bull, v. 69, p. 181–189.

Sarjeant, A. A. S. (ed.), 1983, Terrestrial trace fossils: Benchmark Papers in Geology, v. 76. Hutchinson and Ross, Stroudsburg, Pa., 415 p.

Saxov, S., and J. K. Nieuwenhuis (eds.), 1982, Marine slides and other mass movements: NATO Conference Series IV: Marine Sciences, v. 6, Plenum, New York, 353 p.

Scheltema, R. S., 1977, Dispersal of marine invertebrate organisms: Paleobiogeographic and biostratigraphic implications, in E. G. Kauffman and J. E. Hazel (eds.), Concepts and methods of biostratigraphy: Dowden, Hutchinson and Ross, Stroudsburg, Pa., p. 73–108.

Schidlowski, M., 1982, Content and isotopic composition of reduced carbon in sediments, in H. D. Holland and M. Schidloski (eds.), Mineral deposits and the evolution of the biosphere: Springer-Verlag, New York, p. 103–122.

Schmidt, V., and D. A. McDonald, 1979, The role of secondary porosity in the course of sandstone diagenesis, in P. A. Scholle and P. R. Schluger (eds.), Aspects of diagenesis: Soc. Econ. Paleontologists and Mineralogists Spec. Pub. 26, p. 175–208.

Scholle, P. A., 1978, A color illustrated guide to carbonate rock constituents, textures, cements, and porosities: Am. Assoc. Petroleum Geologists Mem. 27, 241 p.

————1979, A color illustrated guide to constituents, textures, cements, and porosities of sandstones and associated rocks: Am. Assoc. Petroleum Geologists Mem. 28, 201 p.

Scholle, P. A., M. A. Arthur, and A. A. Ekdale, 1983, Pelagic environment, in P. A. Scholle, D. G. Bebout, and C. H. Moore (eds.), Carbonate depositional environments: Am. Assoc. Petroleum Geologists Mem. 33, p. 620–691.

Scholle, P. A., D. G. Bebout, and C. H. Moore (eds.), 1983, Carbonate depositional environments: Am. Assoc. Petroleum Geologists Mem. 33, 708 p.

Scholle, P. A., and P. R. Schluger (eds.), 1979, Aspects of diagenesis: Soc. Econ. Paleontologists and Mineralogists Spec. Pub. 26, 400 p.

Scholle, P. A., and D. Spearing (eds.), 1982, Sandstone depositional environments: Am. Assoc. Petroleum Geologists Mem. 31, 410 p.

Schopf, J. M., 1956, A definition of coal: Econ. Geology, v. 51, p. 521–527.

Schopf, T. J. M., 1977, Patterns of evolution: A summary and discussion, in A. Hallam (ed.), Patterns of evolution as illustrated by the fossil record: Elsevier, New York, p. 547–561.

———1980, Paleoceanography: Harvard University Press, Cambridge, Mass., 341 p.

Schumm, S. A., 1977, The fluvial system: John Wiley & Sons, New York, 338 p.

Schwartz, M. L. (ed.), 1972, Spits and bars: Benchmark Papers in Geology, v. 3, Dowden, Hutchinson and Ross, Stroudsburg, Pa., 452 p.

———1973, Barrier islands: Benchmark Papers in Geology, v. 9, Dowden, Hutchinson and Ross, Stroudsburg, Pa., 451 p.

Schwarzacher, W. J., 1975, Sedimentation models and quantitative stratigraphy: Developments in Sedimentology 19, Elsevier, Amsterdam, 382 p.

Sedimentation Seminar, 1981, Comparison of methods of size analysis for sands of the Amazon-Solimões Rivers, Brazil and Peru: Sedimentology, v. 28, p. 123–128.

Seibold, E., and W. H. Berger, 1982, The sea floor: Springer-Verlag, Berlin, 288 p.

Seilacher, A., 1964, Biogenic sedimentary structures, in J. Imbrie and N. D. Newell (eds.), Approaches to paleoecology: John Wiley & Sons, New York, p. 296–315.

Selley, R. C., 1970, Studies of sequences in sediments using a simple mathematical device: Geol. Soc. London Quart. Jour., v. 125, p. 557–581.

———1978, Ancient sedimentary environments, 2nd ed.: Cornell University Press, Ithaca, N.Y., 287 p.

Sellwood, B. W., 1972, Tidal flat sedimentation in the Lower Jurassic of Bornholm, Denmark: Palaeogeography, Palaeoclimatology and Palaeoecology, v 11, p. 93–106.

———1975, Lower Jurassic tidal flat deposits, Bornholm, Denmark, in R. N. Ginsburg (ed.), Tidal deposits: Springer-Verlag, New York, p. 93 101.

———1978, Shallow-water carbonate environments, in H. G. Reading (ed.), Sedimentary environments and facies: Elsevier, New York, p. 259–313.

Shackleton, N. J., 1967, Oxygen isotope analyses and paleotemperatures reassessed: Nature, v. 215, p. 15–17.

Shaw, A. B., 1964, Time in stratigraphy: McGraw-Hill, New York, 365 p.

Shaw, B. R., 1982, A short note on the correlation of geologic sequences, in J. M. Cubitt and R. A. Reyment (eds.), Quantitative stratigraphic correlation: John Wiley & Sons, New York, p. 7–12.

Shaw, B. R., and J. M. Cubitt, 1978, Stratigraphic correlation of well logs: An automated approach, in D. Gill and D. F. Merriam (eds.), Geomathematical and petrophysical studies in sedimentology: Pergamon, Oxford, p. 127–148.

Shaw, J., 1985, Subglacial and ice marginal environments, in G. M. Ashley, J. Shaw, and N. D. Smith (eds.), Glacial sedimentary environments: Soc. Econ. Paleontologists and Mineralogists Short Course No. 16, p. 7–84.

Shea, J. H., 1982, Twelve fallacies of uniformitarianism: Geology, v. 10, p. 455–460.

Shearman, D. J., 1978, Evaporites of coastal sabkhas, in W. E. Dean and B. C. Schreiber (eds.), Marine evaporites: Soc. Econ. Paleontologists and Mineralogists Short Course Notes No. 4, p. 6–42.

Shepard, F. P., 1932, Sediments on the continental shelves: Geol. Soc. America Bull., v. 43, p. 1017–1039.

———1961, Deep-sea sand: 21st Internat. Geol. Cong. Rept., p. 23, p. 26–42.

———1973, Submarine geology, 3rd ed.: Harper & Row, New York, 551 p.

———1977, Geological oceanography: Crane, Russak, New York, 214 p.

———1979, Currents in submarine canyons and other types of seavalleys, in L. J. Doyle and O. H. Pilkey (eds.), Geology of continental slopes: Soc. Econ. Paleontologists and Mineralogists Spec. Publ. 27, p. 85–94.

Shepard, F. P., and R. F. Dill, 1966, Submarine canyons and other sea valleys: Rand McNally, Chicago, 381 p.

Shepard, F. P., and D. G. Moore, 1955, Central Texas coast sedimentation: Characteristics of sedimentary environment, recent history and diagenesis: Am. Assoc. Petroleum Geologists Bull., v. 39, p. 1463–1593.

Sheriff, R. E., 1980, Seismic stratigraphy: International Human Resources Development Corp., Boston, 227 p.

Sheriff, R. E., and L. P. Geldart, 1982, Exploration seismology: History, theory, and data acquisition: Cambridge University Press, Cambridge, England, 253 p.

Shimp, N. F., J. Witters, P. E. Potter, and J. A. Schleicher, 1969, Distinguishing marine and fresh-water muds: Jour. Geology, v. 77, p. 566–580.

Shinn, E. A., 1971, Holocene submarine sedimentation in the Persian Gulf, in O. P. Bricker (ed.), Carbonate cements: Johns Hopkins University Press, Baltimore, Md., p. 63–65.

———1983, Tidal flat environment, in P. A. Scholle, D. G. Bebout, and C. H. Moore (eds.), Carbonate depositional environments: Am. Assoc. Petroleum Geologists Mem. 33, p. 171–210.

Shinn, E. A., R. B. Halley, H. H. Hudson, and B. H. Lidz, 1977, Limestone compaction: An enigma: Geology, v. 5, p. 21–24.

Shinn, E. A., and D. M. Robbin, 1983, Mechanical and chemical compaction in fine-grained shallow-water limestones: Jour. Sed. Petrology, v. 53, p. 595–618.

Shirley, M. L. and J. A. Ragsdale (eds.), 1966, Deltas in their geologic framework: Houston Geological Society, 251 p.

Sibley, D. F., and H. Blatt, 1976, Intergranular pressure solution and cementation of the Tuscarora Orthoquartzite: Jour. Sed. Petrology, v. 46, p. 881–896.

Siemers, C. T., R. W. Tillman, and C. R. Williamson, 1981, Deep-water clastic sediments: A core workshop: Soc. Econ. Paleontologists and Mineralogists Core Workshop No. 2, 416 p.

Siever, R., 1979, Plate-tectonic controls on diageneis: Jour. Geology, v. 87, p. 127–155.

———1983, Evolution of chert at active and passive continental margins, in A. Iijima, J. R. Hein, and R. Siever (eds.), Siliceous deposits in the Pacific region: Developments in Sedimentology 36, Elsevier, Amsterdam, p. 7–24.

Simons, D. B., and E. V. Richardson, 1961, Forms of bed roughness in alluvial channels: Am. Soc. Civil Engineers Proc., Jour. Hydraulics Div., v. 87 (HY3), p. 87–105.

Simonson, B. M., 1985, Sedimentological constraints on the origins of Precambrian iron-formations: Geol. Soc. America Bull., v. 96, p. 244–252.

Simpson, S., 1975, Classification of trace fossils, in R. W. Frey (ed.), The study of trace fossils: Springer-Verlag, New York, p. 39–54.

Sloss, L. L., 1963, Sequences in the cratonic interior of North America: Geol. Soc. America Bull., v. 74, p. 93–114.

Sly, P. G., 1978, Sedimentation processes in lakes, in A. Lerman (ed.), Lakes—Chemistry, geology, physics: Springer-Verlag, New York, p. 65–89.

Sneed, E. D., and R. L. Folk, 1958, Pebbles in the lower Colorado River, Texas, a study in particle morphogenesis: Jour. Geology, v. 66, p. 114–150.

Sonnenfeld, P., 1984, Brines and evaporites: Academic Press, London, 624 p.

Spencer, A. M., 1975, Late Precambrian glaciation in the North Atlantic region, in A. E. Wright and F. Moseley (eds.), Ice Ages: Ancient and modern: Seel House Press, Liverpool, England, p. 7–42.

Sperling, C. H. B., and R. U. Cooke, 1980, Salt weathering in arid environments: Experimental investigations of the relative importance of hydration and crystallization processes. II. Laboratory studies: Bedford College, London, Papers in Geography 9, 53 p.

Stach, E., 1975, Handbook of coal petrology, 2nd ed.: Gebrüder Borntraeger, Berlin, 428 p.

Stanley, D. J. (ed.), 1969, New concepts of continental margin sedimentation: Am. Geol. Inst. Short Course Notes: Washington, D.C., 400 p.

Stanley, D. J., and G. T. Moore (eds.), 1983, The shelfbreak: Critical interface on continental margins: Soc. Econ. Paleontologists and Mineralogists Spec. Pub. 33, 467 p.

Stanley, D. J., and D. J. P. Swift (eds.), 1976, Marine sediment transport and environmental management: John Wiley & Sons, New York, 602 p.

Stanley, S. M., 1979, Macroevolution, pattern and process: W. H. Freeman, San Francisco, 332 p.

Stewart, F. H., 1963, Marine evaporites, in M. Fleischer (ed.), Data of geochemistry: U.S. Geol. Survey Prof. Paper 440-Y, 54 p.

Stockman, K. W., R. N. Ginsburg, and E. A. Shinn, 1967, The production of lime mud by algae in south Florida: Jour. Sed. Petrology, v. 37, p. 633–648.

Stopes, M. C., 1919, On the four visible ingredients in banded bituminous coal. Studies in the composition of coal: Royal Soc. London Proc., Ser. B, v. 90, p. 470–487.

———1935, On the petrology of banded bituminous coal: Fuel, London, v. 14, p. 4–13.

Stow, D. A. V., and D. J. W. Piper (eds.), 1984, Fine-grained sediments: Deep-water processes and facies: Geol. Soc. Spec. Pub. 15, Blackwell, Oxford, 659 p.

Stride, A. H. (ed.), 1982, Offshore tidal sands: Processes and deposits: Chapman and Hall, London, 222 p.

Stride, A. H., R. H. Belderson, N. H. Kenyon, and M. A. Johnson, 1982, Offshore tidal deposits: Sand sheet and sand bank facies, in A. H. Stride (ed.), Offshore tidal sands: Processes and deposits: Chapman and Hall, London, p. 95–125.

Stuiver, M., S. W. Robinson, and I. C. Yang, 1979, ^{14}C dating to 60,000 years B.P. with proportional counters, in R. Berger and H. E. Suess (eds.), Radiocarbon dating: University of California Press, Berkeley, p. 202–215.

Stumm, W., and J. J. Morgan, 1981, Aquatic chemistry. An introduction emphasizing chemical equilibria in natural waters: John Wiley & Sons, New York, 583 p.

Sugden, D. E., and B. S. John, 1976, Glaciers and landscape: Edward Arnold, London, 376 p.

Sundborg, A., 1956, the River Klarälven, a study of fluvial processes: Geograf. Annaler, v. 38, p. 125–316.

Swift, D. J. P., 1975, Tidal sand ridges and shoal retreat massifs: Marine Geology, v. 18, p. 105–134.

Swift, D. J. P., D. B. Duane, and O. H. Pilkey (eds.), 1972, Shelf sediment transport: Process and pattern: Dowden, Hutchinson and Ross, Stroudsburg, Pa., 656 p.

Swift, D. J. P., J. R. Schubel, and R. E. Sheldon, 1972, Size analysis of fine-grained suspended sediments: A review: Jour. Sed. Petrology, v. 42, p. 122–134.

Swift, D. J. P., D. J. Stanley, and J. R. Curray, 1971, Relict sediments on continental shelves: A recommendation: Jour. Geology, v. 79, p. 322–346.

Sylvester-Bradley, P. C., 1977, Biostratigraphical tests of evolutionary theory, in E. G. Kauffman and J. E. Hazel (eds.), Concepts and methods of biostratigraphy: Dowden, Hutchinson and Ross, Stroudsburg, Pa., p. 41–64.

Tankard, A. J., and J. H. Barwis, 1982, Wave-dominated deltaic sedimentation in the Devonian Bokkeveld Basin of South Africa: Jour. Sed. Petrology, v. 52, p. 959–974.

Tarling, D. H., 1983, Paleomagnetism: Principles and applications in geology, geophysics and archaeology: Chapman and Hall, London, 379 p.

Taylor, J. M., 1950, Pore-space reduction in sandstones: Am. Assoc. Petroleum Geologists Bull., v. 34, p. 701–716.

Teichert, C., 1958, Concepts of facies: Am. Assoc. Petroleum Geologists Bull., v. 42, p. 2718–2744.

Thode, H. G., and J. Monster, 1965, Sulfur-isotope geochemistry of petroleum evaporites and ancient seas: Am. Assoc. Petroleum Geologists Mem. 4, p. 367–377.

Thompson, R. W., 1968, Tidal flat sedimentation on the Colorado River delta, northwest Gulf of California: Geol. Soc. America Mem. 107, 133 p.

Tickell, F. G., 1965, The techniques of sedimentary mineralogy: Elsevier, New York, 220 p.

Tietz, G., and G. Müller, 1971, Recent beachrocks, Fuerteventura, Canary Islands, Spain, in O. P. Bricker (ed.), Carbonate cements: Johns Hopkins University Press, Baltimore, Md., p. 4–8.

Tillman, R. W., and C. T. Siemers (eds.), 1984, Siliciclastic shelf sediments: Soc. Econ. Paleontologists and Mineralogists Spec. Pub. 34, 268 p.

Tissot, B. P., and Welte, D. H., 1978, Petroleum formation and occurrence: Springer-Verlag, Berlin, 538 p.

Toomey, D. F. (ed.), 1981, European fossil reef models: Soc. Econ. Paleontologists and Mineralogists Spec. Pub. 30, 546 p.

Tourtelot, H. A., 1960, Origin and use of the word "shale": Am. Jour. Sci., Bradley Volume, v. 258-A, p. 335–343.

Trask, P. D. (ed.), 1950, Applied sedimentation: John Wiley & Sons, New York, 707 p.

Trendall, A. F. and R. C. Morris (eds.), 1983, Iron-formations facts and problems: Developments in Precambrian Geology 6. Elsevier, Amsterdam, 558 p.

Tucker, R. W., and H. L. Vacher, 1980, Effectiveness of discriminating beach, dune, and river sands by moments and the cumulative weight percentages: Jour. Sed. Petrology, v. 50, p. 165–172.

Twenhofel, W. H., 1950, Principles of sedimentation, 2nd ed.: McGraw-Hill, New York, 673 p.

Twenhofel, W. H., and collaborators, 1926, Treatise on sedimentation: Williams and Wilkins, Baltimore, 661 p.

Udden, J. A., 1898, Mechanical composition of wind deposits: Augustana Library Pub. 1, 69 p.

Uffen, R. J., 1963, Influence of the Earth's core on origin and evolution of life: Nature, v. 198, p. 143.

Vacquier, V., 1972, Geomagnetism in marine geology: Elsevier Oceanography Ser. 6, Elsevier, Amsterdam, 185 p.

Vail, P. R., and R. M. Mitchum, Jr., 1977, Seismic stratigraphy and global changes of sea level, Part I: Overview, in C. E. Payton (ed.), Seismic stratigraphy—Application to hydrocarbon exploration: Am. Assoc. Petroleum Geologists Mem. 26, p. 51–52.

Vail, P. R., R. M. Mitchum, Jr., and S. Thompson, III, 1977a, Seismic stratigraphy and global change of sea level, Part 3: Relative changes of sea level from coastal onlap, in C. E. Payton (ed.), Seismic stratigraphy—Applications to hydrocarbon exploration: Am. Assoc. Petroleum Geologists Mem. 26, p. 63–81.

————1977b, Seismic stratigraphy and global change of sea level, Part 4: Global cycles of relative changes of sea level, in C. E. Payton (ed.), Seismic stratigraphy—Applications to hydrocarbon exploration: Am. Assoc. Petroleum Geologists Mem. 26, p. 83–97.

Vail, P. R., and R. G. Todd, 1981, Northern North Sea Jurassic unconformities, chronostratigraphy, and sea-level changes from seismic stratigraphy, in Petroleum geology of the continental shelf of northwest Europe: Heyden, London, p. 216–235.

Valentine, J. W., 1971, Plate tectonics and shallow marine diversity and endemism, an actualistic model: Systematic Zoology, v. 20, p. 253–264.

————1977a, General patterns of metazoan evolution, in A. Hallam (ed.), Patterns of evolution as illustrated by the fossil record: Elsevier, New York, p. 27–57.

————1977b, Biogeography and biostratigraphy, in E. G. Kauffmann and J. E. Hazel (eds.), Concepts and methods of biostratigraphy: Dowden, Hutchinson and Ross, Stroudsburg, Pa., p. 143–162.

Vandenberghe, N., 1975, An evaluation of CM patterns for grain-size studies of fine grained sediments: Sedimentology, v. 22, p. 615–622.

Van der Leeden, F., 1975, Water resources of the world—selected statistics: Water Information Centre, Point Washington, New York, 568 p.

Van der Linder, G. J. (ed.), 1977, Diagenesis of deep-sea biogenic sediments: Benchmark Papers in Geology, v. 40. Dowden, Hutchinson and Ross, Stroudsburg, Pa., 385 p.

Van Houten, F. B., 1964, Cyclic lacustrine sedimentation, Upper Triassic Lockatong Formation, central New Jersey and adjacent Pennsylvania, in D. F. Merriam, (ed.), Symposium on cyclic sedimentation: Kansas Geol. Survey Bull., v. 169, no. 2, p. 497–531.

Van Houten, F. B. (ed.), 1977, Ancient continental deposits: Benchmark Papers in Geology 43, Dowden, Hutchinson and Ross, Stroudsburg, Pa., 367 p.

Van Houten, F. B., and D. P. Bhattacharyya, 1982, Phanerozoic oolitic ironstone: Geologic record and facies models: Ann. Rev. Earth and Planetary Sci., v. 10, p. 441–457.

Van Straaten, L. M. J. U., 1961, Sedimentation in tidal flat areas: Alberta Soc. Petroleum Geologists Jour., v. 9, p. 203–213, 216–226.

Van Straaten, L. M. J. U. (ed.), 1964, Deltaic and shallow marine deposits: Developments in Sedimentology, v. 1, Elsevier, New York, 464 p.

Van Valen, L., 1973, A new evolutionary law: Evolution Theory, v. 1, p. 1–30.

Varshal, G. M., I. Ya. Koshcheyeva, I. S. Sirotkina, T. K. Velyukhanova, L. N. Intskirveli, and N. S. Zamokina, 1979, Interactions of metal ions with organic matter in surface waters: Trans. from Geokhimiya, no. 4, p. 598–607.

Veizer, J., and R. Demovic, 1974, Strontium as a tool in facies analysis: Jour. Sed. Petrology, v. 44, p. 93–115.

Veizer, J., W. T. Holser, and C. K. Wilgus, 1980, Correlation of $^{13}C/^{12}C$ and $^{34}S/^{32}S$ secular variations: Geochim. et Cosmochim. Acta, v. 44, p. 579–587.

Vine, F. H., and D. H. Matthews, 1963, Magnetic anomalies over oceanic ridges: Nature, v. 199, p. 947–949.

Visher, G. S., 1965, Use of vertical profiles in environmental reconstruction: Am. Assoc. Petroleum Geologists Bull. v. 49, p. 41–61.

————1969, Grain size distributions and depositional processes: Jour. Sed. Petrology, v. 39, p. 1074–1106.

————1984, Exploration stratigraphy: Pennwell, Tulsa, Okla., 334 p.

Wadell, H., 1932, Volume, shape and roundness of rock particles: Jour. Geology, v. 40, p. 443–451.

Walker, R. G., 1978, Deep-water sandstone facies of ancient submarine fans: Models for exploration for stratigraphic traps: Am. Assoc. Petroleum Geologists Bull., v. 62, p. 932–966.

————1979a, Facies and facies models. General introduction, in R. G. Walker (ed.), Facies models: Geoscience Canada Reprint Ser. 1, p. 1–8.

Walker, R. G. (ed.), 1979b, Facies models: Geoscience Canada Reprint Ser. 1, 211 p.

Walker, R. G., 1984a, Shelf and shallow marine sands, in R. G. Walker (ed.), Facies models, 2nd ed.: Geoscience Canada Reprint Ser. 1, p. 141–170.

Walker, R. G. (ed.), 1984b, Facies models, 2nd ed.: Geoscience Canada Reprint Ser. 1, 317 p.

Walker, R. G., 1984c, Turbidites and associated coarse clastic deposits, in R. G. Walker (ed.), Facies models: Geoscience Canada Reprint Ser. 1, p. 171–188.

Walker, R. G., and D. J. Cant, 1979, Facies models 3. Sandy fluvial systems, in R. G. Walker (ed.), Facies models: Geoscience Canada Reprint Ser. 1, p. 23–31.

——1984, Sandy fluvial systems, in R. G. Walker (ed.), Facies models: Geoscience Canada Reprint Ser. 1, p. 71–89.

Walker, R. G., and G. V. Middleton, 1979, Facies models 4. Eolian sands, in R. G. Walker (ed.), Facies models: Geoscience Canada Reprint Ser. 1, p. 33–41.

Walker, T. R., 1962, Reversible nature of chert-carbonate replacement in sedimentary rocks: Geol. Soc. America Bull., v. 73, p. 237–242.

——1967, Formation of red beds in modern and ancient deserts: Geol. Soc. America Bull., v. 78, p. 353–368.

——1984, 1984 SEPM Presidential Address: Diagenetic albitization of potassium feldspars in arkosic sandstones: Jour. Sed. Petrology, v. 54, p. 3–16.

Walker, T. R., and J. C. Harms, 1972, Eolian origin of flagstone beds, Lyons Sandstone (Permian), type area, Boulder County, Colorado: Mountain Geologists, v. 9, p. 279–288.

Walter, M. R. (ed.), 1976, Stromatolites: Elsevier, New York, 790 p.

Ward, C. R. (ed.), 1984, Coal geology and coal technology: Blackwell, Oxford, 345 p.

Warme, J. E., R. G. Douglas, and E. L. Winterer (eds.), 1981, The Deep Sea Drilling Project: A decade of progress: Soc. Econ. Paleontologists and Mineralogists Spec. Pub. 32, 564 p.

Watkins, N. D., 1972, A review of the development of the geomagnetic polarity time scale and discussion of prospects for its finer definition: Geol. Soc. America Bull., v. 83, p. 551–574.

Weaver, M., and S. W. Wise, Jr., 1974, Opaline sediments of the southeastern coastal plain and Horizon A: Biogenic origin: Science, v. 184, p. 899–901.

Weimer, R. J., J. D. Howard, and D. R. Lindsay, 1982, Tidal flats and associated tidal channels, in P. A. Scholle and D. Spearing (eds.), Sandstone depositional environments: Am. Assoc. Petroleum Geologists Mem. 31, p. 191–245.

Weller, J. M., 1958, Stratigraphic facies differentiation and nomenclature: Am. Assoc. Petroleum Geologists Bull., v. 42, p. 609–639.

——1960, Stratigraphic principles and practices: Harper and Brothers, New York, 725 p.

Wentworth, C. K., 1919, A laboratory and field study of cobble abrasion. Jour. Geology, v. 27, p. 507–521.

——1922, A scale of grade and class terms for clastic sediments: Jour. Geology, v. 30, p. 377–392.

Wheeler, H. E., and V. S. Mallory, 1956, Factors in lithostratigraphy: Am. Assoc. Petroleum Geologists Bull., v. 40, p. 2711–2723.

White, D. E., 1965, Saline waters of sedimentary rocks, in A. Young and J. E. Galley (eds.), Fluids in subsurface environments: Am. Assoc. Petroleum Geologists Mem. 4, p. 342–366.

Wiley, M. (ed.), 1976, Estuarine processes, v. II, Circulation, sediments, and transfer of material in the estuary: Academic Press, New York, 428 p.

Wilkinson, B. R., 1982, Cyclic cratonic carbonates and phanerozoic calcite seas: Jour. Geol. Education, v. 30, p. 189–203.

Williams, H., F. J. Turner, and C.M. Gilbert, 1982, Petrography, 2nd ed.: W. H. Freeman, San Francisco, 626 p.

Williams, L. A., and D. A. Crerar, 1985, Silica diagenesis, II. General mechanisms: Jour. Sed. Petrology, v. 55, p. 312–321.

Williams, L. A., G. A. Parks, and D. A. Crerar, 1985, Silica diagenesis, I. Solubility controls: Jour. Sed. Petrology, v. 55, p. 301–311.

Wilson, I. G., 1972, Aeolian bedforms—Their development and origins: Sedimentology, v. 19, p. 173–210.

Wilson, J. L. (1975) Carbonate facies in geologic history: Springer-Verlag, Berlin, 471 p.

Wilson, J. L., and C. Jordan, 1983, Middle shelf, in P. A. Scholle, D. G. Bebout, and C. H. Moore (eds.), Carbonate depositional environments: Am. Assoc. Petroleum Geologists Mem. 33, p. 297–344.

Wolf, K. H., G. V. Chilingar, and F. W. Beales, 1967, Elemental composition of carbonate skeletons, minerals, and sediments, in G. V. Chilingar, H. J. Bissell, and R. W. Fairbridge (eds.), Carbonate rocks: Developments in Sedimentology 9B, Elsevier, New York, p. 23–150.

Wright, A. E., and F. Moseley, 1975, Ice ages: Ancient and modern: Geol. Jour., Spec. Issue 6, Seel House Press, Liverpool, 320 p.

Wright, L. D., 1977, Sediment transport and deposition at river mouths: A synthesis: Geol. Soc. America Bull., v. 88, p. 857–868.

————1978, River deltas, *in* R. A. Davis, Jr. (ed), Coastal sedimentary environments: Springer-Verlag, New York, p. 5–68.

————1985, River deltas, *in* R. A. Davis, Jr. (ed.), Coastal sedimentary environments, 2nd ed.: Springer-Verlag, New York, p. 1–76.

Wyllie, P. J., 1976, The way the Earth works: John Wiley & Sons, New York, 296 p.

Yalin, M. S., 1977, Mechanics of sediment transport, 2nd ed.: Pergamon, New York, 298 p.

Yanov, E. N., 1978, Classification of sandstones and siltstones by composition of grains: Lithology and Mineral Resources, v. 12, p. 466–472.

Yeh, H., and S. Savin, 1977, The mechanisms of burial diagenetic reactions in argillaceous sediments: 3. Oxygen isotope evidence: Geol. Soc. America Bull., v. 88, p. 1321–1330.

Yen, T. F., and G. V. Chilingarian, 1976a, Introduction to oil shales, *in* T. F. Yen and G. V. Chilingarian (eds.), Oil shales: Elsevier, New York, p. 1–12.

Yen, T. F., and G. V. Chilingarian (eds.), 1976b, Oil shales: Elsevier, New York, 292 p.

York, D., and R. M. Farquhar, 1972, The Earth's age and geochronology: Pergamon, New York, 178 p.

Young, F. G., and G. E. Reinson, 1975, Sedimentology of Blood Reserve and adjacent formations (Upper Cretaceous), St. Mary River, southern Alberta, *in* M. S. Shawa (ed.), Guidebook to selected sedimentary environments in southwestern Alberta, Canada: Canadian Soc. Petroleum Geologists, Field Conference, p. 10–20.

Zenger, D. H., and J. B. Dunham, 1980, Concepts and models of dolomitization—An introduction, *in* D. H. Zenger, J. B. Dunham, and R. L. Ethington (eds.), Concepts and models of dolomitization: Soc. Econ. Paleontologists and Mineralogists, Spec. Pub. 28, p. 1–9.

Zenger, D. H., J. B. Dunham, and R. L. Ethington (eds.), 1980, Concepts and models of dolomitization: Soc. Econ. Paleontologists and Mineralogists Spec. Pub. 28, 320 p.

Zenger, D. H., and S. J. Mazzullo (eds.), 1982, Dolomitization: Benchmark Papers in Geology, v. 65. Hutchinson and Ross, Stroudsburg, Pa., 426 p.

Zingg, Th., 1935, Beiträge zur Schotteranalyse: Schweiz. Mineralog. Petrog. Mitt., v. 15, p. 39–140.

Zuffa, G. G., 1980, Hybrid arenites: Their composition and classification: Jour. Sed. Petrology, v. 50, p. 21–29.

Zuffa, G. G. (ed.), 1985, Provenance of arenites: D. Reidel, Dordrecht, Holland, 408 p.

APPENDIX

North American Stratigraphic Code[1]

NORTH AMERICAN COMMISSION ON STRATIGRAPHIC NOMENCLATURE

FOREWORD

This code of recommended procedures for classifying and naming stratigraphic and related units has been prepared during a four-year period, by and for North American earth scientists, under the auspices of the North American Commission on Stratigraphic Nomenclature. It represents the thought and work of scores of persons, and thousands of hours of writing and editing. Opportunities to participate in and review the work have been provided throughout its development, as cited in the Preamble, to a degree unprecedented during preparation of earlier codes.

Publication of the International Stratigraphic Guide in 1976 made evident some insufficiencies of the American Stratigraphic Codes of 1961 and 1970. The Commission considered whether to discard our codes, patch them over, or rewrite them fully, and chose the last. We believe it desirable to sponsor a code of stratigraphic practice for use in North America, for we can adapt to new methods and points of view more rapidly than a worldwide body. A timely example was the recognized need to develop modes of establishing formal nonstratiform (igneous and high-grade metamorphic) rock units, an objective which is met in this Code, but not yet in the Guide.

The ways in which this Code differs from earlier American codes are evident from the Contents. Some categories have disappeared and others are new, but this Code has evolved from earlier codes and from the International Stratigraphic Guide. Some new units have not yet stood the test of long practice, and conceivably may not, but they are introduced toward meeting recognized and defined needs of the profession. Take this Code, use it, but do not condemn it because it contains something new or not of direct interest to you. Innovations that prove unacceptable to the profession will expire without damage to other concepts and procedures, just as did the geologic-climate units of the 1961 Code.

This Code is necessarily somewhat innovative because of: (1) the decision to write a new code, rather than to revise the old; (2) the open invitation to members of the geologic profession to

offer suggestions and ideas, both in writing and orally; and (3) the progress in the earth sciences since completion of previous codes. This report strives to incorporate the strength and acceptance of established practice, with suggestions for meeting future needs perceived by our colleagues; its authors have attempted to bring together the good from the past, the lessons of the Guide, and carefully reasoned provisions for the immediate future.

Participants in preparation of this Code are listed in Appendix I, but many others helped with their suggestions and comments. Major contributions were made by the members, and especially the chairmen, of the named subcommittees and advisory groups under the guidance of the Code Committee, chaired by Steven S. Oriel, who also served as principal, but not sole, editor. Amidst the noteworthy contributions by many, those of James D. Aitken have been outstanding. The work was performed for and supported by the Commission, chaired by Malcolm P. Weiss from 1978 to 1982.

This Code is the product of a truly North American effort. Many former and current commissioners representing not only the ten organizational members of the North American Commission on Stratigraphic Nomenclature (Appendix II), but other institutions as well, generated the product. Endorsement by constituent organizations is anticipated, and scientific communication will be fostered if Canadian, United States, and Mexican scientists, editors, and administrators consult Code recommendations for guidance in scientific reports. The Commission will appreciate reports of formal adoption or endorsement of the Code, and asks that they be transmitted to the Chairman of the Commission (c/o American Association of Petroleum Geologists, Box 979, Tulsa, Oklahoma 74101, U.S.A.).

Any code necessarily represents but a stage in the evolution of scientific communication. Suggestions for future changes of, or additions to, the North American Stratigraphic Code are welcome. Suggested and adopted modifications will be announced to the profession, as in the past, by serial Notes and Reports published in the *Bulletin* of the American Association of Petroleum Geologists. Suggestions may be made to representatives of your association or agency who are current commissioners, or directly to the Commission itself. The Commission meets annually, during the national meetings of the Geological Society of America.

1982 NORTH AMERICAN COMMISSION
ON STRATIGRAPHIC NOMENCLATURE

[1]Reprinted by permission from American Association of Petroleum Geologists Bulletin. v. 67, no. 5 (May 1983), p 841–875.

Copies are available at $1.00 per copy postpaid. Order from American Association of Petroleum Geologists, Box 979, Tulsa, Oklahoma 74101.

CONTENTS

PART I. PREAMBLE

BACKGROUND

PERSPECTIVE

Codes of Stratigraphic Nomenclature prepared by the American Commission on Stratigraphic Nomenclature (ACSN, 1961) and its predecessor (Committee on Stratigraphic Nomenclature, 1933) have been used widely as a basis for stratigraphic terminology. Their formulation was a response to needs recognized during the past century by government surveys (both national and local) and by editors of scientific journals for uniform standards and common procedures in defining and classifying formal rock bodies, their fossils, and the time spans represented by them. The most recent Code (ACSN, 1970) is a slightly revised version of that published in 1961, incorporating some minor amendments adopted by the Commission between 1962 and 1969. The Codes have served the profession admirably and have been drawn upon heavily for codes and guides prepared in other parts of the world (ISSC, 1976, p. 104-106). The principles embodied by any code, however, reflect the state of knowledge at the time of its preparation, and even the most recent code is now in need of revision.

New concepts and techniques developed during the past two decades have revolutionized the earth sciences. Moreover, increasingly evident have been the limitations of previous codes in meeting some needs of Precambrian and Quaternary geology and in classification of plutonic, high-grade metamorphic, volcanic, and intensely deformed rock assemblages. In addition, the important contributions of numerous international stratigraphic organizations associated with both the International Union of Geological Sciences (IUGS) and UNESCO, including working groups of the International Geological Correlation Program (IGCP), merit recognition and incorporation into a North American code.

For these and other reasons, revision of the American Code has been undertaken by committees appointed by the North American Commission on Stratigraphic Nomenclature (NACSN). The Commission, founded as the American Commission on Stratigraphic Nomenclature in 1946 (ACSN, 1947), was renamed the NACSN in 1978 (Weiss, 1979b) to emphasize that delegates from ten organizations in Canada, the United States, and Mexico represent the geological profession throughout North America (Appendix II).

Although many past and current members of the Commission helped prepare this revision of the Code, the participation of all interested geologists has been sought (for example, Weiss, 1979a). Open forums were held at the national meetings of both the Geological Society of America at San Diego in November, 1979, and the American Association of Petroleum Geologists at Denver in June, 1980, at which comments and suggestions were offered by more than 150 geologists. The resulting draft of this report was printed, through the courtesy of the Canadian Society of Petroleum Geologists, on October 1, 1981, and additional comments were invited from the profession for a period of one year before submittal of this report to the Commission for adoption. More than 50 responses were received with sufficient suggestions for improvement to prompt moderate revision of the printed draft (NACSN, 1981). We are particularly indebted to Hollis D. Hedberg and Amos Salvador for their exhaustive and perceptive reviews of early drafts of this Code, as well as to those who responded to the request for comments. Participants in the preparation and revisions of this report, and conferees, are listed in Appendix I.

Some of the expenses incurred in the course of this work were defrayed by National Science Foundation Grant EAR 7919845, for which we express appreciation. Institutions represented by the participants have been especially generous in their support.

SCOPE

The North American Stratigraphic Code seeks to describe explicit practices for classifying and naming all formally defined geologic units. *Stratigraphic procedures* and principles, although developed initially to bring order to strata and the events recorded therein, are applicable to all earth materials, not solely to strata. They promote systematic and rigorous study of the composition, geometry, sequence, history, and genesis of rocks and unconsolidated materials. They provide the framework within which time and space relations among rock bodies that constitute the Earth are ordered systematically. Stratigraphic procedures are used not only to reconstruct the history of the Earth and of extra-terrestrial bodies, but also to define the distribution and geometry of some commodities needed by society. *Stratigraphic classification* systematically arranges and partitions bodies of rock or unconsolidated materials of the Earth's crust into units based on their inherent properties or attributes.

A *stratigraphic code* or guide is a formulation of current views on stratigraphic principles and procedures designed to promote standardized classification and formal nomenclature of rock materials. It provides the basis for formalization of the language used to denote rock units and their spatial and temporal relations. To be effective, a code must be widely accepted and used; geologic organizations and journals may adopt its recommendations for nomenclatural procedure. Because any code embodies only current concepts and principles, it should have the flexibility to provide for both changes and additions to improve its relevance to new scientific problems.

Any system of nomenclature must be sufficiently explicit to enable users to distinguish objects that are embraced in a class from those that are not. This stratigraphic code makes no attempt to systematize structural, petrographic, paleontologic, or physiographic terms. Terms from these other fields that are used as part of formal stratigraphic names should be sufficiently general as to be unaffected by revisions of precise petrographic or other classifications.

The objective of a system of classification is to promote unambiguous communication in a manner not so restrictive as to inhibit scientific progress. To minimize ambiguity, a code must promote recognition of the distinction between observable features (reproducible data) and inferences or interpretations. Moreover, it should be sufficiently adaptable and flexible to promote the further development of science.

Stratigraphic classification promotes understanding of the *geometry* and *sequence* of rock bodies. The development of stratigraphy as a science required formulation of the Law of Superposition to explain sequential stratal relations. Although superposition is not applicable to many igneous, metamorphic, and tectonic rock assemblages, other criteria (such as crosscutting relations and isotopic dating) can be used to determine sequential arrangements among rock bodies.

The term *stratigraphic unit* may be defined in several ways. Etymological emphasis requires that it be a stratum or assemblage of adjacent strata distinguished by any or several of the many properties that rocks may possess (ISSC, 1976, p. 13). The scope of stratigraphic classification and procedures, however, suggests a broader definition: a naturally occurring body of rock or rock material distinguished from adjoining rock on the basis of some stated property or properties. Commonly used properties include composition, texture, included fossils, magnetic signature, radioactivity, seismic velocity, and age. Sufficient care is required in defining the boundaries of a unit to enable others to distinguish the material body from those adjoining it. Units based on one property commonly do not coincide with those based on another and, therefore, distinctive terms are needed to identify the property used in defining each unit.

The adjective *stratigraphic* is used in two ways in the remainder of this report. In discussions of lithic (used here as synonymous with "lithologic") units, a conscious attempt is made to restrict the term to lithostratigraphic or layered rocks and sequences that obey the Law of Superposition. For nonstratiform rocks (of plutonic or tectonic origin, for example), the term *lithodemic* (see Article 27) is used. The adjective *stratigraphic* is also used in a broader sense to refer to those procedures derived from stratigraphy which are now applied to all classes of earth materials.

An assumption made in the material that follows is that the reader has some degree of familiarity with basic principles of stratigraphy as outlined, for example, by Dunbar and Rodgers (1957), Weller (1960), Shaw (1964), Matthews (1974), or the International Stratigraphic Guide (ISSC, 1976).

RELATION OF CODES TO INTERNATIONAL GUIDE

Publication of the International Stratigraphic Guide by the International Subcommission on Stratigraphic Classification (ISSC, 1976), which is being endorsed and adopted throughout the world, played a part in prompting examination of the American Stratigraphic Code and the decision to revise it.

The International Guide embodies principles and procedures that had been adopted by several national and regional stratigraphic committees and commissions. More than two decades of effort by H. D. Hedberg and other members of the Subcommission (ISSC, 1976, p. VI, 1, 3) developed the consensus required for preparation of the Guide. Although the Guide attempts to cover all kinds of rocks and the diverse ways of investigating them, it is necessarily incomplete. Mechanisms are needed to stimulate individual innovations toward promulgating new concepts, principles, and practices which subsequently may be found worthy of inclusion in later editions of the Guide. The flexibility of national and regional committees or commissions enables them to perform this function more readily than an international subcommission, even while they adopt the Guide as the international standard of stratigraphic classification.

A guiding principle in preparing this Code has been to make it as consistent as possible with the International Guide, which was endorsed by the ACSN in 1976, and at the same time to foster further innovations to meet the expanding and changing needs of earth scientists on the North American continent.

OVERVIEW

CATEGORIES RECOGNIZED

An attempt is made in this Code to strike a balance between serving the needs of those in evolving specialties and resisting the proliferation of categories of units. Consequently, more formal categories are recognized here than in previous codes or in the International Guide (ISSC, 1976). On the other hand, no special provision is made for formalizing certain kinds of units (deep oceanic, for example) which may be accommodated by available categories.

Four principal categories of units have previously been used widely in traditional stratigraphic work; these have been termed lithostratigraphic, biostratigraphic, chronostratigraphic, and geochronologic and are distinguished as follows:

1. A *lithostratigraphic unit* is a stratum or body of strata, generally but not invariably layered, generally but not invariably tabular, which conforms to the Law of Superposition and is distinguished and delimited on the basis of lithic characteristics and stratigraphic position. Example: Navajo Sandstone.

2. A *biostratigraphic unit* is a body of rock defined and characterized by its fossil content. Example: *Discoaster multiradiatus* Interval Zone.

3. A *chronostratigraphic unit* is a body of rock established to serve as the material reference for all rocks formed during the same span of time. Example: Devonian System. Each boundary of a chronostratigraphic unit is synchronous. Chronostratigraphy provides a means of organizing strata into units based on their age relations. A chronostratigraphic body also serves as the basis for defining the specific interval of geologic time, or geochronologic unit, represented by the referent.

4. A *geochronologic unit* is a division of time distinguished on the basis of the rock record preserved in a chronostratigraphic unit. Example: Devonian Period.

The first two categories are comparable in that they consist of material units defined on the basis of content. The third category differs from the first two in that it serves primarily as the standard for recognizing and isolating materials of a specific age. The fourth, in contrast, is not a material, but rather a conceptual, unit; it is a division of time. Although a geochronologic unit is not a stratigraphic body, it is so intimately tied to chronostratigraphy that the two are discussed properly together.

Properties and procedures that may be used in distinguishing geologic units are both diverse and numerous (ISSC, 1976, p. 1, 96; Harland, 1977, p. 230), but all may be assigned to the following principal classes of categories used in stratigraphic classification (Table 1), which are discussed below:

I. Material categories based on content, inherent attributes, or physical limits.

II. Categories distinguished by geologic age:
   A. Material categories used to define temporal spans, and
   B. Temporal categories.

Table 1. Categories of Units Defined*

MATERIAL CATEGORIES BASED ON CONTENT OR PHYSICAL LIMITS

Lithostratigraphic (22)
Lithodemic (31)**
Magnetopolarity (44)
Biostratigraphic (48)
Pedostratigraphic (55)
Allostratigraphic (58)

CATEGORIES EXPRESSING OR RELATED TO GEOLOGIC AGE

Material Categories Used to Define Temporal Spans
   Chronostratigraphic (66)
   Polarity-Chronostratigraphic (83)
Temporal (Non-Material) Categories
   Geochronologic (80)
   Polarity-Chronologic (88)
   Diachronic (91)
   Geochronometric (96)

*Numbers in parentheses are the numbers of the Articles where units are defined.

**Italicized categories are those introduced or developed since publication of the previous code (ACSN, 1970).

Material Categories Based on Content or Physical Limits

The basic building blocks for most geologic work are rock bodies defined on the basis of composition and related lithic characteristics, or on their physical, chemical, or biologic content or properties. Emphasis is placed on the relative objectivity and reproducibility of data used in defining units within each category.

Foremost properties of rocks are composition, texture, fabric, structure, and color, which together are designated *lithic characteristics*. These serve as the basis for distinguishing and defining the most fundamental of all formal units. Such units based primarily on composition are divided into two categories (Henderson and others, 1980): lithostratigraphic (Article 22) and lithodemic (defined here in Article 31). A lithostratigraphic unit obeys the Law of Superposition, whereas a lithodemic unit does not. A *lithodemic unit* is a defined body of predominantly intrusive, highly metamorphosed, or intensely deformed rock that, because it is intrusive or has lost primary structure through metamorphism or tectonism, generally does not conform to the Law of Superposition.

Recognition during the past several decades that remanent magnetism in rocks records the Earth's past magnetic characteristics (Cox, Doell, and Dalrymple, 1963) provides a powerful new tool encompassed by magnetostratigraphy (McDougall, 1977; McElhinny, 1978). *Magnetostratigraphy* (Article 43) is the study of remanent magnetism in rocks; it is the record of the Earth's magnetic polarity (or field reversals), dipole-field-pole position (including apparent polar wander), the non-dipole component (secular variation), and field intensity. Polarity is of particular utility and is used to define a *magnetopolarity unit* (Article 44) as a body of rock identified by its remanent magnetic polarity (ACSN, 1976; ISSC, 1979). Empirical demonstration of uniform polarity does not necessarily have direct temporal connotations because the remanent magnetism need not be related to rock deposition or crystallization. Nevertheless, polarity is a physical attribute that may characterize a body of rock.

Biologic remains contained in, or forming, strata are uniquely important in stratigraphic practice. First, they provide the means of defining and recognizing material units based on fossil content (biostratigraphic units). Second, the irreversibility of organic evolution makes it possible to partition enclosing strata temporally. Third, biologic remains provide important data for the reconstruction of ancient environments of deposition.

Composition also is important in distinguishing pedostratigraphic units. A *pedostratigraphic unit* is a body of rock that consists of one or more pedologic horizons developed in one or more lithic units now buried by a formally defined lithostratigraphic or allostratigraphic unit or units. A pedostratigraphic unit is the part of a buried soil characterized by one or more clearly defined soil horizons containing pedogenically formed minerals and organic compounds. Pedostratigraphic terminology is discussed below and in Article 55.

Many upper Cenozoic, especially Quaternary, deposits are distinguished and delineated on the basis of content, for which lithostratigraphic classification is appropriate. However, others are delineated on the basis of criteria other than content. To facilitate the reconstruction of geologic history, some compositionally similar deposits in vertical sequence merit distinction as separate stratigraphic units because they are the products of different processes; others merit distinction because they are of demonstrably different ages. Lithostratigraphic classification of these units is impractical and a new approach, allostratigraphic classification, is introduced here and may prove applicable to older deposits as well. An *allostratigraphic unit* is a mappable stratiform body of sedimentary rock defined and identified on the basis of bounding discontinuities (Article 58 and related Remarks).

Geologic-Climate units, defined in the previous Code (ACSN, 1970, p. 31), are abandoned here because they proved to be of dubious utility. Inferences regarding climate are subjective and too tenuous a basis for the definition of formal geologic units. Such inferences commonly are based on deposits assigned more appropriately to lithostratigraphic or allostratigraphic units and may be expressed in terms of diachronic units (defined below).

Categories Expressing or Related to Geologic Age

Time is a single, irreversible continuum. Nevertheless, various categories of units are used to define intervals of geologic time, just as terms having different bases, such as Paleolithic, Renaissance, and Elizabethan, are used to designate specific periods of human history. Different temporal categories are established to express intervals of time distinguished in different ways.

Major objectives of stratigraphic classification are to provide a basis for systematic ordering of the time and space relations of rock bodies and to establish a time framework for the discussion of geologic history. For such purposes, units of geologic time traditionally have been named to represent the span of time during which a well-described sequence of rock, or a chronostratigraphic unit, was deposited ("time units based on material referents," Fig. 1). This procedure continues, to the exclusion of other possible approaches, to be standard practice in studies of Phanerozoic rocks. Despite admonitions in previous American codes and the International Stratigraphic Guide (ISSC, 1976, p. 81) that similar procedures should be applied to the Precambrian, no comparable chronostratigraphic units, or geochronologic units derived therefrom, proposed for the Precambrian have yet been accepted worldwide. Instead, the IUGS Subcommission on Precambrian Stratigraphy (Sims, 1979) and its Working Groups (Harrison and Peterman, 1980) recommend division of Precambrian time into *geochronometric units* having no material referents.

A distinction is made throughout this report between *isochronous* and *synchronous*, as urged by Cumming, Fuller, and Porter (1959, p. 730), although the terms have been used synonymously by many. *Isochronous* means of equal duration; *synchronous* means simultaneous, or occurring at the same time. Although two rock bodies of very different ages may be formed during equal durations of time, the term *isochronous* is not applied to them in the earth sciences. Rather, isochronous bodies are those bounded by synchronous surfaces and formed during the same span of time. *Isochron*, in contrast, is used for a line connecting points of equal age on a graph representing physical or chemical phenomena; the line represents the same or equal time. The adjective *diachronous* is applied either to a rock unit with one or two bounding surfaces which are not synchronous, or to a boundary which is not synchronous (which "transgresses time").

Two classes of time units based on material referents, or stratotypes, are recognized (Fig. 1). The first is that of the traditional and conceptually isochronous units, and includes *geochronologic units*, which are based on *chronostratigraphic units*, and *polarity-geochronologic units*. These isochronous units have worldwide applicability and may be used even in areas lacking a material record of the named span of time. The second class of time units, newly defined in this Code, consists of

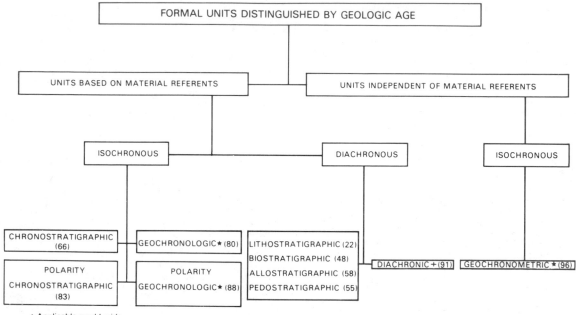

★ Applicable world-wide.
+Applicable only where material referents are present.
( )Number of article in which defined.

FIG. 1.—Relation of geologic time units to the kinds of rock-unit referents on which most are based.

diachronic units (Article 91), which are based on rock bodies known to be diachronous. In contrast to isochronous units, a diachronic term is used only where a material referent is present; a diachronic unit is coextensive with the material body or bodies on which it is based.

A *chronostratigraphic unit*, as defined above and in Article 66, is a body of rock established to serve as the material reference for all rocks formed during the same span of time; its boundaries are synchronous. It is the referent for a *geochronologic unit*, as defined above and in Article 80. Internationally accepted and traditional chronostratigraphic units were based initially on the time spans of lithostratigraphic units, biostratigraphic units, or other features of the rock record that have specific durations. In sum, they form the Standard Global Chronostratigraphic Scale (ISSC, 1976, p. 76-81; Harland, 1978), consisting of established systems and series.

A *polarity-chronostratigraphic unit* is a body of rock that contains a primary magnetopolarity record imposed when the rock was deposited or crystallized (Article 83). It serves as a material standard or referent for a part of geologic time during which the Earth's magnetic field had a characteristic polarity or sequence of polarities; that is, for a *polarity-chronologic unit* (Article 88).

A *diachronic unit* comprises the unequal spans of time represented by one or more specific diachronous rock bodies (Article 91). Such bodies may be lithostratigraphic, biostratigraphic, pedostratigraphic, allostratigraphic, or an assemblage of such units. A diachronic unit is applicable only where its material referent is present.

A *geochronometric* (or chronometric) *unit* is an isochronous direct division of geologic time expressed in years (Article 96). It has no material referent.

Pedostratigraphic Terms

The definition and nomenclature for pedostratigraphic[2] units in this Code differ from those for soil-stratigraphic units in the previous Code (ACSN, 1970, Article 18), by being more specific with regard to content, boundaries, and the basis for determining stratigraphic position.

The term "soil" has different meanings to the geologist, the soil scientist, the engineer, and the layman, and commonly has no stratigraphic significance. The term *paleosol* is currently used in North America for any soil that formed on a landscape of the past; it may be a buried soil, a relict soil, or an exhumed soil (Ruhe, 1965; Valentine and Dalrymple, 1976).

A *pedologic soil* is composed of one or more soil horizons.[3] A *soil horizon* is a layer within a pedologic soil that (1) is approximately parallel to the soil surface, (2) has distinctive physical, chemical, biological, and morphological properties that differ from those of adjacent, genetically related, soil horizons, and (3) is distinguished from other soil horizons by objective compositional properties that can be observed or measured in the field. The physical boundaries of buried pedologic horizons are objective traceable boundaries with stratigraphic significance. A buried pedologic soil provides the material basis for definition of a stratigraphic unit in pedostratigraphic classification (Article 55), but a buried pedologic soil may be somewhat more inclusive than a pedostratigraphic unit. A pedologic soil may contain both an

---

[2]From Greek, *pedon*, ground or soil.
[3]As used in a geological sense, a *horizon* is a surface or line. In pedology, however, it is a body of material, and such usage is continued here.

0-horizon and the entire C-horizon (Fig. 6), whereas the former is excluded and the latter need not be included in a pedostratigraphic unit.

The definition and nomenclature for pedostratigraphic units in this Code differ from those of soil stratigraphic units proposed by the International Union for Quaternary Research and International Society of Soil Science (Parsons, 1981). The pedostratigraphic unit, geosol, also differs from the proposed INQUA-ISSS soil-stratigraphic unit, pedoderm, in several ways, the most important of which are: (1) a geosol may be in any part of the geologic column, whereas a pedoderm is a surficial soil; (2) a geosol is a buried soil, whereas a pedoderm may be a buried, relict, or exhumed soil; (3) the boundaries and stratigraphic position of a geosol are defined and delineated by criteria that differ from those for a pedoderm; and (4) a geosol may be either all or only a part of a buried soil, whereas a pedoderm is the entire soil.

The term *geosol*, as defined by Morrison (1967, p. 3), is a laterally traceable, mappable, geologic weathering profile that has a consistent stratigraphic position. The term is adopted and redefined here as the fundamental and only unit in formal pedostratigraphic classification (Article 56).

FORMAL AND INFORMAL UNITS

Although the emphasis in this Code is necessarily on formal categories of geologic units, informal nomenclature is highly useful in stratigraphic work.

Formally named units are those that are named in accordance with an established scheme of classification; the fact of formality is conveyed by capitalization of the initial letter of the *rank* or *unit* term (for example, Morrison Formation). Informal units, whose unit terms are ordinary nouns, are not protected by the stability provided by proper formalization and recommended classification procedures. Informal terms are devised for both economic and scientific reasons. Formalization is appropriate for those units requiring stability of nomenclature, particularly those likely to be extended far beyond the locality in which they were first recognized. Informal terms are appropriate for casually mentioned, innovative, and most economic units, those defined by unconventional criteria, and those that may be too thin to map at usual scales.

Casually mentioned geologic units not defined in accordance with this Code are informal. For many of these, there may be insufficient need or information, or perhaps an inappropriate basis, for formal designations. Informal designations as beds or lithozones (the pebbly beds, the shaly zone, third coal) are appropriate for many such units.

Most economic units, such as aquifers, oil sands, coal beds, quarry layers, and ore-bearing "reefs," are informal, even though they may be named. Some such units, however, are so significant scientifically and economically that they merit formal recognition as beds, members, or formations.

Innovative approaches in regional stratigraphic studies have resulted in the recognition and definition of units best left as informal, at least for the time being. Units bounded by major regional unconformities on the North American craton were designated "sequences" (example: Sauk sequence) by Sloss (1963). Major unconformity-bounded units also were designated "synthems" by Chang (1975), who recommended that they be treated formally. Marker-defined units that are continuous from one lithofacies to another were designated "formats" by Forgotson (1957). The term "chronosome" was proposed by Schultz (1982) for rocks of diverse facies corresponding to geographic variations in sedimentation during an interval of deposition identified on the basis of bounding stratigraphic markers. Successions of faunal zones containing evolutionarily

related forms, but bounded by non-evolutionary biotic discontinuities, were termed "biomeres" (Palmer, 1965). The foregoing are only a few selected examples to demonstrate how informality provides a continuing avenue for innovation.

The terms *magnafacies* and *parvafacies*, coined by Caster (1934) to emphasize the distinction between lithostratigraphic and chronostratigraphic units in sequences displaying marked facies variation, have remained informal despite their impact on clarifying the concepts involved.

Tephrochronologic studies provide examples of informal units too thin to map at conventional scales but yet invaluable for dating important geologic events. Although some such units are named for physiographic features and places where first recognized (e.g., Guaje pumice bed, where it is not mapped as the Guaje Member of the Bandelier Tuff), others bear the same name as the volcanic vent (e.g., Huckleberry Ridge ash bed of Izett and Wilcox, 1981).

Informal geologic units are designated by ordinary nouns, adjectives or geographic terms and lithic or unit-terms that are not capitalized (chalky formation or beds, St. Francis coal).

No geologic unit should be established and defined, whether formally or informally, unless its recognition serves a clear purpose.

CORRELATION

Correlation is a procedure for demonstrating correspondence between geographically separated parts of a geologic unit. The term is a general one having diverse meanings in different disciplines. Demonstration of temporal correspondence is one of the most important objectives of stratigraphy. The term "correlation" frequently is misused to express the idea that a unit has been identified or recognized.

Correlation is used in this Code as the demonstration of correspondence between two geologic units in both some defined property and relative stratigraphic position. Because correspondence may be based on various properties, three kinds of correlation are best distinguished by more specific terms. *Lithocorrelation* links units of similar lithology and stratigraphic position (or sequential or geometric relation, for lithodemic units). *Biocorrelation* expresses similarity of fossil content and biostratigraphic position. *Chronocorrelation* expresses correspondence in age and in chronostratigraphic position.

Other terms that have been used for the similarity of content and stratal succession are homotaxy and chronotaxy. *Homotaxy* is the similarity in separate regions of the serial arrangement or succession of strata of comparable compositions or of included fossils. The term is derived from *homotaxis*, proposed by Huxley (1862, p. xlvi) to emphasize that similarity in succession does not prove age equivalence of comparable units. The term *chronotaxy* has been applied to similar stratigraphic sequences composed of units which are of equivalent age (Henbest, 1952, p. 310).

Criteria used for ascertaining temporal and other types of correspondence are diverse (ISSC, 1976, p. 86-93) and new criteria will emerge in the future. Evolving statistical tests, as well as isotopic and paleomagnetic techniques, complement the traditional paleontologic and lithologic procedures. Boundaries defined by one set of criteria need not correspond to those defined by others.

PART II. ARTICLES

INTRODUCTION

Article 1.—**Purpose.** This Code describes explicit stratigraphic procedures for classifying and naming geologic units accorded formal status. Such procedures, if widely adopted, assure consistent and uniform usage in classification and terminology and therefore promote unambiguous communication.

Article 2.—**Categories.** Categories of formal stratigraphic units, though diverse, are of three classes (Table 1). The first class is of rock-material categories based on inherent attributes or content and stratigraphic position, and includes lithostratigraphic, lithodemic, magnetopolarity, biostratigraphic, pedostratigraphic, and allostratigraphic units. The second class is of material categories used as standards for defining spans of geologic time, and includes chronostratigraphic and polarity-chronostratigraphic units. The third class is of non-material temporal categories, and includes geochronologic, polarity-chronologic, geochronometric, and diachronic units.

GENERAL PROCEDURES

DEFINITION OF FORMAL UNITS

Article 3.—**Requirements for Formally Named Geologic Units.** Naming, establishing, revising, redefining, and abandoning formal geologic units require publication in a recognized scientific medium of a comprehensive statement which includes: (i) intent to designate or modify a formal unit; (ii) designation of category and rank of unit; (iii) selection and derivation of name; (iv) specification of stratotype (where applicable); (v) description of unit; (vi) definition of boundaries; (vii) historical background; (viii) dimensions, shape, and other regional aspects; (ix) geologic age; (x) correlations; and possibly (xi) genesis (where applicable). These requirements apply to subsurface and offshore, as well as exposed, units.

Article 4.—**Publication.**[4] "Publication in a recognized scientific medium" in conformance with this Code means that a work, when first issued, must (1) be reproduced in ink on paper or by some method that assures numerous identical copies and wide distribution; (2) be issued for the purpose of scientific, public, permanent record; and (3) be readily obtainable by purchase or free distribution.

Remarks. (a) **Inadequate publication.**—The following do not constitute publication within the meaning of the Code: (1) distribution of microfilms, microcards, or matter reproduced by similar methods; (2) distribution to colleagues or students of a note, even if printed, in explanation of an accompanying illustration; (3) distribution of proof sheets; (4) open-file release; (5) theses, dissertations, and dissertation abstracts; (6) mention at a scientific or other meeting; (7) mention in an abstract, map explanation, or figure caption; (8) labeling of a rock specimen in a collection; (9) mere deposit of a document in a library; (10) anonymous publication; or (11) mention in the popular press or in a legal document.
(b). **Guidebooks.**—A guidebook with distribution limited to participants of a field excursion does not meet the test of availability. Some organizations publish and distribute widely large editions of serial guidebooks that include refereed regional papers; although these do meet the tests of scientific purpose and availability, and therefore constitute valid publication, other media are preferable.

Article 5.—**Intent and Utility.** To be valid, a new unit must serve a clear purpose and be duly proposed and duly described, and the intent to establish it must be specified. Casual mention of a unit, such as "the granite exposed near the Middleville

schoolhouse," does not establish a new formal unit, nor does mere use in a table, columnar section, or map.

Remark. (a) **Demonstration of purpose served.**—The initial definition or revision of a named geologic unit constitutes, in essence, a proposal. As such, it lacks status until use by others demonstrates that a clear purpose has been served. A unit becomes established through repeated demonstration of its utility. The decision not to use a newly proposed or a newly revised term requires a full discussion of its unsuitability.

Article 6.—**Category and Rank.** The category and rank of a new or revised unit must be specified.

Remark. (a) **Need for specification.**—Many stratigraphic controversies have arisen from confusion or misinterpretation of the category of a unit (for example, lithostratigraphic vs. chronostratigraphic). Specification and unambiguous description of the category is of paramount importance. Selection and designation of an appropriate rank from the distinctive terminology developed for each category help serve this function (Table 2).

Article 7.—**Name.** The name of a formal geologic unit is compound. For most categories, the name of a unit should consist of a geographic name combined with an appropriate rank (Wasatch Formation) or descriptive term (Viola Limestone). Biostratigraphic units are designated by appropriate biologic forms (*Exus albus* Assemblage Biozone). Worldwide chronostratigraphic units bear long established and generally accepted names of diverse origins (Triassic System). The first letters of all words used in the names of formal geologic units are capitalized (except for the trivial species and subspecies terms in the name of a biostratigraphic unit).

Remarks. (a) **Appropriate geographic terms.**—Geographic names derived from permanent natural or artificial features at or near which the unit is present are preferable to those derived from impermanent features such as farms, schools, stores, churches, crossroads, and small communities. Appropriate names may be selected from those shown on topographic, state, provincial, county, forest service, hydrographic, or comparable maps, particularly those showing names approved by a national board for geographic names. The generic part of a geographic name, e.g., river, lake, village, should be omitted from new terms, unless required to distinguish between two otherwise identical names (e.g., Redstone Formation and Redstone River Formation). Two names should not be derived from the same geographic feature. A unit should not be named for the source of its components; for example, a deposit inferred to have been derived from the Keewatin glaciation center should not be designated the "Keewatin Till."
(b) **Duplication of names.**—Responsibility for avoiding duplication, either in use of the same name for different units (homonymy) or in use of different names for the same unit (synonymy), rests with the proposer. Although the same geographic term has been applied to different categories of units (example: the lithostratigraphic Word Formation and the chronostratigraphic Wordian Stage) now entrenched in the literature, the practice is undesirable. The extensive geologic nomenclature of North America, including not only names but also nomenclatural history of formal units, is recorded in compendia maintained by the Committee on Stratigraphic Nomenclature of the Geological Survey of Canada, Ottawa, Ontario; by the Geologic Names Committee of the United States Geological Survey, Reston, Virginia; by the Instituto de Geología, Ciudad Universitaria, México, D.F.; and by many state and provincial geological surveys. These organizations respond to inquiries regarding the availability of names, and some are prepared to reserve names for units likely to be defined in the next year or two.
(c) **Priority and preservation of established names.**—Stability of nomenclature is maintained by use of the rule of priority and by preservation of well-established names. Names should not be modified without explaining the need. Priority in publication is to be respected, but priority alone does not justify displacing a well-established name by one

---

[4]This article is modified slightly from a statement by the International Commission of Zoological Nomenclature (1964, p. 7-9).

Table 2. Categories and Ranks of Units Defined in This Code*

A. Material Units

LITHOSTRATIGRAPHIC	LITHODEMIC	MAGNETOPOLARITY	BIOSTRATIGRAPHIC	PEDOSTRATIGRAPHIC	ALLOSTRATIGRAPHIC
Supergroup	Supersuite (Complex)				
Group	Suite (Complex)	Polarity Superzone			Allogroup
Formation	*Lithodeme*	*Polarity zone*	*Biozone* (Interval, Assemblage or Abundance)	*Geosol*	*Alloformation*
Member (or Lens, or Tongue)		Polarity Subzone	Subbiozone		Allomember
Bed(s) or Flow(s)					

B. Temporal and Related Chronostratigraphic Units

CHRONO-STRATIGRAPHIC	GEOCHRONOLOGIC GEOCHRONOMETRIC	POLARITY CHRONO-STRATIGRAPHIC	POLARITY CHRONOLOGIC	DIACHRONIC
Eonothem	Eon	Polarity Superchronozone	Polarity Superchron	
Erathem (Supersystem)	Era (Superperiod)			
System (Subsystem)	*Period* (Subperiod)	*Polarity Chronozone*	*Polarity Chron*	*Episode*
Series	Epoch			Diachron: Phase
Stage (Substage)	Age (Subage)	Polarity Subchronozone	Polarity Subchron	Diachron: Span
Chronozone	Chron			Diachron: Cline

*Fundamental units are italicized.

neither well-known nor commonly used; nor should an inadequately established name be preserved merely on the basis of priority. Redefinitions in precise terms are preferable to abandonment of the names of well-established units which may have been defined imprecisely but nonetheless in conformance with older and less stringent standards.

(d) **Differences of spelling and changes in name.**—The geographic component of a well-established stratigraphic name is not changed due to differences in spelling or changes in the name of a geographic feature. The name Bennett Shale, for example, used for more than half a century, need not be altered because the town is named Bennet. Nor should the Mauch Chunk Formation be changed because the town has been renamed Jim Thorpe. Disappearance of an impermanent geographic feature, such as a town, does not affect the name of an established geologic unit.

(e) **Names in different countries and different languages.**—For geologic units that cross local and international boundaries, a single name for each is preferable to several. Spelling of a geographic name commonly conforms to the usage of the country and linguistic group involved. Although geographic names are not translated (Cuchillo is not translated to Knife), lithologic or rank terms are (Edwards Limestone, Caliza Edwards; Formación La Casita, La Casita Formation).

Article 8.—**Stratotypes.** The designation of a unit or boundary stratotype (type section or type locality) is essential in the definition of most formal geologic units. Many kinds of units are best defined by reference to an accessible and specific sequence of rock that may be examined and studied by others. A stratotype is the standard (original or subsequently designated) for a named geologic unit or boundary and constitutes the basis for definition or recognition of that unit or boundary: therefore, it must be illustrative and representative of the concept of the unit or boundary being defined.

Remarks. (a) **Unit stratotypes.**—A unit stratotype is the type section for a stratiform deposit or the type area for a nonstratiform body that serves as the standard for definition and recognition of a geologic unit. The upper and lower limits of a unit stratotype are designated points in a specific sequence or locality and serve as the standards for definition and recognition of a stratigraphic unit's boundaries.

(b) **Boundary stratotype.**—A boundary stratotype is the type locality for the boundary reference point for a stratigraphic unit. Both boundary stratotypes for any unit need not be in the same section or region. Each boundary stratotype serves as the standard for definition and recognition of the base of a stratigraphic unit. The top of a unit may be defined by the boundary stratotype of the next higher stratigraphic unit.

(c) **Type locality.**—A type locality is the specified geographic locality where the stratotype of a formal unit or unit boundary was originally defined and named. A type area is the geographic territory encompassing the type locality. Before the concept of a stratotype was developed, only type localities and areas were designated for many geologic units which are now long- and well-established. Stratotypes, though now mandatory in defining most stratiform units, are impractical in definitions of many large nonstratiform rock bodies whose diverse major components may be best displayed at several reference localities.

(d) **Composite-stratotype.**—A composite-stratotype consists of several reference sections (which may include a type section) required to demonstrate the range or totality of a stratigraphic unit.

(e) **Reference sections.**—Reference sections may serve as invaluable standards in definitions or revisions of formal geologic units. For those well-established stratigraphic units for which a type section never was specified, a principal reference section (lectostratotype of ISSC, 1976, p. 26) may be designated. A principal reference section (neostratotype of ISSC, 1976, p. 26) also may be designated for those units or boundaries whose stratotypes have been destroyed, covered, or otherwise made inaccessible. Supplementary reference sections often are designated to illustrate the diversity or heterogeneity of a defined unit or some critical feature not evident or exposed in the stratotype. Once a unit or boundary stratotype section is designated, it is never abandoned or changed; however, if a stratotype proves inadequate, it may be supplemented by a principal reference section or by several reference sections that may constitute a composite-stratotype.

(f) **Stratotype descriptions.**—Stratotypes should be described both geographically and geologically. Sufficient geographic detail must be included to enable others to find the stratotype in the field, and may consist of maps and/or aerial photographs showing location and access, as well as appropriate coordinates or bearings. Geologic information should include thickness, descriptive criteria appropriate to the recognition of the unit and its boundaries, and discussion of the relation of the unit to other geologic units of the area. A carefully measured and described section provides the best foundation for definition of stratiform units. Graphic profiles, columnar sections, structure-sections, and photographs are useful supplements to a description; a geologic map of the area including the type locality is essential.

Article 9.—**Unit Description.** A unit proposed for formal status should be described and defined so clearly that any subsequent investigator can recognize that unit unequivocally. Distinguishing features that characterize a unit may include any or several of the following: composition, texture, primary structures, structural attitudes, biologic remains, readily apparent mineral composition (e.g., calcite vs. dolomite), geochemistry, geophysical properties (including magnetic signatures), geomorphic expression, unconformable or cross-cutting relations, and age. Although all distinguishing features pertinent to the unit category should be described sufficiently to characterize the unit, those not pertinent to the category (such as age and inferred genesis for lithostratigraphic units, or lithology for biostratigraphic units) should not be made part of the definition.

Article 10.—**Boundaries.** The criteria specified for the recognition of boundaries between adjoining geologic units are of paramount importance because they provide the basis for scientific reproducibility of results. Care is required in describing the criteria, which must be appropriate to the category of unit involved.

Remarks. (a) **Boundaries between intergradational units.**—Contacts between rocks of markedly contrasting composition are appropriate boundaries of lithic units, but some rocks grade into, or intertongue with, others of different lithology. Consequently, some boundaries are necessarily arbitrary as, for example, the top of the uppermost limestone in a sequence of interbedded limestone and shale. Such arbitrary boundaries commonly are diachronous.

(b) **Overlaps and gaps.**—The problem of overlaps and gaps between long-established adjacent chronostratigraphic units is being addressed by international IUGS and IGCP working groups appointed to deal with various parts of the geologic column. The procedure recommended by the Geological Society of London (George and others, 1969; Holland and others, 1978), of defining only the basal boundaries of chronostratigraphic units, has been widely adopted (e.g., McLaren, 1977) to resolve the problem. Such boundaries are defined by a carefully selected and agreed-upon boundary-stratotype (marker-point type section or "golden spike") which becomes the standard for the base of a chronostratigraphic unit. The concept of the mutual-boundary stratotype (ISSC, 1976, p. 84-86), based on the assumption of continuous deposition in selected sequences, also has been used to define chronostratigraphic units.

Although international chronostratigraphic units of series and higher rank are being redefined by IUGS and IGCP working groups, there may be a continuing need for some provincial series. Adoption of the basal boundary-stratotype concept is urged.

Article 11.—**Historical Background.** A proposal for a new name must include a nomenclatorial history of rocks assigned to the proposed unit, describing how they were treated previously and by whom (references), as well as such matters as priorities, possible synonymy, and other pertinent considerations. Consideration of the historical background of an older unit commonly provides the basis for justifying definition of a new unit.

Article 12.—**Dimensions and Regional Relations.** A perspective on the magnitude of a unit should be provided by such information as may be available on the geographic extent of a unit; observed ranges in thickness, composition, and geomorphic expression; relations to other kinds and ranks of stratigraphic units; correlations with other nearby sequences; and the bases for recognizing and extending the unit beyond the type locality. If the unit is not known anywhere but in an area of limited extent, informal designation is recommended.

Article 13.—**Age.** For most formal material geologic units, other than chronostratigraphic and polarity-chronostratigraphic, inferences regarding geologic age play no proper role in their definition. Nevertheless, the age, as well as the basis for its assignment, are important features of the unit and should be stated. For many lithodemic units, the age of the protolith should be distinguished from that of the metamorphism or deformation. If the basis for assigning an age is tenuous, a doubt should be expressed.

Remarks. (a) **Dating.**—The geochronologic ordering of the rock record, whether in terms of radioactive-decay rates or other processes, is generally called "dating." However, the use of the noun "date" to mean "isotopic age" is not recommended. Similarly, the term "absolute age" should be suppressed in favor of "isotopic age" for an age determined on the basis of isotopic ratios. The more inclusive term "numerical age" is recommended for all ages determined from isotopic ratios, fission tracks, and other quantifiable age-related phenomena.

(b) **Calibration**—The dating of chronostratigraphic boundaries in terms of numerical ages is a special form of dating for which the word "calibration" should be used. The geochronologic time-scale now in use has been developed mainly through such calibration of chronostratigraphic sequences.

(c) **Convention and abbreviations.**—The age of a stratigraphic unit or the time of a geologic event, as commonly determined by numerical dating or by reference to a calibrated time-scale, may be expressed in years before the present. The unit of time is the modern year as presently recognized worldwide. Recommended (but not mandatory) abbreviations for such ages are SI (International System of Units) multipliers coupled with "a" for annum: ka, Ma, and Ga[5] for kilo-annum (10^3 years), Mega-annum (10^6 years), and Giga-annum (10^9 years), respectively. Use of these terms after the age value follows the convention established in the field of C-14 dating. The "present" refers to 1950 AD, and such qualifiers as "ago" or "before the present" are omitted after the value because measurement of the duration from the present to the past is implicit in the designation. In contrast, the duration of a remote interval of geologic time, as a number of years, should not be expressed by the same symbols. Abbreviations for numbers of years, without reference to the present, are informal (e.g., y or yr for years; my, m.y., or m.yr. for

---

[5]Note that the initial letters of Mega- and Giga- are capitalized, but that of kilo- is not, by SI convention.

millions of years; and so forth, as preference dictates). For example, boundaries of the Late Cretaceous Epoch currently are calibrated at 63 Ma and 96 Ma, but the interval of time represented by this epoch is 33 m.y.

(d) **Expression of "age" of lithodemic units.**—The adjectives "early," "middle," and "late" should be used with the appropriate geochronologic term to designate the age of lithodemic units. For example, a granite dated isotopically at 510 Ma should be referred to using the geochronologic term "Late Cambrian granite" rather than either the chronostratigraphic term "Upper Cambrian granite" or the more cumbersome designation "granite of Late Cambrian age."

Article 14.—Correlation. Information regarding spatial and temporal counterparts of a newly defined unit beyond the type area provides readers with an enlarged perspective. Discussions of criteria used in correlating a unit with those in other areas should make clear the distinction between data and inferences.

Article 15.—Genesis. Objective data are used to define and classify geologic units and to express their spatial and temporal relations. Although many of the categories defined in this Code (e.g., lithostratigraphic group, plutonic suite) have genetic connotations, inferences regarding geologic history or specific environments of formation may play no proper role in the definition of a unit. However, observations, as well as inferences, that bear on genesis are of great interest to readers and should be discussed.

Article 16.—Subsurface and Subsea Units. The foregoing procedures for establishing formal geologic units apply also to subsurface and offshore or subsea units. Complete lithologic and paleontologic descriptions or logs of the samples or cores are required in written or graphic form, or both. Boundaries and divisions, if any, of the unit should be indicated clearly with their depths from an established datum.

Remarks. (a) **Naming subsurface units.**—A subsurface unit may be named for the borehole (Eagle Mills Formation), oil field (Smackover Limestone), or mine which is intended to serve as the stratotype, or for a nearby geographic feature. The hole or mine should be located precisely, both with map and exact geographic coordinates, and identified fully (operator or company, farm or lease block, dates drilled or mined, surface elevation and total depth, etc).

(b) **Additional recommendations.**—Inclusion of appropriate borehole geophysical logs is urged. Moreover, rock and fossil samples and cores and all pertinent accompanying materials should be stored, and available for examination, at appropriate federal, state, provincial, university, or museum depositories. For offshore or subsea units (Clipperton Formation of Tracey and others, 1971, p. 22; Argo Salt of McIver, 1972, p. 57), the names of the project and vessel, depth of sea floor, and pertinent regional sampling and geophysical data should be added.

(c) **Seismostratigraphic units.**—High-resolution seismic methods now can delineate stratal geometry and continuity at a level of confidence not previously attainable. Accordingly, seismic surveys have come to be the principal adjunct of the drill in subsurface exploration. On the other hand, the method identifies rock types only broadly and by inference. Thus, formalization of units known only from seismic profiles is inappropriate. Once the stratigraphy is calibrated by drilling, the seismic method may provide objective well-to-well correlations.

REVISION AND ABANDONMENT OF FORMAL UNITS

Article 17.—Requirements for Major Changes. Formally defined and named geologic units may be redefined, revised, or abandoned, but revision and abandonment require as much justification as establishment of a new unit.

Remark. (a) **Distinction between redefinition and revision.**—Redefinition of a unit involves changing the view or emphasis on the content of the unit without changing the boundaries or rank, and differs only slightly from redescription. Neither redefinition nor redescription is considered revision. A redescription corrects an inadequate or inaccurate description, whereas a redefinition may change a descriptive (for example, lithologic) designation. Revision involves either minor changes in the definition of one or both boundaries or in the rank of a unit (normally, elevation to a higher rank). Correction of a misidentification of a unit outside its type area is neither redefinition nor revision.

Article 18.—Redefinition. A correction or change in the descriptive term applied to a stratigraphic or lithodemic unit is a redefinition which does not require a new geographic term.

Remarks. (a) **Change in lithic designation.**—Priority should not prevent more exact lithic designation if the original designation is not everywhere applicable; for example, the Niobrara Chalk changes gradually westward to a unit in which shale is prominent, for which the designation "Niobrara Shale" or "Formation" is more appropriate. Many carbonate formations originally designated "limestone" or "dolomite" are found to be geographically inconsistent as to prevailing rock type. The appropriate lithic term or "formation" is again preferable for such units.

(b) **Original lithic designation inappropriate.**—Restudy of some long-established lithostratigraphic units has shown that the original lithic designation was incorrect according to modern criteria; for example, some "shales" have the chemical and mineralogical composition of limestone, and some rocks described as felsic lavas now are understood to be welded tuffs. Such new knowledge is recognized by changing the lithic designation of the unit, while retaining the original geographic term. Similarly, changes in the classification of igneous rocks have resulted in recognition that rocks originally described as quartz monzonite now are more appropriately termed granite. Such lithic designations may be modernized when the new classification is widely adopted. If heterogeneous bodies of plutonic rock have been misleadingly identified with a single compositional term, such as "gabbro," the adoption of a neutral term, such as "intrusion" or "pluton," may be advisable.

Article 19.—Revision. Revision involves either minor changes in the definition of one or both boundaries of a unit, or in the unit's rank.

Remarks. (a) **Boundary change.**—Revision is justifiable if a minor change in boundary or content will make a unit more natural and useful. If revision modifies only a minor part of the content of a previously established unit, the original name may be retained.

(b) **Change in rank.**—Change in rank of a stratigraphic or temporal unit requires neither redefinition of its boundaries nor alteration of the geographic part of its name. A member may become a formation or vice versa, a formation may become a group or vice versa, and a lithodeme may become a suite or vice versa.

(c) **Examples of changes from area to area.**—The Conasauga Shale is recognized as a formation in Georgia and as a group in eastern Tennessee; the Osgood Formation, Laurel Limestone, and Waldron Shale in Indiana are classed as members of the Wayne Formation in a part of Tennessee; the Virgelle Sandstone is a formation in western Montana and a member of the Eagle Sandstone in central Montana; the Skull Creek Shale and the Newcastle Sandstone in North Dakota are members of the Ashville Formation in Manitoba.

(d) **Example of change in single area.**—The rank of a unit may be changed without changing its content. For example, the Madison Limestone of early work in Montana later became the Madison Group, containing several formations.

(e) **Retention of type section.**—When the rank of a geologic unit is changed, the original type section or type locality is retained for the newly ranked unit (see Article 22c).

(f) **Different geographic name for a unit and its parts.**—In changing the rank of a unit, the same name may not be applied both to the unit as a whole and to a part of it. For example, the Astoria Group should not contain an Astoria Sandstone, nor the Washington Formation, a Washington Sandstone Member.

(g) **Undesirable restriction.**—When a unit is divided into two or more of the same rank as the original, the original name should not be used for any of the divisions. Retention of the old name for one of the units precludes use of the name in a term of higher rank. Furthermore, in order to understand an author's meaning, a later reader would have to know about the modification and its date, and whether the author is following the original or the modified usage. For these reasons, the normal practice is to raise the rank of an established unit when units of the same rank are recognized and mapped within it.

Article 20.—Abandonment. An improperly defined or obsolete stratigraphic, lithodemic, or temporal unit may be formally abandoned, provided that (a) sufficient justification is presented to demonstrate a concern for nomenclatural stability, and (b) recommendations are made for the classification and nomenclature to be used in its place.

Remarks. (a) **Reasons for abandonment.**—A formally defined unit may be abandoned by the demonstration of synonymy or homonymy, of assignment to an improper category (for example, definition of a lithostratigraphic unit in a chronostratigraphic sense), or of other direct violations of a stratigraphic code or procedures prevailing at the time of the original definition. Disuse, or the lack of need or useful purpose for a unit, may be a basis for abandonment; so, too, may widespread misuse in diverse ways which compound confusion. A unit also may be abandoned if it proves impracticable, neither recognizable nor mappable elsewhere.

(b) **Abandoned names.**—A name for a lithostratigraphic or lithodemic unit, once applied and then abandoned, is available for some other unit only if the name was introduced casually, or if it has been published only once in the last several decades and is not in current usage, and if its reintroduction will cause no confusion. An explanation of the history of the name and of the new usage should be a part of the designation.

(c) **Obsolete names.**—Authors may refer to national and provincial records of stratigraphic names to determine whether a name is obsolete (see Article 7b).

(d) **Reference to abandoned names.**—When it is useful to refer to an obsolete or abandoned formal name, its status is made clear by some such term as "abandoned" or "obsolete," and by using a phrase such as "La Plata Sandstone of Cross (1898)". (The same phrase also is used to convey that a named unit has not yet been adopted for usage by the organization involved.)

(e) **Reinstatement.**—A name abandoned for reasons that seem valid at the time, but which subsequently are found to be erroneous, may be reinstated. Example: the Washakie Formation, defined in 1869, was abandoned in 1918 and reinstated in 1973.

CODE AMENDMENT

Article 21.—Procedure for Amendment. Additions to, or changes of, this Code may be proposed in writing to the Commission by any geoscientist at any time. If accepted for consideration by a majority vote of the Commission, they may be adopted by a two-thirds vote of the Commission at an annual meeting not less than a year after publication of the proposal.

FORMAL UNITS DISTINGUISHED BY CONTENT, PROPERTIES, OR PHYSICAL LIMITS

LITHOSTRATIGRAPHIC UNITS

Nature and Boundaries

Article 22.—Nature of Lithostratigraphic Units. A lithostratigraphic unit is a defined body of sedimentary, extrusive igneous, metasedimentary, or metavolcanic strata which is distinguished and delimited on the basis of lithic characteristics and stratigraphic position. A lithostratigraphic unit generally conforms to the Law of Superposition and commonly is stratified and tabular in form.

Remarks. (a) **Basic units.**—Lithostratigraphic units are the basic units of general geologic work and serve as the foundation for delineating strata, local and regional structure, economic resources, and geologic history in regions of stratified rocks. They are recognized and defined by observable rock characteristics; boundaries may be placed at clearly distinguished contacts or drawn arbitrarily within a zone of gradation. Lithification or cementation is not a necessary property; clay, gravel, till, and other unconsolidated deposits may constitute valid lithostratigraphic units.

(b) **Type section and locality.**—The definition of a lithostratigraphic unit should be based, if possible, on a stratotype consisting of readily accessible rocks in place, e.g., in outcrops, excavations, and mines, or of rocks accessible only to remote sampling devices, such as those in drill holes and underwater. Even where remote methods are used, definitions must be based on lithic criteria and not on the geophysical characteristics of the rocks, nor the implied age of their contained fossils. Definitions must be based on descriptions of actual rock material. Regional validity must be demonstrated for all such units. In regions where the stratigraphy has been established through studies of surface exposures, the naming of new units in the subsurface is justified only where the subsurface section differs materially from the surface section, or where there is doubt as to the equivalence of a subsurface and a surface unit. The establishment of subsurface reference sections for units originally defined in outcrop is encouraged.

(c) **Type section never changed.**—The definition and name of a lithostratigraphic unit are established at a type section (or locality) that, once specified, must not be changed. If the type section is poorly designated or delimited, it may be redefined subsequently. If the originally specified stratotype is incomplete, poorly exposed, structurally complicated, or unrepresentative of the unit, a principal reference section or several reference sections may be designated to supplement, but not to supplant, the type section (Article 8e).

(d) **Independence from inferred geologic history.**—Inferred geologic history, depositional environment, and biological sequence have no place in the definition of a lithostratigraphic unit, which must be based on composition and other lithic characteristics; nevertheless, considerations of well-documented geologic history properly may influence the choice of vertical and lateral boundaries of a new unit. Fossils may be valuable during mapping in distinguishing between two lithologically similar, non-contiguous lithostratigraphic units. The fossil content of a lithostratigraphic unit is a legitimate lithic characteristic; for example, oyster-rich sandstone, coquina, coral reef, or graptolitic shale. Moreover, otherwise similar units, such as the Formación Mendez and Formación Velasco mudstones, may be distinguished on the basis of coarseness of contained fossils (foraminifera).

(e) **Independence from time concepts.**—The boundaries of most lithostratigraphic units may transgress time horizons, but some may be approximately synchronous. Inferred time-spans, however measured, play no part in differentiating or determining the boundaries of any lithostratigraphic unit. Either relatively short or relatively long intervals of time may be represented by a single unit. The accumulation of material assigned to a particular unit may have begun or ended earlier in some localities than in others; also, removal of rock by erosion, either within the time-span of deposition of the unit or later, may reduce the time-span represented by the unit locally. The body in some places may be entirely younger than in other places. On the other hand, the establishment of formal units that straddle known, identifiable, regional disconformities is to be avoided, if at all possible. Although concepts of time or age play no part in defining lithostratigraphic units nor in determining their boundaries, evidence of age may aid recognition of similar lithostratigraphic units at localities far removed from the type sections or areas.

(f) **Surface form.**—Erosional morphology or secondary surface form may be a factor in the recognition of a lithostratigraphic unit, but properly should play a minor part at most in the definition of such units.

Because the surface expression of lithostratigraphic units is an important aid in mapping, it is commonly advisable, where other factors do not countervail, to define lithostratigraphic boundaries so as to coincide with lithic changes that are expressed in topography.

(g) **Economically exploited units.**—Aquifers, oil sands, coal beds, and quarry layers are, in general, informal units even though named. Some such units, however, may be recognized formally as beds, members, or formations because they are important in the elucidation of regional stratigraphy.

(h) **Instrumentally defined units.**—In subsurface investigations, certain bodies of rock and their boundaries are widely recognized on borehole geophysical logs showing their electrical resistivity, radioactivity, density, or other physical properties. Such bodies and their boundaries may or may not correspond to formal lithostratigraphic units and their boundaries. Where other considerations do not countervail, the boundaries of subsurface units should be defined so as to correspond to useful geophysical markers; nevertheless, units defined exclusively on the basis of remotely sensed physical properties, although commonly useful in stratigraphic analysis, stand completely apart from the hierarchy of formal lithostratigraphic units and are considered informal.

(i) **Zone.**—As applied to the designation of lithostratigraphic units, the term "zone" is informal. Examples are "producing zone," "mineralized zone," "metamorphic zone," and "heavy-mineral zone." A zone may include all or parts of a bed, a member, a formation, or even a group.

(j) **Cyclothems.**—Cyclic or rhythmic sequences of sedimentary rocks, whose repetitive divisions have been named cyclothems, have been recognized in sedimentary basins around the world. Some cyclothems have been identified by geographic names, but such names are considered informal. A clear distinction must be maintained between the division of a stratigraphic column into cyclothems and its division into groups, formations, and members. Where a cyclothem is identified by a geographic name, the word *cyclothem* should be part of the name, and the geographic term should not be the same as that of any formal unit embraced by the cyclothem.

(k) **Soils and paleosols.**—Soils and paleosols are layers composed of the in-situ products of weathering of older rocks which may be of diverse composition and age. Soils and paleosols differ in several respects from lithostratigraphic units, and should not be treated as such (see "Pedostratigraphic Units," Articles 55 et seq).

(l) **Depositional facies.**—Depositional facies are informal units, whether objective (conglomeratic, black shale, graptolitic) or genetic and environmental (platform, turbiditic, fluvial), even when a geographic term has been applied, e.g., Lantz Mills facies. Descriptive designations convey more information than geographic terms and are preferable.

Article 23.—Boundaries.
Boundaries of lithostratigraphic units are placed at positions of lithic change. Boundaries are placed at distinct contacts or may be fixed arbitrarily within zones of gradation (Fig. 2a). Both vertical and lateral boundaries are based on the lithic criteria that provide the greatest unity and utility.

Remarks. (a) **Boundary in a vertically gradational sequence.**—A named lithostratigraphic unit is preferably bounded by a single lower and a single upper surface so that the name does not recur in a normal stratigraphic succession (see Remark b). Where a rock unit passes vertically into another by intergrading or interfingering of two or more kinds of rock, unless the gradational strata are sufficiently thick to warrant designation of a third, independent unit, the boundary is necessarily arbitrary and should be selected on the basis of practicality (Fig. 2b). For example, where a shale unit overlies a unit of interbedded limestone and shale, the boundary commonly is placed at the top of the highest readily traceable limestone bed. Where a sandstone unit grades upward into shale, the boundary may be so gradational as to be difficult to place even arbitrarily; ideally it should be drawn at the level where the rock is composed of one-half of each component. Because of creep in outcrops and caving in boreholes, it is generally best to define such arbitrary boundaries by the highest occurrence of a particular rock type, rather than the lowest.

(b) **Boundaries in lateral lithologic change.**—Where a unit changes laterally through abrupt gradation into, or intertongues with, a markedly different kind of rock, a new unit should be proposed for the different rock type. An arbitrary lateral boundary may be placed between the two equivalent units. Where the area of lateral intergradation or intertonguing is sufficiently extensive, a transitional interval of interbedded rocks may constitute a third independent unit (Fig. 2c). Where tongues (Article 25b) of formations are mapped separately or otherwise set apart without being formally named, the unmodified formation name should not be repeated in a normal stratigraphic sequence, although the modified name may be repeated in such phrases as "lower tongue of Mancos Shale" and "upper tongue of Mancos Shale." To show the order of superposition on maps and cross sections, the unnamed tongues may be distinguished informally (Fig. 2d) by number, letter, or other means. Such relationships may also be dealt with informally through the recognition of depositional facies (Article 22-1).

(c) **Key beds used for boundaries.**—Key beds (Article 26b) may be used as boundaries for a formal lithostratigraphic unit where the internal lithic characteristics of the unit remain relatively constant. Even though bounding key beds may be traceable beyond the area of the diagnostic overall rock type, geographic extension of the lithostratigraphic unit bounded thereby is not necessarily justified. Where the rock between key beds becomes drastically different from that of the type locality, a new name should be applied (Fig. 2e), even though the key beds are continuous (Article 26b). Stratigraphic and sedimentologic studies of stratigraphic units (usually informal) bounded by key beds may be very informative and useful, especially in subsurface work where the key beds may be recognized by their geophysical signatures. Such units, however, may be a kind of chronostratigraphic, rather than lithostratigraphic, unit (Article 75, 75c), although others are diachronous because one, or both, of the key beds are also diachronous.

(d) **Unconformities as boundaries.**—Unconformities, where recognizable objectively on lithic criteria, are ideal boundaries for lithostratigraphic units. However, a sequence of similar rocks may include an obscure unconformity so that separation into two units may be desirable but impracticable. If no lithic distinction adequate to define a widely recognizable boundary can be made, only one unit should be recognized, even though it may include rock that accumulated in different epochs, periods, or eras.

(e) **Correspondence with genetic units.** The boundaries of lithostratigraphic units should be chosen on the basis of lithic changes and, where feasible, to correspond with the boundaries of genetic units, so that subsequent studies of genesis will not have to deal with units that straddle formal boundaries.

Ranks of Lithostratigraphic Units

Article 24.—Formation.
The formation is the fundamental unit in lithostratigraphic classification. A formation is a body of rock identified by lithic characteristics and stratigraphic position; it is prevailingly but not necessarily tabular and is mappable at the Earth's surface or traceable in the subsurface.

Remarks. (a) **Fundamental unit.**—Formations are the basic lithostratigraphic units used in describing and interpreting the geology of a region. The limits of a formation normally are those surfaces of lithic change that give it the greatest practicable unity of constitution. A formation may represent a long or short time interval, may be composed of materials from one or several sources, and may include breaks in deposition (see Article 23d).

(b) **Content.**—A formation should possess some degree of internal lithic homogeneity or distinctive lithic features. It may contain between its upper and lower limits (i) rock of one lithic type, (ii) repetitions of two or more lithic types, or (iii) extreme lithic heterogeneity which in itself may constitute a form of unity when compared to the adjacent rock units.

(c) **Lithic characteristics.**—Distinctive lithic characteristics include chemical and mineralogical composition, texture, and such supplementary features as color, primary sedimentary or volcanic structures, fossils (viewed as rock-forming particles), or other organic content (coal, oil-shale). A unit distinguishable only by the taxonomy of its fossils is not a lithostratigraphic but a biostratigraphic unit (Article 48). Rock type may be distinctively represented by electrical, radioactive, seismic, or other

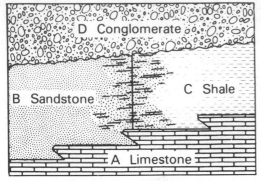

A.--Boundaries at sharp lithologic contacts and in laterally gradational sequence.

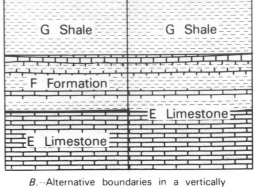

B.--Alternative boundaries in a vertically gradational or interlayered sequence.

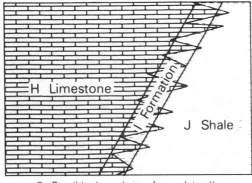

C.--Possible boundaries for a laterally intertonguing sequence.

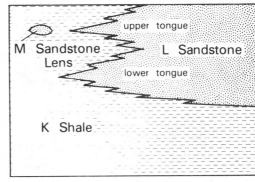

D.--Possible classification of parts of an intertonguing sequence.

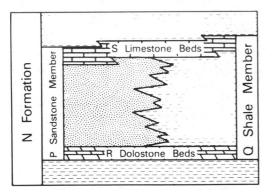

E.--Key beds, here designated the R Dolostone Beds and the S Limestone Beds, are used as boundaries to distinguish the Q Shale Member from the other parts of the N Formation. A lateral change in composition between the key beds requires that another name, P Sandstone Member, be applied. The key beds are part of each member.

EXPLANATION

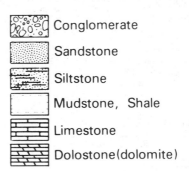

Conglomerate

Sandstone

Siltstone

Mudstone, Shale

Limestone

Dolostone (dolomite)

FIG. 2.—Diagrammatic examples of lithostratigraphic boundaries and classification.

properties (Article 22h), but these properties by themselves do not describe adequately the lithic character of the unit.

(d) **Mappability and thickness.**—The proposal of a new formation must be based on tested mappability. Well-established formations commonly are divisible into several widely recognizable lithostratigraphic units; where formal recognition of these smaller units serves a useful purpose, they may be established as members and beds, for which the requirement of mappability is not mandatory. A unit formally recognized as a formation in one area may be treated elsewhere as a group, or as a member of another formation, without change of name. Example: the Niobrara is mapped at different places as a member of the Mancos Shale, of the Cody Shale, or of the Colorado Shale, and also as the Niobrara Formation, as the Niobrara Limestone, and as the Niobrara Shale.

Thickness is not a determining parameter in dividing a rock succession into formations; the thickness of a formation may range from a feather edge at its depositional or erosional limit to thousands of meters elsewhere. No formation is considered valid that cannot be delineated at the scale of geologic mapping practiced in the region when the formation is proposed. Although representation of a formation on maps and cross sections by a labeled line may be justified, proliferation of such exceptionally thin units is undesirable. The methods of subsurface mapping permit delineation of units much thinner than those usually practicable for surface studies; before such thin units are formalized, consideration should be given to the effect on subsequent surface and subsurface studies.

(e) **Organic reefs and carbonate mounds.**—Organic reefs and carbonate mounds ("buildups") may be distinguished formally, if desirable, as formations distinct from their surrounding, thinner, temporal equivalents. For the requirements of formalization, see Article 30f.

(f) **Interbedded volcanic and sedimentary rock.** Sedimentary rock and volcanic rock that are interbedded may be assembled into a formation under one name which should indicate the predominant or distinguishing lithology, such as Mindego Basalt.

(g) **Volcanic rock.**—Mappable distinguishable sequences of stratified volcanic rock should be treated as formations or lithostratigraphic units of higher or lower rank. A small intrusive component of a dominantly stratiform volcanic assemblage may be treated informally.

(h) **Metamorphic rock.**—Formations composed of low-grade metamorphic rock (defined for this purpose as rock in which primary structures are clearly recognizable) are, like sedimentary formations, distinguished mainly by lithic characteristics. The mineral facies may differ from place to place, but these variations do not require definition of a new formation. High-grade metamorphic rocks whose relation to established formations is uncertain are treated as lithodemic units (see Articles 31 et seq).

Article 25.—Member. A member is the formal lithostratigraphic unit next in rank below a formation and is always a part of some formation. It is recognized as a named entity within a formation because it possesses characteristics distinguishing it from adjacent parts of the formation. A formation need not be divided into members unless a useful purpose is served by doing so. Some formations may be divided completely into members; others may have only certain parts designated as members; still others may have no members. A member may extend laterally from one formation to another.

Remarks. (a) **Mapping of members.**—A member is established when it is advantageous to recognize a particular part of a heterogeneous formation. A member, whether formally or informally designated, need not be mappable at the scale required for formations. Even if all members of a formation are locally mappable, it does not follow that they should be raised to formational rank, because proliferation of formation names may obscure rather than clarify relations with other areas.

(b) **Lens and tongue.**—A geographically restricted member that terminates on all sides within a formation may be called a lens (lentil). A wedging member that extends outward beyond a formation or wedges ("pinches") out within another formation may be called a tongue.

(c) **Organic reefs and carbonate mounds.**—Organic reefs and carbonate mounds may be distinguished formally, if desirable, as members

within a formation. For the requirements of formalization, see Article 30f.

(d) **Division of members.**—A formally or informally recognized division of a member is called a bed or beds, except for volcanic flow-rocks, for which the smallest formal unit is a flow. Members may contain beds or flows, but may never contain other members.

(e) **Laterally equivalent members.**—Although members normally are in vertical sequence, laterally equivalent parts of a formation that differ recognizably may also be considered members.

Article 26.—Bed(s). A bed, or beds, is the smallest formal lithostratigraphic unit of sedimentary rocks.

Remarks. (a) **Limitations.**—The designation of a bed or a unit of beds as a formally named lithostratigraphic unit generally should be limited to certain distinctive beds whose recognition is particularly useful. Coal beds, oil sands, and other beds of economic importance commonly are named, but such units and their names usually are not a part of formal stratigraphic nomenclature (Articles 22g and 30g).

(b) **Key or marker beds.**—A key or marker bed is a thin bed of distinctive rock that is widely distributed. Such beds may be named, but usually are considered informal units. Individual key beds may be traced beyond the lateral limits of a particular formal unit (Article 23c).

Article 27.—Flow. A flow is the smallest formal lithostratigraphic unit of volcanic flow rocks. A flow is a discrete, extrusive, volcanic body distinguishable by texture, composition, order of superposition, paleomagnetism, or other objective criteria. It is part of a member and thus is equivalent in rank to a bed or beds of sedimentary-rock classification. Many flows are informal units. The designation and naming of flows as formal rock-stratigraphic units should be limited to those that are distinctive and widespread.

Article 28.—Group. A group is the lithostratigraphic unit next higher in rank to formation; a group may consist entirely of named formations, or alternatively, need not be composed entirely of named formations.

Remarks. (a) **Use and content.**—Groups are defined to express the natural relationships of associated formations. They are useful in small-scale mapping and regional stratigraphic analysis. In some reconnaissance work, the term "group" has been applied to lithostratigraphic units that appear to be divisible into formations, but have not yet been so divided. In such cases, formations may be erected subsequently for one or all of the practical divisions of the group.

(b) **Change in component formations.**—The formations making up a group need not necessarily be everywhere the same. The Rundle Group, for example, is widespread in western Canada and undergoes several changes in formational content. In southwestern Alberta, it comprises the Livingstone, Mount Head, and Etherington Formations in the Front Ranges, whereas in the foothills and subsurface of the adjacent plains, it comprises the Pekisko, Shunda, Turner Valley, and Mount Head Formations. However, a formation or its parts may not be assigned to two vertically adjacent groups.

(c) **Change in rank.**—The wedge-out of a component formation or formations may justify the reduction of a group to formation rank, retaining the same name. When a group is extended laterally beyond where it is divided into formations, it becomes in effect a formation, even if it is still called a group. When a previously established formation is divided into two or more component units that are given formal formation rank, the old formation, with its old geographic name, should be raised to group status. Raising the rank of the unit is preferable to restricting the old name to a part of its former content, because a change in rank leaves the sense of a well-established unit unchanged (Articles 19b, 19g).

Article 29.—Supergroup. A supergroup is a formal assemblage of related or superposed groups, or of groups and formations. Such units have proved useful in regional and provincial syntheses. Supergroups should be named only where their recognition serves a clear purpose.

Remark. (a) **Misuse of "series" for group or supergroup.**—Although "series" is a useful general term, it is applied formally only to a chronostratigraphic unit and should not be used for a lithostratigraphic unit. The term "series" should no longer be employed for an assemblage of formations or an assemblage of formations and groups, as it has been, especially in studies of the Precambrian. These assemblages are groups or supergroups.

Lithostratigraphic Nomenclature

Article 30.—**Compound Character.** The formal name of a lithostratigraphic unit is compound. It consists of a geographic name combined with a descriptive lithic term or with the appropriate rank term, or both. Initial letters of all words used in forming the names of formal rock-stratigraphic units are capitalized.

Remarks. (a) **Omission of part of a name.**—Where frequent repetition would be cumbersome, the geographic name, the lithic term, or the rank term may be used alone, once the full name has been introduced; as "the Burlington," "the limestone," or "the formation," for the Burlington Limestone.
(b) **Use of simple lithic terms.**—The lithic part of the name should indicate the predominant or diagnostic lithology, even if subordinate lithologies are included. Where a lithic term is used in the name of a lithostratigraphic unit, the simplest generally acceptable term is recommended (for example, limestone, sandstone, shale, tuff, quartzite). Compound terms (for example, clay shale) and terms that are not in common usage (for example, calcirudite, orthoquartzite) should be avoided. Combined terms, such as "sand and clay," should not be used for the lithic part of the names of lithostratigraphic units, nor should an adjective be used between the geographic and the lithic terms, as "Chattanooga Black Shale" and "Biwabik Iron-Bearing Formation."
(c) **Group names.**—A group name combines a geographic name with the term "group," and no lithic designation is included; for example, San Rafael Group.
(d) **Formation names.**—A formation name consists of a geographic name followed by a lithic designation or by the word "formation." Examples: Dakota Sandstone, Mitchell Mesa Rhyolite, Monmouth Formation, Halton Till.
(e) **Member names.**—All member names include a geographic term and the word "member;" some have an intervening lithic designation, if useful; for example, Wedington Sandstone Member of the Fayetteville Shale. Members designated solely by lithic character (for example, siliceous shale member), by position (upper, lower), or by letter or number, are informal.
(f) **Names of reefs.**—Organic reefs identified as formations or members are formal units only where the name combines a geographic name with the appropriate rank term, e.g., Leduc Formation (a name applied to the several reefs enveloped by the Ireton Formation), Rainbow Reef Member.
(g) **Bed and flow names.**—The names of beds or flows combine a geographic term, a lithic term, and the term "bed" or "flow;" for example, Knee Hills Tuff Bed, Ardmore Bentonite Beds, Negus Variolitic Flows.
(h) **Informal units.**—When geographic names are applied to such informal units as oil sands, coal beds, mineralized zones, and informal members (see Articles 22g and 26a), the unit term should not be capitalized. A name is not necessarily formal because it is capitalized, nor does failure to capitalize a name render it informal. Geographic names should be combined with the terms "formation" or "group" only in formal nomenclature.
(i) **Informal usage of identical geographic names.**—The application of identical geographic names to several minor units in one vertical sequence is considered informal nomenclature (lower Mount Savage coal, Mount Savage fireclay, upper Mount Savage coal, Mount Savage rider coal, and Mount Savage sandstone). The application of identical geographic names to the several lithologic units constituting a cyclothem likewise is considered informal.
(j) **Metamorphic rock.**—Metamorphic rock recognized as a normal stratified sequence, commonly low-grade metavolcanic or metasedimen-

tary rocks, should be assigned to named groups, formations, and members, such as the Deception Rhyolite, a formation of the Ash Creek Group, or the Bonner Quartzite, a formation of the Missoula Group. High-grade metamorphic and metasomatic rocks are treated as lithodemes and suites (see Articles 31, 33, 35).
(k) **Misuse of well-known name.**—A name that suggests some well-known locality, region, or political division should not be applied to a unit typically developed in another less well-known locality of the same name. For example, it would be inadvisable to use the name "Chicago Formation" for a unit in California.

LITHODEMIC UNITS

Nature and Boundaries

Article 31.—**Nature of Lithodemic Units.** A lithodemic[6] unit is a defined body of predominantly intrusive, highly deformed, and/or highly metamorphosed rock, distinguished and delimited on the basis of rock characteristics. In contrast to lithostratigraphic units, a lithodemic unit generally does not conform to the Law of Superposition. Its contacts with other rock units may be sedimentary, extrusive, intrusive, tectonic, or metamorphic (Fig. 3).

Remarks. (a) **Recognition and definition.**—Lithodemic units are defined and recognized by observable rock characteristics. They are the practical units of general geological work in terranes in which rocks generally lack primary stratification; in such terranes they serve as the foundation for studying, describing, and delineating lithology, local and regional structure, economic resources, and geologic history.
(b) **Type and reference localities.**—The definition of a lithodemic unit should be based on as full a knowledge as possible of its lateral and vertical variations and its contact relationships. For purposes of nomenclatural stability, a type locality and, wherever appropriate, reference localities should be designated.
(c) **Independence from inferred geologic history.**—Concepts based on inferred geologic history properly play no part in the definition of a lithodemic unit. Nevertheless, where two rock masses are lithically similar but display objective structural relations that preclude the possibility of their being even broadly of the same age, they should be assigned to different lithodemic units.
(d) **Use of "zone."**—As applied to the designation of lithodemic units, the term "zone" is informal. Examples are: "mineralized zone," "contact zone," and "pegmatitic zone."

Article 32.—**Boundaries.** Boundaries of lithodemic units are placed at positions of lithic change. They may be placed at clearly distinguished contacts or within zones of gradation. Boundaries, both vertical and lateral, are based on the lithic criteria that provide the greatest unity and practical utility. Contacts with other lithodemic and lithostratigraphic units may be depositional, intrusive, metamorphic, or tectonic.

Remark. (a) **Boundaries within gradational zones.**—Where a lithodemic unit changes through gradation into, or intertongues with, a rock-mass with markedly different characteristics, it is usually desirable to propose a new unit. It may be necessary to draw an arbitrary boundary within the zone of gradation. Where the area of intergradation or intertonguing is sufficiently extensive, the rocks of mixed character may constitute a third unit.

Ranks of Lithodemic Units

Article 33.—**Lithodeme.** The lithodeme is the fundamental unit in lithodemic classification. A lithodeme is a body of intru-

---

[6]From the Greek *demas, -os:* "living body, frame".

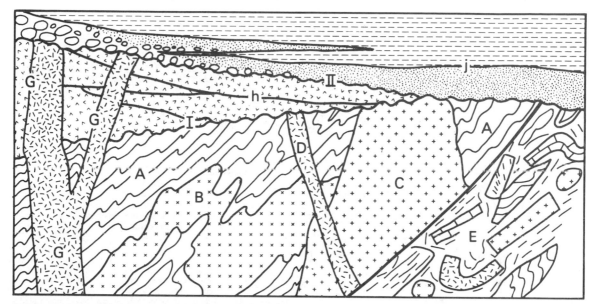

FIG. 3.—Lithodemic (upper case) and lithostratigraphic (lower case) units. A *lithodeme* of *gneiss* (A) contains an *intrusion* of diorite (B) that was deformed with the gneiss. A and B may be treated jointly as a *complex*. A younger *granite* (C) is cut by a dike of *syenite* (D), that is cut in turn by unconformity I. All the foregoing are in fault contact with a *structural complex* (E). A *volcanic complex* (G) is built upon unconformity I, and its feeder dikes cut the unconformity. Laterally equivalent volcanic strata in orderly, mappable succession (h) are treated as lithostratigraphic units. A *gabbro* feeder (G'), to the volcanic complex, where surrounded by gneiss is readily distinguished as a separate lithodeme and named as a *gabbro* or an *intrusion*. All the foregoing are overlain, at unconformity II, by sedimentary rocks (j) divided into formations and members.

sive, pervasively deformed, or highly metamorphosed rock, generally non-tabular and lacking primary depositional structures, and characterized by lithic homogeneity. It is mappable at the Earth's surface and traceable in the subsurface. For cartographic and hierarchical purposes, it is comparable to a formation (see Table 2).

Remarks. (a) **Content.**—A lithodeme should possess distinctive lithic features and some degree of internal lithic homogeneity. It may consist of (i) rock of one type, (ii) a mixture of rocks of two or more types, or (iii) extreme heterogeneity of composition, which may constitute in itself a form of unity when compared to adjoining rock-masses (see also "complex," Article 37).

(b) **Lithic characteristics.**—Distinctive lithic characteristics may include mineralogy, textural features such as grain size, and structural features such as schistose or gneissic structure. A unit distinguishable from its neighbors only by means of chemical analysis is informal.

(c) **Mappability.**—Practicability of surface or subsurface mapping is an essential characteristic of a lithodeme (see Article 24d).

Article 34.—**Division of Lithodemes.** Units below the rank of lithodeme are informal.

Article 35.—**Suite.** A *suite* (metamorphic suite, intrusive suite, plutonic suite) is the lithodemic unit next higher in rank to lithodeme. It comprises two or more associated lithodemes of the same class (e.g., plutonic, metamorphic). For cartographic and hierarchical purposes, suite is comparable to group (see Table 2).

Remarks. (a) **Purpose.**—Suites are recognized for the purpose of expressing the natural relations of associated lithodemes having signifi-

cant lithic features in common, and of depicting geology at compilation scales too small to allow delineation of individual lithodemes. Ideally, a suite consists entirely of named lithodemes, but may contain both named and unnamed units.

(b) **Change in component units.**—The named and unnamed units constituting a suite may change from place to place, so long as the original sense of natural relations and of common lithic features is not violated.

(c) **Change in rank.**—Traced laterally, a suite may lose all of its formally named divisions but remain a recognizable, mappable entity. Under such circumstances, it may be treated as a lithodeme but retain the same name. Conversely, when a previously established lithodeme is divided into two or more mappable divisions, it may be desirable to raise its rank to suite, retaining the original geographic component of the name. To avoid confusion, the original name should not be retained for one of the divisions of the original unit (see Article 19g).

Article 36.—**Supersuite.** A supersuite is the unit next higher in rank to a suite. It comprises two or more suites or complexes having a degree of natural relationship to one another, either in the vertical or the lateral sense. For cartographic and hierarchical purposes, supersuite is similar in rank to supergroup.

Article 37.—**Complex.** An assemblage or mixture of rocks of *two or more genetic classes*, i.e., igneous, sedimentary, or metamorphic, with or without highly complicated structure, may be named a *complex*. The term "complex" takes the place of the lithic or rank term (for example, Boil Mountain Complex, Franciscan Complex) and, although unranked, commonly is comparable to suite or supersuite and is named in the same manner (Articles 41, 42).

Remarks (a) **Use of "complex."**—Identification of an assemblage of diverse rocks as a complex is useful where the mapping of each separate lithic component is impractical at ordinary mapping scales. "Complex" is unranked but commonly comparable to suite or supersuite; therefore, the term may be retained if subsequent, detailed mapping distinguishes some or all of the component lithodemes or lithostratigraphic units.

(b) **Volcanic complex.**—Sites of persistent volcanic activity commonly are characterized by a diverse assemblage of extrusive volcanic rocks, related intrusions, and their weathering products. Such an assemblage may be designated a *volcanic complex.*

(c) **Structural complex.**—In some terranes, tectonic processes (e.g., shearing, faulting) have produced heterogeneous mixtures or disrupted bodies of rock in which some individual components are too small to be mapped. *Where there is no doubt that the mixing or disruption is due to tectonic processes,* such a mixture may be designated as a structural complex, whether it consists of two or more classes of rock, or a single class only. A simpler solution for some mapping purposes is to indicate intense deformation by an overprinted pattern.

(d) **Misuse of "complex".**—Where the rock assemblage to be united under a single, formal name consists of diverse types of a *single class* of rock, as in many terranes that expose a variety of either intrusive igneous or high-grade metamorphic rocks, the term "intrusive suite," "plutonic suite," or "metamorphic suite" should be used, rather than the unmodified term "complex." Exceptions to this rule are the terms *structural complex* and *volcanic complex* (see Remarks c and b, above).

Article 38.—Misuse of "Series" for Suite, Complex, or Supersuite. The term "series" has been employed for an assemblage of lithodemes or an assemblage of lithodemes and suites, especially in studies of the Precambrian. This practice now is regarded as improper; these assemblages are suites, complexes, or supersuites. The term "series" also has been applied to a sequence of rocks resulting from a succession of eruptions or intrusions. In these cases a different term should be used; "group" should replace "series" for volcanic and low-grade metamorphic rocks, and "intrusive suite" or "plutonic suite" should replace "series" for intrusive rocks of group rank.

Lithodemic Nomenclature

Article 39.—General Provisions. The formal name of a lithodemic unit is compound. It consists of a geographic name combined with a descriptive or appropriate rank term. The principles for the selection of the geographic term, concerning suitability, availability, priority, etc, follow those established in Article 7, where the rules for capitalization are also specified.

Article 40.—Lithodeme Names. The name of a lithodeme combines a geographic term with a lithic or descriptive term, e.g., Killarney Granite, Adamant Pluton, Manhattan Schist, Skaergaard Intrusion, Duluth Gabbro. The term *formation* should not be used.

Remarks. (a) **Lithic term.**—The lithic term should be a common and familiar term, such as schist, gneiss, gabbro. Specialized terms and terms not widely used, such as websterite and jacupirangite, and compound terms, such as graphitic schist and augen gneiss, should be avoided.

(b) **Intrusive and plutonic rocks.**—Because many bodies of intrusive rock range in composition from place to place and are difficult to characterize with a single lithic term, and because many bodies of plutonic rock are considered not to be intrusions, latitude is allowed in the choice of a lithic or descriptive term. Thus, the descriptive term should preferably be compositional (e.g., gabbro, granodiorite), but may, if necessary, denote form (e.g., dike, sill), or be neutral (e.g., intrusion, pluton[7]). In any event, specialized compositional terms not widely used are to be avoided, as are form terms that are not widely used, such as bysmalith and chonolith. Terms implying genesis should be avoided as much as possible, because interpretations of genesis may change.

Article 41.—Suite Names. The name of a suite combines a geographic term, the term "suite," and an adjective denoting the fundamental character of the suite; for example, Idaho Springs Metamorphic Suite, Tuolumne Intrusive Suite, Cassiar Plutonic Suite. The geographic name of a suite may not be the same as that of a component lithodeme (see Article 19f). Intrusive assemblages, however, may share the same geographic name if an intrusive lithodeme is representative of the suite.

Article 42.—Supersuite Names. The name of a supersuite combines a geographic term with the term "supersuite."

MAGNETOSTRATIGRAPHIC UNITS

Nature and Boundaries

Article 43.—Nature of Magnetostratigraphic Units. A magnetostratigraphic unit is a body of rock unified by specified remanent-magnetic properties and is distinct from underlying and overlying magnetostratigraphic units having different magnetic properties.

Remarks. (a) **Definition.**—Magnetostratigraphy is defined here as all aspects of stratigraphy based on remanent magnetism (paleomagnetic signatures). Four basic paleomagnetic phenomena can be determined or inferred from remanent magnetism: polarity, dipole-field-pole position (including apparent polar wander), the non-dipole component (secular variation), and field intensity.

(b) **Contemporaneity of rock and remanent magnetism.**—Many paleomagnetic signatures reflect earth magnetism at the time the rock formed. Nevertheless, some rocks have been subjected subsequently to physical and/or chemical processes which altered the magnetic properties. For example, a body of rock may be heated above the blocking temperature or Curie point for one or more minerals, or a ferromagnetic mineral may be produced by low-temperature alteration long after the enclosing rock formed, thus acquiring a component of remanent magnetism reflecting the field at the time of alteration, rather than the time of original rock deposition or crystallization.

(c) **Designations and scope.**—The prefix *magneto* is used with an appropriate term to designate the aspect of remanent magnetism used to define a unit. The terms "magnetointensity" or "magnetosecular-variation" are possible examples. This Code considers only polarity reversals, which now are recognized widely as a stratigraphic tool. However, apparent-polar-wander paths offer increasing promise for correlations within Precambrian rocks.

Article 44.—Definition of Magnetopolarity Unit. A magnetopolarity unit is a body of rock unified by its remanent magnetic polarity and distinguished from adjacent rock that has different polarity.

Remarks. (a) **Nature.**—Magnetopolarity is the record in rocks of the polarity history of the Earth's magnetic-dipole field. Frequent past reversals of the polarity of the Earth's magnetic field provide a basis for magnetopolarity stratigraphy.

(b) **Stratotype.**—A stratotype for a magnetopolarity unit should be designated and the boundaries defined in terms of recognized lithostratigraphic and/or biostratigraphic units in the stratotype. The formal definition of a magnetopolarity unit should meet the applicable specific requirements of Articles 3 to 16.

(c) **Independence from inferred history.**—Definition of a magnetopolarity unit does not require knowledge of the time at which the unit acquired its remanent magnetism; its magnetism may be primary or sec-

---

[7]Pluton—a mappable body of plutonic rock.

ondary. Nevertheless, the unit's present polarity is a property that may be ascertained and confirmed by others.

(d) **Relation to lithostratigraphic and biostratigraphic units.**—Magnetopolarity units resemble lithostratigraphic and biostratigraphic units in that they are defined on the basis of an objective recognizable property, but differ fundamentally in that most magnetopolarity unit boundaries are thought not to be time transgressive. Their boundaries may coincide with those of lithostratigraphic or biostratigraphic units, or be parallel to but displaced from those of such units, or be crossed by them.

(e) **Relation of magnetopolarity units to chronostratigraphic units.**—Although transitions between polarity reversals are of global extent, a magnetopolarity unit does not contain within itself evidence that the polarity is primary, or criteria that permit its unequivocal recognition in chronocorrelative strata of other areas. Other criteria, such as paleontologic or numerical age, are required for both correlation and dating. Although polarity reversals are useful in recognizing chronostratigraphic units, magnetopolarity alone is insufficient for their definition.

Article 45.—Boundaries. The upper and lower limits of a magnetopolarity unit are defined by boundaries marking a change of polarity. Such boundaries may represent either a depositional discontinuity or a magnetic-field transition. The boundaries are either polarity-reversal horizons or polarity transition-zones, respectively.

Remark. (a) **Polarity-reversal horizons and transition-zones.**—A polarity-reversal horizon is either a single, clearly definable surface or a thin body of strata constituting a transitional interval across which a change in magnetic polarity is recorded. Polarity-reversal horizons describe transitional intervals of 1 m or less; where the change in polarity takes place over a stratigraphic interval greater than 1 m, the term "polarity transition-zone" should be used. Polarity-reversal horizons and polarity transition-zones provide the boundaries for polarity zones, although they may also be contained within a polarity zone where they mark an internal change subsidiary in rank to those at its boundaries.

Ranks of Magnetopolarity Units

Article 46.—Fundamental Unit. A polarity zone is the fundamental unit of magnetopolarity classification. A polarity zone is a unit of rock characterized by the polarity of its magnetic signature. Magnetopolarity zone, rather than polarity zone, should be used where there is risk of confusion with other kinds of polarity.

Remarks. (a) **Content.**—A polarity zone should possess some degree of internal homogeneity. It may contain rocks of (1) entirely or predominantly one polarity, or (2) mixed polarity.

(b) **Thickness and duration.**—The thickness of rock of a polarity zone or the amount of time represented should play no part in the definition of the zone. The polarity signature is the essential property for definition.

(c) **Ranks.**—When continued work at the stratotype for a polarity zone, or new work in correlative rocks elsewhere, reveals smaller polarity units, these may be recognized formally as polarity subzones. If it should prove necessary or desirable to group polarity zones, these should be termed polarity superzones. The rank of a polarity unit may be changed when deemed appropriate.

Magnetopolarity Nomenclature

Article 47.—Compound Name. The formal name of a magnetopolarity zone should consist of a geographic name and the term *Polarity Zone*. The term may be modified by *Normal*, *Reversed*, or *Mixed* (example: Deer Park Reversed Polarity Zone). In naming or revising magnetopolarity units, appropriate parts of Articles 7 and 19 apply. The use of informal designations, e.g., numbers or letters, is not precluded.

BIOSTRATIGRAPHIC UNITS

Nature and Boundaries

Article 48.—Nature of Biostratigraphic Units. A biostratigraphic unit is a body of rock defined or characterized by its fossil content. The basic unit in biostratigraphic classification is the biozone, of which there are several kinds.

Remarks. (a) **Enclosing strata.**—Fossils that define or characterize a biostratigraphic unit commonly are contemporaneous with the body of rock that contains them. Some biostratigraphic units, however, may be represented only by their fossils, preserved in normal stratigraphic succession (e.g., on hardgrounds, in lag deposits, in certain types of remanié accumulations), which alone represent the rock of the biostratigraphic unit. In addition, some strata contain fossils derived from older or younger rocks or from essentially coeval materials of different facies; such fossils should not be used to define a biostratigraphic unit.

(b) **Independence from lithostratigraphic units.**—Biostratigraphic units are based on criteria which differ fundamentally from those for lithostratigraphic units. Their boundaries may or may not coincide with the boundaries of lithostratigraphic units, but they bear no inherent relation to them.

(c) **Independence from chronostratigraphic units.**—The boundaries of most biostratigraphic units, unlike the boundaries of chronostratigraphic units, are both characteristically and conceptually diachronous. An exception is an abundance biozone boundary that reflects a mass-mortality event. The vertical and lateral limits of the rock body that constitutes the biostratigraphic unit represent the limits in distribution of the defining biotic elements. The lateral limits never represent, and the vertical limits rarely represent, regionally synchronous events. Nevertheless, biostratigraphic units are effective for interpreting chronostratigraphic relations.

Article 49.—Kinds of Biostratigraphic Units. Three principal kinds of biostratigraphic units are recognized: *interval*, *assemblage*, and *abundance* biozones.

Remark: (a) **Boundary definitions.**—Boundaries of interval zones are defined by lowest and/or highest occurrences of single taxa; boundaries of some kinds of assemblage zones (Oppel or concurrent range zones) are defined by lowest and/or highest occurrences of more than one taxon; and boundaries of abundance zones are defined by marked changes in relative abundances of preserved taxa.

Article 50.—Definition of Interval Zone. An interval zone (or subzone) is the body of strata between two specified, documented lowest and/or highest occurrences of single taxa.

Remarks. (a) **Interval zone types.**—Three basic types of interval zones are recognized (Fig. 4). These include the range zones and interval zones of the International Stratigraphic Guide (ISSC, 1976, p. 53, 60) and are:

1. The interval between the documented lowest and highest occurrences of a single taxon (Fig. 4A). This is the *taxon range zone* of ISSC (1976, p. 53).

2. The interval included between the documented lowest occurrence of one taxon and the documented highest occurrence of another taxon (Fig. 4B). When such occurrences result in stratigraphic overlap of the taxa (Fig. 4B-1), the interval zone is the *concurrent range zone* of ISSC (1976, p. 55), that involves only two taxa. When such occurrences do not result in stratigraphic overlap (Fig. 4B-2), but are used to partition the range of a third taxon, the interval is the *partial range zone* of George and others (1969).

3. The interval between documented successive lowest occurrences or successive highest occurrences of two taxa (Fig. 4C). When the interval is between successive documented lowest occurrences within an evolutionary lineage (Fig. 4C-1), it is the *lineage zone* of ISSC (1976, p. 58). When

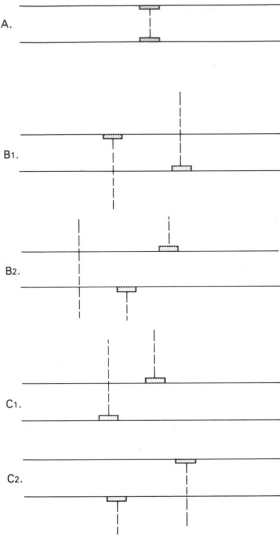

FIG. 4.—Examples of biostratigraphic interval zones.
Vertical broken lines indicate ranges of taxa; bars indicate lowest or highest documented occurrences.

the interval is between successive lowest occurrences of unrelated taxa or between successive highest occurrences of either related or unrelated taxa (Fig. 4C-2), it is a kind of *interval zone* of ISSC (1976, p. 60).

(b) **Unfossiliferous intervals.**—Unfossiliferous intervals between or within biozones are the *barren interzones* and *intrazones* of ISSC (1976, p. 49).

Article 51.—Definition of Assemblage Zone. An assemblage zone is a biozone characterized by the association of three or more taxa. It may be based on all kinds of fossils present, or restricted to only certain kinds of fossils.

Remarks. (a) **Assemblage zone contents.**—An assemblage zone may consist of a geographically or stratigraphically restricted assemblage, or may incorporate two or more contemporaneous assemblages with shared characterizing taxa (*composite assemblage zones* of Kauffman, 1969) (Fig. 5c).

(b) **Assemblage zone types.**—In practice, two assemblage zone concepts are used:

1. The *assemblage zone* (or cenozone) of ISSC (1976, p. 50), which is characterized by taxa without regard to their range limits (Fig. 5a). Recognition of this type of assemblage zone can be aided by using techniques of multivariate analysis. Careful designation of the characterizing taxa is especially important.

2. The *Oppel zone,* or the *concurrent range zone* of ISSC (1976, p. 55, 57), a type of zone characterized by more than two taxa and having boundaries based on two or more documented first and/or last occurrences of the included characterizing taxa (Fig. 5b).

Article 52.—Definition of Abundance Zone. An abundance zone is a biozone characterized by quantitatively distinctive maxima of relative abundance of one or more taxa. This is the *acme zone* of ISSC (1976, p. 59).

Remark. (a) **Ecologic controls.**—The distribution of biotic assemblages used to characterize some assemblage and abundance biozones may reflect strong local ecological control. Biozones based on such assemblages are included within the concept of ecozones (Vella, 1964), and are informal.

Ranks of Biostratigraphic Units

Article 53.—Fundamental Unit. The fundamental unit of biostratigraphic classification is a biozone.

Remarks. (a) **Scope.**—A single body of rock may be divided into various kinds and scales of biozones or subzones, as discussed in the International Stratigraphic Guide (ISSC, 1976, p. 62). Such usage is recommended if it will promote clarity, but only the unmodified term *biozone* is accorded formal status.

(b) **Divisions.**—A biozone may be completely or partly divided into formally designated sub-biozones (subzones), if such divisions serve a useful purpose.

Biostratigraphic Nomenclature

Article 54.—Establishing Formal Units. Formal establishment of a biozone or subzone must meet the requirements of Article 3 and requires a unique name, a description of its content and its boundaries, reference to a stratigraphic sequence in which the zone is characteristically developed, and a discussion of its spatial extent.

Remarks. (a) **Name.**—The name, which is compound and designates the kind of biozone, may be based on:

1. One or two characteristic and common taxa that are restricted to the biozone, reach peak relative abundance within the biozone, or have their total stratigraphic overlap within the biozone. These names most commonly are those of genera or subgenera, binomial designations of species, or trinomial designations of subspecies. If names of the nominate taxa change, names of the zones should be changed accordingly. Generic or subgeneric names may be abbreviated. Trivial species or subspecies names should not be used alone because they may not be unique.

2. Combinations of letters derived from taxa which characterize the biozone. However, alpha-numeric code designations (e.g., N1, N2, N3...) are informal and not recommended because they do not lend themselves readily to subsequent insertions, combinations, or eliminations. Biozonal

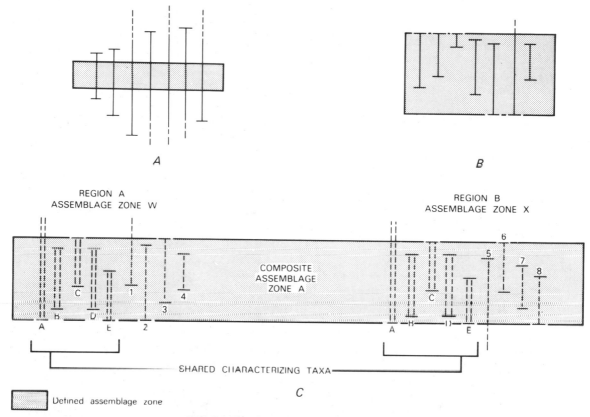

FIG. 5.—Examples of assemblage zone concepts.

systems based *only* on simple progressions of letters or numbers (e.g., A, B, C, or 1, 2, 3) are also not recommended.

(b) **Revision.**—Biozones and subzones are established empirically and may be modified on the basis of new evidence. Positions of established biozone or subzone boundaries may be stratigraphically refined, new characterizing taxa may be recognized, or original characterizing taxa may be superseded. If the concept of a particular biozone or subzone is substantially modified, a new unique designation is required to avoid ambiguity in subsequent citations.

(c) **Specifying kind of zone.**—Initial designation of a formally proposed biozone or subzone as an abundance zone, or as one of the types of interval zones, or assemblage zones (Articles 49-52), is strongly recommended. Once the type of biozone is clearly identified, the designation may be dropped in the remainder of a text (e.g., *Exus albus* taxon range zone to *Exus albus* biozone).

(d) **Defining taxa.**—Initial description or subsequent emendation of a biozone or subzone requires designation of the defining and characteristic taxa, and/or the documented first and last occurrences which mark the biozone or subzone boundaries.

(e) **Stratotypes.**—The geographic and stratigraphic position and boundaries of a formally proposed biozone or subzone should be defined precisely or characterized in one or more designated reference sections. Designation of a stratotype for each new biostratigraphic unit and of reference sections for emended biostratigraphic units is required.

PEDOSTRATIGRAPHIC UNITS

Nature and Boundaries

Article 55.—**Nature of Pedostratigraphic Units.** A pedostratigraphic unit is a body of rock that consists of one or more pedologic horizons developed in one or more lithostratigraphic, allostratigraphic, or lithodemic units (Fig. 6) and is overlain by one or more formally defined lithostratigraphic or allostratigraphic units.

Remarks. (a) **Definition.**—A pedostratigraphic[8] unit is a buried, traceable, three-dimensional body of rock that consists of one or more differentiated pedologic horizons.

(b) **Recognition.**—The distinguishing property of a pedostratigraphic unit is the presence of one or more distinct, differentiated, pedologic horizons. Pedologic horizons are products of soil development (pedogenesis) which occurred subsequent to formation of the lithostrati-

---

[8]Terminology related to pedostratigraphic classification is summarized on page 850.

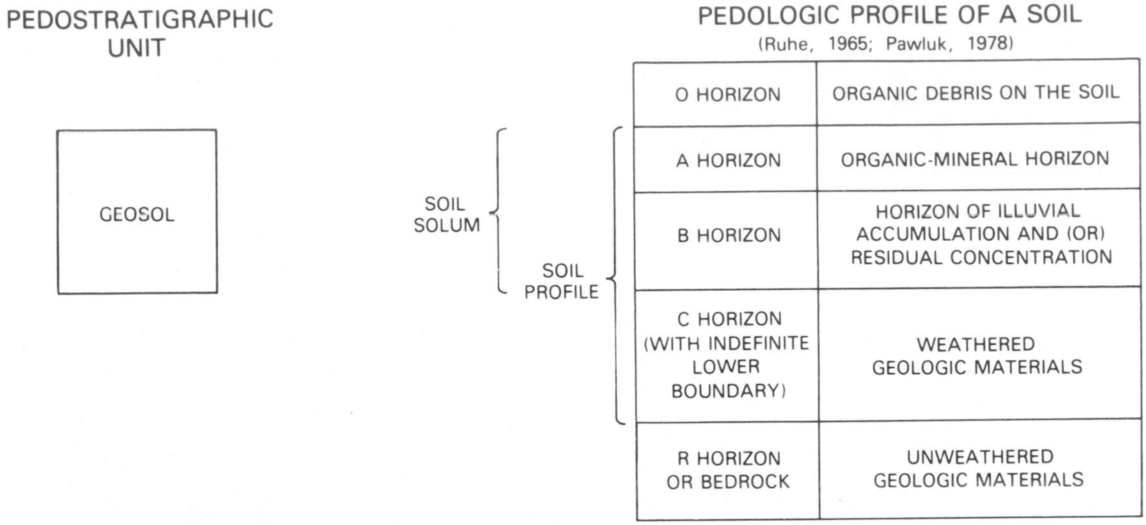

PEDOSTRATIGRAPHIC
UNIT

PEDOLOGIC PROFILE OF A SOIL
(Ruhe, 1965; Pawluk, 1978)

GEOSOL

SOIL
SOLUM

SOIL
PROFILE

O HORIZON	ORGANIC DEBRIS ON THE SOIL
A HORIZON	ORGANIC-MINERAL HORIZON
B HORIZON	HORIZON OF ILLUVIAL ACCUMULATION AND (OR) RESIDUAL CONCENTRATION
C HORIZON (WITH INDEFINITE LOWER BOUNDARY)	WEATHERED GEOLOGIC MATERIALS
R HORIZON OR BEDROCK	UNWEATHERED GEOLOGIC MATERIALS

FIG. 6.—Relationship between pedostratigraphic units and pedologic profiles.
 The base of a geosol is the lowest clearly defined physical boundary of a pedologic horizon in a buried soil profile. In this example it is the lower boundary of the B horizon because the base of the C horizon is not a clearly defined physical boundary. In other profiles the base may be the lower boundary of a C horizon.

graphic, allostratigraphic, or lithodemic unit or units on which the buried soil was formed; these units are the parent materials in which pedogenesis occurred. Pedologic horizons are recognized in the field by diagnostic features such as color, soil structure, organic-matter accumulation, texture, clay coatings, stains, or concretions. Micromorphology, particle size, clay mineralogy, and other properties determined in the laboratory also may be used to identify and distinguish pedostratigraphic units.

(c) **Boundaries and stratigraphic position.**—The upper boundary of a pedostratigraphic unit is the top of the uppermost pedologic horizon formed by pedogenesis in a buried soil profile. The lower boundary of a pedostratigraphic unit is the lowest *definite* physical boundary of a pedologic horizon within a buried soil profile. The stratigraphic position of a pedostratigraphic unit is determined by its relation to overlying and underlying stratigraphic units (see Remark d).

(d) **Traceability.**—Practicability of subsurface tracing of the upper boundary of a buried soil is essential in establishing a pedostratigraphic unit because (1) few buried soils are exposed continuously for great distances, (2) the physical and chemical properties of a specific pedostratigraphic unit may vary greatly, both vertically and laterally, from place to place, and (3) pedostratigraphic units of different stratigraphic significance in the same region generally do not have unique identifying physical and chemical characteristics. Consequently, extension of a pedostratigraphic unit is accomplished by lateral tracing of the contact between a buried soil and an overlying, formally defined lithostratigraphic or allostratigraphic unit, or between a soil and two or more demonstrably correlative stratigraphic units.

(e) **Distinction from pedologic soils.**—Pedologic soils may include organic deposits (e.g., litter zones, peat deposits, or swamp deposits) that overlie or grade laterally into differentiated buried soils. The organic deposits are not products of pedogenesis, and O horizons are not included in a pedostratigraphic unit (Fig. 6); they may be classified as biostratigraphic or lithostratigraphic units. Pedologic soils also include the entire C horizon of a soil. The C horizon in pedology is not rigidly defined; it is merely the part of a soil profile that underlies the B horizon. The base of the C horizon in many soil profiles is gradational or unidentifiable; commonly it is placed arbitrarily. The need for clearly defined and easily recognized physical boundaries for a stratigraphic unit requires that the

lower boundary of a pedostratigraphic unit be defined as the lowest *definite* physical boundary of a pedologic horizon in a buried soil profile, and part or all of the C horizon may be excluded from a pedostratigraphic unit.

(f) **Relation to saprolite and other weathered materials.**—A material derived by in situ weathering of lithostratigraphic, allostratigraphic, and(or) lithodemic units (e.g., saprolite, bauxite, residuum) may be the parent material in which pedologic horizons form, but is not a pedologic soil. A pedostratigraphic unit may be based on the pedologic horizons of a buried soil developed in the product of in-situ weathering, such as saprolite. The parents of such a pedostratigraphic unit are both the saprolite and, indirectly, the rock from which it formed.

(g) **Distinction from other stratigraphic units.**—A pedostratigraphic unit differs from other stratigraphic units in that (1) it is a product of surface alteration of one or more older material units by specific processes (pedogenesis), (2) its lithology and other properties differ markedly from those of the parent material(s), and (3) a single pedostratigraphic unit may be formed in situ in parent material units of diverse compositions and ages.

(h) **Independence from time concepts.**—The boundaries of a pedostratigraphic unit are time-transgressive. Concepts of time spans, however measured, play no part in defining the boundaries of a pedostratigraphic unit. Nonetheless, evidence of age, whether based on fossils, numerical ages, or geometrical or other relationships, may play an important role in distinguishing and identifying non-contiguous pedostratigraphic units at localities away from the type areas. The name of a pedostratigraphic unit should be chosen from a geographic feature in the type area, and not from a time span.

Pedostratigraphic Nomenclature and Unit

 Article 56.—**Fundamental Unit.** The fundamental and only unit in pedostratigraphic classification is a geosol.

 Article 57.—**Nomenclature.**—The formal name of a pedostratigraphic unit consists of a geographic name combined with the term "geosol." Capitalization of the initial letter in each word

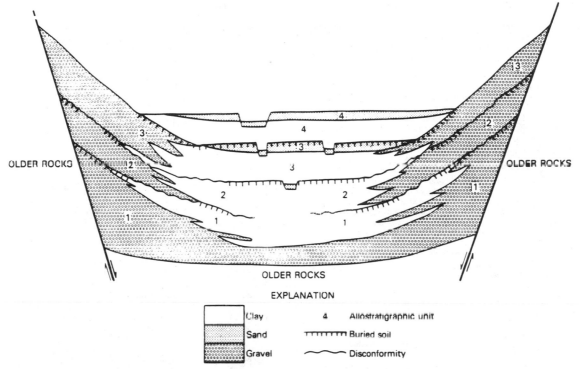

EXPLANATION

Clay   4   Allostratigraphic unit

Sand   ⊤⊤⊤⊤⊤⊤   Buried soil

Gravel   ~~~~   Disconformity

FIG. 7.—Example of allostratigraphic classification of alluvial and lacustrine deposits in a graben.
The alluvial and lacustrine deposits may be included in a single formation, or may be separated laterally into formations distinguished on the basis of contrasting texture (gravel, clay). Textural changes are abrupt and sharp, both vertically and laterally. The gravel deposits and clay deposits, respectively, are lithologically similar and thus cannot be distinguished as members of a formation. Four allostratigraphic units, each including two or three textural facies, may be defined on the basis of laterally traceable discontinuities (buried soils and disconformities).

serves to identify formal usage. The geographic name should be selected in accordance with recommendations in Article 7 and should not duplicate the name of another formal geologic unit. Names based on subjacent and superjacent rock units, for example the super-Wilcox–sub-Claiborne soil, are informal, as are those with time connotations (post-Wilcox–pre-Claiborne soil).

Remarks. (a) **Composite geosols.**—Where the horizons of two or more merged or "welded" buried soils can be distinguished, formal names of pedostratigraphic units based on the horizon boundaries can be retained. Where the horizon boundaries of the respective merged or "welded" soils cannot be distinguished, formal pedostratigraphic classification is abandoned and a combined name such as Hallettville-Jamesville geosol may be used informally.

(b) **Characterization.**—The physical and chemical properties of a pedostratigraphic unit commonly vary vertically and laterally throughout the geographic extent of the unit. A pedostratigraphic unit is characterized by the *range* of physical and chemical properties of the unit in the type area, rather than by "typical" properties exhibited in a type section. Consequently, a pedostratigraphic unit is characterized on the basis of a composite stratotype (Article 8d).

(c) **Procedures for establishing formal pedostratigraphic units.**—A formal pedostratigraphic unit may be established in accordance with the

applicable requirements of Article 3, and additionally by describing major soil horizons in each soil facies.

ALLOSTRATIGRAPHIC UNITS

Nature and Boundaries

Article 58.—**Nature of Allostratigraphic Units.** An allostratigraphic[9] unit is a mappable stratiform body of sedimentary rock that is defined and identified on the basis of its bounding discontinuities.

Remarks. (a) **Purpose.**—Formal allostratigraphic units may be defined to distinguish between different (1) superposed discontinuity-bounded deposits of similar lithology (Figs. 7, 9), (2) contiguous discontinuity-bounded deposits of similar lithology (Fig. 8), or (3) geographically separated discontinuity-bounded units of similar lithology (Fig. 9), or to distinguish as single units discontinuity-bounded deposits characterized by lithic heterogeneity (Fig. 8).

(b) **Internal characteristics.**—Internal characteristics (physical, chemical, and paleontological) may vary laterally and vertically throughout the unit.

(c) **Boundaries.**—Boundaries of allostratigraphic units are laterally traceable discontinuities (Figs. 7, 8, and 9).

(d) **Mappability.**—A formal allostratigraphic unit must be mappable at the scale practiced in the region where the unit is defined.

---

[9]From the Greek *allo*: "other, different."

(e) **Type locality and extent.**—A type locality and type area must be designated; a composite stratotype or a type section and several reference sections are desirable. An allostratigraphic unit may be laterally contiguous with a formally defined lithostratigraphic unit; a vertical cut-off between such units is placed where the units meet.

(f) **Relation to genesis.**—Genetic interpretation is an inappropriate basis for defining an allostratigraphic unit. However, genetic interpretation may influence the choice of its boundaries.

(g) **Relation to geomorphic surfaces.**—A geomorphic surface may be used as a boundary of an allostratigraphic unit, but the unit should not be given the geographic name of the surface.

(h) **Relation to soils and paleosols.**—Soils and paleosols are composed of products of weathering and pedogenesis and differ in many respects from allostratigraphic units, which are depositional units (see "Pedostratigraphic Units," Article 55). The upper boundary of a surface or buried soil may be used as a boundary of an allostratigraphic unit.

(i) **Relation to inferred geologic history.**—Inferred geologic history is not used to define an allostratigraphic unit. However, well-documented geologic history may influence the choice of the unit's boundaries.

(j) **Relation to time concepts.**—Inferred time spans, however measured, are not used to define an allostratigraphic unit. However, age relationships may influence the choice of the unit's boundaries.

(k) **Extension of allostratigraphic units.**—An allostratigraphic unit is extended from its type area by tracing the boundary discontinuities or by tracing or matching the deposits between the discontinuities.

Ranks of Allostratigraphic Units

Article 59.—**Hierarchy.** The hierarchy of allostratigraphic units, in order of decreasing rank, is allogroup, alloformation, and allomember.

Remarks. (a) **Alloformation.**—The alloformation is the fundamental unit in allostratigraphic classification. An alloformation may be completely or only partly divided into allomembers, if some useful purpose is served, or it may have no allomembers.

(b) **Allomember.**—An allomember is the formal allostratigraphic unit next in rank below an alloformation.

(c) **Allogroup.**—An allogroup is the allostratigraphic unit next in rank above an alloformation. An allogroup is established only if a unit of that rank is essential to elucidation of geologic history. An allogroup may consist entirely of named alloformations or, alternatively, may contain one or more named alloformations which jointly do not comprise the entire allogroup.

(d) **Changes in rank.**—The principles and procedures for elevation and reduction in rank of formal allostratigraphic units are the same as those in Articles 19b, 19g, and 28.

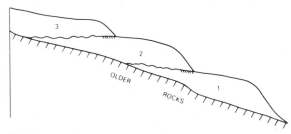

FIG. 8.—Example of allostratigraphic classification of contiguous deposits of similar lithology.
Allostratigraphic units 1, 2, and 3 are physical records of three glaciations. They are lithologically similar, reflecting derivation from the same bedrock, and constitute a single lithostratigraphic unit.

Allostratigraphic Nomenclature

Article 60.—**Nomenclature.** The principles and procedures for naming allostratigraphic units are the same as those for naming of lithostratigraphic units (see Articles 7, 30).

Remark. (a) **Revision.**—Allostratigraphic units may be revised or otherwise modified in accordance with the recommendations in Articles 17 to 20.

FORMAL UNITS DISTINGUISHED BY AGE

GEOLOGIC-TIME UNITS

Nature and Types

Article 61.—**Types.** Geologic-time units are conceptual, rather than material, in nature. Two types are recognized: those based on material standards or referents (specific rock sequences or bodies), and those independent of material referents (Fig. 1).

Units Based on Material Referents

Article 62.—**Types Based on Referents.** Two types of formal geologic-time units based on material referents are recognized: they are isochronous and diachronous units.

Article 63.—**Isochronous Categories.** Isochronous time units and the material bodies from which they are derived are twofold: geochronologic units (Article 80), which are based on corresponding material chronostratigraphic units (Article 66), and polarity-geochronologic units (Article 88), based on corresponding material polarity-chronostratigraphic units (Article 83).

Remark. (a) **Extent.**—Isochronous units are applicable worldwide; they may be referred to even in areas lacking a material record of the named span of time. The duration of the time may be represented by a unit-stratotype referent. The beginning and end of the time are represented by point-boundary-stratotypes either in a single stratigraphic sequence or in separate stratotype sections (Articles 8b, 10b).

Article 64.—**Diachronous Categories.** Diachronic units (Article 91) are time units corresponding to diachronous material allostratigraphic units (Article 58), pedostratigraphic units (Article 55), and most lithostratigraphic (Article 22) and biostratigraphic (Article 48) units.

Remarks. (a) **Diachroneity.**—Some lithostratigraphic and biostratigraphic units are clearly diachronous, whereas others have boundaries which are not demonstrably diachronous within the resolving power of available dating methods. The latter commonly are treated as isochronous and are used for purposes of chronocorrelation (see biochronozone, Article 75). However, the assumption of isochroneity must be tested continually.

(b) **Extent.**—Diachronic units are coextensive with the diachronous material stratigraphic units on which they are based and are not used beyond the extent of their material referents.

Units Independent of Material Referents

Article 65.—**Numerical Divisions of Time.** Isochronous geologic-time units based on numerical divisions of time in years are geochronometric units (Article 96) and have no material referents.

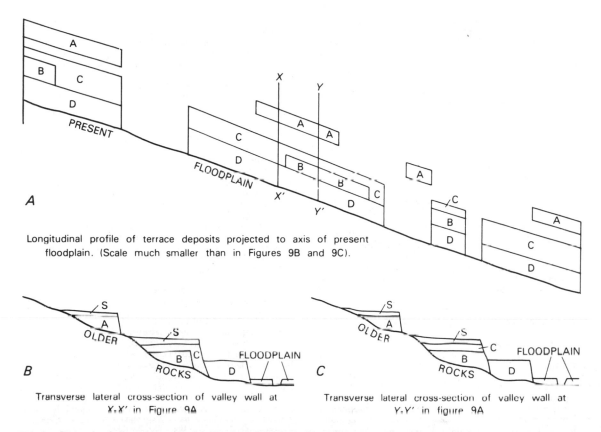

Longitudinal profile of terrace deposits projected to axis of present floodplain. (Scale much smaller than in Figures 9B and 9C).

Transverse lateral cross-section of valley wall at X,X' in Figure 9A

Transverse lateral cross-section of valley wall at Y,Y' in figure 9A

FIG. 9.—Example of allostratigraphic classification of lithologically similar, discontinuous terrace deposits.

A, B, C, and D are terrace gravel units of similar lithology at different topographic positions on a valley wall. The deposits may be defined as separate formal allostratigraphic units if such units are useful and if bounding discontinuities can be traced laterally. Terrace gravels of the same age commonly are separated geographically by exposures of older rocks. Where the bounding discontinuities cannot be traced continuously, they may be extended geographically on the basis of objective correlation of internal properties of the deposits other than lithology (e.g., fossil content, included tephras), topographic position, numerical ages, or relative-age criteria (e.g., soils or other weathering phenomena). The criteria for such extension should be documented. Slope deposits and eolian deposits (S) that mantle terrace surfaces may be of diverse ages and are not included in a terrace-gravel allostratigraphic unit. A single terrace surface may be underlain by more than one allostratigraphic unit (units B and C in sections b and c).

CHRONOSTRATIGRAPHIC UNITS

Nature and Boundaries

Article 66.—**Definition.** A chronostratigraphic unit is a body of rock established to serve as the material reference for all rocks formed during the same span of time. Each of its boundaries is synchronous. The body also serves as the basis for defining the specific interval of time, or geochronologic unit (Article 80), represented by the referent.

Remarks. (a) **Purposes.**—Chronostratigraphic classification provides a means of establishing the temporally sequential order of rock bodies. Principal purposes are to provide a framework for (1) temporal correlation of the rocks in one area with those in another, (2) placing the rocks of the Earth's crust in a systematic sequence and indicating their rel-

ative position and age with respect to earth history as a whole, and (3) constructing an internationally recognized Standard Global Chronostratigraphic Scale.

(b) **Nature.**—A chronostratigraphic unit is a material unit and consists of a body of strata formed during a specific time span. Such a unit represents all rocks, and only those rocks, formed during that time span.

(c) **Content.**—A chronostratigraphic unit may be based upon the time span of a biostratigraphic unit, a lithic unit, a magnetopolarity unit, or any other feature of the rock record that has a time range. Or it may be any arbitrary but specified sequence of rocks, provided it has properties allowing chronocorrelation with rock sequences elsewhere.

Article 67.—**Boundaries.** Boundaries of chronostratigraphic units should be defined in a designated stratotype on the basis of observable paleontological or physical features of the rocks.

Remark. (a) **Emphasis on lower boundaries of chronostratigraphic units.**—Designation of point boundaries for both base and top of chronostratigraphic units is not recommended, because subsequent information on relations between successive units may identify overlaps or gaps. One means of minimizing or eliminating problems of duplication or gaps in chronostratigraphic successions is to define formally as a point-boundary stratotype only the base of the unit. Thus, a chronostratigraphic unit with its base defined at one locality, will have its top defined by the base of an overlying unit at the same, but more commonly another, locality (Article 8b).

Article 68.—**Correlation.** Demonstration of time equivalence is required for geographic extension of a chronostratigraphic unit from its type section or area. Boundaries of chronostratigraphic units can be extended only within the limits of resolution of available means of chronocorrelation, which currently include paleontology, numerical dating, remanent magnetism, thermoluminescence, relative-age criteria (examples are superposition and cross-cutting relations), and such indirect and inferential physical criteria as climatic changes, degree of weathering, and relations to unconformities. Ideally, the boundaries of chronostratigraphic units are independent of lithology, fossil content, or other material bases of stratigraphic division, but, in practice, the correlation or geographic extension of these boundaries relies at least in part on such features. Boundaries of chronostratigraphic units commonly are intersected by boundaries of most other kinds of material units.

Ranks of Chronostratigraphic Units

Article 69.—**Hierarchy.** The hierarchy of chronostratigraphic units, in order of decreasing rank, is eonothem, erathem, system, series, and stage. Of these, system is the primary unit of worldwide major rank; its primacy derives from the history of development of stratigraphic classification. All systems and units of higher rank are divided completely into units of the next lower rank. Chronozones are non-hierarchical and commonly lower-rank chronostratigraphic units. Stages and chronozones in sum do not necessarily equal the units of next higher rank and need not be contiguous. The rank and magnitude of chronostratigraphic units are related to the time interval represented by the units, rather than to the thickness or areal extent of the rocks on which the units are based.

Article 70.—**Eonothem.** The unit highest in rank is eonothem. The Phanerozoic Eonothem encompasses the Paleozoic, Mesozoic, and Cenozoic Erathems. Although older rocks have been assigned heretofore to the Precambrian Eonothem, they also have been assigned recently to other (Archean and Proterozoic) eonothems by the IUGS Precambrian Subcommission. The span of time corresponding to an eonothem is an *eon*.

Article 71.—**Erathem.** An erathem is the formal chronostratigraphic unit of rank next lower to eonothem and consists of several adjacent systems. The span of time corresponding to an erathem is an *era*.

Remark. (a) **Names.**—Names given to traditional Phanerozoic erathems were based upon major stages in the development of life on Earth: Paleozoic (old), Mesozoic (intermediate), and Cenozoic (recent) life. Although somewhat comparable terms have been applied to Precambrian units, the names and ranks of Precambrian divisions are not yet universally agreed upon and are under consideration by the IUGS Subcommission on Precambrian Stratigraphy.

Article 72.—**System.** The unit of rank next lower to erathem is the system. Rocks encompassed by a system represent a time-

span and an episode of Earth history sufficiently great to serve as a worldwide chronostratigraphic reference unit. The temporal equivalent of a system is a *period*.

Remark. (a) **Subsystem and supersystem.**—Some systems initially established in Europe later were divided or grouped elsewhere into units ranked as systems. *Subsystems* (Mississippian Subsystem of the Carboniferous System) and *supersystems* (Karoo Supersystem) are more appropriate.

Article 73.—**Series.** Series is a conventional chronostratigraphic unit that ranks below a system and always is a division of a system. A series commonly constitutes a major unit of chronostratigraphic correlation within a province, between provinces, or between continents. Although many European series are being adopted increasingly for dividing systems on other continents, provincial series of regional scope continue to be useful. The temporal equivalent of a series is an *epoch*.

Article 74.—**Stage.** A stage is a chronostratigraphic unit of smaller scope and rank than a series. It is most commonly of greatest use in intra-continental classification and correlation, although it has the potential for worldwide recognition. The geochronologic equivalent of stage is *age*.

Remark. (a) **Substage.**—Stages may be, but need not be, divided completely into substages.

Article 75.—**Chronozone.** A chronozone is a non-hierarchical, but commonly small, formal chronostratigraphic unit, and its boundaries may be independent of those of ranked units. Although a chronozone is an isochronous unit, it may be based on a biostratigraphic unit (example: *Cardioceras cordatum* Biochronozone), a lithostratigraphic unit (Woodbend Lithochronozone), or a magnetopolarity unit (Gilbert Reversed-Polarity Chronozone). Modifiers (litho-, bio-, polarity) used in formal names of the units need not be repeated in general discussions where the meaning is evident from the context, e.g., *Exus albus* Chronozone.

Remarks. (a) **Boundaries of chronozones.**—The base and top of a *chronozone* correspond in the unit's stratotype to the observed, defining, physical and paleontological features, but they are extended to other areas by any means available for recognition of synchroneity. The temporal equivalent of a chronozone is a chron.

(b) **Scope.**—The scope of the non-hierarchical chronozone may range markedly, depending upon the purpose for which it is defined either formally or informally. The informal "biochronozone of the ammonites," for example, represents a duration of time which is enormous and exceeds that of a system. In contrast, a biochronozone defined by a species of limited range, such as the *Exus albus* Chronozone, may represent a duration equal to or briefer than that of a stage.

(c) **Practical utility.**—Chronozones, especially thin and informal biochronozones and lithochronozones bounded by key beds or other "markers," are the units used most commonly in industry investigations of selected parts of the stratigraphy of economically favorable basins. Such units are useful to define geographic distributions of lithofacies or biofacies, which provide a basis for genetic interpretations and the selection of targets to drill.

Chronostratigraphic Nomenclature

Article 76.—**Requirements.** Requirements for establishing a formal chronostratigraphic unit include: (i) statement of intention to designate such a unit; (ii) selection of name; (iii) statement of kind and rank of unit; (iv) statement of general concept of unit including historical background, synonymy, previous treatment, and reasons for proposed establishment; (v) description of characterizing physical and/or biological features; (vi) designation and description of boundary type sections, stratotypes, or other kinds of units on which it is based; (vii) correlation and age relations; and (viii) publication in a recognized scientific medium as specified in Article 4.

Article 77.—**Nomenclature.** A formal chronostratigraphic unit is given a compound name, and the initial letter of all words, except for trivial taxonomic terms, is capitalized. Except for chronozones (Article 75), names proposed for new chronostratigraphic units should not duplicate those for other stratigraphic units. For example, naming a new chronostratigraphic unit simply by adding "-an" or "-ian" to the name of a lithostratigraphic unit is improper.

Remarks. (a) **Systems and units of higher rank.**—Names that are generally accepted for systems and units of higher rank have diverse origins, and they also have different kinds of endings (Paleozoic, Cambrian, Cretaceous, Jurassic, Quaternary).

(b) **Series and units of lower rank.**—Series and units of lower rank are commonly known either by geographic names (Virgilian Series, Ochoan Series) or by names of their encompassing units modified by the capitalized adjectives Upper, Middle, and Lower (Lower Ordovician). Names of chronozones are derived from the unit on which they are based (Article 75). For series and stage, a geographic name is preferable because it may be related to a type area. For geographic names, the adjectival endings -an or -ian are recommended (Cincinnatian Series), but it is permissible to use the geographic name without any special ending, if more euphonious. Many series and stage names already in use have been based on lithic units (groups, formations, and members) and bear the names of these units (Wolfcampian Series, Claibornian Stage). Nevertheless, a stage preferably should have a geographic name not previously used in stratigraphic nomenclature. Use of internationally accepted (mainly European) stage names is preferable to the proliferation of others.

Article 78.—**Stratotypes.** An ideal stratotype for a chronostratigraphic unit is a completely exposed unbroken and continuous sequence of fossiliferous stratified rocks extending from a well-defined lower boundary to the base of the next higher unit. Unfortunately, few available sequences are sufficiently complete to define stages and units of higher rank, which therefore are best defined by boundary-stratotypes (Article 8b).

Boundary-stratotypes for major chronostratigraphic units ideally should be based on complete sequences of either fossiliferous monofacial marine strata or rocks with other criteria for chronocorrelation to permit widespread tracing of synchronous horizons. Extension of synchronous surfaces should be based on as many indicators of age as possible.

Article 79.—**Revision of units.** Revision of a chronostratigraphic unit without changing its name is allowable but requires as much justification as the establishment of a new unit (Articles 17, 19, and 76). Revision or redefinition of a unit of system or higher rank requires international agreement. If the definition of a chronostratigraphic unit is inadequate, it may be clarified by establishment of boundary stratotypes in a principal reference section.

GEOCHRONOLOGIC UNITS

Nature and Boundaries

Article 80.—**Definition and Basis.** Geochronologic units are divisions of time traditionally distinguished on the basis of the rock record as expressed by chronostratigraphic units. A geochronologic unit is not a stratigraphic unit (i.e., it is not a material unit), but it corresponds to the time span of an established

chronostratigraphic unit (Articles 65 and 66), and its beginning and ending corresponds to the base and top of the referent.

Ranks and Nomenclature of Geochronologic Units

Article 81.—**Hierarchy.** The hierarchy of geochronologic units in order of decreasing rank is *eon, era, period, epoch,* and *age.* Chron is a non-hierarchical, but commonly brief, geochronologic unit. Ages in sum do not necessarily equal epochs and need not form a continuum. An eon is the time represented by the rocks constituting an eonothem; era by an erathem; period by a system; epoch by a series; age by a stage; and chron by a chronozone.

Article 82.—**Nomenclature.** Names for periods and units of lower rank are identical with those of the corresponding chronostratigraphic units; the names of some eras and eons are independently formed. Rules of capitalization for chronostratigraphic units (Article 77) apply to geochronologic units. The adjectives Early, Middle, and Late are used for the geochronologic epochs equivalent to the corresponding chronostratigraphic Lower, Middle, and Upper series, where these are formally established.

POLARITY-CHRONOSTRATIGRAPHIC UNITS

Nature and Boundaries

Article 83.—**Definition.** A polarity-chronostratigraphic unit is a body of rock that contains the primary magnetic-polarity record imposed when the rock was deposited, or crystallized, during a specific interval of geologic time.

Remarks. (a) **Nature.**—Polarity-chronostratigraphic units depend fundamentally for definition on actual sections or sequences, or measurements on individual rock units, and without these standards they are meaningless. They are based on material units, the polarity zones of magnetopolarity classification. Each polarity-chronostratigraphic unit is the record of the time during which the rock formed and the Earth's magnetic field had a designated polarity. Care should be taken to define polarity-chronologic units in terms of polarity-chronostratigraphic units, and not vice versa.

(b) **Principal purposes.**—Two principal purposes are served by polarity-chronostratigraphic classification: (1) correlation of rocks at one place with those of the same age and polarity at other places; and (2) delineation of the polarity history of the Earth's magnetic field.

(c) **Recognition.**—A polarity-chronostratigraphic unit may be extended geographically from its type locality only with the support of physical and/or paleontologic criteria used to confirm its age.

Article 84.—**Boundaries.** The boundaries of a polarity chronozone are placed at polarity-reversal horizons or polarity transition-zones (see Article 45).

Ranks and Nomenclature of Polarity-Chronostratigraphic Units

Article 85.—**Fundamental Unit.** The polarity chronozone consists of rocks of a specified primary polarity and is the fundamental unit of worldwide polarity-chronostratigraphic classification.

Remarks. (a) **Meaning of term.**—A polarity chronozone is the worldwide body of rock strata that is collectively defined as a polarity-chronostratigraphic unit.

(b) **Scope.**—Individual polarity zones are the basic building blocks of polarity chronozones. Recognition and definition of polarity chronozones may thus involve step-by-step assembly of carefully dated or correlated individual polarity zones, especially in work with rocks older than the oldest ocean-floor magnetic anomalies. This procedure is the method by which the Brunhes, Matuyama, Gauss, and Gilbert Chronozones were recognized (Cox, Doell, and Dalrymple, 1963) and defined originally (Cox, Doell, and Dalrymple, 1964).

(c) **Ranks.**—Divisions of polarity chronozones are designated polarity subchronozones. Assemblages of polarity chronozones may be termed polarity superchronozones.

Article 86.—Establishing Formal Units. Requirements for establishing a polarity-chronostratigraphic unit include those specified in Articles 3 and 4, and also (1) definition of boundaries of the unit, with specific references to designated sections and data; (2) distinguishing polarity characteristics, lithologic descriptions, and included fossils; and (3) correlation and age relations.

Article 87.—Name. A formal polarity-chronostratigraphic unit is given a compound name beginning with that for a named geographic feature; the second component indicates the normal, reversed, or mixed polarity of the unit, and the third component is *chronozone*. The initial letter of each term is capitalized. If the same geographic name is used for both a magnetopolarity zone and a polarity-chronostratigraphic unit, the latter should be distinguished by an -an or -ian ending. Example: Tetonian Reversed-Polarity Chronozone.

Remarks: (a) **Preservation of established name.**—A particularly well-established name should not be displaced, either on the basis of priority, as described in Article 7c, or because it was not taken from a geographic feature. Continued use of Brunhes, Matuyama, Gauss, and Gilbert, for example, is endorsed so long as they remain valid units.

(b) **Expression of doubt.**—Doubt in the assignment of polarity zones to polarity-chronostratigraphic units should be made explicit if criteria of time equivalence are inconclusive.

POLARITY-CHRONOLOGIC UNITS

Nature and Boundaries

Article 88.—Definition. Polarity-chronologic units are divisions of geologic time distinguished on the basis of the record of magnetopolarity as embodied in polarity-chronostratigraphic units. No special kind of magnetic time is implied; the designations used are meant to convey the parts of geologic time during which the Earth's magnetic field had a characteristic polarity or sequence of polarities. These units correspond to the time spans represented by polarity chronozones, e.g., Gauss Normal Polarity Chronozone. They are not material units.

Ranks and Nomenclature of Polarity-Chronologic Units

Article 89.—Fundamental Unit. The polarity chron is the fundamental unit of geologic time designating the time span of a polarity chronozone.

Remark. (a) **Hierarchy.**—Polarity-chronologic units of decreasing hierarchical ranks are polarity superchron, polarity chron, and polarity subchron.

Article 90.—Nomenclature. Names for polarity chronologic units are identical with those of corresponding polarity-chronostratigraphic units, except that the term chron (or superchron, etc) is substituted for chronozone (or superchronozone, etc).

DIACHRONIC UNITS

Nature and Boundaries

Article 91.—Definition. A diachronic unit comprises the unequal spans of time represented either by a specific lithostratigraphic, allostratigraphic, biostratigraphic, or pedostratigraphic unit, or by an assemblage of such units.

Remarks. (a) **Purposes.**—Diachronic classification provides (1) a means of comparing the spans of time represented by stratigraphic units with diachronous boundaries at different localities, (2) a basis for broadly establishing in time the beginning and ending of deposition of diachronous stratigraphic units at different sites, (3) a basis for inferring the rate of change in areal extent of depositional processes, (4) a means of determining and comparing rates and durations of deposition at different localities, and (5) a means of comparing temporal and spatial relations of diachronous stratigraphic units (Watson and Wright, 1980).

(b) **Scope.**—The scope of a diachronic unit is related to (1) the relative magnitude of the transgressive division of time represented by the stratigraphic unit or units on which it is based and (2) the areal extent of those units. A diachronic unit is not extended beyond the geographic limits of the stratigraphic unit or units on which it is based.

(c) **Basis.**—The basis for a diachronic unit is the diachronous referent.

(d) **Duration.**—A diachronic unit may be of equal duration at different places despite differences in the times at which it began and ended at those places.

Article 92.—Boundaries. The boundaries of a diachronic unit are the times recorded by the beginning and end of deposition of the material referent at the point under consideration (Figs. 10, 11).

Remark. (a) **Temporal relations.**—One or both of the boundaries of a diachronic unit are demonstrably time-transgressive. The varying time significance of the boundaries is defined by a series of boundary reference sections (Article 8b, 8e). The duration and age of a diachronic unit differ from place to place (Figs. 10, 11).

Ranks and Nomenclature of Diachronic Units

Article 93.—Ranks. A diachron is the fundamental and non-hierarchical diachronic unit. If a hierarchy of diachronic units is needed, the terms episode, phase, span, and cline, in order of decreasing rank, are recommended. The rank of a hierarchical unit is determined by the scope of the unit (Article 91 b), and not by the time span represented by the unit at a particular place.

Remarks. (a) **Diachron.**—Diachrons may differ greatly in magnitude because they are the spans of time represented by individual or grouped lithostratigraphic, allostratigraphic, biostratigraphic, and(or) pedostratigraphic units.

(b) **Hierarchical ordering permissible.**—A hierarchy of diachronic units may be defined if the resolution of spatial and temporal relations of diachronous stratigraphic units is sufficiently precise to make the hierarchy useful (Watson and Wright, 1980). Although all hierarchical units of rank lower than episode are part of a unit next higher in rank, not all parts of an episode, phase, or span need be represented by a unit of lower rank.

(c) **Episode.**—An episode is the unit of highest rank and greatest scope in hierarchical classification. If the "Wisconsinan Age" were to be redefined as a diachronic unit, it would have the rank of episode.

Article 94.—Name. The name for a diachronic unit should be compound, consisting of a geographic name followed by the term diachron or a hierarchical rank term. Both parts of the com-

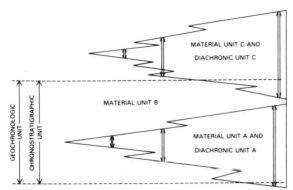

FIG. 10.—Comparison of geochronologic, chronostratigraphic, and diachronic units.

pound name are capitalized to indicate formal status. If the diachronic unit is defined by a single stratigraphic unit, the geographic name of the unit may be applied to the diachronic unit. Otherwise, the geographic name of the diachronic unit should not duplicate that of another formal stratigraphic unit. Genetic terms (e.g., alluvial, marine) or climatic terms (e.g., glacial, interglacial) are not included in the names of diachronic units.

Remarks. (a) **Formal designation of units.**—Diachronic units should be formally defined and named only if such definition is useful.

(b) **Inter-regional extension of geographic names.**—The geographic name of a diachronic unit may be extended from one region to another if the stratigraphic units on which the diachronic unit is based extend across the regions. If different diachronic units in contiguous regions eventually

prove to be based on laterally continuous stratigraphic units, one name should be applied to the unit in both regions. If two names have been applied, one name should be abandoned and the other formally extended. Rules of priority (Article 7d) apply. Priority in publication is to be respected, but priority alone does not justify displacing a well-established name by one not well-known or commonly used.

(c) **Change from geochronologic to diachronic classification.**—Lithostratigraphic units have served as the material basis for widely accepted chronostratigraphic and geochronologic classifications of Quaternary nonmarine deposits, such as the classifications of Frye et al (1968), Willman and Frye (1970), and Dreimanis and Karrow (1972). In practice, time-parallel horizons have been extended from the stratotypes on the basis of markedly time-transgressive lithostratigraphic and pedostratigraphic unit boundaries. The time ("geochronologic") units, defined on the basis of the stratotype sections but extended on the basis of diachronous stratigraphic boundaries, are diachronic units. Geographic names established for such "geochronologic" units may be used in diachronic classification if (1) the chronostratigraphic and geochronologic classifications are formally abandoned and diachronic classifications are proposed to replace the former "geochronologic" classifications, and (2) the units are redefined as formal diachronic units. Preservation of well-established names in these specific circumstances retains the intent and purpose of the names and the units, retains the practical significance of the units, enhances communication, and avoids proliferation of nomenclature.

Article 95.—**Establishing Formal Units.** Requirements for establishing a formal diachronic unit, in addition to those in Article 3, include (1) specification of the nature, stratigraphic relations, and geographic or areal relations of the stratigraphic unit or units that serve as a basis for definition of the unit, and (2) specific designation and description of multiple reference sections that illustrate the temporal and spatial relations of the defining stratigraphic unit or units and the boundaries of the unit or units.

Remark. (a) **Revision or abandonment.**—Revision or abandonment of the stratigraphic unit or units that serve as the material basis for defini

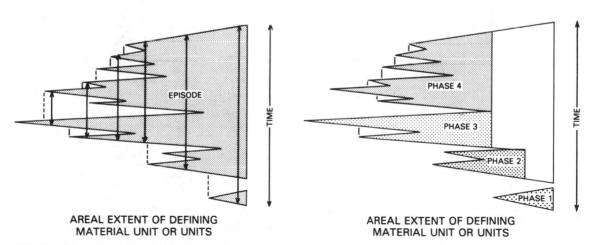

FIG. 11.—Schematic relation of phases to an episode.
Parts of a phase similarly may be divided into spans, and spans into clines. Formal definition of spans and clines is unnecessary in most diachronic unit hierarchies.

tion of a diachronic unit may require revision or abandonment of the diachronic unit. Procedure for revision must follow the requirements for establishing a new diachronic unit.

GEOCHRONOMETRIC UNITS

Nature and Boundaries

Article 96.—**Definition.** Geochronometric units are units established through the direct division of geologic time, expressed in years. Like geochronologic units (Article 80), geochronometric units are abstractions, i.e., they are not material units. Unlike geochronologic units, geochronometric units are not based on the time span of designated chronostratigraphic units (stratotypes), but are simply time divisions of convenient magnitude for the purpose for which they are established, such as the development of a time scale for the Precambrian. Their boundaries are arbitrarily chosen or agreed-upon ages in years.

Ranks and Nomenclature of Geochronometric Units

Article 97.—**Nomenclature.** Geochronologic rank terms (eon, era, period, epoch, age, and chron) may be used for geochronometric units when such terms are formalized. For example, Archean Eon and Proterozoic Eon, as recognized by the IUGS Subcommission on Precambrian Stratigraphy, are formal geochronometric units in the sense of Article 96, distinguished on the basis of an arbitrarily chosen boundary at 2.5 Ga. Geochronometric units are not defined by, but may have, corresponding chronostratigraphic units (eonothem, erathem, system, series, stage, and chronozone).

PART III: REFERENCES[10]

American Commission on Stratigraphic Nomenclature, 1947, Note 1—Organization and objectives of the Stratigraphic Commission: American Association of Petroleum Geologists Bulletin, v. 31, no. 3, p. 513-518.

——— ,1961, Code of Stratigraphic Nomenclature: American Association of Petroleum Geologists Bulletin, v. 45, no. 5, p. 645-665.

——— ,1970, Code of Stratigraphic Nomenclature (2d ed.): American Association of Petroleum Geologists, Tulsa, Okla., 45 p.

——— ,1976, Note 44—Application for addition to code concerning magnetostratigraphic units: American Association of Petroleum Geologists Bulletin, v. 60, no. 2, p. 273-277.

Caster, K. E., 1934, The stratigraphy and paleontology of northwestern Pennsylvania, Part 1, Stratigraphy: Bulletins of American Paleontology, v. 21, 185 p.

Chang, K. H., 1975, Unconformity-bounded stratigraphic units: Geological Society of America Bulletin, v. 86, no. 11, p. 1544-1552.

Committee on Stratigraphic Nomenclature, 1933, Classification and nomenclature of rock units: Geological Society of America Bulletin, v. 44, no. 2, p. 423-459, and American Association of Petroleum Geologists Bulletin, v. 17, no. 7, p. 843-868.

Cox, A. V., R. R. Doell, and G. B. Dalrymple, 1963, Geomagnetic polarity epochs and Pleistocene geochronometry: Nature, v. 198, p. 1049-1051.

——— ,1964, Reversals of the Earth's magnetic field: Science, v. 144, no. 3626, p. 1537-1543.

Cross, C. W., 1898, Geology of the Telluride area: U.S. Geological Survey 18th Annual Report, pt. 3, p. 759.

Cumming, A. D., J. G. C. M. Fuller, and J. W. Porter, 1959, Separation of strata: Paleozoic limestones of the Williston basin: American Journal of Science, v. 257, no. 10, p. 722-733.

Dreimanis, Aleksis, and P. F. Karrow, 1972, Glacial history of the Great Lakes–St. Lawrence region, the classification of the Wisconsin(an) Stage, and its correlatives: International Geologic Congress, 24th Session, Montreal, 1972, Section 12, Quaternary Geology, p. 5-15.

Dunbar, C. O., and John Rodgers, 1957, Principles of stratigraphy: Wiley, New York, 356 p.

Forgotson, J. M., Jr., 1957, Nature, usage and definition of marker-defined vertically segregated rock units: American Association of Petroleum Geologists Bulletin, v. 41, no. 9, p. 2108-2113.

Frye, J. C., H. B. Willman, Meyer Rubin, and R. F. Black, 1968, Definition of Wisconsinan Stage: U.S. Geological Survey Bulletin 1274-E, 22 p.

George, T. N., and others, 1969, Recommendations on stratigraphical usage: Geological Society of London, Proceedings no. 1656, p. 139-166.

Harland, W. B., 1977, Essay review [of] International Stratigraphic Guide, 1976: Geology Magazine, v. 114, no. 3, p. 229-235.

——— ,1978, Geochronologic scales, in G. V. Cohee et al, eds., Contributions to the Geologic Time Scale: American Association of Petroleum Geologists, Studies in Geology, no. 6, p. 9-32.

Harrison, J. E., and Z. E. Peterman, 1980, North American Commission on Stratigraphic Nomenclature Note 52—A preliminary proposal for a chronometric time scale for the Precambrian of the United States and Mexico: Geological Society of America Bulletin, v. 91, no. 6, p. 377-380.

Henbest, L. G., 1952, Significance of evolutionary explosions for diastrophic division of Earth history: Journal of Paleontology, v. 26, p. 299-318.

Henderson, J. B., W. G. E. Caldwell, and J. E. Harrison, 1980, North American Commission on Stratigraphic Nomenclature, Report 8—Amendment of code concerning terminology for igneous and high-grade metamorphic rocks: Geological Society of America Bulletin, v. 91, no. 6, p. 374-376.

Holland, C. H., and others, 1978, A guide to stratigraphical procedure: Geological Society of London, Special Report 10, p. 1-18.

Huxley, T. H., 1862, The anniversary address: Geological Society of London, Quarterly Journal, v. 18, p. xl-liv.

International Commission on Zoological Nomenclature, 1964: International Code of Zoological Nomenclature adopted by the XV International Congress of Zoology: International Trust for Zoological Nomenclature, London, 176 p.

International Subcommission on Stratigraphic Classification (ISSC), 1976, International Stratigraphic Guide (H. D. Hedberg, ed.): John Wiley and Sons, New York, 200 p.

International Subcommission on Stratigraphic Classification, 1979, Magnetostratigraphy polarity units—a supplementary chapter of the ISSC International Stratigraphic Guide: Geology, v. 7, p. 578-583.

Izett, G. A., and R. E. Wilcox, 1981, Map showing the distribution of the Huckleberry Ridge, Mesa Falls, and Lava Creek volcanic ash beds (Pearlette family ash beds) of Pliocene and Pleistocene age in the western United States and southern Canada: U. S. Geological Survey Miscellaneous Geological Investigations Map I-1325.

Kauffman, E. G., 1969, Cretaceous marine cycles of the Western Interior: Mountain Geologist: Rocky Mountain Association of Geologists, v. 6, no. 4, p. 227-245.

Matthews, R. K., 1974, Dynamic stratigraphy—an introduction to sedimentation and stratigraphy: Prentice-Hall, New Jersey, 370 p.

McDougall, Ian, 1977, The present status of the geomagnetic polarity time scale: Research School of Earth Sciences, Australian National University, Publication no. 1288, 34 p.

McElhinny, M. W., 1978, The magnetic polarity time scale; prospects and possibilities in magnetostratigraphy, in G. V. Cohee et al, eds., Contributions to the Geologic Time Scale, American Association of Petroleum Geologists, Studies in Geology, no. 6, p. 57-65.

---

[10]Readers are reminded of the extensive and noteworthy bibliography of contributions to stratigraphic principles, classification, and terminology cited by the International Stratigraphic Guide (ISSC, 1976, p. 111-187).

McIver, N. L., 1972, Cenozoic and Mesozoic stratigraphy of the Nova Scotia shelf: Canadian Journal of Earth Science, v. 9, p. 54-70.

McLaren, D. J., 1977, The Silurian-Devonian Boundary Committee. A final report, in A. Martinsson, ed., The Silurian-Devonian boundary: IUGS Series A, no. 5, p. 1-34.

Morrison, R. B., 1967, Principles of Quaternary soil stratigraphy, in R. B. Morrison and H. E. Wright, Jr., eds., Quaternary soils: Reno, Nevada, Center for Water Resources Research, Desert Research Institute, Univ. Nevada, p. 1-69.

North American Commission on Stratigraphic Nomenclature, 1981, Draft North American Stratigraphic Code: Canadian Society of Petroleum Geologists, Calgary, 63 p.

Palmer, A. R., 1965, Biomere-a new kind of biostratigraphic unit: Journal of Paleontology, v. 39, no. 1, p. 149-153.

Parsons, R. B., 1981, Proposed soil-stratigraphic guide, in International Union for Quaternary Research and International Society of Soil Science: INQUA Commission 6 and ISSS Commission 5 Working Group, Pedology, Report, p. 6-12.

Pawluk, S., 1978, The pedogenic profile in the stratigraphic section, in W. C. Mahaney, ed., Quaternary soils: Norwich, England, GeoAbstracts, Ltd., p. 61-75.

Ruhe, R. V., 1965, Quaternary paleopedology, in H. E. Wright, Jr., and D. G. Frey, eds., The Quaternary of the United States: Princeton, N.J., Princeton University Press, p. 755-764.

Schultz, E. H., 1982, The chronosome and supersome--terms proposed for low-rank chronostratigraphic units: Canadian Petroleum Geology, v. 30, no. 1, p. 29-33.

Shaw, A. B., 1964, Time in stratigraphy: McGraw-Hill, New York, 365 p.

Sims, P. K., 1979, Precambrian subdivided: Geotimes, v. 24, no. 12, p. 15.

Sloss, L. L., 1963, Sequences in the cratonic interior of North America: Geological Society of America Bulletin, v. 74, no. 2, p. 94-114.

Tracey, J. I., Jr., and others, 1971, Initial reports of the Deep Sea Drilling Project, v. 8: U.S. Government Printing Office, Washington, 1037 p.

Valentine, K. W. G., and J. B. Dalrymple, 1976, Quaternary buried paleosols: A critical review: Quaternary Research, v. 6, p. 209-222.

Vella, P., 1964, Biostratigraphic units: New Zealand Journal of Geology and Geophysics, v. 7, no. 3, p. 615-625.

Watson, R. A., and H. E. Wright, Jr., 1980, The end of the Pleistocene: A general critique of chronostratigraphic classification: Boreas, v. 9, p. 153-163.

Weiss, M. P., 1979a, Comments and suggestions invited for revision of American Stratigraphic Code: Geological Society of America, News and Information, v. 1, no. 7, p. 97-99.

——— ,1979b, Stratigraphic Commission Note 50--Proposal to change name of Commission: American Association of Petroleum Geologists Bulletin, v. 63, no. 10, p. 1986.

Weller, J. M., 1960, Stratigraphic principles and practice: Harper and Brothers, New York, 725 p.

Willman, H. B., and J. C. Frye, 1970, Pleistocene stratigraphy of Illinois: Illinois State Geological Survey Bulletin 94, 204 p.

Author Index

Subject Index

772   SUBJECT INDEX